IET POWER AND ENERGY SERIES 58

Lightning Protection

Other volumes in this series:

Volume 1	**Power circuit breaker theory and design** C.H. Flurscheim (Editor)	
Volume 4	**Industrial microwave heating** A.C. Metaxas and R.J. Meredith	
Volume 7	**Insulators for high voltages** J.S.T. Looms	
Volume 8	**Variable frequency AC motor drive systems** D. Finney	
Volume 10	**SF6 switchgear** H.M. Ryan and G.R. Jones	
Volume 11	**Conduction and induction heating** E.J. Davies	
Volume 13	**Statistical techniques for high voltage engineering** W. Hauschild and W. Mosch	
Volume 14	**Uninterruptible power supplies** J. Platts and J.D. St Aubyn (Editors)	
Volume 15	**Digital protection for power systems** A.T. Johns and S.K. Salman	
Volume 16	**Electricity economics and planning** T.W. Berrie	
Volume 18	**Vacuum switchgear** A. Greenwood	
Volume 19	**Electrical safety: a guide to causes and prevention of hazards** J. Maxwell Adams	
Volume 21	**Electricity distribution network design, 2nd edition** E. Lakervi and E.J. Holmes	
Volume 22	**Artificial intelligence techniques in power systems** K. Warwick, A.O. Ekwue and R. Aggarwal (Editors)	
Volume 24	**Power system commissioning and maintenance practice** K. Harker	
Volume 25	**Engineers' handbook of industrial microwave heating** R.J. Meredith	
Volume 26	**Small electric motors** H. Moczala *et al.*	
Volume 27	**AC-DC power system analysis** J. Arrillaga and B.C. Smith	
Volume 29	**High voltage direct current transmission, 2nd edition** J. Arrillaga	
Volume 30	**Flexible AC Transmission Systems (FACTS)** Y-H. Song (Editor)	
Volume 31	**Embedded generation** N. Jenkins *et al.*	
Volume 32	**High voltage engineering and testing, 2nd edition** H.M. Ryan (Editor)	
Volume 33	**Overvoltage protection of low-voltage systems, revised edition** P. Hasse	
Volume 34	**The lightning flash** V. Cooray	
Volume 35	**Control techniques drives and controls handbook** W. Drury (Editor)	
Volume 36	**Voltage quality in electrical power systems** J. Schlabbach *et al.*	
Volume 37	**Electrical steels for rotating machines** P. Beckley	
Volume 38	**The electric car: development and future of battery, hybrid and fuel-cell cars** M. Westbrook	
Volume 39	**Power systems electromagnetic transients simulation** J. Arrillaga and N. Watson	
Volume 40	**Advances in high voltage engineering** M. Haddad and D. Warne	
Volume 41	**Electrical operation of electrostatic precipitators** K. Parker	
Volume 43	**Thermal power plant simulation and control** D. Flynn	
Volume 44	**Economic evaluation of projects in the electricity supply industry** H. Khatib	
Volume 45	**Propulsion systems for hybrid vehicles** J. Miller	
Volume 46	**Distribution switchgear** S. Stewart	
Volume 47	**Protection of electricity distribution networks, 2nd edition** J. Gers and E. Holmes	
Volume 48	**Wood pole overhead lines** B. Wareing	
Volume 49	**Electric fuses, 3rd edition** A. Wright and G. Newbery	
Volume 50	**Wind power integration: connection and system operational aspects** B. Fox *et al.*	
Volume 51	**Short circuit currents** J. Schlabbach	
Volume 52	**Nuclear power** J. Wood	
Volume 53	**Condition assessment of high voltage insulation in power system equipment** R.E. James and Q. Su	
Volume 55	**Local energy: distributed generation of heat and power** J. Wood	
Volume 56	**Condition monitoring of rotating electrical machines** P. Tavner, L. Ran, J. Penman and H. Sedding	
Volume 905	**Power system protection**, 4 volumes	

Lightning Protection

Edited by Vernon Cooray

The Institution of Engineering and Technology

Published by The Institution of Engineering and Technology, London, United Kingdom

© 2010 The Institution of Engineering and Technology

First published 2010

This publication is copyright under the Berne Convention and the Universal Copyright Convention. All rights reserved. Apart from any fair dealing for the purposes of research or private study, or criticism or review, as permitted under the Copyright, Designs and Patents Act 1988, this publication may be reproduced, stored or transmitted, in any form or by any means, only with the prior permission in writing of the publishers, or in the case of reprographic reproduction in accordance with the terms of licences issued by the Copyright Licensing Agency. Enquiries concerning reproduction outside those terms should be sent to the publishers at the undermentioned address:

The Institution of Engineering and Technology
Michael Faraday House
Six Hills Way, Stevenage
Herts SG1 2AY, United Kingdom

www.theiet.org

While the authors and publisher believe that the information and guidance given in this work are correct, all parties must rely upon their own skill and judgement when making use of them. Neither the authors nor publisher assumes any liability to anyone for any loss or damage caused by any error or omission in the work, whether such an error or omission is the result of negligence or any other cause. Any and all such liability is disclaimed.

The moral rights of the authors to be identified as authors of this work have been asserted by them in accordance with the Copyright, Designs and Patents Act 1988.

British Library Cataloguing in Publication Data
A catalogue record for this product is available from the British Library

ISBN 978-0-86341-744-3 (hardback)
ISBN 978-1-84919-106-7 (PDF)

Typeset in India by Techset Composition Ltd, Chennai
Printed in the UK by Athenaeum Press Ltd, Gateshead, Tyne & Wear

In appreciation of the work of a merciful God

Contents

List of contributors		xxvii
Preface		xxix
Acknowledgements		xxxi

1 Benjamin Franklin and lightning rods — 1
E. Philip Krider
1.1 A Philadelphia story — 3
1.2 The French connection — 5
1.3 Experiments in colonial America — 6
1.4 First protection system — 7
1.5 Improvements — 9
1.6 'Snatching lightning from the sky' — 12
Acknowledgement — 13
References — 13

2 Lightning parameters of engineering interest — 15
Vernon Cooray and Mahendra Fernando
2.1 Introduction — 15
2.2 Electric fields generated by thunderclouds — 19
2.3 Thunderstorm days and ground flash density — 23
2.4 Number of strokes and time interval between strokes in ground flashes — 25
   2.4.1 Number of strokes per flash — 28
   2.4.2 Interstroke interval — 30
2.5 Number of channel terminations in ground flashes — 31
2.6 Occurrence of surface flash over — 34
2.7 Lightning leaders — 36
   2.7.1 Speed of stepped leaders — 36
   2.7.2 Speed of dart leaders — 37
   2.7.3 Electric fields generated by stepped leaders — 39
   2.7.4 Electric fields generated by dart leaders — 40
   2.7.5 Speed of connecting leaders — 44
   2.7.6 Currents in connecting leaders — 45

2.8	Current parameters of first and subsequent return strokes		47
	2.8.1	Berger's measurements	50
	2.8.2	Garbagnati and Piparo's measurements	51
	2.8.3	Eriksson's measurements	51
	2.8.4	Analysis of Andersson and Eriksson	53
	2.8.5	Measurements of Takami and Okabe	53
	2.8.6	Measurements of Visacro and colleagues	54
	2.8.7	Summary of current measurements	54
2.9	Statistical representation of lightning current parameters		55
	2.9.1	Correlation between different current parameters	57
	2.9.2	Effect of tower height	60
	2.9.3	Mathematical representation of current waveforms	67
		2.9.3.1 Current waveform recommended by the CIGRE study group	67
		2.9.3.2 Analytical form of the current used in the International Electrotechnical Commission standard	70
		2.9.3.3 Analytical expression of Nucci and colleagues	71
		2.9.3.4 Analytical expression of Diendorfer and Uman	71
		2.9.3.5 Analytical expression of Delfino and colleagues	72
		2.9.3.6 Analytical expression of Cooray and colleagues	72
	2.9.4	Current wave shapes of upward-initiated flashes	72
2.10	Electric fields from first and subsequent strokes		74
2.11	Peak electric radiation fields of first and subsequent strokes		81
2.12	Continuing currents		84
2.13	M-components		88
References			88

3 Rocket-triggered lightning and new insights into lightning protection gained from triggered-lightning experiments   97
V.A. Rakov

3.1	Introduction	97
3.2	Triggering techniques	98
	3.2.1 Classical triggering	98
	3.2.2 Altitude triggering	103
	3.2.3 Triggering facility at Camp Blanding, Florida	105
3.3	Overall current waveforms	107
	3.3.1 Classical triggering	107
	3.3.2 Altitude triggering	109
3.4	Parameters of return-stroke current waveforms	110
3.5	Return-stroke current peak versus grounding conditions	122
3.6	Characterization of the close lightning electromagnetic environment	128
3.7	Studies of interaction of lightning with various objects and systems	131

	3.7.1	Overhead power distribution lines	131
		3.7.1.1 Nearby strikes	131
		3.7.1.2 Direct strikes	135
	3.7.2	Underground cables	143
	3.7.3	Power transmission lines	143
	3.7.4	Residential buildings	144
	3.7.5	Airport runway lighting system	144
	3.7.6	Miscellaneous experiments	149
3.8	Concluding remarks		150
References			150
Bibliography			159

4 Attachment of lightning flashes to grounded structures — 165
Vernon Cooray and Marley Becerra

4.1	Introduction		165
	4.1.1	The protection angle method	166
	4.1.2	The electro-geometrical method	168
	4.1.3	The rolling sphere method	170
	4.1.4	The mesh method	175
4.2	Striking distance to flat ground		176
	4.2.1	Golde	177
	4.2.2	Eriksson	177
	4.2.3	Dellera and Garbagnati	179
	4.2.4	Cooray and colleagues	179
	4.2.5	Armstrong and Whitehead	180
4.3	Striking distance to elevated structures		182
	4.3.1	Positive leader discharges	183
	4.3.2	Leader inception models	186
		4.3.2.1 The critical radius concept	186
		4.3.2.2 Rizk's generalized leader inception equation	190
		4.3.2.3 Critical streamer length concept	194
		4.3.2.4 Bazelyan and Raizer's empirical leader model	195
		4.3.2.5 Lalande's stabilization field equation	196
		4.3.2.6 The self-consistent leader inception model of Becerra and Cooray	197
4.4	The leader progression model		212
	4.4.1	The basic concept of the leader progression model	212
	4.4.2	The leader progression model of Eriksson	212
	4.4.3	The leader progression model of Dellera and Garbagnati	215
	4.4.4	The leader progression model of Rizk	217
	4.4.5	Attempts to validate the existing leader progression models	220
	4.4.6	Critical overview of the assumptions of leader progression models	223
		4.4.6.1 Orientation of the stepped leader	223

			4.4.6.2	Leader inception criterion	224
			4.4.6.3	Parameters and propagation of the upward connecting leader	225
			4.4.6.4	Effects of leader branches and tortuosity	226
			4.4.6.5	Thundercloud electric field	226
		4.4.7	Becerra and Cooray leader progression model		227
			4.4.7.1	Basic theory	227
			4.4.7.2	Self-consistent lightning interception model (SLIM)	231
	4.5	Non-conventional lightning protection systems			239
		4.5.1	The early streamer emission concept		240
			4.5.1.1	Experimental evidence in conflict with the concept of ESE	241
			4.5.1.2	Theoretical evidence in conflict with the concept of ESE	243
		4.5.2	The concept of dissipation array systems (DAS)		251
			4.5.2.1	Experimental evidence against dissipation arrays	253
	References				260

5 Protection against lightning surges — 269
Rajeev Thottappillil and Nelson Theethayi

5.1	Introduction				269
	5.1.1	Direct strike to power lines			269
	5.1.2	Lightning activity in the vicinity of networks			270
5.2	Characteristics of lightning transients and their impact on systems				273
	5.2.1	Parameters of lightning current important for surge protection design			274
			5.2.1.1	Peak current	274
			5.2.1.2	Charge transferred	274
			5.2.1.3	Prospective energy	275
			5.2.1.4	Waveshape	276
	5.2.2	Parameters of lightning electric and magnetic fields important for surge protection design			276
5.3	Philosophy of surge protection				278
	5.3.1	Surge protection as part of achieving electromagnetic compatibility			278
			5.3.1.1	Conductor penetration through a shield	279
	5.3.2	Principle of surge protection			280
			5.3.2.1	Gas discharge tubes (spark gaps)	282
			5.3.2.2	Varistors	285
			5.3.2.3	Diodes and thyristors	286
			5.3.2.4	Current limiters	288
			5.3.2.5	Isolation devices	290
			5.3.2.6	Filters	291

				Contents	xi

		5.3.2.7	Special protection devices used in power distribution networks	293
5.4	Effects of parasitic elements in surge protection and filter components			294
	5.4.1	Capacitors		295
	5.4.2	Inductors		297
	5.4.3	Resistors		298
		5.4.3.1	Behaviour of resistors at various frequencies	299
5.5	Surge protection coordination			302
References				303

6 External lightning protection system — 307
Christian Bouquegneau

6.1	Introduction			307
6.2	Air-termination system			308
	6.2.1	Location of air-terminations on the structure		308
		6.2.1.1	Positioning of the air-termination system utilizing the rolling sphere method	310
		6.2.1.2	Positioning of the air-termination system utilizing the mesh method	311
		6.2.1.3	Positioning of the air-termination system utilizing the protection angle method	313
		6.2.1.4	Comparison of methods for the positioning of the air-termination system	316
	6.2.2	Construction of air-termination systems		316
	6.2.3	Non-conventional air-termination systems		320
6.3	Down-conductor system			320
	6.3.1	Location and positioning of down-conductors on the structure		320
	6.3.2	Construction of down-conductor systems		322
	6.3.3	Structure with a cantilevered part		325
	6.3.4	Joints, connections and test joints in down-conductors		326
	6.3.5	Lightning equipotential bonding		328
	6.3.6	Current distribution in down-conductors		330
6.4	Earth-termination system			330
	6.4.1	General principles		330
	6.4.2	Earthing arrangements in general conditions		332
	6.4.3	Examples of earthing arrangements in common small structures		335
	6.4.4	Special earthing arrangements		341
		6.4.4.1	Earth electrodes in rocky and sandy soils	341
		6.4.4.2	Earth-termination systems in large areas	341
		6.4.4.3	Artificial decrease of earth resistance	342
	6.4.5	Effect of soil ionization and breakdown		343

xii Contents

		6.4.6	Touch voltages and step voltages	343
			6.4.6.1 Touch voltages	343
			6.4.6.2 Step voltages	344
		6.4.8	Measurement of soil resistivity and earth resistances	345
			6.4.8.1 Measurement of soil resistivity	345
			6.4.8.2 Measurement of earth resistances	348
	6.5	Selection of materials		348
	Acknowledgements			352
	References			353

7 Internal lightning protection system — 355
Peter Hasse and Peter Zahlmann

7.1 Damage due to lightning and other surges — 355
   7.1.1 Damage in hazardous areas — 357
   7.1.2 Lightning damage in city areas — 357
   7.1.3 Damage to airports — 360
   7.1.4 Consequences of lightning damage — 361
7.2 Protective measures — 361
   7.2.1 Internal lightning protection — 363
      7.2.1.1 Equipotential bonding in accordance with IEC 60364-4-41 and -5-54 — 363
      7.2.1.2 Equipotential bonding for a low-voltage system — 364
      7.2.1.3 Equipotential bonding for information technology system — 364
   7.2.2 Lightning protection zones concept — 365
   7.2.3 Basic protection measures: earthing magnetic shielding and bonding — 366
      7.2.3.1 Magnetic shielding — 366
      7.2.3.2 Cable shielding — 369
      7.2.3.3 Equipotential bonding network — 371
      7.2.3.4 Equipotential bonding on the boundaries of lightning protection zones — 373
      7.2.3.5 Equipotential bonding at the boundary of LPZ 0_A and LPZ 2 — 380
      7.2.3.6 Equipotential bonding on the boundary of LPZ 1 and LPZ 2 and higher — 383
      7.2.3.7 Coordination of the protective measures at various LPZ boundaries — 385
      7.2.3.8 Inspection and maintenance of LEMP protection — 388
7.3 Surge protection for power systems: power supply systems (within the scope of the lightning protection zones concept according to IEC 62305-4) — 389
   7.3.1 Technical characteristics of SPDs — 391
      7.3.1.1 Maximum continuous voltage U_C — 391

		7.3.1.2	Impulse current I_{imp}	392
		7.3.1.3	Nominal discharge current I_n	392
		7.3.1.4	Voltage protection level U_p	392
		7.3.1.5	Short-circuit withstand capability	392
		7.3.1.6	Follow current extinguishing capability at U_C (I_{fi})	392
		7.3.1.7	Follow current limitation (for spark-gap based SPDs class I)	393
		7.3.1.8	Coordination	393
		7.3.1.9	Temporary overvoltage	393
	7.3.2	Use of SPDs in various systems		393
	7.3.3	Use of SPDs in TN systems		396
	7.3.4	Use of SPDs in TT systems		398
	7.3.5	Use of SPDs in IT systems		401
	7.3.6	Rating the lengths of connecting leads for SPDs		401
		7.3.6.1	Series connection (V-shape) in accordance with IEC 60364-5-53	402
		7.3.6.2	Parallel connection system in accordance with IEC 60364-5-53	403
		7.3.6.3	Design of the connecting lead on the earth side	405
		7.3.6.4	Design of the phase-side connecting cables	406
	7.3.7	Rating of cross-sectional areas and backup protection of SPDs		410
		7.3.7.1	Selectivity to the protection of the installation	416
7.4	Surge protection for telecommunication systems			417
	7.4.1	Procedure for selection and installation of arresters: example BLITZDUCTOR CT		417
		7.4.1.1	Technical data	418
	7.4.2	Measuring and control systems		421
		7.4.2.1	Electrical isolation using optocouplers	422
7.5	Examples for application			423
	7.5.1	Lightning and surge protection of wind turbines		423
		7.5.1.1	Positive prognoses	424
		7.5.1.2	Danger resulting from lightning effects	424
		7.5.1.3	Frequency of lightning strokes	424
		7.5.1.4	Standardization	425
		7.5.1.5	Protection measures	425
		7.5.1.6	Shielding measures	426
		7.5.1.7	Earth-termination system	426
	7.5.2	Lightning and surge protection for photovoltaic systems and solar power plants		428
		7.5.2.1	Lightning and surge protection for photovoltaic systems	428
		7.5.2.2	Lightning and surge protection for solar power plants	434

xiv Contents

        7.5.3 Surge protection for video surveillance systems   438
                7.5.3.1 Video surveillance systems   439
                7.5.3.2 Choice of SPDs   439
    Bibliography   441

8 Risk analysis   443
Z. Flisowski and C. Mazzetti

  8.1 General considerations   443
  8.2 General concept of risk due to lightning   445
  8.3 Number of strikes to a selected location   447
  8.4 Damage probabilities   452
  8.5 Simplified practical approach to damage probability   457
  8.6 Question of relative loss assessment   458
  8.7 Concept of risk components   459
  8.8 Standardized procedure of risk assessment   461
        8.8.1 Basic relations   461
        8.8.2 Evaluation of risk components   464
        8.8.3 Risk reduction criteria   468
  8.9 Meaning of subsequent strokes in a flash   470
  8.10 Final remarks and conclusions   471
  References   472

9 Low-frequency grounding resistance and lightning protection   475
Silverio Visacro

  9.1 Introduction   475
  9.2 Basic considerations about grounding systems   475
  9.3 The concept of grounding resistance   476
  9.4 Grounding resistance of some simple electrode arrangements   479
  9.5 Relation of the grounding resistance to the experimental response of electrodes to lightning currents   482
  9.6 Typical arrangements of grounding electrodes for some relevant applications and their lightning-protection-related requirements   487
        9.6.1 Transmission lines   487
        9.6.2 Substations   490
        9.6.3 Lightning protection systems   494
        9.6.4 Overhead distribution lines   496
  9.7 Conclusion   497
  References   499

10 High-frequency grounding   503
Leonid Grcev

  10.1 Introduction   503
  10.2 Basic circuit concepts   503
  10.3 Basic field considerations   506
  10.4 Frequency-dependent characteristics of the soil   509

	10.5	Grounding modelling for high frequencies	511
	10.6	Frequency-dependent grounding behaviour	514
	10.7	Frequency-dependent dynamic grounding behaviour	521
	10.8	Relation between frequency-dependent and non-linear grounding behaviour	526
	References		527
11	**Soil ionization**		**531**
	Vernon Cooray		
	11.1	Introduction	531
	11.2	Critical electric field necessary for ionization in soil	534
	11.3	Various models used in describing soil ionization	536
		11.3.1 Ionized region as a perfect conductor	536
		11.3.2 Model of Liew and Darveniza	537
		11.3.2.1 Application of the model to a single driven rod	538
		11.3.3 Model of Wang and colleagues	540
		11.3.4 Model of Sekioka and colleagues	543
		11.3.5 Model of Cooray and colleagues	545
		11.3.5.1 Mathematical description	546
		11.3.5.2 Model parameters	548
	11.4	Conclusions	550
	References		550
12	**Lightning protection of low-voltage networks**		**553**
	Alexandre Piantini		
	12.1	Introduction	553
	12.2	Low-voltage networks	554
		12.2.1 Typical configurations and earthing practices	554
		12.2.2 Distribution transformers	558
		12.2.3 Low-voltage power installations	564
	12.3	Lightning surges on low-voltage systems	568
		12.3.1 Direct strikes	568
		12.3.2 Cloud discharges	570
		12.3.3 Indirect strikes	571
		12.3.3.1 Calculation of lightning-induced voltages	573
		12.3.3.2 Sensitivity analysis	578
		12.3.4 Transference from the medium-voltage line	601
		12.3.4.1 Direct strikes	601
		12.3.4.2 Indirect strikes	605
	12.4	Lightning protection of LV networks	611
		12.4.1 Distribution transformers	612
		12.4.2 Low-voltage power installations	614
	12.5	Concluding remarks	624
	References		625

13 Lightning protection of medium voltage lines — 635
C.A. Nucci and F. Rachidi

- 13.1 Introduction — 635
- 13.2 Lightning strike incidence to distribution lines — 636
  - 13.2.1 Expected number of direct lightning strikes — 638
  - 13.2.2 Shielding by nearby objects — 639
- 13.3 Typical overvoltages generated by direct and indirect lightning strikes — 640
  - 13.3.1 Direct overvoltages — 640
  - 13.3.2 Induced overvoltages — 642
- 13.4 Main principles in lightning protection of distribution lines — 645
  - 13.4.1 Basic impulse insulation level and critical impulse flashover voltage — 646
  - 13.4.2 Shield wires — 648
  - 13.4.3 Protective devices — 650
    - 13.4.3.1 General considerations using protective devices — 650
    - 13.4.3.2 Spark gaps — 651
    - 13.4.3.3 Surge arresters — 652
    - 13.4.3.4 Capacitors — 653
- 13.5 Lightning protection of distribution systems — 653
  - 13.5.1 Effect of the shield wire — 653
  - 13.5.2 Effect of surge arresters — 655
  - 13.5.3 Lightning performance of MV distribution lines — 660
    - 13.5.3.1 Effect of soil resistivity — 662
    - 13.5.3.2 Effect of the presence of shield wires — 662
    - 13.5.3.3 Effect of the presence of surge arresters — 666
- Appendix A13 Procedure to calculate the lightning performance of distribution lines according to IEEE Std. 1410-2004 (from Reference 4) — 666
- Appendix B13 The LIOV-Monte Carlo (LIOV–MC) procedure to calculate the lightning performance of distribution lines (from Reference 56) — 668
- Appendix C13 The LIOV code: models and equations — 670
  - C13.1 Agrawal and colleagues field-to-transmission line coupling equations extended to the case of multiconductor lines above a lossy earth — 671
  - C13.2 Lightning-induced voltages on distribution networks: LIOV code interfaced with EMTPrv — 675
- Acknowledgements — 675
- References — 675

Contents   xvii

14 Lightning protection of wind turbines — **681**
Troels Soerensen

- 14.1 Introduction — 681
- 14.2 Nature of the lightning threat to wind turbines — 686
- 14.3 Statistics of lightning damage to wind turbines — 687
- 14.4 Risk assessment and cost–benefit evaluation — 688
- 14.5 Lightning protection zoning concept — 691
- 14.6 Earthing and equipotential bonding — 693
- 14.7 Protection of wind turbine components — 695
  - 14.7.1 Blades — 695
    - 14.7.1.1 Blades with carbon fibre — 699
    - 14.7.1.2 Guidelines, quality assurance and test methods — 700
  - 14.7.2 Hub — 702
  - 14.7.3 Nacelle — 703
  - 14.7.4 Tower — 703
  - 14.7.5 Bearings and gears — 704
  - 14.7.6 Hydraulic systems — 705
  - 14.7.7 Electrical systems, control and communication systems — 706
    - 14.7.7.1 Electrical systems — 707
    - 14.7.7.2 Generator circuit — 707
    - 14.7.7.3 Medium-voltage system — 709
    - 14.7.7.4 Auxiliary power circuit(s) — 709
    - 14.7.7.5 Control and communication systems — 711
- 14.8 Wind farm considerations — 713
- 14.9 Off-shore wind turbines — 714
- 14.10 Lightning sensors and registration methods — 716
- 14.11 Construction phase and personnel safety — 717
- References — 718

15 Lightning protection of telecommunication towers — **723**
G.B. Lo Piparo

- 15.1 Lightning as a source of damage to broadcasting stations — 723
  - 15.1.1 General — 723
  - 15.1.2 Injury to people — 725
  - 15.1.3 Physical damage — 725
    - 15.1.3.1 Thermal effects — 725
    - 15.1.3.2 Electrodynamic effects — 726
  - 15.1.4 Failure of internal electrical and electronic systems — 726
- 15.2 Effects of lightning flashes to the broadcasting station — 726
  - 15.2.1 Effects of lightning flashes to the antenna support structure — 726
    - 15.2.1.1 Lightning current flowing through external conductive parts and lines connected to the station — 729

		15.2.1.2 Potential differences between different parts of the earth-termination system of the station	730
15.3	Lightning flashes affecting the power supply system		731
	15.3.1	Power supply by overhead lines	731
		15.3.1.1 Lightning flashes to an overhead line	731
		15.3.1.2 Lightning flash to ground near an overhead line	733
		15.3.1.3 Surges at the point of entry of an overhead line in the station	733
		15.3.1.4 Overhead line connected directly to the transformer	734
		15.3.1.5 Buried cable connection between the overhead line and the transformer	734
	15.3.2	Power supply by underground cable	735
	15.3.3	Transfer of overvoltages across the transformer	735
15.4	The basic principles of lightning protection		736
	15.4.1	The protection level to be provided	736
	15.4.2	Basic criteria for protection of stations	737
	15.4.3	Protection measures	738
	15.4.4	Procedure for selection of protection measures	739
	15.4.5	Implementation of protection measures	739
15.5	Erection of protection measures to reduce injury of living beings		742
	15.5.1	Protection measures against step voltages	742
	15.5.2	Protection measures against touch-voltages	742
15.6	Erection of the LPS to reduce physical damage		743
	15.6.1	Air-termination system	743
	15.6.2	Down-conductors system	743
	15.6.3	Earth-termination system	745
		15.6.3.1 Earth-termination system for stations with an autonomous power supply	747
		15.6.3.2 Earth-termination system for stations powered by an external source	749
	15.6.4	Protection against corrosion	753
	15.6.5	Earthing improvement	753
	15.6.6	Foundation earth electrode	754
15.7	Potential equalization to reduce failures of electrical and electronic systems		757
	15.7.1	Potential equalization for the earth-termination system of the station	757
	15.7.2	Potential equalization for the antenna support structure	761
	15.7.3	Potential equalization for the equipment within the building	761
	15.7.4	Potential equalization for metallic objects outside the building	763

15.8	Screening to reduce failures of electrical and electronic systems			763
	15.8.1	Screening of circuits within the building		763
	15.8.2	Circuits of the station entering the building		764
15.9	Coordinated SPD protection system to reduce failures of electrical and electronic systems			766
	15.9.1	Selection of SPDs with regard to voltage protection level		766
	15.9.2	Selection of SPD with regard to location and to discharge current		766
	15.9.3	Installation of SPDs in a coordinated SPD protection system		767
		15.9.3.1	Protective distance l_P	767
		15.9.3.2	Induction protective distance l_{Pi}	768
15.10	Protection of lines and services entering the station			768
	15.10.1	Overhead lines		769
	15.10.2	Screened cables		770
15.11	Arrangement of power supply circuits			771
	15.11.1	Stations supplied at high voltage		771
	15.11.2	Stations supplied at low voltage		773
	15.11.3	Stations with self-contained power supplies only		776
Annex A15:	Surge testing of installations			776
	A15.1	Simulation of surge phenomena		776
		A15.1.1	General	776
		A15.1.2	Tests for determining overvoltages due to lightning flashes to the antenna-support structure	777
		A15.1.3	Determination of the transferred overvoltages	777
		A15.1.4	Results	778
	A15.2	Description of a typical test programme		779
		A15.2.1	Introduction	779
		A15.2.2	The power supply arrangements of the station	780
		A15.2.3	The test equipment	781
		A15.2.4	The test procedure and results	781
			A15.2.4.1 Measurement of the overvoltages due to lightning flashes to the antenna-support structure	783
			A15.2.4.2 Measurement of the transferred overvoltages	783
	A15.3	Discussion of the results		784
		A15.3.1	Lightning flashes to the antenna-support structure	786
		A15.3.2	Overvoltages transferred by the power supply line	786

xx Contents

		A15.4	Conclusions	787
	Acknowledgements			787
	References			787

16 Lightning protection of satellite launch pads — 789
Udaya Kumar

- 16.1 Introduction — 789
- 16.2 Structure of a rocket — 790
- 16.3 Launch pad — 791
  - 16.3.1 Launch campaign and duration of exposure — 792
- 16.4 Lightning threat to launch vehicle — 792
  - 16.4.1 Limitations of present-day knowledge in quantifying the risk — 793
- 16.5 Lightning protection systems — 794
  - 16.5.1 External protection — 794
    - 16.5.1.1 Brief description of some of the present protection schemes — 796
  - 16.5.2 Principles used for the design of the external protection system — 798
    - 16.5.2.1 Air termination network — 798
    - 16.5.2.2 Earth termination — 799
    - 16.5.2.3 Down-conductor system — 799
  - 16.5.3 Internal protection — 799
    - 16.5.3.1 Launch vehicle — 799
    - 16.5.3.2 Vehicle on launch pad — 801
- 16.6 Weather launch commit criteria — 802
- 16.7 Review of present status and suggested direction for further work — 803
  - 16.7.1 Attachment process — 803
  - 16.7.2 Lightning surge response — 806
    - 16.7.2.1 Earth termination — 806
    - 16.7.2.2 Down-conductor system — 807
  - 16.7.3 Weather launch commit criteria — 813
- 16.8 Indirect effects — 813
- 16.9 Protection of other supporting systems — 814
- 16.10 On-site measurements — 814
- 16.11 Summary — 815
- References — 816

17 Lightning protection of structures with risk of fire and explosion — 821
Arturo Galván Diego

- 17.1 Introduction — 821
- 17.2 Tanks and vessels containing flammable materials — 822
  - 17.2.1 General — 822
  - 17.2.2 Risk assessment — 824

		17.2.3	Lightning protection measures	825
			17.2.3.1 Air terminations	826
			17.2.3.2 Equipotential bonding	828
			17.2.3.3 A clean environment	829
			17.2.3.4 Self-protecting system	829
			17.2.3.5 Resumé for lightning protection	830
	17.3	Offshore oil platforms		830
		17.3.1	General	830
		17.3.2	Relevant standards	831
		17.3.3	Risk assessment	832
		17.3.4	Lightning protection measures	832
			17.3.4.1 External lightning protection	832
			17.3.4.2 Grounding system and common bonding network	834
			17.3.4.3 Internal grounding system	835
			17.3.4.4 Shielding	838
			17.3.4.5 Location of SPD	839
	References			839

18 Lightning and trees 843
Mahendra Fernando, Jakke Mäkelä and Vernon Cooray

18.1	Introduction			843
18.2	Strike and damage probability of lightning to trees			844
18.3	Types of lightning damage			846
	18.3.1	Microscale damage		846
	18.3.2	Macroscale damage		847
		18.3.2.1	No physical damage	847
		18.3.2.2	Bark-loss damage	849
		18.3.2.3	Wood-loss damage	849
		18.3.2.4	Explosive damage	849
		18.3.2.5	Ignition	850
	18.3.3	Other damage scenarios		852
		18.3.3.1	Long-term propagation of damage	853
		18.3.3.2	Group damage to trees	853
		18.3.3.3	Damage to ground	853
		18.3.3.4	Damage to vegetation	854
18.4	Protection of trees			854
18.5	Conclusions			855
Acknowledgements				855
References				855

19 Lightning warning systems 859
Martin J. Murphy, Kenneth L. Cummins and Ronald L. Holle

19.1	Introduction	859

xxii *Contents*

	19.2	Thunderstorm lifecycle and associated detection methods		860
		19.2.1	Thunderstorm life cycle	860
			19.2.1.1 Convective development and electrification	860
			19.2.1.2 Early stages of lightning activity	861
			19.2.1.3 Late stages of lightning activity	861
		19.2.2	Associated detection methods	862
			19.2.2.1 Detection of initial electrification	862
			19.2.2.2 Single-point lightning detection sensors	863
			19.2.2.3 Lightning detection networks	864
	19.3	Examples of warning systems		864
		19.3.1	Fixed-point warning applications	864
		19.3.2	Storm-following algorithms	867
	19.4	Warning system performance measures		869
		19.4.1	Performance metrics	869
			19.4.1.1 Performance metrics for fixed-point algorithms	870
			19.4.1.2 Performance metrics for storm-following algorithms	873
		19.4.2	Specific challenges at different stages of the warning problem	875
			19.4.2.1 Lightning onset	875
			19.4.2.2 Lightning cessation	876
	19.5	Application of performance measures to cloud-to-ground warning systems		879
		19.5.1	Assessment of a fixed-point warning algorithm	879
			19.5.1.1 Effects of lightning detection technology	879
			19.5.1.2 Effects of algorithm configuration using a single detection technology	884
		19.5.2	Lightning cessation in MCS cases	884
		19.5.3	Radar applications for lightning onset in storm-following algorithms	885
	19.6	Assessing the risks		887
		19.6.1	Decision making	887
		19.6.2	Equipment protection application	891
		19.6.3	Trade-offs between performance and risks for cloud-to-ground warning in safety applications	892
			19.6.3.1 Personal and small-group warning	892
			19.6.3.2 Large venue warning	893
	References			894
20	**Lightning-caused injuries in humans**			**901**
	Vernon Cooray, Charith Cooray and Christopher Andrews			
	20.1	Introduction		901
	20.2	The different ways in which lightning can interact with humans		902

	20.3	Different types of injuries	906
		20.3.1 Injuries to the respiratory and cardiovascular system	906
		20.3.2 Injuries to the eye	909
		20.3.3 Ear	910
		20.3.4 Nervous system	912
		20.3.5 Skin and burn injuries	913
		20.3.6 Psychological	914
		20.3.7 Blunt injuries	914
		20.3.8 Disability caused by lightning	915
		20.3.9 Remote injuries	915
		20.3.10 Lightning electromagnetic fields	917
	20.4	Concluding remarks	920
	References		920
21	**Lightning standards**		**925**
	Fridolin Heidler and E.U. Landers		
	21.1	Introduction	925
	21.2	Standardized lightning currents	926
		21.2.1 Threat parameters of the lightning current	926
		21.2.2 Current waveforms	928
		21.2.3 Requirements for the current tests	930
	21.3	Determination of possible striking points	931
		21.3.1 Rolling sphere method	931
		21.3.2 Mesh method	933
		21.3.3 Protection angle method	933
	21.4	The lightning protection system (LPS)	934
		21.4.1 Air termination system	934
		21.4.2 Down-conductor system	935
		21.4.3 Earth termination system	935
		21.4.4 Lightning equipotential bonding	937
		21.4.5 Separation distance	938
	21.5	The LEMP protection measures system (LPMS)	939
		21.5.1 The lightning protection zones (LPZ) concept	939
		21.5.2 Earthing system and bonding network	941
		21.5.3 Line routing and shielding	942
		21.5.4 Coordinated surge protection device application	942
		21.5.5 Spatial magnetic shielding	942
	21.6	Conclusions	944
	References		946
22	**High-voltage and high-current testing**		**947**
	Wolfgang Zischank		
	22.1	Introduction	947
	22.2	Lightning test equipment	948
		22.2.1 High-voltage impulse test generators	948

xxiv Contents

			22.2.1.1	Single-stage impulse voltage circuits	950
			22.2.1.2	Multistage impulse voltage circuits	953
		22.2.2	High-current test generators		956
			22.2.2.1	Simulation of first return stroke effects	957
			22.2.2.2	Simulation of subsequent return stroke effects	967
			22.2.2.3	Generation of long-duration currents	970
			22.2.2.4	Current injection for direct effects testing	971
		22.2.3	Indirect effects testing		972
	22.3	Measurement techniques			974
		22.3.1	Measurement of impulse voltages		974
		22.3.2	Measurement of impulse currents		975
			22.3.2.1	Resistive shunts	975
			22.3.2.2	Rogowski coils	976
			22.3.2.3	Current monitors	977
	References				978

23 Return stroke models for engineering applications — 981
Vernon Cooray

23.1	Introduction			981
23.2	Current propagation models (CP models)			983
	23.2.1	Basic concept		983
	23.2.2	Most general description		984
23.3	Current generation models (CG models)			986
	23.3.1	Basic concept		986
	23.3.2	Mathematical background		988
		23.3.2.1	Evaluate $I_b(t)$ given $\rho(z)$, $\tau(z)$ and $v(z)$	988
		23.3.2.2	Evaluate $\tau(z)$ given $I_b(t)$, $\rho(z)$ and $v(z)$	989
		23.3.2.3	Evaluate $\rho(z)$ given $I_b(t)$, $\tau(z)$ and $v(z)$	990
		23.3.2.4	Evaluate $v(z)$, given $I_b(t)$, $\rho(z)$ and $\tau(z)$	990
	23.3.3	CG models in practice		990
		23.3.3.1	Model of Wagner	991
		23.3.3.2	Model of Heidler	991
		23.3.3.3	Model of Hubert	992
		23.3.3.4	Model of Cooray	993
		23.3.3.5	Model of Diendofer and Uman	994
		23.3.3.6	First modification of the Diendofer and Uman model by Thottappillil *et al.*	996
		23.3.3.7	Second modification of the Diendofer and Uman model by Thottappillil and Uman	997
		23.3.3.8	Model of Cooray	998
		23.3.3.9	Model of Cooray and Rakov	999
		23.3.3.10	Model of Cooray, Rakov and Montano	1000
23.4	Current dissipation models (CD Models)			1001
	23.4.1	General description		1001

	23.4.2	Mathematical background	1003
	23.4.3	Cooray and Rakov model – a combination of current dissipation and current generation models	1004
23.5		Generalization of any model to the current generation type	1006
23.6		Generalization of any model to the current dissipation type	1008
23.7		Current dissipation models and the modified transmission line models	1009
23.8		Effect of ground conductivity	1010
23.9		Equations necessary to calculate the electric and magnetic fields	1012
23.10		Concluding remarks	1015
References			1016

Index   **1019**

Contributors

Christopher Andrews
Indooroopilly Medical Centre,
Indooroopilly, Queensland,
Australia

Marley Becerra
ABB Corporate Research, Västerås,
Sweden

Christian Bouquegneau
Polytechnical University of Mons,
Belgium

Charith Cooray
Karolinska Institute, Stockholm,
Sweden

Vernon Cooray
Uppsala University, Uppsala,
Sweden

Kenneth L. Cummins
University of Arizona, Tucson,
Arizona, USA

Arturo Galván Diego
Electrical Research Institute of Mexico,
Mexico

Mahendra Fernando
University of Colombo, Colombo,
Sri Lanka

Zdobyslaw Flisowski
Warsaw University of Technology,
Poland

Leonid Grcev
University of Skopje, Skopje,
Macedonia

Peter Hasse
DEHN + SÖHNE, Neumarkt,
Germany

Fridolin Heidler
University of the Federal
Armed Forces, Munich, Germany

Ronald L. Holle
Holle Meteorology & Photography,
Oro Valley, Arizona, USA

Philip Krider
University of Arizona, Tucson,
Arizona, USA

Udaya Kumar
Indian Institute of Science,
Bangalore, India

Giovanni Battista Lo Piparo
RAI – Radiotelevisione Italiana,
Rome, Italy

Jakke Mäkelä
Nokia Devices R&D, Salo, Finland

Carlo Mazzetti
University of Rome 'La Sapienza',
Rome, Italy

Martin J. Murphy
Vaisala, Tucson, Arizona, USA

Carlo Alberto Nucci
University of Bologna, Bologna, Italy

Alexandre Piantini
University of São Paulo,
São Paulo, Brazil

Farhad Rachidi
Swiss Federal Institute of
Technology, Lausanne,
Switzerland

Vladimir A. Rakov
University of Florida, Gainesville,
Florida, USA

Troels Stybe Sorensen
DONG Energy, Copenhagen,
Denmark

Nelson Theethayi
Bombadier Transportation Sweden,
Västerås, Sweden

Rajeev Thottappillil
Royal Institute of Technology (KTH),
Stockholm, Sweden

Silverio Visacro
Federal University of Minas Gerais,
Brazil

Peter Zahlmann
DEHN + SÖHNE, Neumarkt,
Germany

Wolfgang Zischank
University of the Federal
Armed Forces, Munich, Germany

Preface

Lightning protection relies upon the application of some of the principles of electricity and the physics of electrical discharges to mitigate the effects of direct currents and electromagnetic fields generated by lightning discharges. Structures, storage facilities for flammable and explosive materials, power distribution and transmission systems, telecommunication systems and electrical and electronic equipment all require such protection. Since the initial launch of the concept of lightning protection by Benjamin Franklin in 1753, the subject of lightning protection has made significant progress, especially in the last century, thanks to experimental observations of the mechanism and properties of lightning flashes. This book summarises the state of the art of lightning protection as it stands today. The information provided in this book should be of value to professionals who are engaged in the engineering practice of lightning protection as a source of reference and to engineering students as a textbook.

The main goal of the book is not solely to educate the reader in the art of lightning protection, but to provide the necessary scientific background to enable him or her to make appropriate judgments in situations where conventional engineering solutions might be inadequate. Many engineers engaged in lightning protection have learned their work by applying lightning protection standards without the requisite information being provided to them on the reasons why they might select a particular solution to a problem under consideration instead of another one. However, several companies have been introducing fraudulent devices, claiming them to be superior to more conventional protection equipment and procedures, taking advantage of a gap in the knowledge of lightning protection engineers in decision-making positions. It is only through the provision of a thorough education to the engineers examining the basic scientific problems associated with lightning protection that one can remedy this situation. This book is intended to provide such an education to satisfy the needs of those working or studying in the field of lightning protection.

Vernon Cooray

Acknowledgements

I am indebted to the late Prof. James Wait who introduced me as an author to the Institution of Engineering and Technology, London. I wish to thank all my colleagues who have spent a good deal of their free time writing the chapters of this book.

I wish to express my sincere thanks to Ms Lisa Reading of the IET's publications department, who was at all times prepared to listen to my suggestions and accommodated submission delays without any complaints.

My sincere thanks to the fund from the B. John F. and Svea Andersson donation at Uppsala University and to the Swedish Natural Science Research Council (Vätenskapsrådet) for supporting my research work during the past 30 years at Uppsala University; without this support I would not be in a position to be the editor of this book.

Finally, I express my gratitude to my wife Ruby Cooray and two sons, Gerald Cooray and Charith Cooray for their encouragement and patience during the organisation of this book.

Vernon Cooray

Chapter 1
Benjamin Franklin and lightning rods*
E. Philip Krider

Benjamin Franklin's work on electricity and lightning earned him worldwide fame and respect – ideal assets for brokering aid from France during the American Revolution.

On 10 May 1752, as a thunderstorm passed over the village of Marly-la-Ville near Paris, a retired French dragoon, acting on instructions from naturalist Thomas-François Dalibard, drew sparks from a tall iron rod that had been carefully insulated from ground (see Figure 1.1). The sparks showed that thunderclouds are electrified and that lightning is an electrical discharge. In the mid-eighteenth century, such an observation was sensational, and it was soon verified by Delor, Dalibard's collaborator; within weeks many others throughout Europe had successfully repeated the experiment [1,2].

When Dalibard and Delor reported their results to the Académie des Sciences in Paris three days later, they acknowledged that in doing these experiments, they had merely followed a path that Benjamin Franklin had traced for them. In June of 1752, shortly after the experiment at Marly-la-Ville but before he knew about it, Franklin drew sparks himself from a key attached to the conducting string of his famous electrical kite that was insulated from ground by a silk ribbon.

The French results were important because they called attention to Franklin's small pamphlet entitled *Experiments and Observations on Electricity, made at Philadelphia in America* [3], which helped to stimulate other work in electricity and contributed to the beginning of modern physics [4]. The observations also validated the key assumptions that lay behind Franklin's supposition that tall, grounded rods will protect buildings from lightning damage.

---

*Reprinted with permission from E. Philip Krider, Physics Today, January 2006, page 42. Copyright 2006, American Institute of Physics.

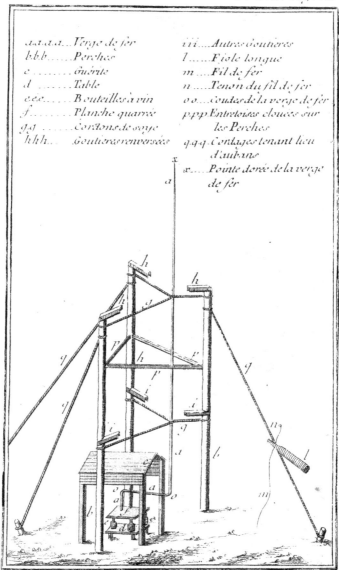

Figure 1.1 This sketch of the 'sentry box' experiment conducted at Marly-la-Ville, France, in 1752 was based on Benjamin Franklin's proposal to determine whether thunderclouds are electrified. Silk ropes (g) and wine bottles (e) insulated a 13 m iron rod from ground, and covers (h) sheltered the ropes from rain. A person standing on the ground could draw sparks from the rod or charge a Leyden jar when a storm was in the area (from Reference 19).

1.1 A Philadelphia story

Franklin performed his initial experiments on electricity in collaboration with friends and neighbours, including Thomas Hopkinson, a lawyer and judge, Ebenezer Kinnersley, a clergyman and teacher, and Philip Syng, Jr, a master silversmith. Franklin described the experiments and their results in five formal letters to Peter Collinson, a fellow of the Royal Society of London, in the years from 1747 to 1750, and Collinson in turn communicated those letters to the Society and published them in April of 1751.

In his first letter [5], Franklin described 'the wonderful Effect of Points, both in *drawing* off and *throwing* off the Electrical Fire'. He showed that points work quickly at 'a considerable Distance', that sharp points work better than blunt ones, that metal points work better than dry wood, and that the pointed object should be touched – that is, grounded – to obtain the maximum draw effect.

Next, Franklin introduced the idea that rubbing glass with wool or silk does not actually create electricity; rather, at the moment of friction, the glass simply takes 'the Electrical Fire' out of the rubbing material; whatever amount is added to the glass, an equal amount is lost by the wool or silk. The terms plus and minus were used to describe those electrical states, and the glass was assumed to be electrified positively and the rubbing material negatively. The idea that electricity is a single fluid that is never created or destroyed, but simply transferred from one place to another, was profound, and it greatly simplified the interpretation of many observations.

In his second letter [5], Franklin was able to describe the behaviour of a Leyden (Leiden) jar capacitor by combining the concept of equal positive and negative states with an assumption that glass is a perfect insulator. 'So wonderfully are these two States of Electricity, the *plus* and *minus* combined and balanced in this miraculous Bottle!' He also made an analogy between electricity and lightning when he described a discharge through the gold trim on the cover of a book that produced 'a vivid Flame, like the sharpest Lightning'.

In his third letter [5], Franklin began to use terms such as 'charging' and 'discharging' when describing how a Leyden jar works, and he noted the importance of grounding when charging and discharging the jar. He also showed that the electricity in such a device resides entirely in the glass and not on the conductors that are inside and outside the jar. Franklin described how several capacitors could be charged in series 'with the same total Labour' as charging one, and he constructed an 'Electrical Battery' – a capacitor bank in today's parlance – using panes of window glass sandwiched between thin lead plates, and then discharged them together so that they provided the 'Force of all the Plates of Glass at once thro' the Body of any Animal forming the Circle with them'. Later, Franklin used discharges from large batteries to simulate the effects of lightning in a variety of materials.

In the fourth letter [5], he applied his knowledge of electricity to lightning by introducing the concept of the sparking or striking distance: If two electrified gun barrels 'will strike at two Inches Distance, and make a loud Snap; to what great a Distance may 10,000 Acres of Electrified Cloud strike and give its Fire, and how loud must be that Crack!' Based on his previous experiments with sharp points, Franklin then postulated that when an electrified cloud passes over a region, it might draw electricity from, or discharge electricity to, high hills and trees, lofty towers, spires, masts of ships, chimneys. That supposition then led to some practical advice against taking shelter under a single, isolated tree during a thunderstorm; crouching in an open field was seen to be less dangerous. Franklin also noted that out in the open during a thunderstorm, clothing tends to become wet, thereby providing a conducting path outside the body. His laboratory analogy was that 'a wet Rat can not be kill'd by the exploding electrical Bottle, when a dry Rat may'.

In the fifth letter [5], Franklin described how discharges between smooth or blunt conductors occur with a 'Stroke and Crack', whereas sharp points discharge silently and produce large effects at greater distances. He then introduced what he viewed to be a 'Law of Electricity, That Points as they are more or less acute, both draw on and throw off the electrical fluid with more or less Power, and at greater or less Distances, and in larger or smaller Quantities in the same Time'. Given his interest in lightning and the effects of metallic points, it was a short step to the lightning rod:

I say, if these Things are so, may not the Knowledge of this Power of Points be of Use to Mankind; in preserving Houses, Churches, Ships, etc. from the Stroke of Lightning; by Directing us to fix on the highest Parts of those Edifices upright Rods of Iron, made sharp as a Needle and gilt to prevent Rusting, and from the Foot of those Rods a Wire down the outside of the Building into the Ground; or down round one of the Shrouds of a Ship and down her Side, till it reaches the Water? Would not these pointed Rods probably draw the Electrical Fire silently out of a Cloud before it came nigh enough to strike, and thereby secure us from that most sudden and terrible Mischief!

Clearly, Franklin supposed that silent discharges from one or more sharp points might reduce or eliminate the electricity in the clouds above and thereby reduce or eliminate the chances of the structure being struck by lightning. From his earlier observations, he knew that point discharges work best when the conductor is grounded, and he also knew that lightning tends to strike tall objects. Therefore, even if the point discharges did not neutralize the cloud, a tall conductor would provide a preferred place for the lightning to strike, and the grounded conductor would provide a safe path for the lightning current to flow into ground. Franklin also stated in his fifth letter [5]:

To determine the Question, whether the Clouds that contain Lightning are electrified or not, I would propose an Experiment to be try'd where it may be done conveniently.

On the Top of some high Tower or Steeple, place a Kind of Sentry Box (see Figure 1.1) big enough to contain a Man and an electrical Stand. From the Middle of the Stand let an Iron Rod rise, and pass bending out of the Door, and then upright 20 or 30 feet, pointed very sharp at the End. If the Electrical Stand be kept clean and dry, a Man standing on it when such Clouds are passing low, might be electrified, and afford Sparks, the Rod drawing Fire to him from the Cloud.

Franklin was not the first person to compare sparks with lightning or to hypothesize that lightning might be an electrical discharge. In fact, almost every experimenter who had previously described electric sparks had, at one time or other, mentioned an analogy to lightning. Franklin's seminal contributions were the suggestions that tall, insulated rods could be used to determine if thunderclouds are electrified and that tall, grounded rods would protect against lightning damage.

1.2 The French connection

Shortly after Collinson published the first edition of *Experiments and Observations*, he sent a copy to the famous French naturalist, the Comte de Buffon, who asked Dalibard to translate it from English into French. While he did this, Dalibard asked Delor to help him repeat many of the Philadelphia experiments. In March of 1752, Buffon arranged for the pair to show the experiments to King Louis XV. The King's delight inspired Dalibard to try the sentry-box experiment at Marly-la-Ville.

At the time of the sentry-box experiment, Abbé Jean-Antoine Nollet was the leading 'electrician' in France and was known throughout Europe for his skill in making apparatus and in performing demonstrations. Unfortunately, because of personal rivalries, Buffon and Dalibard completely ignored Nollet's work in a short history that preceded their translation of Franklin's book. After Dalibard read an account of the sentry-box experiment to the Académie des Sciences on 13 May 1752, Nollet suppressed publication of the results [6]. News reached the Paris newspapers, however, and from there spread very rapidly. After Louis XV saw the experiment, he sent a personal message of congratulations to Franklin, Collinson, and the Royal Society of London for communicating 'the useful Discoveries in Electricity, and Application of Pointed Rods to prevent the terrible Effects of Thunderstorms' [7].

Nollet was both surprised and chagrined by the experiment at Marly-la-Ville. He acknowledged that insulated rods or 'electroscopes' did verify that thunderclouds are electrified, but for the rest of his life he steadfastly opposed the use of grounded rods as 'preservatives'. In 1753, he published a series of letters attacking Franklin's *Experiments and Observations* and suggested other methods of lightning protection. On 6 August 1753, the Swedish scientist Georg Wilhelm Richmann was electrocuted in St. Petersburg while trying to quantify the response of an insulated rod to a nearby storm. This incident, reported worldwide, underscored the dangers inherent in experimenting with insulated rods and in using protective rods with faulty ground connections. Nollet used Richmann's death to heighten the public's fears and to generate opposition to both types of rods [8].

In London, members of the Royal Society were amused when Franklin's letter about lightning conductors was read to the Society, and they did not publish it in their *Philosophical Transactions*. In 1753, however, they awarded Franklin their highest scientific honour, the Copley Gold Medal. In his 1767 history of electricity, Joseph Priestley described the kite experiment as drawing 'lightning from the heavens', and said it was 'the greatest, perhaps, in the whole compass of philosophy since the time of Sir Isaac Newton' [9].

1.3 Experiments in colonial America

After Franklin learned about the success of the sentry-box experiment in France, he installed a tall, insulated rod on the roof of his house to study the characteristics of thunderstorm electricity. The conductor ran down a stairwell to ground but had a gap in the middle, as illustrated on the left side of Figure 1.2. A small ball suspended

Figure 1.2   Modelled after a 1762 painting by Mason Chamberlain, this etching depicts Benjamin Franklin looking at electrostatic bells he used to study cloud electricity. Two chimes, separated from each other by a small gap, are connected to rods that go up through the roof and to ground. A thundercloud charges the right-hand bell, either by induction or point discharge; the bell then alternately attracts or repels a small ball suspended between the chimes on a silk thread. The ball rattles between the bells, ringing an alarm when a storm approaches. The electroscope hanging from the right-hand bell was used to measure the cloud's polarity. A grounded rod of Franklin's 1762 design can be seen through the window on the right (from Reference 20).

between chimes mounted on each end of the gap would ring the chimes whenever an electrified cloud passed overhead. Franklin used this apparatus to compare the properties of atmospheric electricity with the electricity generated by friction and to measure the polarity of thunderclouds.

He found that both types of electricity were the same and 'that the Clouds of a Thunder Gust are *most commonly* in a negative State of Electricity, but *sometimes* in a positive State' [10], a result that was regarded as definitive for the next 170 years. At that time, Franklin thought that all discharges went from positive to negative, so he concluded 'that for the most part in Thunder Strokes, *'tis the Earth that strikes into the Clouds*, and not the Clouds that strike into the Earth'. Judging by his later correspondence, Franklin was fascinated by this discovery, and he postulated that the effects of lightning would be very nearly the same regardless of the direction of the current flow.

1.4 First protection system

In the 1753 issue of *Poor Richard's Almanack*, Franklin published a method for protecting houses from lightning damage:

How to secure Houses, etc. from Lightning

It has pleased God in his Goodness to Mankind, at length to discover to them the Means of securing their Habitations and other Buildings from Mischief by Thunder and Lightning. The Method is this: Provide a small Iron Rod (it may be made of the Rod-iron used by the Nailers) but of such a Length, that one End being three or four Feet in the moist Ground, the other may be six or eight Feet above the highest Part of the Building. To the upper End of the Rod fasten about a Foot of Brass Wire, the Size of a common Knitting-needle, sharpened to a fine Point; the Rod may be secured to the House by a few small Staples. If the House or Barn be long, there may be a Rod and Point at each End, and a middling Wire along the Ridge from one to the other. A House thus furnished will not be damaged by Lightning, it being attracted by the Points, and passing thro the Metal into the Ground without hurting any Thing. Vessels also, having a sharp pointed Rod fix'd on the Top of their Masts, with a Wire from the Foot of the Rod reaching down, round one of the Shrouds, to the Water, will not be hurt by Lightning.

The opening phrase of this description anticipated a religious objection to protective rods that would soon appear in America and Europe. In the late summer or autumn of 1752, grounded conductors were installed on the Academy of Philadelphia (later the University of Pennsylvania) and the Pennsylvania State House (later Independence Hall). Figures 1.3 and 1.4 show fragments of the original grounding conductors that were installed inside the tower of Independence Hall and on the Gloria Dei (Old Swede's) Church in Philadelphia, respectively.

Three key elements made up Franklin's protection system. Metallic rods, or air terminals as they are now called, were mounted on the roof of the structure and connected by horizontal roof conductors and vertical down-conductors to a ground connection. Franklin initially thought point discharges might provide protection, so

8 *Lightning Protection*

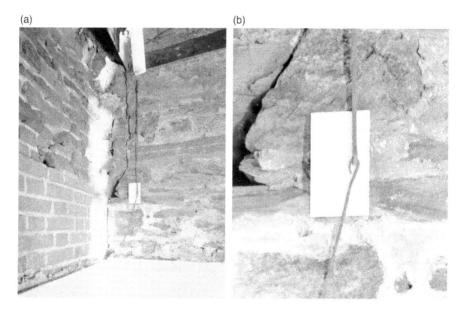

Figure 1.3 Independence Hall, Philadelphia. During a partial restoration, fragments of the original grounding conductor were found under panelling and plaster on the inside wall of the northwest corner of the tower stairwell (from the Independence National Historical Park Collection).

the first air terminals were thin, sharp needles mounted on top of an iron rod. The first down-conductors were chains of iron rods, each several feet long, that were mechanically linked or hooked together, as shown in Figures 1.3 and 1.4. As the current in point discharges is usually less than a few hundred microamperes, the roof and down conductors could be mechanically hooked together and attached to the inside walls of towers and steeples without creating a hazard.

Franklin wanted to verify that lightning would actually follow the path of a metallic conductor and determine what size that conductor should be, so in June of 1753 he published a 'Request for Information on Lightning' in *The Pennsylvania Gazette* and other newspapers:

> Those of our Readers in this and the neighboring Provinces, who may have an Opportunity of observing, during the present Summer, any of the Effects of Lightning on Houses, Ships, Trees, Etc. are requested to take particular Notice of its Course, and Deviation from a strait Line, in the Walls or other Matter affected by it, its different Operations or Effects on Wood, Stone, Bricks, Glass, Metals, Animal Bodies, Etc. and every other Circumstance that may tend to discover the Nature, and compleat the History of that terrible Meteor. Such Observations being put in Writing, and communicated to Benjamin Franklin, in Philadelphia, will be very thankfully accepted and gratefully acknowledged.

Figure 1.4   *David B. Rivers, pastor of the Gloria Dei (Old Swedes') Church in Philadelphia, holds a section of the original iron conductor that protected the church. The upper links in the chain were stapled to the inside of a wooden steeple. The inset shows how a mechanical link may have been ruptured, its hook forced open by an explosive arc during a lightning strike (photographs by the author).*

In the summer of 1753, Dr John Lining, a physician with many scientific interests, verified Franklin's kite experiment in Charleston, South Carolina, but when he tried to install a rod on his house, the local populace objected. They thought the rod was presumptuous – that it would interfere with the will of God – or that it might attract lightning and be dangerous [11]. In April of that year, Franklin commented on that issue [12]:

[Nollet] speaks as if he thought it Presumption in Man to propose guarding himself against Thunders of Heaven! Surely the Thunder of Heaven is no more supernatural than the Rain, Hail, or Sunshine of Heaven, against the Inconvenience of which we guard by Roofs and Shades without Scruple.

But I can now ease the Gentleman of this Apprehension; for by some late Experiments I find, that it is not Lightning from the Clouds that strikes the Earth, but Lightning from the Earth that Strikes the Clouds.

1.5   Improvements

In the following years, Franklin continued to gather information about lightning, and in 1757 he went to London as an agent of the Pennsylvania Assembly. In March of

1761, Kinnersley sent Franklin a detailed description of a lightning flash that struck a Philadelphia house equipped with a protective rod. An observer had reported at the time that 'the Lightning diffused over the Pavement, which was then very wet with Rain, the Distance of two or three Yards from the Foot of the Conductor'. Further investigation showed that the lightning had melted a few inches of the brass air terminal and Kinnersley concluded [12] that 'Surely it will now be thought as expedient to provide Conductors for the Lightning as for the Rain.'

Before Kinnersley's letter, Franklin had received reports of two similar strikes to protected houses in South Carolina. In one case, the points and a length of the brass down-conductor had melted. In the other, three brass points, each about seven inches long and mounted on top of an iron rod, had evaporated. Moreover, several sections of the iron down-conductor, each about a half-inch in diameter and hooked together, had become unhooked by the discharge (see Figure 1.4). Nearly all the staples that held the conductor to the outside of the house had also been loosened. 'Considerable cavities' had been made in the earth near the rod, sunk about three feet underground, and the lightning had produced several furrows in the ground 'some yards in length'. Franklin was pleased by these reports, and replied to Kinnersley that 'a conductor formed of nail rods, not much above a quarter of an inch thick, served well to convey the lightning' but 'when too small, may be destroyed in executing its office'. Franklin sent the reports from South Carolina to Kinnersley with a recommendation to use larger, more substantial conductors and a deeper, more extensive grounding system to protect the foundation of the house against the effects of surface arcs and explosions in the soil.

All reports from North America showed that grounded rods did indeed protect houses from lightning damage, so in January 1762 Franklin sent an improved design for 'the shortest and simplest Method of securing Buildings, Etc. from the Mischiefs of Lightning' together with excerpts from Kinnersley's letter and the reports from South Carolina, to Scottish philosopher David Hume. That letter was subsequently read to the Edinburgh philosophical society, which published it in 1771.

In the letter to Hume, Franklin recommended large, steel air terminals, 5 to 6 ft long and tapered to a sharp point. He said that any building with a dimension greater than ~100 ft should have a pointed rod mounted on each end with a conductor between them. All roof and down-conductors should be at least a half-inch in diameter, continuous, and routed outside the building (the earlier design allowed routing the conductors inside a building's walls). Any links or joints in these conductors should be filled with lead solder to ensure a good connection. The grounding conductor should be a one-inch-diameter iron bar driven 10 to 12 ft into the earth, and if possible, kept at least 10 ft away from the foundation. Franklin also recommended that the ground rods be painted to minimize rust and connected to a well, if one happened to be nearby. Figure 1.5 illustrates an implementation of Franklin's 1762 design.

In the 1769 edition of *Experiments and Observations*, Franklin published his reply to Kinnersley and the reports from South Carolina together with some 'Remarks' on the construction and use of protective rods. After repeating his recommendations for

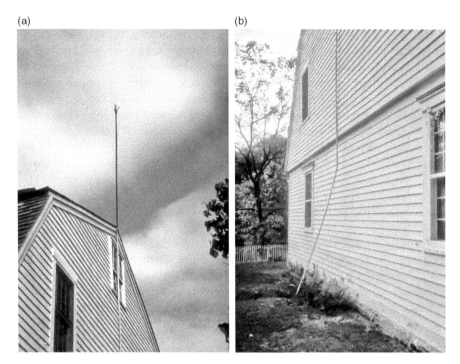

Figure 1.5 An eighteenth-century house with a lightning rod of Franklin's 1762 design. The thick, continuous rod can carry tens of kiloamperes of current to ground without harming the house or its foundation (photograph by the author).

an improved design, he also noted a psychological benefit of having protection against lightning [14]:

> Those who calculate chances may perhaps find that not one death (or the destruction of one house) in a hundred thousand happens from that cause, and that therefore it is scarce worth while to be at any expense to guard against it. But in all countries there are particular situations of buildings more exposed than others to such accidents, and there are minds so strongly impressed with the apprehension of them, as to be very unhappy every time a little thunder is within their hearing; it may therefore be well to render this little piece of new knowledge as general and well understood as possible, since to make us *safe* in not all its advantage, it is some to make us *easy*. And as the stroke it secures us from might have chanced perhaps but once in our lives, while it may relieve us a hundred times from those painful apprehensions, the latter may possibly on the whole contribute more to the happiness of mankind than the former.

Today, most authorities agree that lightning rods define and control the points where lightning will strike the structure and then guide the current safely into ground. As

Franklin noted in 1761, 'Indeed, in the construction of an instrument so new, and of which we could have so little experience, it is rather lucky that we should at first be so near the truth as we seem to be, and commit so few errors.' Franklin was truly lucky: his original 1752 design was based on the low current levels of point discharges, but direct lightning strikes deliver tens of kiloamperes of current, enough to produce explosive arcs across any imperfect mechanical connections; and those arcs can produce momentary over-pressures of several hundred atmospheres and enough heat to ignite flammable materials. The early applications of lightning rods could have been disastrous. Franklin's 1762 design, however, has stood the test of time and remains the basis for all modern lightning protection codes in the world today.

1.6 'Snatching lightning from the sky'

It is difficult for us living in an electrical age to appreciate how important lightning conductors were in the eighteenth century. The discovery that thunderclouds contain electricity and that lightning is an electrical discharge revolutionized human perceptions of the natural world, and the invention of protective rods was a clear example of how basic, curiosity-driven research can lead to significant practical benefits. In his later years, Franklin devoted most of his time to public service, but he continued to follow the work of others and conduct occasional experiments. He also participated on scientific advisory boards and panels that reviewed methods of lightning protection, and made recommendations for protecting cathedrals and facilities for manufacturing and storing gunpowder.

Eventually, Franklin became a leader of the American Revolution. When he embarked for France in November 1776 to seek aid for the newly declared United States of America in the war against Great Britain, he took with him a unique asset – his worldwide fame. By then his work on lightning and electricity had called attention to his other writings in science, politics and moral philosophy [15], and the intellectuals of France and Europe viewed Franklin as one of their own.

In 1811, John Adams, the first Vice-President and second President of the USA who served with Franklin in France in the 1770s (and who actually hated him), summarized Franklin's reputation [16,17]:

> Nothing, perhaps, that ever occurred upon this earth was so well calculated to give any man an extensive and universal celebrity as the discovery of the efficacy of iron points and the invention of lightning rods. The idea was one of the most sublime that ever entered a human imagination, that a mortal should disarm the clouds of heaven, and almost 'snatch from his hand the sceptre and the rod!' The ancients would have enrolled him with Bacchus and Ceres, Hercules and Minerva. His *Paratonnerres* erected their heads in all parts of the world, on temples and palaces no less than on cottages of peasants and the habitations of ordinary citizens. These visible objects reminded all men of the name and character of their inventor; and, in the course of time, have not only tranquilized the minds and dissipated the fears of the tender sex and their timorous children, but have almost annihilated that

panic terror and superstitious horror which was once almost universal in violent storms of thunder and lightning...

His reputation was more universal than that of Leibnitz or Newton, Frederick or Voltaire, and his character more beloved and esteemed than any or all of them. Newton had astonished perhaps forty or fifty men in Europe; for not more than that number, probably, at any one time had read him and understood him by his discoveries and demonstrations. And these being held in admiration in their respective countries as at the head of the philosophers, had spread among scientific people a mysterious wonder at the genius of this perhaps the greatest man that ever lived. But this fame was confined to men of letters. The common people knew little and cared nothing about such a recluse philosopher. Leibnitz's name was more confined still... But Franklin's fame was universal. His name was familiar to government and people, to kings, courtiers, nobility, clergy, and philosophers, as well as plebeians, to such a degree that there was scarcely a peasant or a citizen, a valet de chambre, coachman or footman, a lady's chambermaid or a scullion in a kitchen, who was not familiar with it, and who did not consider him as a friend to human kind. When they spoke of him, they seemed to think he was to restore the golden age.

In June of 1776, the celebrated economist and former comptroller-general of France, Anne-Robert Jacques Turgot, composed a prophetic epigram in Latin that captures Franklin's legacy in a single sentence: 'Eripuit caelo fulmen, sceptrumque tyrannis', ('He snatched lightning from the sky and the scepter from tyrants') [18].

Acknowledgement

The author is grateful to the late Penelope Hartshorne Batcheler for calling his attention to the photograph in Figure 3.

References

1. Portions of this paper are based on the author's presentation at the Inaugural Symposium of the International Commission on History of Meteorology, International Congress of History of Science, Mexico City, 11–12 July 2001, and on Reference 21.
2. Cohen I.B. *Benjamin Franklin's Science*. Cambridge, MA: Harvard University Press; 1990, Chapter 6.
3. Cohen I.B. *Benjamin Franklin's Experiments: A New Edition of Franklin's Experiments and Observations on Electricity*. Cambridge, MA: Harvard University Press; 1941.
4. Heilbron J.L. *Electricity in the 17th and 18th Centuries: A Study of Early Modern Physics*. Berkeley, CA: University California Press; 1979, Part Four.
5. Labaree L.W., Wilcox W.B., Lopez C.A., Oberg B.B., Cohn E.R. *et al.* (eds). *The Papers of Benjamin Franklin*. New Haven, CT: Yale University Press, Vol. I, 1959

to Vol. 37, 2003. The first letter is in The Papers of Benjamin Franklin **3**, pp. 126–35, and the remaining four letters are in **3**, pp. 156–64; **3**, 352–65; **3**, 365–76; and **4**, 9–34.
6. Heilbron J.L. *Electricity in the 17th and 18th Centuries: A Study of Early Modern Physics*. Berkeley, CA: University California Press; 1979, Chapter 15. Benjamin Franklin and Lightning Rods, 13.
7. Labaree L.W., Wilcox W.B., Lopez C.A., Oberg B.B., Cohn E.R. *et al.* (eds). *The Papers of Benjamin Franklin* New Haven, CT: Yale University Press **4**, 1962, pp. 465–67.
8. Cohen I.B. *Benjamin Franklin's Experiments: A New Edition of Franklin's Experiments and Observations on Electricity*. Cambridge, MA: Harvard University Press; 1941, Chapter 8.
9. Priestley J. *History and Present State of Electricity, with Original Experiments*. London: Printed for J. Dodsley, J. Johnson, B. Davenport, and T. Cadell; 1767, pp. 179–80.
10. Labaree L.W., Wilcox W.B., Lopez C.A., Oberg B.B., Cohn E.R. *et al.* (eds). *The Papers of Benjamin Franklin*. New Haven, CT: Yale University Press **5**, 71.
11. Lemay J.A.L. *Ebenezer Kinnersley: Franklin's Friend*. Philadelphia, PA: University of Pennsylvania Press; 1964, p. 78.
12. Labaree L.W., Wilcox W.B., Lopez C.A., Oberg B.B., Cohn E.R. *et al.* (eds). *The Papers of Benjamin Franklin*. New Haven, CT: Yale University Press **4**, p. 463.
13. Labaree L.W., Wilcox W.B., Lopez C.A., Oberg B.B., Cohn E.R. *et al.* (eds). *The Papers of Benjamin Franklin*. New Haven, CT: Yale University Press **4**, p. 293.
14. Labaree L.W., Wilcox W.B., Lopez C.A., Oberg B.B., Cohn E.R. *et al.* (eds). *The Papers of Benjamin Franklin*. New Haven, CT: Yale University Press **10**, p. 52.
15. Lemay J.A.L. (ed.). *Benjamin Franklin Writings*. New York: The Library of America; 1987.
16. Adams C.F. (ed.). *The Works of John Adams. Vol. I*. Boston: Little Brown, and Co.; 1856, pp. 660–62.
17. Dray P. *Stealing God's Thunder: Benjamin Franklin's Lightning Rod and the Invention of America*. New York: Random House; 2005.
18. Schiff S. *A Great Improvisation: Franklin, France, and the Birth of America*. New York: Henry Holt; 2005.
19. Franklin B. *Experiences et Observations sur L'Electricite . . . Trad. De l'Anglais par M. Dalibard*, 2nd edn. Paris, Vol. II, 1756, p. 128.
20. Frontispiece from Oeuvres de M. Franklin, translated by Dubourg J.B. Paris: Chez Quillau; 1773.
21. Krider E.P. in *Benjamin Franklin: In Search of a Better World*. Talbott P. (ed.). New Haven, CT: Yale University Press; 2005, Chapter 5.

Chapter 2
Lightning parameters of engineering interest
Vernon Cooray and Mahendra Fernando

2.1 Introduction

Electrical discharges generated in the Earth's atmosphere by cumulonimbus clouds, volcanic eruptions, dust storms and snow storms are usually referred to as lightning discharges. In this chapter we confine ourselves to the lightning discharges produced by cumulonimbus clouds. Lightning discharges can be separated into two main categories, ground flashes and cloud flashes. Lightning discharges that make contact with ground are referred to as ground flashes and the rest are referred to as cloud flashes. Cloud flashes in turn can be divided into three types: intracloud flashes, air discharges and intercloud discharges. These different categories of lightning flashes are illustrated in Figure 2.1a. A ground flash can be divided into four categories based on the polarity of charge it brings to the ground and its point of initiation. These four categories are illustrated in Figure 2.1b: downward negative ground flashes, downward positive ground flashes, upward positive ground flashes and upward negative ground flashes. The polarity of the flash, i.e. negative or positive, is based on the polarity of the charge brought to the ground from the cloud. Upward lightning flashes are usually initiated by tall objects of heights more than ~ 100 m or structures of moderate heights located on mountain and hill tops. The basic features of the mechanism of lightning ground flashes, summarized next, are given in Reference 1.

Electromagnetic field measurements show that a downward negative ground flash is initiated by an electrical breakdown process in the cloud called the *preliminary breakdown*. This process leads to the creation of a column of charge, called the *stepped leader*, which travels from cloud to ground in a stepped manner. Some researchers use the term preliminary breakdown to refer to both the initial electrical activity inside the cloud and the subsequent stepped leader stage. On its way towards the ground a stepped leader may give rise to several branches. As the stepped leader approaches the ground, the electric field at ground level increases steadily. When the stepped leader reaches a height of about a few hundred metres or less above ground, the electric field at the tip of the grounded structures increases to such a level that electrical discharges are initiated from them. These discharges,

16 Lightning Protection

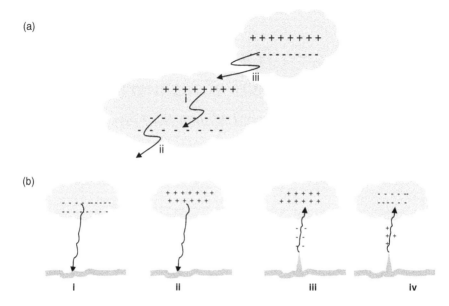

Figure 2.1 (a) Types of cloud flashes: (i) intracloud; (ii) air discharges; (iii) intercloud. (b) Types of ground flashes: (i) downward negative ground flashes; (ii) downward positive ground flashes; (iii) upward positive ground flashes; (iv) upward negative ground flashes.

called *connecting leaders*, travel towards the down-coming stepped leader. One of the connecting leaders may successfully bridge the gap between the ground and the down-coming stepped leader. The object that initiated the successful connecting leader is the one that will be struck by lightning. The distance between the object struck and the tip of the stepped leader at the inception of the connecting leader is called the *striking distance* (see Chapter 4 for a detailed description of this attachment process).

The moment a connection is made between the stepped leader and ground, a wave of near-ground potential travels at a speed close to that of light along the channel towards the cloud. The current associated with this wave heats the channel to several tens of thousands of degrees Kelvin, making the channel luminous. This event is called the *return stroke*. Whenever the upward-moving return stroke front encounters a branch, there is an immediate increase in the luminosity of the channel; such events are called *branch components*. Although the current associated with the return stroke tends to last for a few hundred microseconds, in certain instances the return stroke current may not go to zero within this time, but may continue to flow at a low level for a few to few hundreds of milliseconds. Such long duration currents are called *continuing currents*. The arrival of the first return-stroke front at the cloud end of the return-stroke channel leads to a change of potential in the vicinity of this point. This change in potential may initiate a positive discharge that travels away from the end of the return-stroke channel. Occasionally, a negative *recoil streamer* may be initiated at the outer extremity of this positive discharge channel and propagates along it towards the end of the return-stroke channel. Sometimes, discharges originate at a point several kilometres away from the end of the return-stroke channel and travel towards it. On some occasions these discharges may die out before they make contact with the end of the return-stroke channel. Such events are called *K-changes*. If these discharges make contact with the previous return-stroke channel, the events that follow may depend on the physical state of the return-stroke channel. If the return-stroke channel happens to be carrying a continuing current at the time of the encounter, it will result in a discharge that travels towards the ground. These are called *M-components*. When the M-components reach the ground no return strokes are initiated, but recent analyses of the electric fields generated by M-components show that the current wave associated with them may reflect from the ground. If the return-stroke channel happens to be in a *partially conducting stage* with no current flow during the encounter, it may initiate a *dart leader* that travels towards the ground. Sometimes the lower part of the channel decays to such an extent that the dart leader stops *before actually reaching the* ground. These are termed *attempted leaders*. In other instances, the dart leader may encounter a channel section whose ionization has decayed to such an extent that it cannot support the continuous propagation of the dart leader. In this case the dart leader may start to propagate towards the ground as a stepped leader. Such a leader is called a *dart-stepped leader*. If these leaders travel all the way to ground, then another return stroke, called the *subsequent return stroke*, is initiated. In general, dart leaders travel along the residual channel of the first return strokes, but it is not uncommon for the dart leader to take a different path from that of the first stroke. In this case it ceases to be a dart leader and travels towards the ground as a stepped leader. The point at which this leader terminates may be different from that of the original first leader. Thus, a single flash may generate multiple terminations. Electrical activity similar to that which occurs after the first return strokes may also take place after the subsequent return strokes. However, branch components occur mainly in the first return strokes and occasionally in the first subsequent stroke. This is the case because, in general, dart leaders do not give rise to branches. In the literature on lightning, the electrical activities in the cloud that take place between the strokes and after the final stroke are called, collectively, *Junction processes* or *J processes*.

The mechanisms of downward positive ground flashes have not been studied in detail, but their main features are qualitatively similar to those of downward negative ground flashes with differences in the finer details.

In the case of an upward ground flash an upward leader is initiated from a tall structure under the influence of the background electric field of the cloud. The arrival of this leader, the polarity of which depends on whether the upward initiated flash is negative or positive, at the charge centre in the cloud leads to the initiation of a continuing current that may last for several tens to hundreds of milliseconds. Some flashes end at the cessation of this continuing current, whereas in others this may be followed by a series of dart leader–return-stroke sequences that travel along the already established channel.

18 Lightning Protection

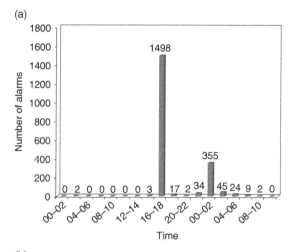

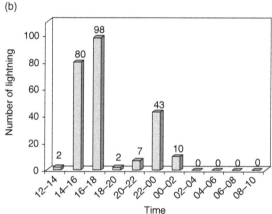

Figure 2.2 (a) Distribution of the number of false alarms caused by a thunderstorm in Malmö and Lund, Sweden on 1 July 1988. (b) The number of lightning flashes observed in the same region by the lightning-location system (modified from Reference 2).

In the protection of structures located at ground it is the ground flash that is of interest. However, the high sensitivity of modern-day electronic devices to electromagnetic disturbances may make them vulnerable even to electromagnetic fields generated by cloud flashes. On the other hand, the electromagnetic environment created by a ground flash striking in the vicinity of a structure is more severe than that of a cloud flash and any protection procedures developed to protect electronic systems from electromagnetic fields of close ground flashes will also mitigate the effects of cloud flashes.

A ground flash can interact with a structure in two ways. First, if the structure is not equipped with a lightning protection system, the direct injection of current at the point of strike may interact with the structure in various ways and cause structural damage and, in some cases, initiate fires. The injected current may also enter into the electrical and other conducting systems of the structure, which again causes damage and destruction in the electrical systems. Second, the electromagnetic field generated by the lightning flash induces large voltages in various electrical systems of the structure, irrespective of whether it is provided with an external lightning protection system or not, and so causes disturbances and damage in sensitive electronics. Indeed, the vulnerability of modern-day electrical and electronic systems to lightning is demonstrated by the effect of a thunderstorm on 1 July 1988 on the civil alarm system of southern Sweden [2]. Figure 2.2a shows the distribution of the number of false alarms caused by the thunderstorm at the alarm centre in Malmö-Lund, Sweden. Note that 1 498 alarms were received in a period of two hours, which completely paralysed the action of the fire-protection services. Figure 2.2b shows the number of lightning flashes registered in the same area by the Swedish lightning-location system. Note the strong correlation between the data in the two diagrams. This typical example illustrates the importance of appropriate protection of modern-day electrical and electronic systems from lightning flashes.

Study of the interaction of ground flashes with structures and other electrical systems can be separated into two parts. The first part deals with the processes that lead to the attachment of the lightning flash to the structure. This part is important in evaluating the point of strike of the lightning flash on the structure. The second part deals with the interaction of the structure and its contents with the injected current and the radiated electromagnetic field. Information concerning various lightning parameters that are of interest in analysing both these effects is essential to mitigate the effects of lightning strikes.

2.2 Electric fields generated by thunderclouds

The electric fields generated by thunderclouds are of interest in lightning protection studies in three ways. First, the electric fields generated by them at ground level are responsible for the initiation of upward flashes from tall structures or moderately tall structures located on mountains. Second, these electric fields cause sharp, grounded tips and pointed leaves of vegetation to go into corona, which generates space charge. The cumulative effect of these could influence the process of lightning attachment. Third, the electric fields generated by thunderclouds can be used in issuing warnings on the threat of lightning strikes.

The distribution of electric fields measured at ground level during thunderstorms in Pretoria, South Africa, by Eriksson [3] is shown in Figure 2.3. Note that the maximum electric fields recorded are less than $\sim 20\,\text{kV}\,\text{m}^{-1}$. However, the measurements conducted by Soula and Chauzy [4] show that thundercloud-generated electric fields at altitudes of 603 m can reach values as high as $60\,\text{kV}\,\text{m}^{-1}$, whereas the electric field at ground level at the same time is clamped below $\sim 10\,\text{kV}\,\text{m}^{-1}$. In another study, Willett and colleagues [5] measured the ambient electric fields below thunderclouds using rockets equipped with field-measuring devices. An example of a measured

20 Lightning Protection

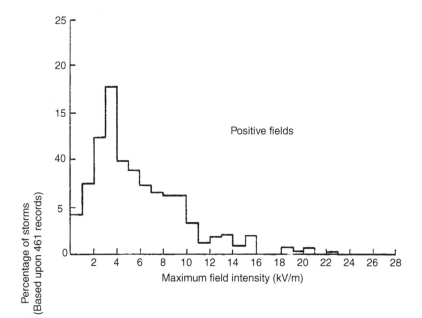

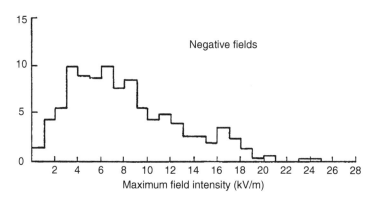

Figure 2.3   *The distribution of electric fields at ground level generated by thunderclouds in Pretoria, South Africa. The polarity of the electrostatic field is assumed to be positive when the dominant positive charge is overhead (from Reference 3).*

electric field as a function of altitude in that study is shown in Figure 2.4. Note how the electric field is clamped to a value less than ~ 10 kV m^{-1} at ground level. The electric field increases with height to reach a steady value within about several hundred metres or so. Why the thundercloud electric field is clamped to a value less than ~ 10–20 kV m^{-1} at ground level, but maintains a value several times larger at higher elevations, is described below.

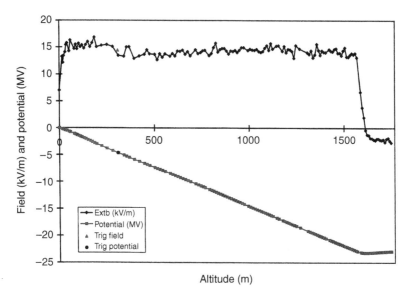

Figure 2.4 Ambient electrostatic field (upper trace) and potential profile (lower trace) as a function of height (from Reference 5)

As the charges in a thundercloud are generated at a rate of about 1 C s^{-1} (i.e. a charging current of $\sim$1 A) the electric field between the negative charge centre in the cloud and the ground increases. When the electric field at ground level reaches about a few kV m^{-1}, small protrusions, sharp points, pointed leaves, and so on, go into corona. These corona discharges give rise to a space charge layer. With increasing time the thickness of the space charge layer increases through the drift of ions upwards in the background electric field. The space charge layer screens the objects at ground level from the background electric field by clamping the electric field at ground level to a value close to the corona threshold. Of course, this clamping action of the space charge has a certain time constant and, as a consequence, it cannot reduce the magnitude of rapidly changing electric fields, such as those generated by return strokes. However, the variation of the thundercloud electric field is slow enough for the space charge effects to be dominant. At a given time the space charge layer has a certain thickness and it cannot influence significantly the fields at heights above this thickness. The thickness of the space charge layer at a given time depends on the drift speed of ions and hence on the aerosol concentration.

The effect of space charge on the electric field generated at ground level by a thundercloud can be taken into account using the model outlined in Reference 6, which is a two-dimensional extension of the model proposed previously by Chauzy and Rennela [7]. We can use this model to illustrate the effect of space charge on the thundercloud electric field at ground level. Assume that the lateral extension of the charge centre in the cloud is large and uniform enough to treat the electric field below the cloud as uniform. Assume also that the rate of generation of charge in the thundercloud is such that the electric field below the thundercloud increases to a peak value of 50 kV m^{-1} over a time of 60 s (Figure 2.5a). Thus, in the absence of corona the electric field at ground level should reach 50 kV m^{-1} in 60 s. Figure 2.5b shows how the electric field below the cloud varies at different times in the presence of corona space charges. First, observe how the space charge reduces the electric field initially at ground level and later at higher elevations. Observe also how the thickness of the space charge layer (the height where the electric field becomes constant) increases with increasing time. In this calculation the aerosol density is assumed to be 1×10^9 mol m^{-3}. With decreasing aerosol density the mobility of small ions increases and therefore the thickness of the space charge layer at a given time increases with decreasing aerosol concentration. In ~50 s the height of the space charge layer increases to ~200 m and therefore small structures become completely immersed in it. Thus, in evaluating the possibility of upward-initiated lightning flashes from a given structure, it is necessary to include the effect of space charge in the analysis

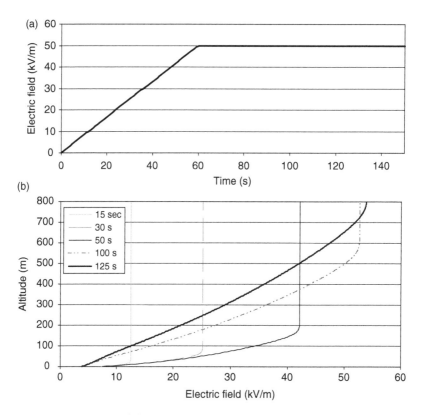

Figure 2.5 (a) Electric field at any point below the thundercloud in the absence of corona at ground level. (b) The electric field at different altitudes at different times in the presence of corona. The aerosol concentration is assumed to be 1×10^9 mol m^{-3}.

(see also Chapter 4). If the action of corona was not present at ground level, we would see the full strength of the electric field at ground level. In fact, space charge generation caused by the corona is very limited on calm water bodies. Not surprisingly, rather high electric fields have been observed over lakes under thunderstorm conditions [8].

2.3 Thunderstorm days and ground flash density

The ground flash density is an important parameter in lightning protection because risk evaluation in lightning protection procedures is based on this parameter. For example, let us represent the attractive radius of a structure of height h (see Chapter 4 for more details of this parameter) for a stepped leader that will give rise to a return stroke with a peak current of i_p (i.e. prospective return-stroke peak current) by $R(i_p, h)$. In other words, any stepped leader with a prospective return-stroke peak current of i_p will be attracted to the structure if it comes within a radial distance of $R(i_p, h)$ from the structure. As shown in Chapter 4 the attractive radius is both a function of i_p and the height h of the structure.

Let us also represent by $f(i_p)di$ the fraction of lightning flashes that have first return-stroke peak currents in the interval between i and $i + di$. Then the number of lightning flashes with first return-stroke current peaks in the above interval that strike the structure in one year is given by

$$dN = N_g \pi [R(i_p, h)]^2 f(i_p) di \qquad (2.1)$$

where N_g is called the ground flash density. It is defined as the number of lightning flashes that strike a unit area in a given region in a year. Then the total number of lightning flashes that strike the structure is given by

$$N = N_g \pi \int_0^\infty [R(i_p, h)]^2 f(i_p) di \qquad (2.2)$$

Different expressions for the attractive radius of structures have been derived by different scientists (see Chapter 4). These expressions differ from each other because of the various assumptions made in the analysis. However, all these expressions predict (and also experience shows) that $R(i_p, h)$ increases with increasing structure height. Therefore, the number of strikes per year on a structure increases with its height.

Ground flash density N_g in a given region can be estimated by counting the number of lightning flashes that strike ground in that region by using lightning flash counters, lightning location systems or using information on lightning strikes provided by satellites. However, lightning-protection engineers still use *thunderstorm days* to extract ground flash density because information concerning this parameter is available worldwide. A thunderstorm day is normally defined as the local calendar day in which thunder is heard by meteorological observers. It is a good source of information about the seasonal and geographical variation in lightning-flash frequency. However, it does not include the intensity of the thunderstorm or the number of times thunder

24   *Lightning Protection*

is heard on one particular day. Figure 2.6a depicts the distribution of the thunderstorm days around the globe. In the absence of better information about ground flash density it can be estimated from thunderstorm days, T_d, using an equation of the form

$$N_g = aT_d^b \text{ flashes km}^{-2} \text{ year}^{-1} \tag{2.3}$$

The parameters of this equation have been derived by many workers from different parts of the globe. The data obtained by these parameters are summarized in Table 2.1. Importantly, in all these studies there is a large scatter in the thunderstorm

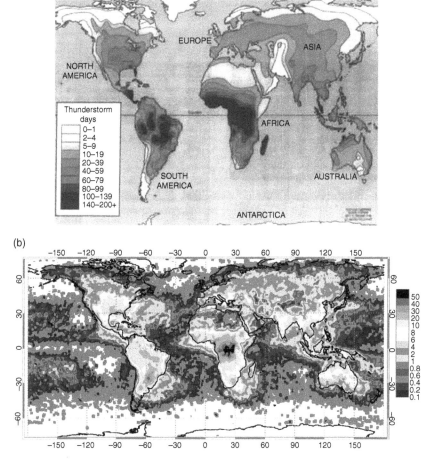

Figure 2.6   (a) *Thunderstorm day map of the world (from the National Lightning Safety Institute web page).* (b) *Total (includes both ground and cloud flashes) lightning-density map of the world (flashes km^{-2} year^{-1}) (from Reference 9).*

Table 2.1 Parameters of equation (2.3) obtained from studies conducted in different regions of the world

Study	Value of a	Value of b	Comments
Mackerras [10]	0.01	1.4	Based on the data obtained from 26 sites in Australia over the period 1965–1977
Anderson and Eriksson [11]	0.023	1.3	Based on 120 observations over two years in South Africa
Anderson et al. [12]	0.04	1.25	Based on data obtained from 62 stations over a period of five years spanning 1976–1980; equation (2.3) is based on the latter values and is generally known as the 'CIGRE* formula'
Kuleshov and Jayaratne [13]	0.012	1.4	Obtained using long-term lightning-flash counter registrations and thunderstorm day observations in Australia
Chen et al. [14]	0.0054	1.537	Data obtained in China from 82 stations
de la Rosa et al. [15]	0.024	1.12	Study conducted in tropical Mexico
Torres [16]	0.003	1.12	Study conducted in tropical Brazil
Younes [17]	0.0017	1.56	Study conducted in tropical Colombia

*International Council on Large Electrical Systems.

day versus ground flash density plots. Moreover, there is a large spread in the best estimates of a and b obtained in different parts of the world.

Figure 2.6b shows the variation of lightning-flash density obtained in various parts of the world using satellite data. In this estimation it is difficult to separate ground and cloud flashes and therefore the data in Figure 2.6b gives the total lightning-flash density in different regions of the world.

Fortunately, the recent deployment of lightning-detection systems in many parts of the world has led to the development of ground flash density maps from the direct measurement of lightning ground flashes in different parts of the world. Some of these maps are depicted in Figure 2.7. At present the data being gathered by lightning-location systems are also being used to convert the satellite data into ground and cloud flashes.

2.4 Number of strokes and time interval between strokes in ground flashes

The number of strokes in a lightning flash and the time separation between them are important parameters in lightning protection. For example, this information is essential to the coordination of circuit breakers in power distribution systems. Moreover,

26 *Lightning Protection*

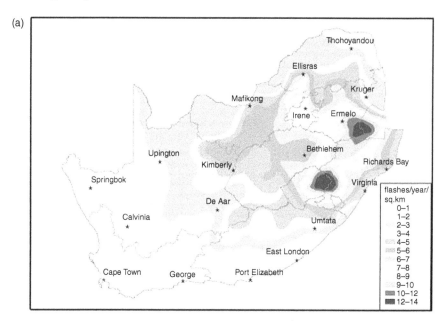

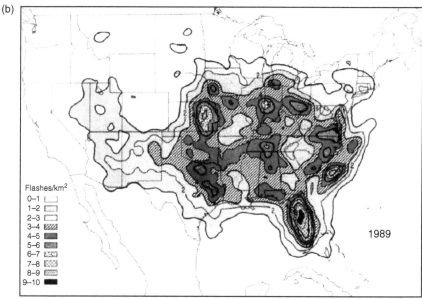

Figure 2.7 Average lightning-flash densities as measured in different regions of the world. (a) Lightning ground flash density map of South Africa [18]. (b) Lightning ground flash density map of the United States (from Reference 19).

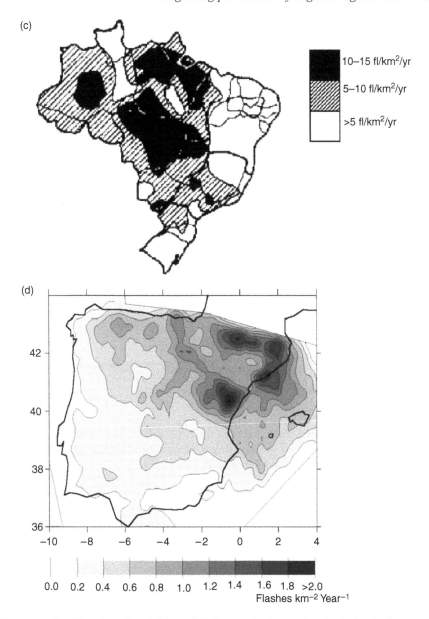

Figure 2.7 (Continued) (c) Map of lightning-flash density (includes both ground and cloud flashes) of Brazil obtained from the lightning imaging sensor (LIS) on board the tropical rainfall measuring mission (TRMM) satellite from December 1997 to December 1999 [20]. (d) Average lightning ground flash density distribution over the Iberian Peninsula, Spain, for the period 1992–2001. The numbers on the x axis are longitudes (negative longitudes are western longitudes) and numbers on the y axis are latitudes (from Reference 21).

28 *Lightning Protection*

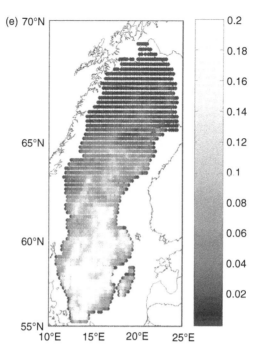

Figure 2.7 *(Continued) (e) The lightning ground flash density map of Sweden (from Reference 22).*

studies conducted by Darveniza and colleagues [23] show that the failure modes of surge-protective devices deployed in power systems depend on the stroke multiplicity of and the time interval between lightning flashes.

2.4.1 Number of strokes per flash

A study conducted by Thomson [24] showed that the number of strokes per flash does not vary significantly from one geographical region to another. In fact, the data obtained from widely different regions around the globe show similar characteristics. The average number of strokes per flash, percentage of single flashes and the maximum multiplicity in different regions of the world are tabulated in Table 2.2. The actual distributions of the number of strokes measured in Brazil, Sri Lanka and Sweden are depicted in Figure 2.8.

In the case of positive ground flashes, subsequent strokes are observed only seldomly, and almost all flashes are single-stroke flashes. For example, in the analysis carried out by Heidler and Hofp [30], out of 45 positive ground flashes, 33 were single, eight had two strokes and two flashes had three strokes. According to this study about 75 per cent of positive flashes are single-stroke flashes. In the study conducted by Saba and colleagues [31], out of 39 positive ground flashes 11 had two or

Table 2.2 Number of strokes per flash and percentage of single-stroke flashes in negative ground flashes in different regions of the world (from Reference 25)

Study	Location and latitude	Number and type of storm	Total number of flashes	Maximum multiplicity	Percentage of single-stroke flashes	Average multiplicity
Kitagawa et al. [26]	Socorro, New Mexico, USA (34 N)	Three summer night thunderstorms	193	26	14	6.4
Rakov et al. [27]	Tampa, Florida, USA (27.4 N)	Three convective summer thunderstorms	76	18	17	4.6
Cooray and Perez [28]	Uppsala, Sweden (60 N)	Two frontal summer thunderstorms	137	10	18	3.4
Cooray and Jayaratne [29]	Colombo, Sri Lanka (6.55 N)	Two convective thunderstorms	81	12	21	4.5
Saba et al. [25]	Sao Jose dos Campos and Cahoeira, Sao Paulo, Brazil (23.2 and 22.6 S)	Twenty-seven frontal and convective summer thunderstorms	233	16	20	3.8

Lightning parameters of engineering interest   29

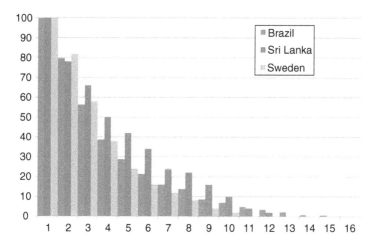

Figure 2.8 Percentage of flashes having a certain number of strokes as measured in Brazil, Sri Lanka and Sweden

more strokes, making the percentage of single-stroke positive ground flashes equal to 73 per cent.

2.4.2 Interstroke interval

In his statistical analysis Thomson [24] investigated whether there is a significant difference between the interstroke intervals measured in studies conducted in different geographical regions. As for the number of strokes per flash, he concluded that there is no statistical difference between different datasets.

The interstroke time intervals of negative ground flashes observed in Germany, Florida (USA), Sweden, Sri Lanka, China and Brazil are given in Table 2.3. Note the similarity between the interstroke time intervals of lightning flashes in different regions. Both Saba and colleagues [25] and Rakov and colleagues [27] analysed the interstroke time intervals associated with strokes that created new terminations at ground (see Section 2.5). The results gave the geometric mean values of 68 and 92 ms, respectively. These values have to be compared, respectively, with 61 and 60 ms, the interstroke time intervals obtained for all strokes in these two studies. In the Brazilian study the difference is not significant, whereas in the Florida study the mean interstroke time interval associated with strokes that created new terminations is larger than those associated with normal strokes.

Information concerning the interstroke intervals of positive return strokes is scarce in the literature. To the best of our knowledge only three studies are reported, conducted in Sweden, Germany and Brazil. The results obtained in these studies are presented in Table 2.4. The interstroke interval distributions observed in Brazil and Sweden are shown in Figure 2.9. Note that these intervals are significantly larger than those that correspond to negative ground flashes.

Table 2.3 Interstroke time intervals observed in different studies conducted at different geographical regions

Study	Total number of flashes	Total number of strokes	Arithmetic mean (ms)	Geometric mean (ms)
Heidler and Hopf, Germany 1986 [30]		116	87	
Heidler and Hopf, Germany 1988 [30]		414	87	96
Thottapillil et al., Florida 1992 [32]	46	199		57
Cooray and Perez, Sweden 1994 [28]	271	568	65	48
Cooray and Jayaratne, Sri Lanka 1994 [29]	81	284	82.8	56.5
Rakov et al., Florida 1994 [27]		270		60
Qie et al., China 2002 [33]	50	238	64.3	46.6
Miranda et al., Brazil 2003 [34]	26	131	69.0	49.6
Saba et al., Brazil 2006 [25]	186	608	83	61

2.5 Number of channel terminations in ground flashes

In the classical lightning literature, ground flashes that have two terminations are known as fork lightning. There are two mechanisms by which a ground flash may generate two strike points at ground. In the first, two branches of a single stepped leader may approach the ground more or less simultaneously and, if they both touch ground within a few microseconds (i.e. before the return stroke initiated by the first touch could neutralize the charge on the second branch), two return strokes are initiated

Table 2.4 Interstroke time intervals of positive ground flashes (from Reference 25)

Study	Number of flashes	Arithmetic mean	Geometric mean
Heidler and Hopf, Germany 1993 [30]	16	120	101
Cooray and Perez, Sweden 1994 [28]	29	64	92
Saba et al., Brazil 2006 [31]	13	168	117

32 Lightning Protection

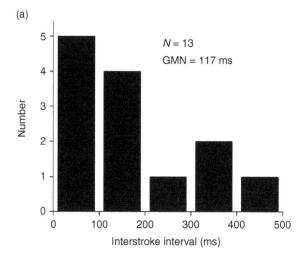

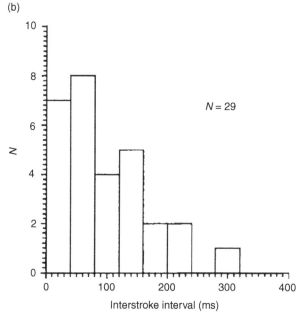

Figure 2.9 Interstroke interval of positive ground flashes as observed in (a) Brazil [31] and (b) Sweden (from Reference 28)

from the two branches that touch the ground. However, in this case, the two branches have to touch the ground within 10 μs or so of each other, and the probability of this is extremely small. In the second scenario multiple terminations are created when the down-coming dart leader deviates from the previous channel and takes a new path to ground. Optical observations show that this is the most common process by which multiple terminations are created. Davis [35] (as referenced in Rakov and Uman [36]) identified that the jump from the previous to the new channel takes place at a height of around 0.7–3.4 km. According to the observations presented by Thottappillil and colleagues [32] the separation between strike points in multiple terminations varies from 0.3 to 7.3 km with a geometric mean of 1.7 km.

Information about multiple terminations in lightning flashes has been gathered by Rakov and colleagues [27], Kitagawa and colleagues [26], Valine and Krider [37] and Saba and colleagues [25]. The data reported in these studies show that the percentage of flashes with multiple strike points is 50 per cent in Florida, 49 per cent in New Mexico, 35 per cent in Tuscan, Arizona and 51 per cent in Brazil. The average number of strike points per flash is 1.7 and 1.67 in Brazil and Florida, respectively. Figure 2.10 shows the number of flashes that produced a given number of strike points in the studies conducted in Florida and Brazil – a few flashes produced four terminations at ground. Figure 2.11 gives the probability of the creation of new paths to ground by strokes of different order. It clearly shows that the probability of creating a new termination is highest in the earlier strokes. For example, the second stroke has the highest probability to create a new termination and this probability decreases with increasing stroke order. This indicates that each stroke preconditions the channel in such a way that the probability of creating a new channel decreases with increasing stroke order. The reason why some strokes deviate from the previous path and create new paths to ground is unknown at present. It is possible that the location on the channel where the new path deviates from the previous one corresponds to the points where the stepped leader has generated a branch point on its way towards ground. Only the strongest of these branches would be visible in

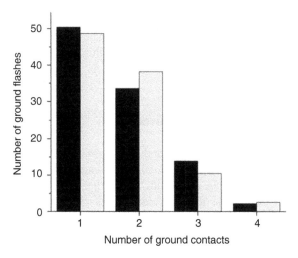

Figure 2.10 *Number of ground flashes that produce the given number of ground terminations: black, Brazil [25]; grey, Florida [27] (from Reference 25)*

34 *Lightning Protection*

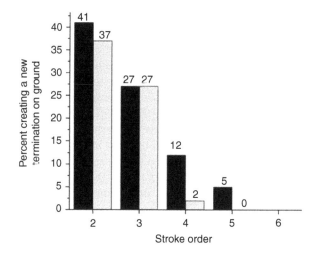

Figure 2.11 Percentage of strokes that create a new termination at ground as a function of stroke order. Numbers on the bars give the actual percentage: black, Brazil [25]; grey, Florida [27] (from Reference 25).

photographic records. The charges that reside on branches will be neutralized by the first return stroke and, therefore, at branch points the dart leader may encounter two paths of elevated temperatures whose density is less than ambient. Such encounters may promote the dart leader to take a new path to ground. However, if the second stroke follows the same path as the first one, the third stroke may also prefer the same path. This is because the second stroke does not usually travel along the branches and therefore it will not reheat the branch channels. Thus the third stroke may prefer the path taken by the second stroke because at branch points it will find a preferred path along the channel traversed by the second stroke. In this way, higher-order strokes may find it more difficult to create new terminations.

2.6 Occurrence of surface flash over

A lightning flash that strikes a finitely conducting ground generates a horizontal electric field. This electric field has its maximum strength at the strike point, which decreases with distance. If the magnitude of this horizontal electric field is larger than a certain critical value then an electrical discharge propagating along the ground (i.e. a surface discharge) is generated. Figure 2.12 shows a surface discharge created at the point of strike by a triggered lightning flash at Fort McClellen [39]. The length of the surface discharge depends on the magnitude of the horizontal electric field and how fast it decreases when moving away from the strike point. The properties of these surface flashovers and the conditions under which they are generated are of interest in the design of grounding systems in lightning protection.

Lightning parameters of engineering interest 35

Figure 2.12 Photograph of the surface arcs emanating from the ground rod conducting a 29.6 kA rocket-triggered lightning strike to ground at Fort McClellan, Alabama [38] (from Reference 39)

Experimental data on the formation of surface discharges at the point of strike have been reported by Fisher and colleagues [38] and Uman and colleagues [40]. The length of surface discharges can reach up to 20 m and a current of ∼1 kA has been measured in one case. Figure 2.13 shows the percentage of return strokes that produced optically detectable surface arcing as observed in the study conducted by Rakov and colleagues [42]. In this figure the percentage of strokes that generate surface discharges increases with increasing peak current. This is in accordance with theory because the surface electric field increases with increasing return-stroke current, and hence the probability of creating a surface arc increases with increasing current.

According to the experiment conducted by Liew [43] (as referenced in Wang and colleagues [44]), injection of laboratory currents up to 20 kA into loamy sand in the presence of water sprays, to imitate rain, resulted in surface arcing. This indicates that wetting the soil increases the probability of surface arcing. Let us consider the reason for this behaviour.

The critical electric field necessary to cause electrical breakdown in a liquid or solid material is usually one to two orders of magnitude larger than that of gaseous substances. However, if the solid or the liquid contains gaseous material in the form of gaseous cavities or bubbles, then there will be a drastic reduction in the breakdown electric field. In this case, the discharge process is initiated in the cavity or the bubble (see also Chapter 11). In a similar manner, when a lightning flash strikes the

36 *Lightning Protection*

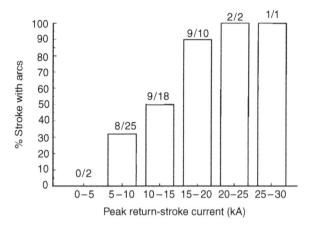

Figure 2.13 *Percentages of return strokes that produced optically detectable surface arcing as a function of return stroke peak current (Fort McClellan, 1993 and 1995). Numbers above each histogram column indicate the number of strokes that produced optically detectable arcing (numerator) and the total number of strokes in that return-stroke current range (denominator) (from Reference 42).*

ground, ionization processes start in the air gaps in the soil and lead to the formation of discharge channels that create a low resistive path for the current flow in soil. The sparking inside the soil reduces the impedance at the strike point, which in turn leads to a reduction in the surface electric field and so decreases the probability of surface discharges. If the soil is wet and the air pockets are filled with water, the soil behaves as a solid or a liquid material without air pockets. This increases the critical electric field necessary for breakdown inside the soil. This, in turn, leads to a larger surface electric field, which increases the probability of surface flashover.

2.7 Lightning leaders

2.7.1 *Speed of stepped leaders*

The speed of a stepped leader is important in lightning protection in two ways. First, it determines the rate of change of electric field produced at ground level by the down-coming stepped leader. Studies conducted by Becerra and Cooray [45] show that the inception of connecting leaders from grounded structures depends not only on the amplitude, but also on the rate of change of the electric field generated by the stepped leader (see also Chapter 4). Thus, to evaluate the conditions under which upward leaders are initiated from grounded structures it is necessary to know the speed of the down-coming leader. Second, once a connecting leader is incepted, whether it will make a successful connection with the down-coming stepped leader or not is determined by the relative speed of the two leaders. Thus, numerical simulation of the attachment process in lightning-protection studies requires statistics concerning the speed of stepped leaders.

The speeds of stepped leaders have been measured and reported by Schonland and Collens [46], Schonland [47,48], Schonland and colleagues [49], McEachron [50], Orville and Idone [51], Berger and Vogelsanger [52] and Saba and colleagues [53]. The stepped leader speed distributions obtained by Schonland (combined with the data of McEachron) [48] and Saba and colleagues [53] are shown in Figure 2.14. The average leader speeds observed in the two studies were 1.3×10^5 m s^{-1} and 3.36×10^5 m s^{-1}. The minimum leader speed observed by Schonland was 8×10^4 m s^{-1} and that by Saba and colleagues was 9×10^4 m s^{-1} [53]. The minimum value observed by McEachron [50] was 6×10^4 m s^{-1}. The maximum values observed by Schonland [48] and Saba and colleagues [53] were 2.6×10^6 m s^{-1} and 1.98×10^6 m s^{-1}, respectively. Note that the two studies, one the oldest and the other the latest, generated more or less similar results. The speed of stepped leaders observed by Orville and Idone [51], Cheng and colleagues [54] and Berger and Vogelsanger [52] also fall within these ranges.

2.7.2 Speed of dart leaders

There is direct evidence to show the existence of upward-connecting leaders that make contact with down-coming dart leaders. For example, in a recent study Wang and

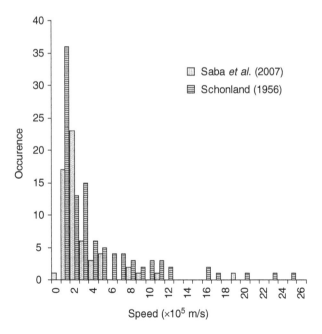

Figure 2.14 Distribution of the speed of stepped leaders as observed by Schonland [48] and Saba and colleagues [53]

38   Lightning Protection

colleagues [55] observed the upward-connecting leader lengths in two dart leader-return stroke sequences to be 7–11 m and 4–7 m. The conditions necessary for the inception of these upward leaders are determined by the amplitude and the rate of change of the electric fields produced by dart leaders. Simulation of the attachment process in dart leader-return stroke sequences requires, therefore, the speed of dart leaders.

The first observations of the speed of dart leaders were probably made by Schonland and colleagues [49]. However, these are probably biased towards smaller values because of the limited resolution of the cameras they used. Since then, several studies have been conducted to measure the speed of dart leaders. A summary of the dart-leader speeds as obtained by different workers is presented in Table 2.5. Importantly, the speed of the dart leader may also vary as it propagates towards the ground. For example, Wang and colleagues [55] found two dart leaders that exhibited a speed increase as they approached the ground. In one case the speed increased from 8×10^6 to 13×10^6 m s^{-1} during its downward propagation from 350 to 40 m and in the other case the leader speed increased from 2×10^6 to 8×10^6 m s^{-1} during its propagation from 200 to 40 m. However, Orville and Idone [51] reported that several

Table 2.5   Dart leader speeds observed in different studies

Study	Stroke number	Previous interstroke interval	Speed (m μs^{-1})
Jordan et al., Florida [56]	3	138	12*
	6	38	15
	2	25	11
	3	30	17
	2	56	5.4*
	3	44	24
	4	20	9.2
	7	75	7.9
	2	44	16
	3	30	18
	4†	34	17
Jordan et al., Florida [57]	3	30	17
	2	44	16
	3	31	18
Wang et al., Florida, for triggered lightning [58]			30 40
Saba et al., Brazil [53]		Median of partial speed	30
		Median of average speed	40

*Leaders showing stepping near ground.
†Second stroke in a new channel created by the third stroke of the flash.

dart leaders they observed showed a decrease in propagation speed as they approached the ground while four dart leaders showed the opposite tendency. Also, Schonland and colleagues [49] reported that dart leaders have more or less constant speed, but a few slowed down as they approached the ground.

2.7.3 Electric fields generated by stepped leaders

The electric field produced by stepped leaders at ground level as they travel towards the ground is the source that drives the inception and propagation of upward-connecting leaders. Thus, information concerning the electric fields generated by stepped leaders is essential to finding out the possible points of attachment of lightning flashes on structures (see also Chapter 4). Even though data for electric fields generated by leaders at distances larger than several hundreds of metres are available in the literature, no information is available on the electric fields generated by stepped leaders directly below them at ground. Thus, we are confined to theoretical investigations to obtain information about electric fields at ground level directly beneath the downcoming stepped leaders. Such theoretical investigations require both the speed of stepped leaders, which we present in Section 2.7.1, and the charge distribution along the stepped leader channel as a function of height. Let us now consider the charge distribution along the leader channel.

Based on experimental observations, Schonland [59] ascertained that the charge on the stepped leader is distributed uniformly. However, Golde [60] assumed that the line-charge density ρ_S on the stepped leader channel decreases exponentially with increasing height above ground:

$$\rho_S = \rho_{S0} e^{-z/\lambda} \tag{2.4}$$

where z is the height, ρ_{S0} the charge density at ground level and λ the charge decay height constant. In the calculations, Golde [60] used $\lambda = 1\,000$ m. Eriksson [3], in turn, assumed that the charge is distributed linearly along a vertical leader channel with the maximum charge density at ground level (features of these distributions are discussed in detail in Chapter 4).

As the stepped leader extends towards the ground, its charge distribution is determined by the background electric field generated by cloud charges and any field enhancement caused by the ground (e.g. the proximity effect). During the return stroke of a negative ground flash, positive charge is transported from ground into the stepped leader channel. Part of this positive charge neutralizes the negative leader charge, while the rest supplies the positive charge induced on the channel to maintain it at ground potential in the background electric field of the cloud. If the total positive charge injected into the return-stroke channel during the return-stroke stage is known, it can be combined with theory to give the distribution of the charge on the leader channel. Such a study was conducted by Cooray and colleagues [61]. They measured the charge transported by the first 100 μs of the first return strokes and combined it with theory to generate the charge distribution along the leader channel (see Section 2.9.1). Their results show that the variation of the charge per unit length of the stepped leader with height can be approximated by the following analytical expressions:

$$\rho(\zeta) = a_0\left(1 - \frac{\zeta}{H - z_0}\right)G(z_0)I_p + \frac{I_p(a + b\zeta)}{1 + c\zeta + d\zeta^2}J(z_0) \quad (2.5)$$

$$G(z_0) = 1 - \frac{z_0}{H} \quad (2.6)$$

$$J(z_0) = 0.3\alpha + 0.7\beta \quad (2.7)$$

$$\alpha = e^{-(z_0 - 10)}/75 \quad (2.8)$$

$$\beta = \left(1 - \frac{z_0}{H}\right) \quad (2.9)$$

where z_0 is the height (m) of the leader tip above ground, H the total length (m) of the stepped leader channel, $\rho(\zeta)$ the charge per unit length (C m^{-1}), ζ the length (m) along the stepped leader channel (with $\zeta = 0$ at the tip of the leader), I_p the return-stroke peak current (kA), $a_0 = 1.476 \times 10^{-5}$, $a = 4.857 \times 10^{-5}$, $b = 3.9097 \times 10^{-6}$, $c = 0.522$ and $d = 3.73 \times 10^{-3}$. Note that equations (2.5) to (2.9) are valid for $z_0 \geq 10$ m.

One can use this equation to calculate how the electric field at ground level directly below the path of a stepped leader varies as a function of time. For example, Figure 2.15 shows such electric fields for stepped leaders with prospective return-stroke currents of 15, 30, 50, 80 and 120 kA. In the calculation the height of the charge centre is assumed to be 4 km and the speed of propagation of the stepped leader is assumed to be 5×10^5 m s^{-1}. The electric field variation is very different to that present in a uniform laboratory gap excited by switching impulse voltage waveforms, which is used in testing electrical equipment. In Chapter 4 it is shown that, because of this dissimilarity, it is not possible to use switching impulse voltages to test the performance of lightning rods.

2.7.4 Electric fields generated by dart leaders

Even though data on the electric fields from natural dart leaders measured at the vicinity of the striking point is scarce, such fields are available from triggered lightning flashes. Because the features of return strokes in triggered lightning flashes are similar to those of natural subsequent strokes, the electric fields of dart leaders of triggered lightning return strokes also provide information about the electric fields of natural dart leaders in the vicinity of the strike point. Knowledge of these fields is important in analysing the final attachment of the dart leader to grounded objects. In addition, recent studies show that dart leaders can induce significant voltages in overhead power lines, and in this connection the electric fields generated by dart leaders close to the channel are of interest.

Typical examples of dart-leader electric fields measured close to the strike point of triggered lightning flashes are shown in Figure 2.16. The left half of the V shape is produced by the dart leader and the right half is produced by the return stroke. The

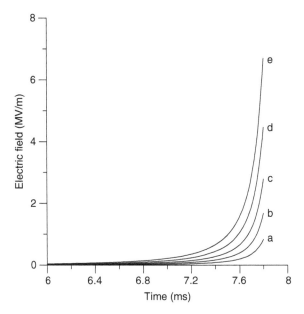

Figure 2.15 Electric field at ground level at a point directly below the path of a down-coming stepped leader with speed 5×10^5 m s^{-1}. The height of the charge centre is taken to be 4 km. The charge distribution in the stepped leader is given by equations (2.5) to (2.9). The calculation stops when the tip of the stepped leader is 100 m above ground. Calculations are given for the prospective first return-stroke currents of (a) 15 kA, (b) 30 kA, (c) 50 kA, (d) 80 kA and (e) 120 kA. The polarity of the electric field is assumed to be positive when it is produced by negative charge overhead.

reason for this shape becomes apparent when one recalls that the return stroke neutralizes and removes the charge deposited by the dart leader on the channel. Figure 2.17 shows the relationship observed between the return-stroke peak current and the peak dart-leader field at two different distances from triggered lightning flashes. The relationship between these parameters is approximately linear. The data given in Figure 2.17 shows that the peak electric field of dart leaders can reach values as high as 100 kV m^{-1} within 10 m of the strike point. To evaluate the electric fields directly below the down-coming dart leaders it is necessary to know the charge distribution along the dart-leader channel.

The charge distribution along the dart-leader channel was evaluated by Cooray and colleagues [61] using a technique similar to the one they used to evaluate the charge distribution along the stepped leader channel (see Section 2.9.1). The results are based on the charge brought to ground by subsequent return strokes over the first 50 μs. The results obtained can also be represented by equations (2.5) to (2.9) using the constants $a_0 = 5.09 \times 10^{-6}$, $a = 1.325 \times 10^{-5}$, $b = 7.06 \times 10^{-6}$, $c = 2.089$ and $d = 1.492 \times 10^{-2}$. Again, these are valid for $z_0 \geq 10$ m.

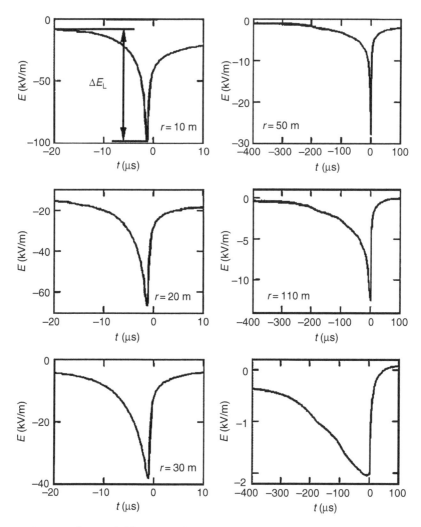

Figure 2.16 Electric field measured at 10, 20, 30, 50, 110 and 500 m from the lightning channel for dart leader–subsequent return stroke. Note that the definition of the positive electric field is opposite to that used in Figure 2.15 (from Reference 62).

Once the charge distribution along the leader channel is known, the close electric field produced by the dart leader at a given point at ground level can be calculated and compared with measurements. For a vertical dart-leader channel of length H the electric field E_z at distance D from the ground strike point is given by

$$E_z = \int_0^H \rho(\zeta) \frac{\zeta d\zeta}{2\pi\varepsilon_0 (D^2 + \zeta^2)^{3/2}} \qquad (2.10)$$

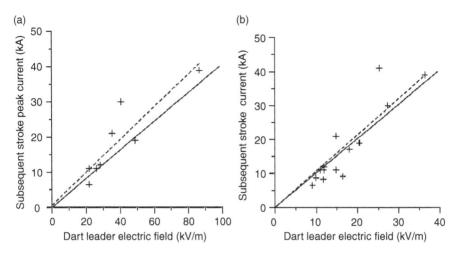

Figure 2.17 Variation of dart-leader electric field at (a) 10 m and (b) 110 m from the lightning channel base as a function of return-stroke peak current. Crosses show the experimental values obtained by Crawford et al. [62] and the dotted line shows the best fit for this data. The solid line shows the calculated values by Cooray et al. [61] using equations (2.5) to (2.9) with parameters appropriate to dart leaders (from Reference 61).

where $\rho(\zeta)$ is given by equations (2.5) to (2.9) with parameters appropriate to dart leaders and where ε_0 is the permittivity of free space. The electric fields of dart leaders at 10 and 110 m as a function of the ensuing return-stroke peak current, calculated using (2.10), are represented by solid lines in Figure 2.17. Figure 2.17 shows the experimental data (crosses and broken lines) for triggered lightning in Florida (1997, 1998 and 1999) as reported by Crawford and colleagues [62]. The calculated fields agree with the measurements to within $\sim$20 per cent. This supports the procedure used by Cooray and colleagues [61] to obtain the charge distribution on the dart-leader channel as a function of peak current.

Based on the charge distribution described above, the electric field at ground level directly below a dart-leader channel is evaluated for several prospective subsequent return-stroke peak currents. The results are shown in Figure 2.18. Let us consider a 10-m-tall grounded conductor of radius 0.28 m. The tip of the conductor is shaped as a hemisphere. As shown in Chapter 4, the background electric field necessary to launch a stable connecting leader from this conductor is $\sim$110 kV m^{-1}. A downward-moving dart leader that supports a prospective current of 12 kA produces a field of this magnitude at ground level when the tip of the dart leader is $\sim$150 m from ground. If the speed of the connecting leader issued from the conductor is $\sim 1 \times 10^5$ m s^{-1}, a typical value observed for connecting leaders, a connection between the two leaders takes place at a height of $\sim$15 m from the tip of the conductor, assuming that the speed of the dart leader is 1×10^7 m s^{-1}. However, this distance is more a minimum value for several reasons. First, the critical electric field necessary to

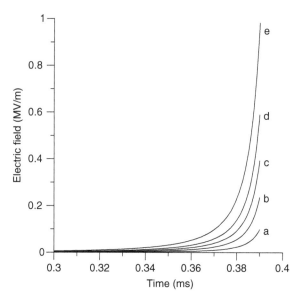

Figure 2.18 Electric field at ground level at a point directly below the path of a down-coming dart leader with speed 1×10^7 m s^{-1}. The height of the charge centre is taken to be 4 km. The charge distribution in the dart-leader channel is given by equations (2.5) to (2.9), with constants appropriate to the dart leaders. The calculation stops when the tip of the dart leader is 100 m above ground. Calculations are given for the prospective subsequent return-stroke currents of (a) 5 kA, (b) 12 kA, (c) 20 kA, (d) 30 kA and (e) 50 kA.

launch a connecting leader decreases with the rate of change of the electric field. The critical field of 110 kV m^{-1} is based on a static background electric field and therefore the critical electric field pertinent to the situation under consideration could be smaller than 110 kV m^{-1}. Second, the latter value is valid for standard atmospheric conditions. However, the density of the gas in a defunct return-stroke channel is less than that at normal atmospheric conditions, which means the critical electric field necessary to launch a connecting leader is less than the above value. The experimental evidence for the existence connecting leaders associated with dart leaders is discussed in Section 2.7.2.

2.7.5 Speed of connecting leaders

Once a connecting leader is issued from a grounded body under the influence of the down-coming stepped leader, whether a connection between these two leaders is made or not depends on the relative speed of the stepped leader and the connecting leader. Thus, the speed of connecting leaders is a necessary parameter to simulate the lightning attachment to structures.

The majority of the speeds of upward-moving leaders reported in the literature are for those in either rocket-triggered lightning or upward-initiated lightning flashes from tall structures. In both these cases the leaders move under the influence of a more or less static background electric field generated by an overhead thundercloud. The reported values of leader speeds vary between 1×10^4 and 1.4×10^6 m s^{-1} [63–67]. However, in studies related to lightning attachment what is needed is the speed of upward-moving connecting leaders propagated under the influence of the electric field created by down-coming stepped leaders. Yokoyama and colleagues [67] studied lightning attachment to an 80-m-tall tower and managed to obtain a few samples of the speed of upward-moving connecting leaders propagated in this way. In the four examples presented by Yokoyama and colleagues [67], the speed of stepped leaders and connecting leaders at the moment of connection was estimated as 5.9×10^6 and 1.3×10^6 m s^{-1}, 2.7×10^5 and 14×10^5 m s^{-1}, 2.7×10^6 and 2.9×10^6 m s^{-1}, and 6.9×10^5 and 5×10^5 m s^{-1}, respectively. The average propagation speed of the upward-connecting leaders ranged from 0.8×10^5 to 2.7×10^5 m s^{-1}. Unfortunately, it is not possible to distinguish between the positive and negative upward-connecting leaders in this experiment, because neither the polarity nor the return-stroke peak current were reported.

Recently, Becerra and Cooray [68] evaluated the time development of upward leaders from a tall structure under the influence of the electric field created by down-coming stepped leaders. The charge distribution of the stepped leader used in the calculation is identical to that represented by equations (2.5) to (2.9). Figure 2.19 shows the simulated streak image of the upward-connecting leader. The results show that the speed of the upward-moving leader depends on the speed of the down-coming stepped leader. Simulations also show that the connecting leader reaches speeds close to 1×10^5 m s^{-1} before the final connection. Note how both the upward-leader speed and the location of the point of connection are decided by the speed of the down-coming leader.

2.7.6 Currents in connecting leaders

As the stepped leader approaches the ground, several connecting leaders may be issued from grounded objects, but only one will be successful in bridging the gap between the ground and the stepped leader. However, the currents in these aborted connecting leaders may still be high enough to damage sensitive electronic components, cause ignition in flammable vapours and injure humans. Because upward leaders could be generated by any structure located in the vicinity of lightning flashes, they may pose a threat in various ways, even if the threat of a direct lightning strike is reduced to a minimum by a lightning-protection system. For example, a vent of a flammable liquid storage located close to a tall tower could very well be ignited by a connecting leader.

Direct measurement of currents in upward-connecting leaders of natural lightning flashes is complicated because, in addition to the current measurements, we need to have simultaneous time-resolved records of the development of the discharge to separate the measured current into contributions from the connecting leader and the return

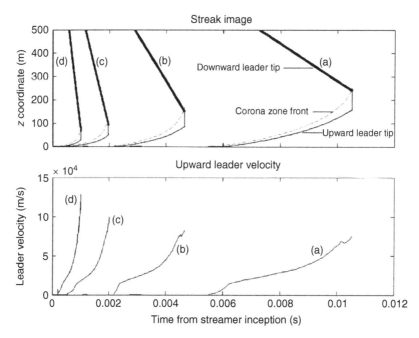

Figure 2.19 Simulated streak image and variation of the upward-connecting leader speed for different values of the downward leader velocities: (a) 8×10^4 m s^{-1}, (b) 2×10^5 m s^{-1}, (c) 5×10^5 m s^{-1} and (d) 1×10^6 m s^{-1} (from Reference 67)

stroke. However, a triggered-lightning technique at altitude has provided a possible way to measure the currents associated with the development of upward-connecting leaders in the presence of a down-coming negative leader (see Chapter 3 for a description of rocket-triggered lightning). Using this technique, Lalande and colleagues [69] managed to record the currents in upward leaders. An example of a current measured by them is shown in Figure 2.20. Note that the current can reach values as large as 100 A before the connection is made.

The results shown in Figure 2.20 are for a connecting leader that made a successful connection with a down-coming negative leader. For the reasons mentioned earlier, the currents in aborted connecting leaders (connecting leaders that do not attach to the stepped leader) are also of interest in lightning-protection studies. The peak currents in aborted connecting leaders may be lower than those in connecting leaders that succeed in bridging the gap between the ground and the stepped leader. This is so because the former cannot enter into the high-field region or the streamer region of the stepped leader, but the latter can. However, the currents in the aborted leaders cannot be smaller than 10 A or so, because such current amplitudes occur even in leaders in laboratory sparks. More information about the currents in connecting leaders can be obtained from theory. Modelling of positive connecting leaders has been carried out by Becerra and Cooray [68,70,71] and Aleksandrov and colleagues

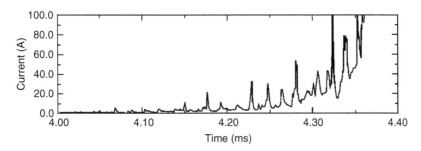

Figure 2.20 Plots of lightning triggered at altitude 9 516 showing the current produced by the upward-connecting positive leader from the grounded 50-m-tall wire (from Reference 69)

[72] (see also Chapter 4). These studies also show that positive leaders, a few metres in length, can support currents of the order of 10 A. Because the aborted leaders can be up to several metres in length and their speeds lie in the range 1×10^4 to 1×10^5 m s^{-1}, this current may flow for several tens to hundreds of microseconds. Because the temperature of the leader channel can be several thousands of degrees kelvin, aborted connecting leaders can easily cause explosions if they occur in environments with flammable vapours. Moreover, during the return-stroke stage, the high electric field that had driven the aborted connecting leader will be removed within a few microseconds, a field change that may cause a back-discharge current to flow along the aborted connecting leader and so neutralize the charge deposited on it. Depending on how fast the neutralization takes place, the amplitude of the back-discharge currents may reach values much larger than the magnitude of the currents generated during the growth of the connecting leaders. However, the duration of the back-discharge currents would be much shorter than the duration of the currents associated with the growth of the connecting leader (see also Chapter 20).

2.8 Current parameters of first and subsequent return strokes

The parameters of currents generated by return strokes at the point of strike are of utmost importance in lightning-protection studies. The peak current of the return stroke decides the voltage drop developed across a resistive load during a lightning strike. It is this voltage that leads to side flashes from (or along) struck objects. For example, if a resistance of a struck object is 10 Ω, a 30 kA peak return-stroke current will induce a 3 MV peak voltage across the object.

The peak-current derivative governs the magnitude of over-voltage produced across an inductive load during a lightning strike. For example, a lightning current with a 30 kA μs^{-1} current derivative will produce a voltage of $\sim$30 kV across an inductive load of 1 μH. Moreover, the magnetic field time derivative in the vicinity of the lightning channel is proportional to the current derivative. The magnetic field time derivative, in turn, decides the voltage induced in a loop as the magnetic field penetrates it.

For example, the peak voltage induced in a loop of 1 m² area placed 10 m away from a return stroke with a current derivative of 30 kA μs^{-1} is ~600 V.

The charge transported to ground by the lightning discharge determines the heating effects at the strike point. Therefore, the amount of metal melted and the probability of burn through when the lightning strike point happens to be on a metal surface (e.g. a metal roof or the surface of an air plane) is determined by this parameter. When a lightning arc is attached to a metal surface, energy is supplied to the metal by the acceleration of electrons across the anode potential V_a or by the acceleration of ions across the cathode potential V_c. These potential drops are independent of the magnitude of the current and therefore the energy liberated by a current pulse $i(t)$ is $\int i(t)V_a\,dt$ or $\int i(t)V_c\,dt$. Thus, the energy liberated is proportional to the charge transported by the current. Of course, there is another supply of energy from the thermal velocities of the particles, but this is not strongly dependent on current. Thus the main supply of energy is proportional to the charge.

If a lightning current is passed through a resistive load, for example a surge-protection device, of resistance R, the total heat generated in the load is given by $R\int[i(t)]^2 dt$. The term containing the integral is called the action integral. In this case it determines the amount of heat dissipated in the resistive load. Moreover, this parameter also governs the magnitude of the force between conductors located in the vicinity of each other or the forces experienced by a bend in a conductor carrying a lightning channel. Assume that the lightning current is divided into two parts at the strike point and it propagates along two conductors located parallel to each other and separated by a distance r. The impulse experienced by the two conductors as a result of the current flow along them is given by:

$$I_{im} = \frac{\mu_0}{2\pi r} \int_0^\infty [i(t)]^2 dt \qquad (2.11)$$

which is proportional to the action integral.

The risk assessment in lightning protection is based on the distribution of various lightning-current parameters, some of which were mentioned earlier. For practical purposes these distributions have to be either measured or evaluated indirectly from other measured parameters. There are several ways to measure the lightning-current parameters: (i) using instrumented towers, (ii) using triggered lightning and (iii) remote sensing using electromagnetic fields. The first is the best procedure because the inherent nature of triggering procedure, triggered lightning flashes do not contain the first return stroke, which are mediated by stepped leaders in natural lightning flashes. The parameters of first return strokes are those that are used for risk evaluation in lightning protection. Unfortunately, the third procedure is not yet developed enough to make accurate estimations of the first return-stroke currents. Consequently, at present, lightning protection standards are based on the data gathered at instrumented towers and in this section we mainly deal with such data.

The measurements of lightning currents with instrumented towers were conducted by Berger [63] in Switzerland, Ericsson [3] in South Africa, Heidler and colleagues

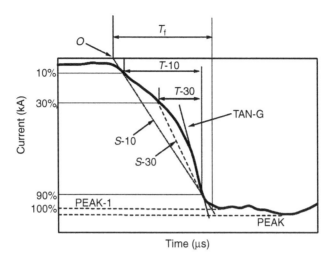

Figure 2.21   Lightning-current parameters (from Reference 76)

[73] in Germany, Pichler and co-workers in Austria [74], Garbagnati and Piparo [75] in Italy, Takami and Okabe [76] in Japan and Visacro and colleagues [77] in Brazil. We restrict the data presentation to short towers because the lightning strikes on tall towers consist mainly of upward-initiated flashes. The important features of the current in upward-initiated lightning flashes are described later. Before presenting the data, however, it is necessary to identify various parameters that are of interest in lightning-current waveforms. These parameters are identified in Figure 2.21 and additional information concerning these parameters is given in Table 2.6.

Table 2.6   Definition of different lightning-current parameters

PEAK-1 (kA)	**Maximum amplitude of the first peak**
PEAK (kA)	The highest current peak
T_f (μs)	Front duration
T-10 (μs)	Time between the 10% and 90% values of PEAK-1 at the wave front
T-30 (μs)	Time between the 30% and 90% values of PEAK-1 at the wave front
T_t (μs)	Stroke duration; time from the virtual zero time to the half-peak value at the wave tail
TAN-10 (kA μs^{-1})	Rate of rise at the 10% point of PEAK-1
TAN-G (kA μs^{-1})	Maximum rate of rise
S-10 (kA μs^{-1})	Average rate of rise between the 10% and 90% values of PEAK-1
S-30 (kA μs^{-1})	Average rate of rise between the 30% and 90% values of PEAK-1

2.8.1 Berger's measurements

The most complete and extensively analysed dataset of tower measurements was made by Berger [63]. He conducted measurements at a tower located on top of Mount San Salvatore at an elevation of 915 m from sea level. The height of the tower was 70 m. The current shunt was located at the top of the tower and the signal from the shunt was connected to oscilloscopes at the tower base through screened cables. The oscilloscope was triggered by the incoming signal and the screen was photographed by a moving film camera. The oscilloscope did not have pre-trigger capabilities and therefore events that took place before the trigger could not be captured. The maximum current recordable was 200 kA. Even though the current shunt had a rather high time sensitivity (response time 16 ns), the time resolution was limited by the accuracy with which the oscilloscope records could be read. The lightning-current parameters obtained by Berger are summarized in Table 2.7.

Table 2.7   Current parameters reported by Berger et al. [78]

Parameter	Sample	Percentage of cases exceeding the tabulated values		
		95%	50%	5%
PEAK (kA)				
Negative first RS	101	14	30	80
Negative subsequent RS	135	4.6	12	30
Positive RS	26	4.6	35	250
Front duration, T_f (μs)				
Negative first RS	89	1.8	5.5	18
Negative subsequent RS	118	0.22	1.1	4.5
Positive RS	19	3.5	22	200
TAN-G (kA μs^{-1})				
Negative first RS	92	5.5	12	32
Negative subsequent RS	122	12	40	120
Positive RS	21	0.20	2.4	32
Stroke duration (μs)				
Negative first RS	90	30	75	200
Negative subsequent RS	115	6.5	32	140
Positive RS	16	25	230	2 000
Impulse charge (C)				
Negative first RS	90	1.1	4.5	20
Negative subsequent RS	117	0.22	0.95	4.0
Positive RS	25	2.0	16	150

RS, return strokes.

Table 2.8   Lightning-current parameters measured by Grabagnati and Piparo [75]

Parameter	Downward lightning		Upward lightning	
	First strokes	Subsequent strokes	First strokes	Subsequent strokes
Number of events	42	33	61	142
PEAK (kA)	33	18	61	142
Maximum rate of current rise (kA μs^{-1})	14	33	5	13
Front duration (μs)	9	1.1	4	1.3
Back duration (μs)	56	28	35	31
Impulse charge (C)	2.8	1.4		

2.8.2   Garbagnati and Piparo's measurements

Grabagnati and Piparo [75] conducted current measurements from 1969 to 1979 at a tower in Northern Italy. The tower, which was 40 m tall, was located on flat ground at an altitude of 900 m. The current was measured by a shunt resistor at the top of the tower. The response time of the shunt was less than 55 ns. The data were fed using a shielded cable to a wideband digital oscilloscope located at the tower base. No pre-trigger facility was available in the oscilloscope. The oscilloscope screen was photographed using a moving film camera. The maximum current recordable was 140 kA. The median values of lightning-current parameters measured by Garbagnati and Piparo [75] are summarized in Table 2.8.

2.8.3   Eriksson's measurements

Eriksson and colleagues [79] conducted current measurements from 1972 to 1987 at a tower in South Africa. The tower, which was 60 m tall, was located on flat ground at an altitude of 1 400 m. The current was measured by a current transformer with a bandwidth of 1 Hz to 10 MHz. The current transformer was located at the base of the tower. The data from the current transformer was fed to a wideband digital oscilloscope through a shielded cable. The sampling time of the oscilloscope varied from 50 to 200 ns and the minimum trigger level was set to 1–2 kA. No pre-trigger facility was available in the oscilloscope. The number of lightning flashes collected by Eriksson in his study is limited and therefore a detailed summery of various parameters pertinent to that study is not given. However, various features of the dataset obtained by Eriksson are summarized in Table 2.9. In this data set one subsequent return stroke had a peak current derivative of 170 kA μs^{-1}. Also, in this dataset the log-normal distribution of the peak current amplitude is characterized by a mode of 44 kA and a standard deviation of 0.3.

Table 2.9 Summary of current-waveform parameters measured by Eriksson et al. [79]

Event	First (F) or subsequent (S)	$\hat{\imath}$ (kA)	Q (C)	dI/dt (kA s^{-1})	$I_2^2 dt$ (A^{2}S)	Direction of propagation*
721129/1	F	−35	—	—	—	↓
731126/7	F	−73	—	—	—	↓ ?
731216/9	F	−10	>1.6	—	>1.2 × 10^4	↓
740104/10/1	F	−41	>2	—	>3 × 10^4	↓ ?
740104/10/2	S	>−12	—	—	—	
740104/10/3	S	>−4	—	—	—	
740118/11	F	−50	6.5	>11.0	3.2 × 10^5	↓ ?
740118/12/1	F	−58	12.4	>4.0	7.2 × 10^5	↓ ?
740118/12/2	S	−55	10.4	>5.0	5.7 × 10^5	—
740118/12/3	S	−22	5.8	>7.0	1.3 × 10^5	—
741224/16	F	−53	1.5	>8.0	3.7 × 10^4	→
750227/19	F	−26	—	—	—	→
751102/21	F	−19	0.8	24	8.4 × 10^3	→
751114/22/1	F	−10	0.9	3.1	7.3 × 10^3	←
751114/22/2	S	−16	0.9	22	9.0 × 10^3	—
751114/22/3	S	−15	0.8	29	7.0 × 10^3	—
760207/23	F	>−100	>9	>12	>9.0 × 10^5	↓ ?
770222/25/1	F	−83	>3	20	>2.0 × 10^5	→
770222/25/2	S	−82	>2	170	>1.0 × 10^5	
770222/25/3	S	−13	>0.3	27	>2.0 × 10^4	
771009/27	F	−87	15	12	6.8 × 10^5	→

* ↓, downward; ↓?, presumed downward; ↑, upward.

2.8.4 Analysis of Andersson and Eriksson

The dataset obtained by Berger was digitized and re-analysed by Anderson and Eriksson [80]. The results obtained in the study are summarized in Table 2.10. CIGRE have recommended that these parameters be used in engineering applications.

2.8.5 Measurements of Takami and Okabe

Takami and Okabe [76] conducted measurements on 60 transmission-line towers from 1994 to 2004 in the Kanto area of Japan. The tower heights varied from 60 to 140 m. The currents were measured by Rogowski coils located at the tower tops. The bandwidth of the sensors was 10 Hz to 1 MHz. The signals were recorded digitally by equipment located at the top of the tower. The trigger level was 9 kA and the sampling time was 100 ns. The maximum current that could be recorded by the equipment was 300 kA. The lightning-current parameters pertinent to the negative first return strokes measured by Takami and Okabe are summarized in Table 2.11.

Table 2.10 Lightning-current parameters as reported Anderson and Eriksson [80]

Parameter	Sample	Percentage of cases exceeding the tabulated values		
		95%	50%	5%
First stroke				
T-30 (μs)	80	0.91	2.3	5.76
T-10 (μs)	80	1.76	4.46	11.32
T-30 (μs)	80	0.91	2.3	5.76
TAN-10 (kA μs^{-1})	75	0.58	2.61	11.80
S-10 (kA μs^{-1})	75	1.74	4.95	14.09
S-30 (kA μs^{-1})	73	2.62	7.23	19.95
TAN-G (kA μs^{-1})	75	9.06	24.27	64.97
PEAK-1 (kA)	75	12.87	27.67	59.47
PEAK (kA)	80	14.07	31.05	68.52
Subsequent strokes				
T-10 (μs)	114	0.13	0.61	2.79
T-30 (μs)	114	0.07	0.35	1.83
TAN-10 (kA μs^{-1})	108	1.90	18.88	187.44
S-10 (kA μs^{-1})	114	3.30	15.42	72.02
S-30 (kA μs^{-1})	114	4.12	20.14	98.46
TAN-G (kA μs^{-1})	113	7.54	37.84	190.01
PEAK-1 (kA)	114	4.86	11.80	28.64
PEAK (kA)	114	5.19	12.30	29.18

Table 2.11 Negative first return-stroke lightning-current parameters (logarithmic normalized) measured by Takami and Okabe [76]

Parameter	95% value	50% value	5% value
PEAK (kA)	10.1	29.3	84.9
PEAK-1 (kA)	9.8	27.7	78.7
T_f (μs)	2.5	4.8	9.0
T-10/0.8 (μs)	2.5	4.8	9.0
T-30/0.6 (μs)	1.6	3.2	6.2
TAN-G (kA μs^{-1})	7.0	18.9	51.2
S-10 (kA μs^{-1})	2.0	5.8	17.0
S-30 (kA μs^{-1})	3.2	8.8	23.7
TAN-10 (kA μs^{-1})	0.6	2.1	7.5
T_t (μs)	9.5	36.5	139.7

2.8.6 Measurements of Visacro and colleagues

Visacro and colleagues [77] conducted lightning-current measurements in Brazil, the first time that various features of return-stroke currents were measured in a tropical region. The research mast used in the measurements was at Morro do Cachimbo Hill, 1 430 m above sea level. It is located in the outskirts of Belo Horizonte (43° 58′ W, 20° 00′ S). The height of the tower was 60 m and the current was measured at the base of the tower using two Pearson coils. The frequency bandwidth of the Pearson coils extended from 100 Hz to 100 MHz. The lightning-current parameters measured in this study are summarized in Table 2.12.

2.8.7 Summary of current measurements

The median values of peak current and peak-current derivatives measured in different studies are summarized in Table 2.13. In the same table we give the median values of the same parameters of the subsequent return strokes of triggered lightning. Triggered lightning flashes lack the first return stroke because they are initiated by a positive leader that travels towards cloud either from the top of the grounded rocket in triggered lightning or from the top of the tower in the case of upward-initiated lightning flashes from tall towers.

The median peak current in all measurements lies in the range 30–45 kA, with the largest value reported from the tropical region. Usually, it is assumed that Berger's data on dI/dt are distorted by instrument response times, but the first return-stroke current derivative extracted from Berger's measurement by Anderson and Eriksson, 24.3 kA μs^{-1}, is close to the median values of this parameter obtained in other studies with equipment of better time resolution. From this we can conclude that the median value of the peak-current derivative of first strokes lies close to 20–30 kA μs^{-1}. The median value of the subsequent return stroke currents of all measurements lies in the range 10–18 kA. The median value of the peak-current

Table 2.12 Lightning-current parameters observed by Visacro et al. [77] in Brazil

Parameter	Sample	Percentage of cases not exceeding tabulated values		
		5%	50%	95%
First stroke				
PEAK-1 (kA)	31	73	40	22
PEAK (kA)	31	85	45	24
T-10 (µs)	31	3.1	5.6	9.9
T-30 (µs)	31	1.4	2.9	5.9
S-10 (kA µs^{-1})	31	3.5	5.8	9.6
S-30 (kA µs^{-1})	31	5.1	8.4	13.7
TAN-G (kA µs^{-1})	31	11.9	19.4	31.4
T_{50} (µs)	31	19.7	53.5	145.2
Subsequent strokes				
PEAK (kA)	59	7.0	16.3	37.7
T-10 (µs)	59	0.2	0.7	2.3
T-30 (µs)	59	0.12	0.4	1.2
S-10 (kA µs^{-1})	59	5.6	18.7	62.7
S-30 (kA µs^{-1})	59	8.1	24.7	75.0
TAN-G (kA µs^{-1})	59	10.1	29.9	88.6
T_{50} (µs)	59	2.2	16.4	122.3

derivative of subsequent return strokes observed by Berger is also similar to the median value of this parameter observed both in the Brazilian study and in subsequent strokes that struck the Peissenberg tower. However, in the case of triggered lightning flashes, the experiments conducted at the Kennedy Space Centre (KSC) gave a median value of 91.4 kA µs^{-1}. At present, the reason for this discrepancy is not known.

2.9 Statistical representation of lightning current parameters

It is a general consensus among lightning researchers that lightning-current parameters can be approximated by log-normal distributions. According to this distribution the logarithm of the random variable x follows the normal or Gaussian distribution. This distribution is characterized by a median value x_m and a standard deviation σ. The probability density function $p(x)$ of x of this distribution is given by

$$p(x) = \frac{1}{\sqrt{2\pi}x\sigma} e^{-0.5\left[\frac{\ln(x)-\ln(x_m)}{\sigma}\right]^2} \qquad (2.12)$$

Table 2.13 Median values of peak current and peak-current derivatives measured in different studies

Study	First stroke		Subsequent stroke	
	Peak current (kA)	Peak-current derivative (kA μs^{-1})	Peak current (kA)	Peak-current derivative (kA μs^{-1})
Anderson and Eriksson [81]	31.1 (0.21)	24.3 (0.26)	12.3	39.9
Eriksson [3]	44	>13	18	>43
Fisher et al. [82] for triggered lightning				
Alabama			11	
Florida			15	
Total			12	
Leteinturier et al. [83] for triggered lightning			15	
Berger et al. [78]	30	12	12	40
Berger et al. [78] for positive flashes	35	2.4		
Garbagnati et al. [84] as given in [85]	33 (0.25)	14 (0.35)		
Saba et al. [86]	13.5		28.3	
Depasee [87] for triggered lightning				
Florida			12 (9.0)	91.4 (97.1)
France			9.9 (4.6)	37.1 (18.6)
Visacro et al. [77]	45 (0.37)	19.4 (0.29)	16.3 (0.51)	29.9 (0.66)
Takami and Okabe [76]	29.3	18.9		

Standard deviation of the logarithm of the variations are given in parentheses.

The probability that the value of the parameter exceeds x_0 is given by:

$$P_c(x) = \frac{1}{\sqrt{\pi}} \int_{u_0}^{\infty} e^{-u^2} du = 0.5 erfc(u_0) \quad (2.13)$$

where $u = \ln(x) - \ln(x_m)/\sqrt{2}\sigma$ and $u_0 = \ln(x_0) - \ln(x_m)/\sqrt{2}\sigma$. In many studies the lightning parameter distributions are used in simulations of the Monte Carlo type to evaluate the effect of lightning flashes on various systems. In such studies it is necessary to calculate the joint probability distribution of two parameters correlated with each other. The joint probability density function of two parameters x and y is given by

$$p(x, y) = \frac{1}{2\pi x y \sigma_x \sigma_y \sqrt{1-\rho^2}} e^{-0.5\left[\frac{(f_1 - f_2 + f_3)}{1-\rho^2}\right]} \qquad (2.14)$$

where ρ is the coefficient of correlation between the two parameters. In equation (2.14),

$$f_1 = \left(\frac{\ln x - \ln x_m}{\sigma_x}\right)^2 \qquad (2.15)$$

$$f_2 = 2\rho \left(\frac{\ln x - \ln x_m}{\sigma_x}\right)\left(\frac{\ln y - \ln y_m}{\sigma_y}\right) \qquad (2.16)$$

$$f_3 = \left(\frac{\ln y - \ln y_m}{\sigma_y}\right)^2 \qquad (2.17)$$

If the two parameters are independently distributed then $\rho = 0$ and $p(x, y) = p(x)p(y)$. The cumulative probability that $x > x_0$ and $y > y_0$ is given in this case by

$$P(x > x_0, y > y_0) = \lfloor 0.5 erfc(u_{x_0}) \rfloor \lfloor 0.5 erfc(u_{y_0}) \rfloor \qquad (2.18)$$

Similarly, the probability that $x < x_0$ and $y > y_0$ is given by

$$P(x < x_0, y > y_0) = \lfloor 1 - 0.5 erfc(u_{x_0}) \rfloor \lfloor 0.5 erfc(u_{y_0}) \rfloor \qquad (2.19)$$

As mentioned earlier, the log-normal distribution is completely defined by the median or mode (50 per cent value) and the standard deviation. These values for the distributions obtained by Takami and Okabe [76] and Visacro and colleagues [77] have already been given in Tables 2.11 and 2.12. The data gathered by Berger and colleagues [78] were scrutinized thoroughly by CIGRE and the Institute of Electrical and Electronic Engineers (IEEE) study committees and the median and standard deviation of distributions of different lightning parameters derived by them are given in Table 2.14. The statistical parameters for positive strokes pertinent to Berger's data are given in Table 2.15. IEEE suggests that the peak-current distribution of negative ground flashes obtained by Berger can be approximated by two straight lines (when plotted by probability) intersecting at 20 kA. According to this description, for $I_p \leq 20$ kA the median value is 61.1 kA and the standard deviation is $\sigma_{\ln I p} = 1.33$. For $I_p > 20$ kA, the median value is 33.3 kA and the standard deviation is $\sigma_{\ln I p} = 0.605$.

2.9.1 Correlation between different current parameters

The correlation coefficients between different lightning parameters as observed in the measurements conducted by Berger and colleagues [78], Takami and Okabe [76] and Visacro and colleagues [77] are tabulated in Tables 2.16, 2.17 and 2.18. In the same table the parameters that provide the best fit if the correlated quantities are represented by the equation $y = ax^d$ are also given. There is a strong correlation between the peak

58 Lightning Protection

Table 2.14 The median and standard deviation of log normal distributions of downward negative lightning-current parameters based on Berger's data (from Reference 88)

Parameter	First stroke		Second stroke	
	Median	Logarithmic standard deviation, σ	Median	Logarithmic standard deviation, σ
$T_{d10/90} = T_{10/90}/0.8$ (μs)	5.63	0.576	0.75	0.921
$T_{d30/90} = T_{30/90}/0.6$ (μs)	3.83	0.553	0.67	1.013
$t_m = I_F/S_m$ (μs)	1.28	0.611	0.308	0.708
S_m maximum (kA μs^{-1})	24.3	0.599	39.9	0.853
S_{10} at 10% (kA μs^{-1})	2.6	0.921	18.9	1.404
$S_{10/90}$ (kA μs^{-1})	5.0	0.645	15.4	0.944
$S_{30/90}$ (kA μs^{-1})	7.0	0.622	20.1	0.967
PEAK-1 (kA)	27.7	0.461	11.8	0.530
PEAK (kA)	31.1	0.484	12.3	0.530

current and the impulse charge of the first return strokes in Berger's study. The data of Takami and Okabe [76] show a strong correlation between the peak current and the peak-current derivative of the first strokes. Their results are depicted in Figure 2.22. However, in the datasets of Berger and colleagues [78] and Visacro and colleagues [77] only a weak correlation is observed between the peak current and the peak-current derivative. The data of Berger and colleagues [78] are shown in Figure 2.23.

As mentioned previously, Berger and colleagues [78] found a strong correlation between the charge brought to ground by negative first return strokes during the first 2 ms (called the impulse charge) and the peak return-stroke current. The corresponding dataset is shown in Figure 2.24a. For comparison a plot of the impulse charge of positive first return strokes as a function of peak current observed in the same study is given in Figure 2.24b. Note that the correlation is not as strong as that of the negative counterpart. Using the same dataset of Berger and colleagues

Table 2.15 Statistical parameters of positive strokes (from Reference 89)

Parameter	Sample size	Median	Σ
PEAK (kA)	26	35	1.21
T_f, front time (μs)	19	22	1.23
T_h, stroke duration (μs)	16	230	1.33
S_m (kA μ^{-1}s μs^{-1})	21	2.4	1.54

Table 2.16 Correlation coefficients (ρ_c) and derived functions; conditional median, $y_{mc}/x = ax^d$ from data presented by Anderson and Eriksson [81] and Berger et al. [78] (from Reference 89)

Conditional median	ρ_c	a	d	$\sigma_{\ln(y/x)}$
First negative stroke				
t_f/I_p (μs) Anderson and Eriksson [81]	0.47	0.61	0.535	0.4855
I_p/t_f (kA) Anderson and Eriksson [81]	0.47	17.857	0.4132	0.4268
S_m/I_p (kA μs^{-1}) Anderson and Eriksson [81]	0.38	4.805	0.472	0.5550
I_p/S_m (kA) Anderson and Eriksson [81]	0.38	11.708	0.3062	0.4472
Q_{flash}/I_p (C) Berger et al. [78]	0.54	0.149	1.1392	0.8585
I_p/Q_{flash} (kA) Berger et al. [78]	0.54	18.568	0.2560	0.4069
Q_{imp}/I_p (C) Berger et al. [78]	0.77	3.20	1.4811	0.5934
I_p/Q_{imp} (kA) Berger et al. [78]	0.77	16.074	0.40	0.3085
E_{flash}/I_p (A^{2}s) Berger et al. [78]	0.88	8.643	2.5481	0.6650
I_p/E_{flash} (kA) Berger et al. [78]	0.88	1.127	0.3039	0.2296
Subsequent negative strokes				
S_m/I_p (kA μs^{-1}) Berger et al. [78]	0.11	25.618	0.1765	0.8448
I_p/S_m (kA) Berger et al. [78]	0.11	9.554	0.0685	0.5264
Q_{stroke}/I_p (C) Berger et al. [78]	0.43	0.11	1.0149	1.1285
I_p/Q_{stroke} (kA) Berger et al. [78]	0.43	11.569	0.1822	0.4781
Positive stroke				
t_f/I_p (μs) Berger et al. [78]	0.07	17.083	0.0712	1.2270
I_p/t_f (kA) Berger et al. [78]	0.07	28.290	0.0689	1.2070
S_m/I_p (kA μs^{-1}) Berger et al. [78]	0.49	0.261	0.6236	1.3425
I_p/S_m (kA) Berger et al. [78]	0.49	24.985	0.3850	1.0548
Q_{flash}/I_p (C) Berger et al. [78]	0.62	15.525	0.4612	0.7061
I_p/Q_{flash} (kA) Berger et al. [78]	0.62	0.907	0.8336	0.9494
E_{flash}/I_p (A^{2}s) Berger et al. [78]	0.84	5 828.489	1.326	1.0363
I_p/E_{flash} (kA) Berger et al. [78]	0.84	2.823	0.5321	0.6565

Table 2.17 Correlation coefficients between different current parameters of first strokes as observed in the study conducted by Takami and Okabe [76]

	PEAK	PEAK-1	T_f	T-10	T-30	TAN-G	S-10	S-30
PEAK	1							
PEAK-1	0.988	1						
T_f	0.257	0.238	1					
T-10	0.241	0.223	0.995	1				
T-30	0.368	0.345	0.802	0.815	1			
TAN-G	0.819	0.846	−0.030	−0.042	−0.040	1		
S-10	0.804	0.827	−0.350	−0.361	−0.132	0.831	1	
S-30	0.758	0.787	−0.297	−0.307	−0.293	0.889	0.931	1

Table 2.18 Correlation coefficients between different current parameters of first and subsequent strokes as observed in the study conducted by Visacro et al. [77]

	PEAK-1	PEAK	TAN-G	S-10	S-30	T-10	T_d-10	T-30	T_d-30
				First strokes					
PEAK-1	1								
PEAK	0.944	1							
TAN-G	0.160	0.127	1						
S-10	0.290	0.306	0.178	1					
S-30	0.104	0.145	0.134	0.817	1				
T-10	0.736	0.651	0.071	−0.365	−0.488	1			
T_d-10	0.734	0.650	0.068	−0.367	−0.490	1	1		
T-30	0.711	0.612	0.140	−0.217	−0.503	0.939	0.939	1	
T_d-30	0.709	0.611	0.140	−0.218	−0.503	0.939	0.939	1	1
				Subsequent strokes					
PEAK-1	1								
PEAK	–	–							
TAN-G	0.383	–	1						
S-10	0.239	–	0.747	1					
S-30	0.309	–	0.904	0.822	1				
T-10	0.212	–	−0.365	−0.574	−0.443	1			
T_d-10	0.209	–	−0.364	−0.569	−0.444	1	1		
T-30	0.300	–	−0.531	−0.577	−0.605	0.878	0.879	1	
T_d-30	0.309	–	−0.532	−0.566	−0.594	0.878	0.879	0.999	1

[78], Cooray and colleagues [61] analysed the charge dissipated within the first 100 μs of first strokes and within the first 50 μs of subsequent strokes. They found that these charge magnitudes strongly correlate to the peak return-stroke current. The data obtained together with the best-fit line that passes through the origin are shown in Figures 2.25a and b. The analysis shows that the charge and the peak current are linearly correlated with a correlation coefficient of 0.98 for first strokes and 0.95 for subsequent strokes. This strong correlation between the charge and the peak current is used by Cooray and colleagues [61] to extract the charge distribution along the leader channels as a function of prospective return-stroke currents.

Figure 2.26 shows the relationship between current parameters for subsequent return strokes of triggered lightning measured in KSC, Florida, and Fort McClellan, Alabama. Note the reasonable correlation between the peak current and the current derivatives, S-10 and S-30 (see Figure 2.21 and Table 2.6 for definitions of S-10 and S-30).

2.9.2 Effect of tower height

Because most of the data pertinent to lightning flashes have been obtained from instrumented towers, it is important to know the effects, if any, of the tower on the measured

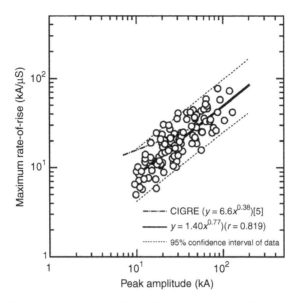

Figure 2.22   Correlation between the peak current and peak-current derivatives as observed by Takami and Okabe [76]

parameters. Consider a tower of a certain height and a stepped leader that comes down vertically at a horizontal distance y from the tower. Depending on the charge distribution on the leader channel, there is a certain critical value of y below which the stepped leader will be attracted to the tower. This critical distance depends on the

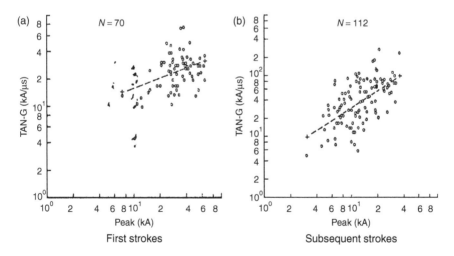

Figure 2.23   Correlation between the peak current and peak-current derivatives for first and subsequent strokes as observed by Anderson and Eriksson for the data of Berger et al. [78]

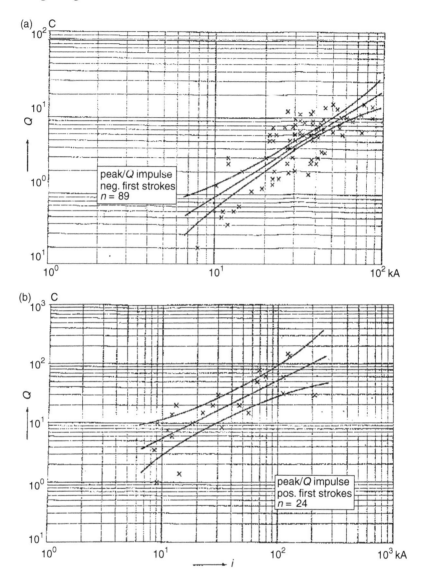

Figure 2.24 (a) Variation of impulse charge of negative first return strokes (charge brought to ground during the first 2 ms of the return stroke) as a function of peak current. The results were obtained by Berger et al. [78] using the current waveforms recorded at Mount San Salvatore. The relationship between these two parameters can be described by $Q_{imp} = 3.2 I_p^{1.48}$. (b) Variation of impulse charge of positive first return strokes (charge brought to ground during the first 2 ms of the return stroke) as a function of peak current. The results are obtained by Berger et al. [78] using the current waveforms recorded at Mount San Salvatore.

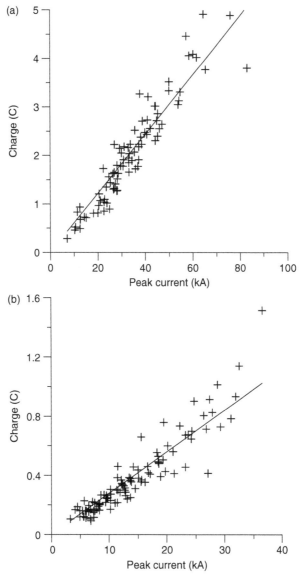

Figure 2.25 (a) Plot of the charge brought to ground by first return strokes in the first 100 μs ($Q_{f,100\mu s}$) as a function of the peak current (I_{pf}). The relationship between the two parameters can be represented by a straight line (also shown in the plot) with a correlation coefficient of 0.98 ($Q_{f,100\mu s} = 0.061 I_{pf}$). Results obtained by Cooray et al. [61] using the dataset of Berger et al. [78]. (b) Plot of the charge brought to ground by subsequent return strokes in the first 50 μs ($Q_{S,50\mu s}$) as a function of the peak current (I_{ps}). The relationship between the two parameters can be represented by a straight line (also shown in the plot) with a correlation coefficient of 0.95 ($Q_{S,50\mu s} = 0.028 I_{ps}$). Results obtained by Cooray et al. [61] using the dataset of Berger et al. [78].

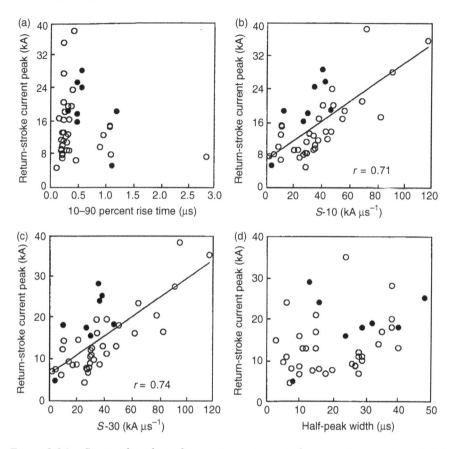

Figure 2.26 Scatterplots that relate various return-stroke current parameters. Solid circles represent 1990 data from KSC, Florida, and the open circles represent 1991 data from Fort McClellan, Alabama. (a) Current peak versus 10–90 per cent rise time; (b) current peak versus S-10; (c) current peak versus S-30; (d) current peak versus half-peak width (from Reference 36).

interaction between the down-coming leader and the connecting leader issued from the tower. As mentioned previously, this critical distance is called the attractive radius of the tower for that particular leader charge. (It is not equal to the striking distance for the reasons given in Chapter 4.) Because there is a strong correlation between the charge on the leader channel and the return-stroke peak current (see Section 2.9.1 and Chapter 4), this attractive radius for a given tower height can be expressed as a function of the peak of the prospective return-stroke current. The attractive radius of the tower increases with increasing charge density on the leader and hence on the return-stroke peak current. Thus, the area over which a lightning flash with a particular first return-stroke peak current is attracted to the tower increases with increasing peak current. In other words, the tower attracts lightning flashes that have larger first return-stroke currents from a large area, whereas flashes with small first return-stroke currents are attracted from a smaller area. Consequently, if the ground flash density in the region is uniform, the current distribution measured at the tower is biased towards higher currents. If the attractive radius of the tower as a function of the first return-stroke current is known, we can correct for this bias and obtain the unbiased distribution, i.e. a distribution as seen from a point located at plane ground. This correction can be performed as follows. Let us represent the probability density function of the peak current I_p of an elevated structure by P_{ds} and the probability density of strokes for flat ground by P_{dg}. Then we can show that [90,91]:

$$P_{ds} = \frac{r^v(h, I_p)P_{dg}(I_p)}{\int_0^\infty r^v(h, I_p)P_{dg}(I_p)dI_p} \quad (2.20)$$

where h is the height of the structure, r is the attractive radius of the structure (which, of course, is a function of the peak current) and $v = 2$ if the structure is a tower and $k = 1$ if it is a horizontal conductor. Borghetti and colleagues [85] studied this problem in detail and evaluated the corrected distributions for different expressions for the attractive radii available in the literature. As summarized in Reference 85, several expressions have been proposed in the literature to evaluate the attractive radius. Those based on electrogeometrical models (see Chapter 4) can be expressed as:

$$r = \sqrt{r_s^2 - (r_g - h)^2} \quad \text{for } h < r_g \quad (2.21)$$
$$r = r_s \quad \text{for } h \geq r_g$$

where r_s and r_g are the striking distance to the structure and to the ground, respectively.

The striking distance is connected to the return-stroke peak current by equation (2.22):

$$r_s = \alpha I_p^\beta \quad r_g = kr_s \quad (2.22)$$

where the values of α, β and k are independent of I_p. Attractive radii obtained from more complex models can be expressed as:

$$r = c + aI_p^\beta \quad (2.23)$$

where the values of a, b and c are independent of I_p. Tables 2.19 and 2.20 give the values of parameters in the above equations as proposed by different authors (as given by Borghetti and colleagues [85]). Based on these parameters, the mean and the standard deviation of current parameters at ground level, as obtained by Borghetti and colleagues [85], are listed in Tables 2.21 and 2.22. Note, for example, that in the case of Berger's study when moving from tower to ground distribution the median peak current of first strokes may decrease from 30 kA to a value in the

Table 2.19 Constants pertinent to equation (2.22) as proposed by different authors (from Reference 1)

	Striking distance model	α	β	k
1	Armstrong and Whitehead [92]	6.7	0.8	0.9
2	Adopted by IEEE Std. 1243, [93]	10	0.65	0.55
3	Adopted from Golde [94]	3.3	0.78	1

Table 2.20 Constants pertinent to equation (2.23) as proposed by different authors (from Reference 1)

	Attractive radius expression	c	a	b
4	Eriksson [96]	0	$0.84h^{0.6}$	$0.7h^{0.02}$
5	Rizk [90]	0	$4.27h^{0.41}$	0.55
6	Dellera and Gabagnati [97] – as given in [85]	$3h^{0.6}$	$0.028h$	1

Table 2.21 Median and standard deviations of current parameters at ground obtained using equation (2.20) with the constants given in Tables 2.19 and 2.20 for tower measurements published by Berger et al. [78] and Anderson and Eriksson [81] (from Reference 1)

Parameter	Berger et al. [78]						Anderson and Eriksson [81]					
	Attractive radius expression						Attractive radius expression					
	1	2	3	4	5	6	1	2	3	4	5	6
PEAK (kA)	17.5	18.9	17.3	17.4	20.3	20.8	22.1	23.0	21.7	21.8	24.1	24.1
PEAK-1 (kA)							20.2	21.1	20.0	20.1	22.0	22.2
t_f (μs)	4.3	4.5	4.3	4.3	4.6	4.7	3.2	3.3	3.1	3.2	3.3	3.3
TAN-G (maximum)	9.9	10.2	9.9	9.9	10.4	10.5	20.8	21.1	20.6	20.7	21.6	21.6
Time to half value	55.3	57.6	54.8	55.2	59.9	61.0	61.4	63.2	60.7	60.9	65.0	65.3
Impulse charge (C)	2.4	2.6	2.4	2.4	2.9	2.9	2.9	3.1	2.8	2.9	3.3	3.3

Table 2.22 Median and standard deviations of current parameters at ground obtained using equation (2.20) with the constants given in Tables 2.19 and 2.20 for the tower measurements of Garbagnati et al. [84] (from Reference 1)

Parameter	Garbagnati et al. [84]					
	Attractive radius expression					
	1	2	3	4	5	6
PEAK (kA)	21.6	21.8	20.5	20.1	23.0	23.8
PEAK-1 (kA)						
t_f (μs)	5.7	5.7	5.3	5.2	6.1	6.3
TAN-G (maximum)	12.5	12.5	12.3	12.2	12.7	12.8
Time to half value	46.0	46.2	45.0	44.5	47.2	48.3
Impulse charge (C)	1.7	1.8	1.6	1.6	1.9	1.9

range 17–21 kA, depending on the expression for the attractive radius used in the calculation.

2.9.3 Mathematical representation of current waveforms

The mean current waveforms of first and subsequent return strokes as measured by Berger are shown in Figure 2.27. These current waveforms were constructed by Berger and colleagues [98] first by normalizing each individual waveform to a common amplitude (i.e. setting the peak value to unity) and then aligning the peaks and averaging.

The first 10 μs of the median current waveform of first and subsequent return strokes observed in several other studies together with that extracted by Berger and colleagues [98] (marked in the figure as San Salvatore) are shown in Figures 2.28 and 2.29.

In theoretical analysis, either in connection with the mechanical and thermal effects of lightning currents or in evaluating the electromagnetic fields generated by lightning flashes, it is necessary to represent the current waveform using analytical expressions. Several analytical expressions that represent the first and subsequent return-stroke currents are available in the literature and a few of these expressions are given below.

2.9.3.1 Current waveform recommended by the CIGRE study group

According to CIGRE [99] the initial rising part of the first return-stroke current waveform, including the peak, can be represented by

$$I(t) = At + Bt^x \qquad (2.24)$$

68 *Lightning Protection*

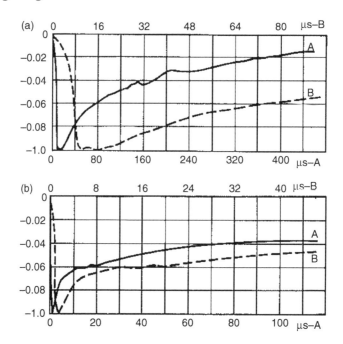

Figure 2.27 The mean current wave shapes of (a) first and (b) subsequent return-stroke current waveforms based on the experimental data of Berger et al. [98]

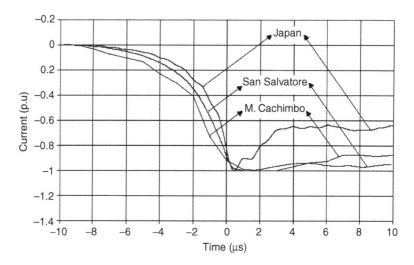

Figure 2.28 The mean curve representing the first 10 μs of the first return-stroke current waveform. The peak current is normalized to unity.

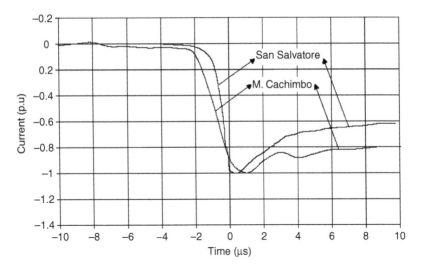

Figure 2.29 The mean curve representing the first 10 µs of the subsequent return-stroke current waveform. The peak current is normalized to unity.

where A and B are constants and t is the time. The rising part of the subsequent return-stroke current, including the peak, is represented by

$$I(t) = s_m t \tag{2.25}$$

where s_m is the maximum steepness. According to equation (2.24) the current shape reaches its maximum steepness (90 per cent of amplitude) at a time t_n that depends on the exponent χ. Both variables (i.e. s_m and t_n) have to be evaluated by an iterative solution of the generalized equation:

$$\left(\frac{1-3x}{2S_N}\right)(1-x)^\chi = \frac{x(\chi-1)}{2S_N} + \left(\frac{1-3x\chi}{2S_N}\right)(1-x) \tag{2.26}$$

with

$$S_N = s_m \frac{t_f}{I}; \quad X_N = 0.6 \frac{t_f}{t_n} \tag{2.27}$$

where t_f is the duration of the front. Sufficiently accurate solutions can be obtained using equations (2.28) and (2.29) for χ and t_n:

$$\chi = 1 + 2(S_N - 1)\left(2 + \frac{1}{S_N}\right) \tag{2.28}$$

and

$$t_n = 0.6 t_f \left[\frac{3S_N^2}{(1+S_N^2)}\right] \tag{2.29}$$

The constants A and B are given by

$$A = \frac{1}{\chi - 1}\left[0.9\frac{I}{t_n}\cdot \chi - S_m\right] \quad (2.30)$$

$$B = \frac{1}{t_n^\chi(\chi - 1)}[S_m \cdot t_n - 0.9I] \quad (2.31)$$

The tail of the current waveform is represented by

$$I - I_1 e^{-(t-t_n)/t_1} - I_2 e^{-(t-t_n)/t_2} \quad (2.32)$$

where I_1 and I_2 are constants, and t_1 and t_2 are time constants. The time constants are given by:

$$t_1 = \frac{(t_h - t_n)}{\ln 2} \quad (2.33)$$

$$t_2 = 0.1\frac{I}{S_m} \quad (2.34)$$

where t_h is the time-to-half value. The constants I_1 and I_2 are given by

$$I_1 = \frac{t_1 \cdot t_2}{t_1 - t_2}\left[S_m + 0.9\frac{I}{t_2}\right] \quad (2.35)$$

$$I_2 = \frac{t_1 \cdot t_2}{t_1 - t_2}\left[S_m + 0.9\frac{I}{t_1}\right] \quad (2.36)$$

2.9.3.2 Analytical form of the current used in the International Electrotechnical Commission standard

In the International Electrotechnical Commission (IEC) lightning protection standard the Heidler function [100] is used to represent the return-stroke current waveform. According to this, the current waveform at the channel base can be represented by

$$i(t) = \frac{I_P}{\eta}\frac{k_s^n}{1 + k_s^n}e^{\frac{t}{\tau_2}} \quad (2.37)$$

$$k_s = \frac{t}{\tau_1} \quad (2.38)$$

In these equations I_p is the peak current, η is a correlation factor of the peak current, n is the current steepness factor (assumed to be 10) and τ_1 and τ_2 are time constants that determine the current rise time and decay time. The parameters of the current waveforms to be used for different protection levels (actually, only the peak current varies from one level of protection to another; see also Chapter 4) are tabulated in Table 2.23.

Table 2.23 Parameters of equation (2.37)

Parameters	First stroke current			Subsequent stroke current		
	LPL			LPL		
	I	II	III–IV	I	II	III–IV
I (kA)	200	150	100	50	37.5	25
η	0.93	0.93	0.93	0.993	0.993	0.993
τ_1 (μs)	19.0	19.0	19.0	0.454	0.454	0.454
τ_2 (μs)	485	485	485	143	143	143

LPL, lightning protection level.

2.9.3.3 Analytical expression of Nucci and colleagues

Nucci and colleagues [101] represented the current at the channel base of subsequent return strokes by the expression

$$i(t) = \frac{I_{01}}{\eta} \frac{(t/\tau_1)^2}{(t/\tau_1)^2+1} e^{-t/\tau_2} + I_{02}(e^{-t/\tau_3} + e^{-t/\tau_4}) \quad (2.39)$$

Using this equation, we can independently vary the peak current and peak-current derivative by changing I_{01} and τ_1. The parameters used to represent a typical subsequent return-stroke current are $I_{01} = 9.9$ kA, $\eta = 0.845$, $\tau_1 = 0.072$ μs, $\tau_2 = 5$ μs, $I_{02} = 7.5$ kA, $\tau_3 = 100$ μs and $\tau_4 = 6$ μs.

2.9.3.4 Analytical expression of Diendorfer and Uman

Diendorfer and Uman [102] represented the first and subsequent return-stroke currents as a sum of two Heidler functions. Their expression is given by

$$i(t) = \frac{I_{01}}{\eta_1} \frac{(t/\tau_{11})^2}{(t/\tau_{11})^2+1} e^{-t/\tau_{21}} + \frac{I_{02}}{\eta_2} \frac{(t/\tau_{12})^2}{(t/\tau_{12})^2+1} e^{-t/\tau_{22}} \quad (2.40)$$

To represent the first return strokes, the authors recommended the parameters $I_{01} = 28$ kA, $\eta_1 = 0.73$, $\tau_{11} = 0.3$ μs, $\tau_{21} = 6$ μs, $I_{02} = 16$ kA, $\eta_2 = 0.53$, $\tau_{12} = 10$ μs and $\tau_{22} = 50$ μs. The parameters recommended for the subsequent strokes are $I_{01} = 13$ kA, $\eta_1 = 0.77$, $\tau_{11} = 0.15$ μs, $\tau_{21} = 3$ μs, $I_{02} = 7$ kA, $\eta_2 = 0.64$, $\tau_{12} = 5$ μs and $\tau_{22} = 50$ μs. With these parameters the peak current and peak-current derivative of first return strokes become 30 kA and 80 kA μs^{-1}, respectively. For subsequent strokes the corresponding parameters are 14 kA and 75 kA μs^{-1}.

2.9.3.5 Analytical expression of Delfino and colleagues

Delfino and colleagues [103] used an expression similar to that used by Diendorfer and Uman [102] to represent the current of first and subsequent return strokes. The current is given by

$$i(t) = I_{01} \frac{(t/\tau_{11})^2}{(t/\tau_{11})^2 + 1} e^{-t/\tau_{21}} + I_{02} \frac{(t/\tau_{12})^2}{(t/\tau_{12})^2 + 1} e^{-t/\tau_{22}} \quad (2.41)$$

In the case of first return strokes only the first term of equation (2.41) is used together with the parameters $I_{01} = 28$ kA, $\tau_{11} = 1.8$ μs and $\tau_{21} = 95$ μs. For subsequent return strokes both terms of (2.41) are used with the parameters $I_{01} = 10.7$ kA, $\tau_{11} = 0.25$ μs, $\tau_{21} = 2.5$ μs, $I_{02} = 6.5$ kA, $\tau_{12} = 2$ μs and $\tau_{22} = 230$ μs.

2.9.3.6 Analytical expression of Cooray and colleagues

Cooray and colleagues [104] constructed the analytical expression (2.42) to represent the first return-stroke currents that contain a slow front followed by a fast transition:

$$i(t) = I_{01} \frac{(t/\tau_1)^n}{(t/\tau_1)^n + 1} + I_{02} \left[1 - e^{-(t/\tau_1)^3}\right](ae^{-t/\tau_2} + be^{-t/\tau_3}) \quad (2.42)$$

For a typical first return-stroke current these authors suggested the parameters: $I_{01} = 7.8$ kA, $\tau_1 = 5$ μs, $n = 100$, $I_{02} = 32.5$ kA, $\tau_2 = 4$ μs, $\tau_3 = 100$ μs, $a = 0.2$ and $b = 0.8$. This waveform has a slow front duration of ~5 μs, 10 to 90 per cent rise time of 4.5 μs, total charge of 3 C, action integral of 4.5×10^4 A^{2}s^{-1} and a peak current derivative of 37 kA μs^{-1}.

2.9.4 Current wave shapes of upward-initiated flashes

If a grounded structure such as a mast or a tower is taller than ~100 m or a structure of moderate height is located on a hill or a mountain, it might launch an upward lightning flash provided that the background electric field is high enough for it to do so (see Chapter 4). In the case of upward-initiated negative ground flashes, a positive upward-moving leader is initiated from the grounded structure and travels upwards, bridging the gap between the ground and the charge centre in the cloud.

Several researchers have studied the parameters of lightning flashes initiated by upward-moving leaders [63,105,107,109]. In these flashes, when the upward-moving leader bridges the gap, an initial continuing current that flows for hundreds of milliseconds and whose amplitude can be as high as tens to thousands of amperes is established. A typical current waveform of an upward negative ground flash measured at the Peissenberg tower is shown in Figure 2.30. In some cases this initial continuing current contain pulses known as continuing current pulses (CCPs). After cessation of the continuing current the channel may be traversed by several dart leader–return

Lightning parameters of engineering interest   73

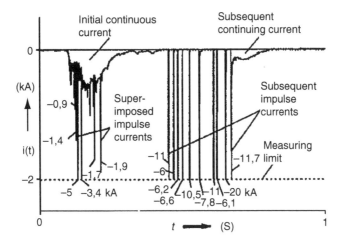

Figure 2.30   Current waveform of a negative upward ground flash as recorded at the Peissenberg Tower (from Reference 73)

stroke sequences that generate current pulses at the channel base. These pulses are marked as subsequent impulse currents in Figure 2.30. They are similar to the subsequent return-stroke current pulses as recorded in instrumented towers and triggered lightning flashes and therefore we concentrate here on the initial continuing current.

In the current waveforms recorded at Peissenberg tower, approximately two-thirds of the flashes consist of only an initial continuing current without any superimposed pulses. The remaining one-third contains pulses that ride on the continuing current. If the initial continuing current is followed by subsequent impulse currents, then the probability of having pulses on the continuing current is also increased. Statistics concerning the duration of the initial stage, average current of the initial stage, the integral of the initial stage current and its action integral are given in Table 2.24. In Reference 109, a comparison is made between the initial stages of tower-initiated currents (natural upward) and triggered lightning currents. The authors concluded that the characteristics of pulses superimposed on the initial continuing current in tower-initiated lightning in different geographical regions are similar within a factor of two, but differ more significantly from their counterpart in rocket-triggered lightning. For example, the ICC pulses in tower-initiated lightning exhibit larger peaks, shorter rise times and shorter half-peak widths than do the ICC pulses in rocket-triggered lightning.

A typical current waveform of a positive upward-initiated lightning flash recorded at Gaisberg tower is shown in Figure 2.31. In the first few milliseconds of the positive upward-initiated flashes, there is a steady increase in the current ramp with superimposed pulses, as shown in Figure 2.32. The peak current and duration of these pulses are ~ 5 kA and ~ 30 μs, respectively. Parameters of these pulses together with the transferred charge, the peak current and the duration of these pulses are tabulated in Table 2.25.

74   Lightning Protection

Table 2.24   Average characteristics of the initial stage of rocket triggered and natural upward lightning (from Reference 109)

Study	Sample size	Duration (ms)	Charge transfer (C)	Average current (A)	Action integral, (10^3 A^{-2}s)
Rocket-triggered lightning, Florida	45	305	30.4	99.6	8.5
Gaisberg tower, Austria	74	231	29.1	126	1.5
Peissenberg tower, Germany	21	290	38.5	133	3.5
Fukui chimney, Japan	36	82.5	38.3	465	40

2.10 Electric fields from first and subsequent strokes

For electromagnetic fields generated by first and subsequent return strokes, the most important parameters are the peak values of the electric field and the peak time derivative of the magnetic field. For distances larger than ∼1 km from the lightning channel,

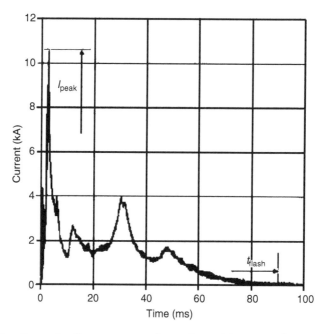

Figure 2.31   Example of a current waveform of a positive upward-initiated lightning discharge (from Reference 110)

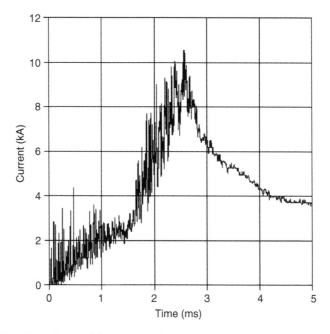

Figure 2.32   First 5 ms of the current shown in Figure 2.31 (from Reference 110)

the electric field peak is radiation. Thus, if the peak value of the electric field at any given distance larger than ~1 km is known, the peak value that corresponds to any other distance can be obtained by using the inverse distance relationships appropriate for radiation fields. The data on the peak values of radiation fields normalized to 100 km (using inverse distance relationships) as observed by different authors are tabulated in Table 2.26.

The values given in Table 2.26 cannot be normalized to distances closer than ~1 km and unfortunately we do not have many statistics concerning how the peak of the first return-stroke electric field varies with distance for distances less than ~1 km. However, we can study how the first return-stroke electric field varies within 1 km by simulations based on different return-stroke models available in the literature. For example, Figure 2.33 shows how the electric fields of first return strokes associated with 30 and 120 kA peak currents vary as a function of distance as calculated by using the return-stroke model of Cooray and Rakov [115]. In the model the channel base current is represented by the analytical expression developed by Delfino and colleagues [103] (also given in Section 2.9.3.5) and the charge neutralized by the return stroke is based on the stepped leader charge distribution given by equations (2.6) to (2.9), but also includes the positive charge induced on the channel during the return-stroke stage. The corona current is represented by an exponential function with a decay-time constant that increases with height. The value of the decay-time constant is assumed to be 10 ns at ground level and its rate of increase with height is assumed to be 1 μs km^{-1}.

76 Lightning Protection

Table 2.25 Overall features and parameters of pulses superimposed on the first few milliseconds of the current waveforms of positive upward-initiated lightning flashes at Gaisberg tower (from Reference 110)

Number of flashes	N	Δt_{tot} (ms)	$I_{max,rel}$ (kA)	Δt_{puls} (µs)	Q_{puls} (As)	$Q_{puls,tot}$ (As)	Q_{corona} (As)	Q_{tot} (As)	Q_{flash} (As)	I_{peak} (kA)	t_{flash} (ms)
98	89	2.8	1.6	28.1	0.017	1.5	9.0	10.5	125	10.6	100
112	68	2.5	5.0	35.7	0.069	4.7	19.7	24.3	356	18.0	70
139	88	3.8	5.0	40	0.082	7.2	45.9	53.1	162	25.0	45
161	63	2.4	3.0	34.6	0.036	2.3	11.4	13.8	220	11.9	200
209	41	1.9	13.7	44.5	0.321	13.2	22.8	35.9	62	39.3	5
210	48	1.9	13.1	37.7	0.238	11.4	19.0	30.4	55	39.3	16
214	72	2.2	3.0	30.6	0.036	2.6	9.3	11.9	63	13.9	32
270	25	0.9	1.8	27.2	0.013	0.4	0.7	1.1	20	2.0	12
298	48	1.6	2.8	31.5	0.030	1.4	4.9	6.3	90	7.0	75
Min	25	0.9	1.6	27.2	0.013	0.4	0.7	1.1	20	2.0	5
Mean	60	2.2	5.5	34.4	0.093	5.0	15.8	20.8	128	18.5	62
Max	89	3.8	13.7	44.5	0.321	13.2	45.9	53.1	356	39.9	200

N, number of pulses of the flash; Δt_{tot}, total time duration of pulsing section; $I_{max,rel}$, mean peak current of pulses of the flash; Δt_{puls}, mean pulse duration; Q_{puls}, mean charge of all pulses within the flash; $Q_{puls,tot}$, accumulated pulse charge from 0 to Δt_{tot}; Q_{corona}, total charge transfer by the continuous current; Q_{tot}, sum of $Q_{puls,tot}$ and Q_{corona}; Q_{flash}, charge transferred by the entire flash; I_{peak}, peak current of the flash; t_{flash}, total time of current flow in the lightning channel.

Lightning parameters of engineering interest 77

Table 2.26 Peak value of the electric field (normalized to 100 km) observed by different authors (from Reference 30)

Study	Number	Mean (V m^{-1})	Standard deviation (V m^{-1})
Heidler and Hopf [30]	*Period 1988–1989*		
	Negative – 19	6.5	4.0
	Positive – 03	−21.6	6.3
	Period 1988–1993	5.3	3.2
	Negative – 148	3.6	2.0
	Negative subsequent – 302	−10.8	6.8
	Positive – 45		
Willett *et al.* [111]	125	8.6	4.4
Krider *et al.* [112]	65	8.5	2.5
Willett and Krider [113]	76	7.9	3.6
Rakov *et al.* [27]	All first strokes – 76	5.9	0.22 (log)
	First stroke in multiple-stroke flashes – 63	6.2	0.23
	First strokes in single-stroke flashes – 13	4.7	0.12
	All subsequent strokes – 270	2.9	0.30
	Subsequent strokes creating a new termination – 38	4.1	0.23
	Subsequent strikes in previously formed channel – 232	2.7	0.30
Filho *et al.* [114]	All first strokes	6.35	
	First strokes in single-stroke flashes	6.04	
	First stroke in multiple-stroke flashes	6.53	

For subsequent return strokes the close electric fields can be evaluated from the data collected by triggered-lightning experiments. The peak values of the electric fields obtained from such experiments are tabulated in Table 2.27.

For electric and magnetic field time derivatives, the distance data can be interpolated using inverse distance relationships down to distances as small as 50 m. The reason is that, even at these distances, the radiation field dominates in the peak electric field time derivative. Moreover, because the fields are radiation the magnetic field time derivative can be obtained from the electric field time derivative because these two field components are related by $E = Bc$, where c is the speed of light in free space.

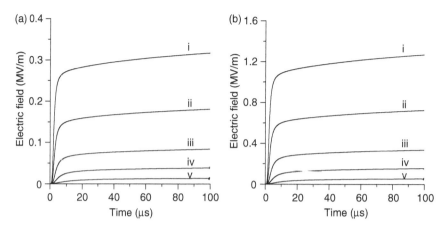

Figure 2.33 *Electric field at different distances from a first return stroke obtained using the return-stroke model of Cooray and Rakov [115] for (a) 30 kA and (b) 120 kA. The input parameters of the return-stroke models are the channel base current, distribution of the charge deposited by the return stroke along the channel and the corona discharge time constant. The current at channel base is represented by the analytical expression given by Delfino et al. [103]. The charge distribution is obtained by appealing to the bi-directional leader concept assuming that the background electric field produced by the cloud remains the same during the leader and return-stroke stages. The corona decay time constant, $\tau(s)$, is assumed to vary along the channel according to the equation $\tau = 10^{-8} + 10^{-6} * (z/1\,000.0)$, where z is a coordinate directed along the channel. The curves i, ii, iii, iv and v represent electric fields at 30, 50, 100, 200 and 500 m, respectively.*

Measurements at coastal stations of the peak-time derivatives of the electric field of return strokes in lightning flashes that strike sea have been reported by Weidman and Krider [118], Willett and colleagues [111] and Cooray and colleagues [119]. The results obtained by these authors are summarized in Table 2.28. The largest value for an electric field peak-time derivative normalized to 100 km observed in these studies is ~ 100 V m^{-1}μs^{-1}. Importantly, the large content of high frequencies in the electric field time derivative means this field component attenuates very rapidly when it propagates along finitely conducting ground. Such attenuation does not affect the electromagnetic fields of lightning flashes significantly when they propagate over the sea because of the high conductivity of sea water. Both Willett and Krider [113] and Murray and colleagues [120] observed a strong linear correlation between the peaks of electric field and electric field time derivatives. Their data are shown in Figure 2.34. Measurements conducted by Heidler and Hopf [30] over land show that the peak electric field time derivatives at different distances are smaller than

Table 2.27 Peak electric field measured for triggered lightning at distances less than 1 km

Study	Distance (m)	Sample	Average (kV m^{-1})
Crawford [116]	10	8	97
	20	3	48
	30	10	35
	50	4	28
	110	4	13
	500	4	2.0
Leteinturier et al. [117]	50	40	119

Table 2.28 Peak values of electric field time derivatives of return strokes observed in different studies (corrected values after removing propagation effects are given in parentheses)

Reference	Number of observations	Mean (V m^{-1}μs^{-1})	Standard deviation (V m^{-1}μs^{-1})	Comments
Willett et al. [111]	131	37 (42)	12	First strokes
Krider et al. [112]	63	39 (46)	11 (13)	First strokes
Weidman and Krider [118]	97	29	12	First and subsequent
Cooray et al. [119]	40	25	11.6	Positive first strokes

those observed in Florida when extrapolated to the same distance. The data obtained by Heidler and Hopf [30] are given in Table 2.29. Based on this comparison, the latter authors suggested that the electric field peak derivatives of return strokes are different in Florida and Germany. Fernando and Cooray [121] suggested that this difference probably results from the propagation effects. However, the studies of Cooray and Rakov [122] show that ground conductivity can also affect the current derivative of the return stroke at the channel base. According to their study the current derivative decreases with decreasing conductivity. If this is correct, we can expect the current derivatives in lightning flashes that strike salt water to be higher than those in lightning flashes that strike ground. Because a smaller current derivative can lead to a smaller peak electric field time derivative the differences in the two datasets probably result from both propagation and source effects.

In analysing the effects of electromagnetic fields on electrical and telecommunication systems it is necessary to know the amplitude of electromagnetic fields from

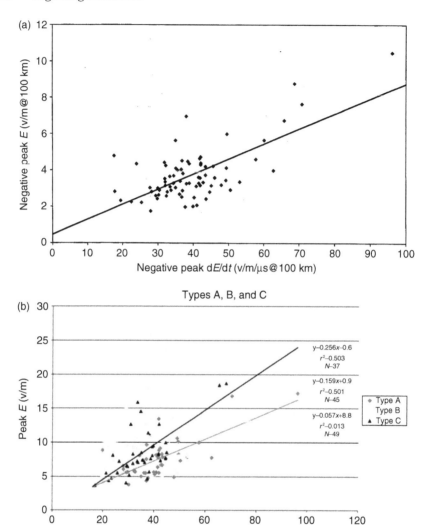

Figure 2.34 Relationship between electric field and electric field time derivative as observed in the studies of (a) Willett and Krider [113] and (b) Murray et al. [120]. Type A strokes have a single, dominant peak in dE/dt. The electric field waveforms for Type B strokes tend to have an inflection point (or shoulder) within or near the fast (negative-going) transition in the E field or, if the additional dE/dt pulse contains a (positive-going) zero crossing, multiple peaks in E. Type C strokes contain one or more large pulses in dE/dt near the beginning or during the slow front, and the corresponding E signatures have a structure that depends on the amplitude and duration of these pulses (or bursts) and when they occur relative to the dominant peak.

Table 2.29   Peak amplitudes of electric field time derivative normalized to 100 km (from Reference 30)

Year	Return stroke	Sample	Mean ($V\,m^{-1}\mu s^{-1}$)	Standard deviation ($V\,m^{-1}\mu s^{-1}$)	Maximum ($V\,m^{-1}\mu s^{-1}$)
1984–1985	Negative	19	6.5	4.5	17
	Positive	3	−12.6	9.6	−24
1986	Negative first	39	3.0	1.5	9
	Negative subsequent	76	3.0	1.3	7
1988–1993	Negative first	148	5.4	3.4	20
	Negative subsequent	302	4.4	2.2	20
	Positive	45	−7.1	2.4	−18

lightning flashes at different frequencies, i.e. the electromagnetic spectrum. The spectrum of electromagnetic fields can be calculated from the measured broadband electric or magnetic field, or it can be obtained by conducting narrow-band measurements at a given frequency. Figure 2.35a shows the spectrum of a return-stroke electromagnetic field normalized to 100 km as obtained from Fourier transformation of broadband data by Willett and colleagues [123]. Figure 2.35b depicts the electric field spectrum constructed from data pertinent to narrow-band measurements.

2.11   Peak electric radiation fields of first and subsequent strokes

In lightning-protection studies it is usually assumed that the first stroke has the largest peak current. Although this assumption is true on average, there is a certain percentage of flashes in which at least one subsequent return-stroke current peak is larger than that of the first. Analysis of the current waveforms of Berger [125] by Thottappillil and colleagues [32] shows that in ∼15 per cent of the negative downward flashes there was at least one subsequent return stroke with a current amplitude larger than the that of the first. A similar tendency is also seen in the electromagnetic fields generated by first and subsequent return strokes. For example, the percentage of flashes in which at least one subsequent stroke had an amplitude larger than the first was 33 per cent in Florida, 24 per cent in Sweden, 35 per cent in Sri Lanka and 38.2 per cent in Brazil. In the Brazilian study [114], one flash in which all the subsequent strokes were larger than the first was found. How the average peak field varies with stroke order as obtained in that study is shown in Figure 2.36. The authors of that study conjectured that strokes 2 to 5 were larger than strokes 6 to 11 because the former can make multiple terminations.

82 Lightning Protection

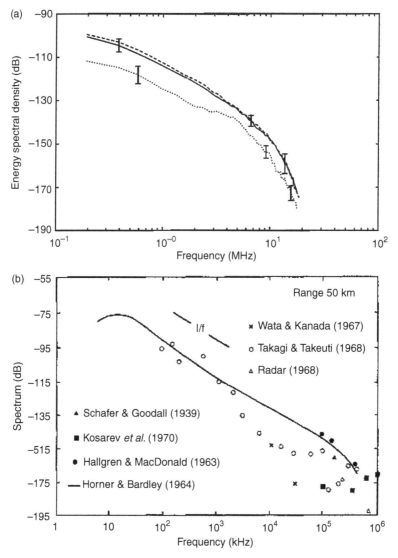

Figure 2.35 *Spectrum of lightning electromagnetic radiation obtained (a) using Fourier transformation of broadband return-stroke electric fields (normalized to 50 km) [123] and (b) using narrow-band receivers (normalized to 50 km) [124] (reference in the figure are taken from Reference 124). Note that in the case of narrow-band receivers all the events in the flash contribute to the spectrum.*

Lightning parameters of engineering interest 83

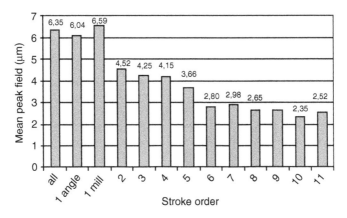

Figure 2.36 Mean peak value ($V\,m^{-1}$) of return-stroke fields normalized to 100 km versus stroke order. Results are based on negative cloud-to-ground flashes observed in Brazil (from Reference 114).

If the transmission-line model is used to interpret the electric field data, assuming that the return-stroke speed is more or less the same in first and subsequent strokes, we could conclude that in 30 per cent of the flashes at least one subsequent stroke may have a current larger than the first. It is important to take this into account in a risk evaluation for lightning protection studies because risk evaluation for lightning protection usually assumes that the first return stroke has the highest peak current amplitude. Based on this the external lightning protection system is designed in such a way that a first return-stroke current that could bypass it will have an amplitude smaller than a critical value. Given the above information we can infer that there is a risk that, once the system is bypassed by a weak first return stroke, a heavy subsequent stroke with a peak current larger than the critical value may follow the same channel and terminate on the structure.

Table 2.30 Characteristics of continuing currents in downward negative ground flashes observed in Germany (from Reference 30)

	Number of strokes	Duration of continuing currents (ms)
Mean	4.79	221
Standard deviation	2.08	103
Maximum	11	551
Minimum	1	67
5% value	8.54	391
95% value	2	82.7

84 *Lightning Protection*

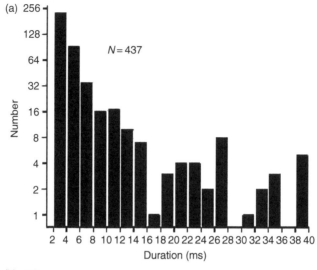

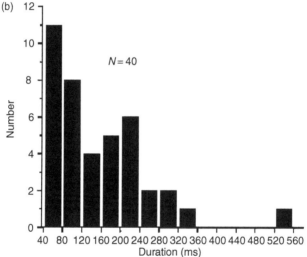

Figure 2.37 Histogram of very short continuing current durations observed by Saba et al. [25]: (a) duration below 40 ms and (b) duration above 40 ms (from Reference 24)

2.12 Continuing currents

Although the currents associated with return strokes usually reach zero within several hundred microseconds, in some return strokes the current amplitude decreases to ~100 A or so within this time but then, instead of decreasing to zero, maintains this amplitude for a few milliseconds to a few hundreds of milliseconds. Such currents are known in the literature as continuing currents.

The existence and the duration of continuing currents can be identified by the close electric fields produced by lightning flashes. Based on the results obtained from such records, the continuing current can be divided into several categories. Kitagawa and colleagues [26] and Brook and colleagues [126] defined continuing currents longer than 40 ms as *long continuing currents*, whereas Shindo and Uman [127] defined the continuing currents of duration between 10–40 ms as *short continuing currents*. They also found examples in which the continuing current duration was 1–10 ms. Saba and colleagues [25] defined these as *very short continuing currents*. However, this division, although it helps the bookkeeping, does not have a physical basis. That is, it could be the same physical process that gives rise to continuing currents, irrespective of their duration.

Heidler and Hopf [30] studied continuing currents in lightning flashes in Germany using electric-field records. The mean, maximum and minimum durations observed in that study are tabulated in Table 2.30. According to Heidler and Hopf [30], 48 per cent of negative flashes were hybrid flashes that contained at least one continuing current (these are long continuing currents). In a study conducted by Thompson [128] it was found that in 34 multiple-stroke flashes, 47 per cent had at least one continuing current. In a study conducted by Shonland [48], this value was 20 per cent. According to the observations of Livingston and Krider [129] the frequency of flashes that have continuing currents range from 29 to 46 per cent. In a study conducted by Shindo and Uman [127] 22 out of 90 negative flashes contained long continuing currents and 11 contained short continuing currents.

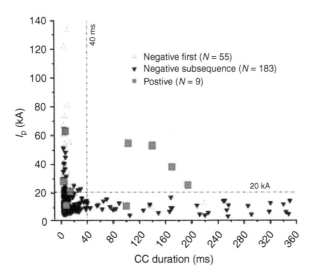

Figure 2.38   *Peak current versus continuing current duration for 248 negative strokes and nine positive strokes (from Reference 132)*

86 Lightning Protection

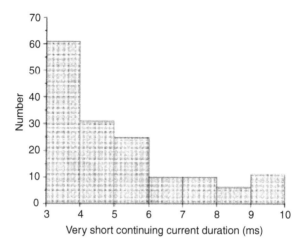

Figure 2.39 Histogram of very short continuing current durations as observed by Ballarotti et al. [131]

Saba and colleagues [25], using high-speed video cameras, managed to obtain data on continuing currents in lightning flashes in Brazil. Analysing 233 negative ground flashes that contained 608 strokes, they found that 50 per cent of the strokes supported continuing currents longer than ~1 ms and 35.6 per cent of the strokes were followed by short or long continuing currents. The distribution of the duration of continuing currents observed in the study is given in Figure 2.37. Saba and colleagues [132] analysed 454 negative strokes followed by continuing currents and observed that

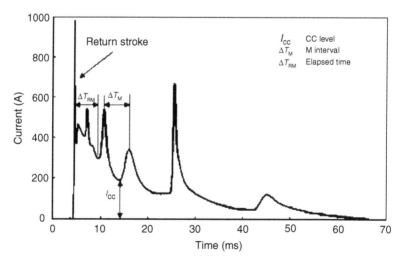

Figure 2.40 A current record showing one return stroke followed by several M-components (I_{CC}, continuing current level; ΔT_M, M-interval; ΔT_{RM}, elapsed time) (from Reference 133)

Lightning parameters of engineering interest 87

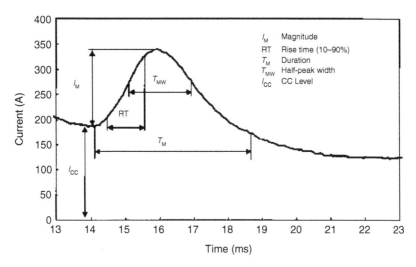

Figure 2.41 An expanded portion of the continuing current current (I_{CC}) record of Figure 2.40 showing the definitions used in the measurement of M-current magnitude (I_M), 10–90 per cent rise time (RT), duration (T_M), and half-peak width (T_{MW})

combinations of stroke amplitudes greater than 20 kA and continuing currents longer than 40 ms are highly unlikely to occur (as shown in Figure 2.38). However, such a restriction was not observed for positive ground flashes. They also observed that the peak currents of the strokes (estimated from electric-field records) that supported long continuing currents are smaller, on average, than those of other strokes. This observation is similar those made by Rakov and Uman [130]. Ballarotti and colleagues

Table 2.31 Summary of statistics of M-components (from Reference 133)

Parameter	Sample	GM	Standard deviation $\log_{10}(x)$	Cases exceeding tabulated value		
				95%	50%	5%
Magnitude (A)	124	117	0.50	20	121	757
Rise time (µs)	124	422	0.42	102	415	1 785
Duration (ms)	114	2.1	0.37	0.6	2.0	7.6
Half-peak width (µs)	113	816	0.41	192	800	3 580
Charge (mC)	104	129	0.32	33	131	377
Continuing current level (A)	140	177	0.45	34	183	991
M-interval (ms)	107	4.9	0.47	0.8	4.9	23
Elapsed time (ms)	158	9.1	0.73	0.7	7.7	156

[131] analysed 890 strokes of 233 negative ground flashes and found that the geometric mean duration of continuing currents was 5.3 ms. The histogram of very short continuing-current durations observed by Ballarotti and colleagues [130] is shown in Figure 2.39. According to the study, about 28 per cent of all negative strokes observed were followed by continuing currents longer than 3 ms.

2.13 M-components

M-components are discharge events travelling from cloud to ground along the lightning channels that support a continuing current 1. The statistics concerning these currents are important in the study of ageing effects and failure modes of surge-protection devices. The M-components measured at the channel base typically have some hundreds of amperes of current and a rise time of some hundreds of microseconds. A current record that shows several M-components is given in Figure 2.40, and a typical M-component current is depicted in Figure 2.41. The statistics of M-components are summarized in Table 2.31. Thottappillil and colleagues [133] did not find much correlation between current magnitudes and other current parameters of M-components.

References

1. Cooray V. *The Lightning Flash*. IEE Power and Energy Series, no. 34; 2003.
2. Aronsson I.-L. *Det Sårbara Informationssamhället – En Studie Avåskans Skadeverkningar*. UURIE:220-89, Division for Electricity, Uppsala University, Sweden; 1989.
3. Eriksson A.J. *The Lightning Ground Flash – an Engineering Study*. PhD Thesis, 1979.
4. Soula S., Chauzy S. 'Multilevel measurement of the electric field underneath a thundercloud 2. Dynamical evolution of a ground space charge layer'. *Journal of Geophysical Research*, 1991;**96**(D12):22327–36.
5. Willett J.C., Davis D.A., Larochec P. 'An experimental study of positive leaders initiating rocket-triggered lightning'. *Atmospheric Research*, 1999;**51**:189–219.
6. Becerra M., Cooray V., Soula S., Chauzy S. 'Effect of the space charge layer created by corona at ground level on the inception of upward lightning leaders from tall towers'. *Journal of Geophysical Research*, 2007;**112**: doi: 10.1029/2006JD008308, D 12205.
7. Chauzy S., Rennela C. 'Computed response of the space charge layer created by corona at ground level to external electric field variations beneath a thundercloud'. *Journal of Geophysical Research*, 1985;**90**(D4):6051–57.
8. Moore P.K., Orville R.E. 'Lightning characteristics in lake-effect thunderstorms'. *Monthly Weather Review*, 1990;**118**:1767–82.
9. Christian H.J., Blakeslee R.J., Boccippio D.J., Boeck W.L., Buechler D.E., Driscoll K.T. et al. 'Global frequency and distribution of lightning as observed from space by the Optical Transient Detector'. *Journal of Geophysical Research*, 2003;**108**: doi: 10.1029/2002JD002347.

10. Mackerras D. 'Prediction of lightning incidence and effects in electrical systems'. *Electr. Eng. Trans. Int. Eng. Aust. EE*, 1978;**14**:73–77.
11. Anderson R.B., Eriksson A.J. 'Lightning parameters for engineering application'. *Electra*, 1980;**69**:65–102.
12. Anderson R.B., Eriksson A.J., Kroninger H., Meal D.V., Smith M.A. 'Lightning and thunderstorm parameters'. Presented at International Conference on Lightning and Power Systems, Institute of Electrical and Electronic Engineers; London, 1984.
13. Kuleshov Y., Jayaratne E.R. 'Estimates of lightning ground flash density in Australia and its relationship to thunder-days'. *Australian Meteorological Magazine*, 2004;**53**:189–96.
14. Shui Ming Chen, Du Y., Ling Meng Fan. 'Lightning data observed with lightning location system in Guang-Dong Province, China'. *IEEE Transactions on Power Delivery*, 2004;**19**(3):1148–53.
15. Rosa F., de la Cummins K., Dellera L., Diendorfer G., Galván A., Husse J. *et al.* 'Characterization of lightning for applications in Electric Power Systems'. Technical Brochure no. 172, CIGRE WG. 33.01.02; December 2000.
16. Torres H. *Ground Flash Density: Definition of the Appropriate Grid Size and a Proposal of Relationship N_g vs. T_d for Tropical Zones*. Activity report of TF C4.01.02-B, Working Group C4.01 'Lightning', CIGRE Dallas, TX, USA, September 2003.
17. Younes C. *Evaluation of Lightning Parameters in Colombia from LLS and Satellite Data*. MSc. Thesis, National University of Colombia, Bogotá, 2002.
18. www.weathersa.co.za/Pressroom/2006/2006Jan13LightningDetectNetwork.pdf.
19. Orville R.E. 'Lightning ground flash density in the contiguous United States – 1989'. *Monthly Weather Review*, 1991;**119**(2):573–77.
20. Pinto I.R.C.A., Pinto O. Jr. 'Cloud-to-ground lightning distribution in Brazil'. *Journal of Atmospheric and Solar-Terrestrial Physics*, 2003;**65**:733–37.
21. Soriano L.R., Pablo de F., Tomas C. 'Ten-year study of cloud-to-ground lightning activity in the Iberian Peninsula'. *Journal of Atmospheric and Solar-Terrestrial Physics*, 2005;**67**:1632–39.
22. Upul S., Cooray V., Götschl T. 'Characteristics of cloud-to-ground lightning flashes over Sweden'. *Physical Scripta*, 2006;**74**:541–8; doi: 10.1088/0031-8949/74/5/010.
23. Darveniza M., Tumma L.R., Richter B., Roby D.A. 'Multipulse lightning currents metal-oxide arresters'. *IEEE Transactions on Power Delivery*, 1997;**12**(3):1168–75.
24. Thomson E.M. 'The dependence of lightning return stroke characteristics on latitude'. *Journal of Geophysical Research*, 1980;**85**:1050–56.
25. Saba M.M.F., Ballarotti M.G., Pinto O. Jr. 'Negative cloud-to-ground lightning properties from high-speed video observations'. *Journal of Geophysical Research*, 2006;**111**:D03101.
26. Kitagawa N., Brook M., Workman E.J. 'Continuing currents in cloud-to-ground lightning discharges'. *Journal of Geophysical Research*, 1962;**67**:637–47.

27. Rakov V.A., Uman M.A., Thottappillil R. 'Review of lightning properties from electric field and TV observations'. *Journal of Geophysical Research*, 1994;**99**:10745–50.
28. Cooray V., Perez H. 'Some features of lightning flashes observed in Sweden'. *Journal of Geophysical Research*, 1994;**99**(D5):10683–88.
29. Cooray V., Jayaratne K.P.S.C. 'Characteristics of lightning flashes observed in Sri Lanka in the tropics'. *Journal of Geophysical Research*, 1994;**99**(D10):21051–65.
30. Heidler F., Hopf C. 'Measurement results of the electric fields in cloud-to-ground lightning in nearby Munich, Germany'. *IEEE Transactions on Electromagnetic Compatibility*, 1998;**40**(4):436–43.
31. Saba M.M.F., Ballarotti G.M, Campos L.Z.S., Pinto O. Jr. 'High-speed video observations of positive lightning'. *IX International Symposium on Lightning Protection*; Foz do Iguaçu, Brazil, 26–30 November 2007.
32. Thottappillil R., Rakov V.A., Uman M.A., Beasley W.H., Master M.J., Shelukhin D.V. 'Lightning subsequent-stroke electric field peak greater than the first stroke peak and multiple ground terminations'. *Journal of Geophysical Research*, 1992;**97**(D7):7503–509.
33. Qie Xiushu, Ye Yu, Daohong Wang, Huibin Wang, Rongzhong Chu. 'Characteristics of cloud-to-ground lightning in Chinese Inland Plateau notes and correspondence'. *Journal of the Meteorological Society of Japan*, 2002;**80**(4):745–54.
34. Miranda F.J., de Pinto O. Jr., Saba M.M.F. 'A study of the time interval between return strokes and K-changes of negative cloud-to-ground lightning flashes in Brazil'. *Journal of Atmospheric and Solar-Terrestrial Physics*, 2003;**65**:293–97.
35. Davis S.M. *Properties of Lightning Discharges from Multiple-Station Wideband Electric Field Measurements*. PhD Thesis, University of Florida, Gainesville, p. 228, 1999.
36. Rakov V.A., Uman M. *Lightning Physics and Effects*. Cambridge: Cambridge University Press; 2003. p. 286.
37. Valine W.C., Krider E.P. 'Statistics and characteristics of cloud-to-ground lightning with multiple ground contacts'. *Journal of Geophysical Research*, 2002; **107**(D20):4441; doi:10.1029/2001jd001360.
38. Fisher R.J., Schnetzer G.H. 'Triggered lightning test program: environments within 20 meters of the lightning channel and small area temporary lightning protection concepts'. Sandia National Laboratories Report SAND93-0311, Albuquerque, NM, 1994.
39. More C.B., Aulich G.D., Rison W. *An Examination of Lightning-Strike-Grounding Physics* [online]. Available from http://www.lightningsafety.com/nlsi_lhm/Radials.pdf.
40. Uman M.A., Rakov V.A., Rambo K.J., Vaught T.W., Fernandez M.I., Cordier D.J. et al. 'Triggered lightning experiments at Camp Blanding, Florida (1993–1995)'. *Transactions of the Institute of Electrical Engineers of Japan*, 1997;**117-B**:446–52.

41. Uman M.A., Cordier D.J., Chandler R.M., Rakov V.A., Bernstein R., Barker P.P. 'Fulgurites produced by triggered lightning'. *Eos Transactions, American Geophysical Union*, 1994;**75**(44):99, Fall Meeting Suppl.
42. Rakov V.A., Uman M.A., Rambo K.J., Fernandez M.I., Fisher R.J., Schnetzer G.H. *et al.* 'New insights into lightning processes gained from triggered-lightning experiments in Florida and Alabama'. *Journal of Geophysical Research*, 1998;**103**(D12):14117–30.
43. Liew A.C. *Calculation of the Lightning Performance of Transmission Lines*. PhD Thesis, University of Queensland, Australia, 1972.
44. Wang Junping, Ah Choy Liew, Darveniza M. 'Extension of dynamic model of impulse behavior of concentrated grounds at high currents'. *IEEE Transactions on Power Delivery*, 2005;**20**(3):2160–65.
45. Becerra M., Cooray V. 'A simplified physical model to determine the lightning upward connecting leader inception'. *IEEE Transactions on Power Delivery*, 2006;**21**(2):897–908.
46. Schonland B.F.J., Collens H. 'Progressive lightning'. *Proceedings of the Royal Society London, Series A*, 1934;**143**:654–74.
47. Schonland B.F.J. 'Progressive lightning, Part. 4, The discharge mechanism'. *Proceedings of the Royal Society London, Series A*, 1938;**164**:132–50.
48. Schonland B.F.J. 'The lightning discharge', in *Handbuch der Physik*, Vol. 22. Berlin: Springer-Verlag; 1956. pp. 576–628.
49. Schonland B.F.J., Malan D.J., Collens H. 'Progressive lightning, 2'. *Proceedings of the Royal Society London, Series A*, 1935;**152**:595–625.
50. McEachron K.B. 'Lightning to the Empire State Building'. *Journal of the Franklin Institute*, 1939;**227**:149–217.
51. Orville R.E., Idone V.P. 'Lightning leader characteristics in the thunderstorm research international program (TRIP)'. *Journal of Geophysical Research*, 1982;**87**(C13):11, 177–92.
52. Berger K., Vogelsanger E. 'Photographische Blitzuntersuchungender Jahre 1955–1965 auf dem Monte San Salvatore'. *Bulletin SEV*, 1966; **57**:1–22.
53. Sab M.M.F., Campos L.Z.S., Ballarotti M.G., Pinto O. Jr. 'Measurements of cloud to ground and spider leader speeds with high speed video observations'. *Presented at 13th International Conference on Atmospheric Electricity*, Beijing, China, August 13–18, 2007.
54. Chen Mingli, Nobuyuki Takagi, Teiji Watanabe, Daohong Wang, Zen-Ichiro Kawasaki, Xinsheng Liu. 'Spatial and temporal properties of optical radiation produced by stepped leaders'. *Journal of Geophysical Research*, 1999; **104**(D22):27573–84.
55. Wang D., Rakov V.A., Uman M.A., Takagil N., Watanabe T., Crawford D.E. *et al.* 'Attachment process in rocket-triggered lightning strokes'. *Journal of Geophysical Research*, 1999;**104**(D2):2143–50.
56. Jordan D.M., Idone V.P., Rakov V.A., Uman M.A., Beasley W.H., Jurenka H. 'Observed dart leader speed in natural and triggered lightning'. *Journal of Geophysical Research*, 1992;**97**(D9):9951–57.

57. Jordan D.M., Rakov V.A., Beasley W.H., Uman M.A. 'Luminosity characteristics of dart leaders return strokes in natural lightning'. *Journal of Geophysical Research*, 1997;**102**(D18):22025–32.
58. Wang D., Takagi N., Watanabe T., Rakov V.A., Uman M.A. 'Observed leader and return-stroke propagation characteristics in the bottom 400 in of a rocket-triggered lightning channel'. *Journal of Geophysical Research*, 1999; **104**(D12):14369–76.
59. Schonland B.F.J. 'The pilot streamer in the lightning and long spark'. *Proceedings of the Royal Society of London Series A*, 1953;**220**:25–38.
60. Golde R.H. 'The frequency of occurrence and their distribution of lightning flashes to transmission lines'. *AIEE Transactions*, 1945;**64**:902–10.
61. Cooray V., Rakov V., Theethayi N. 'The lightning striking distance – revisited'. *Journal of Electrostatics*, 2007;**65**(5–6):296–306.
62. Crawford D.E., Rakov V.A., Uman M.A, Schnetzer G.H., Rambo K.J., Stapleton M.V. et al. 'The close lightning electromagnetic environment: Dart-leader electric field change versus distance'. *Journal of Geophysical Research*, 2001;**106**(D14):14909–17.
63. Berger K. 'Novel observations on lightning discharges: Result of research on Mount San Salvatore'. *Journal of the Franklin Institute*, 1967;**283**:478–525.
64. Kito Y., Horii K., Higashiyama Y., Nakamura K. 'Optical aspects of winter lightning discharges triggered by the rocket-wire technique in Hokuriku District of Japan'. *Journal of Geophysical Research*, 1985;**90**(D4): 6147–57.
65. Asakawa A., Miyake K., Yokoyama S., Shindo T., Yokota T., Sakai T. 'Two types of lightning discharges to a high stack on the coast of the sea of Japan in winter'. *IEEE Transactions on Power Delivery*, 1997;**12**(3):1222–31.
66. Wada A., Asakawa A.T., Shindo A., Yokoyama S. 'Leader and return stroke speed of upward-initiated lightning'. *Proceedings of the 12th International Conference on Atmospheric Electricity*; pp. 553–56.
67. Yokoyama S., Miyake K., Suzuki T., Kanao S. 'Winter lightning on Japan Sea coast-development of measuring system on *progressing feature of lightning discharge*'. *IEEE Transactions on Power Delivery*, 1990;**5**(3):1418.
68. Becerra M., Cooray V. 'On the velocity of positive connecting leaders associated with negative downward lightning leaders'. *Geophysical Research Letters*, 2008;**35**:L02801; doi:10.1029/2007gl032506.
69. Lalande P., Bondiou-Clergerie A., Laroche P., Eybert-Berard A., Betlandis J.-P., Bador B. et al. 'Leader properties determined with triggered lightning techniques'. *Journal of Geophysical Research*, 1998;**103**(D12):14109–15.
70. Becerra M., Cooray V. 'A self-consistent upward leader propagation model'. *Journal of Physics D: Applied Physics*, 2006;**39**:3708–15; doi:10.1088/0022-3727/39/16/028.
71. Becerra M., Cooray V. 'A simplified physical model to determine the lightning upward connecting leader inception'. *IEEE Transactions on Power Delivery*, 2006;**21**(2):897–908.

72. Aleksandrov N.L., Bazelyan E.M., Raizer Yu. P. 'The effect of a corona discharge on a lightning attachment'. *Plasma Physics Reports*, 2005;**31**(1):75–91.
73. Heidler F., Zischank W., Weisinger J. 'Current parameters of lightning measured at Piessenberg Telecommunication Tower'. Presented at the *19th International Aerospace and Ground Conference on Lightning and Static Electricity*, Seattle, USA. Report 2879, September 11–13, 2001.
74. Pichler H., Diendorfer G., Mair M. 'Statistics of lightning current parameters measured at the Gaisberg Tower'. Presented at the *18th International Lightning Detection Conference (ILDC)*, Helsinki 7–9 June, 2004.
75. Garbagnati E., Lo Piparo G.B. 'Parameter von Blitzstromen'. *Elektrotechnische Zeitschrift A*, 1982;**103**(2):61–65.
76. Takami J., Okabe S. 'Observational results of lightning current on transmission towers'. *IEEE Transactions on Power Delivery*, 2007;**22**(1):547–56.
77. Visacro S., Soares A., Jr., Aurélio M., Schroeder O., Cherchiglia L.C.L., de Sousa V.J. 'Statistical analysis of lightning current parameters: measurements at Morro do Cachimbo Station'. *Journal of Geophysical Research*, 2004;**109**, D01105; doi:10.1029/2003jd003662.
78. Berger K., Anderson R.B., Kroninger H. 'Parameters of lightning flashes'. *Electra*, 1975;**41**:23–37.
79. Eriksson A.J., Geldenhuys H.J., Bourn G.W. 'Fifteen year's data of lightning measurements on a 60 m mast'. *Transactions of the South African Institute of Electrical Engineers*, 1989;**80**(1):98–103.
80. Anderson R.B., Erikkson A.J. 'Lightning parameters for engineering applications'. ELEK 170, CIGRE Study Committee 33, June 1979.
81. Anderson R.B., Eriksson A.J. 'Lightning parameters for engineering applications'. *Electra*, 1980;**69**:65–102.
82. Fisher R.J., Schnetzer G.H., Thottappillil R., Rakov V.A., Uman M.A., Golberg J.D. 'Parameters of triggered lightning flashes in Florida and Alabama'. *Journal of Geophysical Research*, 1993;**98**(D12):22887–902.
83. Leteinturier C., Hamelin J.H., Eybert-Berard A. 'Submicrosecond characteristics of lightning return-stroke currents'. *IEEE Transactions on Electromagnetic Compatibility*, 1991;**33**:351–57.
84. Garbagnati E., Giudice E., Lo Piparo G.B. 'Messung von Blitzströmen in Italien – Ergebnisse einer statistischen Auswertung'. *Elektrotechnische Zeitschrift A*, 1978;**11**:664–68.
85. Borghetti A., Nucci C.A., Paolone M. 'Estimation of the statistical distributions of lightning current parameters at ground level from the data recorded by instrumented towers'. *IEEE Transactions on Power Delivery*, 2004;**19**(3):1400–09.
86. Saba M.M.F., Pinto O., Jr., Ballarotti M.G. 'Relation between lightning return stroke peak current and following continuing current'. *Geophysical Research Letters*, 2006;**33**, L23807; doi:10.1029/2006gl027455.
87. Depasse P. 'Statistics on artificially triggered lightning'. *Journal of Geophysical Research*, 1994;**99**(18):515–22.

88. 'Draft guide for improving the lightning performance of electrical power overhead distribution lines'. IEEE Power Engineering Society.
89. Chowdhuri P., Anderson J.G., Chisholm W.A., Field T.E., Ishii M., Martinez J.A. et al. 'Parameters of lightning strokes: A review'. *IEEE Transactions on Power Delivery*, 2005;**20**(1):346–58.
90. Rizk F.A.M. 'Modeling of lightning incidence to tall structures. Part I: Theory, Part II: Application'. *IEEE Transactions on Power Delivery*, 1994;**9**:162–93.
91. Brown G.W. 'Joint frequency distribution of stroke current rates of rise and crest magnitude to transmission lines'. *IEEE Transactions on Power Apparatus and Systems*, 1978;**97**:53–58.
92. Armstrong H.R., Whitehead E.R. 'Field and analytical studies of transmission lines shielding'. *IEEE Transactions on Power Apparatus and Systems*, 1968;**87**:270–381.
93. IEEE guide for improving the lightning performance of transmission lines, IEEE Standard 1243–1997, December 1997.
94. Golde R.H. 'Lightning and tall structures'. *Proceedings of the Institute of Electrical Engineers*, 1978;**125**(4):347–51.
95. Golde R.H. 'The frequency of occurrence and the distribution of lightning flashes to transmission lines'. *AIEE Transactions*, 1945;**64**:902–10.
96. Eriksson A.J. 'An improved electrogeometric model for transmission line shielding analysis'. *IEEE Transactions on Power Delivery*, 1987;**2**:871–86.
97. Dellera L., Garbagnati E. 'Lightning stroke simulation by means of the leader progression model, Parts I and II'. *IEEE Transactions on Power Delivery*, 1990;**5**:2009–29.
98. Berger K., Anderson R.B., Kroninger H. 'Parameters of lightning flashes'. *Electra*, 1975;**40**:101–19.
99. Guide to procedures for estimating the lightning performance of transmission lines, Working Group 01 (Lightning) of Study Committee 33, October 1991.
100. Heidler F., Svetic J.M., Stanic B.V. 'Calculation of lightning current parameters'. *IEEE Transactions on Power Delivery*, 1997;**14**(2):399–404.
101. Nucci C.A., Diendorfer G., Uman M.A., Rachidi F., Ianoz M., Mazzetti C. 'Lightning return stroke current model with specified channel-base current: a review and comparison'. *Journal of Geophysical Research*, 1990;**95**:20395–408.
102. Diendorfer G., Uman M.A. 'An improved return stroke model with specified channel base current'. *Journal of Geophysical Research*, 1990;**95**:13621–44.
103. Delfino F., Procopio R., Rossi M., Rachidi F., Nucci C.A. 'An algorithm for the exact evaluation of the underground lightning electromagnetic fields'. *IEEE Transactions on Electromagnetic Compatibility*, 2007;**49**:401–11.
104. Cooray V., Montano R., Rakov V. 'A model to represent negative and positive lightning first strokes with connecting leaders'. *Journal of Electrostatics*, 2004;**60**:97–109.
105. McEachron K.B. 'Lightning to the Empire State Building'. *Journal of the Franklin Institute*, 1939;**227**:1149–217.

106. McEachron K.B. 'Lightning to the Empire State Building II'. *Transactions of the AIEE*, 1941;**60**:885–89.
107. Hagenguth J.H., Anderson J.G. 'Lightning to the Empire State Building, Part III'. *Transactions of the AIEE*, 1952;**71**:641–49.
108. Berger K. 'The Earth Flash', in R.H. Golde (ed.). *Lightning*. New York: Elsevier; 1977. pp. 119–90.
109. Miki M., Rakov V.A., Shindo T., Diendorfer G., Mair M., Heidler F. 'Initial stage in lightning initiated from tall objects and in rocket-triggered lightning'. *Journal of Geophysical Research*, 2005;**110**:D02109; doi:10.1029/2003jd004474.
110. Diendorfer G. 'Characteristics of positive upward lightning measured on an instrumented tower'. Presented at the *28th International Conference on Lightning Protection P*, Japan, 2006.
111. Willett J.C., Krider E.P., Leteinturier C. 'Submicrosecond field variations during the onset of first return strokes in cloud-to-ground lightning'. *Journal of Geophysical Research*, 1998;**103**(D8):9027–34.
112. Krider E.P., Leteinturier C., Willett J.C. 'Submicrosecond fields radiated during the onset of first return strokes in cloud-to-ground lightning'. *Journal of Geophysical Research*, 1996;**101**(D1):1589–97.
113. Willett J.C., Krider E.P. 'Rise times of impulsive high-current processes in cloud-to-ground lightning'. *IEEE Transactions on Antenna and Propagation*, 2000;**48**(9):1442–45.
114. Filho A.O., Schulz W., Saba M.M.F., Pinto O., Jr., Ballarotti M.G. 'First and subsequent stroke electric field peaks in negative cloud-to-ground lightning'. Presented at the *IX International Symposium on Lightning Protection*, 26–30 November 2007, Foz do Iguaçu, Brazil.
115. Cooray V., Rakov V. A current generation type return stroke model that predicts the return stroke velocity. *Journal of Lightning Research*, 2007;**1**:32–39.
116. Crawford D.E. 'Multiple station measurements of triggered lightning electric and magnetic fields'. Master Thesis, Department of Electrical and Computer Engineering, University of Florida, Gainsville, USA, 1998.
117. Leteinturier C., Weidman C., Hamelin J. 'Current and electric field derivatives in triggered lightning return strokes'. *Journal of Geophysical Research*, 1990;**95**(D1):811–28.
118. Weidman C., Krider E.P. 'Submicrosecond risetimes in lightning return stroke fields'. *Journal of Geophysical Research Letters*, 1980;**7**:955–58.
119. Cooray V., Fernando M., Gomes C., Sorensen T. 'The fine structure of positive return stroke radiation fields'. *IEEE Transactions on Electromagnetic Compatibility*, 2004;**46**:87–95.
120. Murray M.D., Krider E.P., Willett J.C. 'Multiple pulses in dE/dt and the fine-structure of E during the onset of first return strokes in cloud-to-ocean lightning'. *Atmospheric Research*, 2005;**76**:455–80.
121. Mahendra F., Cooray V. 'Propagation effects on the electric field time derivatives generated by return strokes in lightning flashes'. *Journal of Atmospheric and Solar-Terrestrial Physics*, 2007;**69**(12):1388–96.

122. Cooray V., Rakov V. 'A 'hybrid current source' lightning return stroke model'. Presented at the *IX International Symposium on Lightning Protection*, 26–30 November 2007, Foz do Iguaçu, Brazil.
123. Willett J.C., Bailey J.C., Leteinturier C., Krider E.P., 'Lightning electromagnetic radiation field spectra in the interval from 0.2 to 20 MHz'. *Journal of Geophysical Research*, 1990;**95**(D12):20367–87.
124. LeVine D.M. 'Review of measurements of the RF spectrum of radiation from lightning'. *Meteorology and Atmospheric Physics*, 1987;**37**:195–204.
125. Berger K. 'Methoden und Resultate derBlitzforschung auf dem Mounte San Salvatorebei Lugano in dem Jahren 1963–1971'. *Bulletin SEV*, 1972;**63**: 1403–22.
126. Brook M., Kitagawa N., Workman E.J. 'Quantitative study of strokes and continuing currents in lightning discharges to ground'. *Journal of Geophysical Research*, 1962;**67**:649–59.
127. Shindo T., Uman M.A. 'Continuing current in negative cloud-to-ground lightning'. *Journal of Geophysical Research*, 1989;**94**(D4):5189–98.
128. Thomson E.M. 'Characteristics of Port Moresby ground flashes'. *Journal of Geophysical Research*, 1980;**85**(C2):1027–36.
129. Livingston J.M., Krider E.P. 'Electric fields produced by Florida thunderstorms' *Journal of Geophysical Research*, 1978;**83**(C1):385–401.
130. Rakov V.A., Uman M.A. 'Some properties of negative cloud-to-ground lightning flashes versus stroke order'. *Journal of Geophysical Research*, 1990;**95**: 5447–53.
131. Ballarotti M.G., Saba M.M.F., Pinto O., Jr. 'High-speed camera observations of negative ground flashes on a millisecond-scale'. *Geophysical Research Letters*, 2005;**32**: L23802; doi:10.1029/2005GL023889.
132. Saba M.M.F., Pinto O., Jr., Ballarotti M.G. 'Relation between lightning return stroke peak current and following continuing current'. *Geophysical Research Letters*, 2006;**33**: l23807; doi:10.1029/2006gl027455.
133. Thottappillil R., Goldberg J.D., Rakov V.A., Uman M.A., Fisher R.J., Schnetzer G.H. 'Properties of M components from currents measured at triggered lightning channel base'. *Journal of Geophysical Research*, 1995;**100**(D12):25711–20.

Chapter 3

Rocket-triggered lightning and new insights into lightning protection gained from triggered-lightning experiments

V.A. Rakov

3.1 Introduction

An understanding of the physical properties and deleterious effects of lightning is critical to the adequate protection of power and communication lines, aircraft, spacecraft, and other objects and systems. Many aspects of lightning are not yet well understood and are in need of research that often requires the termination of lightning channel on an instrumented object or in the immediate vicinity of various sensors. The probability of natural lightning striking a given point on the earth's surface or an object or structure of interest is very low, even in areas of relatively high lightning activity. Simulation of the lightning channel in a high-voltage laboratory has very limited applications, because it does not allow the reproduction of the many features of lightning important for lightning protection, and it does not allow the testing of large distributed systems such as overhead power lines. One promising tool for studying both the direct and induced effects of lightning is an artificially initiated (or triggered) lightning discharge from a thunderstorm cloud to a designated point on the ground. In most respects the triggered lightning is a controllable analogue of natural lightning. The most effective technique for artificial lightning initiation is the so-called rocket-and-wire technique. This technique involves the launching of a small rocket extending a thin wire (either grounded or ungrounded) into the gap between the ground and a charged cloud overhead.

The possibility of artificially initiating lightning by ground-based activity was apparently first discussed by Newman [1] and by Brook and colleagues [2]. Brook and colleagues [2] showed that, in the laboratory, a spark discharge could be triggered by the rapid introduction of a thin wire into an electric field, while the steady presence of the wire did not result in a spark. They suggested that the corona discharge from a stationary conductor acts to shield this conductor so that the high fields necessary to initiate electrical breakdown are not obtained, whereas the field enhancement due to the rapid introduction of a conductor is not significantly reduced by the corona, because there is insufficient time for its development.

The first triggered lightning discharges were produced in 1960 by launching small rockets trailing thin grounded wires from a research vessel off the west coast of Florida [1,3,4]. The first triggering over land was accomplished in 1973, at Saint-Privat d'Allier in France [5,6]. In the following decades, a number of triggered-lightning programmes have been developed in different countries, as summarized in Table 3.1. Rocket-triggered lightning experiments in France have been reviewed by Fieux and colleagues [6], in Japan by Horii [7], Kito and colleagues [8], Nakamura and colleagues [9,10] and Horii and colleagues [11], in New Mexico by Hubert and colleagues [12], at the Kennedy Space Center, Florida, by Willett [13], at Camp Blanding, Florida, by Uman *et al.* [14] and Rakov *et al.* [15–17], in China by Liu *et al.* [18] and Liu and Zhang [19], and in Brazil by Pinto *et al.* [20]. Triggered-lightning experiments conducted in different countries have been reviewed by Uman [21], Horii and Nakano [22], Rakov [23], and Rakov and Uman [24].

In all published experiments, the triggering wires were made of either steel or copper with a diameter of typically ~ 0.2 mm, wound on a spool located either on the ground or on the rocket. Various rockets made of plastic and of steel have been used, with the rocket length being typically ~ 1 m. Most of the experiments in Japan were conducted in the winter, the several attempts made to trigger in the summer months being unsuccessful. At Camp Blanding, Florida, lightning has been triggered in both summer and winter storms. All other triggering sites have apparently been operated only during the summer. The results from these programmes have made possible a number of new insights into the various lightning processes and effects.

Descriptions of the classical and altitude rocket and wire triggering techniques are given in Sections 3.2.1 and 3.2.2, respectively. Probably close to a thousand lightning discharges have been triggered using these techniques to date. An overview of lightning-triggering facilities is found in Table 3.1, with a description of the Camp Blanding facility being given in Section 3.2.3. Over 300 lightning flashes have been triggered to date at the Camp Blanding site. The properties of rocket-triggered lightning (including its close electromagnetic environment) are reviewed in Sections 3.3 to 3.6. The use of rocket-triggered lightning for testing various objects and systems is described in Section 3.7.

3.2 Triggering techniques

Two techniques for triggering lightning with a small rocket that extends a thin wire in the gap between a thundercloud and the ground are discussed here. 'Classical' triggering is described in Section 3.2.1 and 'altitude' triggering in Section 3.2.2. These descriptions primarily apply to triggering negative lightning.

3.2.1 Classical triggering

This triggering method involves the launching of a small rocket trailing a thin grounded wire toward a charged cloud overhead, as illustrated in Figure 3.1. Still

Table 3.1 An overview of major triggered-lightning programmes

Experimental site	Height above sea level (m)	Years of operation*	Wire material	Location of wire spool	Selected references
Saint Privat d'Allier, France	1 100	1973–1996	Steel or copper	Ground or rocket	[6,124]
Kahokugata, Hokuriku coast, Japan	0	1977–1985	Steel	Ground	[7,8]
Langmuir Laboratory, New Mexico	3 230	1979–present	Steel	Ground	[12,69]
KSC, Florida (south of Melbourne, Florida in 1983)	0	1983–1991	Copper	Rocket	[13,59,60]
Okushishiku, Japan	930	1986–1998	Steel	Ground or rocket	[9,10]
Four sites in northern and southeastern China	Various	1989–present	Steel or copper	Ground or rocket	[18,19]
Fort McClellan, Alabama	190	1991–1995	Copper	Rocket	[68,119]
Camp Blanding, Florida	20–25	1993–present	Copper	Rocket	[14,15,17]
Cachoeira Paulista, Brazil	570	1999–2007	Copper	Rocket	[20,125]

*As of this writing and not necessarily continuous.
Additionally, triggered-lightning experiments have been conducted in Germany [127], in Indonesia [128] and in Russia [129].

100   *Lightning Protection*

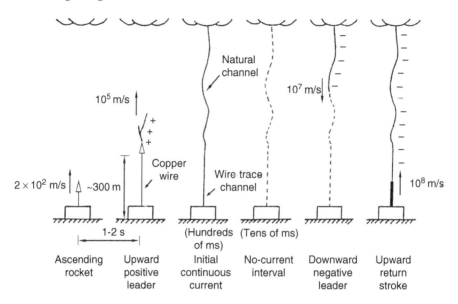

Figure 3.1   *Sequence of events (except for the attachment process [25]) in classical triggered lightning. The upward positive leader (UPL) and initial continuous current (ICC) constitute the initial stage (IS) (adapted from Reference 15).*

photographs of classical triggered lightning flashes are shown in Figure 3.2. To decide when to launch a triggering rocket, the cloud charge is indirectly sensed by measuring the electric field at ground, with absolute values of 4–10 kV m^{-1} generally being good indicators of favourable conditions for negative lightning initiation in Florida, as seen in Figure 3.3. However, other factors, such as the general trend of the electric field and the frequency of occurrence of natural lightning discharges, are usually taken into account in making the decision to launch a rocket. The triggering success rate is generally relatively low during very active periods of thunderstorms, one reason being that during such periods the electric field is more likely to be reduced by a natural lightning discharge before the rocket rises to a height sufficient for triggering.

When the rocket, ascending at ∼200 m s^{-1}, is about 200 to 300 m high, the field enhancement near the rocket tip launches a positively charged leader that propagates upwards towards the cloud. This upward positive leader (UPL) vaporizes the trailing wire, bridges the gap between the cloud charge source and ground, and establishes an initial continuous current (ICC) with a duration of some hundreds of milliseconds that transports negative charge from the cloud charge source to the triggering facility. The ICC can be viewed as a continuation of the UPL when the latter has reached the main negative charge region in the cloud. At that time the upper extremity of the UPL is likely to become heavily branched. The UPL and ICC constitute the initial stage (IS) of a classical triggered lightning discharge. After the cessation of the ICC, one or more downward dart leader/upward return stroke sequences may traverse the

Rocket-triggered lightning and new insights 101

Figure 3.2 Photographs of lightning flashes triggered at Camp Blanding, Florida. Top, a distant view of a strike to the test runway; middle, a strike to the test power system initiated from the tower launcher; bottom, a strike initiated from the underground launcher at the centre of a $70 \times 70\ m^2$ buried metallic grid.

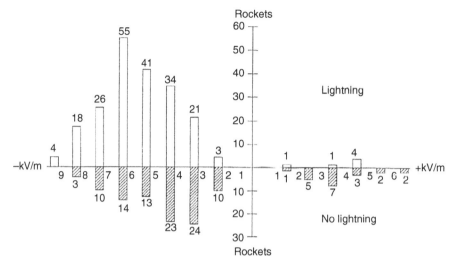

Figure 3.3 Histograms of successful (above the horizontal axis) and unsuccessful (below the horizontal axis) classical triggering attempts in 1983 to 1991 at the NASA Kennedy Space Center. Individual histogram bins correspond to different positive and negative values of surface electric field at the time of rocket launch. The upward-directed field is considered negative (atmospheric electricity sign convention) (adapted from Reference 26).

same path to the triggering facility. The dart leaders and the following return strokes in triggered lightning are similar to dart leader/return stroke sequences in natural lightning, although the initial processes in natural downward and classical triggered lightning are distinctly different.

In summer, the triggering success rate for positive lightning is apparently lower than for negative lightning [6], one known exception being the triggered lightning experiment in northern China [18,19], although all discharges triggered there were composed of an IS only; that is, no leader/return stroke sequences occurred.

There is contradictory information regarding whether the height H of the rocket at the time of lightning triggering depends on the electric field intensity E at ground at the time of launching the rocket. Hubert and colleagues [12] found a strong correlation (correlation coefficient $= -0.82$) between H and E for triggered lightning in New Mexico. They gave the following equation between H (in metres) and E (in kV m^{-1})

$$H = 3\,900\, E^{-1.33} \qquad (3.1)$$

In the study by Hubert and colleagues [12], E varied from ~ 5 to 13 kV m^{-1} and H from ~ 100 to 600 m, with a mean value of 216 m. On the other hand, in winter triggered-lightning studies at the Kahokugata site in Japan (Table 3.1), no clear relation was observed between H and E for either sign of E [22; figure 6.2.3].

Willett and colleagues [27], who used electric field sounding rockets in Florida, studied ambient-field conditions that are sufficient to initiate and sustain the propagation of upward positive leaders in triggered lightning. It was found that lightning can be initiated with grounded triggering wires ~ 400 m long when the ambient fields aloft are as small as 13 kV m^{-1}. When lightning occurred, ambient potentials with respect to earth at the triggering-rocket altitude were 3.6 MV (negative with respect to earth). These potentials were referred to as triggering potentials by Willett and colleagues [27].

3.2.2 Altitude triggering

A stepped leader followed by a first return stroke in natural downward lightning can be reproduced to some degree by triggering lightning via a metallic wire that is not attached to the ground. This ungrounded-wire technique is usually called altitude triggering and is illustrated in Figure 3.4, which shows that a bidirectional (positive charge up and negative charge down) leader process is involved in the initiation of the first return stroke from ground. Note that the 'gap' (in this case, the length of the insulating kevlar cable) between the bottom end of the upper (triggering) wire and the top end of the grounded (intercepting) wire is some hundreds of metres. Altitude triggering can also be accomplished without using an intercepting wire, whose only function is to increase the probability of lightning attachment to the instrumented rocket-launching facility. In some triggered-lightning experiments, the bottom end of the triggering wire has been attached to an air gap of up to 10 m in length [10]. Such triggering is not considered as being of the altitude type, because it was not intended to simulate the downward stepped leader (discussed below) from the bottom of the triggering wire. On the other hand, altitude triggering may also occur as a result of the accidental breakage of the wire during classical triggering, so that the wire connection to ground is unintentionally lost. Additionally, altitude triggering has been accomplished using a two-stage rocket system in which the two rockets separated in the air with the triggering wire extending between them [10]. The properties of altitude triggered lightning are discussed in papers by Laroche *et al.* [28], Lalande *et al.* [29,30], Uman *et al.* [31], Rakov *et al.* [15,32], Wang *et al.* [33], Chen *et al.* [34] and Saba *et al.* [35].

In the following, we briefly discuss the sequence of processes involved in altitude triggered lightning, as illustrated in Figure 3.4. A downward negative leader is usually launched from the lower end of the elevated triggering wire some milliseconds after the initiation of the upward positive leader from the upper end of the wire [30; figure 6]. The downward negative leader shown in Figure 6 of a paper by Lalande *et al.* [30] was apparently initiated after two unsuccessful attempts. As the negative downward leader approaches the triggering facility, an upward connecting

104   Lightning Protection

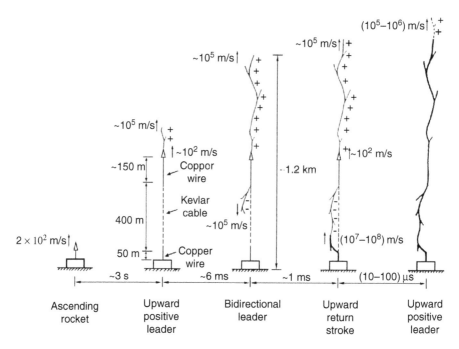

Figure 3.4 *Sequence of events in altitude triggered lightning leading to the establishment of a relatively low-resistance connection between the upward-moving positive leader tip and the ground (except for the attachment process [30]), based on the event described by Laroche et al. [28]. The processes that follow the sequence of events shown, the ICC and downward leader/upward return-stroke sequences, are similar to their counterparts in classical triggered lightning (see Figure 3.1) (adapted from Reference 15).*

leader (not shown in Figure 3.4) is initiated from the grounded intercepting wire. Once the attachment between the two leaders is made, the return stroke is initiated. Because (i) the length of the channel available for the propagation of the first return stroke in altitude triggered lightning is relatively small (of the order of 1 km) and (ii) the return-stroke speed is two to three orders of magnitude higher than that of the leader, the return stroke catches up with the tip of the upward leader within 10 μs or so. As a result, the upward leader becomes strongly intensified. The processes that follow, the ICC and downward leader/upward return-stroke sequences, are probably similar to those in classical triggered lightning (see Figure 3.1). Thus the downward-moving negative leader of the bidirectional leader system and the resulting return stroke in altitude triggered lightning serve to provide a relatively low-resistance connection between the upward-moving positive leader tip and the ground. The IS of altitude triggered lightning can be viewed as composed of an initial upward leader, a bidirectional leader (part of which is a continuation of the initial upward leader), an attachment process, an IS return stroke, an intensified upward leader and an ICC.

Wang and colleagues [33] reported on a positive flash that was initiated using the altitude triggering technique from a summer thunderstorm in China. This is the first documented triggering of a positive lightning using the altitude triggering technique. For this flash, the length of grounded intercepting wire was 35 m and the length of insulating cable was 86 m. The flash was apparently initiated when the rocket was at an altitude of 550 m, so that the length of the ungrounded triggering wire was 429 m.

3.2.3 Triggering facility at Camp Blanding, Florida

The lightning-triggering facility at Camp Blanding, Florida, was established in 1993 by the Electric Power Research Institute (EPRI) and Power Technologies, Inc. (PTI). Since September 1994, the facility has been operated by the University of Florida (UF). Over 40 researchers (excluding UF faculty, students and staff) from 15 countries representing 4 continents have performed experiments at Camp Blanding concerned with various aspects of atmospheric electricity, lightning and lightning protection. Since 1995, the Camp Blanding facility has been referred to as the International Center for Lightning Research and Testing (ICLRT) at Camp Blanding, Florida. A summary of the lightning triggering operations conducted for various experiments from 1993 to 2005 is presented in Table 3.2. Over the 13-year period, the total number of triggered flashes was 315, that is, on average about 24 per year, with about 17 ($\sim$70 per cent) of them containing return strokes. Of the total of 315 flashes in Table 3.2, 312 transported negative charge and 3 either positive or both negative and positive charge to ground.

The principal results obtained from 1993 through 2005 at the ICLRT include

- characterization of the close lightning electromagnetic environment [15,36–40]
- first lightning return-stroke speed profiles within 400 m of ground [41,42]
- new insights into the mechanism of the dart-stepped (and by inference stepped) leader [15,41]
- identification of the M-component mode of charge transfer to ground [15,36,43]
- first optical image of the UCL in lightning strokes developing in previously conditioned channels [25]
- electric fields in the immediate vicinity of the lightning channel core, inside the radial corona sheath [44]
- inferences on the interaction of lightning with ground and with grounding electrodes [15,45–47]
- discovery of X-rays produced by triggered-lightning strokes [48–52]
- new insights into the mechanism of cutoff and reestablishment of current in the lightning channel [53,54]
- first direct measurements of NO_x production by lightning [55]
- direct estimates of lightning input energy [56]

106 Lightning Protection

Table 3.2 Triggered-lightning experiments at the ICLRT at Camp Blanding, Florida, 1993–2005

Year(s)	Rocket launchers used	Total flashes triggered	Flashes with return strokes	Positive or bipolar flashes	Time period
1993	1	32	22	–	7 June–21 September
1994	2	15	11	–	4 August–September
1995	2	14	13	–	25 June–19 August
1996	2	30	25	–	20 June–11 September
1997	4	48	28	1	24 May–26 September
1998	3	34	27	–	15 May, 24 July–30 September
1999	2	30	22	1	23 January, 26 June–27 September
2000	2	30	27	–	12 June–6 September
2001	2	23	11	–	13 July–5 September
2002	2	19	14	–	9 July–13 September
2003	2	24	12	1	30 June–15 August
2004	1	5	3	–	23 June–24 July
2005	2	11	8	–	2 July–5 August
1993–2005 (13 years)		315	223	3	

3.3 Overall current waveforms

In this Section, we discuss currents measured at the rocket launcher. For both classical and altitude triggered lightning, the emphasis will be placed on the IS, with the characterization of current waveforms due to return strokes (primarily from classical triggered lightning) being presented in Section 3.4. Initial-stage return strokes in altitude triggered lightning are discussed in Section 3.3.2. For classical triggered lightning, IS current initially flows through the triggering wire until the wire is destroyed and replaced by a plasma channel [53,54]. For altitude triggered lightning, current exceeding some amperes is first measured when an UCL (not shown in Figure 3.4) emanates from the launcher (or from a grounded intercepting wire) in response to the approaching downward-extending, negative part of the bidirectional leader system.

3.3.1 Classical triggering

The overall current record for a typical negative classical triggered lightning flash is presented in Figure 3.5a, and portions of this record are shown on expanded timescales in Figure 3.5b and c. The record is intentionally clipped at the 2-kA level in order to accentuate the current components in the hundreds of amperes range. Other researchers [59,60] used recorders with a logarithmic scale in order to be able to view both small currents and large currents on the same record. Median values of the overall flash duration from triggered lightning experiments in France and New Mexico are 350 and 470 ms [61], respectively. The median flash charges from the same studies are 50 and 35 C, respectively. Both the flash duration and charge transferred are comparable, within a factor of 2 to 4, to their counterparts in object-initiated lightning and in natural downward lightning.

We first consider the overall characteristics, that is, the duration, the charge transfer, and the average current, of the IS, and then discuss (i) the current variation at the beginning of the IS, termed the initial current variation (ICV), and (ii) the current pulses superimposed on the later part of the IS current, referred to as ICC pulses. Parameters of the return-stroke current pulses (three pulses are shown in Figure 3.5a) that often follow the IS current are discussed in Section 3.4.

Miki and colleagues [62], based on data from Camp Blanding, Florida (see Table 3.1), reported that the IS had a geometric mean (GM) duration of 305 ms and lowered to ground a GM charge of 30 C. The average IS current in an individual lightning discharge had a GM value of 100 A.

In many cases the initial current variation includes a current drop, as illustrated in Figure 3.5b where it is labelled ABC. This current drop is associated with the disintegration of the copper triggering wire (abrupt current decrease from A to B in Figure 3.5b) and the following current re-establishment (abrupt current increase from B to C in Figure 3.5b). The processes of current cutoff and re-establishment were studied in detail by Rakov and colleagues [53] and Olsen *et al.* [54].

The ICC usually includes impulsive processes, illustrated in Figure 3.5c, that resemble the M processes observed during the continuing currents that often follow return strokes in both natural and triggered lightning [63–65]. Wang and colleagues

108 Lightning Protection

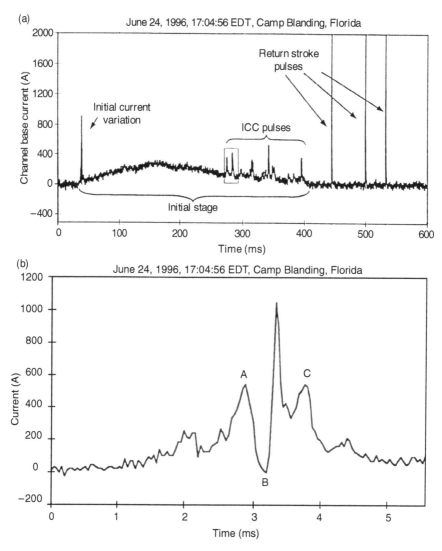

Figure 3.5 (a) Example of the overall current record of a triggered lightning at Camp Blanding, Florida, containing an IS and three return strokes. The initial tens of milliseconds of IS are due to the upward positive leader (UPL), and the rest of the IS is due to the ICC. The record is intentionally clipped at ∼2 kA (adapted from Reference 57). (b) Initial current variation (ICV) shown in Figure 3.5a but on an expanded timescale (adapted from Reference 57). (c) First two ICC pulses of Figure 3.5a on an expanded timescale. This figure illustrates the definitions of the ICC pulse magnitude I_M, 10–90 per cent rise time R_T, duration T_D, half-peak width T_H, interpulse interval T_I and preceding continuous current level I_{CC}. All these parameters have been found

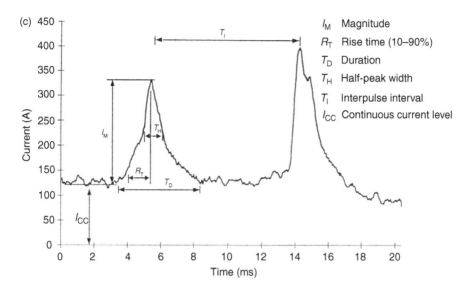

Figure 3.5 (*Continued*) *to be similar to the corresponding parameters of M-component current pulses analysed by Thottappillil et al. [58] (adapted from Reference 57).*

[57], from a comparison of various characteristics of the ICC pulses with the characteristics of the M-component current pulses analysed by Thottappillil and colleagues [58], concluded that these two types of pulses are similar and hence likely due to similar lightning processes. Like M component pulses, the ICC pulses sometimes have amplitudes in the kiloamperes range.

3.3.2 Altitude triggering

As noted in Section 3.2.2, the IS of altitude triggered lightning includes an initial upward leader, a bidirectional leader (which includes a continuation of the initial upward leader), an attachment process, an initial-stage return stroke, an intensified upward leader and an ICC. Because the triggering wire is ungrounded, no current can be directly measured at ground during the initial upward leader and bidirectional leader stages. Shown in Figure 3.6b is the current associated with an upward positive connecting leader initiated in response to the approaching downward negative leader of the bidirectional leader system (Figure 3.4), with the corresponding electric field measured 50 m from the lightning attachment point being shown in Figure 3.6a. This current record, reported by Lalande and colleagues [30], suggests that the upward positive connecting leader is stepped, with the interstep interval being 20 µs or so. When contact is established between the downward leader and the UCL, the IS return stroke begins. The current waveform of this return stroke differs appreciably from a typical return-stroke current waveform in that the former appears to be chopped soon after reaching its peak value (see, e.g. figure 7c of Reference 61).

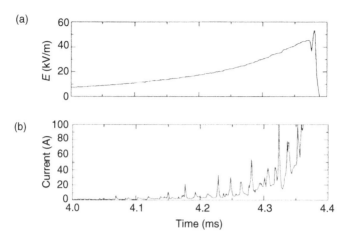

Figure 3.6 (a) *Electric field measured 50 m from the lightning attachment point and* (b) *current produced by the upward connecting positive leader from the grounded 50 m wire in altitude triggered lightning 9516 at Camp Blanding, Florida (adapted from Reference 30)*

As a result, the width of the current waveform produced by the IS return stroke is appreciably smaller than that of the following return strokes in the same flash. As discussed in Section 3.2.2, the IS return stroke front catches up with the upward-moving leader tip after 10 μs or so. This is likely to produce an opposite polarity downward-moving reflected current wave that is presumably responsible for the chopped shape of both the channel-base current and the close magnetic field waveforms. Examples of the latter are shown, along with waveforms produced by 'normal' return strokes, in Figure 3.7. The IS characteristics of altitude triggered lightning, after the return stroke has established a relatively low-resistance connection between the upward-moving positive leader tip and ground (see Figure 3.4), are apparently similar to their counterparts in classical triggered lightning [66]. Further, the downward leader/upward return-stroke sequences that follow the IS in altitude triggered lightning are thought to be similar to those in classical triggered lightning (see Figure 3.1).

3.4 Parameters of return-stroke current waveforms

In this Section, we discuss return-stroke current peak and current waveform parameters such as rise time, rate of rise (steepness) and half-peak width. We will additionally consider interstroke intervals and characteristics that may involve both the return-stroke current component and the following continuing current component, such as the total stroke duration, the total stroke charge, $\int I(t)\,dt$, and the total stroke action integral, $\int I^2(t)\,dt$. The action integral is measured in A s², which is the same as JΩ^{-1}, and represents the joule or ohmic heating energy dissipated per unit resistance

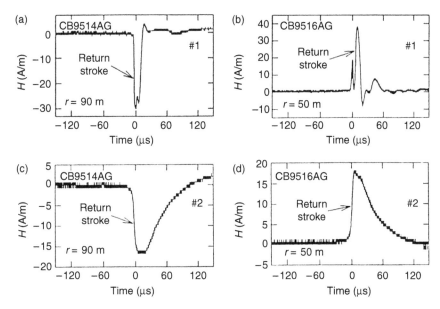

Figure 3.7 The magnetic fields produced by the first two strokes of the Camp Blanding altitude triggered lightning flashes 9514 (a, first stroke; c, second stroke; four strokes total) and 9516 (b, first stroke; d, second stroke; four strokes total). In each case, the waveshapes of all the higher-order strokes are similar to the second-stroke waveshape. The measuring system's decay time constant was ~ 120 μs. The difference in polarity of the waveforms is due to different positions of the lightning channel with respect to the magnetic field antenna, all strokes lowering negative charge to ground. Note that the first-stroke magnetic field pulses in (a) and (b) are appreciably shorter than the corresponding second-stroke magnetic field pulses in (c) and (d), respectively (adapted from Reference 15).

at the lightning attachment point. The action integral is also called the specific energy. We will additionally discuss correlations among the various parameters listed above. The characterization of the return-stroke current waveforms presented in this Section is based primarily on data for classical triggered lightning. It is possible that some of the samples on which the statistics presented here are based contain a small number of IS return strokes from altitude triggered lightning, but their exclusion would have essentially no effect on the statistics.

Some researchers [61,67], in presenting statistics on triggered lightning currents, do not distinguish between current pulses associated with return strokes and those produced by other lightning processes such as M components and processes giving rise to the initial current variation and ICC pulses described in Section 3.3.1. In this Section, we consider only return-stroke current pulses. These can usually be distinguished from other types of pulses by the absence of a steady current immediately prior to a pulse [68]. Further, we do not consider here three unusual New Mexico triggered lightning flashes, each of which contained 24 return strokes [69]. For these three flashes, the geometric means of the return-stroke current peak and inter-stroke interval are 5.6 kA and 8.5 ms, respectively, each considerably smaller than its counterpart in either natural lightning or other triggered lightning discussed below.

We first review measurements of the peak values of current and current derivative. Summaries of the statistical characteristics of measured return-stroke currents, I, and derivatives of current with respect to time, dI/dt, taken from a paper by Schoene and colleagues [40], are given in Tables 3.3 and 3.4, respectively. As seen in Table 3.3, the geometric mean values of current peak range from ~ 12 to 15 kA. These values are similar to the median value of 12 kA reported by Anderson and Eriksson [70] for subsequent strokes in natural lightning. The geometric mean values of dI/dt peak based on data from two studies presented in Table 3.4 are 73 and 97 kA μs^{-1}.

Scatter plots of dI/dt peak vs. I peak from the triggered lightning experiments in Florida (1985, 1987 and 1988) and in France (1986) are shown in Figure 3.8. Correlation coefficients are 0.87, 0.80 and 0.70 for the 1985, 1987 and 1988 Florida data, respectively, and 0.78 for the 1986 data from France. The largest measured value of dI/dt is 411 kA μs^{-1}, as reported from Florida (KSC) studies by Leteinturier and colleagues [71]. The corresponding measured peak current is greater than 60 kA, the largest value of this parameter reported for summer triggered lightning to date. Also shown in Figure 3.8 are the linear regression line and the regression equation for each of the four subsets of the data. Note that the correlation coefficients between the logarithms of dI/dt and I for the same data were found to be lower: 0.79, 0.56 and 0.60 for the 1985, 1987 and 1988 Florida data, respectively, and 0.71 for the 1986 data from France [72; table 10].

Fisher and colleagues [68] compared a number of return-stroke current parameters for classical triggered-lightning strokes from Florida and Alabama with the corresponding parameters for natural lightning in Switzerland reported by Berger *et al.* [73] and Anderson and Eriksson [70]. This comparison is given in Figures 3.9 to 3.17. Recall that triggered-lightning strokes are considered to be similar to subsequent strokes in natural lightning. Therefore, the comparison in Figures 3.9 to 3.17 applies only to subsequent strokes that are usually initiated by leaders that follow the path of the previous stroke. Both Berger and colleagues [73] and Anderson and Eriksson [70] fitted a straight line representing a lognormal approximation to the experimental statistical distribution in order to determine the percentages (95, 50 and 5 per cent) of cases exceeding the tabulated values, while Fisher and colleagues [68] used the nearest experimental point instead. Distributions of peak currents are very similar, with median values being 13 and 12 kA for triggered and natural lightning, respectively. On the other hand, there appear to be appreciable differences between the triggered-lightning data of Fisher and colleagues [68] and the natural-lightning data of Berger and colleagues [73] and Anderson and Eriksson [70] in terms of current wavefront parameters, half-peak width and stroke charge. The shorter rise time and higher average slope (steepness) in the triggered-lightning data may be explained by the better time resolution of the measuring systems used in the triggered-lightning

Table 3.3 Current waveform parameters for negative rocket-triggered lightning (adapted from Reference 40)

Location/year	n	Min.	Max.	Arithmetic mean	σ	Geometric mean	$\sigma_{\log}$	Reference
Current peak (kA)								
Kennedy Space Center, Florida; 1985–1991	305	2.5	60.0	14.3	9.0	–	–	[72]
Saint-Privat d'Allier, France; 1986, 1990–1991	54	4.5	49.9	11.0	5.6	–	–	[72]
Kennedy Space Center, Florida and Fort McClellan, Alabama; 1990, 1991	45	–	–	–	–	12	0.28	[68]
Camp Blanding, Florida; 1993	37	5.3	44.4	15.1	–	13.3	0.23	[15]
Camp Blanding, Florida; 1997	11	5.3	22.6	12.8	5.6	11.7	0.20	[126]
Camp Blanding, Florida; 1998	25	5.9	33.2	14.8	7.0	13.5	0.19	[37]
Present study	64	5	36.8	16.2	7.6	14.5	0.21	
Current 10–90% rise time (ns)								
Kennedy Space Center, Florida and Fort McClellan, Alabama; 1990, 1991	43	–	–	–	–	370	0.29	[68]
Saint-Privat d'Allier, France; 1990–1991	37	250	4 900	1 140	1 100	–	–	[72]
Camp Blanding, Florida; 1997	11	300	4 000	900	1 200	600	0.39	[126]

(Continues)

Table 3.3 Continued

Location/year	n	Min.	Max.	Arithmetic mean	σ	Geometric mean	$\sigma_{\log}$	Reference
				Current 30–90% rise time (ns)				
Kennedy Space Center, Florida and Fort McClellan, Alabama; 1990, 1991	43	–	–	–	190	280	0.28	[68]
Present study	65	54	1 751	260	316	191	0.29	
				Current half-peak width (μs)				
Saint-Privat d'Allier, France; 1990–1991	24	14.7	103.2	49.8	22.4	–	–	[72,87]
Kennedy Space Center, Florida and Fort McClellan, Alabama; 1990, 1991	41	–	–	–	–	18	0.30	[68]
Camp Blanding, Florida; 1997	11	6.5	100	35.7	24.6	29.4	0.29	[126]
Present study	64	2.4	37.2	13.2	8.5	10.5	0.32	

The polarity of the peak values is ignored.

Table 3.4 Current derivative waveform parameters for negative rocket-triggered lightning* (adapted from Reference 40)

Location/year	n	Min.	Max.	Arithmetic mean	σ	Geometric mean	$\sigma_{\log}$
			dI/dt peak (kA/μs)				
Kennedy Space Center, Florida; 1985–1998[†]	134	5	411	118	97	—	—
Saint-Privat d'Allier, France; 1986, 1990–1991[†]	47	13	139	43	25	—	—
Camp Blanding, Florida; 1998[‡]	15	45	152	80	35	73	0.17
Present study[§]	64	8	292	117	65	97	0.31
			dI/dt 30–90% rise time (ns)				
Present study	29	17	69	32	13	30	0.16
			dI/dt 10–10% width (ns)				
Saint-Privat d'Allier, France; 1990–1991[†]	17	70	2 010	400	210	—	—
			dI/dt half-peak width (ns)				
Present study	29	49	149	92	25	89	0.12

*The polarity of the peak values is ignored.
[†]Reference 72.
[‡]Reference 37.
[§]Fifteen dI/dt peaks obtained from differentiation of I.

116   Lightning Protection

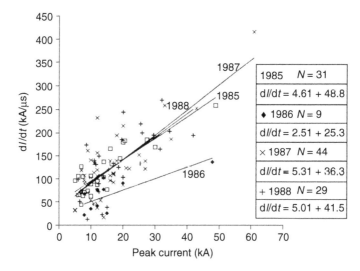

Figure 3.8   Relation between peak current rate of rise, dI/dt, and peak current I, from triggered-lightning experiments conducted at the NASA Kennedy Space Center, Florida, in 1985, 1987 and 1988 and in France in 1986. The regression line for each year is shown, and the sample size and the regression equation are given (adapted from Reference 71).

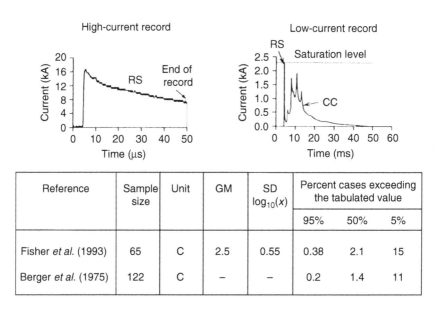

Figure 3.9   Total stroke charge. RS is the return stroke, CC is the continuing current, GM the geometric mean and SD the standard deviation of the logarithm (base 10) of the parameter (adapted from Reference 68).

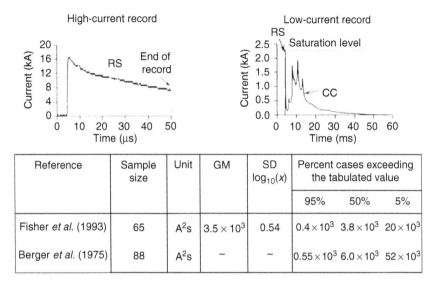

Figure 3.10 Total stroke action integral. RS is the return stroke, CC the continuing current, GM the geometric mean and SD the standard deviation of the logarithm (base 10) of the parameter (adapted from reference 68).

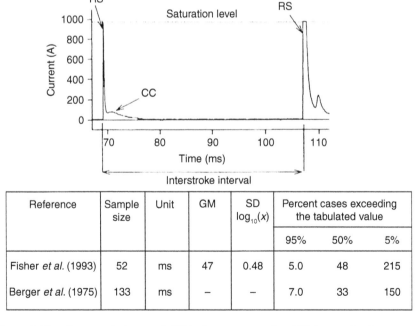

Figure 3.11 Interstroke interval. RS is the return stroke, CC the continuing current, GM the geometric mean and SD the standard deviation of the logarithm (base 10) of the parameter (adapted from Reference 68).

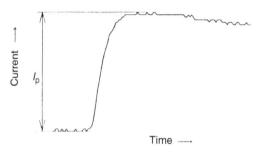

Symbol	Reference	Sample size	Unit	GM	SD $\log_{10}(x)$	Percent cases exceeding the tabulated value		
						95%	50%	5%
I_p	Fisher et al. (1993)	45	kA	12	0.28	4.7	13	29
	Anderson and Eriksson (1980)	114	kA	–	–	4.9	12	29

Figure 3.12   Peak current. GM is the geometric mean and SD is the standard deviation of the logarithm (base 10) of the parameter (adapted from Reference 68).

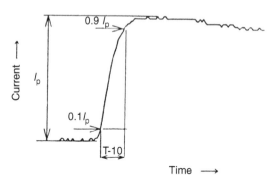

Symbol	Reference	Sample size	Unit	GM	SD $\log_{10}(x)$	Percent cases exceeding the tabulated value		
						95%	50%	5%
T-10	Fisher et al. (1993)	43	µs	0.37	0.29	0.20	0.32	1.1
	Anderson and Eriksson (1980)	114	µs	–	–	0.1	0.6	2.8

Figure 3.13   The 10–90 per cent rise time. GM is the geometric mean and SD the standard deviation of the logarithm (base 10) of the parameter (adapted from Reference 68).

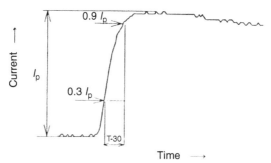

Symbol	Reference	Sample size	Unit	GM	SD $\log_{10}(x)$	Percent cases exceeding the tabulated value		
						95%	50%	5%
T-30	Fisher et al. (1993)	43	µs	0.28	0.28	0.14	0.24	0.96
	Anderson and Eriksson (1980)	114	µs	–	–	0.1	0.4	1.8

Figure 3.14 The 30–90 per cent rise time. GM is the geometric mean and SD the standard deviation of the logarithm (base 10) of the parameter (adapted from Reference 68).

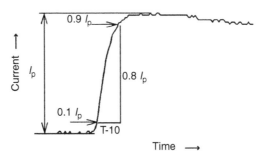

Symbol	Reference	Sample size	Unit	GM	SD $\log_{10}(x)$	Percent cases exceeding the tabulated value		
						95%	50%	5%
S-10	Fisher et al. (1993)	43	kA/µs	28	0.37	5.4	34	83
	Anderson and Eriksson (1980)	114	kA/µs	–	–	3.3	15	72

Figure 3.15 The 10–90 per cent average slope (steepness). GM is the geometric mean and SD the standard deviation of the logarithm (base 10) of the parameter. S-10 = $0.8I_p/T\text{-}10$ (adapted from Reference 68).

120 Lightning Protection

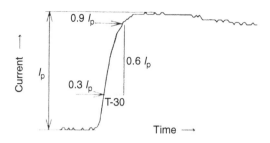

Symbol	Reference	Sample size	Unit	GM	SD log₁₀(x)	Percent cases exceeding the tabulated value		
						95%	50%	5%
S-30	Fisher et al. (1993)	43	kA/μs	28	0.36	4.6	31	91
	Anderson and Eriksson (1980)	114	kA/μs	–	–	4.1	20	99

Figure 3.16 The 30–90 per cent average slope (steepness). GM is the geometric mean and SD the standard deviation of the logarithm (base 10) of the parameter. $S\text{-}30 = 0.6 I_p / T\text{-}30$ (adapted from Reference 68).

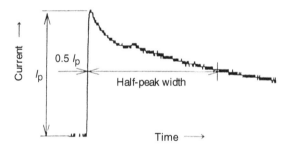

Reference	Sample size	Unit	GM	SD log₁₀(x)	Percent cases exceeding the tabulated value		
					95%	50%	5%
Fisher et al. (1993)	41	μs	18	0.30	5.0	20	40
Anderson and Eriksson (1980)	115	μs	–	–	6.5	32	140

Figure 3.17 Half-peak width. GM is the geometric mean and SD the standard deviation of the logarithm (base 10) of the parameter (adapted from Reference 68).

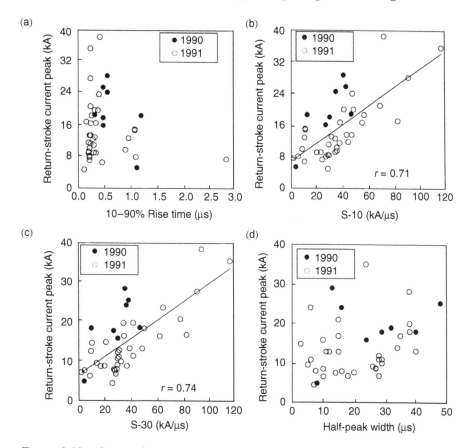

Figure 3.18 Scatterplots relating various return stroke parameters. Solid circles represent 1990 data from KSC, Florida, and open circles represent 1991 data from Fort McClellan, Alabama. (a) Current peak versus 10–90 per cent rise time. (b) Current peak versus S-10. (c) Current peak versus S-30. (d) Current peak versus half-peak width. Regression lines and correlation coefficients (r) are given in (b) and (c) (adapted from Reference 68).

studies. The Swiss data were recorded as oscilloscopic traces with the smallest measurable time being 0.5 μs [74].

Fisher and colleagues [68] also studied relations among some return-stroke parameters, the results being shown in Figure 3.18. They found a relatively strong positive correlation between the 10–90 per cent average steepness (S-10) and current peak (correlation coefficient = 0.71) and between the 30–90 per cent average steepness (S-30) and current peak (correlation coefficient = 0.74). As seen in Figure 3.18a, there is essentially no linear correlation between current peak and 10–90 per cent rise time.

3.5 Return-stroke current peak versus grounding conditions

In examining the lightning current flowing from the bottom of the channel into the ground, it is convenient to approximate lightning by a Norton equivalent circuit [75], i.e. by a current source equal to the lightning current that would be injected into the ground if that ground were perfectly conducting (a short-circuit current) in parallel with a lightning-channel equivalent impedance Z_{ch} assumed to be constant. The lightning grounding impedance Z_{gr} is a load connected in parallel with the lightning Norton equivalent. Thus the 'short-circuit' lightning current I effectively splits between Z_{gr} and Z_{ch} so that the current measured at the lightning-channel base is found as $I_{meas} = I\, Z_{ch}/(Z_{ch} + Z_{gr})$. Both source characteristics, I and Z_{ch}, vary from stroke to stroke, and Z_{ch} is a function of channel current, the latter non-linearity being in violation of the linearity requirement necessary for obtaining the Norton equivalent circuit. Nevertheless, if we are concerned only with the peak value of current and assume that for a large number of strokes the average peak value of I and the average value of Z_{ch} at current peak are each more or less constant, the Norton equivalent becomes a useful tool for studying the relation between lightning current peak and the corresponding values of Z_{ch} and Z_{gr}. For example, if the measured channel-base current peak statistics are similar under a variety of grounding conditions, then Z_{ch} must always be much larger than Z_{gr} at the time of the current peak. In the following, we will compare the geometric mean current peaks from triggered lightning experiments in which similar rocket launchers having a relatively small height of 4–5 m were used, but grounding conditions differed considerably. All the information needed for this comparison is given in Table 3.5.

As seen in Table 3.5, Camp Blanding measurements of lightning currents that entered sandy soil with a relatively poor conductivity of 2.5×10^{-4} S m^{-1} without any grounding electrode resulted in a value of the geometric mean return-stroke current peak (13 kA) that is similar to the geometric mean value (14 kA) estimated from measurements at KSC made in 1987 using a launcher of the same geometry that was much better grounded into salt water with a conductivity of 3–6 S m^{-1} via underwater braided metallic cables. Additionally, fairly similar geometric mean values were found from the Fort McClellan, Alabama, measurements using a poorly grounded launcher (10 kA) and the same launcher well grounded (11 kA) in 1993 and 1991, respectively. Also, Ben Rhouma and colleagues [76] give arithmetic mean values of return-stroke current peaks in the range from 15 to 16 kA for the Florida triggered-lightning experiments at Camp Blanding in 1993 and at KSC in 1987, 1989 and 1991.

The values of grounding resistance (probably the dominant component of Z_{gr}) given in Table 3.5 should be understood as the initial values encountered by a lightning downward leader before the onset of any breakdown processes in the soil or along the ground surface associated with the return stroke. Note from Table 3.5 that the grounding resistance varies from 0.1 Ω to 64 kΩ, whereas Z_{ch}, assumed to be a real number, was estimated from the analysis of the current waves travelling along the 540-m-high tower to be in the range from hundreds of ohms to some kΩ [77,78]. The observation that the average return-stroke current is not much influenced by the level of manmade

Table 3.5 Geometric mean peak current versus grounding conditions from different triggered-lightning experiments (adapted from Reference 15)

Experiment	Reference	Trigger threshold (kA)	Sample size	GM peak current (kA)	Soil	Artificial grounding	Grounding resistance (Ω)
KSC, Florida; 1987	[60,68,71]	5	36	14	0.5 m deep salt water (3–6 S m^{-1})	1.2×1.2 m^2 square metal plane connected through three 0.5-m-long wires at the four corners to salt water	0.1
Fort McClellan, Alabama; 1991	[68]	2 (two strokes below 2 kA from continuous tape record included)	37	11	Clay (3×10^{-3} S m^{-1})	Rebar framework of the munition storage bunker interconnected with lightning protection system including air terminals, down-conductors and buried counterpoise	Presumably low

(Continues)

124 Lightning Protection

Table 3.5 Continued

Experiment	Reference	Trigger threshold (kA)	Sample size	GM peak current (kA)	Soil	Artificial grounding	Grounding resistance (Ω)
Camp Blanding, Florida; 1993	[14,90]	3.3 and 4.2	37	13	Sand (2.5×10^{-4} S m^{-1})	None. Launcher was based on two parallel 15-m long, 2-m spaced concrete slabs above three unenergized power cables buried 1 m deep and 5 m apart	64×10^3 (assuming that the contact surface between the channel and ground was a hemisphere with 1 cm radius)
Fort McClellan, Alabama; 1993	[82]	~4	31	10	Heavy red clay (1.8×10^{-3} S m^{-1})	Single 0.3- or 1.3-m-long vertical grounding rod	260

KSC = Kennedy Space Center. The values of grounding resistance are determined by the geometry of the grounding electrode (or the geometry of the contact surface between the channel and the ground in the absence of grounding electrode) and soil conductivity. They are measured under low-frequency, low-current conditions and should be understood as the initial values of resistance encountered by lightning before the onset of any breakdown processes in the soil or along the ground surface.

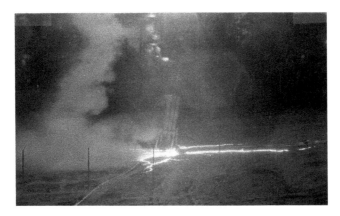

Figure 3.19 *Photograph of surface arcing associated with the second stroke (current peak of 30 kA) of flash 9312 triggered at Fort McClellan, Alabama. The lightning channel is outside the field of view. One of the surface arcs approached the right edge of the photograph, a distance of 10 m from the rocket launcher (adapted from Reference 68).*

grounding, ranging from excellent to none, implies that lightning is capable of lowering the grounding impedance it initially encounters (Table 3.5) to a value that is always much lower than the equivalent impedance of the main channel. On the basis of (i) the evidence of the formation of plasma channels (fulgurites) in the sandy soil at Camp Blanding [14,79–81] and (ii) optical records showing arcing along the ground surface at both Camp Blanding and Fort McClellan [15,82], it can be inferred that surface and underground plasma channels are important means of lowering the

Figure 3.20 *Evidence of surface arcing on a golf course green in Arizona (courtesy of E.P. Krider)*

126 *Lightning Protection*

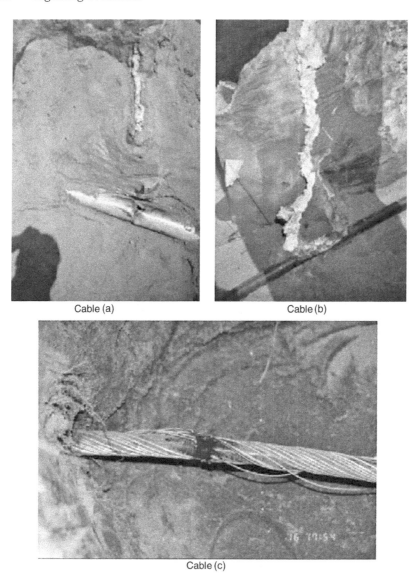

Figure 3.21 Lightning damage to underground power cables. (a) Coaxial cable in an insulating jacket inside a PVC conduit; note the section of vertical fulgurite in the upper part of the picture (the lower portion of this fulgurite was destroyed during excavation) and the hole melted through the PVC conduit. (b) Coaxial cable in an insulating jacket, directly buried; note the fulgurite attached to the cable. (c) Coaxial cable for which the neutral (shield) was in contact with earth; note that many strands of the neutral are melted through. The cables were tested at Camp Blanding, Florida in 1993. (Photos in (a) and (b) courtesy of V.A. Rakov and in (c) of P.P. Barker.)

lightning grounding impedance, at least for the types of soil at the lightning triggering sites in Florida and Alabama (sand and clay, respectively). A photograph of surface arcing during a triggered-lightning flash from Fort McClellan, Alabama, is shown in Figure 3.19, and evidence of surface arcing in natural lightning is presented in Figure 3.20. Injection of laboratory currents up to 20 kA into loamy sand in the presence of water sprays simulating rain resulted in surface arcing that significantly reduced the grounding resistance at the current peak (M. Darveniza, personal communication, 1995). The fulgurites (glassy tubes produced by lightning in sand; Figures 3.21 and 3.22) found at Camp Blanding usually show that the in-soil plasma channels tend to develop towards the better conducting layers of soil or towards buried metallic objects that, when contacted, serve to further lower the grounding resistance. The percentages of return strokes producing optically detectable surface arcing versus return stroke peak current, from the 1993 and 1995 Fort McClellan experiments, are shown in Figure 3.23. The surface arcing appears to be random in direction and often leaves little if any evidence on the ground. Even within the same flash, individual strokes can produce arcs developing in different directions. In one case, it was possible to estimate the current carried by one arc branch which contacted the instrumentation. That current was approximately 1 kA,

Figure 3.22 *A Florida fulgurite of* ~5 *m length excavated by the University of Florida lightning research group*

128 Lightning Protection

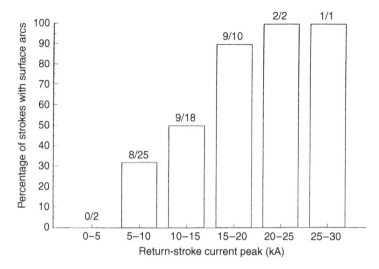

Figure 3.23   *Percentages of return strokes producing optically detectable surface arcing as a function of return-stroke current peak (Fort McClellan, Alabama, 1993 and 1995). The numbers above each histogram column indicate the number of strokes producing optically detectable arcing (numerator) and the total number of strokes in that current peak range (denominator) (adapted from Reference 15).*

or 5 per cent of the total current peak in that stroke. The observed horizontal extent of surface arcs was up to 20 m, which was the limit of the photographic coverage during the 1993 Fort McClellan experiment. No fulgurites were found in the soil (red clay) at Fort McClellan, only concentrated current exit points at several spots along the 0.3 or 1.3 m steel earthing rod (Table 3.5). It is likely that the uniform ionization of soil, usually postulated in studies of the behaviour of grounding electrodes subjected to lightning surges, is not an adequate assumption, at least not in the southeastern United States, where distinct plasma channels in the soil and on the ground surface appear to contribute considerably to lowering the grounding resistance.

3.6 Characterization of the close lightning electromagnetic environment

A knowledge of close lightning electric and magnetic fields is needed for the evaluation of lightning-induced effects in various electric circuits and systems [83] and for the testing of the validity of lightning models [84,85]. The close (within tens to hundreds of metres) lightning electromagnetic environment is most easily studied using rocket-triggered lightning for which the termination point on ground is known [15,36–40,63,86–88].

Rubinstein and colleagues [88,89] measured and analysed electric field waveforms at 500 m for 31 leader/return-stroke sequences and at 30 m for two leader/return-stroke sequences in lightning flashes triggered at the Kennedy Space Center, Florida, in 1986 and 1991, respectively. They found that, at tens to hundreds of metres from the lightning channel, leader/return-stroke vertical electric field waveforms appear as asymmetrical V-shaped pulses, the negative slope of the leading edge being lower than the positive slope of the trailing edge. The bottom of the V is associated with the transition from the leader (the leading edge of the pulse) to the return stroke (the trailing edge of the pulse). The first multiple-station electric field measurements within a few hundred metres of the triggered-lightning channel were performed in 1993 at Camp Blanding, Florida [90] and at Fort McClellan, Alabama [82]. Detailed analyses of these data have been presented by Rakov and colleagues [15]. From the 1993 experiment, the geometric mean width of the V at half of peak value is 3.2 µs at 30 m, 7.3 µs at 50 m and 13 µs at 110 m, a distance dependence close to linear.

In 1997, the multiple-station field measuring experiment at Camp Blanding, Florida, was extended to include seven stations at distances of 5, 10, 20, 30, 50, 110 and 500 m from the triggered-lightning channel [91]. Most of the data obtained at 5 m appeared to be corrupted, possibly due to ground surface arcs (see Section 3.5) and are not considered here. Leader/return-stroke electric field waveforms in one flash (S9721) simultaneously measured at 10, 20, 30, 50, 110 and 500 m are shown in Figure 3.24. The evolution of the leader/return-stroke electric field waveform as distance increases is consistent with previous measurements [15,85] and reflects an increasing contribution to the field from progressively higher channel sections.

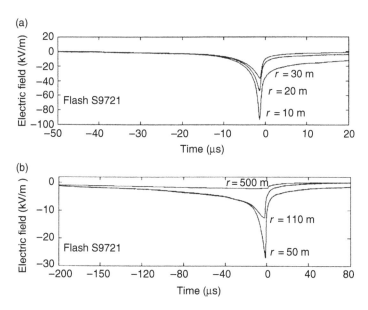

Figure 3.24   Electric field waveforms of the first leader/return-stroke sequence of flash S9721 as recorded in 1997 at distances (a) 10, 20 and 30 m and (b) 50, 110 and 500 m at Camp Blanding, Florida. The initial downward-going portion of the waveform is due to the dart leader, and the upward-going portion is due to the return stroke (adapted from Reference 91).

Crawford and colleagues [39] analysed net electric field changes due to dart leaders in triggered lightning from experiments conducted in 1993, 1997, 1998 and 1999 at Camp Blanding, Florida, and in 1993 at Fort McClellan, Alabama. In 1997 to 1999, the fields were measured at 2 to 10 stations with distances from the lightning channel ranging from 10 to 621 m, while in 1993 the fields were measured at three distances, 30, 50 and 110 m, in Florida, and at two distances, ~10 and 20 m, in Alabama.

Table 3.6 Dart-leader electric field change as a function of distance from the lightning channel for events recorded at the ICLRT in 1993–1999 (adapted from Reference 39)

Year	Flash	Stroke	Number of stations	I_p (kA)	$\Delta E_L = f(r)$ (kV m^{-1})	Distances (m)
1993	9313	2	3	9.7	$61 r^{-0.28}$	30/50/110
		3	3	11	$69 r^{-0.30}$	30/50/110
		4	3	13	$76 r^{-0.30}$	30/50/110
		5	3	11	$56 r^{-0.25}$	30/50/110
	9320	1	3	9.6	$1.7 \times 10^2 r^{-0.51}$	30/50/110
		2	3	8.4	$1.0 \times 10^2 r^{-0.42}$	30/50/110
1997	S9711	1	3	6.5	$1.6 \times 10^3 r^{-1.1}$	50/110/500
	S9712	1	3	5.3	$1.4 \times 10^2 r^{-0.59}$	10/20/30
	S9718	1	5	12	$2.1 \times 10^3 r^{-1.1}$	20–500
			3		$1.4 \times 10^3 r^{-1.0}$	30/50/110
	S9720	1	4	21	$2.6 \times 10^3 r^{-1.1}$	30–500
			3		$1.7 \times 10^3 r^{-0.99}$	30/50/110
	S9721	1	6	11	$1.3 \times 10^3 r^{-1.0}$	10–500
			3		$9.9 \times 10^2 r^{-0.93}$	30/50/110
			3		$7.1 \times 10^2 r^{-0.84}$	10/20/30
1998	U9801	1	10	8.7	$2.8 \times 10^3 r^{-1.2}$	102–410
	U9822	1	10	11	$2.6 \times 10^3 r^{-1.1}$	92–380
	U9824	1	10	17	$5.1 \times 10^3 r^{-1.2}$	102–410
	U9825	1	10	NR	$5.8 \times 10^3 r^{-1.2}$	102–410
	U9827	1	9	41	$7.1 \times 10^3 r^{-1.2}$	92–380
	S9806	1	10	9.1	$1.5 \times 10^3 r^{-0.96}$	67–619
1999	U9901	1	10	8.2	$3.3 \times 10^3 r^{-1.2}$	91–380
	U9902	1	10	12	$2.1 \times 10^3 r^{-1.1}$	91–380
	S9915	1	9	11	$1.0 \times 10^3 r^{-0.98}$	15–621
	S9918	1	9	26*	$5.3 \times 10^3 r^{-1.2}$	15–621
	S9930	1	3	39	$4.0 \times 10^3 r^{-1.0}$	15–507
	S9932	1	4	19	$3.6 \times 10^3 r^{-1.1}$	15–507
	S9934	1	4	30	$3.0 \times 10^3 r^{-1.0}$	15–507
	S9935	1	3	21*	$2.1 \times 10^3 r^{-1.0}$	15–507

NR = not recorded. I_p = return-stroke peak current. *Peak current estimated from peak magnetic field recorded at 15 m from the channel using Ampere's law for magnetostatics.

The data on the leader electric field change as a function of distance for Florida are presented in Table 3.6. With a few exceptions, the 1997 to 1999 data indicate that the distance dependence of the leader electric field change is close to an inverse proportionality (r^{-1}), in contrast with the 1993 data (from both Florida, shown in Table 3.6, and Alabama, not shown) in which a somewhat weaker distance dependence was observed. The typically observed r^{-1} dependence is consistent with a uniform distribution of leader charge along the bottom kilometre or so of the channel. This observation simply indicates that for such a relatively short channel section a non-uniform charge density distribution will appear approximately uniform. Cooray and colleagues (2004) compared Crawford and colleagues' (2001) experimental results with theoretical predictions for a vertical conductor in an external electric field and found a fairly good agreement. A variation of ΔE_L with distance slower than r^{-1} dependence implies a decrease of leader charge density with decreasing height.

3.7 Studies of interaction of lightning with various objects and systems

In Sections 3.7.1 to 3.7.3 we consider the triggered-lightning testing of overhead power distribution lines, underground cables, and power transmission lines, respectively. Lightning interaction with lightning protective systems of a residential building and an airport runway lighting system is discussed in Sections 3.7.4 and 3.7.5, respectively. In Section 3.7.6, we briefly review the use of triggered lightning for testing components of power systems, different types of lightning rods, and other objects, and also for measuring step voltages and for making fulgurites.

3.7.1 Overhead power distribution lines

Most of the published studies concerned with the responses of power distribution lines to direct and nearby triggered-lightning strikes have been conducted in Japan and in Florida.

3.7.1.1 Nearby strikes

From 1977 to 1985, a test power distribution line at the Kahokugata site in Japan (see Table 3.1) was used for studying the induced effects of close triggered-lightning strikes to ground [7]. Both negative and positive polarity flashes were triggered. The wire simulating the phase conductor was 9 m above ground, and the minimum distance between the test line and the rocket launcher was 77 m. The peak value of induced voltage was found to be linearly related to the peak value of lightning current, with 25–30 kV corresponding to a 10-kA stroke. Installation of a grounded wire 1 m above the phase conductor resulted in a reduction of the induced voltage peak by ∼40 per cent. Horii and Nakano [22; figure 6.4.2] show a photograph of the test distribution line being struck directly during the induced-effect experiments. All triggered-lightning experiments in Japan were performed in winter.

In 1986, the University of Florida lightning research group studied the interaction of triggered lightning with an unenergized, three-phase 448-m overhead test line at the NASA Kennedy Space Center. Lightning was triggered 20 m from one end of the line, and acquired data included induced voltages on the top phase (10 m above ground) and fields at a distance of 500 m from the lightning channel [92]. Two types of induced-voltage waveforms were recorded: oscillatory and impulsive. The former exhibit peak values that range from tens of kilovolts to ~100 kV, while the latter show peak voltages nearly an order of magnitude larger. The oscillatory nature of the waveforms is due to multiple reflections at the ends of the line. Both types of voltage waveforms were observed to occur for different strokes within a single flash. The time domain technique of Agrawal and colleagues [93] as adopted by Master and Uman [94], Rubinstein and colleagues [95] and Georgiadis and colleagues [96] was used to model the observed voltages. Some success was achieved in the modelling of the oscillatory voltage waveforms, whereas all attempts to model the impulsive waveforms failed, probably because these measurements had been affected by a flashover in the measuring system. Rubinstein and colleagues [92] used only the return-stroke electric field as the source in their modelling, assuming that the contribution from the leader was negligible. In a later analysis of the same data, Rachidi and colleagues [97] found that the overall agreement between calculated and measured voltages of the oscillatory type was appreciably improved by taking into account the electric field of the dart leader.

From 1993 to 2004, studies of the interaction of triggered and natural lightning with power distribution systems were conducted at Camp Blanding, Florida. An overview of the Camp Blanding facility in 1997 is given in Figure 3.25.

During the 1993 experiment at Camp Blanding, the voltages induced on the overhead distribution line shown in Figure 3.25 were measured at poles 1, 9 and 15. The line had a length of ~730 m. The distance between the line and the triggered lightning strikes was 145 m. The line was terminated at both ends with a resistance of 500 Ω, and its neutral (the bottom conductor; see Figure 3.25) was grounded at poles 1, 9 and 15. The results of this experiment have been reported by Barker and colleagues [98] and are briefly reviewed next. Waveforms of the induced voltage and of the total lightning current were obtained for 63 return strokes in 30 triggered flashes. The typical induced voltage waveform at pole 9 and corresponding lightning return-stroke current waveform are shown in Figure 3.26. A strong correlation was observed between the peak values of the return-stroke current, ranging from 4 to 44 kA, and the voltage, ranging from 8 to 100 kV, induced at pole 9, with a correlation coefficient of 0.97 (see Figure 3.27). Voltages induced at the terminal poles were typically half the value of the voltage induced at pole 9.

In the period 1994 to 1997, the test distribution system at Camp Blanding shown in Figure 3.25 was subjected to both direct (see Section 3.7.1.2) and nearby triggered-lightning strikes. A large number of system configurations were tested, and several important results were obtained. It was observed, for example, that when lightning strikes earth at tens of metres from the system's grounds, an appreciable fraction of the total lightning current enters the system from earth [99–101]. The observed peak values of current entering the system from earth, in per cent of the total lightning current peak, were (for three different events) 10 per cent at 60 m (see Figure 3.28), 5 per cent at 40 m and 18 per cent at 19 m from the ground strike point. These observations have important implications for modelling of lightning-induced effects on power lines.

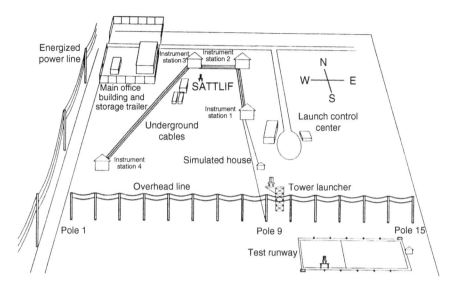

Figure 3.25 Overview of the International Center for Lightning Research and Testing (ICLRT) at Camp Blanding, Florida, 1997 (artwork by C.T. Mata)

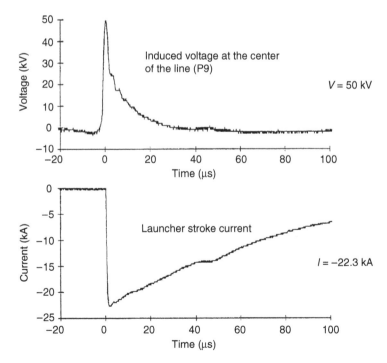

Figure 3.26 Typical induced voltage at pole 9 and corresponding lightning return-stroke current (93-05) reported by Barker and colleagues [98]

134   Lightning Protection

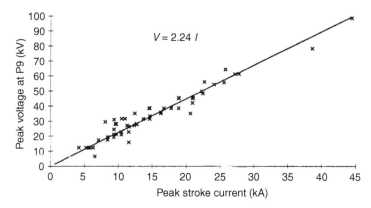

Figure 3.27   Peak induced voltage (8 to 100 kV) at pole 9 versus return-stroke peak current (4 to 44 kA), n = 63 (adapted from Reference 98)

In 2003, the vertical-configuration, three-phase plus neutral, power distribution line (see Figure 3.29) at Camp Blanding, Florida, was subjected to induced effects of triggered-lightning strikes to ground 7 and 15 m from the centre of the line and 11 m from one of its termination poles. The line was equipped with surge arresters (at 4 out of 15 poles), and its neutral was grounded at 6 poles (4 poles with arresters

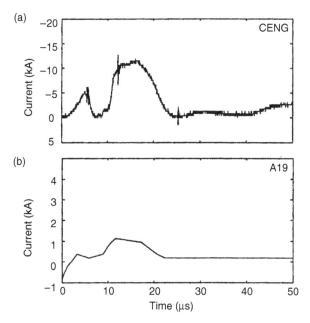

Figure 3.28   Current versus time waveforms for flash 9516, displayed on a 50 μs scale: (a) total lightning current, CENG; (b) ground-rod current, A19, measured 60 m from the lightning strike point (adapted from Reference 46)

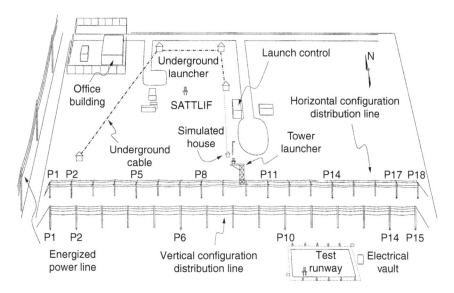

Figure 3.29   Overview of the ICLRT at Camp Blanding, Florida, 2000–2003

and 2 termination poles). At the termination poles, 500-Ω resistors were connected between each phase conductor and the neutral. Paolone and colleagues [102] compared measured lightning-induced currents along the line with those predicted by the LIOV-EMTP96 Code [103] and found a reasonably good agreement for most, although not all, current measurement locations.

3.7.1.2 Direct strikes

As noted above, various configurations of distribution system at Camp Blanding (see Figure 3.25) were tested in the period 1994 to 1997. In 1996, the responses of MOV arresters in the system, composed of an overhead line, underground cable, and padmount transformer with a resistive load, were measured during very close, direct lightning strikes to the overhead line. Arresters were installed on the overhead line at two locations 50 m apart (on either side of the strike point) and at the primary of the padmount transformer, which was connected to the line via the underground cable. Simultaneously recorded arrester discharge current and voltage waveforms were obtained. Additionally, the energy absorbed by an arrester on the line as a function of time for the first 4 ms for one lightning event was estimated. The total energy absorbed by the arrester was 25 kJ (∼60 per cent of its maximum energy capability). The energy absorbed during the initial 200 μs was ∼8 kJ.

Mata and colleagues [104] used EMTP to model a direct lightning strike to the overhead power line shown in Figure 3.25. Overall, measured voltages and currents have been fairly well reproduced by EMTP simulations.

More details on findings from the 1994 to 1997 experiments at Camp Blanding are found in References 14, 99 and 104–106.

Presented below are results of triggered-lightning experiments conducted in 2000, 2001 and 2002 at the ICLRT at Camp Blanding, Florida, to study the responses of four-conductor (three-phase plus neutral) overhead distribution lines to direct lightning strikes (see Figure 3.29). Presented first are direct-strike results for the line with horizontally configured phase conductors obtained in 2000 and then for the line with vertically configured phase conductors obtained in 2001 and 2002.

Horizontal configuration distribution line
The horizontal configuration, 856-m line was subjected to eight lightning flashes containing return strokes between 11 July and 6 August 2000 [107]. The line was additionally subjected to two flashes without return strokes that are not considered here. The lightning current was injected into the phase C conductor in the middle of the line. Six of the eight flashes with return strokes produced damage to the phase C arrester at pole 8. Of the two that did not, one had a triggering wire over the line and the other produced a flashover at the current injection point. The eight triggered flashes contained 34 recorded return strokes. These return strokes were characterized by submicrosecond current rise times and by peak currents having geometric and arithmetic means between 15 and 20 kA with a maximum peak current of 57 kA. Each triggered flash also contained an ICC of the order of hundreds of amperes, which flowed for a time of the order of hundreds of milliseconds, and some flashes contained a similar continuing current after subsequent strokes. The placement of conductors and arresters on the test distribution line is illustrated in Figure 3.30. A total of six three-phase sets of arresters were installed on the line, at poles 2, 5, 8, 11, 14 and 17, the arresters being connected between the phase conductors and the neutral conductor. The neutral of the line was grounded at these poles and at the two line-terminating poles, 1 and 18. The 856-m three-phase line was terminated at each end in an impedance of $\sim 500\ \Omega$. The distance between poles of the line varied from 47 to 73 m.

The grounding of the neutral at each arrester-equipped pole and at each of the two terminating poles was accomplished by means of 24 m vertically driven ground rods. The low-frequency, low-current grounding resistance of each pole ground was measured on several occasions using the fall-of-potential method. The measured grounding resistances in September 2000 were 41, 47, 28, 52, 55, 46, 37 and 22 Ω for the ground rods at poles 1, 2, 5, 8, 11, 14, 17 and 18, respectively. Two different brands of 18-kV MOV arresters were used in the experiment: arresters installed at poles 2, 5, 14, 17 were from manufacturer A and those installed at poles 8 and 11 were from manufacturer B. Polymer insulators were used at the terminating poles and ceramic insulators on all other poles, all 35-kV rated.

Arrester currents, line currents, and neutral currents were measured with current transformers (CTs), and currents through the terminating resistors at pole 1 and at each pole ground location with 1-mΩ current viewing resistors (shunts). The current signals were recorded on Lecroy digitizing oscilloscopes at a sampling rate of 20 MHz. The total triggered-lightning current was measured at the rocket launching unit with a 1-mΩ shunt and recorded with a Lecroy digitizing oscilloscope having a sampling rate of 25 MHz.

The focus of the study was on the paths of return stroke current and charge transfer from the current injection point on one phase, C, between poles 9 and 10, to the eight grounds. This current division was examined in detail only for the case, flash 0036, in

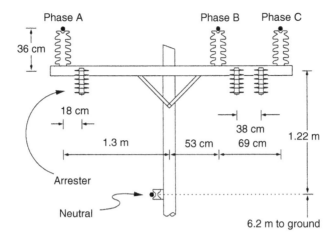

Figure 3.30 Placement of conductors and arresters on the test distribution line (adapted from Reference 107)

which arrester failure did not occur or had not yet occurred in the flash, except for Figure 3.34 where all strokes recorded in 2000 without severe saturation were included. In flash 0036, an ICC and five return strokes were injected into phase C between poles 9 and 10 prior to the arrester failure at pole 8. The arrester on pole 8 failed following the fifth stroke, perhaps from the accumulation of energy from the ICC and the five strokes or from those currents and any following unrecorded continuing current and additional strokes. As an example, Figure 3.31 depicts the division of the incident current for the first stroke of flash 0036. This stroke had a peak current of ∼26 kA. Note that the arrester current at pole 8 was lost due to instrumentation (fibre-optic link) malfunction, but it was likely similar to the arrester current at pole 11, given the symmetry of the other currents on the line. Also, the current through the terminating resistor at pole 18 was not measured.

Figure 3.32 shows the arrester and terminating-resistor peak currents recorded for all five strokes of flash 0036, while Figure 3.33 gives the peak currents entering all eight pole grounds for the five return strokes. It is evident from Figures 3.31 to 3.33 that the bulk of the peak current injected into phase C passed through the arrester at pole 11, and by inference at pole 8, and also went to ground mostly at poles 8 and 11.

Figure 3.34 shows the measured distribution of peak current to ground for all strokes triggered to the horizontal configuration line in 2000. In many of these events there were line flashovers. It is evident that all strokes show a behaviour similar to that in the example above for flash 0036.

Figure 3.31 shows current waveforms only to 100 μs, although the total duration of current records is 10 ms. Figure 3.35 shows percentages of charge transfer through arresters and terminating resistor at pole 1, and Figure 3.36 percentages of charge transfer through ground rods, at 100 μs, 500 μs and 1 ms.

It is clear from Figure 3.31, an observation also illustrated in Figure 3.36, that after 25 μs or so the current from the neutral to ground no longer flows primarily through

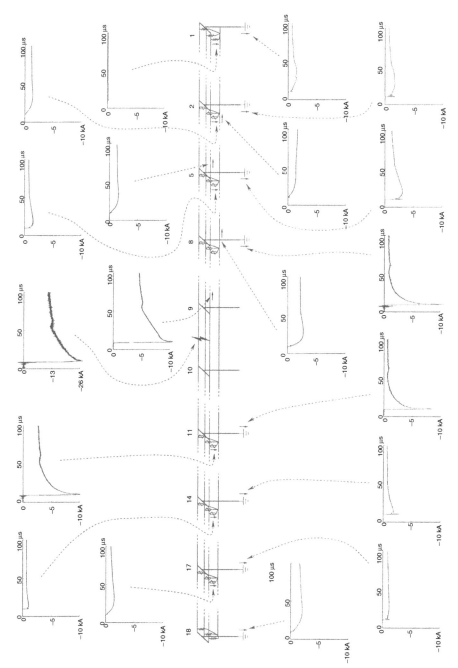

Figure 3.31 Current distribution for flash 0036, stroke 1 (adapted from Reference 107)

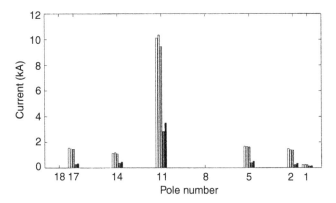

Figure 3.32   *Measured peak currents through arresters and terminating resistor at pole 1 for strokes 1 through 5 (in ascending order from left to right) of flash 0036. Arrester currents at pole 8 were lost due to instrumentation malfunction. Currents through the terminating resistor at pole 18 were not measured (adapted from Reference 107).*

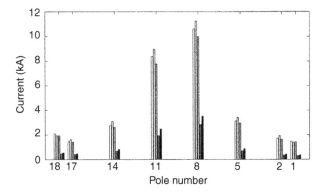

Figure 3.33   *Measured peak currents to ground for strokes 1 through 5 (in ascending order from left to right) of flash 0036 (adapted from Reference 107)*

the grounds closest to the strike point but is more uniformly distributed among the eight grounds. In fact, the currents after 25 μs are distributed roughly inversely to the measured low-frequency, low-current grounding resistance. Figure 3.36 shows that the percentage of charge transferred to a given ground rod in the first 100 μs is not much different from that transferred in the first millisecond.

As seen in Figure 3.31, there are considerable differences among the waveshapes of currents measured in different parts of the test system. As a result, the division of peak current to ground (Figure 3.33) is very different from the division of associated charge transfer (Figure 3.36). It appears that the higher-frequency current components that are

140   Lightning Protection

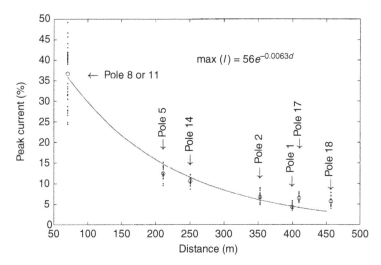

Figure 3.34   *Measured peak current to ground in per cent of the total lightning peak current as a function of distance from the strike point. The dots represent measured peak current to ground for all strokes triggered in 2000 with no severe saturation, circles indicate mean values, and the solid line is the exponential function that fits the mean values (adapted from Reference 107).*

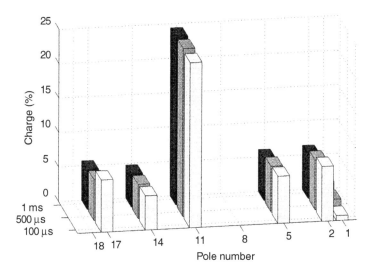

Figure 3.35   *Percentage of total charge transferred through phase C arresters at different poles and terminating resistor at pole 1, calculated at three different instants of time (100 µs, 500 µs and 1 ms from the beginning of the return stroke) for stroke 1 of flash 0036. No measurements are available at pole 8 and pole 18 (adapted from Reference 107).*

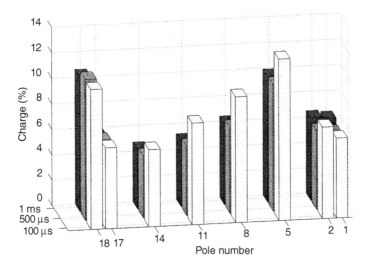

Figure 3.36 *Percentage of total charge transferred to ground at different poles, calculated at three different instants of time (100 μs, 500 μs and 1 ms from the beginning of the return stroke) for stroke 1 of flash 0036 (adapted from Reference 107)*

associated with the initial current peak tend to flow from the struck phase to ground through the arresters and ground rods at the two poles closest to the current injection point (see also Figure 3.34). The low-frequency, low-current grounding resistances of the ground rods apparently have little or no effect on determining the paths for these current components. The lower-frequency current components that are associated with the tail of current waveforms are distributed more evenly among the multiple ground rods of the test system and appear to be significantly influenced by the low-frequency, low-current grounding resistances of the ground rods. In fact, the distribution of charge transfer in Figure 3.36 is very similar to the distribution of the inverse of the low-frequency, low-current grounding resistances of the ground rods, with poles 5 and 18 having the largest charge transfer and the lowest grounding resistances. Because the current waveshapes may differ considerably throughout the system, charge transfer is apparently a better quantity than the peak current for studying the division of lightning current among the various paths in the system.

In summary, for the 856-m, horizontally configured four-conductor, unenergized test power distribution line equipped with six sets of MOV arresters and each phase terminated at each end in a 500-Ω resistor, the following statements can be made.

- There are considerable differences among the waveshapes of currents flowing from the struck phase to neutral and from the neutral to ground at different poles of the line.
- The higher-frequency current components that are associated with the initial current peak tend to flow from the struck phase to neutral and then to ground at the two poles adjacent to the lightning current injection point.

142 *Lightning Protection*

- The division of lightning charge among the multiple paths between the struck phase and neutral is different from the division of charge among the multiple ground rods. The charge transfer from the struck phase to neutral tends to occur at the two poles adjacent to the lightning current injection point, while the charge transfer from the neutral to ground is apparently determined by the low-frequency low-current grounding resistances of the ground rods.

Vertical configuration distribution line

The vertical configuration, 812-m line was subjected to four lightning flashes containing return strokes (also to four flashes without return strokes) between 26 July and 5 September 2001 and to ten flashes with return strokes between 27 June and 13 September 2002 [108,109]. In 2001, return-stroke peak currents ranged from 6 to 28 kA and in 2002 from 6 to 34 kA. Arresters were installed at poles 2, 6, 10 and 14. Lightning current was injected into the top conductor near the centre of the line.

In 2001, for one of the flashes having return strokes, an arrester failed early in the flash, probably during the IS. The three other flashes with return strokes were triggered with failed arresters already on the line, those failures being caused by previous flashes without return strokes and by the one flash that likely caused an arrester failure during its IS. Two flashes without return strokes did not damage arresters. One flash with return strokes was triggered when the line contained two damaged arresters, resulting in the failure of a third arrester. Note that the charge transfer associated with the IS current is of the order of tens of coulombs, more than an order of magnitude larger than the charge transfer associated with triggered-lightning return strokes.

In 2002, in order to reduce arrester damage during the IS of rocket-triggered lightning, a different configuration of the tower launching system was used. This new configuration allowed the diversion of most of the IS current to ground at the tower base. Additionally, two arresters were installed in parallel on the struck (top) phase conductor. In 2002, arresters failed on three storm days out of a total of five (60 per cent), compared with two out of three storm days (67 per cent) in 2001. Flashovers on the line were very frequent during the direct strike tests. Significant currents were detected in phase B, which was not directly struck by lightning, with the waveshape of phase B currents being similar to that of the corresponding current in phase A that was directly struck.

Overall, the results presented in this section suggest that many direct lightning strikes to power distribution lines are capable of damaging MOV arresters, unless alternative current paths (flashovers, transformers, underground cable connections, and so on) are available to allow the lightning current to bypass the arrester.

In 2003, the vertical configuration line was equipped with a pole-mounted transformer. With the transformer on the line, the bulk of the return-stroke current injected into the line after ~ 1 ms flowed from the struck phase to the neutral through the transformer primary protected by an MOV arrester. Very little lightning current was passing through the transformer primary during the first few hundred microseconds.

3.7.2 Underground cables

In 1993, an experiment was conducted at Camp Blanding to study the effects of lightning on underground power distribution systems. All three cables shown in Figure 3.25 were used in this experiment. The cables were 15-kV coaxial cables with polyethylene insulation between the centre conductor and the outer concentric shield (neutral). One of the cables (Cable A) had an insulating jacket and was placed in a PVC conduit, another one (Cable B) had an insulating jacket and was directly buried, and the third one (Cable C) had no jacket and was directly buried. The three cables were buried 5 m apart at a depth of 1 m. Thirty lightning flashes were triggered, and lightning current was injected into the ground directly above the cables, with the current injection point being approximately equidistant from instrument stations 1 and 2 (see Figure 3.25) but at different positions with respect to the cables. The cables were unenergized. Transformers at instrument stations 1, 2, 3 and 4 were connected to Cable A. More details on this test system configuration are found in Reference 105.

Barker and Short [110–112] reported the following results from the underground power cables experiment. After lightning attachment to ground, a substantial fraction of the lightning current flowed into the neutral conductor of the cable, with ∼15 to 25 per cent of the total lightning current (measured at the rocket launcher) being detected 70 m in either direction from the strike point at instrument stations 1 and 2. The largest voltage measured between the centre conductor and the concentric neutral of the cable was 17 kV, which is below the cable's basic insulation level (BIL) rating. Voltages measured at the transformer secondary were up to 4 kV. These could pose a threat to residential appliances. The underground power cables were excavated by the University of Florida research team in 1994. Lightning damage to these three cables is illustrated in Figure 3.21.

Paolone and colleagues [113] measured, at Camp Blanding, Florida, currents induced by triggered (and natural) lightning events at the end of a buried coaxial cable, both in the cable shield and in the inner conductor. The horizontal magnetic field above the ground surface was also measured. The obtained experimental data have been used to test the theoretical models and the developed time- and frequency-domain computer codes. In general, a reasonably good agreement has been found between numerical simulations and experimentally recorded waveforms.

3.7.3 Power transmission lines

Extensive studies of the interaction of triggered lightning with an unenergized power transmission line, the Okushishiku test line, were performed in Japan. The line was designed to operate at a voltage of 275 kV and had six conductors and one ground wire suspended on seven steel 60-m towers. The total length of the test line was 2 km. All experiments were conducted in winter, primarily using the altitude triggering technique.

The distribution of triggered lightning current injected into the tower top among the four tower legs and the overhead ground wire was studied. The currents through the four legs were not equal, presumably because of the differences among the grounding impedances of the individual legs. It was observed that the higher-frequency components of current tended to flow to ground through the struck tower, and the lower-frequency components appeared to travel to other towers along the ground wire. Currents in the phase conductors and voltages between each phase conductor and the tower were also measured.

3.7.4 Residential buildings

In 1997, the grounding system of a test house (labelled 'simulated house' in Figure 3.25) at Camp Blanding was subjected to triggered-lightning discharges for three different configurations, with the house's electrical circuit being connected to the secondary of a transformer in IS1, about 50 m distant. The primary of the transformer was connected to the underground cable which was open-circuited at IS4. The cable's neutral was grounded at IS1 and IS4. The test system was unenergized. The division of lightning current injected into the grounding system of the test house among the various paths in the overall test system was analysed. The waveshapes of currents in the ground rods of the test house differed markedly from the current waveshapes in other parts of the overall system. The ground rods at the test house appeared to filter out the higher-frequency components of the lightning current, allowing the lower-frequency components to enter the house's electrical circuit. In other words, the ground rods exhibited a capacitive rather than the often expected and usually modelled resistive behaviour. This effect was observed for d.c. resistances of the ground rods (in typical Florida sandy soil) ranging from more than a thousand ohms to some tens of ohms. The peak value of the current entering the test house's electrical circuit was found to be over 80 per cent of the injected lightning current peak, in contrast with the 25 or 50 per cent assumed in two IEC-suggested scenarios, illustrated in Figure 3.37. Similarly, the percentages of current flowing (i) to the transformer secondary neutral and (ii) through the SPDs were observed to be approximately a factor of two to four greater than those expected in the IEC hypothetical scenario shown in Figure 3.37a. Selected current waveforms for one of the configurations tested are presented in Figure 3.38. Because the current waveshapes may differ considerably throughout the system, charge transfer is apparently a better quantity than the peak current for studying the division of lightning current among the various paths in the system.

3.7.5 Airport runway lighting system

In 1997 to 1998, the University of Florida conducted a major experiment to study the interaction of lightning with an airport lighting system, shown in Figure 3.39. The experiment was conducted at Camp Blanding, Florida (see Figure 3.25). The test airport lighting system was subjected to a total of 16 lightning strikes, 12 of which contained one or more return strokes. The total number of return strokes was 47 (24 in 1997 and 23 in 1998). Lightning current injection points were (i) the pavement, (ii) one of the stake-mounted lights, (iii) the counterpoise, and (iv) the ground directly above the counterpoise or between the counterpoise and the edge of the pavement.

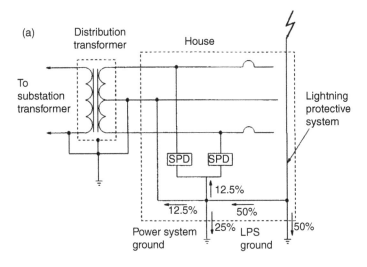

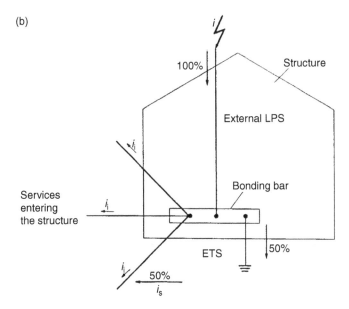

Figure 3.37 (a) Currents in different parts of the electrical circuit of a house when it is struck by lightning, in per cent of the injected lightning current, as hypothesized by the International Electrotechnical Commission (J.L. Koepfinger, personal communication, 1998). SPD, surge protective device; LPS, lightning protective system. (b) Division of lightning current between the structure's earth termination (grounding) system, ETS, and services entering the structure, as assumed by IEC 61 312-1 (1995). LPS, lightning protective system.

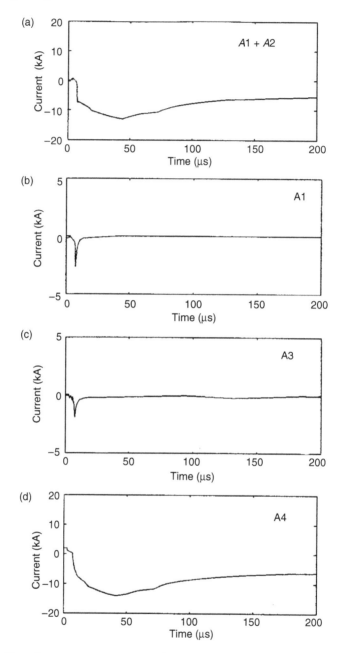

Figure 3.38 Current versus time waveforms for Flash 9706, displayed on a 200 μs scale: (a) injected lightning current; (b) current to ground at node A (lightning protective system ground, 1 550 Ω); (c) current to ground at node B (power system ground, 590 Ω); (d) current entering the test house's electrical circuit (adapted from Reference 114)

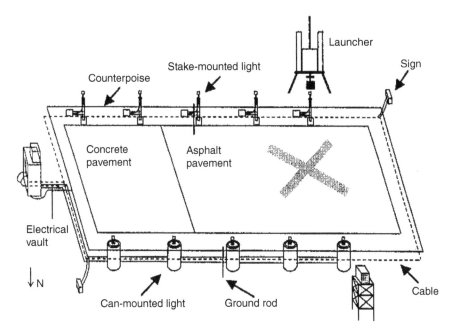

Figure 3.39 *Schematic representation of the test runway and its lighting system. The horizontal dimensions of the lighting system are ~ 106 m $\times$ 31 m. The cable is buried at a depth of 0.4 m with the counterpoise placed in the same trench 0.1 m or so above the cable. The counterpoise was connected to the light stakes and cans (adapted from Reference 47).*

The system was energized using a generator and a current regulator for some of the tests and unenergized for others. The total lightning current and the currents and voltages at various points on the lighting system were measured.

The results of these experiments are presented by Bejleri and colleagues [47]. They include the first measurements of the responses of an underground bare conductor (counterpoise) to direct lightning strikes. These measurements can serve as ground truth for the testing of the validity of various counterpoise models. Overall results of the experiments can be summarized as follows.

Current decay along the counterpoise
When lightning struck a stake-mounted light or directly struck the counterpoise, 10 to 30 per cent of the total lightning current was dissipated locally, within 3 m of the strike point (from measurements made at a distance of 3 m on either side of the strike point), while 70 to 90 per cent was carried by the counterpoise further away from the strike point. Measurements of the counterpoise current at four different locations (two on each side of the strike point) made it possible to estimate that $\sim$63 per cent of the current detected 3 m from the strike point was dissipated in the ground after propagating along 50 m of the counterpoise, and $\sim$73 per cent of the current detected 3 m on the other side from the strike point was dissipated in the ground after propagating

along 67 m of the counterpoise. The average per cent current decay rate is ~1 per cent per metre, independent of the peak current at the origin (peak current measured 3 m from the strike point). The current waveshape changes as the current wave propagates along the counterpoise; however, the rise time remains more or less the same, a plateau or a broad maximum, not seen in the total lightning current waveform, is observed at distances of 50 and 67 m. The plateau duration is approximately between 10 and 50 μs. In some cases, when the lightning current is smaller than 10 kA, current waveforms do not exhibit the plateau.

Currents in vertical ground rods

During experiments with configurations 1 and 2 (a total of four configurations were tested) the entry point of current in the counterpoise was ~12 m from the north ground rod (see Figure 3.39). In this case, the current through the ground rod was as high as 1 to 2 kA, accounting for 10 to 15 per cent of the total lightning current.

During experiments with configurations 3 and 4 the entry point of current in the counterpoise was ~36 m from the south ground rod. Shown in Figure 3.40 are the waveforms of the injected lightning current and the current through the ground rod for the first stroke of triggered lightning flash U9841. In this particular case the

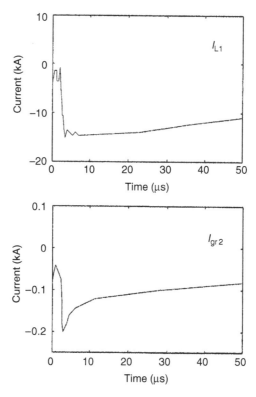

Figure 3.40 Lightning channel current, I_{L1}, and ground rod current, I_{gr2}, for flash U9841, first return stroke (adapted from Reference 47)

current through the ground rod accounted for ~1.3 per cent of the lightning channel current (peak values). For all the lightning strikes 36 m from the south ground rod, the maximum value of current leaving the system through this ground rod was ~300 A, which was less than 5 per cent of the total lightning current. The ground rod current waveform had approximately the same rise time as the lightning current but shorter duration. This suggests that the ground rod is a better path than the counterpoise for the higher-frequency current components.

Cable currents
From the data recorded, it appears that the current flowing in the counterpoise induced current in the cable. The largest currents in the cable were observed near the current injection point. No evidence of direct lightning current injection into the cable or flashover to the cable from the counterpoise was found, but they definitely cannot be ruled out. Voltage pulses between the cable and the counterpoise had magnitudes of some tens of kilovolts (likely underestimated due to insufficiently short sampling interval of 50 ns) and very short durations, ranging from a few hundred nanoseconds to a few microseconds.

Lightning damage to the system
Several elements of the test airport runway lighting system sustained damage caused by one or more lightning strikes. The damage included (i) failure of one of the electronic boards of the current regulator (CCR), (ii) minor damage to the light fixture and to the glass cover of the light bulb of the stake-mounted light under the launcher, (iii) multiple burn marks on the surface of the secondary cable of the current transformers (at the strike point and at a distance of 36 m from it), (iv) pinholes on the secondary cable of the current transformer, and (v) melting of the counterpoise conductor at the point where the lightning attached to the system.

3.7.6 Miscellaneous experiments

Besides the tests described above, triggered-lightning experiments have been performed in order to study the interaction of lightning with a number of miscellaneous objects and systems and for a variety of other reasons. Some of those studies are briefly reviewed below. Triggered lightning has been used to test power transformers [7], lightning arresters [22,115,116], overhead ground wires [22], lightning rods including so-called early-streamer-emission rods [117] and high-resistance (tens to hundreds of kΩ), current-limiting rods [118], explosive materials [6] and explosives storage facilities [119]. Various aspects of lightning safety have been studied using a mannequin with a hairpin on the top of its head and a metal-roof car with a live rabbit inside [7,11]. The car was confirmed to be a lightning-safe enclosure. Step voltages have been measured within a few tens of metres of the triggered-lightning strike point [7,82,120]. Voltages have been measured across a single overhead power line tower and between the tower footing and remote ground (over a distance of 60 m), along with the lightning current injected into the tower [121]. Additionally, triggered lightning has been used to make fulgurites [79,80, 122,123]. Photographs of fulgurites are found in Figures 3.21 and 3.22. Oxide reduction during triggered-lightning fulgurite formation has been examined by Jones and colleagues [81].

3.8 Concluding remarks

The rocket-and-wire technique has been routinely used since the 1970s to artificially initiate (trigger) lightning from natural thunderclouds for purposes of research and testing. Leader/return stroke sequences in triggered lightning are similar in most (if not all) respects to subsequent leader/return-stroke sequences in natural downward lightning and to all such sequences in object-initiated lightning. The initial processes in triggered lightning are similar to those in object-initiated (upward) lightning and are distinctly different from the first leader/return-stroke sequence in natural downward lightning. The results of triggered-lightning experiments have provided considerable insight into natural lightning processes that would not have been possible from studies of natural lightning due to its random occurrence in space and time. Among such findings are the observation of an UCL in a dart leader/return-stroke sequence, identification of the M-component mode of charge transfer to ground, the observation of a lack of dependence of return-stroke current peak on grounding conditions, discovery of X-rays produced by lightning dart leaders, new insights into the mechanism of cutoff and reestablishment of current in the lightning channel, direct measurements of NO_x production by an isolated lightning channel section and the characterization of the electromagnetic environment within tens to hundreds of metres of the lightning channel. Triggered-lightning experiments have contributed significantly to testing the validity of various lightning models and to providing ground-truth data for the US National Lightning Detection Network (NLDN). Triggered lightning has proved to be a very useful tool to study the interaction of lightning with various objects and systems, particularly in view of the fact that simulation of the lightning channel in a high-voltage laboratory does not allow the reproduction of many lightning features important for lightning protection and it does not allow the testing of large distributed systems such as overhead power lines.

References

1. Newman M.M. 'Use of triggered lightning to study the discharge channel'. *Problems of Atmospheric and Space Electricity*. New York: Elsevier; 1965, pp. 482–90.
2. Brook M., Armstrong G., Winder R.P.H., Vonnegut B. and Moore C.B. 'Artificial initiation of lightning discharges'. *J. Geophys. Res.* 1961;**66**:3967–69.
3. Newman M.M., Stahmann J.R., Robb J.D., Lewis E.A., Martin S.G. and Zinn S.V. 'Triggered lightning strokes at very close range'. *J. Geophys. Res.* 1967;**72**:4761–64.
4. Newman M.M. and Robb J.D. 'Protection of aircraft', in R.H. Golde (ed.). *Lightning, vol. 2: Lightning Protection*. London: Academic Press; 1977, pp. 659–96.

5. Fieux R., Gary C. and Hubert P. 'Artificially triggered lightning above land'. *Nature* 1975;**257**:212–14.
6. Fieux R.P., Gary C.H., Hutzler B.P., Eybert-Berard A.R., Hubert P.L., Meesters A.C. *et al.* 'Research on artificially triggered lightning in France'. *IEEE Trans. Power Apparat. Syst.* 1978;**PAS-97**:725–33.
7. Horii K. 'Experiment of artificial lightning triggered with rocket'. *Memoirs of the Faculty of Engineering, Nagoya Univ. Japan* 1982;**34**:77–112.
8. Kito Y., Horii K., Higasbiyama Y. and Nakamura K. 'Optical aspects of winter lightning discharges triggered by the rocket-wire technique in Hokuriku district of Japan'. *J. Geophys. Res.* 1985;**90**:6147–57.
9. Nakamura K., Horii K., Kito Y., Wada A., Ikeda G., Sumi S. *et al.* 'Artificially triggered lightning experiments to an EHV transmission line'. *IEEE Trans. Power Deliv.* 1991;**6**:1311–18.
10. Nakamura K., Horii K., Nakano M. and Sumi S. 'Experiments on rocket triggered lightning'. *Res. Lett. Atmos. Electr.* 1992;**12**:29–35.
11. Horii K., Nakamura K. and Sumi S.I. 'Review of the experiment of triggered lightning by rocket in Japan'. *Proceedings of the 28th International Conference on Lightning Protection*, Kanazawa, Japan, 2006.
12. Hubert P., Laroche P., Eybert-Berard A. and Barret L. 'Triggered lightning in New Mexico'. *J. Geophys. Res.* 1984;**89**:2511–21.
13. Willett J.C. 'Rocket-triggered-lightning experiments in Florida'. *Res. Lett. Atmos. Electr.* 1992;**12**:37–45.
14. Uman M.A., Rakov V.A., Rambo K.J., Vaught T.W., Fernandez M.I., Cordier D.J. *et al.* 'Triggered-lightning experiments at Camp Blanding, Florida (1993–1995)'. *Trans. IEE Japan* 1997;**117-B**:446–52.
15. Rakov V.A., Uman M.A., Rambo K.J., Fernandez M.I., Fisher R.J., Schnetzer G.H. *et al.* 'New insights into lightning processes gained from triggered-lightning experiments in Florida and Alabama'. *J. Geophys. Res.* 1998;**103**: 14117–30.
16. Rakov V.A., Uman M.A., Wang D., Rambo K.J., Crawford D.E. and Schnetzer G.H. 'Lightning properties from triggered-lightning experiments at Camp Blanding, Florida (1997–1999)'. *Proceedings of the 25th International Conference on Lightning Protection*; Rhodes, Greece, 2000, pp. 54–59.
17. Rakov V.A., Uman M.A. and Rambo K.J. 'A review of ten years of triggered-lightning experiments at Camp Blanding, Florida'. *Atmos. Res.* 2005;**76**(1–4):504–18.
18. Liu X.-S., Wang C., Zhang Y., Xiao Q., Wang D., Zhou Z. and Guo C. 'Experiment of artificially triggering lightning in China'. *J. Geophys. Res.* 1994;**99**:10727–31.
19. Liu X. and Zhang Y. 'Review of artificially lightning study in China'. *Trans. IEE Japan* 1998;**118**-B(2):170–75.
20. Pinto O. Jr., Pinto I.R.C.A., Saba M.M.F., Solorzano N.N. and Guedes D. 'Return stroke peak current observations of negative natural and triggered lightning in Brazil'. *Atmos. Res.* 2005;**76**:493–502.

21. Uman M.A. *The Lightning Discharge*. San Diego, California: Academic Press; 1987, p. 377.
22. Horii K. and Nakano M. 'Artificially triggered lightning' in H. Volland (ed.). *Handbook of Atmospheric Electrodynamics*. Boca Raton, Florida: CRC Press, 1995, vol. 1, pp. 151–66.
23. Rakov V.A. 'Lightning discharges triggered using rocket-and-wire techniques'. *Recent Res. Devel. Geophys.* 1999;**2**:141–71.
24. Rakov V.A. and Uman M.A. *Lightning: Physics and Effects*. Cambridge Univ. Press; 2003, 687 p., ISBN 0521583276.
25. Wang D., Rakov V.A., Uman M.A., Takagi N., Watanabe T., Crawford D.E. *et al.* 'Attachment process in rocket-triggered lightning strokes'. *J. Geophys. Res.* 1999;**104**:2141–50.
26. Jafferis W. 'Rocket triggered lightning – Kennedy Space Center and beyond'. *Proceedings of the 1995 International Conference on Lightning and Static Electricity*; Williamsburg, Virginia, 1995, pp. 57/1–57/20.
27. Willett J.C., Davis D.A. and Laroche P. 'An experimental study of positive leaders initiating rocket-triggered lightning'. *Atmos. Res.* 1999;**51**: 189–219.
28. Laroche P., Idone V., Eybert-Berard A. and Barret L. 'Observations of bi-directional leader development in a triggered lightning flash'. *Proceedings of the 1991 International Conference on Lightning and Static Electricity*; Cocoa Beach, Florida, 1991, pp. 57/1–57/10.
29. Lalande P., Bondiou-Clergerie A., Laroche P., Eybert-Berard A., Berlandis J.P., Bador B. *et al.* 'Connection to ground of an artificially triggered negative downward stepped leader'. *Proceedings of the 10th International Conference on Atmospheric Electricity 1996*; Osaka, Japan, 1996, pp. 668–71.
30. Lalande P., Bondiou-Clergerie A., Laroche P., Eybert-Berard A., Berlandis J.-P., Bador B. *et al.* 'Leader properties determined with triggered lightning techniques'. *J. Geophys. Res.* 1998;**103**:14109–15.
31. Uman M.A., Rakov V.A., Rambo K.J., Vaught T.W., Fernandez M.I., Bach J.A. *et al.* '1995 triggered lightning experiment in Florida'. *Proceedings of the 10th International Conference on Atmospheric Electricity*; Osaka, Japan, 1996, pp. 644–47.
32. Rakov V.A., Uman M.A., Fernandez M.I., Thottappillil R., Eybert-Berard A., Berlandis J.P. *et al.* 'Observed electromagnetic environment close to the lightning channel'. *Proceedings of the 23rd International Conference on Lightning Protection*; Florence, Italy, 1996, pp. 30–35.
33. Wang C., Yan M., Liu X., Zhang Y., Dong W. and Zhang C. 'Bidirectional propagation of lightning leader'. *Chinese Sci. Bull.* 1999;**44**(2):163–66.
34. Chen M., Watanabe T., Takagi N., Du Y., Wang D. and Liu X. 'Simultaneous observations of optical and electrical signals in altitude-triggered negative lightning flashes'. *J. Geophys. Res.* 2003;**108**(D8):4240, doi: 10.1029/2002JD002676.
35. Saba M.M.F., Pinto O. Jr., Solorzano N.N. and Eybert-Berard A. 'Lightning current observation of an altitude-triggered flash'. *Atmos. Res.* 2005;**76**:402–11.

36. Rakov V.A., Crawford D.E., Rambo K.J., Schnetzer G.H., Uman M.A. and Thottappillil R. 'M-component mode of charge transfer to ground in lightning discharges'. *J. Geophys. Res.* 2001;**106**:22817–31.
37. Uman M.A., Rakov V.A., Schnetzer G.H., Rambo K.J., Crawford D.E. and Fisher R.J. 'Time derivative of the electric field 10, 14 and 30 m from triggered lightning strokes'. *J. Geophys. Res.* 2000;**105**:15577–95.
38. Uman M.A., Schoene J., Rakov V.A., Rambo K.J. and Schnetzer G.H. 'Correlated time derivatives of current, electric field intensity and magnetic flux density for triggered lightning at 15 m'. *J. Geophys. Res.* 2002;**107**(D13), doi: 10.1029/2000JD000249, 11 p.
39. Crawford D.E., Rakov V.A., Uman M.A., Schnetzer G.H., Rambo K.J. and Stapleton M.V. et al. 'The close lightning electromagnetic environment: Dart-leader electric field change versus distance'. *J. Geophys. Res.* 2001;**106**: 14909–17.
40. Schoene J., Uman M.A., Rakov V.A., Kodali V., Rambo K.J. and Schnetzer G.H. 'Statistical characteristics of the electric and magnetic fields and their time derivatives 15 m and 30 m from triggered lightning'. *J. Geophys. Res.* 2003;**108**(D6):4192, doi: 10.1029/2002JD002698.
41. Wang D., Takagi N., Watanabe T., Rakov V.A. and Uman M.A. 'Observed leader and return-stroke propagation characteristics in the bottom 400 m of the rocket triggered lightning channel'. *J. Geophys. Res.* 1999;**104**:14369–76.
42. Olsen R.C., Jordan D.M., Rakov V.A., Uman M.A. and Grimes N. 'Observed one-dimensional return stroke propagation speeds in the bottom 170 m of a rocket-triggered lightning channel'. *Geophys. Res. Lett.* 2004;**31**:L16107, doi: 10.1029/2004GL020187.
43. Rakov V.A., Thottappillil R., Uman M.A. and Barker P.P. 'Mechanism of the lightning M component'. *J. Geophys. Res.* 1995;**100**:25701–10.
44. Miki M., Rakov V.A., Rambo K.J., Schnetzer G.H. and Uman M.A. 'Electric fields near triggered lightning channels measured with Pockels sensors'. *J. Geophys. Res.* 2002;**107**(D16):4277, doi: 10.1029/2001JD001087.
45. Rakov V.A. 'Review of non-conventional approaches to triggering lightning discharges'. *Recent Res. Devel. Geophys.* 2002;**4**:1–8.
46. Rakov V.A., Mata C.T., Uman M.A., Rambo K.J. and Mata A.G. 'Review of triggered-lightning experiments at the ICLRT at Camp Blanding, Florida'. *Proceedings of the 5th IEEE Power Tech Conference*; Bologna, Italy, 2003a, Paper 381, 8 p.
47. Bejleri M., Rakov V.A., Uman M.A., Rambo K.J., Mata C.T. and Fernandez M.I. 'Triggered lightning testing of an airport runway lighting system'. *IEEE Trans. on Electromagnetic Compatibility* 2004;**46**:96–101.
48. Dwyer J.R., Al-Dayeh M., Rassoul H.K., Uman M.A., Rakov V.A., Jerauld J. et al. 'Observations of energetic radiation from triggered lightning'. *2002 Fall AGU Meeting*; San Francisco, California, 2002.
49. Dwyer J.R., Uman M.A., Rassoul H.K., Al-Dayeh M., Caraway L., Jerauld J. et al. 'Energetic radiation produced by rocket-triggered lightning'. *Science* 2003;**299**:694–97.

50. Al-Dayeh M., Dwyer J.R., Rassoul H.K., Uman M.A., Rakov V.A., Jerauld J. et al. 'A new instrument for measuring energetic radiation from triggered lightning'. *2002 Fall AGU Meeting*; San Francisco, California.
51. Dwyer J.R., Rassoul H.K., Al-Dayeh M., Caraway L., Wright B., Chrest A. et al. 'Measurements of X-ray emission from rocket-triggered lightning'. *Geophys. Res. Lett.* 2004;**31**:L05118, doi: 10.1029/2003GL018770, 4 p.
52. Dwyer J.R., Rassoul H.K., Al-Dayeh M., Caraway L., Wright B., Chrest A. et al. 'A ground level gamma-ray burst observed in association with rocket-triggered lightning'. *Geophys. Res. Lett.* 2004;**31**:L05119, doi: 10.1029/2003GL018771, 4 p.
53. Rakov V.A., Crawford D.E., Kodali V., Idone V.P., Uman M.A. and Schnetzer G.H. et al. 'Cutoff and re-establishment of current in rocket-triggered lightning'. *J. Geophys. Res.* 2003;**108**(D23):4747, doi: 10.1029/2003JD003694.
54. Olsen R.C., Rakov V.A., Jordan D.M., Jerauld J., Uman M.A. and Rambo K.J. 'Leader/return-stroke-like processes in the initial stage of rocket-triggered lightning'. *J. Geophys. Res.* 2006;**111**:D13202, doi: 10.1029/2005JD006790, 11 p.
55. Rahman M., Cooray V., Rakov V.A., Uman M.A., Liyanage P., DeCarlo B.A., Jerauld J. and Olsen III R.C. 'Measurements of NO_x produced by rocket-triggered lightning'. *Geophys. Res. Lett.* 2007;34:L03816, doi: 10.1029/2006GL027956.
56. Jayakumar V., Rakov V.A., Miki M., Uman M.A., Schnetzer G.H. and Rambo K.J. 'Estimation of input energy in rocket-triggered lightning'. *Geophys. Res. Lett.* 2006;**33**:L05702, doi: 10.1029/2005GL025141.
57. Wang D., Rakov V.A., Uman M.A., Fernandez M.I., Rambo K.J., Schnetzer G.H. et al. 'Characterization of the initial stage of negative rocket-triggered lightning'. *J. Geophys. Res.* 1999;**104**:4213–22.
58. Thottappillil R., Goldberg J.D., Rakov V.A. and Uman M.A. 'Properties of M components from currents measured at triggered lightning channel base'. *J. Geophys. Res.* 1995;**100**:25711–20.
59. Eybert-Berard A., Barret L. and Berlandis J.P. 'Campagne foudre aux ETATS-UNIS Kennedy Space Center (Florida), Programme RTLP 85* (in French). STT/ASP 86-01'. Cent. D'Etud. Nucl. de Grenoble, Grenoble, France, 1986.
60. Eybert-Berard A., Barret L. and Berlandis J.P. 'Campagne d'experimentations foudre RTLP 87, NASA Kennedy Space Center, Florida, USA (in French). STT/LASP 88-21/AEB/JPB-pD', Cent. D'Etud. Nucl. de Grenoble, Grenoble, France, 1988.
61. Hubert P. 'Triggered lightning in France and New Mexico'. *Endeavour* 1984;**8**:85–89.
62. Miki M., Rakov V.A., Shindo T., Diendorfer G., Mair M., Heidler F. et al. 'Initial stage in lightning initiated from tall objects and in rocket-triggered lightning'. *J. Geophys. Res.* 2005;**110**:D02109, doi: 10.1029/2003JD004474.
63. Rakov V.A., Uman M.A. and Thottappillil R. 'Review of recent lightning research at the University of Florida'. *Electrotechnik und Informationstechnik* 1995;**112**:262–65.

64. Rakov V.A. 'Characterization of lightning electromagnetic fields and their modeling'. *Proceedings of the 14th International Zurich Symposium on Electromagnetic Compatibility*; Zurich, Switzerland, 2001, pp. 3–16.
65. Rakov V.A. 'Positive Blitzentladungen'. *ETZ Elektrotech. Autom.* 2001; **122**(5):26–29.
66. Rakov V.A., Uman M.A., Rambo K.J., Fernandez M.I., Eybert-Berard A., Berlandis J.P. *et al.* 'Initial processes in triggered lightning'. *Eos Trans. AGU* 1996;**77**:F86.
67. Horii K. and Ikeda G. 'A consideration on success conditions of triggered lightning'. *Proceedings of the 18th International Conference on Lightning Protection*; Munich, Germany, 1985, paper 1–3, 6 p.
68. Fisher R.J., Schnetzer G.H., Thottappillil R., Rakov V.A., Uman M.A. and Goldberg J.D. 'Parameters of triggered-lightning flashes in Florida and Alabama'. *J. Geophys. Res.* 1993;**98**:22887–902.
69. Idone V.P., Orville R.E., Hubert P., Barret L. and Eybert-Berard A. 'Correlated observations of three triggered lightning flashes'. *J. Geophys. Res.* 1984;**89**: 1385–94.
70. Anderson R.B. and Eriksson A.J. 'Lightning parameters for engineering application'. *Electra* 1980;**69**:65–102.
71. Leteinturier C., Hamelin J.H. and Eybert-Berard A. 'Submicrosecond characteristics of lightning return-stroke currents'. *IEEE Trans. Electromagn. Compat.* 1991;**33**:351–57.
72. Depasse P. 'Statistics on artificially triggered lightning'. *J. Geophys. Res.* 1994;**99**: 18515–22.
73. Berger K., Anderson R.B. and Knoninger H. 'Parameters of lightning flashes'. *Electra* 1975;**41**:23–37.
74. Berger K. and Garabagnati E. 'Lightning current parameters. Results obtained in Switzerland and in Italy'. *URSI Conference*, Florence, Italy, 1984, 13 p.
75. Carlson A.B. *Circuits*. New York: John Wiley & Sons; 1996, pp. 838.
76. Ben Rhouma A., Auriol A.P., Eybert-Berard A., Berlandis J.-P. and Bador B. 'Nearby lightning electromagnetic fields'. *Proceedings of the 11th International Zurich Symposium on Electromagnetic Compatibility*; Zurich, Switzerland, 1995, pp. 423–428.
77. Gorin B.N., Levitov V.I. and Shkilev A.V. 'Lightning strikes to the Ostankino tower'. *Elektrichestvo* 1977;**8**:19–23.
78. Gorin B.N. and Shkilev A.V. 'Measurements of lightning currents at the Ostankino tower'. *Elektrichestvo* 1984;**8**:64–65.
79. Uman M.A., Cordier D.J., Chandler R.M., Rakov V.A., Bernstein R. and Barker P.P. 'Fulgurites produced by triggered lightning'. *Eos Trans. AGU* 1994; **75**(44):99.
80. Rakov V.A. 'Lightning makes glass'. *Journal of the Glass Art Society* 1999; 45–50.
81. Jones B.E., Jones K.S., Rambo K.J., Rakov V.A., Jerauld J. and Uman M.A. 'Oxide reduction during triggered-lightning fulgurite formation'. *J. Atmospheric and Solar-Terrestrial Physics* 2005;**67**:427–28.

82. Fisher R.J., Schnetzer G.H. and Morris M.E. 'Measured fields and earth potentials at 10 and 20 m from the base of triggered-lightning channels'. *Proceedings of the 22nd International Conference on Lightning Protection*; Budapest, Hungary, 1994, Paper R 1c–10, 6 p.
83. Nucci C.A. and Rachidi F. 'On the contribution of the electromagnetic field components in the field-to-transmission line interaction'. *IEEE Trans. Electromagn. Compat.* 1995;**37**:505–08.
84. Rakov V.A. and Uman M.A. 'Review and evaluation of lightning return stroke models including some aspects of their application'. *IEEE Trans. Electromagn. Compat.* 1998;**40**:403–26.
85. Schoene J., Uman M.A., Rakov V.A., Rambo K.J., Jerauld J. and Schnetzer G.H. 'Test of the transmission line model and the traveling current source model with triggered lightning return strokes at very close range'. *J. Geophys. Res.* 2003;**108**(D23):4737, doi: 10.1029/2003JD003683.
86. Leteinturier C., Weidman C. and Hamelin J. 'Current and electric field derivatives in triggered lightning return strokes'. *J. Geophys. Res.* 1990; **95**:811–28.
87. Depasse P. 'Lightning acoustic signature'. *J. Geophys. Res.* 1994;**99**:25933–40.
88. Rubinstein M., Rachidi F., Uman M.A., Thottappillil R., Rakov V.A. and Nucci C.A. 'Characterization of vertical electric fields 500 m and 30 m from triggered lightning'. *J. Geophys. Res.* 1995;**100**:8863–72.
89. Rubinstein M., Uman M.A., Thomson E.M., Medelius P. and Rachidi F. 'Measurements and characterization of ground level vertical electric fields 500 m and 30 m from triggered-lightning'. *Proceedings of the 9th International Conference on Atmospheric Electricity*; St. Petersburg, Russia, 1992, pp. 276–78.
90. Uman M.A., Rakov V.A., Versaggi J.A., Thottappillil R., Eybert-Berard A., Barret L. *et al.* 'Electric fields close to triggered lightning'. *Proceedings of the International Symposium on Electromagnetic Compatibility (EMC'94 ROMA)*; Rome, Italy, 1994, pp. 33–37.
91. Crawford D.E., Rakov V.A., Uman M.A., Schnetzer G.H., Rambo K.J. and Stapleton M.V. 1999. 'Multiple-station measurements of triggered-lightning electric and magnetic fields'. *Proceedings of the 11th International Conference on Atmospheric Electricity*; Guntersville, Alabama, 1999, pp. 154–157.
92. Rubinstein M., Uman M.A., Medelius P.J. and Thomson E.M. 'Measurements of the voltage induced on an overhead power line 20 m from triggered lightning'. *IEEE Trans. Electromagn. Compat.* 1994;**36**(2):134–40.
93. Agrawal A.K., Price H.J. and Gurbaxani S.H. 'Transient response of multiconductor transmission lines excited by a nonuniform electromagnetic field'. *IEEE Trans. Electromagn. Compat.* 1980;**22**:119–29.
94. Master M.J. and Uman M.A. 'Lightning induced voltages on power lines: Theory'. *IEEE Trans. Power Apparat. Syst.* 1984;**103**:2505–17.
95. Rubinstein M., Tzeng A.Y., Uman M.A., Medelius P.J. and Thomson E.M. 'An experimental test of a theory of lightning-induced voltages on an overhead wire'. *IEEE Trans. Electromagn. Compat.* 1989;**31**:376–83.

96. Georgiadis N., Rubinstein M., Uman M.A., Medelius P.J. and Thomson E.M. 'Lightning-induced voltages at both ends of a 450-m distribution line'. *IEEE Trans. Electromagn. Compat.* 1992;**34**:451–60.
97. Rachidi F., Rubinstein M., Guerrieri S. and Nucci C.A. 'Voltages induced on overhead lines by dart leaders and subsequent return strokes in natural and rocket-triggered lightning'. *IEEE Trans. Electromagn. Compat.* 1997;**39**(2):160–66.
98. Barker P.P., Short T.A., Eybert-Berard A.R. and Berlandis J.P. 'Induced voltage measurements on an experimental distribution line during nearby rocket triggered lightning flashes'. *IEEE Trans. Power Deliv.* 1996;**11**:980–95.
99. Fernandez M.I. 'Responses of an unenergized test power distribution system to direct and nearby lightning strikes'. M.S. Thesis, University of Florida, Gainesville, 1997, 249 p.
100. Fernandez M.I., Rambo K.J., Stapleton M.V., Rakov V.A. and Uman M.A. 'Review of triggered lightning experiments performed on a power distribution system at Camp Blanding, Florida, during 1996 and 1997'. *Proceedings of the 24th International Conference on Lightning Protection*; Birmingham, United Kingdom, 1998, pp. 29–35.
101. Fernandez M.I., Rakov V.A. and Uman M.A. 'Transient currents and voltages in a power distribution system due to natural lightning'. *Proceedings of the 24th International Conference on Lightning Protection*; Birmingham, United Kingdom, 1998, pp. 622–29.
102. Paolone M., Schoene J., Uman M., Rakov V., Jordan D., Rambo K. *et al.* 'Testing of the LIOV-EMTP96 code for computing lightning-induced currents on real distribution lines: Triggered-lightning experiments'. *Proceedings of the 27th International Conference on Lightning Protection*; Avignon, France, 13–16 September 2004, pp. 286–90.
103. Rachidi F., Nucci C.A., Ianoz M. and Mazzetti C. 'Response of multiconductor power lines to nearby lightning return stroke electromagnetic fields'. *IEEE Trans. Power Deliv.* 1997;**12**(3):1404–11.
104. Mata C.T., Fernandez M.I., Rakov V.A. and Uman M.A. 'EMTP modeling of a triggered-lightning strike to the phase conductor of an overhead distribution line'. *IEEE Trans. Power Deliv.* 2000;**15**(4):1175–81.
105. Fernandez M.I., Mata C.T., Rakov V.A., Uman M.A., Rambo K.J., Stapleton M.V. and Bejleri M. 'Improved lightning arrester protection results, final results'. Technical report, TR-109670-R1 (Addendum AD-109670-R1), EPRI, 3412 Hillview Avenue, Palo Alto, California 94304, 1998.
106. Fernandez M.I., Rambo K.J., Rakov V.A. and Uman M.A. 'Performance of MOV arresters during very close, direct lightning strikes to a power distribution system'. *IEEE Trans. Power Deliv.* 1999;**14**(2):411–18.
107. Mata C.T., Rakov V.A., Rambo K.J., Diaz P., Rey R. and Uman M.A. 'Measurement of the division of lightning return stroke current among the multiple arresters and grounds of a power distribution line'. *IEEE Trans. Power Deliv.* 2003;**18**(4):1203–208.

108. Mata A.G., Rakov V.A., Rambo K.J., Stapleton M.V. and Uman M.A. 'UF/FPL study of triggered lightning strikes to FPL distribution lines: 2001 experiments'. Phase III Report. University of Florida, December 2001, 25 p.
109. Mata A.G., Mata C.T., Rakov V.A., Uman M.A., Schoene J.D., Rambo K.J. *et al*. 'Study of triggered lightning strikes to FPL distribution lines'. Phase IV Report. University of Florida, December 2002, 258 p.
110. Barker P.P. and Short T.A. 'Lightning effects studied: The underground cable program'. *Transmission and Distribution World* May 1996a;24–33.
111. Barker P.P. and Short T.A. 'Lightning measurements lead to an improved understanding of lightning problems on utility power systems'. *Proceedings of the 11th CEPSI*; Kuala Lumpur, Malaysia, 1996b, vol. 2, pp. 74–83.
112. Barker P. and Short T. 'Findings of recent experiments involving natural and triggered lightning. Panel Session Paper presented at' *1996 Transmission and Distribution Conference*; Los Angeles, California, 16–20 September 1996.
113. Paolone M., Petrache E., Rachidi F., Nucci C.A., Rakov V.A., Uman M.A. *et al*. 'Lightning induced disturbances in buried cables – Part 2: Experiment and model validation'. *IEEE Trans. Electromagn. Compat.* 2005;**47**(3):509–20.
114. Rakov V.A., Uman M.A., Fernandez M.I., Mata C.T., Rambo K.T., Stapleton M.V. and Sutil R.R. 'Direct lightning strikes to the lightning protective system of a residential building: Triggered-lightning experiments'. *IEEE Trans. Power Deliv.* 2002;**17**(2):575–86.
115. Kobayashi M., Sasaki H. and Nakamura K. 'Rocket-triggered lightning experiment and consideration for metal oxide surge arresters'. *Proceedings of the 10th International Symposium on High Voltage Engineering*; Montreal, Quebec, Canada, 1997, 4 p.
116. Barker P., Short T., Mercure H., Cyr S. and O'Brien J. 'Surge arrester energy duty considerations following from triggered lightning experiments'. *IEEE PES, 1998 Winter Meeting, Panel Session on Transmission Line Surge Arrester Application Experience*, 1998, 19 p.
117. Eybert-Berard A., Lefort A. and Thirion B. 'On-site tests'. *Proceedings of the 24th International Conference on Lightning Protection*; Birmingham, United Kingdom, 1998, pp. 425–35.
118. Teramoto M., Yamada T., Nakamura K., Matsuoka R., Sumi S. and Horii K. 'Triggered lightning to a new type lightning rod with high resistance'. *Proceedings of the 10th International Conference on Atmospheric Electricity*; Osaka, Japan, 1996, pp. 341–44.
119. Morris M.E., Fisher R.J., Schnetzer G.H., Merewether K.O. and Jorgenson R.E. 'Rocket-triggered lightning studies for the protection of critical assets'. *IEEE Trans. Indust. Appl.* 1994; **30**:791–804.
120. Schnetzer G.H. and Fisher R.J. 'Earth potential distributions within 20 m of triggered lightning strike points'. *Proceedings of the 24th International Conference on Lightning Protection*; Birmingham, United Kingdom, 1998, pp. 501–505.
121. Gary C., Cimador A. and Fieux R. 'La foudre: Étude du phénomène. Applications à la protection des lignes de transport'. *Revue Générale de l'Électricité* 1975;**84**:24–62.

122. Kumazaki K., Nakamura K., Naito K. and Horii K. 'Production of artificial fulgurite by utilizing rocket triggered lightning'. *Proceedings of the 8th International Symposium on High Voltage Engineering*; Yokohama, Japan, 1993, pp. 269–72.
123. Davis D.A., Murray W.C., Winn W.P., Mo Q., Buseck P.R. and Hibbs B.D. 'Fulgurites from triggered lightning'. *Eos Trans. AGU* 1993;**74**(43): Fall Meet. Suppl. 165.
124. SPARG (1982): Saint-Privat-d'Allier Research Group (SPARG). 'Eight years of lightning experiments at Saint-Privat-d'Allier'. *Extrait de la Revue Générale de l'Electricité* September 1982, Paris.
125. Saba M.M.F., Pinto O., Pinto I.R.C.A., Pissolato F.J., Eybert-Berard A., Lefort A. et al. 'An international center for triggered and natural lightning research in Brazil'. *Proceedings of the 2000 International Lightning Detection Conference*; GAI, 2705 East Medina Road, Tucson, Arizona 85706-7155, 2000, paper 40, 7 p.
126. Crawford D.E. 'Multiple-station measurements of triggered lightning electric and magnetic fields'. M.S. Thesis, University of Florida, Gainesville, 1998, 282 p.
127. Hierl A. 'Strommessungen der Blitztriggerstation Steingaden'. *Proceedings of the 16th International Conference on Lightning Protection*; Szeged, Hungary, 1981, Paper R-1.04, 10 p.
128. Horii K., Wada A., Nakamura K., Yoda M., Kawasaki Z., Sirait K.T. et al. 'Experiment of rocket-triggered lightning in Indonesia'. *Trans. IEE Japan* 1990;**110-B**:1068–69.
129. Beituganov M.N. and Zashakuev T.Z. 'The experience of initiation of lightning and spark discharges'. *Proceedings of the 9th International Conference on Atmospheric Electricity*; St. Petersburg, Russia, 1992, pp. 283–286.

Bibliography

Baum C.E., O'Neill J.P., Breen E.L., Hall D.L. and Moore C.B. 'Electromagnetic measurement of and location of lightning'. *Electromagnetics* 1987;**7**:395–422.
Bazelyan E.M., Gorin B.N. and Levitov V.I. *Physical and Engineering Foundations of Lightning Protection.* Leningrad: Gidrometeoizdat; 1978, 223 p.
Bazelyan E.M. and Raizer Yu.P. *Lightning Physics and Lightning Protection*, Bristol: IOP Publishing; 2000a, 325 p.
Bazelyan E.M. and Raizer Yu.P. 'Lightning attraction mechanism and the problem of lightning initiation by lasers'. *UFN* 2000b;**170**(7):753–69.
Bejleri M. 'Triggered-lightning testing of an airport runway lightning system'. M.S. Thesis, University of Florida, Gainesville, 1999, 307 p.
Bejleri M., Rakov V.A., Uman M.A., Rambo K.J., Mata C.T. and Fernandez M.I. 'Triggered lightning testing of an airport runway lighting system'. *Proceedings of the 25th International Conference on Lightning Protection*; Rhodes, Greece, 2000, pp. 825–830.

Berger K. 'Novel observations on lightning discharges: Results of research on Mount San Salvatore'. *J. Franklin Inst.* 1967;**283**:478–525.

Berger K. 'The Earth Flash', in R.H. Golde (ed.). *Lightning, Vol. 1, Physics of Lightning*. New York: Academic Press; 1977, pp. 119–190.

Berger K. and Vogelsanger E. 'New results of lightning observations' in S.C. Coroniti and J. Hughes (eds.). *Planetary Electrodynamics*. New York: Gordon and Breach; 1969, pp. 489–510.

Bondiou-Clergerie A., Lalande P., Laroche P., Willett J.C., Davis D. and Gallimberti I. 'The inception phase of positive leaders in triggered lightning: comparison of modeling with expcrimental data'. *Proceedings of the 11th International Conference on Atmospheric Electricity*; Guntersville, Alabama, 1999, pp. 22–25.

Chauzy S., Medale J.-C., Prieur S. and Soula S. 'Multilevel measurement of the electric field underneath a thundercloud, 1. A new system and the associated data processing'. *J. Geophys. Res.* 1991;**96**:22319–26.

Chen H. and Liu X. 'Triggered lightning flashes at Beijing Lightning Trigger Laboratory'. *Proceedings of the International Aerospace and Ground Conference on Lightning and Static Electricity*; Atlantic City, New Jersey, 1992, pp. 54/1–54/7.

Cooray V., Fernando M., Sörensen T., Götschl T. and Pedersen Aa. 'Propagation of lightning generated transient electromagnetic fields over finitely conducting ground'. *J. Atmos. Solar-Terr. Phys.* 2000;**62**:583–600.

Cooray V., Rakov V. and Theethayi N. 'The lightning striking distance-revisited'. *J. Electrostatics* 2007;**65**:296–306.

Cooray V., Montano R. and Rakov V. 'A model to represent negative and positive lightning first return strokes with connecting leaders'. *J. Electrostatics* 2004;**60**:97–109.

Davis D.A. and Laroche P. 'Positive leaders in rocket-triggered lightning: Propagation velocity from measured current and electric field derivative at ground'. *Proceedings of the 11th International Conference of Atmospheric Electricity*; Guntersville, Alabama, 1999, pp. 158–161.

Djebari B., Hamelin J., Leteinturier C. and Fontaine J. 'Comparison between experimental measurements of the electromagnetic field emitted by lightning and different theoretical models – influence of the upward velocity of the return stroke'. *Proceedings of the 4th International Zurich Symposium on Electromagnetic Compatibility*; Zurich, Switzerland, 1981, pp. 511–16.

Fieux R. and Hubert P. 'Triggered lightning hazards'. *Nature* 1976;**260**:188.

Gardner R.L., Baum C.E. and Rison W. 'Resistive leaders for triggered lightning'. Kiva Memos, Memo 7, November 1987, 10 p.

Hagenguth J.H. and Anderson J.G. 'Lightning to the Empire State Building – Part III'. *AIEE Trans.* 1952;**71**(Part III):641–49.

Hamelin J. 'Sources of natural noise' in P. Deguauque and J. Hamelin (eds.). *Electromagnetic Compatibility*. New York: Oxford Univ Press, 1993, pp. 652.

Hamelin J., Karczewsky J.F. and Sene F.X. 'Sonde de mesure du champ magnétique dû a une décharge orageuse'. *Ann. des Télécommun.* 1978;**33**:198–205.

Hamelin J., Leteinturier C., Weidman C., Eybert-Berard A. and Barret L. 'Current and current-derivative in triggered lightning flashes – Florida' 1985. *Proceedings of the International Conference on Lightning and Static Electricity*; NASA, Dayton, Ohio, 1986, 10 p.

Horii K. and Sakurano H. 'Observation on final jump of the discharge in the experiment of artificially triggered lightning'. *IEEE Trans. Power Apparat. Syst.* 1985; **PAS-104**:2910–17.

Hubert P. 'Triggered lightning at Langmuir Laboratory during TRIP-8l'. *Service d'Electronique Physique*, 1981; DPH/EP/81/66.

Hubert P. 'A new model of lightning subsequent stroke – confrontation with triggered lightning observations'. *Proceedings of the 10th International Conference on Lightning and Static Electricity*, Paris, France, 1985, Paper 4B4.

Hubert P. and Mouget G. 'Return stroke velocity measurements in two triggered lightning flashes'. *J. Geophys. Res.* 1981;**86**:5253–61.

Idone V.P. 'The luminous development of Florida triggered lightning'. *Res. Lett. Atmos. Electr.* 1992;**12**:23–28.

Idone V.P. 'Microscale tortuosity and its variation as observed in triggered lightning channels'. *J. Geophys. Res.* 1995;**100**:22943–56.

Idone V.P. and Orville R.E. 'Three unusual strokes in a triggered lightning flash'. *J. Geophys. Res.* 1984;**89**:7311–16.

Idone V.P. and Orville R.E. 'Correlated peak relative light intensity and peak current in triggered lightning subsequent return strokes'. *J. Geophys. Res.* 1985;**90**:6159–64.

Idone V.P. and Orville R.E. 'Channel tortuosity variation in Florida triggered lightning'. *Geophys. Res. Lett.* 1988;**15**:645–48.

IEC 61024-1. *Protection of Structures Against Lightning, Part 1: General Principles*, IEP, 1990.

Jerauld J., Rakov V.A., Uman M.A., Rambo K.J., Jordan D.M. and Cummins K.L. et al. 'An evaluation of the performance characteristics of the U.S. National Lightning Detection Network in Florida using rocket-triggered lightning'. *J. Geophys. Res.* 2005;**110**:D19106, doi: 10.1029/2005JD005924.

Jordan D.M., Idone V.P., Rakov V.A., Uman M.A., Beasley W.H. and Jurenka H. 'Observed dart leader speed in natural and triggered lightning'. *J. Geophys. Res.* 1992;**97**:9951–57.

Kawasaki Z.-I., Kanao T., Matsuura K., Nakano M., Horii K. and Nakamura K.-I. 'The electric field changes and UHF radiations caused by the triggered lightning in Japan'. *Geophys. Res. Lett.* 1991;**18**:1711–14.

Kawasaki Z.-I. and Mazur V. 'Common physical processes in natural and triggered lightning in winter storms in Japan'. *Res. Lett. Atmos. Electr.* 1992;**12**:61–70.

Kawasaki Z.-I. and Mazur V. 'Common physical processes in natural and triggered lightning in winter storms in Japan'. *J. Geophys. Res.* 1992;**97**:12935–45.

Kikuchi H. 'Overview of triggered lightning EMP' in H. Kikuchi (ed.). *Nonlinear and Environmental Electromagnetics*. Amsterdam: Elsevier; 1985. pp. 347–350.

Kodali V., Rakov V.A., Uman M.A., Rambo K.J., Schnetzer G.H. and Schoene J. et al. 'Triggered-lightning properties inferred from measured currents and very close electric fields'. *Atmos. Res.* 2005;**76**:355–76.

Krider E.P., Noogle R.C., Uman M.A. and Orville R.E. 'Lightning and the Apollo/ Saturn V exhaust plume'. *J. Spacecraft Rockets* 1974;**11**:72–75.

Lalande P., Bondiou-Clergerie A., Bacchiega G. and Gallimberti I. 'Observations and modeling of lightning leaders'. *C.R. Physique* 2002;**3**:1375–92.

Laroche P., Eybert-Berard A. and Barret L. 'Triggered lightning flash characteristics'. *Proceedings of the 10th International Aerospace and Ground Conference on Lightning and Static Electricity*; Paris, France, 1985, pp. 231–39.

Laroche P., Eybert-Berard A., Barret L. and Berlandis J.P. 'Observations of preliminary discharges initiating flashes triggered by the rocket and wire technique'. *Proceedings of the 8th International Conference on Atmospheric Electricity*; Uppsala, Sweden, 1988, pp. 327–33.

Leteinturier C. and Hamelin J. 'Rayonnement électromagnétique des décharges orageuses. Analyse submicroseconde'. *Revue Phys. Appl.* 1990;**25**:139–46.

Le Vine D.M., Willett J.C. and Bailey J.C. 'Comparison of fast electric field changes from subsequent return strokes of natural and triggered lightning'. *J. Geophys. Res.* 1989;**94**:13259–65.

Lhermitte R. 'Doppler radar observations of triggered lightning'. *Geophys. Res. Lett.* 1982;**9**:712–15.

Liaw Y.P., Cook D.R. and Sisterson D.L. 'Estimation of lightning stroke peak current as a function of peak electric field and the normalized amplitude of signal strength: Corrections and improvements'. *J. Atmos. Ocean. Tech.* 1996;**13**: 769–73.

Mach D.M. and Rust W.D. 'Two-dimensional speed and optical risetime estimates for natural and triggered dart leaders'. *J. Geophys. Res.* 1997;**102**:13673–84.

Mach D.M. and Rust W.D. 'Photoelectric return-stroke velocity and peak current estimates in natural and triggered lightning'. *J. Geophys. Res.* 1989;**94**: 13237–47.

Mata C.T. 'Interaction of lightning with power distribution lines'. Ph.D. Dissertation, University of Florida, Gainesville, 2000, 388 p.

Mata C.T., Fernandez M.I., Rakov V.A., Uman M.A., Bejleri M. and Rambo K.J. *et al.* 'Overvoltages in underground systems, phase 2 results'. Technical report, TR-109669-R1 (Addendum AD-109669-R1), EPRI, 3412 Hillview Avenue, Palo Alto, California 94304, 1998.

Matsumoto Y., Sakuma O., Shinjo K., Saiki M., Wakai T., Sakai T. *et al.* 'Measurement of lightning surges on test transmission line equipped with arresters struck by natural and triggered lightning'. *IEEE Trans. Power Deliv.* 1996;**11**:996–1002.

Mazur V. and Ruhnke L. 'Common physical processes in natural and artificially triggered lightning'. *J. Geophys. Res.* 1993;**98**:12913–30.

McEachron K.B. 'Lightning to the Empire State Building'. *J. Franklin Inst.* 1939;**227**:149–217.

Merlino R.L. and Goree J.A. 'Dusty plasmas in the laboratory, industry and space'. *Physics Today* 2004; July: 32–38.

Miki M., Shindo T., Rakov V.A., Uman M.A., Rambo K.J., Schnetzer G.H. *et al.* 'Die Anfangsphase von Aufwaertsblitzen'. *ETZ Elektrotech. Autom.* 2003;**124**(3–4): 50–55.

Miyachi I., Horii K., Muto S., Ikeda G. and Aiba S. 'Experiment of long gap discharge by artificially triggered lightning with rocket'. *Proceedings of the 3rd International Symposium on High Voltage Engineering*; Milan, Italy, 1979, paper 51.10, 4 p.

Miyake K., Suzuki T., Takashima M., Takuma M. and Tada T. 'Winter lightning on Japan Sea coast – lightning striking frequency to tall structures'. *IEEE Trans. Power Deliv.* 1990;**5**:1370–76.

Moore C.B., Winn W.P., Hall D.L. and Cobb J.W. 'An investigation into the use of detonating fuses to create conducting paths in the atmosphere'. *Measurement Notes* 1984; March: Note 30.

Motoyama H., Shinjo K., Matsumoto Y. and Itamoto N. 'Observation and analysis of multiphase back flashover on the Okushishiku test transmission line caused by winter lightning'. *IEEE Trans. Power Deliv.* 1998;**13**:1391–98.

Nakano M., Takagi N., Kawasaki Z. and Takeuti T. 'Return strokes of triggered lightning flashes'. *Res. Lett. Atmos. Electr.* 1983;**3**:73–78.

Newman M.M. 'Lightning discharge channel characteristics and related atmospherics', in L.G. Smith (ed.). *Recent Advances in Atmospheric Electricity*. New York: Pergamon Press; 1958, pp. 475–84.

Orville R.E. 'Calibration of a magnetic direction finding network using measured triggered lightning return stroke peak currents'. *J. Geophys. Res.* 1991;**96**:17135–42.

Orville R.E. and Idone V.P. 'Lightning leader characteristics in the Thunderstorm Research International Program (TRIP)'. *J. Geophys. Res.* 1982;**87**:11177–92.

Petrache E., Rachidi F., Paolone M., Nucci C.A., Rakov V. and Uman M. 'Lightning-induced disturbances on buried cables – Part I: Theory'. *IEEE Trans. Electromagnetic Compatability* 2005;**47**(3):498–508.

Pierce E.T. 'Triggered lightning and some unsuspected lightning hazards'. Stanford Research Institute, Menlo Park, California, 1971, 20 p.

Pierce E.T. 'Triggered lightning and some unsuspected lightning hazards'. *Naval Research Reviews* 1972;**25**:14–28.

Pierce E.T. 'Atmospheric electricity – some themes'. *Bull. Am. Meteor. Soc.* 1974;**55**:1186–94.

Rakov V.A. 'Rocket-triggered lightning experiments at Camp Blanding, Florida'. *Proceedings of the 1999 International Conference on Lightning and Static Electricity*; Toulouse, France, 1999, pp. 469–81.

Rakov V.A. 'Evaluation of the performance characteristics of lightning locating systems using rocket-triggered lightning'. *Proceedings of the International Symposium on Lightning Protection (VIII SIPDA)*; Sao Paulo, Brazil, 2005, pp. 697–715.

Rakov V.A., Kodali V., Crawford D.E., Schoene J., Uman M.A. and Rambo K.J. et al. 'Close electric field signatures of dart leader/return stroke sequences in rocket-triggered lightning showing residual fields'. *J. Geophys. Res.* 2005;**110**:D07205, doi: 10.1029/2004JD0054.

Rakov V.A., Uman M.A., Fernandez M.I., Mata C.T., Rambo K.T. and Stapleton M.V. et al. 'Direct lightning strikes to the lightning protective system of a residential building: triggered-lightning experiments'. *IEEE Trans. Power Deliv.* 2002; **17**(2):575–86.

Rakov V.A., Uman M.A., Rambo K.J., Schnetzer G.H. and Miki M. 'Triggered-lightning experiments conducted in 2000 at Camp Blanding, Florida (abstract)'. *Eos Trans. Suppl. AGU* 2000;**81**(48):F90.

Rakov V.A., Uman M.A., Wang D., Rambo K.J., Crawford D.E. and Schnetzer G.H. *et al.* 'Some results from recent experiments at the International Center for Lightning Research and Testing at Camp Blanding, Florida (abstract)'. *Eos Trans. Suppl. AGU* 1999;**80**(46):F203.

Rühling F. 'Gezielte Blitzentladung mittels Raketen'. *Umschau in Wissenschaft und Technik* 1974;**74**:520–21.

Rühling F. 'Raketengetriggerte Blitze im Dienste des Freileitungsschutzes vor Gewitterüberspannungen'. *Bull. SEV* 1974;**65**:1893–98.

Sadiku M.N.O. *Elements of Electromagnetics*. Orlando, Florida: Sounders College; 1994, 821 p.

Schnetzer G.H., Fisher R.J., Rakov V.A. and Uman M.A. 'The magnetic field environment of nearby lightning'. *Proceedings of the 24th International Conference on Lightning Protection*; Birminghan, United Kingdom; 1998, pp. 346–49.

Schonland B.F.J., Malan D.J. and Collens H. 'Progressive Lightning II'. *Proc. Roy. Soc. (London)* 1935;**A152**:595–625.

Shindo T. and Ishii M. 'Japanese research on triggering lightning'. *Elektrotechnik und Informationstechnik* 1995;**112**(6):265–68.

Soula S. and Chauzy S. 'Multilevel measurement of the electric field underneath a thundercloud 2. Dynamical evolution of a ground space charge layer'. *J. Geophys. Res.* 1991;**96**:22327–36.

Uman M.A. *The Lightning Discharge*. Mineola, New York: Dover; 2001, p. 377.

Wada A. 'Discussion on lightning protection by artificial triggered lightning'. *Proceedings of the 22nd International Conference on Lightning Protection*; Budapest, Hungary, 1994, paper R2–06, 5 p.

Waldteufel P., Metzger P., Boulay J.L., Laroche P. and Hubert P. 'Triggered lightning strokes originating in clear air'. *J. Geophys. Res.* 1980;**85**:2861–68.

Wang D., Takagi N., Watanabe T., Rakov V.A., Uman M.A. and Rambo K.J. *et al.* 'A comparison of channel-base currents and optical signals for rocket-triggered lightning strokes'. *Atmos. Res.* 2005;**76**:412–22.

Willett J.C., Davis D.A. and Laroche P. 'Positive leaders in rocket-triggered lightning: propagation velocity from measured current and ambient-field profile'. *Proceedings of the 11th International Conference on Atmospheric Electricity*; Guntersville, Alabama, 1999, pp. 58–61.

Yoda M., Miyachi I., Kawashima T. and Katsuragi Y. 'Lightning current protection of equipments in winter'. *Res. Lett. Atmos. Electr.* 1992;**12**:117–21.

Chapter 4
Attachment of lightning flashes to grounded structures
Vernon Cooray and Marley Becerra

4.1 Introduction

As a stepped leader approaches the ground, the electric field at the extremities of grounded structures increases to such a level that some of these structures or different parts of the same structure may launch connecting leaders towards the down-coming stepped leader. The first return stroke is initiated at the instant contact is made between the down-coming stepped leader and one of these connecting leaders. The strike point of the lightning flash is the place from which the connecting leader that made the successful connection to the stepped leader was initiated.

An exact evaluation of the point of lightning strike of a structure should take into account the development of streamers from the extremities of the structure, the subsequent streamer-to-leader transition, the inception of a stable propagating leader and the final encounter between the upward-moving connecting leader and the down-coming stepped leader. However, current international standards on lightning protection of structures and power transmission and distribution lines are based on different concepts and models, namely the protective angle method and the electro-geometrical method (of which the rolling sphere method was a derivative); these neglect most of the physics associated with the attachment process of lightning flashes with structures. However, lightning research has progressed significantly over the last several decades, resulting in a deeper understanding of the physics of the process of attachment and the possibility of representing this physics in computer simulation procedures. Today, the possibility exists of simulating the inception and propagation of leaders from grounded structures under the influence of down-coming stepped leaders, so that the point of lightning strike of any complex structure may be predicted.

The goal of this chapter is to present the current state of the art of lightning interception, and to show how the computer simulation programs that accommodate the physics of lightning interaction could be used to complement the protection procedures based on either the electro-geometrical model or the rolling sphere method. First, let us explain the basics of the simple procedures used by engineers to protect structures from lightning flashes. Some of these procedures are explained also in Chapters 6 and 21. However, for the sake of completeness they are described here too.

4.1.1 The protection angle method

Determining the volume protected by a vertical conductor has been the subject of discussion since the time of Benjamin Franklin. In 1823, Gay-Lussac stated that a lightning rod protects effectively against a lightning strike in a circular space around it, the radius of the space being twice the height of the rod [1]. Subsequent modifications to the definition of the zone of protection were later published by Lodge [2], and reproduced and discussed by Golde [3]. The results presented by Golde are shown in Figure 4.1. Note that with the exception of the work by Preece [4], the protection zone of a vertical conductor is viewed as a cone. The apex angle of the cone is known as the angle of protection of the vertical conductor. For a protective ratio (the ratio of the base radius of the cone to the rod height) of 1:1, the angle between the vertical rod and the lateral surface of the cone is 45°. A protective of ratio 2:1 corresponds to an angle of 60°. Until recently, the protection angle method was the one recommended by lightning protection standards [5–7]. The concept of the cone of protection can be used to locate lightning conductors on a building, as illustrated in Figure 4.2. Note that the smaller the angle of protection assumed in the analysis, the smaller is the separation between the adjacent lightning conductors located on the structure (and the more reliable the protection offered by the lightning conductors).

During the early decades of the twentieth century, the problem of lightning protection of grounded structures was brought to the fore by the construction of power transmission lines that extended over long distances. Because of the increased exposure of the power grid to lightning, shielding ground conductors were used to protect the phase conductors in transmission lines. The location of the shield wires was based on the concept of protective angle. The volume of space protected by a horizontal conductor according to this concept is shown in Figure 4.3. In designing the power lines, the phase conductors were placed inside the volume of protection offered by

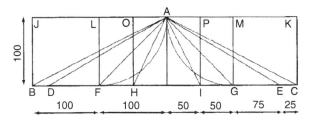

Figure 4.1 Zones of protection of a vertical rod, a JBCK cylinder, Gay Lussac (1823); a BAC cone, DeFonville (1874); a DAE cone, Paris Commission (1875); an LFGM cylinder, Chapman (1875); an FAG cone, Adams (1881); an OHIP cylinder hypothesis and FAG cone, Preece (1880), an HAI cone, Melsens (adapted from Reference 3)

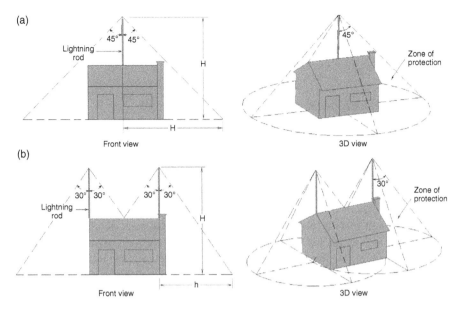

Figure 4.2 *Protective angle method for the protection of a common house for an angle of (a) 45° and (b) 30°*

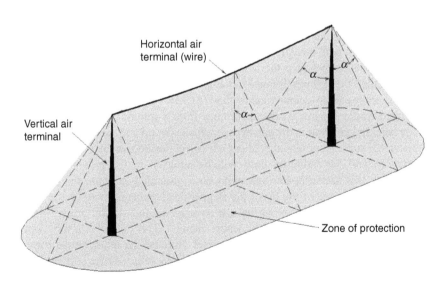

Figure 4.3 *The volume of space protected according to the angle-of-protection method by a horizontal conductor connected to two vertical conductors at the ends*

the shield wire. Initially, based on empirical criteria and scaled laboratory models, the apex angle of the cone was assumed to range between 20 and 75° [8]. Values of the protective angle were formalized by Wagner and colleagues, and protective angles of 30–45° are used in the design of power line geometry [8–10].

4.1.2 The electro-geometrical method

In long laboratory sparks created by switching voltage impulses, the breakdown process is mediated by a leader travelling from the high-voltage electrode towards ground. The propagation of the leader is facilitated by a streamer system emanating from the tip of the leader channel (see Section 4.3.1). This streamer system supplies the current necessary to heat the air and consequently extends the leader channel in the gap. However, if the voltage is suddenly removed, the propagation of the leader will be arrested and no electrical breakdown will take place in the gap. As the leader continues to propagate towards the grounded electrode, a situation will be reached in which the streamers will bridge the gap between the leader tip and the ground. This situation is known as the final jump condition. When the final jump condition is reached, the discharge cannot be arrested by removing the voltage and the breakdown of the gap follows immediately.

Experimental observations and theoretical calculations show that the streamers maintain a constant potential gradient along their axes. Thus, during the final jump condition the average potential gradient between the tip of the leader and the grounded electrode is equal to the potential gradient of the streamers. For positive streamers this potential gradient is $\sim 450-500$ kV m^{-1} and for negatives it is $1\,000-1\,500$ kV m^{-1} [11,12]. In many applications 500 and $1\,000$ kV m^{-1} are taken as typical values. Based on this laboratory observation one can hypothesize that when the average potential gradient between the leader tip and the ground is equal to the potential gradient of the streamers, electrical breakdown takes place between the leader tip and the ground. The critical distance between the leader tip and the ground when this condition is reached is called the striking distance to flat ground. This concept of striking distance is the basis of the electro-geometrical method (EGM). This method assumes that when the stepped leader reaches a critical distance from a grounded structure where the average potential gradient in the gap between the leader tip and the grounded structure is equal to the streamer potential gradient, electrical breakdown takes place in the gap immediately, and the lightning flash will be attracted to the grounded structure. According to this scenario, the first point on a grounded structure that will come within striking distance of the tip of the stepped leader channel will be the point of strike of the lightning flash. Note that the striking distance to flat ground depends only on the potential of the tip of the leader channel. This potential in turn depends on the charge distribution of the leader channel. Because the return stroke current is a result of the neutralization of leader charge, the peak return stroke current that will result when the stepped leader makes ground contact (i.e. prospective return stroke current) depends on the charge distribution of the leader channel. The larger the charge on the leader, the larger will be the prospective return stroke current. This connection between the leader potential, leader charge and the prospective return stroke current makes it possible to express the striking distance either as a function of leader charge or as a function of the peak of the prospective return stroke current. Because the striking distance decreases with decreasing leader charge, it also decreases with decreasing return stroke current. Consequently, a leader channel with a smaller prospective return stroke current has to come much closer to a structure than a leader channel associated with a larger prospective return stroke current before the leader becomes attached to the structure.

Electro-geometrical theory as used today for designing the shielding of power transmission lines was first proposed by Armstrong and Whitehead [13]. In order to illustrate how this concept is applied in power transmission lines, consider an infinitely long horizontal conductor located at height h above ground, as shown in Figure 4.4. Let us denote the striking distance r_s. According to the electro-geometrical concept, if the down-coming stepped leader intercepts the horizontal planes generated by the projection of line segments AB and CD in a horizontal direction perpendicular to ABCD, it will be attracted to ground. However, if the stepped leader intercepts the semicircular arc BC or the surface generated by projecting the semicircular arc in a horizontal direction perpendicular to ABCD, it will be attracted to the horizontal conductor. Now consider a power transmission line as shown in Figure 4.5. According to the EGM, if the tip of the stepped leader reaches any point on the arc CD it will be attracted to the shielding conductor. If it reaches arc BC it will be attracted to the phase conductor and a shielding failure would occur (Figure 4.5a). If it reaches any

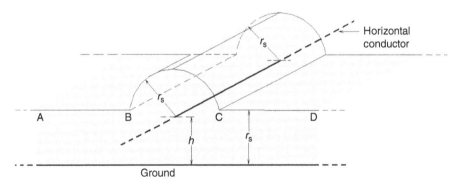

Figure 4.4 *Attachment of a lightning flash to a horizontal conductor according to the electro-geometrical method (EGM). If the down-coming stepped leader intercepts the horizontal planes generated by the projection of line segments AB and CD in a horizontal direction perpendicular to ABCD, it will be attracted to ground. On the other hand, if the stepped leader intercepts the semicircular arc BC or the surface generated by projecting the semicircular arc in a horizontal direction perpendicular to ABCD it will be attracted to the horizontal conductor.*

170 Lightning Protection

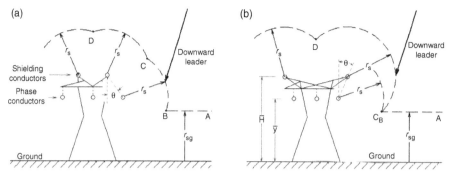

Figure 4.5 Sketch of a power transmission line and the lightning exposure arcs of the conductors according to the electro-geometrical method (EGM). This example shows a transmission line with (a) expected shielding failures and (b) perfect lightning performance.

other point on the line AB it will be attracted to ground. One can see from this figure that by changing the angle θ one can reduce the length of the exposure arc CD of the phase conductor. Power engineers use this concept to locate the overhead ground conductors in such a way as to screen the phase conductors from lightning flashes, as shown in Figure 4.5b. In practical applications the striking distance associated with overhead conductors is assumed to be slightly higher (about 10 per cent) than that corresponding to flat ground. The reason why a longer striking distance is selected for the overhead conductors will be explained in Section 4.3.

4.1.3 The rolling sphere method

In the early 1960s, based on the concept of protected spaces boarded by circular arcs as introduced by Schwaiger [14], Horvath [15] proposed the use of a fictitious sphere for the location of lightning conductors on structures; this was soon introduced into the Hungarian standard. The term 'rolling sphere' originates from the studies conducted in the United States by Lee [16,17]. The concept of the rolling sphere is directly related to the electro-geometrical models in that it is based on the assumption that a stepped leader has to approach to a critical distance, i.e. striking distance, before it will be attracted to the structure. In other words, this concept assumes that there is a spherical region with radius equal to the striking distance and located around the tip of the stepped leader, with the property such that the first point of a grounded structure that enters into this spherical volume will be the point of attachment of the stepped leader. In layman terms one can consider this region as the 'visual region' of the stepped leader. Based on this concept, the air terminals of a grounded structure are located in such a way that when a sphere with a given radius (i.e. striking distance) is rolled around the structure, it should touch only the conductors of the lightning

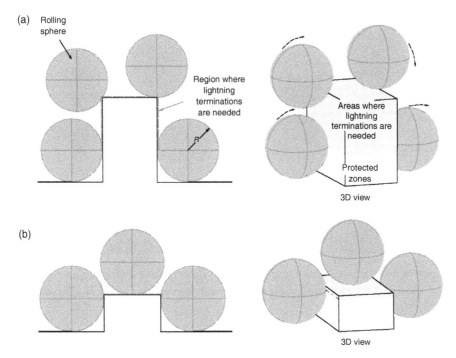

Figure 4.6 *In the protection of structures using the rolling sphere method, when a sphere of radius R (the magnitude of which depends on the peak return stroke current) is rolled over the surface of the structure it should touch only the external lightning protection system. Note that if the height of the building is larger than the radius of the sphere (diagram (a)) the lightning protection system should also cover the sides of the vertical walls. This is the case because the rolling sphere method predicts the possibility of lightning strikes to the sides of the building. However, if the height of the structure is smaller than the radius of the sphere (diagram (b)) this precaution is unnecessary.*

protection system. Two examples of the application of the rolling sphere method are shown in Figure 4.6. Note that the rolling sphere predicts the possibility of lightning strikes below the top of the structure as have been observed in field observations. Figure 4.7 shows an example of a lightning strike below the top of the CN tower in Toronto, Canada [18].

Consider a lightning protection system designed using a rolling sphere of a given radius. This radius, being the striking distance, is associated with a certain peak return stroke current. Let us denote this I_{crit}. Any stepped leader associated with a prospective peak return stroke current larger than I_{crit} will be associated with a rolling sphere of larger radius; such a stepped leader will not be able to penetrate the lightning protection system. On the other hand, a stepped leader associated with a current

172 *Lightning Protection*

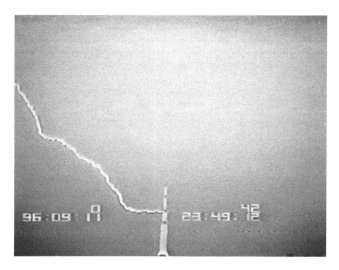

Figure 4.7 Photograph showing a lightning flash striking a point below the top of CN tower in Canada (photograph courtesy of Prof. A.M. Hussein)

smaller than I_{crit} will have a smaller rolling sphere radius and such strokes may be able to penetrate through the lightning protection system and strike the structure. Thus, for a more sensitive structure for lightning strikes a smaller sphere radius should be used in creating the lightning protection system.

In lightning protection standards such as the IEC standards [5–7], the protection levels are defined as I, II, III and IV. A structure lightning protection system based on level I would not allow a return stroke peak current larger than 2.9 kA to penetrate the lightning protection system. The corresponding currents for levels II, III, IV are 5.4, 10.1 and 15.7 kA, respectively. The rolling sphere radii associated with these protection levels are given in Table 4.1. Based on the statistical distribution of the first return stroke peak currents, one can estimate that level I provides complete protection against 99 per cent of all ground flashes and level IV provides protection against 84 per cent of all ground flashes. Note that the statement 'protection against 84 per cent of all ground flashes' does not mean that the remaining 16 per cent of the lightning flashes will strike the structure. In other words, not all the return strokes with peak currents less than 15.7 kA will penetrate through the protection system and strike the structure. Depending on the location of the down-coming stepped leader, some of these low-current flashes will also be captured by the lightning protection system.

One can utilize the rolling sphere method to determine the volume of space protected by a structure and the equivalent protective angle for the cone of protection. This is illustrated in Figure 4.8a. The shaded region shows the volume of space protected by the conductor (only one side of the protected space is shown). Note that for a given rolling sphere radius, the equivalent cone angle varies with the height of the structure. For a rolling sphere with a 20-m radius (corresponding to level I protection), the equivalent cone angle is about 45° for a 10-m-high structure and 25° for a

Table 4.1 Positioning of air terminals according to the protection levels defined by the IEC standards [5–7]

Protection level	Critical minimum prospective return stroke peak current (kA)	Efficiency of protection (%)	Rolling sphere method: sphere radius, R (m)	Protective angle method for different heights of terminals: protective angle, α (°)				Mesh method: maximum distance, D (m)
				20	30	45	60	
I	3	99	20	25	*	*	*	5
II	8	97	30	35	25	*	*	10
III	10	91	45	45	35	25	*	15
IV	16	84	60	55	45	35	25	20

*Not defined.

174   Lightning Protection

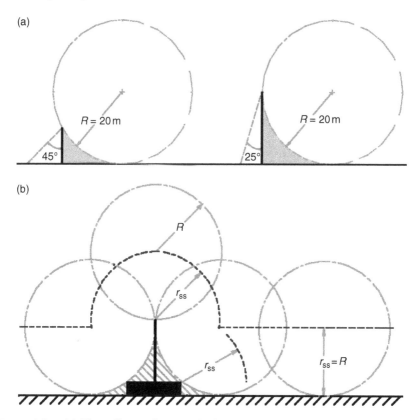

Figure 4.8 (a) The rolling sphere method can be used to estimate the angle of protection offered by a vertical conductor. (b) The critical contour in space that will decide whether the down-coming leader will strike the vertical lightning rod or ground. If the down-coming leader intercepts the semicircular arc marked by a dotted line symmetrically located around and over the lightning rod, then the leader will be intercepted by the rod. If the down-coming stepped leader intercepts the horizontal lines (again marked by dotted lines) then it will be attracted to ground.

20-m-high structure. As one can see from Table 4.1, for a given height of the structure this angle increases when moving from level I to IV. In addition to the volume of space protected by a vertical conductor as dictated by the rolling sphere method, Figure 4.8b also shows the critical contour in space that will decide whether the down-coming leader will strike the conductor or the ground. If the down-coming leader intercepts the semicircular arc marked by a dotted line symmetrically located around and over the conductor, then the leader will be intercepted by the conductor. If the down-coming stepped leader intercepts the horizontal lines (again marked by dotted lines), then it will be attracted to the ground.

Despite the widespread use of the rolling sphere method, there are several factors that call for modifications to its present form [19]. First, the rolling sphere method predicts that the probability of a lightning strike is the same irrespective of whether it strikes flat ground or a flat surface, a sharp point, an edge or a corner of a structure. However, field observations of buildings struck by lightning in Malaysia and Singapore [20,21] have shown that nearly all observed strikes, i.e. more than 90 per cent of the observed cases, terminate on sharp points or protruding corners. Only a few lightning strikes occurred to exposed horizontal or slanting edges (less than 5 per cent) and to elevated vertical edges (less than 2 per cent). Second, the radius of the rolling sphere used in the standard is obtained from a gross oversimplification of the physical nature of the lightning discharge. The magnitude of the radius in current use is actually a product of different compromises made by standardization committees [22]. At the time when the rolling sphere method as it is being used today was created [16,17], the radius of the sphere was taken directly from the striking distance derived and 'calibrated' for power transmission lines according to EGM theory [13]. Owing to the lack of data available at that time, this extension of EGM theory to other structures was done without any further validation. However, the leader progression models proposed to study the interception of lightning flashes with structures (introduced in Section 4.4) have shown that the attachment of lightning flashes to grounded structures depends not only on the prospective return stroke peak current but also on the geometry of the structure to be protected. Thus the assumption that the radius of the rolling sphere is only a function of the prospective return stroke peak current and is independent of the geometry of the structure may lead to serious errors in some situations.

4.1.4 The mesh method

In 1838 Maxwell suggested that the installation of a lightning protection system increases the occurrence of lightning strikes to a building, and the best procedure to

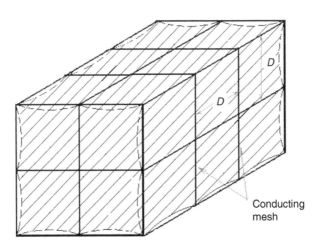

Figure 4.9 Sketch of the mesh method for the design of external lightning protection systems

176 Lightning Protection

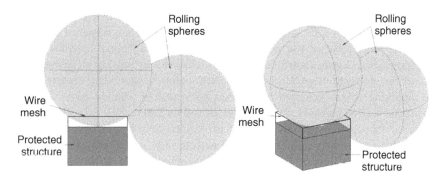

Figure 4.10   *The rolling sphere method predicts that the lightning can strike the building if the mesh is laid directly on the building. In order to avoid such incidences there should be a certain clearance distance between the mesh and the protected structure, as shown.*

protect a building from lightning strikes is to encase the building in a Faraday cage [3]. Even though the statement that the lightning protection system will increase the number of strikes to a building is not correct, the Faraday cage concept is the best procedure to protect a building from lightning strikes. However, because encasing a building completely with a metal cage is not practical, one can encase the building within a conducting mesh to achieve a practical Faraday cage. This method is called a mesh method and is illustrated in Figure 4.9. The protected zone given by the mesh is shown by the dashed zone. The mesh width D suggested by the IEC standards for the four protection levels is documented in Table 4.1. Note that the rolling sphere method predicts that the lightning can strike the building if the mesh is laid directly on the building. In order to avoid such incidences there should be a certain clearance distance between the mesh and the protected structure, as shown in Figure 4.10. This clearance distance decreases with increasing level of protection.

4.2   Striking distance to flat ground

Based on the information gathered from long sparks, the striking distance of a stepped leader to flat ground can be obtained by estimating the separation between the leader tip and ground when the average potential gradient between them is equal to the potential gradient of the streamer channels. As mentioned in Section 4.1.2 for negative streamers, this potential gradient is about $1\,000\,\text{kV}\,\text{m}^{-1}$, and $500\,\text{kV}\,\text{m}^{-1}$ for positive streamers. However, in the case of a negative stepped leader the streamer system may consists of streamer channels of both polarities, negative emanating from the tip of the leader and positive from a space stem [23]. Thus, the average potential gradient of a streamer region of negative stepped leaders may lie somewhere between 500 and $1\,000\,\text{kV}\,\text{m}^{-1}$.

Once the average potential gradient between the tip of the leader and the ground at striking distance is specified, the only other parameter necessary to estimate the

4.2.1 Golde

In his derivation, Golde [24] assumed that the line charge density ρ_s on the vertical stepped leader channel decreases exponentially with increasing height above ground,

$$\rho_s = \rho_{s0} e^{-z/\lambda} \tag{4.1}$$

where ρ_{s0} is the value of ρ_s at $z = 0$ and λ is the decay height constant. Golde estimated that $\lambda = 1\,000$ m. The total charge on the leader channel is given by

$$Q = \rho_{s0}\lambda[1 - e^{-H/\lambda}] \tag{4.2}$$

where H is the total length of the channel. In calculations, Golde assumed that $H = 2.5 \times 10^3$ m. Moreover, making several arguments concerning the possible length of striking distance corresponding to a typical stepped leader, Golde concluded that a stepped leader that gives rise to a return stroke of 25 kA is associated with a stepped leader charge of about 1 C. Furthermore, he assumed that the return stroke peak current increases linearly with increasing leader charge,

$$I_p = kQ \tag{4.3}$$

where I_p is the return stroke peak current in kA, Q is in C and $k = 25$ kA C^{-1}. (Note that in another study Golde [25] suggested $k = 20$ kA C^{-1}.) Golde did not give any justification for the assumed linear relationship in equation (4.3) between the return stroke peak current and stepped leader charge. Combining equations (4.1), (4.2) and (4.3) one obtains

$$\rho_{s0} = 4.36 \times 10^{-5} I_p \tag{4.4}$$

Equations (4.1) and (4.4) completely define the charge distribution along a stepped leader channel associated with a prospective return stroke peak current of I_p. The striking distance to flat ground obtained from this charge distribution is shown by curve a in Figure 4.11. In this calculation the average potential gradient between the leader tip and ground at striking distance is assumed to be 500 kV m^{-1}.

4.2.2 Eriksson [26]

As outlined in Chapter 2, Berger [27] found a relatively strong correlation between the first return stroke current peak I_p and the charge brought to ground within 2 ms from the beginning of the return stroke (termed the impulse charge), Q_{im}. The relation can

178 Lightning Protection

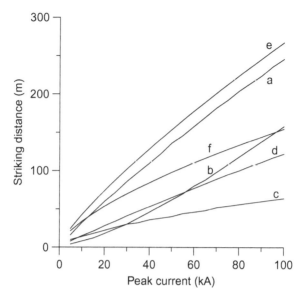

Figure 4.11 *Striking distance to flat ground proposed by different authors: (a) Golde [3], (b) Eriksson [26], (c) Dellera and Garbagnati [28], (d) Cooray and colleagues [31], (e) Armstrong and Whitehead [13] and (f) equation (4.18)*

be represented by the equation

$$I_p = 10.6 Q_{im}^{0.7} \quad (4.5)$$

According to equation (4.5), a 25-kA peak current corresponds to a stepped-leader charge of 3.3 C. Based on Golde's [25] suggestion that only the charge located on the lower portions of the leader channel is related to the peak current (i.e. a 25 kA peak current corresponds to a stepped leader charge of 1 C) and after comparing some of the measured striking distances with analytical results, Eriksson modified the above relationship to

$$I_p = 29.4 Q^{0.7} \quad (4.6)$$

where I_p is in kA and Q in C. Eriksson assumed that the charge is distributed linearly along a vertical leader channel of length 5 km. When this assumption is combined with equation (4.6) one obtains the following relationship between the charge per unit length on the leader channel at ground level:

$$\rho_{s0} = 3.2 \times 10^{-6} I_p^{1.43} \quad (4.7)$$

Recall that the charge density decreases linearly along the channel and it decreases to zero at a height of 5 km. The striking distance obtained from this charge distribution by assuming 500 kV m^{-1} for the potential gradient is shown by curve *b* in Figure 4.11.

4.2.3 Dellera and Garbagnati [28]

In some of the first return stroke currents measured by Berger [27,29] and Vogelsanger one can observe a secondary peak (or a change in slope) appearing in the waveform after a few tens of microseconds from the beginning of the waveform. The time of occurrence of this secondary peak may change from one stroke to another. Dellera and Garbagnati assumed that this subsidiary peak is associated with a return stroke current reflection from the upper end of the leader channel. It is important to note that this subsidiary peak is probably generated by the neutralization of a branch in the stepped leader channel (i.e. a branch component) and may not have any connection to the termination of the lightning channel in the cloud. They selected current waveforms exhibiting this secondary peak from different studies and calculated the charge associated with these current waveforms up to this subsidiary peak (or the change in slope). They assumed that this is the charge originally located on the leader channel. In order to evaluate the charge distribution along the leader channel they needed information concerning the time taken by the return stroke to reach the assumed point of reflection and the form of the charge distribution. The former was obtained by assuming that the return stroke speed is a function of peak current and is given by the equation derived by Wagner [30]. The charge distribution along the leader channel was assumed to be uniform. From their analysis, Dellera and Garbagnati obtained the following relationship between ρ_{s0} and I_p:

$$\rho_{s0} = 3.8 \times 10^{-5} I_p^{0.68} \tag{4.8}$$

where ρ_{s0} is in C m^{-1} and I_p is in kA. The striking distance obtained from this charge distribution by assuming 500 kV m^{-1} for the potential gradient is given by curve *c* in Figure 4.11.

4.2.4 Cooray and colleagues

Cooray and colleagues [31] measured the charge brought down by first return strokes during the first 100 μs. They assumed that this charge is the sum of the positive charge that is necessary to neutralize the negative charge stored on the leader channel and the additional positive charge induced on the channel after the return stroke. Utilizing this information in a model that treats the leader channel as a finitely conducting channel, they studied how the charge distribution on the stepped leader channel varies as it propagates towards the ground. According to this study, the charge distribution along the stepped leader channel when its tip is z_0 metres above the ground is given by

$$\rho(\xi) = a_0\left(1 - \frac{\xi}{H-z_0}\right)G(z_0)I_p + I_p\frac{(a+b\xi)}{(1+c\xi+d\xi^2)}J(z_0), \quad z_0 \geq 10\,\text{m} \quad (4.9)$$

$$G(z_0) = 1 - (z_0/H) \quad (4.10)$$

$$J(z_0) = 0.3\alpha + 0.7\beta \quad (4.11)$$

$$\alpha = e^{-(z_0-10)/75} \quad (4.12)$$

$$\beta = \left(1 - \frac{z_0}{H}\right) \quad (4.13)$$

In these equations H is the height of the channel, $\rho(\xi)$ is the charge per unit length (in C/m), ξ is the distance along the leader channel (in metres) with origin at the leader tip, I_p is the return stroke peak current in kA, $a_0 = 1.48 \times 10^{-5}$, $a = 4.86 \times 10^{-5}$, $b = 3.91 \times 10^{-6}$, $c = 0.52$ and $d = 3.73 \times 10^{-3}$. The striking distance obtained from this charge distribution, again by assuming that the potential gradient in the streamer region is 500 kV m^{-1}, is shown by curve d in Figure 4.11.

4.2.5 Armstrong and Whitehead

Instead of utilizing the charge distribution on the leader channel, Armstrong and Whitehead [13] utilized experimental data obtained from the laboratory together with theory to derive an expression for the striking distance. First they assumed that the striking distance r_s can be expressed in the form

$$r_s = aI_p^b \quad (4.14)$$

where I_p is the return stroke peak current in kA and a and b are constants. Second, they used the return stroke model of Wagner [30] to derive the potential of the downward leader channel V_s in MV as a function of the prospective return stroke current I_p in kA. From the analysis they obtained

$$V_s = 3.7I_p^{0.66} \quad (4.15)$$

Analysis of the laboratory data of the rod–rod gap configuration of lengths up to 5 m available to the authors showed that the breakdown voltage of the gap is related to the gap length s by the equation

$$s = 1.4V_s^{1.2} \quad (4.16)$$

They assumed that the striking distance is related to the voltage of the stepped leader by an identical relationship. Utilizing this relationship in equation (4.15), they obtained the following well known relationship between the striking distance and the prospective return stroke current:

$$r_s = 6.72 I_p^{0.8} \tag{4.17}$$

This relationship is depicted in Figure 4.11 by curve e.

Small changes to coefficients a and b in equation (4.14) have also been suggested by Brown and Whitehead [32], Gilman and Whitehead [33], Love [34] and Whitehead [35], based on the refinement of equation (4.15). Other values for these coefficients have also been proposed in different IEEE standards [36,37]. The values of the parameters a and b in equation (4.14) proposed by the different authors are shown in Table 4.2.

Note that according to these authors equation (4.14) defines the striking distance associated with the structure (or power line conductors). The striking distance to flat ground is obtained by multiplying the right-hand side of equation (4.14) by the proportionality constant K_{sg} (also shown in Table 4.2).

Note that in the case of rod–rod gaps the final jump condition is achieved when the streamers from leaders of opposite polarity emanating from the two rods meet each other. The distance s in equation (4.16) is therefore somewhat larger than the final jump condition (i.e. the distance between the two tips of leaders when the two streamer fronts meet each other). However, because the results are based on gap lengths of a few metres, where the leader development is not significant, equation (4.17) may still be used to approximate the striking distance to flat ground (i.e. $K_{sg} = 1$) provided that Wagner's return stroke model is capable of generating the correct relationship between the peak return stroke current and leader potential.

On the other hand, the potential of the leader channel obtained in equation (4.15) can be directly converted to striking distance to flat ground by finding the height of the leader tip from the ground when the average potential gradient in the gap is

Table 4.2 Coefficients of the striking distance according to expression (4.14)

Source	a	b	$K_{sg}^\dagger$
Armstrong and Whitehead [13]	6.7	0.8	0.9
Brown and Whitehead [32]	7.1	0.75	0.9
Gilman and Whitehead [33]	6.7	0.8	1
	6.0*	0.8*	1*
Love [34]	10	0.65	1
Whitehead CIGRE survey [35]	9.4	0.67	1
	8.5*	0.67*	1*
IEEE Working Group [36]	8	0.65	0.64–1
IEEE Working Group [37]	10	0.65	0.55–1

*Recommended for design of new lines.
†K_{sg}, proportionality constant.

500 kV m^{-1}. The result of this exercise will be

$$r_s = 7.4 I_p^{0.66} \tag{4.18}$$

This curve is also shown by curve f in Figure 4.11. Note that the exponents in equations (4.17) and (4.18) are different.

4.3 Striking distance to elevated structures

According to the definition of striking distance, it is the maximum distance at which the conditions necessary for electrical breakdown are established between the stepped leader and the grounded structure. In the case of flat ground this can be estimated without much difficulty as illustrated in the last section. However, in the case of structures, the final attachment of the stepped leader is mediated by upward-connecting leaders emanating from the structures. To take this into account, Golde [3] defined the striking distance of tall structures as the distance of the leader tip from the structure when a connecting leader is initiated from the structure. This definition assumes that once an upward leader is initiated the conditions necessary for the attachment of the stepped leader to the structure are fulfilled. Of course there are situations in which a launch of an upward-connecting leader may not result in an attachment. For example, in the presence of a down-coming stepped leader, several connecting leaders may be issued either from several structures at ground level or from different parts of the same structure. Only one of these leaders may succeed in making the connection with the down-coming stepped leader.

The striking distance of elevated structures defined above (note that the above definition differs to the way in which the striking distance to flat ground is defined) is a function not only of the charge on the leader channel, and hence the peak return stroke current, but also of the dimension of the structure and the angle of approach of the stepped leader with respect to the structure. Several researchers have utilized the concept of striking distance of tall structures as defined above to evaluate how this parameter varies as a function of height of the structure. It is important to note that when calculating the striking distance using this concept, one has to estimate when a connecting leader is issued from the structure under the influence of the electric field generated by the down-coming stepped leader. As we will see in Section 4.3.2, there are many models that attempt to predict the background electric field necessary for the initiation of upward leaders. Different models predict different conditions for the inception of upward leaders. Thus, depending on the connecting leader inception model used, different researchers may obtain different values for striking distances. Indeed, the differences in the results obtained by different researchers are partly due to the differences in these inception models and partly due to the different assumptions made concerning the charge distribution of the stepped leader channel. It is also important to point out that the inception and subsequent propagation of the connecting leader is mediated by the electric field configuration in the vicinity of the top of the structure. In reality, this electric field configuration depends not only on the height of the structure, but also on the other dimensions of the structure. However, many researchers plot their results as a function of structure height only.

As one can infer directly, the reliability of the inception models of connecting leaders is vital for the accurate determination of the striking distance of tall structures. In Section 4.4 this point will be considered. However, before doing that it is necessary to provide a brief review of the characteristic of positive leaders as observed in long laboratory sparks and in nature.

4.3.1 Positive leader discharges

Les Renardières group [11] have identified the main features of positive leader discharges by using electrical measurements and time-resolved photography. The basic features of long sparks as documented by them are schematically shown in Figure 4.12. When the electric field on the surface of the positive electrode is high enough to initiate streamers, a condition known as streamer inception, a first corona burst is created (stage *a*). In the first stage of development of the corona burst, a streamer starts propagating from the electrode and then splits into many branches, forming a conical volume [11]. These branched streamers usually develop from a common stem. For small-diameter electrodes, the space charge injected into the gap by the first corona distorts and reduces the electric field close to the electrode, giving rise to a dark period where no streamers are created. The duration of the dark period depends upon the injected charge and the rate of increase of the voltage applied to the electrode.

As the external applied voltage grows, the total electric field on the surface of the electrode increases until a second corona burst is initiated (stage *b*). Depending upon the energy supplied by the streamers, the temperature of the stem of the second corona burst can reach a critical value around 1 500 K [38], which leads to the creation of the first leader segment. This transition from streamer to leader (stage *c*), called the unstable leader inception, takes place if the total charge in the second or successive corona bursts is equal to or larger than about 1 μC [38]. This value corresponds to the critical charge required to thermalize the stem of a corona discharge, after at

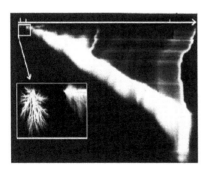 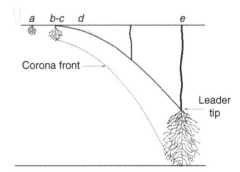

Figure 4.12   Streak image and sketch of the development of positive leaders in laboratory long air gaps (adapted from Reference 71)

least one corona burst (first corona) has occurred. However, this condition is not sufficient to guarantee the stable propagation of the newly created leader. Only when the energy that is being supplied by the streamer discharges emanating from the leader tip is high enough to sustain the thermalization and the creation of new leader segments does the leader start to propagate continuously (stage d in Figure 4.12), with corona streamers developing at its tip (Figure 4.13). This condition defines the stable leader inception.

In the laboratory, it has been observed that positive leaders propagate continuously with an approximately constant velocity [11]. The estimated velocity of positive leaders in the laboratory ranges between 1×10^4 and 3×10^4 m s^{-1}. The leader velocity has been correlated to the leader current through a proportionality term, which represents the charge per unit length required to thermalize a leader segment [11,39,40]. This parameter, which depends mainly on the rise time of the applied voltage (and hence on the rise time of the electric field) and the absolute humidity, has been estimated to lie somewhere between 20 and 50 μC m^{-1} [38].

The last stage of the leader propagation is the 'final jump' (stage e). This takes place when the streamers of the leader corona reach the opposite electrode and is characterized by the creation of a conducting path in air that short circuits the gap, leading to collapse of the voltage and a rapid increase in the current.

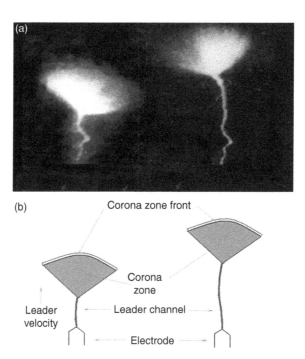

Figure 4.13 Details of the structure of positive leader discharges. (a) Frame image (from Reference 71); (b) sketch of the leader channel and the corona zone at its tip.

Even though positive leaders in nature are not easily detectable with streak photographs [41], optical measurements of upward leaders triggered by tall towers (without a descending downward leader) are available in the literature. The estimated velocity of the observed upward lightning leaders, which sometimes exhibit a kind of stepping, range between 4×10^4 and 1×10^6 m s^{-1} [42,43]. Later measurements [44,45] gave estimates of the velocity of upward leaders ranging between 6×10^4 and 1.4×10^6 m s^{-1}. However, all the upward leaders that were observed propagated continuously without any stepped motion [45].

The first measurements of upward connecting leaders generated under the influence of downward stepped leaders reported in the literature appeared in 1990 [46]. The average propagation velocity of the observed upward connecting leaders initiated from a tall tower ranged from 0.8×10^5 to 2.7×10^5 m s^{-1}. Unfortunately, it is not possible to distinguish between positive or negative upward connecting leaders in the dataset reported in Reference 46, because the polarity of the current was not reported. Recently, the propagation of an upward connecting leader initiated from a tower under the influence of a branched descending stepped leader was detected using a high-speed camera in the United States (see Saba M. High speed video measurements of an upward connecting positive leader, personal communication, 2007). The minimum detected velocity of the upward connecting leader was 2.7×10^4 m s^{-1} after inception and then the velocity gradually increased to a value close than 2.5×10^5 m s^{-1} just before the connection with the downward leader. The descending stepped leader propagated with an average velocity of 2.5×10^5 m s^{-1}. No record of the upward leader current was available for this example.

Rocket-triggered lightning experiments have also been a good source of information on upward positive leaders generated under natural conditions [47]. Experiments with triggered lightning can provide information concerning upward leaders initiated either by the thundercloud electric field in classical triggered lightning or by the descent of a triggered downward stepped leader in altitude-triggered lightning (see also Chapter 3). In the case of classical triggering, an upward positive leader is launched from the tip of a rocket trailing a thin grounded wire under an active thundercloud [48]. The inception of the self-propagating upward leader usually takes place when the rocket is at an altitude of 200 to 300 m [47]. As the rocket ascends, several current pulses are usually measured before the inception of the self-propagating upward leader [49–51]. These current pulses are attributed to unstable aborted leaders (precursors), which stop propagating after a few metres, launched from the tip of the triggering rocket. The charge associated with these individual pulses has been estimated to be in the order of several tens of microcoulombs [49]. After these pulses, the current gradually damps out and merges into a slowly varying current of a few amperes [51]. As the upward leader keeps moving towards the thundercloud, the measured current can reach a few hundred amperes [49,52]. Streak images in classical rocket-triggered lightning experiments report upward leader velocities ranging between 2×10^4 and 1×10^5 m s^{-1} [50]. In some experiments [50,53], the triggered upward leader appears to propagate with discontinuous luminosity of its tip, which has been interpreted as stepping [48]. In some other experiments [52], the upward positive leaders appear to propagate continuously.

In the triggering technique called altitude triggering [48], the ascending rocket trails a thin wire that is not grounded. Usually, this floating wire is connected to an insulating wire followed by a grounded wire. Similar to classical triggering, an upward positive leader is launched from the rocket tip by the ambient thundercloud electric field. Some microseconds later, an upward connecting positive leader is also initiated from the grounded wire under the influence of a downward negative leader triggered from the bottom end of the floating wire. To date, few experiments with altitude rocket triggered lightning have been reported in the literature [49,50]. It has been reported that a small current of a few amperes with superimposed pulses starts to flow in the ground wire when the upward connecting leader is incepted [49]. However, as the descent of the triggered downward stepped leader continues, the upward leader current increases continuously with superimposed pulses [49]. The presence of these pulses suggests some stepping behaviour in the upward moving leader. Because upward connecting leaders created in altitude rocket-triggered lightning are very faint, no streak image or velocity estimations are available at present.

4.3.2 Leader inception models

Because leader inception from the high-voltage electrode is a necessary condition for the breakdown of long gaps when stressed by switching impulses, different models have been proposed to predict the voltages and hence the electric fields necessary for leader inception in a given electrode configuration and applied voltage waveshape. These models were soon used and extrapolated to study the initiation of upward lightning leaders under natural conditions. Several other models were also developed based on either the physics of the streamer-to-leader transition or the observations made in laboratory long sparks for the sole purpose of tackling the problem of leader inception from grounded structures. In the sections to follow, the assumptions of these models used in lightning attachment studies are presented and discussed.

4.3.2.1 The critical radius concept

Over the last few decades, many experiments have been conducted in high-voltage laboratories to extract information to understand the physics of the streamer-to-leader transition and the subsequent electrical breakdown in long laboratory gaps. Because positive electrical breakdown is of more industrial relevance than negative breakdown, most of the experiments have been conducted with different types of gap configurations where the high-voltage electrode is of positive polarity. These experiments have managed to collect a wealth of information concerning the inception and subsequent propagation in the gap of leader discharges and how electrical breakdown is mediated by it in long gaps. One of the most important results relevant to lightning inception came from careful analysis of breakdown in rod–plane gaps by Carrara and Thione [54] – the critical radius concept. In order to illustrate this concept, consider the data obtained by Carrara and Thione [54], which is reproduced in Figure 4.14. In this experiment positive switching voltages of critical front times were applied to rod–plane gaps and the 50 per cent breakdown voltage of the gap was studied as a function

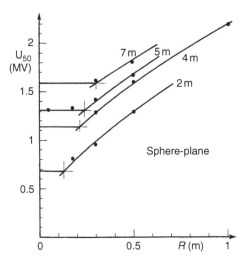

Figure 4.14 Fifty per cent breakdown voltage as a function of the radius of the electrode for different gap spacing as experimentally determined by Carrara and Thione [54]

of the tip radius of the high-voltage electrode. The results show that for a given gap length the breakdown voltage remained the same with increasing electrode radius until a critical radius is reached. Further increase of the radius led to an increase of the breakdown voltage. The radius at which the breakdown voltage starts to increase is named the critical radius. These experiments show that for radii below the critical value the inception of the leader is preceded by one or more corona bursts followed by a dark period. However, at the critical radius, the inception of the leader is immediately followed by the inception of the corona. As one can see from the data, the critical radius is not a constant but increases with increasing gap length. The way in which the critical radius varies as a function of gap length as observed by Carrara and Thione [54] is shown in Figure 4.15. Note that the critical radius initially increases with gap length, but reaches a more or less asymptotic value for large gap lengths. These results show that the initiation of a stable leader requires the attainment of an ionization field, i.e. 3×10^6 V m^{-1}, over a volume bounded by the critical radius. In other words, corona inception alone is not a sufficient criterion for leader inception, but the volume of corona discharge around the conductor should increase to a critical size before a leader is incepted. This study was later extended by Les Renardieres group [11,40] for other voltage wave shapes. It was found that the value of the critical radius changes for different time to crest of the applied voltage.

In the configuration used by Carrara and Thione [54] the rod is located above the ground plane and raised to a high voltage. The positive leader is incepted from the high-voltage electrode. However, in the case of lightning attachment the positive leader is incepted from a grounded rod under the influence of the down-coming stepped leader. In order to make the laboratory experiments more realistic to the

188 Lightning Protection

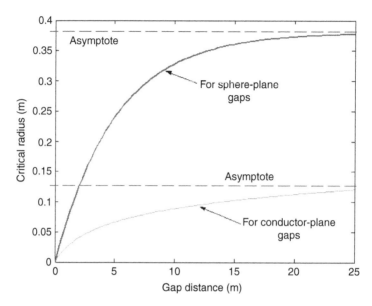

Figure 4.15 The variation of critical radius for sphere–plane and conductor–plane gaps as a function of gap length [54]

problem under consideration, researchers have conducted experiments with inverted rod–plane gaps. Here, the rod is grounded and the plane located above it is raised to a negative high voltage so that a positive leader is initiated from the grounded rod. From such experiments Gary and colleagues [55] found 0.1 m to be the critical radius for 3-m gaps, whereas Carrara and Thione came up with 0.17 m for the same gap length. A similar experiment was conducted by Bernardi and colleagues [56] in which two electrode configurations were tested: a sphere and a horizontal conductor with different radii and height from ground. In this case, the asymptotic value of the critical radius was ~0.28 m for both configurations. A similar experiment was conducted recently by D'Alessandro and colleagues [57] with 5-m gaps and obtained 0.27 m as the critical radius. Indeed, it is difficult to understand why the leader inception should differ in the rod–plane gaps depending on whether the rod is at high voltage, as in the case of Carrara and Thione [54], or when the rod is grounded, as in the case of Bernardi and colleagues [56]. If the electric field distribution is the same in both configurations one should observe the same results in the two configurations. The differences observed are probably caused by the different electric field configurations in the different experiments.

Eriksson was the first to use the critical radius concept in lightning studies [26]. Since then, the critical radius concept has been widely used in the literature to compute the leader inception conditions in rods, masts, power lines [28,58–60] and buildings [61]. In these studies any sharp point on a structure such as the tips of lightning rods or corners and edges of a building are rounded off to the critical radius and it is assumed that a stable leader is initiated when the electric field on the surface is equal to the critical corona inception electric field, which is about 3 MV m^{-1} at atmospheric pressure.

It is important to note that the critical radius depends both on the gap length and the voltage waveshape. Moreover, the documented critical radius is based on 50 per cent breakdown voltage. That means that there is a large spread in the breakdown voltage of individual application of waveforms and hence in the observed critical radius. Furthermore, in the case of lightning, the electric field to which a grounded rod is exposed during the earthward progression of the stepped leader is very different to a rod exposed to an electric field produced by a switching impulse. This may also produce a change in the critical corona volume necessary for the launch of a connecting leader. In addition the application of the technique itself may lead to different results depending on the shape of the rod. In order to illustrate these facts, the magnitude of the background electric field (assumed to be uniform) necessary for the initiation of leaders from two grounded rods of different shapes is computed for two different values of critical radii. The values selected for the critical radii are 0.36 m [54] and 0.28 m [56]. Both values are utilized in lightning studies. One of the rods had a cylindrical shape with a hemispherical tip and the other rod had the form of a semi-ellipsoid. The results obtained for different rod heights are shown in Figure 4.16. For a given rod, the change in the critical radius from 0.28 to 0.36 m resulted in about 25 per cent difference in the background electric field necessary for the launch of connecting leaders. For a given critical radius, the differences in the rod shapes led to variations as large as 30 to 80 per cent in the background electric field necessary for leader inception. Not withstanding these differences both rod

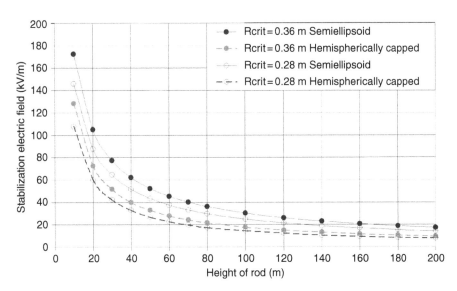

Figure 4.16   *Background leader inception electric fields of a lightning rod computed with the critical radius concept for a semi-ellipsoidal rod and for a hemispherically capped rod*

4.3.2.2 Rizk's generalized leader inception equation

Initially Rizk [62] analysed the leader inception criterion for the case of rod–plane gaps in the laboratory. Subsequently, the results were modified for the case of leader inception from grounded structures under the influence of thunderstorm electric fields. First, let us consider the rod–plane gap in the laboratory. The geometry relevant for the problem under consideration is shown in Figure 4.17.

Rizk assumed that in the configurations of interest the corona inception voltage was smaller than the leader inception voltage from the rod. He also assumed that for the initiation of a stable leader from the rod, the difference between the applied electric field E_{lc} and the electric field generated by the streamer space charge E_{in} at the tip of the streamer stem zone must equal or exceed a certain critical voltage gradient E_c. That is

$$E_{lc} - E_{in} = E_c \tag{4.19}$$

These electric fields were then converted to potentials by assuming that the applied electric field can be represented by that due to a point charge located at a distance r_{eq} from the tip of the streamer zone. Then

$$E_{lc} = U_{lc}/r_{eq} \tag{4.20}$$
$$E_{in} = \alpha U_{in}/r_{eq} \tag{4.21}$$
$$E_c = U_c/r_{eq} \tag{4.22}$$

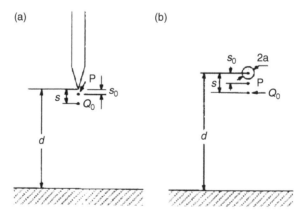

Figure 4.17 Geometry relevant to the calculation of positive leader inception according to Rizk [62] for (a) a rod–plane gap and (b) a conductor–plane gap

where α is the proportionality constant between r_{eq} and the distance from the streamer stem zone tip to the equivalent centre of the space charge region. Equation (4.19) can now be written in terms of potentials as

$$U_{lc} - \alpha U_{in} = U_c \tag{4.23}$$

where U_{lc} is the connecting leader inception voltage, U_{in} the voltage at the tip of the streamer stem zone generated by the space charge under critical condition and U_c the critical potential. The potential created by the space charge can now be divided into two parts as follows:

$$U_{in} = U_{ia} - U_{ib} \tag{4.24}$$

where U_{ia} is the potential produced by a point charge Q_0 located at a distance s from the streamer stem zone and the other part due to induced charges in the ground plane and other conductors. Based on this definition

$$U_{ia} = Q_0/4\pi\varepsilon_0(s - s_0) \tag{4.25}$$

where s_0 is the distance to the streamer tip zone from the tip of the electrode. Now, U_{ib} can be written as

$$U_{ib} = Q_0/4\pi\varepsilon_0 R \tag{4.26}$$

where R is a function that depends on geometry. This parameter can be estimated using numerical techniques such as charge simulation procedure. Using these new definitions equation (4.23) can be written as

$$U_{lc} - \frac{\alpha Q_0}{4\pi\varepsilon_0(s - s_0)} + \frac{\alpha Q_0}{4\pi\varepsilon_0 R} = U_c \tag{4.27}$$

Now it is assumed that the critical charge in the streamer zone at the leader inception is proportional to the voltage necessary for leader inception. Thus

$$Q_0 = cU_{lc} \tag{4.28}$$

where c is a constant which characterizes the discharge. It has dimensions of capacitance and varies with the high voltage electrode geometry.

Combining this with the previous equations one obtains

$$U_{lc} = \frac{U_{c\infty}}{1 + A/R} \tag{4.29}$$

$$A = \frac{\alpha c}{4\pi\varepsilon_0} \bigg/ \left[1 - \frac{\alpha c}{4\pi\varepsilon_0(s - s_0)}\right] \tag{4.30}$$

$$U_{c\infty} = U_c \bigg/ \left[1 - \frac{\alpha c}{4\pi\varepsilon_0(s - s_0)}\right] \tag{4.31}$$

Note that A and $U_{c\infty}$ depend on the configuration of the high-voltage arrangement. In the case of rod–plane gaps

$$U_{ib} = \frac{Q_0}{4\pi\varepsilon_0} \frac{1}{2d} \tag{4.32}$$

and therefore $R = 2d$. Substituting this in the above equation one obtains

$$U_{lc} = \frac{U_{c\infty}}{1 + \frac{A}{2d}} \tag{4.33}$$

Rizk made a comparison of this equation with the experimental data obtained in the laboratory and obtained $U_{c\infty} = 1\,556$ kV and $A = 7.78$ m. Similar analysis conducted for the conductor plane gap resulted in

$$U_{lc} = 2247 \bigg/ \left[1 + \frac{5.15 - 5.49 \ln a}{d \ln(2d/a)}\right] \quad \text{(kV, m)} \tag{4.34}$$

where d is the gap spacing and a is the conductor radius. After this derivation, Rizk proceeded to adapt the equations for the case of leader initiation from a grounded conductor under the influence of a stepped leader. The relevant geometry is shown in Figure 4.18. The main differences between the two cases, as pointed out by Rizk, are the following:

1. The tip of the streamer stem zone is practically at ground potential and not at the applied voltage as in the case of rod–plane gaps.
2. The gap distance d and the height h above ground of the equivalent streamer space charge Q_0 are completely different.

In this case Rizk redefined equation (4.19) by replacing E_{lc} with the induced electric field E_{ic} necessary for the positive leader inception:

$$E_{ic} - E_{in} = E_c \tag{4.35}$$

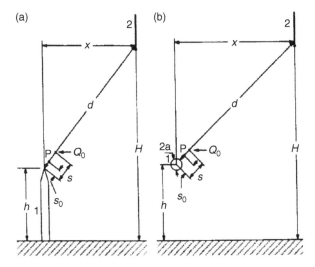

Figure 4.18 Geometry relevant to the calculation of positive leader inception according to Rizk [96], for (a) a lightning rod and (b) a horizontal conductor

The electric field E_{ic} is related to the induced potential U_{ic} in the absence of the lightning conductor through an equivalent radius r_{eq} as in the case for the laboratory gaps. From this it follows that

$$U_{ic} - \alpha U_{in} = U_c \tag{4.36}$$

while the expression for U_{ia} remains the same. Moreover, at the positive leader inception the distance to the descending leader from the tip of the conductor will have practically no influence on the image charge of the equivalent streamer charge. Thus the distance R in the equations corresponding to the laboratory gap will be approximately equal to $2h$, where h is the height of the conductor. Thus the potential generated at the tip of the vertical conductor by the stepped leader at leader inception is given by

$$U_{ic} = \frac{U_{c\infty}}{1 + \frac{A}{2h}} \tag{4.37}$$

where $U_{c\infty}$ and A have the same values as in laboratory discharges. For a conductor–plane gap configuration, Rizk derived the equation

$$U_{ic} = 2247 \bigg/ \left[1 + \frac{5.15 - 5.49 \ln(a)}{h \ln(2h/a)}\right] \quad \text{(kV, m)} \tag{4.38}$$

First, it is important to note that the equations are 'calibrated' using the data on leader inception characteristics of long gaps excited by switching impulses. For this reason it is doubtful whether the results are applicable to lightning conductors exposed to the electric fields generated by stepped leaders. Second, it is necessary to justify the assumption involved in converting the electric field at the tip of the rod to the background electric potential in the absence of the rod using the same r_{eq} as that used in the case of leader inception from high-voltage electrodes in the laboratory.

4.3.2.3 Critical streamer length concept

According to Petrov and Waters [63], the streamer initiated from a given point on a structure under the influence of the electric field generated by the stepped leader must extend to a critical length before an upward leader is initiated from that point. Moreover, the total electric field, i.e. the sum of the ambient electric field and the electric field generated by the structure due to the induced charges on it by the ambient electric field, must exceed a critical electric field over the streamer zone. Based on the results given by Chernov and colleagues [64], the critical streamer length is estimated to be 0.7 m. The critical electric field is taken to be 500 kV m^{-1} for positive streamers and 1 000 kV m^{-1} for negative.

This study was extended by Petrov and D'Alessandro [65] to include the conditions necessary for the continuous propagation of the leader once incepted. Because there is a potential drop along the leader channel, these authors assumed that the external voltage drop generated by the down-coming stepped leader along the leader length should not be less than the drop of the internal potential along the leader channel. Based on this the condition necessary for the propagation of the leader is evaluated as

$$\frac{dU}{dt} = E_l v \tag{4.39}$$

where U is the external voltage drop produced by the background electric field along the leader channel, E_l is the internal electric field of the leader and v is the speed of the leader. Taking an average leader speed of 6×10^4 m s^{-1} and $E_l = 100$ kV m^{-1}, the condition necessary for leader propagation is estimated to be

$$\frac{dU}{dt} > 6 \, \text{kV} \, \mu\text{s}^{-1} \tag{4.40}$$

Under these assumptions, the critical background electric field E_0 required to initiate an upward leader from a structure of height h was obtained as

$$E_0 \approx 697/h^{0.68} \quad (\text{kV m}^{-1}) \tag{4.41}$$

Even though the critical length of the streamer extension necessary for leader inception was assumed to be 0.7 m by Petrov and Waters [63], experiments conducted by the Les Renardières group show that the critical streamer length necessary for leader inception increases with gap length, reaching an asymptotic value of ~3 m for long gaps. For example, Figure 4.19 shows how the critical streamer length

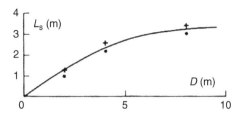

Figure 4.19 The critical streamer length necessary for leader inception as a function of gap length [40]

varies as a function of the gap length as observed in the Les Renardières study. In a study conducted by Akyuz and Cooray [66] the critical streamer length necessary for leader inception was assumed to be 3 m. In their analysis they assumed that the positive streamers extend to the periphery of the region where the background electric field (sum of ambient and the induced by the structure) is larger than 500 kV m^{-1}.

4.3.2.4 Bazelyan and Raizer's empirical leader model

Bazelyan and Raizer [67] proposed in 1996 a model to determine the conditions required to initiate upward leaders from grounded conductors based on semi-empirical expressions derived from experiments in laboratory long gaps. This model considers that the unstable leader inception takes place when the potential drop between the electrode tip and a point about 1 m from it is equal to a critical value ΔU_{cr}, which is assumed to be approximately equal to 400 kV. Because 400 kV m^{-1} is approximately equal to the potential gradient of positive streamers, this criterion assumes that the energy injected in the gap is high enough to thermalize the streamer stem if the streamers produced in front of the electrode extend up to 1 m. After the inception of the first leader segment, it is assumed that a leader with length l_L propagates as far as the potential difference $\Delta U(l_L)$ between the potential of the leader tip $U_L(l_L)$ and the potential produced by the external field $U_0(l_L)$ as the leader tip position continuously increases. In the model, the leader potential $U(l_L)$ is iteratively solved as the leader propagates by using the following set of empirical equations for the leader velocity v_L, current i_L and potential gradient E_L:

$$v_L = a\sqrt{\Delta U(l_L)} \tag{4.42}$$

$$i_L = \left(\frac{2 \cdot \pi \cdot \varepsilon_0}{\ln(l_L/R_{cov})}\right) \cdot \Delta U(l_L) \cdot v_L \tag{4.43}$$

$$E_L = \frac{b}{i_L} \tag{4.44}$$

$$U_L(l_L) = E_L l_L \tag{4.45}$$

where a and b are empirical constants equal to $15 \text{ m s}^{-1} \text{ V}^{-0.5}$ and $3 \times 10^4 \text{ V m}^{-1} \text{ A}^{-1}$, respectively, and ε_0 is the permittivity of vacuum. The parameter R_{cov} is the effective radius of the space charge cover around the leader channel, which is assumed to be approximately equal to the length of the streamer zone. Thus, the effective radius R_{cov} is computed by the ratio between the potential difference $\Delta U(l_L)$ and the streamer potential gradient E_{sc} (assumed to be $\sim$500 kV m^{-1}) [68].

Based on this set of equations, the background external electric field E_0 that is necessary to sustain the propagation of a leader from a grounded electrode of height h can be computed as

$$E_0 \approx 3\,700/h^{0.6} \quad (\text{kV m}^{-1}) \tag{4.46}$$

Because the model is based on laboratory data, it is not clear whether the model can be used to study the inception of positive leaders from structures under the influence of the electric field generated by down-coming stepped leaders.

4.3.2.5 Lalande's stabilization field equation

Lalande [69] used a physical model for leader propagation in long gaps as proposed by Goelian and colleagues [70] and combined it with the thermo-hydrodynamic model of the leader channel as proposed by Gallimberti [38] in order to compute the leader inception condition. In constructing the model he also assumed that there is a constant relationship between the leader velocity and current. Using this model, an analytical expression for the leader inception fields pertinent to lightning rods and horizontal conductors was derived. However, no further analyses were performed for other configurations other than for geometries relevant to rocket-triggered lightning [71]. Although the model, which also takes into account the effect of space charge, could be of use in analysing leader inception from complex structures, several problems and complexities arise when practical cases are analysed following the procedure as presented by Goelian and colleagues [70]. For example, the procedure of calculating the corona charge, which is required in quantifying the streamer-to-leader transition by this method, is only valid for structures with axial symmetry. Moreover, the model requires knowing the number of streamer channels in the streamer zone, which is not known, as an input.

Now let us consider the results obtained by Lalande using this model. Lalande applied the model to vertical rods and horizontal conductors of different height in order to derive an equation for the critical background external field that leads to the stable propagation of upward leaders (called the stabilization field). In the model it was assumed that the background electric field does not vary during the inception process of the leader. This assumption is valid only in the case of upward lightning flashes initiated either from tall towers or from elevated rockets used in lightning triggering under the influence of steady or slowly varying thundercloud electric fields. Based on this analysis, Lalande obtained the following equation for the background electric field necessary to initiate leaders from grounded structures (i.e. the stabilization electric field E_0):

$$E_0 \geq \frac{240}{1 + \frac{h}{10}} + 12 \quad (\text{kV m}^{-1}) \tag{4.47}$$

where h is the height of the structure. In a later study, however, Lalande [71] proposed the following equation, which is different from the above, for the stabilization electric field:

$$E_0 \geq \frac{306.7}{1 + \frac{h}{6.1}} + \frac{21.6}{1 + \frac{h}{132.7}} \quad (\text{kV m}^{-1}) \tag{4.48}$$

Unfortunately, no details were given in Reference 71 as to the modifications necessary both in physics and in mathematics to change the results from equations (4.47) to (4.48).

4.3.2.6 The self-consistent leader inception model of Becerra and Cooray

The model introduced by Becerra and Cooray [72,73] consists of two parts, namely static and dynamic leader inception models. In the static leader inception model, the background electric field is assumed to remain constant during the streamer-to-leader transition process, whereas in the dynamic leader inception model the effect of the variation of the background electric field on the streamer-to-leader transition process is included. Thus the static model is suitable for cases in which the background electric field remains constant or is changing slowly. One such typical situation is the upward leader inception from tall grounded structures under the influence of the electric fields produced by thunderclouds. The dynamic model should be used in studies related to the inception of upward leaders driven by the background electric field generated by down-coming stepped leaders. The models can also be applied to study leader inception in laboratory sparks with the static leader inception model applicable to d.c. breakdown and the dynamic one to switching impulse breakdown.

The main steps that are included in the model are the following:

- formation of a streamer corona discharge at the tip of a grounded object (first, second or third corona inception)
- transformation of the stem of the streamer into a thermalized leader channel (unstable leader inception)
- extension of the positive leader and its self-sustained propagation (stable leader inception)

Corona inception is evaluated using the well known streamer inception criterion [38], whereas the transition from streamer to leader is assumed to take place if the total charge in the second or successive corona bursts is equal to or larger than about

198 Lightning Protection

1 μC [38]. The condition for self-propagation of the leader, i.e. stable leader inception, is assumed to be satisfied if the leader continues to accelerate in the background electric field for a distance of at least a few meters.

Static leader inception evaluation
In cases in which the static leader inception model is applicable, the streamer-to-leader transition, i.e. unstable leader inception, is usually accompanied either by the second or third corona burst. This is the case, for example, in towers with tip radii smaller than some tens of centimetres. In these cases, model simulations start with the corona burst that leads to the inception of the unstable leader (second or third corona, and so on). In other words, the simulation neglects any space charge left behind by corona bursts that did not result in the inception of an unstable leader. The procedure proposed by Becerra and Cooray [72] to evaluate the leader inception condition is as follows:

1. For a given background electric field, the background potential distribution U_1 along a vertical line from the tip of a grounded object is computed and it is approximated by a straight line (see Figure 4.20) with slope E_1 and intercept U_0' such that

$$U_1^{(0)}(z) \approx E_1 z + U_0' \qquad (4.49)$$

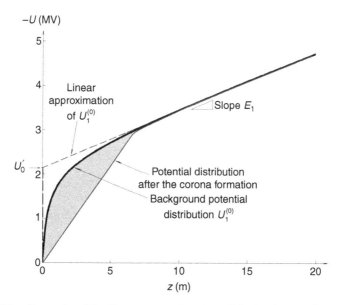

Figure 4.20 Example of the linear approximation of the background potential distribution along a vertical line from the tip of a 100-m-tall tower under the influence of a thundercloud electric field of 20 kV m^{-1}

2. The second (or third) corona charge $\Delta Q^{(0)}$ and length $l_s^{(0)}$ are computed by assuming that the streamers have a constant potential gradient E_{str} across the corona zone. This assumption leads to the following equations:

$$\Delta Q^{(0)} \approx K_Q \cdot \frac{U_0'^2}{2 \cdot (E_{str} - E_1)} \tag{4.50}$$

$$l_s^{(0)} = \frac{U_0}{E_{str} - E_1} \tag{4.51}$$

where K_Q is a geometrical factor that connects the total corona charge ΔQ with the area defined by the potential distribution before and after corona formation. This area is defined in Figure 4.20. If the charge of the second (or third) corona $\Delta Q^{(0)}$ is lower than 1 µC, the unstable leader inception condition is not fulfilled and the analysis stops. Otherwise, an iterative geometrical analysis of the leader propagation starts with $i = 1$ and an assumed initial leader length of $l_L^{(1)}$ as inputs.

3. The potential at the tip of the leader $U_{tip}^{(i)}$ during the current simulation step i is computed with the semi-empirical equation derived by Rizk [62]:

$$U_{tip}^{(i)} = l_L^{(i)} \cdot E_\infty + x_0 \cdot E_\infty \cdot \ln\left[\frac{E_{str}}{E_\infty} - \frac{E_{str} - E_\infty}{E_\infty} \cdot e^{-l_L^{(i)}/x_0}\right] \tag{4.52}$$

where $l_L^{(i)}$ is the leader length at the current simulation step, E_∞ is the final quasi-stationary leader gradient, and x_0 is a constant parameter given by the product $v\theta$, where v is the ascending positive leader speed and θ is the leader time constant.

4. The position and charge of the corona zone in front of the leader tip are calculated as

$$l_s^{(i)} = l_s^{(0)} + \frac{E_{str} \cdot l_L^{(i)} - U_{tip}^{(i)}}{E_{str} - E_1} \tag{4.53}$$

$$\Delta Q^{(i)} \approx K_Q \cdot \left\{\left(E_{str} \cdot \left(l_L^{(i)} - l_L^{(i-1)}\right) + U_{tip}^{(i-1)} - U_{tip}^{(i)}\right) \cdot \left(l_s^{(i-1)} - l_L^{(i)}\right)\right\} \tag{4.54}$$

5. By using the relation between the leader velocity and the current proposed by Gallimberti [38], the leader advancement distance $\Delta l_L^{(i)}$ and the new leader length $l_L^{(i+1)}$ are evaluated as follows:

$$\Delta l_L^{(i)} = \Delta Q^{(i)}/q_L \tag{4.55}$$

$$l_L^{(i+1)} = l_L^{(i)} + \Delta l_L^{(i)} \tag{4.56}$$

where q_L is the charge per unit length necessary to realize the transformation of the streamer corona stem located in the active region in front of the already formed leader channel into a new leader segment.

6. If the leader length $l_L^{(i+1)}$ reaches a maximum value l_{max}, the leader inception condition is fulfilled for the thundercloud electric field under consideration. If the leader advancement distance $l_L^{(i)}$ starts decreasing after several steps, this suggests that the leader will stop propagating eventually and therefore the stable leader inception condition is not reached. Otherwise, the simulation of the leader propagation continues by going back to step 3. A typical value of l_{max} equal to 2 m was observed to be long enough to define the stable propagation of an upward leader when space charge pockets are not considered [72].

The values of the parameters used in equations (4.49) to (4.56) are tabulated in Table 4.3 and a detailed discussion of them can be found in Reference 72.

The predictions of the model for static leader inception are in excellent agreement with the results of the triggered lightning experiment of Willett and colleagues [51] when the space charge left behind the rocket by aborted leaders is included.

Because the model takes into account the effect of space charge, it has also been implemented to evaluate the influence of the space charge layer on the initiation of upward leaders from grounded tall objects under thunderstorms. As discussed in Chapter 2, the space charge layer created from irregularities on the ground surface distorts the ambient electric field and therefore may influence the conditions necessary for the initiation of upward leaders [74].

To make a thorough analysis of the effect of the space charge layer on the inception of upward leaders, Becerra and Cooray [75] considered the electric field measurements performed by Soula and colleagues [76] during a thunderstorm at the Kennedy Space Center in 1989. Using the electric field measured by these authors at 600 m above ground as input, Becerra and Cooray [75] simulated the development of the space charge layer at ground level during the growth of the thunderstorm. Using the simulation, the space charge profiles at the moment of four triggered lightning flashes to ground are obtained, and these profiles in turn are used to estimate the thundercloud

Table 4.3 Parameters used for the evaluation of the leader inception condition

Sym	Description	Value	Units
$l_L^{(1)}$	Initial leader length	5×10^{-2}	m
E_{str}	Positive streamer gradient	450	kV m^{-1}
E_∞	Final quasi-stationary leader gradient	30	kV m^{-1}
x_0	Constant given by the ascending positive leader speed and the leader time constant	0.75	m
q_L	Charge per unit length necessary to thermal transition	65×10^{-6}	C m^{-1}
K_Q	Geometrical constant that correlates the potential distribution and the charge in the corona zone	4×10^{-11}	C V^{-1} m^{-1}

electric fields required to initiate upward lightning leaders from tall towers in the presence of the space charge layers. In the calculation, different values for the neutral aerosol concentration and the corona current density of the site were used to study the effect of these parameters on the space charge layer profiles and the thundercloud electric fields necessary for leader inception. The aerosol concentration is important in the study because it controls the mobility of ions in the background electric field. It is important to note that the neutral aerosol particle concentration changes from typical values of about 5×10^9 particles m^{-3} for clean rural zones to values exceeding 1×10^{11} particles m^{-3} for urban, polluted places [77]. The corona current density at ground level depends on the type of vegetation and surface roughness of the place under consideration [78].

The results of the simulation show that the space charge can shield the grounded towers from the background electric field, thus affecting the background electric fields necessary for the initiation of upward positive leaders from these towers. In particular, the neutral aerosol density is the factor that predominantly influences the space charge layer shielding and, consequently, the leader inception thundercloud electric field. Figure 4.21 shows an example of the estimated critical thundercloud electric field required to initiate upward leaders from tall towers as a function of their height, taking into account different neutral aerosol particle concentrations. For comparison purposes, the critical thundercloud fields necessary for leader inception

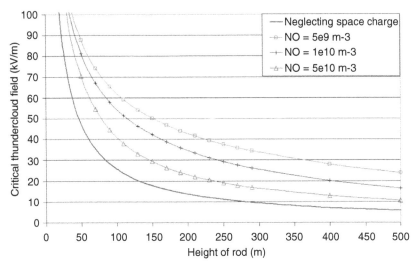

Figure 4.21   *Critical thundercloud electric field $E_{cloud}^{(crit)}$ required to initiate upward leaders as a function of the tower height, for different initial aerosol particle concentrations: 5×10^9 m^{-3} (squares), 1×10^{10} m^{-3} (crosses), 5×10^{10} m^{-3} (triangles). The critical leader inception electric field computed by neglecting the space charge layer is also shown (line with no symbols).*

without the space charge layer are also shown. As one can see from this figure, the critical thundercloud electric field necessary for leader inception is significantly influenced by the neutral aerosol particle concentration at the site. In the case shown in Figure 4.21, observe that the critical cloud electric fields required to initiate upward leaders from a grounded tower of a given height are larger when the neutral aerosol particle concentration of the site is low. For instance, the critical cloud electric field computed for a 200-m-high tower in a site with a neutral aerosol particle concentration of 5×10^9 particles m^{-3} (corresponding to a clean rural zone) is $\sim$43 kV m^{-1}, but for a location with a high aerosol particle concentration of 5×10^{10} particles m^{-3} (moderately polluted urban area), the critical field is only $\sim$23 kV m^{-1}. In the same way, the minimum height of a tower from which an upward lightning leader is initiated under the influence of a given thundercloud electric field decreases as the aerosol particle concentration of the site increases. For example, a thundercloud electric field of 50 kV m^{-1} can initiate upward leaders from towers higher than 150, 115 and 78 m, respectively, in sites with neutral aerosol particle concentrations of 5×10^9, 1×10^{10} and 5×10^{10} particles m^{-3}, respectively. These results clearly show that the initiation of upward lightning leaders from towers (or other tall grounded structures such as transmission lines, wind mills, and so on) depends not only on the height of the tower and the thundercloud electric field, but also on the neutral aerosol particle concentration of the site.

These results show that relatively lower thundercloud electric fields are required to initiate upward leaders from tall towers in polluted sites (with high neutral aerosol particle concentration) than in clean rural zones (with low neutral aerosol particle concentration). This result also applies to coastal areas where aerosol particles generated over the sea from wave breaking and so on are carried inland by wind. In coastal regions one can expect the height of towers that can launch upward flashes in a given thundercloud electric field to be lower than the height of towers in other regions capable of launching upward leaders under the same thundercloud electric field. This is the case because the aerosol particles will retard the upward growth of the space charge layer. This indeed seems to be the case at sites located on the coast of Japan [79]. Field observations in Japan suggest that upward flashes occur even from structures of moderate height (lower than 50 m height and located on flat terrain) in winter [80]. This result is due partly, of course, to the high electric fields produced by winter thunderclouds, for which the charge centres are located closer to the ground than in those in summer thunderstorms, and partly to the high aerosol particle concentration [75] at the coast of Japan during winter thunderstorms.

Dynamic leader inception evaluation
For the case of upward leaders initiated under the influence of a changing electric field produced by the descent of a downward stepped leader, the leader inception model of Becerra and Cooray [72] was modified and extended to account for the time variation of the electric field as well as for the space charge left by streamers and aborted leaders produced before the stable leader inception takes place [73]. In this case, the simulation is initiated by computing the height of the downward leader tip when streamers are incepted from the analysed object. Once the first streamer is initiated, its charge is

computed by using the same representation of the corona zone as in the static case. Owing to the fact that the streamer corona charge shields the electric field at the rod tip, no streamers are produced (dark period) until the streamer inception condition is satisfied again by the increase of the electric field caused by the descent of the downward leader. This analysis is repeated until the total charge ΔQ of any subsequent corona burst is equal to or larger than 1 µC and the first leader segment is created. Because the background electric field changes in time because of the approach of the downward leader, it is not possible to use the same set of equations described in the previous section to compute the corona zone total charge. For this reason, the charge simulation method [81] or a geometric analysis of the potential distribution similar to the one implemented in Reference 72 can be used to estimate the charge and length of the streamer corona zones as well as the dark period time [73].

Once the unstable leader inception condition is reached, the leader starts propagating, with corona streamers developing at its tip as the downward stepped leader moves towards the ground. The simulation of the leader propagation is then started by evaluating the potential at the tip of the first leader segment ($i = 1$). In order to improve the calculation of the leader tip potential used in the static case, the thermohydrodynamical model of the leader channel proposed by Gallimberti [38] is used. Gallimberti's model relates the gradient along the leader channel directly to the injected charge through it. Based on this theory, the radius $a_{L(i)}$ and electric field $E_{L(i)}$ of each ith leader segment produced during each simulation time step t are computed as follows:

$$a_{L(i)}(t) = \sqrt{a_{L(i)}^2(t - \Delta t) + \frac{(\gamma - 1) \cdot E_{L(i)}(t - \Delta t) \cdot \Delta Q(t - \Delta t)}{\pi \cdot \gamma \cdot p_0}} \quad (4.57)$$

$$E_{L(i)}(t) = E_{L(i)}(t - \Delta t) \cdot \frac{a_{L(i)}^2(t - \Delta t)}{a_{L(i)}^2(t)} \quad (4.58)$$

where p_0 is the atmospheric pressure and γ is the ratio between the specific heats at constant volume and constant pressure. The potential at the tip of the leader channel containing n segments is then evaluated as

$$U_{tip}(t) = \sum_{i=1}^{n} l_{L(i)} \cdot E_{L(i)}(t) \quad (4.59)$$

The total corona charge $\Delta Q(t)$ in front of the leader segment is computed with the charge simulation method utilizing the geometrical analysis of the potential distribution before and after the corona formation, updating the value of the electric field for the new position of the downward leader channel. The leader advancement distance $\Delta l_L^{(i)}$ and the new leader length $l_L^{(i+1)}$ are evaluated using equations (4.55) and (4.56). In contrast to the static case, the charge per unit length q_L changes as the upward leader propagates towards the downward moving leader [73]. The

204 Lightning Protection

variation of this parameter is computed using the thermodynamic analysis of the transition region proposed by Gallimberti [38].

Figure 4.22 shows an example of the computed streak image and the upward leader velocity and current of a leader initiated from 10-m-high tower of 0.05 m radius when exposed to the electric field of a stepped leader moving down with a speed of 2×10^5 m s^{-1}. The prospective return stroke current associated with the stepped leader is 10 kA. Note that the model can predict the development of aborted streamer and leaders as observed in field experiments [51]. The unstable leader inception takes place at time T_i'. The stable leader inception condition is assumed to be satisfied by Becerra and Cooray [73] at time T_1 when the leader starts to accelerate uniformly. The time of inception of the stable leader is estimated as the crossing point of a line that connects the leader velocity between 2×10^4 and 4×10^4 m s^{-1} and the time axis.

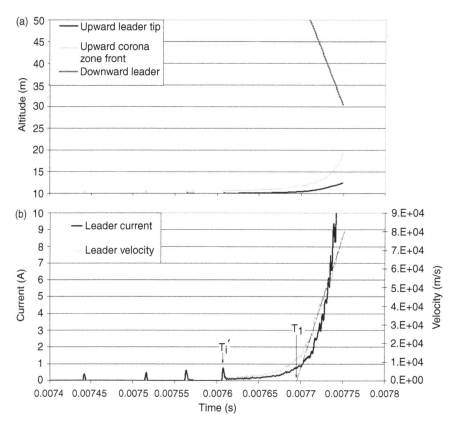

Figure 4.22 Computed discharge characteristics during leader inception for a 10-m-high, 0.05-m tip radius rod in the presence of a downward moving leader with 10 kA prospective return stroke current: (a) streak image; (b) leader velocity and current (adapted from Reference 73)

Comparison of different models
The question of the validity of leader inception models when applied to natural lightning dates back to the moment when the first laboratory-based model was used to evaluate the initiation of lightning positive leaders [26]. Owing to the fact that the experimental measurement of lightning strikes to towers [26,42,43] did not provide enough data related to the conditions under which upward leaders were initiated from them under natural conditions, scientists suggested different laboratory experiments to validate leader inception models.

Unfortunately, laboratory leader discharges are apparently not long enough to become fully thermalized and therefore require larger background electric fields to propagate in comparison with upward lightning leaders [51]. In addition, it appears that the switching voltage waveforms used in the laboratory do not resemble the electric field produced by downward moving stepped leaders. For these reasons leaders in the laboratory have different features when compared with upward leaders initiated under natural conditions. This was demonstrated theoretically by Becerra and Cooray [82]. Because it has been found that laboratory experiments cannot adequately simulate the conditions relevant to lightning flashes [41,74], they cannot be used to validate the leader inception models used for lightning studies.

For the same reasons outlined above, empirical data obtained from laboratory experiments cannot be used to directly model upward leaders of lightning flashes [74]. This raises serious doubts about the validity of the results obtained by applying leader inception models based on empirical equations or data gathered from laboratory leaders to explore the nature of lightning leaders. This is the case for the critical radius concept [54] or the leader inception models of Rizk [62], Petrov and Waters [63] and Bazelyan and Raizer [67] for studying lightning attachment.

For the case of the critical radius concept, laboratory experiments show that critical radius varies with the rise time of the applied voltage [40] and electrode geometry [40,56]. These experiments also show that the critical radius concept is strongly geometry dependent. Therefore the leader inception conditions obtained using this concept depend strongly on the value of critical radius used in the analysis [72,73]. On the other hand, the assumptions used to derive the leader inception model of Rizk [62] are only valid for a voltage waveform with a critical time to rise of $\sim$500 μs [40] (see also Section 4.3.2.1). The temporal variation of this voltage impulse does not give rise to an electric field in the gap similar to that generated by a downward moving stepped leader.

With the development of rocket-triggered lightning techniques, a better source of experimental data under natural conditions became available. One of the first measurements of the electric field at ground level at the moment of initiation of upward leaders as a function of the rocket height was published in 1985 [83]. This experiment reported that upward leaders were initiated from the rocket at a height of about $\sim$100–200 m above ground when the background electric fields ranged between 5 and 10 kV m^{-1}. These data were later used to justify the validity of the leader inception model of Rizk to evaluate upward connecting leaders under natural conditions [98]. However, contrary to earlier investigations, later experiments showed that there was no clear relationship between the altitude of the rocket at the moment of upward leader inception and the electric field at ground [84]. The reason for this is probably the shielding of the ambient electric field by the space charge layers created by the corona at ground level. For this reason, the electric field at ground level during the initiation of upward leaders from triggered rockets or tall towers does not give a correct description of the electric field aloft, and it cannot therefore be used to validate leader inception models.

Although several triggered lightning experiments were performed to gather information on different aspects of lightning flashes [49,50,85], the first measurement of the effective background electric field required to initiate upward leaders from a triggering rocket was conducted in 1999 by Willett and colleagues [51]. In this experiment, the vertical ambient electric field profile beneath thunderstorms was measured with a measuring rocket fired 1 s before the triggering rocket. The experiment made it possible to obtain the space charge modified electric field profile between the rocket and ground, providing the means to test different theories of leader inception. Let us now see how different theories fair against these experimental data.

Figure 4.23 shows the background leader inception (static) electric fields computed with different models and the average electric field necessary for leader inception as measured in the rocket-triggered experiment [51]. Note that the predictions of the Rizk model [96] and the first equation of Lalande [69] do not agree with the results of the triggered lightning experiment [51]. The other leader inception criteria, namely the Petrov and Waters model [63], the Bazelyan and Raizer model [67], the

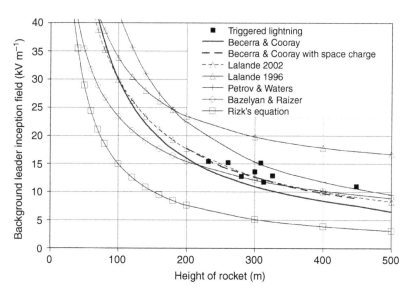

Figure 4.23 Comparison of the average background electric fields required to initiate stable upward positive leaders from a rocket, calculated with different leader inception models and measured in a classical rocket triggered lightning experiment [51]

second equation of Lalande [50] and the proposed static leader inception model of Becerra and Cooray [72] are in rather good agreement with the experimental values. In particular, there is a good agreement for the model presented by Becerra and Cooray [72] with the measurements when the space charge left behind the rocket by aborted streamers and leaders is taken into account. This also shows the importance of taking into account the space charge effects when dealing with leader inception.

Note that leader inception takes place in the experiment when the height of the rocket (or the grounded conductor) is more than 200 m. Thus the above comparison validates model results for (slim) structure heights exceeding this value. Data in Figure 4.23 show that there is considerable disagreement between model predictions for heights lower than ~100 m. Of course, it is generally assumed that structures lower than 100 m or so do not normally initiate upward lightning [86], but recent reports indicate that structures as short as 50 m on flat ground could initiate upward lightning flashes [80]. Because of the large differences between model predictions for heights lower than 100 m, further experiments with instrumented towers and rocket-triggered lightning are required to measure the effective background electric fields required to initiate upward leaders from shorter structures.

In an interesting study, Lalande and colleagues [49] performed an altitude-triggered lightning experiment with simultaneous measurements of current, electric field and luminosity during the initiation and propagation of an upward connecting leader. In this experiment, the rocket first spooled out 50 m of grounded wire, followed by 400 m of insulating Kevlar, and from it to the rocket tail a second (floating) copper wire [49]. During the ascent of the rocket, an upward positive leader was initiated from the top end of the floating wire, and a negative stepped leader was initiated from its bottom end. In response to this downward moving negative leader, an upward connecting positive leader was initiated from the top end of the grounded wire. The upward leader current, the electric field change produced by the descending negative leader (measured at 50 m from the wire), and the leader luminosity (with still and streak photography) were measured simultaneously during the experiment.

In order to reconstruct all the physical parameters during the inception of the upward connecting leader in this experiment [49], the time-dependent leader inception model presented by Becerra and Cooray [73] is used. Figure 4.24 shows the predictions of the model for the main physical parameters of the upward connecting leader in the experiment before and during its initiation. The model predicts that several streamers and aborted leaders are launched before the inception of a successful upward connecting leader at around 4 ms. This calculated leader inception time is in good agreement with the experimentally estimated value of 4.02 ms [49]. As the downward stepped leader approaches to ground, the total corona charge in front of the upward connecting leader tip augments and its channel potential gradient decreases (Figure 4.24d). Consequently, the injected current and the velocity increase and the upward leader starts accelerating continuously, reaching the stable propagation condition. Note the good agreement between the computed leader current and the main component of the measured current (Figure 4.24c).

Based on the reconstructed physical parameters of the upward connecting leader in the experiment, let us consider the validity of the assumptions made in the Petrov and

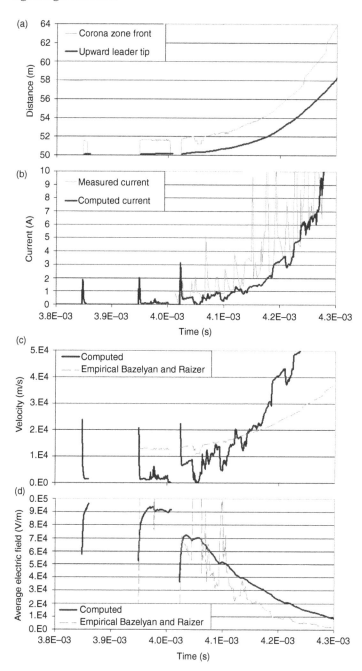

Figure 4.24 Predictions of the leader inception model of Becerra and Cooray [72,73] for the upward connecting leader in an altitude rocket-triggered lightning experiment [51]: (a) streak image; (b) predicted

Waters [63,65] and Bazelyan and Raizer [67] models. No further discussion of the validity of the critical radius concept [54] and the generalized leader model of Rizk [96] is made here, given the inaccuracy of these models when applied to lightning. In addition, the leader inception equations of Lalande [71] are not discussed further because they only consider the initiation of leaders under static electric fields (in the absence of a downward stepped leader). Note first that the length of the streamer zone during the creation of the stable leader in the experiment is ~ 1.7 m (Figure 4.24a). This value is larger than the critical value of 0.7 m assumed by Petrov and Waters [63]. Moreover, as pointed out by Becerra and Cooray [73], the critical length of streamers at the moment of unstable leader inception can vary between 0.7 and 2 m due to the effect of the space charge produced by aborted streamers. In laboratory experiments the streamer length at unstable leader inception is also observed to vary with the rise time of the applied voltage [40]. This raises some doubts as to the validity of using a fixed streamer length as a leader inception criterion.

As to the assumptions relevant to the extension of the Petrov and Waters model [65], note that the velocity of an upward connecting leader can be as low as 1×10^4 m s^{-1}, significantly slower than the minimum upward leader velocity of 6×10^4 m s^{-1} considered in the model [65]. This speed was used by Petrov and Waters to estimate the critical rate of change of the induced potential at the tip of the rod (or at the point of initiation) that would lead to stable propagation of an upward connecting leader [65]. Because the velocity of upward connecting leaders can change from one flash to another, it is not possible to define a fixed critical rate of change in the induced potential as a measure of stable leader propagation.

Consider the leader inception model of Bazelyan and Raizer [67]. It uses a set of empirical equations derived from laboratory experiments to estimate the upward leader velocity and the average leader channel electric field. For the sake of comparison, the leader velocity and channel average field computed with these empirical equations are also shown in Figures 4.24c and d. Note that the empirical equations give a higher velocity, particularly soon after unstable leader inception, than the value obtained from the experiment. The empirical equation derived by Bazelyan and Raizer [67] gives a velocity of 1.2×10^4 m s^{-1} for any newly created leader. This overestimates the propagation velocity of newly created leaders. Moreover, the empirical equation of Bazelyan and Raizer for the average internal electric field of the leader channel gives lower values than the ones estimated from measurements. The combined effect of higher leader velocity and lower internal electric field in the leader channel immediately after unstable leader inception would favour the propagation of leaders that would have been aborted otherwise. This artifact of the model may lead to erroneous estimation of the stable leader inception time.

Figure 4.24 *(Continued) upward leader velocity and values computed with an empirical equation [67]; (c) predicted and measured upward leader currents [51]; (d) predicted average electric field in the leader channel and values computed with an empirical equation [67]*

The effect of the shape of the lightning conductor on lightning attachment

One important question in lightning protection is whether the shape of the lightning conductor can influence the process of attachment between the down-coming stepped leader and the upward moving connecting leader. Moore [87,88] suggested that lightning rods with blunt tips might be more effective than sharp ones in generating connecting leaders that bridge the gap between the stepped leader and ground. A seven-year field study to determine the effect of the radius of the rod on lightning attachment was conducted by Moore and colleagues [87,88]. In this experiment air terminals of different radii were placed on a ridge near the 3 288-m-high summit of South Badly Peak in the Magdalena Mountains of Central New Mexico. Some air terminals were sharp-pointed and others blunt. The diameters of the blunt rods varied from 9.5 to 51 mm. The rods competed with each other to attract lightning flashes occurring in the vicinity.

Over the seven-year study, none of the sharp rods were struck by lightning, but 12 of the blunt rods were. All of the strikes were to blunt rods with tip diameters ranging between 12.7 and 25.4 mm, although most of the flashes struck the 19-mm diameter blunt rods. None of the adjacent blunt rods with diameters 9.5 or 51 mm were struck. Moore and colleagues [87,88] concluded that moderately blunt rods are more likely to generate successful connecting leaders than sharp rods or extremely blunt rods.

In a recent study Becerra and Cooray [73] utilized their model described in Section 4.3.2.6 to study how lightning rods of different radii differ in their ability to launch connecting leaders when exposed to the electric fields of downward moving leaders. For a stepped leader with a prospective return stroke current of 10 kA, they computed the height of the downward leader tip above ground when unstable and stable leaders were incepted from lightning rods of different radii. The results are shown in Figure 4.25. First, note that the height of the downward moving leader tip when an unstable leader is incepted is about twice the height of the leader tip when a stable leader is incepted. Second, note that the tip radius of the lightning rod slightly affects the height of the downward leader when stable leader inception takes place. In other words, the background electric field necessary for stable leader inception is affected by the radius of the lightning rod. However, these differences are not more than ~ 10 per cent for the range of tip radius considered. However, observe that there is an optimum rod radius for which the height of the tip of the downward moving stepped leader that can generate a stable (and unstable) upward leader from the rod is a maximum. In other words, this rod radius is slightly more efficient that other rod radii in generating both unstable and stable upward leaders. For tip radii lower than the optimum, several burst of streamers with low charge (lower than 1 μC) are initiated as the stepped leader proceeds downwards, leading to a reduction of charge in subsequent streamer bursts and to retardation of leader inception. On the other hand, blunter rods produce fewer streamer bursts, but the dark periods are larger, leading to retardation of leader inception. In fact, the optimum radius for efficient stable leader inception is slightly smaller than the one that was most efficient in generating an unstable leader, because the former depends also on the most favourable conditions for the leader propagation.

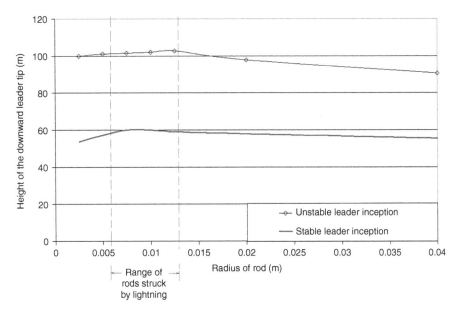

Figure 4.25 *Height of the downward leader tip where unstable and stable upward leader inception takes place from a 10-m-high rod as function of its radius. The downward stepped leader is assumed to have a prospective return stroke peak current of 10 kA and an average velocity of 5×10^5 m s^{-1} [73]. The range of radii of the blunt lightning rods struck by lightning in the experiment of Moore and colleagues [87,88] is also shown.*

These predicted differences in rods of different radii by Becerra and Cooray [73] are in agreement with the field observations of Moore and colleagues [87,88]. However, differences in the efficiency of different rods, as measured by the height of the leader tip at which a stable leader is generated from the rod, are not more than 10 per cent according to Figure 4.25. The question then is why all the lightning flashes became attached to the rods with optimum radii in the experiment of Moore and colleagues, whereas the others did not get a single strike. According to the opinion of the authors, the differences in the efficiency of lightning attachment between rods found by Moore and colleagues [87,88] were influenced by the setup of the experiment. Because the competing rods were placed rather close to each other (about 6 m apart), the small advantage of the 'optimum' rods caused them to initiate upward leaders before the others. Although the presence of the adjacent rods according to Moore and colleagues [87] produced a change of only ~ 1 per cent in the electric field at the tip of each rod, a much larger shielding of the electric field was produced when the corona zone and leader channel were formed from any rod. Thus, once an upward leader was initiated from a rod, it screened the other rods and inhibited the development of stable upward leaders from them. In this way, the 'optimum' rods

212   *Lightning Protection*

managed to become attached to the stepped leaders all the time. Thus, caution should be applied in interpreting the results of this experiment. For example, based on the results one should not conclude that rods of other diameters are not effective strike receptors [73].

In addition, it is noteworthy that the 'optimum' rod radii given in Figure 4.25 apply only for free-standing rods of height less than ~ 10 m. For taller rods the influence of the radius of the rod on lightning attachment becomes less significant. This is because the frequency and magnitude of the streamer corona generated due to the field enhancement at the tip of the rod become less dependent on its radius as the rod height increases [73].

4.4 The leader progression model

4.4.1 The basic concept of the leader progression model

As mentioned previously, the final attachment of the stepped leader to a grounded structure takes place through the interaction of the down-coming stepped leader with the upward moving connecting leader. Leader progression models attempt to simulate both the propagation of the downward leader and the upward connecting leader until the meeting point at which a return stroke is initiated. The goal is to characterize and quantify the process of attachment and to understand how different physical parameters affect the process of attachment. Moreover, these models will be able to estimate the attractive radius of a given structure for a stepped leader with a given prospective return stroke current. The attractive radius of a structure is defined as follows. Consider a stepped leader travelling down along a straight line from cloud to ground. If the lateral or horizontal distance to the vertical path of the stepped leader from the structure is less than a certain critical value, then the stepped leader will be attracted to the structure. This critical lateral (horizontal) distance is called the attractive radius.

Because the exact mechanism of the propagation of stepped leaders is not known, a large number of simplifying assumptions are made in creating leader progression models. First, we will describe the basic assumptions of leader progression models and then we will describe the latest research work that could be utilized to improve the state of the art of these models.

4.4.2 The leader progression model of Eriksson

The first leader progression model was introduced by Eriksson [58,59] and is based on the following assumptions.

1. It is assumed that the grounded structure concerned may be regarded as free-standing and has approximately axial symmetry.
2. The downward moving stepped leader is represented by a vertically descending linearly charged channel. The charge distribution along the leader channel and

its relationship to the peak current of the first return stroke are given by equations (4.5) to (4.7).
3. The electric field enhancement at the structure top caused through its protrusion above ground level is expressed in terms of a field enhancement factor, which in turn is derived from the structure dimensional ratio (i.e. height in relation to the cylindrical radius).
4. The criterion for the initiation of an upward leader from a particular structure is taken as the attainment of critical field intensity, 3×10^6 V m^{-1}, over the critical radius at the structure extremity (i.e. critical radius concept, see Section 4.3.2.1).
5. The stepped leader is assumed to take a straight path to ground and this path is not affected by the presence of a connecting leader.
6. The connecting leader travels in space in such a way that it will find the closest path for the connection with the stepped leader.
7. The ratio between the speed of propagation of the downward stepped leader and the upward moving stepped leader is assumed to be 1.
8. When dealing with the striking distance to flat ground in Section 4.2 we defined the final jump condition. As mentioned previously, in the case of a stepped leader approaching flat ground, the final jump condition is reached when the streamers of the down-coming leader meet the ground plane. One can also define the final jump condition for the case of an encounter between the upward moving connecting leader and the downward moving stepped leader. In this case the final jump condition is reached when the streamer zone of one discharge meets the streamer zone of the other. Once the final jump condition is reached, the electrical breakdown between the two discharges is inevitable. However, in the analysis Eriksson neglected this final jump condition and assumed that for attachment to take place the two tips of the leader channel have to meet one another.

The model predicts that when a stepped leader with a given charge enters into a certain volume in space it will be captured by the structure. This volume is called the 'collective volume' of the structure. It depends on the charge on the leader channel, field enhancement of the structure and the velocity ratio of the two leaders. The collection volume around the cylindrical structure for different leader charges as calculated by Eriksson is shown in Figure 4.26. Note in this figure that for a given charge on the leader channel there is a critical horizontal radial distance (attractive radius) within which the leader is attached to the structure. Because the charge on the leader channel can be expressed as a function of the return stroke peak current, the attractive radius can be expressed as a function of peak return stroke current as follows:

$$R = I^a 0.84 h^{0.4} \quad \text{(m, kA)} \tag{4.60}$$

where $a = 0.7 h^{0.02}$.

D'Alessandro and Gumley [89] applied the collection volume concept proposed by Eriksson [58,59] to the analysis of lightning strikes to buildings. As outlined above, the analysis requires field enhancement at different points on the structure. They

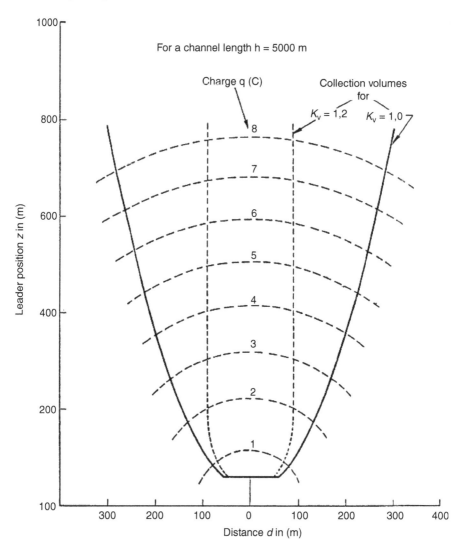

Figure 4.26 Collection volume around a tall rod for different leader charges according to Eriksson [58]

made use of the 'field intensification method' described in Reference 61 to obtain this parameter. The model was then used to compute the striking distance relevant to any given point on a building. Once the striking distance was known, the collection volume of lightning flashes pertinent to any given point on the structure could be evaluated. Based on the analysis they proposed the use of the collection volume concept for the optimum placement of air terminals on grounded structures to be protected. This method is known as the collective volume/field intensification method (CVM/FIM) [90].

The predictions of this method are claimed to be in close agreement with statistics of actual lightning strikes captured by lightning protection systems designed according to this concept in Hong Kong [91] and Malaysia [92]. The procedure assumes that the ground flash density of the site is known and the collective volume of the air terminals on complex structures is identical to the collection volume of a free-standing rod of similar height. The ground flash density is computed from the number of thunder days using three different empirical relationships that connect these two parameters. The attraction zones of terminals are assumed to be circular regions with a radius equal to the attractive radius obtained from equation (4.60). From the computed attraction zones the equivalent average lightning exposure areas of air terminals are computed by taking into account the probability distribution of the prospective return stroke peak current. Finally, using the ground flash density, the expected number of strikes in the lightning protection system per year is obtained. Because the ground flash densities obtained from different equations differ significantly, in the analysis they selected the value that produced a better agreement between the computed expected number of strikes per year and the actual number of strikes captured by the lightning protection system. Not withstanding the simplifying assumptions made in the analysis and the fact that the ground flash density is selected to fit the theory with experiment, the authors of the experiment claimed that the experimental data prove the validity of the CVM/FIM method.

However, serious doubts exist in the lightning research community on the validity of the CVM/FIM method [20,90,93]. Interestingly, Becerra and Cooray [94] have recently shown that the lightning attractiveness of air terminals on complex structures cannot be evaluated using parameters valid for free-standing rods. This is the case because the lightning attraction zones of corners and short air terminals on buildings do not in reality define symmetrical and circular areas as do the ones assumed by the CVM/FIM. They also found that the lightning attractive zones predicted by the CVM/FIM are excessively large and unrealistically circular in comparison to the predictions based on self-consistent physical leader inception models. The study shows that the lightning exposure areas predicted by the CVM are larger than in reality and that the good agreement between the number of strikes per year estimated by the CVM/FIM and the observed number of strikes [91,92] is a result of conveniently selecting the ground flash density to fit experiment and theory.

4.4.3 The leader progression model of Dellera and Garbagnati

These authors [28,60] introduced a more sophisticated and dynamic model of the downward leader and its subsequent interception with the connecting leader issued from a structure. The main assumptions of this model are the following.

1. The charge per unit length on the stepped leader is the same along the whole length of the channel, and its value is given by equation (4.8) as a function of peak current. However, the lower part (the last tens of metres) of the leader channel is assumed to have a charge of $100\ \mu C\ m^{-1}$.

2. The charge per unit length of the positive connecting leader is equal to 50 μC m^{-1}.
3. The streamer zone of the stepped leader extends from the tip of the leader up to a distance where the electric field is 300 kV m^{-1}.
4. The length of the streamer zone of the positive connecting leader is assumed to be equal to the length between the leader tip and the point defined by the intersection of the actual potential distribution curve with a straight line of slope equal to 500 kV m^{-1}, both computed along the maximum field strength line.
5. The direction of propagation of the leaders, both the down-coming stepped leader and the upward moving connecting leader, is determined by the direction of the maximum electric field along an equipotential line at a distance from the leader tip equal to the streamer extension.
6. The inception of the connecting leader is based on the critical radius concept. The value of the critical radius was assumed to be 0.36 m [54]. Based on the results obtained later by Bernardi and colleagues [56] a value of 0.28 m was adopted in later studies. In calculating the electric field at the point of interest of the structure the method of charge simulation was used [81].
7. The model requires the velocity ratio between the negative and positive leaders as an input. In the simulation it is assumed to be 4 at times close to the initiation of the positive leader but it decreases to 1 just before the connection of two leaders.
8. In the model the propagation of the two leaders towards each other is continued until the final jump condition is reached between them. In other words the simulation continues until the streamer regions of the two leaders touch each other.
9. The model takes into account the background electric field generated by the cloud in the analysis. The cloud, assumed to be 10 km in diameter, is simulated with four charge rings of uniform charge density. The rings are placed at equal distances from each other and placed in such a way to cover the whole area of the cloud. It is doubtful, however, whether the electric field generated by the cloud can influence the attachment process.

A sketch of the model is shown in Figure 4.27a. Figure 4.27b and c show examples of application of the model to study lightning attachment to a free-standing structure of 220 m height on flat ground and a 420 kV line in different orographic conditions, respectively. Based on such simulations the lateral distance (having the same definition as the attractive radius of Eriksson's model described earlier) has been evaluated corresponding to free-standing structures of different heights. The results obtained from the model are depicted in Figure 4.27d.

Recently, Ait-Amar and Berger [95] used a leader progression model similar to the one proposed by Dellera and Garbagnati to study the lightning incidence to buildings. In their analysis, the critical radius concept [54] was used as a condition for the initiation of upward connecting leaders from buildings.

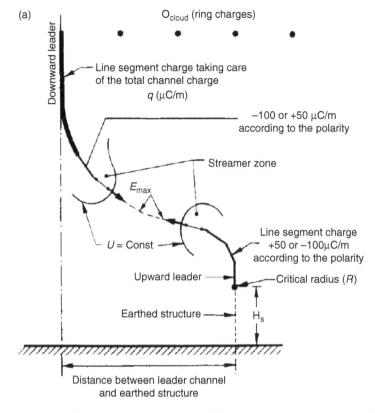

Figure 4.27 (a) Sketch of the propagation of lightning according to the leader progression model of Dellera and Garbagnati [28]. (b) Example of application of the leader progression model of Dellera and Garbagnati for a free-standing structure 220 m in height [28]. (c) Example of the computed leader paths of a 420 kV power line in different orographic conditions [28]. (d) Computed lateral distance as a function of the lightning current for free-standing structures of different heights according to the leader progression model of Dellera and Garbagnati [60].

4.4.4 The leader progression model of Rizk

The basic idea behind the leader progression model of Rizk [96,97] is similar to that of Dellera and Garbagnati [28]. However, there are several fundamental differences in details. The following are the assumptions of the model.

1. The charge per unit length on the stepped leader decreases linearly with height, becoming zero at cloud level and maximum at the downward leader tip height.

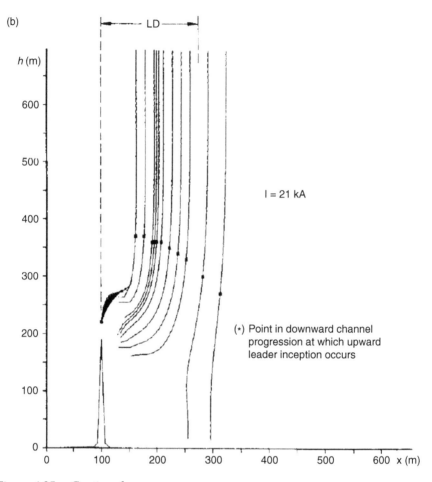

Figure 4.27 Continued

The magnitude of the charge per unit length as a function of peak return stroke current is evaluated by assuming that the total charge satisfies equation (4.5).
2. Rizk's generalized leader inception equation (described in Section 4.3.2.2) is utilized to evaluate the height of the downward leader tip when the upward connecting leader is initiated.
3. The negative downward channel continues its downward motion unperturbed by any object on the ground. It maintains this unperturbed motion until the final jump condition is reached between it and the upward moving connecting leader. At the final jump condition it turns towards the positive leader.
4. At the final jump condition the mean potential gradient across the gap between the two tips of the leaders is assumed to be 500 kV m^{-1}. In order to make a decision whether the final jump condition is reached, Rizk represented the upward connecting leader as a finitely conducting channel. In the model the potential of

Attachment of lightning flashes to grounded structures 219

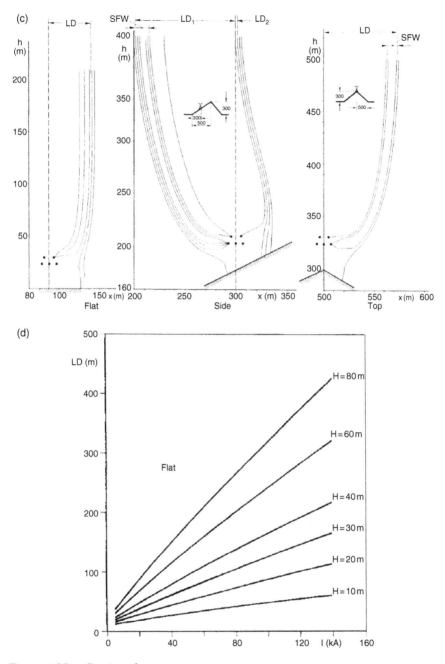

Figure 4.27 Continued

the connecting leader is evaluated by the expression

$$\Delta U_l = lE_\infty + x_0 E_\infty \ln \left[\frac{E_i}{E_\infty} - \frac{E_i - E_\infty}{E_\infty} \exp(-l/x_0) \right] \quad (4.61)$$

where l is the length of the leader channel, E_i is the minimum positive streamer gradient, E_∞ is the final quasi-stationary streamer gradient, $x_0 = v\theta$, θ is the arc (leader) time constant and v is the speed of the ascending positive leader. In the calculations it is assumed that $E_i = 400$ kV m^{-1}, $E_\infty = 3$ kV m^{-1}, $\theta = 50$ μs and $v = 1 \times 10^5$ m s^{-1}. Note that the potential of the downward moving stepped leader is already fixed once the distribution of charge along its channel is specified.

5. The vector motion of the positive upward leader is such that at any instant it seeks the negative leader tip.
6. The ratio of the speeds of the negative and positive leaders is taken to be 1.

Figure 4.28a and b show the model-predicted variation of attractive radius for different mast heights in flat terrain and on a mountain, respectively, for different values of return stroke peak currents. The simulations in Figure 4.28a corresponds to a background electric field (i.e. the field produced by the cloud) of zero and Figure 4.28b for 3 kV m^{-1} [98].

4.4.5 Attempts to validate the existing leader progression models

Although the three leader progression models discussed above describe the main stages of the attachment of lightning flashes to grounded structures, they introduce different simplifying assumptions in each phase. Because of a lack of knowledge at the time when the models were proposed, different models made different assumptions concerning the distribution of the charge on the stepped leader, the leader inception criterion, the velocity ratio and the properties of the upward connecting leader [104]. For this reason, the results obtained for a particular structure using different leader progression models may differ considerably from each other [99]. This is illustrated in Figure 4.29.

Unfortunately, sufficient field data to fully validate the leader progression models are not available at present. Owing to the inherent difficulties associated with performing controlled experiments to evaluate the lightning attractiveness of grounded objects, no direct estimates of lateral attractive distances of towers or structures are available today. The first attempt to estimate indirectly the lightning attractive distances of tall objects by analysing field observations was made by Eriksson [59]. He gathered data from ~3 000 observed lightning flashes to a variety of free-standing structures with heights ranging between 22 and ~540 m in various regions in the world. Because the structures were located in places with different thunderstorm days he normalized all data to an arbitrary flash density of 1 flash km^{-2} yr^{-1}. By taking the normalized number of lightning strikes to the structures per year as a function of their height and considering a uniform lightning flash density of 1 flash km^{-2} yr^{-1} for all the data worldwide, the lateral attractive distance of the structures was estimated. The estimations are compared in Figure 4.30 with the predictions of

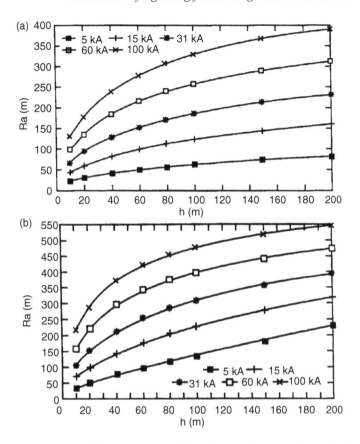

Figure 4.28 Variation of the attractive radius (a) as a function of mast height for flat terrain for different return stroke peak currents according to the leader progression model of Rizk [97] and (b) of a mast on a hemispherical mountain top for different return stroke peak currents according to the leader progression model of Rizk [97]

three leader progression models. Notice the broad scatter of the attractive distances estimated from the field observations, which is caused by various inadequacies in the collection, classification and normalization of the data. One major cause of scatter is due to the large uncertainty in the relationship between the flash density and the thunderstorm days used by Eriksson [100]. The large uncertainty in this relationship is such that, for instance, areas reporting a keraunic level of 20 thunderstorm days per year would have an estimated lightning density of 2 flashes km^{-2} yr^{-1} even though they could in fact experience a flash density varying between 0.4 and 4 flashes km^{-2} yr^{-1} [59]. Such errors in the calculated lightning flash density lead to large errors in the estimated lateral attractive distances. Although the results of Eriksson have been used in validating leader progression models [59,74,97], for the reasons mentioned above, the values of attractive distances estimated from the

222 *Lightning Protection*

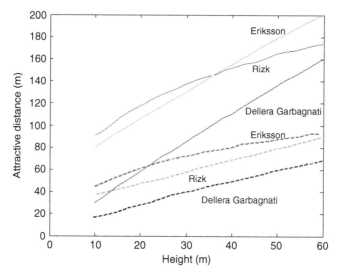

Figure 4.29 Comparison of the attractive distance of a free-standing structure predicted by the Eriksson [58,59], Dellera and Garbagnati [28,60] and Rizk models [96,98] for prospective return stroke peak currents of 60 kA (solid lines) and 20 kA (dashed lines) (adapted from Reference 99)

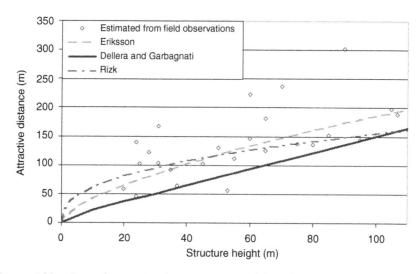

Figure 4.30 Lateral attractive distance estimated from field observations [59] and predicted by the Eriksson [58,59], Dellera and Garbagnati [28,60] and Rizk [96,98] models for free-standing structures as a function of height. A prospective return stroke peak current of 31 kA was assumed for the models.

observed strike incidence to various structures by Eriksson [59] are inaccurate, and they should be used with caution in validating leader progression models.

Other attempts to indirectly evaluate the accuracy of leader progression models have been made by comparing the predicted and experimentally derived shielding angles required to protect phase conductors in power transmission lines [58,101]. The field observations are based on the same dataset used by Whitehead and colleagues [102] to calibrate EGM theory. It corresponds to the observed effective shielding angles of high-performance power transmission lines operating in the United States. However, a comparison of the predictions of leader progression models with this dataset will constitute only a crude validation of the models, because the shielding performance of power transmission lines is affected by several other factors such as environmental shielding (by trees), variations in the topography of the terrain, sag of the line, and so on [34,102,103]. Because these factors are either difficult to evaluate or are unknown for the transmission lines in the existing dataset, it is not possible to make an accurate prediction using leader progression models of the shielding angles that will provide proper lightning protection in these lines.

4.4.6 Critical overview of the assumptions of leader progression models

When considering the complexity of lightning phenomena, it will always be necessary to make a large number of assumptions and simplifications in order to formulate a usable lightning strike model. These simplifications, which on the one hand make the model feasible for calculations within a reasonable time, on the other hand involve a number of limitations that influence the accuracy of the results of the calculations. However, the creation of leader progression models has been a major step forward and these models are capable of predicting several phenomena observed in the field. Moreover, the models seem to be well suited for sensitivity analysis where the effects of various parameters on the efficiency of lightning protection procedures are being studied by changing one parameter at a time. However, the models described above are based on several simplifying assumptions and therefore their ability to describe nature is limited. These assumptions are discussed as follows.

4.4.6.1 Orientation of the stepped leader

In the leader progression models two assumptions are made concerning the path of propagation of the stepped leader. In the model of Dellera and Garbagnati it is assumed that the leaders travel along the maximum electric field direction. In the Rizk model it is assumed that the background electric field does not influence the path of propagation of the stepped leader. Of course the first assumption makes some physical sense, but it is not that clear how this condition should be applied in practice. It is difficult, if not impossible, to determine the direction of the maximum electric field ahead of the leader channel for the following reasons. The space in front of the leader channel is occupied by streamer discharges that supply the current necessary for the propagation of the leader. The electric field configuration in front of the leader channel is determined by the spatial distribution of the space charge of the streamer system, which in reality is not uniform. Thus the exact distribution of the electric field in space in front of the leader cannot be determined with any certainty. Moreover, the direction of the stem of the streamer system, which becomes the new leader segment, may to some extent be controlled by the space charge lying ahead of it. This space charge reduces the electric field at the stem and therefore the direction of the next section of the leader channel may form in a direction away from the main concentration of the space charge. This process may introduce some tortuosity into the leader channel. This tortuosity may also complicate the spatial distribution of the electric field ahead of the leader channel.

The situation is even more complicated in the negative stepped leader. In this case the high electric field at the outer edge of the streamer space charge leads to the creation of a space stem. The space stem generates a positive streamer system towards the tip of the already formed leader channel. The interaction of this positive streamer system coming out of the space stem and the negative streamer system emanating from the tip of the leader channel gives rise to the next step of the leader. Thus the location of the space stem with respect to the tip of the leader channel will decide the direction of the next leader step. The theory available at present cannot be used to predict the exact location of the space stem with respect to the electric field configuration.

The next problem is our lack of knowledge concerning the external electric field necessary to divert the direction of propagation of a leader channel. A newly created section of either positive or negative leader is immersed in the background electric field of the streamer discharges. In negative leaders this is $\sim 1\,000$ kV m^{-1} and in positives it is ~ 500 kV m^{-1}. In order to divert the direction of propagation of the leader channel it is necessary to have a background electric field that is comparable to these streamer electric fields. As long as the background electric field is much less than the electric field in the streamer region, the direction of the leader is determined more by the random nature of the space charge distribution in the streamer volume than by the background electric field. However, background electric fields comparable to streamer electric fields may occur in practice as the stepped leader approaches a structure or when two leaders (i.e. stepped leader and the connecting leader) approach each other. A physically reasonable assumption to be made in leader progression models, therefore, is to assume that that the leaders move without much influence from the background electric fields unless this field becomes comparable to the streamer field, say about one-fifth of the streamer field. This amounts to ~ 200 and 100 kV m^{-1} for negative and positive streamers, respectively.

4.4.6.2 Leader inception criterion

The theoretical results as well as experimental observations gathered from natural and triggered lightning show that the leader formation in the laboratory under switching impulses and under natural lightning conditions are considerably different (see also Section 4.5.1). In the laboratory, the formation of a positive leader takes place under a high external electric field as a cold leader. The internal electric field of a cold leader is $\sim 100–200$ kV m^{-1} with a channel temperature of $\sim 4\,000$ K. It may support a current of about 1 A. Under actual lightning conditions in nature, the positive leader starts as an unstable/cold leader but converts into a stable/hot leader with an internal electric field strength of ~ 10–50 kV m^{-1}, a temperature of $\sim 6\,000$ K, and a current of 10 A or more. The processes in leader formation are strongly nonlinear and therefore the laboratory results should be used in lightning studies only as a vehicle to extract basic parameters that are of interest in the formulation of theory of electrical breakdown in air.

The leader inception criterion utilized by Eriksson [58], Dellera and Garbagnati [28] and Rizk [97] are based on the breakdown characteristics of long gaps obtained in the laboratory. For example, Eriksson [58] and Dellera and Garbagnati [28] utilized the critical radius concept [54], which is extracted from information on the breakdown of long gaps in the laboratory. The same applied to Rizk's leader inception model [62], which is calibrated using the data relevant to the breakdown of long gaps. The laboratory data may depend both on electrode geometry and the type of voltage waveform used in the study. For example, as mentioned previously, the critical radius depends on the gap length and the waveform of the voltage impulse used in the evaluation. Could one use, for example, the critical radius observed in the laboratory for horizontal and vertical lightning conductors placed for instance on a building? The field distribution around a conductor located on a building depends not only on the shape of the conductor but also on the height and shape of the building. Therefore, the leader inception conditions for such conductors could be different in comparison to similar conductors located in free space without the effect of the building. This creates a need for using different values of critical radii for conductors of similar shape but located on different parts of a building. Other problems associated with the critical radius concept have already been discussed in Section 4.3.2.1.

Another important simplification of these models is the use of static conditions to evaluate the inception of leaders. In the evaluation the background electric field is kept constant while ascertaining whether the leader inception criterion is satisfied. In reality, the background electric field produced by a descending stepped leader increases continuously and this variation has to be taken into account in the evaluation of leader inception. For example, the leader inception model of Becerra and Cooray [73], which takes this into account, shows that the leader inception depends on the rate of change of the background electric field (see the subsection 'Comparison of different models' in Section 4.3.2.6 for a comparison of different inception models) The results of leader progression models should be interpreted keeping in mind all these simplifying assumptions.

4.4.6.3 Parameters and propagation of the upward connecting leader

Neither of the leader progression models of Dellera and Garbagnati or Rizk utilizes physics to ascertain whether the conditions necessary for propagation are satisfied continuously as the connecting leader propagates towards the stepped leader. In these models once a leader is incepted it is assumed to propagate continuously in the available background electric field.

Although some predictions of leader progression models agree qualitatively with field observations [74], these models are based on some assumptions that are not in line with current knowledge on the physics of leader discharges and lightning. Furthermore, some input parameters utilized for the modelling of the upward connecting leader are unknown or difficult to estimate. Hence, most of the parameters necessary for the model have been extracted from laboratory experiments pertinent to long sparks. However, the physical properties of upward leaders in nature have been found to be different to those of leaders in the laboratory [104,105]. For instance, it has been observed that the temporal evolution of the upward leader velocity in triggered lightning experiments does not follow a well defined pattern as in the laboratory, where the external parameters are somewhat controlled, but instead changes from flash to flash [13]. A better leader progression model should be capable of self-consistently estimating the physical properties of upward leaders.

One of the most important parameters in leader progression models is the speed of propagation of the connecting leader. In available leader progression models this speed is assumed to be constant and its value is selected somewhat arbitrarily. In reality, the speed of the connecting leader may vary continuously while propagating. Moreover, different experiments with natural and triggered lightning suggest that it is not possible to generalize behaviour, including speed, of upward connecting leaders because it changes from one flash to another [26,44]. However, the existing leader progression models assume that leaders of all lightning flashes behave in an identical manner. A more physically oriented leader progression model should be able to take into account in a self-consistent manner the variation of lightning parameters from one flash to another.

4.4.6.4 Effects of leader branches and tortuosity

All the leader progression models available today represent the leader by a single channel without including branches or tortuosity. The leader deposits charge not only on the main channel but also on branches. Thus the electric field configuration in space in front of the leader channel is controlled to some extent by the branched nature of the channel. Moreover, the final attachment process could also be influenced by branches if several branches approach the ground more or less simultaneously. A more realistic leader progression model should take the branched nature of the lightning channel into account.

In a leader progression model the length of the streamer zone is evaluated by taking into account the electric field ahead of the leader tip. This electric field depends both on the spatial distribution of charge behind the leader tip and the geometry of the channel. Moreover, the charge distribution of the leader channel itself may depend on the geometry of the channel. Thus, the tortuosity of the leader channel should be included in leader progression models to represent the lightning leaders faithfully.

4.4.6.5 Thundercloud electric field

In general, the magnitude of the thunder cloud electric field is no more than a few tens of $kV\ m^{-1}$, and as far as the propagation of the leader is concerned one may neglect the effects of this electric field. However, the thundercloud electric field may affect leader propagation in an indirect manner. The electric field in front of the leader tip, which determines the length of the streamer region and hence the propagation characteristics of the leader, depends on the charge distribution along the leader channel. This charge distribution is intimately related to the background electric field through which the leader is propagating. The background electric field concentrates electric charges at the tip of the leader. This charge in turn generates a high electric field, which is orders of magnitude larger than the background electric field, in front of the leader channel that propels the leader forward. This is the reason why the general path of propagation of leaders is directed along the background electric field. A leader may travel a short distance perpendicular to the background electric field, but soon the charge concentration at the leader tip will be exhausted, forcing the leader to stop propagating. Of course, if the charge distribution along the leader channel is arbitrarily selected, this connection between the background electric field and the charge distribution along the leader channel is lost and one may erroneously come to the conclusion that the thundercloud electric field is unimportant in leader progression models. However, to make leader progression models more realistic it is necessary to connect the charge distribution of the leader channel to the background electric field produced by the thundercloud.

4.4.7 Becerra and Cooray leader progression model

4.4.7.1 Basic theory

Owing to the limitations of the existing models, it is necessary that a leader progression model self-consistently estimates the physical properties of upward connecting leaders in order to reduce the uncertainties of the calculations. These properties include the charge per unit length, the injected current, the leader channel potential gradient and the velocity of the upward connecting leader during its propagation towards the downward stepped leader. With this idea in mind, a self-consistent physical model to simulate the initiation and propagation of upward connecting positive leaders from grounded structures was introduced by Becerra and Cooray [104]. The model takes the time-dependent inception model presented in Section 4.3.2.6 (subsection 'Dynamic leader inception evaluation') as a base, and extends the analysis of the upward leader propagation until the connection with the downward stepped leader. The model self-consistently estimates the leader physical properties during its propagation towards the downward stepped leader. The model uses a thermohydrodynamical model as proposed by Gallimberti [38] to estimate the leader channel properties together with a thermodynamic analysis of the transition zone where the corona converges to the leader tip. In the first step of the model simulation, the radius and electric field of each leader segment produced during each simulation time step are computed with equations given in Section 4.3.2.6 (subsection 'Dynamic leader inception evaluation'). In the second step, the charge per unit length required to thermalize a new leader segment is computed by estimating the specific power input at the tip of the leader channel. This charge per unit length is required to properly estimate the leader advancement caused by the charge in the streamer corona zone at the tip of the leader channel. The thermodynamic analysis of the transition region where the corona converges to the leader tip proposed for Gallimberti [38] is used to estimate the charge per unit length required for the creation of a new leader segment. This theory assumes that the creation of a new leader segment takes place when the temperature rise in the transition zone is high enough to produce thermal detachment of negative ions. Accordingly to this theory, the charge per unit length required to achieve the transition to a new leader segment can be estimated as

$$q_L = \frac{I_L}{K \cdot \left(f_{ert} + f_v \cdot \left(\frac{\tau_1}{\tau_1 + \tau_{vt}}\right)\right) \cdot \int_{l_L}^{l_t} (J \cdot E) \, dl} \quad (4.62)$$

where I_L is the leader current, f_{ert} is the fraction of the energy transferred into electronic, rotational and translational excitation and f_v is the fraction of the collision energy transferred into the vibrational reservoir. The term $\tau_1/(\tau_1 + \tau_{vt})$ represents the fraction of the vibrational energy which can be relaxed into thermal energy during the leader transition time, τ_1. This fraction depends on the vibrational relaxation time τ_{vt}. The integral term in the denominator corresponds to the specific power available in the transition zone. It is defined by the product of the current density J and the average electric field E across the transition zone Δl_1 defined as the separation between the leader tip l_L and a point l_t where the specific power becomes negligible. In this way, the thermal energy is released in the transition zone during the leader transit time $\tau_1 = \Delta l_1/v_L$, where v_L is the leader velocity. This leads to an increase in the temperature in front of the leader until the transition from the corona to the new leader channel segment takes place. The parameter K is a constant that depends on the critical temperature required for the transition and the density of neutrals in the transition zone.

By representing the leader channel with the 'charge simulation method' as shown in Figure 4.31, the specific power input at the tip of the leader channel is computed and equation (4.62) evaluated. In this way, it is possible to estimate the charge per unit length required to thermalize a new leader segment. During the evaluation of equation (4.62) it is assumed that all the energy is transferred into vibrational excitation ($f_{ert} = 0, f_v = 1$) and that the current density J is approximately equal to the ratio of the leader current I_L and the surface area of the transition zone at each radial distance from the leader tip. The vibrational relaxation time τ_{vt} is taken as 100 μs [38] and the value of the constant K is set in such a way that the value of q_L computed with (4.62) is equal to 65 μC m^{-1} when the leader velocity reaches 2×10^4 m s^{-1} [50].

Becerra and Cooray [104] applied the model to predict the features of the upward connecting leader observed in altitude rocket-triggered lightning experiment reported in Reference 49. The details of this experiment are described in Section 4.3.2.6 (subsection 'Comparison of different models'), but for the convenience of the reader the description is reproduced here. In this experiment, a rocket was launched toward the cloud overhead, spooling 50 m of ground wire, followed by 400 m of insulating Kevlar and from it to the rocket tail a second (floating) copper wire. An upward leader was initiated and propagated upward from the top end of the floating wire,

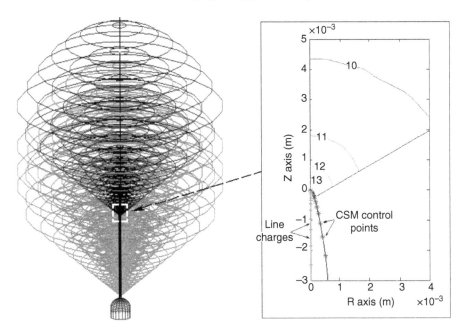

Figure 4.31 Detail of the representation of the leader channel tip and the specific power in the corona zone in the self-consistent leader progression model of Becerra and Cooray [104]. The contours correspond to the logarithm of the specific power evaluated taking into account the effect of space charge produced during previous steps.

leading to the initiation of a downward stepped negative leader at the bottom end of the floating wire. As a consequence, a positive upward connecting leader was initiated from the tip of the grounded wire in response to the triggered downward moving negative leader. The current of the connecting leader, the background ground electric field change at 50 m from the grounded wire and the leader luminosity (with static and streak photography) were measured simultaneously during the experiment [49].

In order to reproduce the experimental conditions, the background electric field necessary for model simulation is calculated using the charge of the downward moving negative leader. This charge was inferred from the ground-level electric field measured at 50 m from the ground wire [49]. Good agreement between the predictions of the proposed self-consistent leader progression model and the experimental data is found. The connecting leader in the experiment started its continuous propagation around 4.02 ms, which is in excellent agreement with the leader inception time of 4 ms calculated by the model (see Figure 4.24). There is also good agreement between the simulated final jump time (at 4.33 ms) and the value observed in the experiment (at 4.37 ms). Furthermore, good agreement was found between the predicted leader current and the continuous component of the current measured in the experiment.

230   Lightning Protection

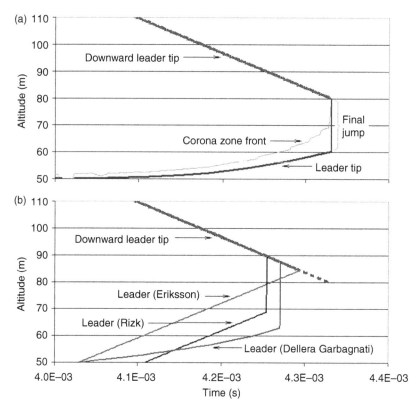

Figure 4.32   Comparison of the streak image computed for the rocket-triggered experiment reported in [49] with (a) the self-consistent model presented in Reference 104 and (b) the leader progression models of Eriksson [58,59], Dellera and Garbagnati [28,60] and Rizk [96,98]

Figure 4.32a shows the streak image of the altitude rocket-triggered experiment [49] simulated with the self-consistent leader progression model presented in References 104 and 106. For the sake of comparison, the predictions of the leader progression models of Eriksson [58,59], Dellera and Garbagnati [28,60] and Rizk [96,98] are included in Figure 4.32b. First, notice that different leader progression models predict different times of leader inception. The reason for this is the different leader inception criteria used by these models. In their models Eriksson and Dellera and Garbagnati use the critical radius concept [54], whereas Rizk uses his own leader inception model [62]. Second, note that the height of the downward leader tip, the connecting leader length and the time at the moment of interception estimated by the existing leader progression models differ considerably from the values computed by the model of Becerra and Cooray [104]. Because the charge of the downward moving leader was the same in all cases, these differences can only be attributed to the manner in which the connecting leader is represented in different models.

Consequently, the differences in the predictions of different leader progression models as shown in Figure 4.32b are caused by a combination of errors introduced by different assumptions made in these models regarding the inception, velocity and channel properties of the connecting leader [106].

In order to test the validity of the assumptions concerning the properties of upward connecting leaders made in the leader progression models of Eriksson, Dellera and Garbagnati, and Rizk, Becerra and Cooray [104] used the self-consistent leader progression model to extract different features of the upward connecting leader pertinent to the triggered lightning experiment describe above. The computed values of the velocity, average channel electric field and charge per unit length of the upward connecting leader are shown in Figure 4.33. Notice that the constant upward leader velocities assumed by Eriksson [58,59] and Rizk [96,98] exceed by several times the computed values, particularly immediately after inception (Figure 4.33a). In the same way, the model of Dellera and Garbagnati [28,60] also overestimates the connecting leader velocity, although to a lesser degree. The direct effect of this is the overestimation of the upward leader length, which in turn leads to a larger interception distance. Regarding the properties of the leader channel, note that the average electric fields computed with the semi-empirical equation [62] assumed by Rizk [96,98] in his model are lower than the self-consistently calculated values (Figure 4.33b). An underestimation of the leader electric field leads to a larger average electric field between the tips of both leaders, resulting in a larger final jump distance. This in turn gives rise to an earlier final jump time [106].

As for the charge per unit length of the connecting leader, the constant value assumed by Dellera and Garbagnati [28,60] disagrees with the values computed with the self-consistent model (Figure 4.33c). Moreover, the leader charge per unit length was found to increase as the connecting leader speeds up, which disagrees with the assumption of constant charge density along the connecting leader channel. It is also worth mentioning that in their model Dellera and Garbagnati represented the upward leader channel by a line charge. This simplification causes large errors in the evaluation of the potential of the leader channel. Moreover, the total charge of the upward leader is partly located within the corona sheath and partly on the channel and the streamer corona zone of the connecting leader [106]. Thus, it is not appropriate to concentrate all of it only along the leader channel. This assumption also leads to an overestimation of the average electric field between the tips of the leaders affecting the correct evaluation of the final jump condition in the model of Dellera and Garbagnati.

4.4.7.2 Self-consistent lightning interception model (SLIM)

The detailed comparison of model predictions with experiment given in the previous section shows that the self-consistent leader progression model of Becerra and Cooray is better suited to evaluate the attachment of a stepped leader to a grounded structure. Moreover, it can be used to a perform sensitivity analysis to investigate how different parameters of stepped leaders and the geometry of the grounded structures influence the attachment process. However, in order to facilitate the implementation of

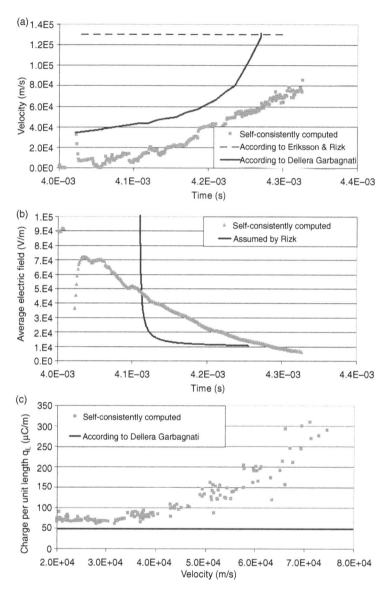

Figure 4.33 Comparison between the upward connecting leader parameters self-consistently computed with the model presented in Reference 104 and the values assumed by the leader progression models for the rocket-triggered experiment reported in Reference 49: (a) leader velocity, (b) average electric field along the leader channel and (c) charge per unit length of the leader as a function of its velocity (adapted from Reference 106)

the model for such an analysis, a simplified procedure to evaluate the charge of the corona zone in front of the leader tip needs to be developed. This is based on the assumption that the total corona charge in front of the leader tip can be determined from the difference between the background geometrical potential distribution and the potential distribution after corona formation. This involves a numerical analysis of the background potential distribution as the leader propagates. The basic procedure of calculation is similar to that developed in Reference 72. In addition, the charge distribution along the downward moving stepped leader is evaluated with the equation proposed by Cooray and colleagues [31].

With these simplifications the model presented in the previous section can be easily used to self-consistently evaluate the initiation and propagation of upward connecting leaders in the presence of downward moving lightning stepped leaders. This version of the model is called the self-consistent lightning interception model (SLIM). An example of the predictive power of SLIM when applied to study the attachment of a lightning flash to a 60-m-tall tower is shown in Figure 4.34a and b. Figure 4.34a shows the simulated streak image and the variation of the velocity of the upward connecting leader under the influence of a stepped leader approaching with different average velocities [105]. Note that, as mentioned in Reference 73, the height of the leader tip when the connecting leader is incepted increases with increasing leader velocity. The reason for this is that, in contrast to a slowly moving stepped leader, a fast-moving one generates a rapidly changing electric field, which facilitates rapid inception and fast propagation of the upward leader [105]. This effect leads to a longer striking (inception) distance. The predicted final value of the upward leader velocity under the influence of a slowly moving downward leader ($V_{\text{down}} = 8 \times 10^4$ m s^{-1}) with a prospective return stroke current of 87 kA is close to 8×10^4 m s^{-1}. For a downward moving leader with the same prospective return stroke current but moving down at 1×10^6 m s^{-1} [105] the final speed of the connecting leader is 1.2×10^5 m s^{-1}.

However, when a downward leader approaches ground with high velocity, the time available for the development of the connecting leader is drastically reduced. As a consequence, the length of the upward connecting leader at the moment of the final jump (i.e. the connection of the streamer zones of both leaders) decreases considerably for fast-moving downward leaders. For the case considered here, the predicted length of the upward connecting leader for a slowly moving downward leader ($V_{\text{down}} = 8 \times 10^4$ m s^{-1}) is ~172 m, but is only 30 m for a fast-moving downward leader ($V_{\text{down}} = 1 \times 10^6$ m s^{-1}) [105].

Figure 4.34b presents the predictions of the model for the case of a 60-m-tall tower under the influence of stepped leaders associated with different prospective return stroke peak currents (i.e. different charge densities). In the simulations, the downward leader is assumed to be directly over the tower. As one may expect, the striking distance, the length of the upward connecting leader and the final jump distance increase with increasing prospective return stroke current. Interestingly, the final velocity of the upward connecting leader also increases with the prospective return stroke current. The estimated upward leader velocity at the moment of connection with a downward leader moving with average velocity V_{down} equal to 2×10^5 m s^{-1} is ~4.5×10^4 m s^{-1} for a

234 *Lightning Protection*

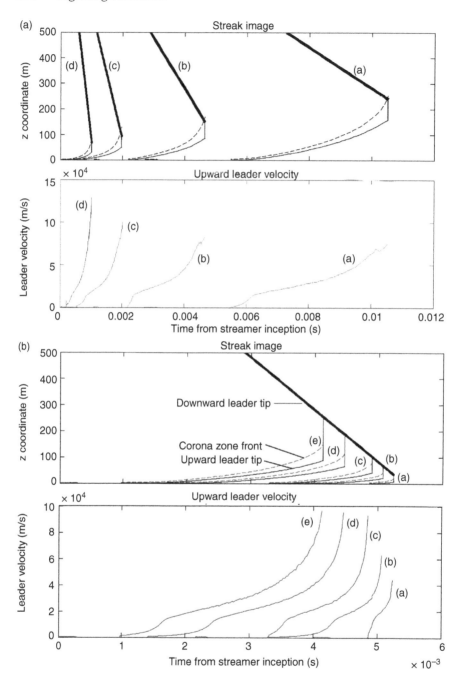

Figure 4.34 *(a) Simulated streak image and leader velocity variation of the upward connecting leader for different values of the downward leader velocity for a lightning flash with 87 kA return stroke peak current striking.*

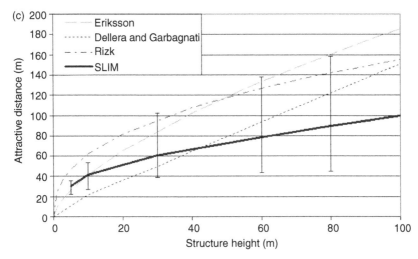

Figure 4.34 *(Continued) a 60-m-tall mast on a mountain: (a) 8×10^4 m s^{-1}, (b) 2×10^5 m s^{-1}, (c) 5×10^5 m s^{-1}, (d) 1×10^6 m s^{-1}. (b) Simulated streak image and leader velocity variation of the upward leader for different prospective return stroke currents for a lightning flash striking a 60-m-tall mast: (a) 10 kA, (b) 20 kA, (c) 40 kA, (d) 70 kA, (e) 110 kA. The downward leader is located directly over the tower and propagates with an average velocity v_{down} of 2×10^5 m s^{-1}. (c) Attractive distance computed with SLIM for free-standing structures corresponding to a downward leader with a prospective return stroke current of 31 kA and an average velocity of 2×10^5 m s^{-1}. The error bars show the variation of the attractive distances due to the dispersion of the observed downward leader velocity probability distribution. The predictions of the existing leader progression models are shown as reference.*

prospective return stroke current of 10 kA, while it close to 9.5×10^4 m s^{-1} for a prospective peak current of 40 kA. However, the final upward leader velocity increases only slightly for prospective return stroke peak currents larger than ∼40 kA. This is the case because the corona zone in front of the upward connecting leader extends over a great distance in the case of downward leaders with high prospective return stroke currents, leading to an early final jump condition before the upward leader reaches a higher velocity. A similar analysis made with SLIM also shows that parameters such as the lateral position of the downward leader channel with respect to the tower axis and the ambient electric field do not significantly affect the final value of the connecting leader velocity, but could influence its time of development. These results clearly show that the velocity of connecting leaders changes from one flash to another due to the variations of these parameters. Thus, it is not appropriate to use generalized ratios between the velocity of the downward and upward leaders as is assumed in existing leader progression models [28,59–61,96,98]. Instead, the upward leader velocity has to be self-consistently computed for each case.

Owing to the strong effect of the upward connecting leader velocity on the point of attachment between connecting leader and stepped leader, the aforementioned parameters also influence the attractiveness of any grounded structure to lightning flashes. Figure 4.34c shows the predictions of SLIM for the lateral attractive distance of a free-standing slender structure under the influence of a downward leader moving with an average velocity of 2×10^5 m s^{-1}. In order to illustrate the effect of the downward leader velocity on the attractive distance, calculations are also performed for the lower and upper limits of the measured values of the average stepped leader velocity [49]. Recent measurements with high-speed cameras show that the two-dimensional average velocity of downward negative stepped leaders is distributed between $\sim 9 \times 10^4$ and 2×10^6 m s^{-1}, with a median of 2.2×10^5 m s^{-1} [107]. The variation of the computed attractive distances for downward leader velocities ranging between those limits is shown with bars in Figure 4.34c. The predictions of the existing leader progression models of the Eriksson [58,59], Dellera and Garbagnati [28,60] and Rizk models [96,98] are also shown for comparison purposes.

The large spread of the downward leader velocities observed in nature results in a rather wide range for the attractive distances estimated by the model. For this reason, the attractive distances of free-standing objects range between the limits shown with the bars in Figure 4.34c. In estimating these limits the probability distribution function of the downward leader velocity is considered. This demonstrates that the attractiveness of a free-standing object to lightning does not depend on the prospective return stroke current or the height of the structure alone, but also depends on the downward leader average velocity. This result suggests that in the analysis of the lightning attractive distances of grounded objects one has to take into account also the downward leader velocity to make a better estimate of this parameter. However, the attractive distance computed with SLIM for isolated slender structures for a downward leader velocity of 2×10^5 m s^{-1} can be qualitatively averaged by the following equation, which expresses the attractive distance in terms of return stroke peak current and structure height:

$$R = 1.86 \cdot I^a h^{0.1746} \quad \text{(m, kA)} \quad (4.63)$$

where $a = -1.617 \times 10^{-3} h + 0.6417 h^{0.0932}$. Note that this expression applies only for thin structures with axial symmetry and cannot be applied to evaluate the attractive distance of other objects without such symmetry, such as buildings or complex structures. Furthermore, Figure 4.35 shows an example of the lightning attractive zones computed with SLIM for an isolated tall air terminal (or mast), an air terminal on the roof, a corner and an edge of a simple building. In all the cases, the point of interest is located 30 m above ground. In the simulations, a stepped leader propagating vertically downwards with an average velocity of 2×10^5 m s^{-1} and associated with a prospective return stroke current of 16 kA is considered. The ambient electric field is assumed to be equal to 20 kV m^{-1}.

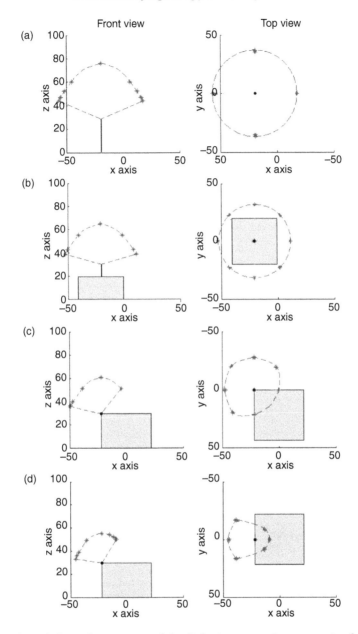

Figure 4.35 Side and top views of the lightning attractive zones simulated with SLIM for (a) a 30-m-tall free-standing air terminal, (b) a 10-m air terminal at the centre of the roof of a $40 \times 40 \times 20$ m^3 building, (c) a corner of a $40 \times 40 \times 30$ m^3 building and (d) an edge of a $40 \times 40 \times 30$ m^3 building. A downward leader with an average velocity of 2×10^5 m s^{-1} and prospective return stroke current of 16 kA were considered.

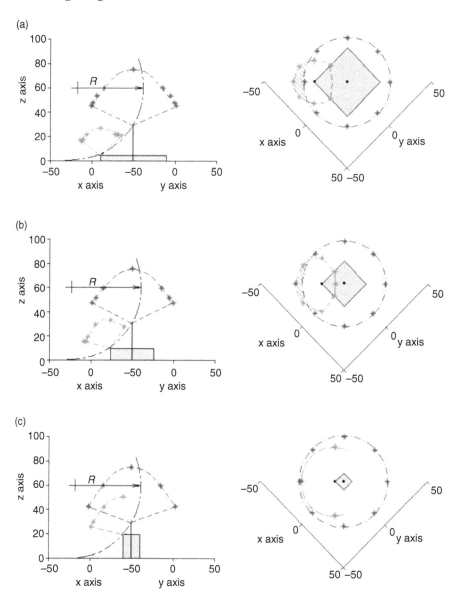

Figure 4.36 Side and top views of a square structure (a) 5 m, (b) 10 m and (c) 20 m tall protected according to the rolling sphere method (R = 60 m) with an air terminal at the centre of its roof. In all cases, the air terminal tip is located 30 m above ground [108]. The figure depicts the attractive zones corresponding to one corner and the air terminal simulated with SLIM. The dashed area in the top view corresponds to the exposure area of the corner to vertical downward stepped leaders.

Note that the attractive zone in each case depends not only on the return stroke peak current, as the EGM predicts, but also on the geometry of the point of interest [108]. Observe also that the lightning attractive zones of corners and edges of structures are not symmetric and circular as is usually assumed [16,17,21,89]. Moreover, it is noteworthy that the maximum attractive distances of an air terminal located either on the roof, at a corner or on an edge of a structure are smaller (29.1, 24.5 and 21.8 m, respectively) than the attractive distance of a free-standing air terminal with the same height (33.8 m). This result clearly confirms that the analysis of the lightning attractiveness of air terminals, corners and edges of buildings cannot be studied on the basis of the attractive zones of free-standing air terminals or masts as is done in the CVM/FIM method [94,109].

Because the effective lightning attraction zones of any structure can be easily calculated with SLIM, it is also possible to make a quantitative comparison of its predictions with the results of the rolling sphere method. Figure 4.36 shows an example of the lightning attractive zones predicted by SLIM for a simple square structure protected according to the rolling sphere method with an air terminal at the centre of the roof. Because protection level IV is considered in the calculations, the structures (including their corners) are inside the protected zone given by the rolling sphere with a radius of 60 m (corresponding to a critical return stroke peak current of 16 kA). The analysis is based on stepped leaders moving vertically downward with an average speed of 2×10^5 m s^{-1}. Owing to the symmetry of the geometry, only the attractive zones of a single corner are computed.

First, note that in the analysis the upper end of the air terminal is kept at the same place while the height of the building is changed from case to case. Notice that the lightning attractive zones of the corners of the structure are not entirely covered by the attractive zone of the air terminal in some cases (Figure 4.36a and b). In these cases the corners are exposed to direct lightning strikes with a prospective return stroke current of 16 kA, even though they are in the 'protected' zone as predicted by the rolling sphere method. However, in other cases the attractive zones of the rolling sphere method agree with the predictions of SLIM (as in Figure 4.36c). This means that in some cases the corners of structures protected by the rolling sphere method are vulnerable to lightning strikes.

As shown above, the study of the lightning attractive zones of structures by SLIM can give valuable information about the conditions under which the rolling sphere method succeeds or fails to properly identify the protected and unprotected zones of a structure. Moreover, such analysis can also provide information as to the correct radius of the rolling sphere that should be used in determining the location of terminals on structures. Further studies pertinent to other cases using SLIM and a preliminary discussion of the validity of the rolling sphere method to locate air terminals on simple structures is given in Reference 108.

4.5 Non-conventional lightning protection systems

External lightning protection systems used by engineers in different countries can be divided into two categories: conventional and non-conventional lightning protection systems. Conventional systems use Franklin rods, the performance of which has been validated in a large number of studies conducted around the globe over many decades. Early streamer emission rods and dissipation arrays (sometimes called charge transfer systems) belong to the category of non-conventional lightning protection systems. The latter systems have been introduced into several lightning protection standards without testing them over a long period of time in the field to assess and validate their performances. Unfortunately, recent studies raise doubts on the specified performances of these systems [123,124]. In the sections to follow we will summarize the results of studies pertinent to these systems as reported in the scientific literature.

4.5.1 The early streamer emission concept

Since the middle of the twentieth century, laboratory experiments in long air gaps have been a source of information in understanding some of the basic physical mechanisms of lightning [11,12,110]. According to this information, in a rod-to-rod configuration stressed by a switching impulse voltage, leader discharges of opposite polarity may develop both from the high-voltage and earthed electrodes. When these two discharges meet somewhere in the middle of the gap, the conditions necessary for the final breakdown process are achieved. The similarity of this process to the final stage of the lightning flash where the down-coming stepped leader is met by an upward moving connecting leader makes it possible to relate the final stage in the development of the lightning stroke to the phenomenon observed in the laboratory [111]. As a result, some of the physical properties of both the negative downward lightning leader that propagates from the cloud towards the ground and of the upward connecting positive leaders initiated from grounded objects were first interpreted based on the leaders observed in the laboratory [11,12]. As mentioned several times previously, it is important to understand that laboratory experiments cannot fully simulate the conditions under natural lightning [25]. This is the case because most laboratory leaders are not long enough to become fully 'thermalized', and therefore leaders in the laboratory require larger background electric fields to propagate in comparison with the lightning leaders [51]. Hence there are reasons to be concerned about the validity of the procedures in which experimental results obtained from leaders in laboratory long air gaps are utilized and extrapolated to obtain information relevant to lightning [112]. Notwithstanding these concerns, long gap laboratory experiments are currently used to simulate the conditions under which upward positive leaders are initiated from lightning rods under natural conditions [113–118]. The continuation of this practice is fuelled by the recent use of laboratory experiments to assess the efficiency of early streamer emission (ESE) terminals to attract lightning as stipulated in some national standards [119,120]. The proponents of ESE devices claim that these terminals have a larger lightning protection zone than the ones offered by a conventional Franklin rods under similar conditions [119–121]. These claims are usually substantiated by the fact that an earlier initiation of streamers in an air gap in the laboratory under switching voltages leads to the reduction of the leader initiation time and therefore to a shorter time to breakdown. This reduction of the leader initiation time observed in the laboratory has been arbitrarily extrapolated to the natural case by

ESE supporters. The main assumption behind this extrapolation is that the switching electric fields applied in the laboratory 'fairly approximate' the electric fields produced by the descent of a negative downward moving leader [113–115].

The ESE terminals used in practice are equipped with a discharge triggering device to initiate streamers from the terminal in an attempt to increase the probability of inception of a connecting leader from the terminal during the approach of a downward lightning leader [119–121]. According to the proponents of ESE, the time advantage realized by the early inception of the connecting leader from an ESE terminal in comparison to a normal Franklin rod would provide a possibility for the connecting leader generated by an ESE terminal to travel a longer distance in comparison to that from a Franklin rod. Consequently, it is claimed that under similar circumstances an ESE terminal will have a larger protection area than a Franklin rod of similar dimensions.

Notwithstanding these claims, the discussion of the efficiency of such air terminals has been the subject of much controversy recently. This is due to the reasonable doubts that exist on the validity of laboratory experiments to assess the efficiency of air terminals and on the procedures used to evaluate the performance of ESE devices [123–125]. Although the best way to evaluate the efficiency of air terminals is to test them in the field under natural conditions, there are several practical limitations that make it difficult to gather conclusive experimental evidence from such tests. Hence, until recently there was a lack of scientific and technical evidence either to reject or to accept these devices [124]. Fortunately, advances in both field observations and theoretical studies made in recent years have led to a growing body of evidence that clearly suggests that these devices do not have superior performance compared to conventional Franklin rods. Let us briefly present the results of experimental and theoretical studies that are in conflict with the claimed performance of ESE devices.

4.5.1.1 Experimental evidence in conflict with the concept of ESE

As mentioned above, the proponents of ESE suggest that the attractive distance of an ESE terminal is larger than that of a Franklin rod. This claimed advantage is taken into account when placing ESE terminals on grounded structures. However, case studies conducted by Hartono and colleagues [20,126] in Malaysia provide undisputable evidence that lightning does bypass the ESE terminals and strike the protected structures well within the claimed protective region of the ESE devices. Two examples provided by Hartono and colleagues are shown in Figure 4.37. The same study showed that no damages were observed on structures equipped with Franklin rods installed according to the international lightning protection standard to cover the vulnerable points such as edges or corners of the structure. However, in structures where Franklin rods were installed without consideration of these high-risk interception points, lightning strikes have been observed at these points.

In another study conducted in New Mexico [87,88], ESE lightning rods were allowed to compete with symmetrically spaced Franklin rods to validate the enhanced attractive zone of ESE devices claimed by its proponents. If, as claimed, ESE rods can initiate an upward leader before the Franklin rods and if they have a larger attractive zone, then one would expect ESE rods to be the preferential point of attachment of

242 *Lightning Protection*

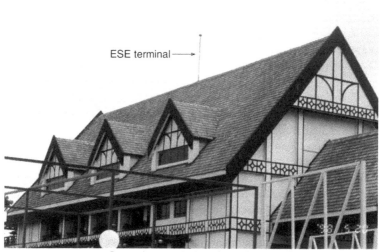

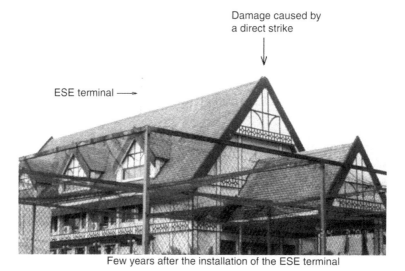

Figure 4.37   Photographs showing the effects of a lightning strike within the claimed protective space of an ESE terminal [20,126] (photograph courtesy of Dr Z.A. Hartono)

the lightning strikes. However, according to the observations, all the lightning strikes attached to the Franklin conductors and not a single one terminated on the ESE devices. This experiment conclusively proves that ESE terminals do not have an advantage over Franklin rods, and the claimed enhanced protective range does not exist.

Proponents of ESE sometimes refer to an experiment conducted in France using triggered lightning [127] to support the action of ESE terminals. In this experiment an ESE terminal was put in competition with a Franklin rod to attach to a downcoming leader created in an altitude triggered lightning experiment. The downward moving leader attached to the ESE terminal, and the proponents of ESE claim that this proves the superior action of ESE terminals in comparison to Franklin rods. However, it is important to note that in the experiment the ESE terminal was located closer to the rocket launcher than the conventional one. The reason for the attachment of the lightning flash to the ESE rod could simply be due to the spatial advantage it had with respect to the conventional rod. Unfortunately, the positions of the rods were not interchanged to validate the claimed enhanced attractive range of the ESE terminal. Thus, one has to conclude that this experiment does not provide any evidence for the claimed superiority of the ESE terminals over conventional ones.

4.5.1.2 Theoretical evidence in conflict with the concept of ESE

The whole concept of ESE is based on the observed fact that by artificial triggering of streamers from the tip of a lightning terminal (i.e. ESE rod) stressed by a switching impulse, one can cause the terminal to initiate a leader earlier than from a lightning terminal placed under identical circumstances but without the action of artificial streamers (i.e. Franklin rod) [124]. In the laboratory, it was found that the time advantage (i.e. the time interval between the initiation of leaders from ESE and Franklin rods) Δt of an ESE terminal is ~ 75 μs. Proponents of ESE terminals have taken this laboratory observation and extended it to natural conditions, claiming that a 75-μs advantage will give rise to a length advantage equal to the product $v\,\Delta t$ where v is the speed of the upward moving leader. Assuming a leader speed of 1×10^6 m s^{-1}, they claim that an ESE terminal would have a length advantage of ~ 75 m over a conventional rod. Thus, the whole concept of the ESE device is based on two assumptions:

1. The early initiation of leaders from ESE terminals observed in the laboratory also takes place under natural conditions. In other words, an ESE terminal can launch a connecting leader long before a conventional rod under natural conditions.
2. The time advantage observed will translate to a length advantage $v\,\Delta t$ over a conventional terminal.

Let us discuss these assumptions separately.

Can one extrapolate the action of early streamer emission rods in the laboratory to natural conditions?
As mentioned above, the claimed action of ESE devices is based on the fact that an artificial triggering of streamers at the tips of ESE rods stressed with switching voltages leads to the reduction of the leader initiation time and therefore to a shorter

time to breakdown [113]. The supporters of ESE arbitrarily extrapolate this reduction in the leader initiation time observed in the laboratory to natural conditions.

The first question that needs to be solved in order to evaluate the efficiency of ESE devices is whether the time advantage observed in the laboratory exists also under natural conditions. In order to answer this question, it is necessary to understand the effect of artificial initiation of streamers from a lightning rod and its effects on leader inception. Moreover one has to understand how this interaction is controlled by the time-varying background electric field. The physics behind these processes can be understood only through a careful analysis of the temporal variation of the background electric field in combination with the statistical time lags associated with leader initiation. Unfortunately, the problem of statistical time lags is very complex and has been avoided by most existing models of leader discharges [124].

Recently, Becerra and Cooray [82] utilized a self-consistent leader inception model, described previously in Section 4.3.2.6 (subsection 'Dynamic leader inception evaluation'), to investigate this problem. Using this model they simulated the initiation and development of positive leaders under the influence of time-varying electric fields used in the laboratory as well as the time-varying electric fields generated at ground level by the descent of the downward leaders. In the sections to follow we will describe the results of this investigation.

The early streamer emission concept under switching impulse voltages. In order to reproduce the conditions under which the early streamer principle was discovered in the laboratory, Becerra and Cooray [82] performed their simulations using an electrode configuration similar to the one used in References 113–115. This consisted of a 3.5-m-tall grounded air terminal placed under an energized plane electrode located 13 m above the ground plane. In the simulations, a switching voltage impulse waveform with 3.2 MV peak value and 350 μs rise time was chosen to roughly reproduce the conditions reported in References 113–115. As in the experiment, this voltage impulse was superimposed on a d.c. voltage equal to 130 kV to reproduce the thundercloud electric field of 10 kV m^{-1}.

Figure 4.38 shows the simulated streak image of a positive leader propagating in the gap under the influence of the switching voltage impulse as simulated by Becerra and Cooray [82]. In this simulation, in order to consider the statistical time lag relevant for streamer inception and its effect on the leader initiation time, two extreme cases for streamer inception times are considered. The lower extreme (Figure 4.38a) corresponds to the minimum possible streamer inception time $t_i^{(min)}$ given by the well known streamer criterion [39]. The upper limit (Figure 4.38b) is the probabilistic maximum streamer inception time $t_i^{(max)}$, where the probability to find a free electron to initiate the streamer is close to one [11,39].

As can be seen in Figure 4.38, the simulated unstable and stable leader inception times t_i' and t_1, as well as the time to breakdown t_B, decrease when the streamer inception t_i takes place earlier. Thus, if a streamer is 'triggered' earlier, a reduction of the leader inception and breakdown times is obtained. This predicted improvement of the leader inception time in the laboratory by reducing the streamer initiation time

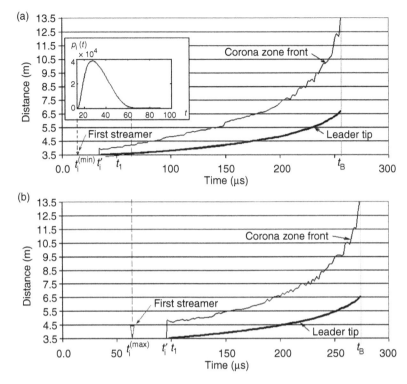

Figure 4.38 Simulated streak image of the propagation of a positive leader in a 9.5-m-long air gap under a switching voltage impulse for different streamer inception times: (a) the minimum possible streamer inception time, (b) the probabilistic maximum streamer inception time (adapted from Reference 82)

agrees with the experimental results presented by Berger [113–115]. Their laboratory tests showed that the leaders initiated from a terminal with a streamer triggering unit starts very early, well below the inception times of leaders from the Franklin rod. Based on streak images obtained during the experiment, the ESE device tested by Berger [113–115] showed a time advantage of ∼75 μs in the leader inception time compared with the control Franklin rod [113]. Consequently, the mean value of the time-to-breakdown probability distributions of the tested ESE terminal were also lower compared with those of the control Franklin rod [113]. Note that the simulation shown in Figure 4.38 also predicts that the time to breakdown t_B in the air gap is reduced when the streamer initiation occurs earlier.

Theoretical analysis confirms that a time advantage on leader initiation can be obtained in the laboratory under switching impulses by triggering an early streamer, as reported in References 113–116. The next main question is then 'Can one extrapolate the results to natural lightning conditions?' The proponents of ESE claim that because a switching electric field produced in the gap 'fairly approximates' the rising electric field produced by the downward lightning leader, it is possible to assume that the same results would also be obtained under natural conditions. Let us consider this assumption now.

The early streamer emission concept under lightning-like electric fields. In order to evaluate how well the switching voltage waveform approximates the lightning electric fields, the simulations are repeated by using the electric field produced by the descent of the downward moving leader as an input. In the analysis, the potential of the upper plane electrode is defined in such a way that the electric field in the gap is identical to that produced by a down-coming stepped leader propagating at 2×10^5 m s^{-1} to ground directly over the rod. The charge on the leader is such that it can give rise to a return stroke peak current of 5 kA. Figure 4.39 shows the simulated streak image of a positive leader propagating in a laboratory air gap under the influence of lightning-like electric fields for the two extreme conditions of streamer inception $t_i^{(\min)}$ and

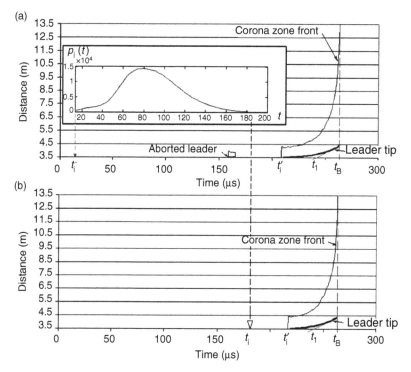

Figure 4.39 Simulated streak image of leader propagation in the laboratory under an electric field similar to that produced at ground level by a downward leader with a prospective return stroke current of 5 kA for different streamer inception times: (a) the minimum possible streamer inception time, (b) the probabilistic maximum streamer inception time (adapted from Reference 82)

$t_i^{(max)}$. Observe that the time to stable leader inception t_1 and the time to breakdown t_B are not affected by the streamer inception time t_i. In other words, the simulation shows that the time advantage that was present when the gap is excited by the switching impulse disappears when the gap is excited by electric fields similar to those produced by down-coming stepped leaders. Note that this time advantage is not present even in the extreme case in which a streamer is triggered 180 μs earlier (at $t_i^{(min)}$) compared with a late streamer onset (at $t_i^{(max)}$).

The reason why the time advantage found under switching waveforms is not present under lightning-like electric fields is due to the differences in the rate of change of the electric field in the two cases (Figure 4.40). The rate of increase of the lightning-like electric field changes from slow to fast as the downward leader approaches, but the rate of change of the switching electric fields applied in laboratory changes from fast to slow with increasing time. Because of this difference, initiation and propagation of positive leaders under lightning-like electric fields are different to

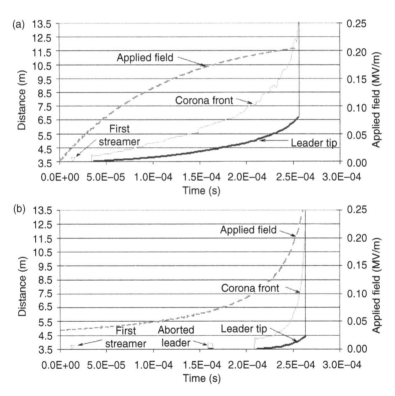

Figure 4.40 Simulated streak images of the initiation and propagation of a leader discharge (a) under a laboratory switching electric field and (b) under the influence of an electric field produced by the descent of a downward leader with prospective return stroke current of 5 kA and average velocity of 2×10^5 m s^{-1} (adapted from Reference 82)

those of leaders present under the switching waveform (Figure 4.40). First, observe that the length of the leader in the laboratory is shorter under the lightning-like electric field than under the switching waveform. In the case considered here, the leader simulated under the influence of the lightning-like electric field is about three times shorter than the one simulated under the switching waveform. Second, the unstable leader inception time t_i' takes place a long time after the inception of the first streamer t_i for lightning-like electric fields. For the switching waveform, this time difference $t_i' - t_i$ is only ~ 25 μs, several times shorter compared with the estimated 200 μs for the lightning-like electric field. In addition, notice that more than one streamer burst can be produced before the initiation of the leader in the case of lightning-like electric fields. Thus, if a streamer is triggered earlier from a rod exposed to lightning-like electric fields, further bursts of streamers and aborted leaders would be produced, without any significant change in the stable leader inception time t_1. Third, the time difference between the breakdown t_B and stable leader inception t_1 in the gap is significantly shorter when lightning-like electric fields are applied. For the lightning-like waveform, the time span $t_B - t_1$ is shorter than 40 μs for the considered case, whereas this time span is more than 150 μs for the switching impulse.

The results presented above clearly show that the switching voltage impulses used in the laboratory do not 'fairly approximate' the electric fields produced by the descent of a downward leader, as claimed in References 113–115. Consequently, the 'time advantage' in the initiation of leaders from terminals observed under switching impulses is not present in lightning-like electric fields. Hence, it is not appropriate to use laboratory experiments conducted with switching impulses to evaluate the efficiency of lightning terminals to attract lightning, as recommended by several national standards [119,120]. Such experiments do not have the capacity to expose the physics of leader discharges generated under lightning-like electric fields. The results presented above show conclusively that the conditions necessary for initiation and propagation of leaders in lightning flashes cannot be extracted from experimental results relevant to leaders created using switching impulses. The same applies to models created using information gathered from laboratory sparks created by switching impulses.

The results obtained from lightning-like electric fields in the laboratory cannot elucidate completely how air terminals will perform in nature when exposed to down-coming stepped leaders. This is the case because in the laboratory the leaders and their associated streamer regions do not have enough space to grow because of the limited space available. Becerra and Cooray [82] therefore performed simulations to study the initiation and propagation of leaders in free space when lightning terminals are exposed to the electric fields of down-coming stepped leaders. From that study they also concluded that the early streamer emission principle does not produce any improvement in lightning attachment. For example, Figure 4.41 shows the predictions pertinent to the development of an upward positive leader connecting a downward moving negative leader with prospective return stroke current of 10 kA. In this case, features similar to the ones obtained in the laboratory when the exciting electric field is lightning-like are obtained. Therefore, there is no any change in the length of the upward leader at

Attachment of lightning flashes to grounded structures 249

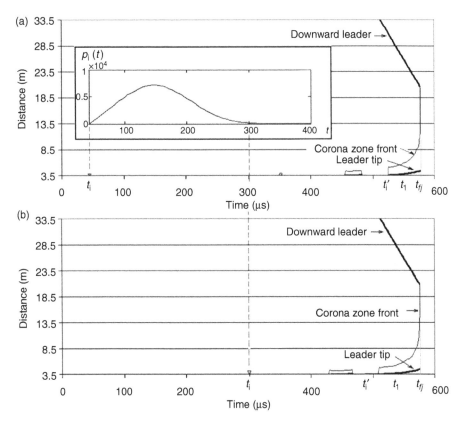

Figure 4.41 Simulated streak image of leader propagation under the electric field produced by a downward leader with prospective return stroke current of 10 kA for different streamer inception times: (a) the minimum possible streamer inception time, (b) the probabilistic maximum streamer inception time (adapted from Reference 82)

the moment of the connection with the downward leader when a streamer is initiated early. In this case, the connection of both leaders takes place at the same instant, regardless of the time of streamer inception. Even if the time difference of the streamer inception times evaluated in Figure 4.41 is ∼300 μs, there is no 'gain' in upward leader length by triggering an early streamer. This result clearly shows that even if ESE terminals increase the probability of streamer inception [119–121], they would not affect the initiation or the length of the self-propagating upward connecting leaders. Based on this theoretical evaluation one can conclude that the ESE principle does not work under natural conditions.

Influence of the amplitude of the voltage pulses applied to the ESE terminal on the propagation of connecting leaders and final jump. Because most commercial ESE devices operate by applying a voltage pulse to the tip of the terminal [122], it is

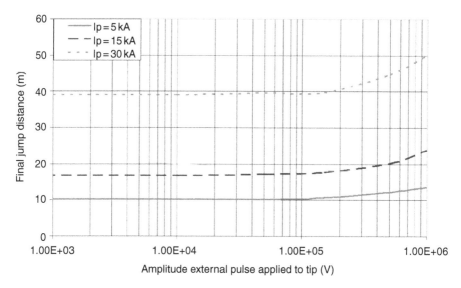

Figure 4.42 Simulated final jump (interception) distance as a function of the amplitude of the external square voltage pulse applied to the tip of a lightning rod. Prospective return stroke peak currents of 5, 10 and 30 kA are considered (adapted from Reference 82).

relevant to investigate the effect of the magnitude of that voltage pulse on the results presented in the previous section. In the simulation the shape of the applied voltage was assumed to be a square. Figure 4.42 shows the predicted distance between the downward leader tip and the ESE rod at the moment of the connection between the upward and the downward leaders (i.e. final jump) for different voltage amplitudes. Simulations are performed for three leader charges corresponding to prospective return stroke peak currents of 5, 10 and 30 kA. As one can see in this figure, the final jump distance is not influenced by the magnitude of the external voltage applied to the terminal unless the peak value of the voltage pulse is larger than $\sim$500 kV. Because the voltage pulses applied to the tip of most ESE terminals are generated from the energy supplied by the ambient electric field, the peak value of such pulses is not larger than a few tens of kilovolts [113,122,124]. Such values are far below the voltage magnitudes required to make any change to the length of the upward connecting leader at the moment of connection between it and the stepped leader. Hence, one can conclude that, contrary to the claims of ESE manufacturers, the external voltage applied to the tip of ESE terminals does not influence the propagation of the upward leader.

The claimed time advantage
The experiments conducted by Berger showed that an ESE terminal, when tested against a switching voltage, has a time advantage of $\sim$50–75 μs [113]. This time advantage was converted to a length advantage of $\sim$50–75 m over a

conventional rod by assuming a leader speed of $\sim 1 \times 10^6$ m s^{-1}. First, the experimental results on leader properties reviewed in Chapter 2 show that the speed of upward leaders immediately after initiation is close to 1×10^4 m s^{-1} and may increase as the leader length increases to values close to 1×10^5 m s^{-1} (see also Saba M., High speed video measurements of an upward connecting positive leader, personal communication, 2007). These values are one to two orders of magnitude smaller than the 1×10^6 m s^{-1} assumed by ESE proponents. If this experimentally observed value of leader speed is used in the conversion of time advantage to distance, the resulting length advantage would be of no use in many practical situations. Second, this conversion of time advantage to a length advantage is not correct, because the eventual length advantage depends on the ratio of the speeds of both downward and upward leaders. If this is taken into account the assumed length advantage will be less than the value calculated by just multiplying Δt by the speed of the leader. Third, according to the proponents of ESE the earlier initiation of a connecting leader from an ESE device occurs in a smaller electric field than is required for the initiation of a leader by a conventional rod. However, for a successful propagation of a connecting leader a certain background electric field is needed. If the background electric field is not large enough, the initiated leader could be aborted. The proponents of the ESE do not consider the requirements for the propagation of a leader and they do not consider the possibility that the initiated leaders could be aborted if the background electric field requirements are not satisfied.

4.5.2 The concept of dissipation array systems (DAS)

Benjamin Franklin conducted static experiments with blunt and sharp conductors. He observed that if he approached a charged conductor with a blunt rod then there was a spark, whereas if it was approached with a sharp conductor the charge was silently discharged without a spark. Extending this static laboratory analogue to dynamic lightning discharges, Franklin hypothesized erroneously that it may be possible to prevent lightning strikes by installing grounded sharp conductors on structures. There was no evidence that sharp points could prevent lightning strikes, but scientists and engineers soon realized that the conductors provided a preferential path for the lightning current without damaging the structure. Unfortunately, this old and incorrect idea of Franklin was resurrected recently in the form of lightning eliminators or dissipation arrays.

The original idea of lightning eliminators or dissipation arrays was to utilize the space charge generated by one or several grounded arrays of sharp points to dissipate the charge in thunderclouds and thus prevent lightning strikes to a structure to be protected. The proponents of this system claimed that the space charge generated by the array would silently discharge the thundercloud. Scientists demonstrated conclusively that this would not be the case, using following arguments. First, a thundercloud generates charge at a rate of about a Coulomb of charge per second, and the charge production rate from dissipation arrays is not large enough to compete with this charging process. The maximum currents from arrays, as claimed by their proponents, are in the range of 500 µA. However, no details of the measurements are given, nor whether this refers to the maximum current or the average is not clear. Even if it is true, it is still not strong enough to neutralize the charge in the thundercloud. Second, the mobility of small ions at ground level is ~ 1 to 3×10^{-4} m^2 V^{-1} s^{-1} and in background electric fields of 10–50 kV m^{-1} the drift velocity may reach 1–15 m s^{-1}. Even if the array can generate charge of sufficient quantities, in the time of regeneration of charge in the thundercloud of ~ 10 s the space charge can move only a distance of ~ 10–150 m. Thus, the space charge would not be able to reach the cloud in time to prevent the occurrence of lightning. Facing this challenging and convincing opposition from lightning researchers the proponents of lightning eliminators accepted that arrays are not capable of neutralizing the cloud charge. In turn they suggested that the function of the dissipation array is to neutralize the charge on down-coming stepped leaders.

A stepped leader may consist of ~ 5 C of charge, and the dissipation array has to generate this charge in ~ 10 s. The proponents of dissipation arrays made the following argument to show the effectiveness of the array in neutralizing the stepped leader. A 10-point dissipation array can produce ~ 1 mA of current. Thus the number of points needed to generate a current that is capable of neutralizing the leader charge is 4 000. In making this claim they have assumed that the current generated by a multipoint array is equal to the current generated by a single point multiplied by the number of points. As one can show (see Section 4.5.2.1) a larger number of points does not necessarily mean a larger current than a single-point array.

More recently, proponents of the dissipation arrays claimed that they work by suppressing the initiation of upward leaders by screening the top of the structure with space charge. This claim was based on the study conducted by Aleksandrov and colleagues [68,128–130]. In that study it was shown that the electric field redistribution due to space charge released long corona discharges near the top of a high object, hindering the initiation and development of an upward leader from an object in a thunderstorm electric field. The finding is in line with the results of Becerra and Cooray [75], who showed that the corona generated at ground level could reduce the probability of upward initiated lightning flashes from tall structures under the influence of electric fields generated by thunderclouds.

The proponents of dissipation arrays claim that according to the anecdotal evidence of the users there is a reduction in the cases of lightning damage after the installation of arrays. However, this does not necessarily mean that the array has prevented any lightning strikes. First, because the array is well grounded it provides a preferential path for the lightning current to go to ground. This itself will reduce the damage due to lightning strikes, even if it does not actually prevent a lightning strike. Second, if the array is connected to a tall mast, due to the geometry itself, the presence of the array can reduce the number of upward initiated flashes. This is the case because the background electric field necessary to initiate upward leaders from a given tower increases with increasing radius of the tip. For example Figure 4.43 shows the background electric fields necessary for streamer inception and stable leader inception for a 60-m-tall tower as a function of its radius. Note how the background electric field necessary

Attachment of lightning flashes to grounded structures 253

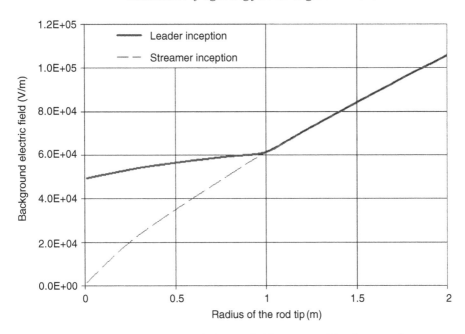

Figure 4.43 Critical thundercloud electric fields at ground level required to initiate streamers and stable upward moving leaders from a 60-m-tall mast as a function of its tip radius

for stable leader inception increases with increasing radius of the tower. Connection of a dissipation array at the top of the mast will increase the effective radius of the mast and, therefore, will require a higher background electric field to launch an upward moving leader. This may lead to a reduction in the number of upward initiated flashes from the tower. However, as noted by Mousa [131], upward flashes are of interest in the case of towers of heights larger than $\sim$100 m and any benefit can be obtained only for these cases.

4.5.2.1 Experimental evidence against dissipation arrays

There are several well documented cases in which lightning has been observed to strike dissipation arrays. The best procedure to conduct such a study is to compare two similar structures, one with a DAS and the other without. Several such studies have been conducted [132–134]. All the studies show that DAS systems were struck by lightning as well as the control structure. No reduction in the frequency of lightning strikes to structures was observed.

Additional experimental and theoretical evidence against some of the principles used by the proponents of dissipation arrays
Effect of rate of change of electric field on corona screening. It is a well documented fact that grounded objects with sharp points, plants and trees go into corona when the

254 *Lightning Protection*

ambient electric field increases above $\sim 1-5$ kV m^{-1}. Depending on the polarity of the ambient electric field, the corona discharge will introduce positive or negative space charge into the surrounding space, which in effect decreases the electric field at ground level. This basic physics associated with the screening of the background electric field from the corona is used by the proponents of dissipation arrays to claim that the space charge produced by these arrays will screen the underlying structure from electric fields generated by down-coming leaders and hence prevent the occurrence of the connecting leaders that mediate lightning attachment to the structure. In claiming this, the proponents of dissipation arrays have completely neglected the basic physics associated with the space charge mediated screening process. The fact that they have neglected is that the space charge can screen an underlying structure from electric fields if the electric field is changing slower than the time constant associated with space charge generation and drift. If the rate of change of the electric field is faster than this time constant, then the space charge will not be able to screen the underlying structure from such field changes.

In order to illustrate this, consider a time-varying electric field produced by a thundercloud and how this field will be modified by ground corona. We will assume that the electric field generated by the charges in a thundercloud increases linearly with time and reaches a steady value after a certain time, t_{ramp}. For simplicity we assume that the electric field is uniform below the cloud. We consider three examples with different values of t_{ramp}. These electric fields are depicted in Figure 4.44a. Figure 4.44b shows how the space charge density produced by the ground corona varies as a function of height. The situation depicted corresponds to a time equal to t_{ramp}. Figure 4.44c depicts the electric field as a function of height at the same time. Observe that when the electric field is changing slowly the corona can completely screen the electric field at ground level. However, the ability of the corona to screen the electric field decreases as the rate of change of the electric field increases. If the rate of change of the electric field is very fast it will not be affected by the corona at all. In a similar manner the electric field generated by the charges in the cloud can be completely screened by the corona charge, whereas the electric field generated by down-coming stepped leaders are not affected at all by the corona space charge.

The corona current generated by a cluster of needles. Cooray and Zitnik [136] conducted experiments to investigate how the corona current produced by an array of sharp points or needles vary as a function of the number of needles in the array. The experimental setup consisted of a parallel plate gap of length 0.3 m with 1.0-m-diameter Rogowski profiled electrodes. The bottom electrode of the gap was prepared in such a way that a cluster of needles could be fixed onto it. The needles used in the experiment were pointed, 2 cm long and 1 mm in diameter. The needles were arranged at the corners of 2×2 cm^2 adjacent squares. A d.c. voltage was applied to the electrode gap and the corona current generated by the needles was measured as a function of the background electric field and the number of needles in the cluster using a micro-ammeter. The lower limit of the corona current that could be measured was ~ 1 μA. The results obtained are shown in Figure 4.45.

Attachment of lightning flashes to grounded structures 255

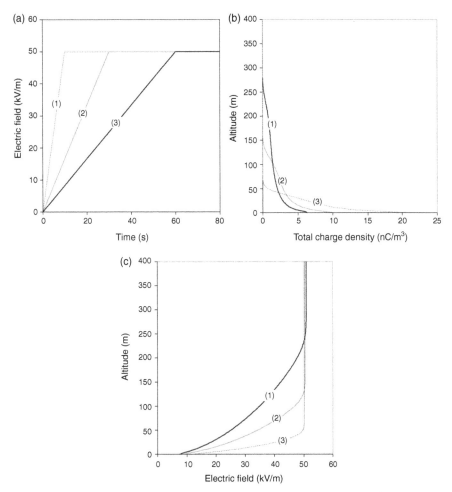

Figure 4.44 The effect of ground corona on the spatial and temporal distribution of the electric field generated by cloud. (a) The electric field below the cloud as a function of time. The electric field is assumed to uniform below the cloud in the absence of corona. (b) The space charge distribution as a function of height at the times when the electric field reaches a steady value. Note that this time is different for different waveforms. (c) Electric field at different heights at the time the electric field reaches a steady value.

Observe first that the corona current increases with increasing electric field and for a given electric field the corona current increases with increasing number of needles. Note, however, that for a given electric field the corona current does not increase linearly with the number of needles. This is probably caused by the electrical screening of the needles by the adjacent ones.

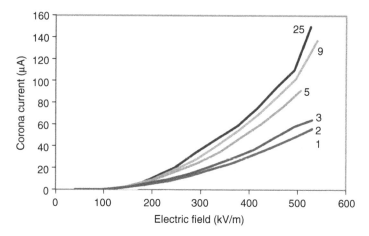

Figure 4.45 *Corona current as a function of the background electric field from clusters of needles. The number of needles in the cluster is indicated (adapted from Reference 136).*

Figure 4.46 depicts the corona current (in µA) at 500 kV m^{-1} (close to the maximum value of the background field achieved in the experiment) as a function of the number of needles. Note that the corona current seems to reach an asymptotic value with increasing number of needles. For example when the number of needles is increased from 1 to 25 the corona current increases only by a factor three. This clearly demonstrates that the assumption made by the proponents of the dissipation arrays that the current from an array increases linearly with the number of needles is not correct.

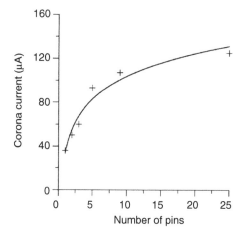

Figure 4.46 *Corona current at 500 kV m^{-1} as a function of the number of needles in the cluster (adapted from Reference 136)*

Effect of space charge on the initiation of connecting leaders. Proponents of dissipation arrays claim that the space charge generated by the array can prevent the formation of connecting leaders and hence the attachment of lightning flashes either to the array or the protected structure. Let us investigate this point now. It is not an easy task to evaluate the screening effect of the space charge generated by a set of needles because the space charge is continuously generated at the needles and it is drifting in a complex manner under the influence of the total (background plus space charge generated) electric field. Cooray and Zitnik [136] simplified the calculating procedure by adopting a particular distribution for the set of needles as follows. Consider a grounded semi-ellipsoid with the base at ground level and immersed in a uniform background electric field. Let $E(\eta_0,\theta)$ be the electric field normal to the surface of the ellipsoid, where η_0 and θ are two of the ellipsoidal coordinates. The surface charge density on the surface of the ellipsoid is proportional to this normal electric field. This surface charge distribution creates a uniform electric field equal in magnitude but opposite in sign to the background electric field inside the ellipsoid. The maximum electric field on the ellipsoid is reached at the location $\theta = 0$. Let us denote that by $E(\eta_0,0)$. Assume that corona needles are distributed over the whole surface of the ellipsoid, and the density of needles $N(\eta_0,\theta)$ on the surface of the ellipsoid is given by

$$N(\eta_0, \theta) = K \frac{E(\eta_0, 0)}{E(\eta_0, \theta)} \tag{4.64}$$

where K is the density of needles at the top of the ellipsoid (i.e. $\theta = 0$). Thus the distribution of the surface charge density in the space charge layer generated by the needles during the time interval between t and $t + \delta t$ is given by

$$\sigma(\eta_0, \theta, t) = c_0 N(\eta_0, \theta)[E(\eta_0, \theta, t)]^2 \delta t \tag{4.65}$$

where $E(\eta_0, \theta, t)$ is the electric field on the surface of the ellipsoid at time t. Substituting from equation (4.64) one obtains

$$\sigma(\eta_0, \theta, t) = c_0 K E(\eta_0, 0, t) E(\eta_0, \theta, t) \delta t \tag{4.66}$$

In the calculations it is assumed that $K = 1\,000$ m^{-2}. Equation (4.66) shows that the charge density of the space charge layer varies in a manner identical to the surface charge density induced on an ellipsoid immersed in a uniform electric field. In our calculations we also assume that as the space charge layer expands due to ion drift it will maintain the shape of an ellipsoid, which is also a simplifying approximation. The electric field, both inside and outside the space charge layer, produced by a space charge layer having an ellipsoidal shape and having a space charge density variation identical to that of (4.66) is known [137]. Now we are ready to investigate the effect of needles.

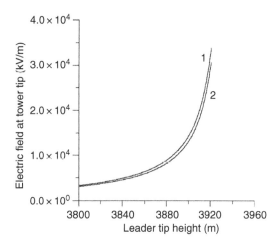

Figure 4.47 *Evolution of the electric field at the top of a 60-m tower located below the stepped leader trajectory as the stepped leader propagates towards the ground. Curve (1) shows the evolution of the electric field in the absence of space charge and curve (2) the electric field in the presence of space charge. The radius of the tower tip is 2 m (adapted from Reference 136).*

Figure 4.47 shows the evolution of the electric field at the top of a 60-m tower during the descent of a stepped leader associated with an 80-kA prospective return stroke current. Results are given both in the presence and absence of needles. In the calculation it is assumed that the leader trajectory, assumed to be vertical, is located directly over the tower. Observe the effect of the space charge in reducing the electric field at the top of the tower. A tower with a tip radius of $\sim$2 m will launch a connecting leader when the electric field at the top of the tower is $\sim 3 \times 10^6$ V m^{-1}. Thus, a tower without the space charge will launch a connecting leader before a tower with similar geometry but with space charge at the tower top. However, the space charge controlled field does not lag far behind the field that would be present in the absence of the space charge. For example, the difference in the leader tip height from the tower top when the electric field at the tower top reaches the critical electric field of 3×10^6 V m^{-1} in the presence and in the absence of space charge is no more than 2 m. Thus, the reduction in the striking distance caused by the space charge is no more than a few metres.

In a recent work Bazelyan and colleagues [138–140] studied lightning attachment to grounded structures taking into account the effect of corona space charge near structures. In their paper they also discussed the question of lightning attachments to dissipation arrays. They conclude that connecting leaders will not be issued from dissipation arrays when exposed to the electric fields of down-coming leaders and therefore they will be struck only in the case in which the leader comes down directly above the dissipation array. They used the following arguments to justify this conclusion:

The starting point for the leader development of a connecting leader from a dissipation array requires the inception of streamers. The inception of streamers from the dissipation array in the presence of glow corona can take place only if the glow corona current is larger than a critical value of 1.9 mA. In a dissipation array the total corona current is distributed over all points and in order for the corona current through a single point to be higher than 1.9 mA the current through all the other points should also be increased beyond 1.9 mA. This means an array of 5 000 points should generate close to 10 A before streamers could be initiated from the array. Because corona discharges cannot sustain such large currents, no connecting leaders could be issued from dissipation arrays when exposed to the electric fields of down-coming stepped leaders.

The main fault of this argument is the assumption that corona discharges will be distributed perfectly uniformly over the dissipation array during the descent of the stepped leader. Even if the dissipation array is perfectly uniform and the electric field is distributed over it uniformly, the turbulent and random nature of the electrical discharges will always cause some points on the array to enhance their current at the expense of the others. The current flowing through such points will increase dramatically under the influence of the intense electric field generated by the down-coming stepped leader. However, in practice, not all the points are identical to each other in an array as far as electrical discharges are concerned, and therefore some points will generate currents that are larger than the others when exposed to electric fields. Streamer discharges may develop from such points, ultimately leading to the inception of leaders when exposed to the electric field of stepped leaders. To look at it from another angle, consider a large sphere (this represent the case $N \rightarrow \infty$, where N is the number of needles in the array) raised to a high voltage. If the arguments raised by Bazelyan and colleagues [138–140] are correct, then as the voltage increases the whole sphere should go into glow corona and no streamers should be issued from it. However, in practice, streamers will be issued from some points on the sphere due to space charge irregularities. Going back to the dissipation arrays, one also has to consider leaders displaced laterally from the array and those approaching it at an angle. In such cases, the tips of all needles on the array will not be exposed to the same electric field. Moreover, the displacement of space charge due to wind and rain drops falling on the array during thunderstorms will also destabilize any symmetry if present. Furthermore, one cannot disregard the possibility that a connecting leader could be issued from the edges of the array. All these considerations show that the argument raised by Bazelyan and colleagues in support of the action of dissipation arrays may not be valid in practice.

References

1. Gay-Lussac F. and Pouillet C. *Introduction sur les paratonnères*. Adopteè par l'Acadèmie des Sciences, Paris, France, 1823.
2. Lodge O.J. *Lightning Conductors and Lightning Guards*. London: Whittaker and Co.; 1892.
3. Golde R.H. 'The lightning conductor' in R.H. Golde (ed.). *Lightning, vol. 2: Lightning Protection*. New York: Academic Press; 1977. pp. 545–76.
4. Preece W.H. 'On the space protected by a lightning conductor'. *Phyl. Mag.* 1880;**9**:427–30.
5. IEC. *Protection of Structures Against Lightning, Part 1: General Principles*. IEC 61024-1, 1990.
6. IEC. *Protection of Structures Against Lightning, Part 1: General Principles*. IEC 61024-1, 1995.
7. IEC. *Protection Against Lightning – Part 1: General Principles*. IEC 62305-1, Ed. 1.0b, 2006.
8. Wagner C.F., McCann G.D. and MacLane G.L. 'Shielding of transmission lines'. *AIEE Trans.* 1941;**60**:313–28.
9. Wagner C.F., McCann G.D. and Lear C.M. 'Shielding of substations'. *AIEE Trans.* 1942;**61**:96–100.
10. IEEE. *IEEE Guide for Direct Lightning Stroke Shielding of Substations*. The Institute of Electrical and Electronics Engineers, Inc., IEEE Standard 998-1996, 1996.
11. Les Renardiéres Group. 'Research on long air gap discharges – 1973 results'. *Electra* 1974;**35**:47–155.
12. Les Renardiéres Group. 'Negative discharges in long air gaps'. *Electra* 1981;**74**:67–216.
13. Armstrong H.R. and Whitehead E.R. 'Field and analytical studies of transmission line shielding'. *IEEE Trans.* 1968;**PAS-87**(1):270–79.
14. Schwaiger A. *Der Schutzbereich von Blitzableitern*. Munich: R. Oldenbourg; 1938.
15. Szedenik N. 'Rolling sphere – method or theory?'. *J. Electrostatics* 2001;**51–52**: 345–50.
16. Lee R.H. 'Protection zone for buildings against lightning strokes using transmission line protection practice'. *IEEE. Trans. Ind. Appl.* 1978;**14**: 465–70.
17. Lee R.H. 'Lightning protection of buildings'. *IEEE Trans. Ind. Appl.* 1979;**IA-15**(3):236–40.
18. Hussein A.M., Milewski M., Janichewskyj W., Noor F. and Jabbar F. 'Characteristics of lightning flashes striking the CN Tower below its tip'. *J. Electrostatics* 2007;**65**:307–15.
19. Darveniza M. 'A modification to the rolling sphere method for positioning air terminals for lightning protection of buildings'. *Proceedings of the 25th International Conference on Lightning Protection*; Rhodes, Greece, 2000. ICLP; 2000. pp. 904–908.

20. Hartono Z.A., Robiah I. and Darveniza M. 'A database of lightning damage caused by bypasses of air terminals on buildings in Kuala Lumpur, Malaysia'. *Proceedings of VI International Symposium on Lightning Protection*; Santos, Brazil, 2001. SIPDA; 2001. pp. 211–16.
21. Hartono Z.A. and Robiah I. 'A method of identifying the lightning strike location on a structure'. *International Conference on Electromagnetic Compatibility*; 1995. ICEMC; Kuala Lumpur, April 11–13, 1995. pp. 112–17.
22. Horvath T. 'Standarization of lightning protection based on the physics or on the tradition'. *J. Electrostatics* 2004;**60**:265–75.
23. Cooray V. 'Mechanism of electrical discharges' in V. Cooray (ed.). *The Lightning Flash*. IEE Publishers: London; 2003.
24. Golde R.H. 'The frequency of occurrence and the distribution of lightning flashes to transmission lines'. *AIEE Trans.* 1945;**64**:902–10.
25. Golde R.H. *Lightning Protection*. Edward Arnold: London; 1973.
26. Eriksson A.J. 'The lightning ground flash – an engineering study'. PhD thesis, Faculty of Engineering, University of Natal, Pretoria, 1979.
27. Berger K. 'Methods and results of lightning records at Monte San Salvatore from 1963–1971'. *Bull. Schweiz. Elektrotech.* 1972;**63**:21403–22 (in German).
28. Dellera L. and Garbagnati E. 'Lightning strike simulation by means of the Leader Progression Model, Part I: Description of the model and evaluation of free-standing structures'. *IEEE Trans.* 1990;**PWRD-5**:2009–23.
29. Berger K. 'Vogelsanger, Measurement and results of lightning records at Monte San Salvatore from 1955–1963'. *Bull. Schweiz. Elektrotech.* 1965;**56**:2–22 (in German).
30. Wagner C.F. 'Relation between stroke current and the velocity of the return stroke'. *IEEE Trans. Power App. Syst.* 1963;**82**:609–17.
31. Cooray V., Rakov V. and Theethayi N. 'The lightning striking distance – revisited'. *J. Electrostatics* 2007;**65**(5–6):296–306.
32. Brown G.W. and Whitehead E.R. 'Field and analytical studies of transmission line shielding: Part II'. *IEEE. Trans.* 1969;**PAS-88**(N.5):617–25.
33. Gilman D.W. and Whitehead E.R. 'The mechanism of lightning flashover on high-voltage and extra high voltage transmission lines'. *Electra* 1973;**27**:65–96.
34. Whitehead E.R. 'Protection of transmission lines' in R.H. Golde (ed.). *Lightning*. Academic Press: London; 1977. Vol. 2.
35. Whitehead E.R. 'Survey of the lightning performance of EHV transmission lines'. *Electra* 1979;**27**:63–89.
36. IEEE Working Group on Lightning Performance of Transmission Lines. 'A simplified method for estimating lightning performance of transmission lines'. *IEEE Transactions on Power Apparatus and Systems PAS* 1985;**104**(4):919–27.
37. IEEE. *Guide for Improving the Lightning Performance of Transmission Lines*. IEEE Std 1243-1997, 1997.
38. Gallimberti I. 'The mechanism of long spark formation'. *J. Physique Coll.* 1972;**40**(C7, Suppl. 7):193–250.
39. Les Renardiéres Group. 'Research on long air gap discharges at Les Renardieres'. *Electra* 1972;**23**:53–157.

40. Les Renardiéres Group. 'Positive discharges in long air gaps – 1975 results and conclusions'. *Electra* 53, 1977, pp. 31–132.
41. Berger K. 'The earth flash' in R.H. Golde (ed.). *Lightning*. Academic Press: London; 1977. Vol. 1, Chapter 5.
42. McEachron K.B. 'Lightning to the Empire State Building'. *J. Franklin Inst.* 1939;**227**:149–217.
43. Berger K. 'Novel observations of lightning discharges: results of research on Mount San Salvatore'. *J. Franklin Inst.* 1967;**283**:478–525.
44. Kito Y., Horii K., Higashiyama Y. and Nakamura K. 'Optical aspects of winter lightning discharges triggered by the rocket-wire technique in Hukuriku district of Japan'. *J. Geophys. Res.* 1985;**90**:6147–57.
45. Wada A., Asakawa A., Shindo T. and Yokoyama S. 'Leader and return stroke speed of upward-initiated lightning'. *Proceedings of International Conference on Atmospheric Electricity*; 2003. ICAE, Versailles, June 9–13, 2003, paper C3-20.
46. Yokoyama S., Miyake K., Suzuki T. and Kanao S. 'Winter lightning on Japan sea coast – development of measuring system on progressing feature of lightning discharge'. *IEEE Trans. Power Deliv.* 1990;**5**(3):1418–25.
47. Uman M.A. *The Lightning Discharge*. Academic Press: London; 1987.
48. Rakov V.A. and Uman M.A. *Lightning: Physics and Effects*. Cambridge, UK: Cambridge University Press; 2003.
49. Lalande P., Bondiou A., Laroche P., Eybert A., Berlandis J., Bador B., Bonamy A., Uman M. and Rakov V. 'Leader properties determined with triggered lightning techniques'. *J. Geophys. Res.* 1998;**103**:14109–15.
50. Laroche P., Eybert-Berard A., Barret L. and Berlandis J.P. 'Observations of preliminary discharge initiating flashes triggered by the rocket and wire technique'. *Proceedings of the International Conference on Atmospheric Electricity*; Uppsala, Sweden, 1988. pp. 327–33.
51. Willet J.C., Davis D.A. and Laroche P. 'An experimental study of positive leaders initiating rocket-triggered lightning'. *Atmos. Res.* 1999;**51**:189–219.
52. Hoori K., Nakamura K. and Ichi Sumi S. 'Review of the experiment of triggered lightning by rocket in Japan'. *Proceedings of the 28th International Conference on Lightning Protection*; 2006. ICLP, Kanazawa, Japan, 2006, pp. 44–51.
53. Idone V.P. 'The luminous development of Florida triggered lightning'. *Res. Lett. Atmos. Elect.* 1992;**12**:23–28.
54. Carrara G. and Thione L. 'Switching surge strength of large air gaps: a physical approach'. *IEEE Trans.* 1979;**PAS-95**(2):512–24.
55. Gary C., Hutzler B., Cristescu D., Dragan G., Enache R. and Popa V. 'Laboratory aspects regarding the upward positive discharges due to negative lightning'. *Rev. Roum. Sci. Technol., Electrotechnol. Energ.*, 1989;**34**:363–77.
56. Bernardi M., Dellera L., Garbagnati E. and Sartorio G. 'Leader progression model of lightning: updating of the model on the basis of recent test results'. *Proceedings of the 23rd International Conference on Lightning Protection*; Florence, Italy, 1996. ICLP, 1996. pp. 399–407.

57. Alessandro F.D., Kossmann C.J., Gaivoronsky A.S. and Ovsyannikov A.G. 'Experimental study of lightning rods using long sparks in air'. *IEEE Trans. Dielectrics and Electrical Insulation* 2004;**11**(4):638–49.
58. Eriksson A.J. 'An improved electrogeometric model for transmission line shielding analysis'. *IEEE Trans.* 1987;**PWDR-2**:871–77.
59. Eriksson A.J. 'The incidence of lightning strikes to power lines'. *IEEE Trans.* 1987;**PWDR-2**:859–70.
60. Dellera L. and Garbagnati E. 'Lightning strike simulation by means of the Leader Progression Model: II. Exposure and shielding failure evaluation of overhead lines with assessment of application graphs'. *IEEE Trans. Power Deliv.* 1990; **PWRD-5**:2023–29.
61. D'Alessandro F. 'The use of "Field Intensification Factors" in calculations for lightning protection of structures'. *J. Electrostatics* 2003;**58**:17–43.
62. Rizk F. 'A model for switching impulse leader inception and breakdown of long air-gaps'. *IEEE Trans. Power Deliv.* 1989;**4**(1):596–603.
63. Petrov N.I. and Waters R.T. 'Determination of the striking distance of lightning to earthed structures'. *Proc. R. Soc. A* 1995;**450**:589–601.
64. Chernov E.N., Lupeiko A.V. and Petrov N.I. 'Investigation of spark discharge in long air gaps using Pockel's device'. In *Proceedings of the 7th International Symposium on High Voltage Engineering ISH*; Dresden, Germany, 1991. pp. 141–44.
65. Petrov N.I. and D'Alessandro F. 'Theoretical analysis of the processes involved in lightning attachment to earthed structures'. *J. Phys. D: Appl. Phys.* 2002;**35**:1788–95.
66. Akyuz M. and Cooray V. 'The Franklin lightning conductor: conditions necessary for the initiation of a connecting leader'. *J. Electrostatics*, 2001;**51–52**:319–25.
67. Bazelyan E.M. and Raizer Y.P. *Lightning Physics and Lightning Protection*. Bristol: Institute of Physics; 2000.
68. Alekandrov N.L., Bazelyan E.M., Carpenter R.B., Drabkin M.M. and Raizer Y. 'The effect of coronae on leader inception and development under thunderstorm conditions and in long air gaps'. *J. Phys. D: Appl. Phys.* 2001;**34**:3256–66.
69. Lalande P. 'Study of the lightning stroke conditions on a grounded structure'. Doctoral Thesis, Office National d'Etudes et de Recherches Aerospatiales ONERA, 1996.
70. Goelian N., Lalande P., Bondiou-Clergerie A., Bacchiega G.L., Gazzani A. and Gallimberti I. 'A simplified model for the simulation of positive-spark development in long air gaps'. *J. Phys. D: Appl. Phys.* 1997;**30**:2441–52.
71. Lalande P., Bondiou-Clergerie A., Bacchiega G. and Gallimberti I. 'Observations and modeling of lightning leaders'. *C.R. Physique* 2002;**3**:1375–92.
72. Becerra M. and Cooray V. 'A simplified physical model to determine the lightning upward connecting leader inception'. *IEEE Trans. Power Deliv.* 2006;**21**(2):897–908.
73. Becerra M. and Cooray V. 'Time dependent evaluation of the lightning upward connecting leader inception'. *J. Phys. D: Appl. Phys.* 2006;**39**:4695–4702.

74. Dellera L., Garbagnati E., Bernardi M., Bondiou A., Cooray V., Gallimberti I., Pedersen A. and Ruhling F. *Lightning Exposure of Structures and Interception Efficiency of Air Terminals*. CIGRE Report, Task Force 33.01.03, 1997.
75. Becerra M., Cooray V., Soula S. and Chauzy S. 'Effect of the space charge layer created by corona at ground level on the inception of upward lightning leaders from tall towers'. *J. Geophys. Res.* 2007;**112**:D12205.
76. Soula S. and Chauzy S. 'Multilevel measurement of the electric field underneath a thundercloud 1: A new system and the associated data processing'. *J. Geophys. Res.* 1991;**96**(D12):22319–26.
77. Twomey S. *Atmospheric Aerosols*. Amsterdam: Elsevier; 1977.
78. Soula S. 'Transfer of electrical space charge from corona and thundercloud: measurements and modeling'. *J. Geophys. Res.* 1994;**99**(D5):10759–65.
79. Sunaga Y. and Shindo T. 'Influence of space charge on upward leader initiation'. *IEEE Trans.* 2005;**PE125**:789–96.
80. Miki M. 'Observation of current and leader development characteristics of winter lightning'. *Proceedings of the 28th International Conference on Lightning Protection*; 2006. ICLP, Kanazawa, Japan, pp. 14–19.
81. Singer H., Steinbigler H. and Weiss P. 'A charge simulation method for the calculation of high voltage fields'. *IEEE Trans.* 1974; **PAS-93**:1660–68.
82. Becerra M. and Cooray V. 'Laboratory experiments cannot be utilized to justify the action of Early Streamer Emission terminals'. *J. Phys. D: Appl. Phys.* 2008;**41**(8):085204.
83. Horii K. and Sakurano H. 'Observation on final jump of the discharge in the experiment of artificially triggered lightning'. *IEEE Trans. Power App. Syst.* 1985;**PAS104**(10):2910–15.
84. Horii K. and Nakano M. 'Artificially triggered lightning' in H. Volland (ed.). *Handbook of Atmospheric Electrodynamics*. CRC Press, 1st Edition, Vol. 1, Chapter 6, pp. 151–66.
85. Rakov V.A., Uman M.A., Rambo K.J., Fernandez M.I., Fisher R.J., Schnetzer G.H., Thottappillil R., Eybert-Berand A., Berlandis J.P., Lalance P., Bonamy A., Laroche P. and Bondiou-Clergerie A. 'New insights into lightning processes gained from triggered lightning experiments in Florida and Alabama'. *J. Geophys. Res.* 1998;**103**(D12):14117–30.
86. Eriksson A.J. 'Lightning and tall structures'. *Trans. SA Inst. Elect. Eng.* 1978;August:1–16.
87. Moore C.B., Aulich G.D. and Rison W. 'Measurements of lightning rod responses to nearby strikes'. *Geophys. Res. Lett* 2000;**27**(10):1487–90.
88. Moore C.B., Rison W., Mathis J. and Aulich G. 'Lightning rod improvements'. *J. Appl. Meteorol.* 2000;**39**:593–609.
89. D'Alessandro F. and Gumley J.R. 'A collection volume method for the placement of air terminals for the protection of structures against lightning'. *J. Electrostatics*, 2001; **50**(4):279–302.
90. Mousa A.M. 'Proposed research on the collection volume method/field intensification method for the placement of air terminals on structures'. *Power Engineering Society General Meeting* 2003;**1**:301–305.

91. Petrov N.I. and D'Alessandro F. 'Assessment of protection system positioning and models using observations of lightning strikes to structures'. *Proc. R. Soc. A, Math. Phys. Eng. Sci.* 2002;**458**:723–42.
92. D'Alessandro F. and Petrov N.I. 'Field study on the interception efficiency of lightning protection systems and comparison with models'. *Proc. R. Soc. A: Math. Phys. Eng. Sci.* 2006;**462**:1365–86.
93. Hartono Z.A. and Robiah I. *The Field Intensification Method: An Assessment Based on Observed Bypass Data on Real Buildings in Malaysia*. Report submitted to Standards Australia Committee EL-024, September 2002.
94. Becerra M., Cooray V. and Roman F. 'Lightning striking distance of complex structures'. *IET Generation, Transmission and Distribution* 2008;**2**(1):131–38.
95. Ait-Amar S. and Berger G. 'A 3-D numerical model of negative lightning leader inception: Application to the collection volume construction'. *Proceedings of the 25th International Conference on Lightning Protection*; Avignon, France. ICLP, 2004.
96. Rizk F. 'Modeling of transmission lines: exposure to direct lightning strokes'. *IEEE Trans. Power Deliv.* 1990;**PWRD-5**:1983–89.
97. Rizk F. 'Modeling of lightning incidence to tall structures Part I: Theory'. *IEEE Trans. Power Deliv.* 1994;**PWRD-9**:162–71.
98. Rizk F. 'Modeling of lightning incidence to tall structures Part II: Application'. *IEEE Trans. Power Deliv.* 1994;**PWRD-9**:172–93.
99. Baldo G. 'Lightning protection and the physics of discharge'. *Proceedings of the High Voltage Engineering Symposium*; 1999. ISH, London, UK, Vol. 2, 1999. pp. 169–176.
100. Mousa A.M. 'Discussion of A. J. Eriksson, The incidence of lightning strikes to power lines'. *IEEE Trans.* 1987;**PWDR-2**:867.
101. CIGRE Working Group 01 on Lightning. *Guide to Procedures for Estimating the Lightning Performance of Transmission Lines*. CIGRE document, no. 63, October 1991.
102. Whitehead E.R. Final Report of Edison Electric Institute – Mechanism of Lightning Flashover Research Project. EEI; EEI Project RP50, 1972, pp. 72–900.
103. Whitehead J.T. 'Lightning performance of TVA-s 500 kV and 161 kV transmission lines'. *IEEE Trans. Power App. Syst.* 1983;**PAS102**(3):752–68.
104. Becerra M. and Cooray V. 'A self-consistent upward leader propagation model'. *J. Phys. D: Appl. Phys.* 2006;**39**:3708–15.
105. Becerra M. and Cooray V. 'On the velocity of positive connecting leaders associated with negative downward lightning leaders'. *Geophys. Res. Lett.* 2008;**35**:L02801.
106. Becerra M. 'On the attachment of lightning flashes to grounded structures, digital comprehensive summaries of Uppsala dissertations of the Faculty of Science and Technology', 438, Uppsala University, PhD Thesis, 2008.
107. Saba M., Campos Z.L., Ballarotti M. and Pinto O. 'Measurements of cloud-to-ground and spider leader speeds with high-speed video observations'.

Proceedings of the 13th International Conference on Atmospheric Electricity; Beijing, China, August 2007. pp. 1–4.
108. Becerra M., Roman F. and Cooray V. 'Lightning attachment to common structures: Is the rolling sphere really adequate?' Presented at the *29th International Conference of Lightning Protection ICLP 2008*, Uppsala, Sweden, 2008, paper 4-2.
109. Becerra M., Cooray V. and Hartono Z.A. 'Identification of lightning vulnerability points on complex grounded structures'. *J. Electrostatics* 2007;**65**: 562–70.
110. Schonland B.F.J., Malan D.J. and Collens H. 'Progressive Lightning II'. *Proc. Roy. Soc. London. Ser. A containing papers of a mathematical and physical character (1905–1934)* 1935;**152**:595–625.
111. Golde R.H. 'The lightning conductor'. *J. Franklin Inst.* 1967;**283**(6):451–77.
112. Lowke J. 'On the physics of lightning'. *IEEE Trans. Plasma Sci.* 2004;**32**: 4–17.
113. Berger G. 'The early streamer emission lightning rod conductor'. *Proceedings of the 15th International Conference on Aerospace and Ground*; ICOLSE, Atlanta City, NJ, 1992, paper 38, pp. 1–9.
114. Berger G. 'Leader inception field from a vertical rod conductor – efficiency of electrical triggering techniques'. *1996 IEEE International Symposium on Electrical Insulation*; Montreal, Canada, 1996. pp. 308–11.
115. Berger G. 'Determination of the inception electric fields of the lightning upward leader'. *Proceedings of the International Symposium on High Voltage Engineering*; ISH, Yokoyama, Japan, 1993. pp. 225–29.
116. Lee J.B., Myung S.H., Cho Y.G., Chang S.H., Kim J.S. and Kil G.S. 'Experimental study on lightning protection performance of air terminals'. *Proceedings of Power System Technology Conference*; PowerCon 2002. Vol. 4, pp. 2222–26.
117. Heary K.P., Chaberski A.Z., Richens F. and Moran J.H. 'Early streamer emission enhanced air terminal performance and zone of protection'. *Conference Record of the Industrial and Commercial Power Systems Technical Conference*; 1993. pp. 26–31.
118. Allen N.L., Evans J.C., Faircloth D.C., Siew W.H. and Chalmers I.D. 'Simulation of an early streamer emission air terminal for application to lightning protection'. *Proceedings of the High Voltage Engineering Symposium*; ISH, 1999, Vol. 2, pp. 208–11.
119. French Standard, *Protection of Structures and Open Areas Against Lightning Using ESE Air Terminals*; French Standard, NF C, 1995.17, 102.
120. Spanish Standard, *Protection of Structure and of Open Areas Against Lightning Using Early Streamer Emission Air Terminals*; UNE, 1996. pp. 21186.
121. NFPA, *Draft Standard NFPA 781-F93TCR, Proposal for Lightning Protection Systems Using Early Streamer Emission Air Terminals*; 1997.
122. Allen N.L., Cornick K.J., Faircloth D.C. and Kouzis C.M. 'Test of the early streamer emission principle for protection against lightning'. *IEE Proc. Sci. Meas. Technol.* 1998;**145**(5):200–206.

123. Uman M.A. and Rakov V. *Bulletin of the American Meteorological Society*; 2002. pp. 1809–1820.
124. Van Brunt R.J., Nelson T.L. and Stricklett K.L. 'Early streamer emission lightning protection systems: An overview'. *IEEE Electrical Insulation Magazine* 2000;**16**(1):5–24.
125. Chalmers L.D., Evans J.C. and Siew W.H. 'Considerations for the assessment of early streamer emission lightning protection'. *IEE Proc. Sci. Meas. Technol.* 1999;**146**(2):57–63.
126. Hartono Z.A. and Robiah I. 'A study of non-conventional air terminals and striken points in a high thunderstorm region'. *Proceedings of the 25th International Conference on Lightning Protection*; ICLP Rhodes, Greece, 2000. pp. 357–361.
127. Eybert-Berard A., Lefort A. and Thirion B. 'Onsite tests'. *Proceedings of the 24th International Conference on Lightning Protection*; Birmingham, England, Staffordshire University, 1998. pp. 425–435.
128. Aleksandrov N.L., Bazelyan E.M., Drabkin M.M., Carpenter R.B. and Raizer Yu. P. 'Corona discharge at the tip of a tall object in the electric field of a thundercloud'. *Plasma Physics Reports* 2002;**28**:1032–1045.
129. Aleksandrov N.L., Bazelyan E.M. and Raizer Y.P. 'The effect of a corona discharge on a lightning attachment'. *Plasma Physics Reports* 2005;**31**(1):75–91.
130. Aleksandrov N.L., Bazelyan E.M. and Raizer Y.P. 'Initiation and development of first lightning leader: the effects of coronae and position of lightning origin'. *Atmospheric Research* 2005;**76**(1–4):307–29.
131. Mousa A.M. 'The applicability of lightning elimination devices to substations and power lines'. *IEEE Trans. Power Delivery* 1998;**13**:1120–1127.
132. FAA, *1989 Lightning Protection Multipoint Discharge Systems Tests: Orlando, Sarasota, and Tampa, Florida*. Federal Aviation Administration, FAATC T16 Power Systems Program, Final Rep. ACN-210, 1990. pp. 48.
133. Durrett W.R. *Dissipation Arrays at Kennedy Space Center. Review of Lightning Protection Technology for Tall Structures*, J. Hughes (ed.). Publ. AD-075 449, Office of Naval Research; 1977. pp. 24–52.
134. Bent R.B. and Llewellyn S.K. *An Investigation of the Lightning Elimination and Strike Reduction Properties of Dissipation Arrays*. Review of Lightning Protection Technology for Tall Structures, J. Hughes (ed.). Publ. AD-A075 449, Office of Naval Research; 1977. pp. 149–241.
135. Rourke C. 'A review of lightning-related operating events at nuclear power plants'. *IEEE Trans. Energy Conversion* 1994;**9**:636–41.
136. Cooray V. and Zitnik M. 'On attempts to protect a structure from lightning strikes by enhanced charge generation'. *Proceedings International Conference Lightning Protection*; ICLP, France, 2004.
137. Moon P. and Spencer D.E. *Field Theory for Engineers*. Princeton, New Jersey: D. Van Nostrand Company, Inc.; 1961.
138. Bazelyan E.M. and Drabkin M.M. 'Scientific and technical basis for preventing lightning strikes to earthbound objects'. *IEEE Power Engineering Society General Meeting*, July 2003, Vol. 4, pp. 2208.

139. Bazelyan E.M., Raizer Y.P. and Aleksandrov N.L. 'A scientific study on the performance of charge transfer technology and the dissipation array system'. Unpublished, 2009.
140. Bazelyan E.M., Raizer Y.P. and Aleksandrov N.L. 'Corona initiated from grounded objects under thunderstorms conditions and its influence on lightning attachment'. *Plasma Sources Sci. Technol.* 2008;**17**:024015.

Chapter 5
Protection against lightning surges
Rajeev Thottappillil and Nelson Theethayi

The term 'surge' denotes a state of electrical overstress that lasts less than a few milliseconds, a duration much less than that of a power frequency cycle. The brief nature of the surge is emphasized by adding the word 'transient' before it. To distinguish from other types of electrical overstresses, some authors prefer the term 'transient overvoltages' [1,2]. Sometimes transients may not exceed the normal operating voltage, but they may still be of concern because of their high-frequency content. The most common sources of transients in power and telecommunication systems are lightning and switching events. Current and voltage transients are part of what is known as conducted electromagnetic interference (EMI). Here, we consider only lightning transients. In this chapter we first give a brief overview of the characteristics of lightning and provide examples of the nature of lightning transients measured in low-voltage networks. This is followed by a discussion on transient protection methods and components.

5.1 Introduction

The most common transients in low-voltage electrical installations are a result of lightning, switch operations in power networks, switching of local loads and residual voltage from the operation of surge protective devices. We will concentrate on the issues associated with lightning protection in low-voltage networks. Throughout this chapter the phrases 'lightning overvoltages', 'lightning transients' and 'lightning surges' are used synonymously. Lightning can create overvoltages in electrical installations either by direct attachment of the lightning (direct strike) or as a result of the coupling of electromagnetic fields from remote lightning (indirect strike). Examples of both these cases are described in the following.

5.1.1 Direct strike to power lines

In 1994 an experiment was carried out by the University of Florida, Gainesville, using triggered lightning [3], in which lightning was allowed to directly strike a test power distribution system (13 kV) [4]. In that study a portion of the lightning current in the phase line was led to an underground cable and from there to the primary of a pad-mounted distribution transformer protected by a metal oxide varistor (MOV) surge arrester; the residual current pulses coming out of the secondary side of the transformer were measured [5]. In Figure 5.1, the top trace shows the lightning current measured at the bottom of the lightning channel, a portion of which is led to the transformer. The middle trace shows the primary voltage of the transformer, clamped by the surge arrester. The bottom trace shows the currents in the secondary windings (220/110 V) of the transformer. There are three important observations to be made.

1. Although there were only four return strokes in the flash, there were 15 voltage surge pulses clamped by the arrester to near 20 kV.
2. All 15 surge voltage pulses produced voltage and current surges on the secondary. Even small current pulses of a hundred amperes (M current pulses [3,5]) in lightning, which are not return strokes, can produce voltage surges of the order of 20 kV in a line of 400 Ω surge impedance in the event of a direct strike. Studies show that in a negative cloud-to-ground lightning there may be up to 20 such current pulses separated in time, on average, by 5 ms [6].
3. The transformer turns ratio is not applicable for fast transients because surges on the primary side are capacitively coupled to the secondary side. This coupling is influenced by transformer type, circuit and load [7,8]. The response of protective systems can be quite different under a multipulse transient environment when compared to the response to single transient pulses. In general, energy-absorbing protective components tend to fail in a multipulse environment where they would not have failed with a single pulse [9].

5.1.2 Lightning activity in the vicinity of networks

A low-voltage power installation (LVPI) network of a single-storey residential building in Uppsala has been extensively studied [10–12] for its response to lightning. In one experiment performed in 1995 the network was exposed to electromagnetic fields from natural lightning occurring at a distance of many kilometres. The induced common mode (CM) voltages in a power outlet of the network were measured simultaneously with the vertical component of the electric field near the installation. The LVPI network was disconnected from the distribution network to avoid the conducted transients entering through the mains and hence the measured induced voltages are due to the direct interaction of lightning electromagnetic fields with the LVPI network. Generally, only the return strokes in a cloud-to-ground flash are considered as important in determining the transient environment of devices connected to the LVPI network. However, this study shows that electric field pulse trains associated with the initiation of both cloud-to-ground lightning and cloud lightning can cause induced CM voltage pulses in LVPI networks that may pose a threat to sensitive devices connected to the power network. The transient environment of sensitive electronic devices connected to LVPI networks is more complex than it would be by considering the return strokes alone as being the determining source for interference in low-voltage electrical systems.

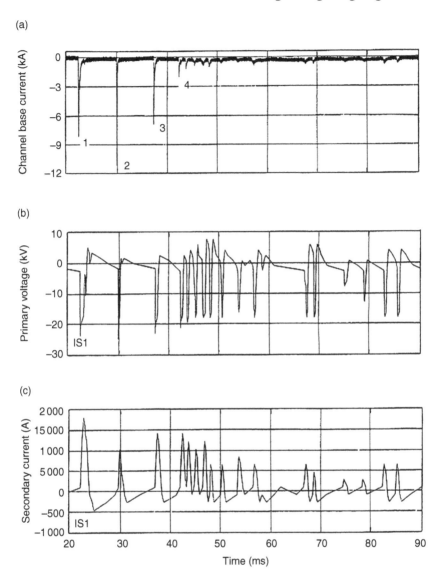

Figure 5.1 (a) *Current at the channel base of a four-stroke triggered-lightning flash. (b, c) Selected responses to this flash of the test power distribution system at Camp Blanding, Florida, in September 1994. The lightning current was injected into the top conductor of the overhead line. The numbered pulses in (a) are due to return strokes, and smaller pulses after pulse 4 are due to M components. The voltage in (b) was measured across the arrester in the transformer primary. The current in (c) was measured in the phase conductor at the transformer secondary (adapted from Reference 5).*

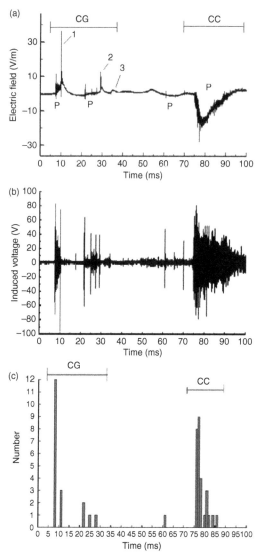

Figure 5.2 *A cloud-to-ground (CG) flash followed by a cloud flash (CC) recorded in Uppsala on 21 June 1995. (a) Vertical component of the electric field. 1, 2 and 3 are return strokes, and P indicates major groups of microsecond-scale pulses. (b) Induced common-mode (CM) voltages in a power outlet of a residential wooden-frame house simultaneously measured with the electric field. The low-voltage power installation network of the house was completely isolated from the external distribution line. (c) Histogram of the induced CM voltages exceeding 80 V peak-to-peak due to the CG flash, followed by CC flash. Each bin is 1 ms wide (from Reference 12).*

Figure 5.2 shows a lightning flash recorded on 21 June 1995 and the corresponding induced CM voltage in the LVPI network. According to the information from the lightning location network, this flash occurred at a distance of 24 km from the measuring station. Figure 5.2a shows the vertical component of the electric field measured near the house. The first half of the record is a cloud-to-ground lightning flash consisting of three return strokes numbered 1, 2 and 3. Towards the end of the record there is a prominent static field change marked CC, with many pulses superimposed on it, indicative of a discharge that has developed within the cloud. The record also contains many pulses, the most prominent groups of them being marked P. The voltage measured between the phase and local ground (heating system) is given in Figure 5.2b. Note from this overall record that there is a one-to-one correspondence between the electric field pulse activity and the induced voltage events.

In order to compare the severity of induced CM voltages during the various stages of a lightning flash, the occurrence of induced voltage events that exceeded a given value were compared [12]. The results are presented in the form of histograms in Figure 5.2c. During the cloud-to-ground flash the maximum induced CM voltage was 160 V peak-to-peak, corresponding to the first return stroke. There were 19 induced voltage events that exceeded 80 V peak-to-peak, half of the maximum induced voltage. Of those, 12 happened within a time period of 0.8 ms during the preliminary breakdown stage of the flash starting at 7 ms, and 7 happened between the first and second return strokes within a time period of 18 ms. In the cloud flash, starting at $\sim$72 ms, most of the big induced voltage pulses were concentrated within the first 3 ms of the flash. During this 3 ms period, there were 21 induced voltage events that exceeded 80 V peak-to-peak.

The flash in the example of Figure 5.2 was at a distance of 24 km. If it were at a distance of 6 km, it is possible to make a crude estimate that the response of the system due to direct interaction with the electromagnetic fields would be at least four times larger than the vertical scale shown in Figure 5.2b.

Having seen the examples of induced transient voltages and currents due to lightning direct and indirect strikes we will now move on to the lightning current and field parameters that would be useful for the design of lightning protection systems.

5.2 Characteristics of lightning transients and their impact on systems

Cloud-to-ground lightning can cause damage to an object on earth by directly attaching to it, or it can cause damage by induction effects while striking somewhere near the object. Sometimes lightning may strike far from the object, but the surge is conducted to the object via power lines or other conducting systems, causing damage. The extent and nature of damage depend both on the characteristics of the lightning and the characteristics of the object [3,13]. The physical properties of lightning that are important in causing damage are the current and electromagnetic fields. In this section, we consider the characteristics of lightning current and electromagnetic fields important in producing damage to earth-bound systems. Summaries of lightning current and field parameters are given in References 3 and 13; these are based on actual measurements on lightning by different research groups around the world.

5.2.1 Parameters of lightning current important for surge protection design

The most important properties of lightning current that cause damage are peak current, the maximum rate of change of current, the integral of the current over time (charge) and the integral of the square of current over time (action integral).

5.2.1.1 Peak current

In cloud-to-ground (CG) lightning, the largest currents are produced by the return strokes. Peak return stroke currents are important in cases where the struck object essentially presents a resistive load, for example the surge impedance of a long power line, a tree, ground rods driven into earth, and so on. As an example, when a lightning return stroke with a peak current of 30 kA strikes a power line with a surge impedance of 400 Ω, it can produce a prospective overvoltage of 600 kV, assuming division of current. This large voltage can cause flashover across insulators, from line to ground, to adjacent lines and to other objects nearby. The magnetic forces produced by the peak current can cause wires to be pulled out of walls and electrical machines, and metal tubes to be crushed. A 30 kA current entering earth through a grounding impedance of 10 Ω causes a potential rise of 300 kV and may also cause surface arcing.

Available evidence indicates that the average value of peak lightning current is not affected by the conductivity of the soil. However, the same peak value of the current will have more adverse effects in low conductivity soil compared to high conductivity soil. Soil conductivity varies widely from region to region; for example, in most parts of Sweden soil has poor soil conductivity in the range 0.2–1 mS m^{-1}, so more surface arcing can be expected. In objects that present essentially inductive impedance such as wires in electronic systems, earth leads and so on, the maximum overvoltage produced is proportional to the maximum rate of change of current. Maximum di/dt occurs at the return stroke current wavefront. Assume that 10 per cent of the 30 kA peak current (i.e. 3 kA) with front time 0.3 μs finds its way to the wiring of an electronic apparatus. For an inductance of 1×10^{-6} H m^{-1}, the inductive voltage produced in a 10-cm-long wire is 1 kV, enough to destroy most electronics unless there is adequate protection. In negative return strokes the average value of di/dt is 110 kA μs^{-1}. In positive return strokes these values are much smaller [13,14].

5.2.1.2 Charge transferred

To a first approximation, the heating and burn through of metal sheets (e.g. metal roofs, airplane wings) is proportional to the amount of charge transferred, and depends also on the current at which this charge is delivered. Charge is the integral of the current over time. The power delivered to the lightning attachment point is the product of the current and the voltage drop (5 to 10 V) at the arc–metal interface. Most of the charge in lightning is due to the long continuing current that follows some of the return strokes. Even a big return stroke that lasts perhaps a few tens of microseconds may not transfer as much charge as a low-level (100–1 000 A) continuing current that lasts a few hundred milliseconds. In 1990, Sandia National Laboratories conducted an experiment in which lightning was allowed to strike stationary metallic samples [15]. Three strokes, each from a different flash, that lowered 5.5, 7.6 and 13.6 C of charge produced significant damage and partial penetration in the 2 mm aluminium sample, but did not burn a hole through. In another flash three consecutive strokes striking a 0.9 mm steel sample produced three significant damage spots corresponding to 5.8, 7.8 and 49 C of stroke charge, one of which producing a burn through, possibly corresponding to the 49 C charge. It seems unlikely that the different strokes in a negative flash attach to the same spot. Therefore in negative lightning burn through may be more correlated to stroke charge than to total flash charge. If the arc could be fixed to a spot on the plate, then due to heat concentration only a lesser amount of charge is required to burn a hole. Laboratory experiments with short-gap metal arcs show that a charge of only 10 C delivered by a current of 500 A is required to burn a hole through 2-mm-thick aluminium plate [16], possibly because arc root does not wander as in lightning. The same value of charge may produce more damage in a less conducting material than in a more conducting material.

Charge lowered by a typical stroke followed by a continuing current is 11–15 C in a negative flash and 80 C in a positive flash [13,17].

Lightning to very tall towers and buildings is often initiated by upward leaders from these objects. Such lightning begins with a long continuous current, some hundreds of amperes in amplitude and several tens of milliseconds in duration, which lowers several coulombs of charge to ground before the onset of a regular sequence of leader and return strokes. Sometimes this lightning will only have initial continuous current without any following strokes. Even those lightning events can do damage associated with the large charges.

5.2.1.3 Prospective energy

Action integral is a measure of the ability of lightning current to generate heat in the resistive impedance of the struck object. This represents the prospective energy that would have been dissipated in a 1 Ω resistor due to joule heating if the entire current of the return stroke were to flow through it and is represented as the time integral of the square of the current. The rapid heating of materials and the resulting explosion of non-conducting materials are, to a first approximation, due to the value of the action integral. A doubling of the return stroke current tends to quadruple the action integral, for similar wave shape and duration of the return stroke. An action integral of 2.0×10^6 A^2 s would create a temperature rise in excess of 200 °C in a copper strap of 10 mm^2 cross-sectional area [16], creating an explosion hazard where flammable materials or vapours may exist. Much thinner wires or straps may melt and vaporize when subjected to the above value of action integral. Action integral is an important parameter that has to be considered in the dimensioning of conductors directly subject to lightning strikes. Typical values of the action integral are 5.5×10^4 and $6.5 \times 10^5 A^2$ s for the negative first return stroke and positive return stroke, respectively.

5.2.1.4 Waveshape

The lightning return stroke current wave shape is highly variable even within the same flash. The rise time can vary from 0.1 μs to several microseconds and the half-peak width from a few microseconds to a few hundreds of microseconds [13,18]. Current wave shapes very rarely follow exactly the 1.2/50 μs, 8/20 μs or 10/350 μs wave shape or any other specified wave shape. These are test wave shapes adopted by various standards for simulating the effects of lightning in the laboratory [13]. Longer-duration waveshapes (e.g. 10/350 μs) are used to simulate the effects of large energy input for the same peak current.

5.2.2 Parameters of lightning electric and magnetic fields important for surge protection design

The most important of the field parameters are the peak electric field and the maximum time rate of change of the electric or magnetic field (Table 5.1). Peak voltages on exposed metallic surfaces in the lightning field are proportional to the peak electric field, and peak voltages produced in a loop of wire are proportional to the rate of change of magnetic field. For example, a typical return stroke striking 100 m away may induce an overvoltage in excess of 200 V per m^2 of loop area formed by the equipment and its cables, for certain orientations of the loop. The degree of penetration of fields inside shielded enclosures through apertures is largely proportional to the rate of change of the magnetic and electric fields. The magnitude of the peak fields and the rate of change of fields are important parameters in overvoltages caused in above-ground wires and underground cables.

The finite conductivity of the ground creates a horizontal component of electric field on the surface of the earth. This component of the field is large if soil conductivity is low. Typically, the peak value of the horizontal component of the field can be 10–20 per cent of the peak vertical component of the field at ground if the ground conductivity is of the order of 1 mS m^{-1} [19]. This field is oriented radially from the lightning channel and induces overvoltages in overhead lines and cables on the ground. A peak vertical electric field of 2 kV m^{-1} at a distance of 1 km from the lightning channel may be accompanied by a horizontal field component of 200–400 V m^{-1}. The effect of this horizontal field may be seen as series voltage sources distributed along the conductors, each source turned on in sequence as the field sweeps along the conductor. These series voltage sources will drive a CM current in the conductors.

Measurements of electric and magnetic fields closer than 1 000 m from the lightning channel are limited. The available data come from the triggered lightning experiments in Florida and are applicable to subsequent return strokes in lightning [20–23]. No data are available for first return strokes. However, the following assumptions can be made on the relationship between the average parameters of negative subsequent return strokes and negative first return strokes:

1. The first return stroke peak electric and magnetic fields are about twice the corresponding values for subsequent return strokes.

Table 5.1 Estimated values of the field parameters of the negative first return stroke close to the lightning channel [13,14]

Distance (m)	Expected typical value				Expected maximum value			
	E (kV m^{-1})	dE/dt (kV m^{-1} μs^{-1})	H (A m^{-1})	dH/dt (A m^{-1} μs^{-1})	E (kV m^{-1})	dE/dt (kV m^{-1} μs^{-1})	H (A m^{-1})	dH/dt (A m^{-1} μs^{-1})
10	300	600	500	1 800	650	1 500	1 300	6 500
50	60	120	100	350	130	300	250	1 300
100	30	60	50	180	65	150	130	650
500	6	12	10	35	13	30	25	130

2. The peak electric field and magnetic field derivative values are approximately the same for both first and subsequent return strokes.
3. The magnetic fields very close to the lightning channel are related to the current in the channel by Ampères law.

Making these assumptions, the electric and magnetic field parameters of the negative first return stroke can be estimated as follows.

It has been demonstrated that the induced voltages due to lightning, whether direct or indirect, are significantly affected in terms of magnitudes and shapes. Recently, a number of research papers have been published on the subject of lightning-induced voltages in the presence of finitely conducting ground [24].

5.3 Philosophy of surge protection

The term 'surge protection' is used usually to denote the protection of circuits and devices from the effects of wire-bound or conducted transients. Surge protection is only one of the measures for controlling the effects of transients or electromagnetic interference (EMI) and applied as part of a strategy to achieve electromagnetic compatibility (EMC).

5.3.1 Surge protection as part of achieving electromagnetic compatibility

One of the most cost-effective methods for controlling EMI in a system is through the proper layout of subsystems, and surge protection forms part of this at conducted interfaces. This can be explained using the following example.

A two-layer shielding topology (geometry) for controlling internal and external interference sources is shown in Figure 5.3. All the sensitive (critical) circuits are physically grouped together as far as possible, and are provided with a shield that prevents fields external to them from having an effect. Similarly, strong internal sources are grouped together and are provided with a shield that confines the emission to within the enclosed shield volume. The remaining weak internal sources and non-critical components are physically grouped together without a special enclosing shield. All connections from sensitive circuits and strong internal sources are controlled by interference diverters, such as filters, surge protectors and equipotential bonds. All the subsystems are surrounded by an external shield that excludes external electromagnetic disturbance (e.g. lightning). All the connections penetrating the external shield (e.g. power, data, telephone, pipeline) are provided with interference diverters.

Despite the sensitive layout and shielding, interference fields and currents may penetrate inside (1) along insulated conductors passing through the shields, (2) through openings or imperfections in the shields, and (3) by diffusion through imperfectly conducting shields. The objective of system hardening (making coupling path inefficient) is to control these interference penetrations at each shield, so that interference reaching the sensitive circuit is within the tolerance of the circuit. We will explain next the interference diverters shown in Figure 5.3.

Protection against lightning surges 279

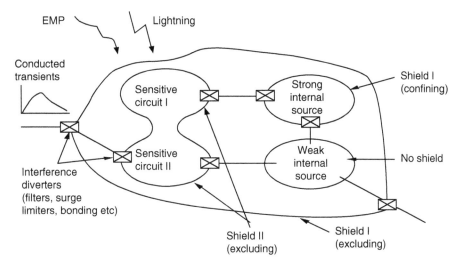

Figure 5.3   A two-layer shielding topology and the role of surge protection

5.3.1.1   Conductor penetration through a shield

A shield has two surfaces, external and internal. If a conductor carrying interference currents is connected to the outside of the shield, interference currents are confined to the outside of the shield and have very little influence on the shield volume enclosed. If the conductor is connected to the inside surface of the shield, all interference currents are available inside the shield and may couple with circuits inside the shield volume, which could potentially cause EMI.

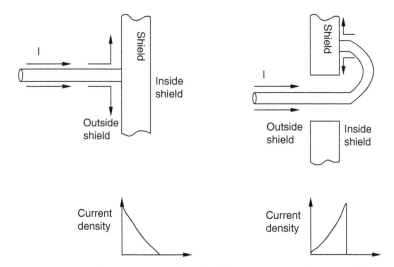

Figure 5.4   Current density across the shield cross-section for two different conditions: (a) confinement of conductor current to the outside surface by the skin effect; (b) conductor current injected on the inside of a shield

280 *Lightning Protection*

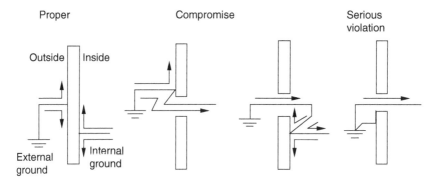

Figure 5.5   Grounding conductors (e.g. grounding of a cabinet)

In short, to preserve the integrity of the shield, interference current of external origin must be diverted to the outside surface of the shield as shown in Figure 5.4. Several examples of the proper application of this principle, together with some common compromises and serious violations are given in Figures 5.5, 5.6 and 5.7 [25].

5.3.2 *Principle of surge protection*

Surges can cause damage or upset in sensitive electronic circuits. Damage is the failure of the hardware requiring replacement of the defective components or modules. Upset is a temporary malfunction of a circuit or system. Recovery from an upset does not require replacement of defective components, but may require an operator's intervention.

A logical approach to transient overvoltage protection would be (1) to determine the threshold at which damage would occur, (2) to determine the worst-case overvoltage that would arrive at a particular device, and (3) to design and install a

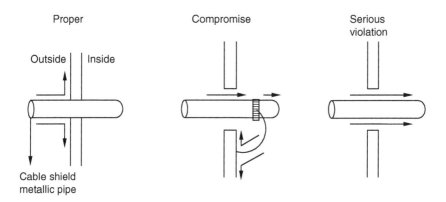

Figure 5.6   Conductors that can be grounded (e.g. metallic water pipe)

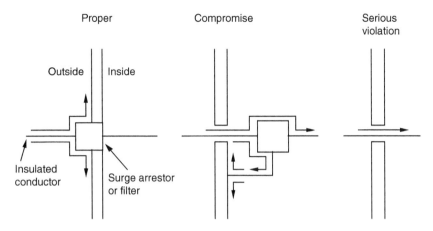

Figure 5.7   Insulated conductors or conductors that cannot be grounded (e.g. power, communication and data cables)

protective circuit that would limit the worst-case overvoltage to less than the damage threshold.

Protection from transients is achieved either by blocking the transients using a large series impedance or by diverting them using a small shunt impedance. Sometimes both methods are used together. A general surge protection circuit is shown in Figure 5.8.

A surge protection circuit should not influence the normal operation of the protected system. That is, series impedance should be very small ($Z_1 \ll Z_2$) and shunt impedance should be very large ($Z_2 \gg Z_L$) for normal signal voltage and frequencies. Let Z_L be the load impedance. During abnormal conditions (during a surge) the series impedance should be very large ($Z_1 \gg Z_L$) to limit the surge current, and shunt impedance should be very small ($Z_2 \ll Z_L$) to divert the surge current. Note that Z_1, Z_2 and Z_L may be functions of frequency, voltage or current. Also, surge protection circuits should not be damaged by the surge themselves.

During electrical overstress, the voltage of the surge is larger than the normal system voltage. Therefore, shunt elements with non-linear voltage–current (V–I) characteristics can provide very low impedance during overvoltage conditions and very high

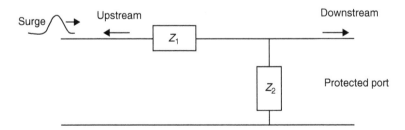

Figure 5.8   General surge protection circuit

282 *Lightning Protection*

impedance during normal system voltage. The overvoltage waveforms will have very fast rising or falling portions (very high rate of change of voltage or current). Therefore the shunt elements are required to respond very fast, in nanoseconds.

Surge diversion to the reference conductor or earth plane has its disadvantages. When large surge currents are allowed to spread over the reference network in an uncontrolled manner, this causes interference in other healthy systems. Therefore, series protection seems to be more desirable. However, to date there is no robust, fast and reliable non-linear series protection devices that can replace shunt protection.

Clearly, from the abovementioned requirements, protection devices (surge protectors) should be non-linear. Non-linear components can be classified into three groups:

- devices that have an approximately constant voltage across them during the conduction of surge (clamp);
- devices that change their state from insulator to good conductor during the conduction of surge (crowbar) (note that clamps absorb energy and crowbars do not absorb much; both are shunt devices);
- devices that offer large series impedance to CM voltages (isolators) (isolators are inserted in series, e.g. CM filters, isolation transformers, opto-isolators; other series surge protection or limiting devices include fuses, circuit breakers, inductors and temperature-dependent resistors).

The advantages and disadvantages of common surge protection components are given in Table 5.2.

Spark gaps enclosed in a ceramic tube filled with inert gas (gas discharge tube) and metal oxide varistors are very popular devices in low-voltage installation protection. Electrical characteristics of spark gaps and varistors will be inspected in detail in the next sections.

5.3.2.1 Gas discharge tubes (spark gaps)

One of the earliest transient surge protectors was the spark gap in air between two carbon blocks. A serious disadvantage of it was that the carbon blocks eroded after conducting high-energy transients. The widening gap due to eroded blocks changed the electrical characteristics of the gap with time. The modern spark gap lies between metal electrodes in a sealed tube containing a mixture of noble gases (neon, argon, and so on). Miniature low-voltage spark gaps sealed in ceramic tubes can conduct transient current pulses of 5–20 kA for 10 μs without appreciable damage to the spark gap. Of all the non-linear shunt protective devices, spark gaps have the lowest parasitic capacitance, typically between 0.5 and 2 pF. Therefore sparkgaps can be used even in applications where the signal frequencies are in excess of 50 MHz.

The operation of a spark gap (gas tube) can be explained with the help of Figure 5.9, which shows the typical response of a low-voltage gas discharge tube to an applied sinusoidal overvoltage. Figure 5.9a shows the applied sinusoidal

Table 5.2   Properties of common protection components (adapted from Reference 1)

	Characteristics
Clamps	
Metal oxide varistor (MOV)	Very fast response (<0.5 ns)
	Large energy absorption
	Can safely conduct large currents (from a few amperes to many kA)
	Available in a wide range of voltages (from a few volts to hundreds of kV)
	Large parasitic capacitance (in nF)
Avalanche diode	Very fast response (<0.1 ns)
	Good control over clamping voltage (~6–200 V)
	Small maximum current (<100 A)
	Large parasitic capacitance (in nF)
Diode	Small clamping voltage (0.7–2 V)
	Small parasitic capacitance
Crowbars	
Spark gap	Slow to conduct
	Can conduct large currents (from a few amperes to many kA)
	Low voltage in arc mode
	Small parasitic capacitance (in pF)
	Possible follow current
Silicon controlled rectifier (SCR) and Triac	Slow to turn on or turn off
	Small voltage across conducting switch (0.7–2 V)
	Possible follow current
	Can tolerate sustained large currents

overvoltage and the voltage across the spark gap, Figure 5.9b shows the current flow through the spark gap. As the voltage across the spark gap is slowly increased (Figure 5.9a), the gap fires at voltage V_s, bringing down the voltage. That is, at V_s the gap switches from the insulating state (resistance >10 GΩ) to the conducting state (resistance <0.1 Ω). The change of state can happen within a fraction of a microsecond. The voltage V_s (90–300 V) is called the d.c. firing voltage of the gap. Later we will see that the actual spark overvoltage and the response time of the spark gap depend upon the rate of increase of the applied voltage across the gap.

During the drop in voltage from V_s, the incremental resistance dV/dI is negative; i.e. this is a negative resistance region. The current through the gap increases (Figure 5.9b) and the gap voltage increases slightly to V_{gl}, the glow voltage. This region is called the glow region. The glow is produced by a thin layer of excited gas atoms covering part of the cathode surface and later extending to the whole cathode surface. Maximum current during the glow region is between 0.1 and 1.5 A and the glow voltage is between 70 and

284 *Lightning Protection*

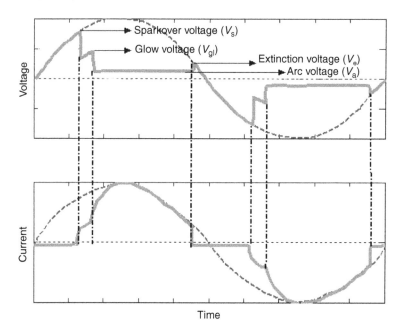

Figure 5.9   *Typical response of a low-voltage gas discharge tube to an applied sinusoidal overvoltage: (a) voltage across the gas discharge tube; (b) current through the gas discharge tube*

150 V. This is positive in the resistance region. Electron–ion pairs are produced in the intense electric field that exists between the electrodes when the spark gap is being operated in the glow region. When they obtain sufficient energy they accelerate and collide with neutral atoms or ions, producing more electron–ion pairs, and finally leading to a general breakdown of the gap. During the arc phase the voltage across the gap reduces to V_a (10–25 V) and becomes virtually independent of the current. The arc current can be very high and is limited mostly by the magnitude of the overvoltage and the parameters of the circuit containing the spark gap. With decreasing overvoltage, as in the second quarter of the applied sine wave, the current through the spark gap decreases until it drops below the minimum value (0.01–0.1 A) necessary to maintain the arc. The arc is extinguished at voltage V_e, after passing through the glow mode again. The above processes are repeated during the negative half cycle of the sine wave. The gas discharge tube is a bipolar device. That is, its characteristics do not depend upon the polarity of the applied voltage.

The gas tube can operate either in the glow regime or in the arc regime while giving protection from overvoltages. Both regimes are associated with a power follow current because the tube do not extinguish unless the voltage across the tube fall below the glow voltage or arc voltage, as the case may be. Sometimes thermionic emission from hot electrodes may maintain the arc even during the brief zero crossings of a sinusoidal voltage. Prolonged follow currents can destroy a gas tube by shattering the case or by melting the electrodes. Therefore it is essential to prevent the follow current after a surge. Follow current is prevented by putting a varistor in series with the spark gap. More on this method will be discussed while considering the varistor in Section 5.3.2.2.

The spark overvoltage (V_s) and the response time (time to conduct) of spark gaps are functions of the rate of rise of transient voltage. The spark overvoltage increases and the response time decreases with increase in dV/dt of the transient. For example, if the static spark overvoltage (or d.c. spark overvoltage), usually determined by applying a low rate of rise transient ($dV/dt \approx 100$ V s^{-1}), of a gas discharge tube is 350 V, the impulse spark overvoltage, usually determined with a fast rate of rise transient ($dV/dt = 1$ kV µs^{-1}), can be 750 V. The response time of the gas tube can be ~4 s at 100 V s^{-1} rate of rise of voltage, whereas it can be as small as 0.8 µs at 1 kV s^{-1}. Gas tubes may conduct within a few nanoseconds if the applied transient has rate of rise times about 1 MV µs^{-1}.

5.3.2.2 Varistors

Varistors are non-linear semiconductor devices whose resistance decreases as the magnitude of the voltage increases. Modern varistors are fabricated from metal oxides, with zinc oxide the primary ingredient. A typical V–I curve of a metal oxide varistor is shown in Figure 5.10. Under normal voltages there is a small leakage current of less than 0.1 mA and the varistor behaves like a simple high value resistor R_{leak}. During overvoltage the current through the varistor increases and the voltage is clamped at a level close to the normal voltage. This is the operating region of the varistor (Figure 5.10) and the voltage–current relationship in this region is given by

$$I = kV^\alpha \tag{5.1}$$

In equation (5.1) α is a coefficient with values between 25 and 60. The parameter k in (5.1) has a value extremely small ($<10^{-100}$), therefore I is expressed in terms of logarithms as in equation (5.2):

$$\log |I| = \log (k) + \alpha \log |V| \tag{5.2}$$

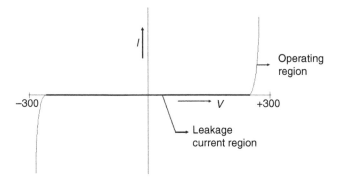

Figure 5.10 V–I *curve of a metal oxide varistor*

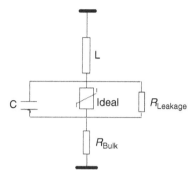

Figure 5.11   Equivalent circuit model of a varistor

If (V_1, I_1) and (V_2, I_2) are two measured data points in the operating region of the varistor, the value of α can be determined as

$$\alpha = \frac{\log(I_1/I_2)}{\log(V_1/V_2)} \qquad (5.3)$$

At very large currents, usually more than 100 A, the varistor characteristics are dominated by the low value bulk resistance R_{bulk} of the device. Usually, varistors are fabricated in the form of discs and hence have large parasitic capacitance values in the range of 0.2–10 nF. Including the inductance of the varistor leads will complete the equivalent circuit of the varistor, which is shown in Figure 5.11.

Varistors are fast acting devices with response times less than 0.5 ns, if parasitic inductance due to leads can be avoided. The performance of varistors is affected by temperature. Excessive leakage currents can raise the temperature of the varistor. Because the varistor has a negative temperature coefficient, the current will increase as it become hotter, which will increase the current even further, resulting finally in a thermal runaway. Varistors are usually used in protecting electronic systems from transient overvoltages that propagate on the mains. There are various varistor models that have been developed in the recent past that are being used for various applications and depend upon the type of varistor used.

Energy absorbed in the ceramic of a varistor is distributed throughout the ceramic at numerous grain boundaries rather than a single junction as in other semiconductor voltage clamping devices. Varistors can withstand single pulse transients up to 150 per cent of their rated current, but may fail at multipulse transients at 75 per cent of peak rated current. When energized at power system operating voltage, they could only withstand 40 per cent of the rated current in a multipulse environment [26].

5.3.2.3   Diodes and thyristors

The Zener or avalanche diode creates a constant voltage clamp. It contains a pn junction with a larger cross-section, proportional to its surge power rating. It works in response to a fast rising voltage potential and is available for wide range of clamping voltages (from less than 10 V up to several hundred volts); the response time is in the range of a few picoseconds [1]. The V–I characteristics of the diode are similar to equation (5.1), but the value of α can be between 7 and 700 depending upon the rating. The diode is placed in parallel with the circuit to be protected and will not operate until a surge exceeds the diode's breakdown voltage. The surge causing the diode to conduct will be clamped to the diode's rated voltage.

Note that these diodes are good protectors for circuits operating typically at low voltages. They are used for protection of data lines on telecommunication and computer systems. They are sometimes referred to as transient voltage suppressers (TVSs). Their large junction is designed specifically for surge protection. TVS diodes are rated for higher current surges than conventional Zener diodes and can carry currents for periods of 2–10 μs. They are also known as avalanche breakdown diodes. Among the prominent advantages of the application of TVS devices in surge suppression is that they have lower clamping ratios and stronger resistance to surges compared to conventional diodes. TVS diodes can be either unidirectional or bidirectional. The peak

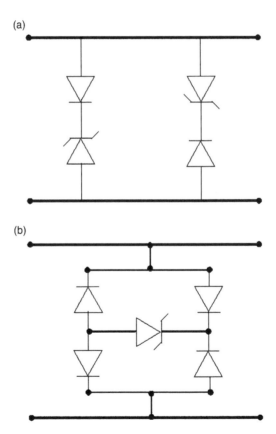

Figure 5.12   *Two techniques for reducing the effective shunt capacitance of the avalanche diode*

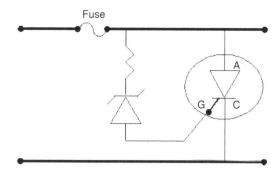

Figure 5.13   Basic thyristor crowbar circuit

power of a TVS diode can be a few watts for small signal circuit protection to several thousand watts for power panel protection. A TVS that handles large currents will also have large capacitance due to geometry, which could be 500 pF to 10 nF, limiting their application for high-frequency signals [1]. One method to reduce this capacitance would be to use suppressor diodes in series with forward-biased switching diodes to effectively reduce the shunt capacitance. Preserving bipolar clamping, two techniques for reducing the capacitance are shown in Figure 5.12.

In addition to diodes, thyristors are also used for transient suppression (Figure 5.13). They are typically four-layer (pnpn) semiconductor devices for unidirectional and five-layer devices for single-chip bidirectional use. Thyristors are turned to the on state by a voltage trigger as shown in Figure 5.10. In the turn on state, the voltage drops across the device is only a few volts, allowing large surge current conduction through it. Operating voltages range from 20 V up to 250 V with current ratings of 50 to 200 A for 10/100 μs.

5.3.2.4   Current limiters

In this section we discuss components such as fuses, circuit breakers, chokes and ferrite. These are current limiting devices that are in series with the lines. Series devices provide high impedance during a surge and that way limit the surge current in the circuit.

Fuses and circuit breakers
Fuses and/or circuit breakers are usually included in the output of d.c. power supplies and also to isolate defective loads from an a.c. power line. The main difference between a fuse and a circuit breaker is that a fuse becomes a permanent open circuit when it faces large fault currents, but a circuit breaker opens the circuit to be protected but can be reset manually or automatically to restore the normal operation of the system. The fuse is faster in action when the current through it is larger. The fuse acts in a time range of 10 ms to 10 μs. Fuses or circuit breakers are always placed in series with the line and sometimes used in conjunction with a surge protective device (SPD) as shown in Figure 5.14.

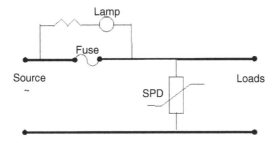

Figure 5.14 Use of a fuse in conjunction with an SPD

Note that when a fuse is in action (during its operation) the voltage across it will appear on the load terminals, which could sometimes be high due to resistance and parasitic inductance of the fuse itself. Thus the overvoltage due to this should be taken into account in the design.

Inductors (chokes and ferrites)
At high frequencies the inductor offers a larger voltage drop across itself (i.e. due to its capability to offer high impedance it can limit the current to a greater extent). It should have sufficient insulation to prevent breakdown and should also have minimum parasitic capacitance across it. It should also be mechanically strong to prevent stress under surge conditions. Inductors are largely used as series devices for power line transient protection rather than in low-voltage signal or data lines. To attenuate transients and associated noise, ceramic materials called ferrites are sometimes used; these are representative of a series circuit with resistance and inductance. The resistance will damp any kind of oscillations that could have resulted from interaction of the inductance and capacitance combination in the system. For this reason, ferrite beads are used in experiments to clean up electromagnetic interference problems associated with measurements.

Magnetic fields tend to concentrate in high-permeability materials. Ferromagnetic rings are very useful in suppressing the unwanted CM noise in cables. The wires are wound through the core in such a way that the fluxes due to the CM currents add in the core, whereas the fluxes due to differential mode (DM) currents or signal currents subtract in the core. Because almost all the flux is confined to the core, the self and mutual inductance of the windings are the same (Figure 5.15).

The DM currents produce fluxes in opposite directions and in the ideal case they exactly cancel each other; the mutual inductance is therefore negative and cancels the self inductance part. Therefore the choke does not present any impedance to DM currents. In the case of CM currents, the fluxes set up by the individual currents add up and therefore the mutual inductance is positive. The choke presents a series impedance of $Z_{CM} = j\omega(L + M) = j\omega 2L$ per winding to CM currents. This type of choke is called CM choke.

Generally, DM currents are much higher than CM currents. The fluxes due to large DM currents cancel in the core. Therefore the core is usually not driven into saturation.

290 *Lightning Protection*

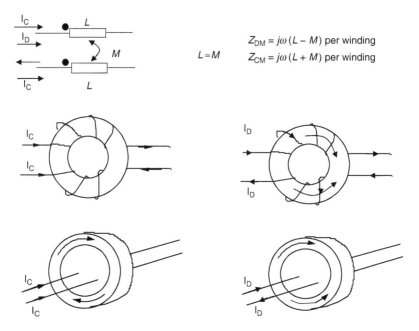

Figure 5.15 *Differential- and common-mode operation of chokes*

The core of the choke is made of high-permeability materials. Permeability is a function of frequency. Depending on the frequency of application, proper materials for the ferrite core have to be selected accordingly.

Positive temperature coefficient (PTC) devices
These devices have characteristic such that their resistance increases as the voltage across them increases and hence they can be used in series for transient protection. They should not break down when a large voltage appears across them. PTC resistors have constant low resistance (fractions of milliohms) at temperatures below the temperature where it switches state. Beyond this critical temperature, when the voltage across them increases, the resistance increases dramatically to a few tens of kV. They can therefore be used as reset-type fuses. When the voltage across them increases, the resistance increases dramatically to a few tens of kΩ; they can therefore be used as reset-type fuses for transient protection machines.

5.3.2.5 Isolation devices

An isolation device has no conductive path between the input and output ports, hence the name isolation. Such devices are mainly used to block the CM voltage from appearing across the loads that are usually/normally working with differential mode voltages. There are two ways in which a signal couples from one port to the other of an isolation device: one through the magnetic field (isolation transformer) and the other through optical signals (optical isolators).

In the case of an isolation transformer there exists in general an unavoidable parasitic capacitance between the input and output terminals. This capacitance is not desirable for CM transient voltages at the input of the isolation device. Usually, two or more conducting screens are inserted between the input and output coils of the transformer, thereby almost eliminating the parasitic capacitance. A proper bonding of the shields to clean ground is needed. Sometimes a pair of capacitors in series with their midpoint grounded are connected across the output port of the transformer for further elimination of the CM surges.

An optical isolator is an electronic component that contains a light source (infrared light-emitting diode) and a photodetector (silicon phototransistor with response time of 1 μs) with no electrical connection between the two. The electrical insulation between the two is a piece of plastic or glass with a dielectric strength of several kilovolts. Unlike isolation transformers, optical isolators can transfer d.c. signals.

5.3.2.6 Filters

Power supply filters are low-pass filters commonly connected in series with the power cord of electronic equipment to attenuate the high-frequency noise that is generated inside the chassis and conducted on the power cord out of the chassis into the mains and environment (Figure 5.16). Low-pass filters may also protect equipment from conducted high-frequency noise on the mains. The high-frequency noise is usually below the normal operating voltage and hence will not trigger the non-linear surge protective devices. Filters are not used as standalone devices to protect against transient overvoltages. However, filters are very useful in attenuating high-frequency noise downstream of a non-linear protective device (spark gap, varistor). This high-frequency noise is partly due to the remnants of the transient overvoltage, and partly due to the action of the non-linear device itself.

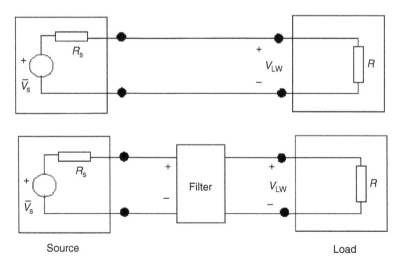

Figure 5.16 System comprising of source and load with and without filter

A simple low-pass filter consists of an inductance in series or a capacitance in parallel or a combination of both. A simple high-pass filter consists of a capacitance in series, or an inductance in parallel, or a combination of both. Inductance has high impedance at high frequencies and low impedance at low frequencies and capacitance has low impedance at high frequencies and high impedance at low frequencies. There will be some attenuation at all frequencies. Therefore filters are characterized by their insertion loss (IL), which is typically stated in decibels. If V_{Lwo} is the magnitude of load voltage without the filter and V_{Lw} is the load voltage with the filter inserted, then the insertion loss is defined as in equation (5.3) with reference to Figure 5.12:

$$\text{IL (dB)} = 20 \log_{10}(V_{Lwo}/V_{Lw}) \quad (5.3)$$

The load voltage without the filter is given by

$$V_{Lwo} = R_L/(R_S + R_L) \quad (5.4)$$

The load voltage with filter is given by equation (5.5), where Z_F is the series impedance of the filter ($Z_F = j\omega L$ in the case of the simple low-pass filter above):

$$V_{Lwo} = R_L/(R_S + Z_F + R_L) \times V_S \quad (5.5)$$

The insertion loss is the ratio of V_{Lwo} and V_{Lw} and is given by

$$\text{IL} = 20 \log_{10} |1 + [Z_F/(R_S + R_L)]| \quad (5.6)$$

From equation (5.6) it is evident that the insertion loss of a filter depends on the source and load impedance, and therefore cannot be stated independently of the terminal impedance. Usually, insertion loss is specified assuming a terminal impedance of 50 Ω.

A power supply filter should give protection to both (CM) and (DM) noise currents. The filter consists of a CM choke, which comprises two identical windings wound over the same ferrite core, two capacitors C_{D1} and C_{D2} between line and neutral on either side of the choke, and two capacitors C_{C1} and C_{C2} between phase/neutral and ground (chassis) on either side of the choke (Figure 5.17). Capacitors C_{D1} and C_{D2} divert the DM noise currents and capacitors C_{C1} and C_{C2} divert the CM current. These are shown in Figure 5.17. Usually $C_{C1} = C_{C2}$, and is kept low

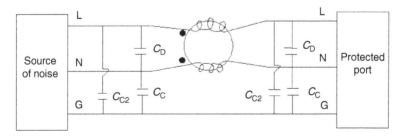

Figure 5.17   A general power supply filter

($\sim$2 nF) to limit the leakage current to below 1 mA for safety reasons. Otherwise there may be a shock hazard if the filter chassis is not earthed properly. Typical values for C_{D1} and C_{D2} are in the range of 0.1–0.5 µF.

5.3.2.7 Special protection devices used in power distribution networks

About 90 per cent of all outages affecting customers originate on the utility distribution system due to lightning and line faults. Voltage spikes, voltage sags and short interruptions cost electric utility customers millions of dollars each year. Most of today's mechanical autoreclosures require six power cycles to react to a line fault caused by a lightning transient or other causes. In most cases there are no satisfactory solutions to voltage sag problems. Low-power customers ride through these difficulties using uninterrupted power supply (UPS) systems, but such solutions are not feasible for high-power consumers. Many solid-state power controller devices have been developed recently that are based on new-generation power electronic components such as gate turn-off thyristors (GTOs) and insulated gate bipolar transistors (IGBTs). A brief description of them is given below and is based on the information from Reference 27.

Solid-state breaker (SSB)
The first SSB for use on a 13 kV system was developed in 1995 by Westinghouse Electric Corporation, with the support of the Electric Power Research Institute (EPRI), United States. The SSB can clear a fault at sub-cycle speed and is based on GTOs, which do not require power cycle zero for turn-off.

Solid-state transfer switch (SSTS)
Mechanical transfer switches that transfer load from one feeder to the other takes 2 to 10 s, whereas an SSTS can transfer the load within half a cycle. The first 15 kV class SSB was developed in 1995 by Silicon Power Networks. The key components of the SSTS are two SCR switches connected back-to-back, controlling primary and secondary feeder.

Dynamic voltage restorer (DVR)
This is a solid-state controller that protects a critical load from power line disturbances other than outages. DVRs are connected in series with the distribution feeder through three single-phase injection transformers, and restore the original voltage waveform by injecting compensating voltages in real time. DVRs are effective against voltage sags, swells, transients and harmonics. The DVR consists of a d.c. to a.c. inverter based on an IGBT. The IGBT switches convert the regulated d.c. source into a synchronous a.c. voltage of controllable amplitude, phase angle and frequency. For example, during voltage sag the DVR supplies a compensating voltage in phase with that on the line to make up the difference, while during harmonics the DVR will generate a complex waveform to cancel them out. The DVR can also limit fault currents by injecting a voltage that leads the line current by 90°, increasing the apparent reactive impedance of the line. Westinghouse Electric Corporation makes DVRs in ratings

from 2 to 10 MVA. The response time of the DVR is less than one 1 ms, which is a small fraction of a power cycle. Westinghouse Electric Corporation is now developing a DVR that can be mounted on distribution poles. The first pole-mounted DVR is expected to be ready this year and will have a rating of 300 kVA, for use on a 15 kV class distribution system. It can be visualized that in the future low-cost series types of devices that can maintain the quality of the voltage waveshape even in the presence of fast transients may be developed, for use in low-voltage applications.

Distribution static compensator (D-STATCOM)
This protects the distribution system against power pollution caused by certain customer loads found in steel plants, saw mills and so on. It replaces the conventional tap changing transformer, voltage regulator and switched capacitors. Like the DVR, a D-STATCOM also consists of an IGBT-based d.c. to a.c. power inverter. However, the D-STATCOM is connected in shunt and usually supplies only reactive power to the line through a coupling transformer. This is also developed by Westinghouse.

Fault current limiters using superconductors
High-temperature superconductors (HTS), which are superconducting around 77 K and maintained by liquid nitrogen, are finding applications in fault-current limiters in power systems [28]. The HTS fault-current limiter is a series device. A 2.4 kV, 3 kA prototype was successfully tested by Lockheed Martin in 1995.

In a screened-core fault-current limiter, an iron core is surrounded by a superconducting cylinder over which there is a conventional copper winding. During normal operation, shielding currents induced in the HTS do not allow magnetic field penetration into the iron core, resulting in low series impedance. During fault current, the magnetic field penetrates into the iron core, resulting in high series impedance. This device can respond within one power cycle and can affect 80 per cent reduction in fault current. Multiple faults within a period of 15 s can be successfully handled. Associated with fast electronics and further technical improvements, HTS-based fault-current limiters may develop into dynamic series impedance that can block any transient surges in distribution lines.

5.4 Effects of parasitic elements in surge protection and filter components

Physical components such as conductors, resistors, inductors and capacitors are used in filters. The symbols used in circuit diagrams to represent these components show only their ideal properties. For example, conducting wires used between components or in component leads have some resistance, however small it may be. While carrying high-frequency currents, the charge carriers may crowd toward the periphery of the conductor, increasing the resistance of the wire. Associated with the currents and charges in the wire there are magnetic fields and electric fields, and therefore inductance and capacitance. We know that at high frequencies inductance has high impedance and capacitance has low impedance. Therefore the behaviour of the

Protection against lightning surges 295

conductors and circuit components used in filters may depart from ideal behaviour at high frequencies. Surge protection components such as varistors, gas discharge tubes, diodes and so on, also come with various packages and connection leads. These introduce parasitic elements and modify the performance of the component.

5.4.1 Capacitors

A capacitor is a discrete component used in filters, ideally providing an impedance inversely proportional to the frequency (Figure 5.18):

$$Z(j\omega) = \frac{1}{j\omega C} = \frac{1}{\omega C} \angle -90°$$

$$|Z| = \frac{1}{2\pi C} \cdot \frac{1}{f}$$

$$20\log|Z| = 20\log\left(\frac{1}{2\pi C}\right) - 20\log(f)$$

The magnitude of the impedance of an ideal capacitor decreases by 20 dB per decade increase in frequency. Capacitors are often used to provide a short-circuit path (very low impedance) at frequencies beyond a certain value. A capacitor uses conducting wires and dielectric in its construction. Therefore the high-frequency behaviour of capacitors is far from ideal. The general high-frequency equivalent circuit of a capacitor is shown in Figure 5.19, where C is the nominal value of the capacitance, R is the resistance of the capacitor plates and connecting leads, and L is the inductance of the external and internal connecting leads. In short, the equivalent circuit for a capacitor is an inductance, capacitance and resistance connected in series, i.e. $Z(j\omega) = R + j\omega L + 1/j\omega C$.

At series resonance, $\omega L = 1/\omega C$, so the resonance frequency f_r is equal to $1/[2\pi\sqrt{(LC)}]$ and the impedance at resonance is equal to R. Below resonance

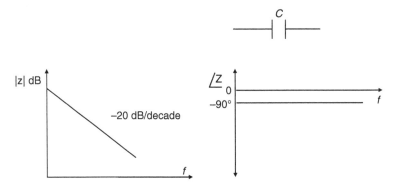

Figure 5.18 Impedance magnitude and phase of an ideal capacitor

296 Lightning Protection

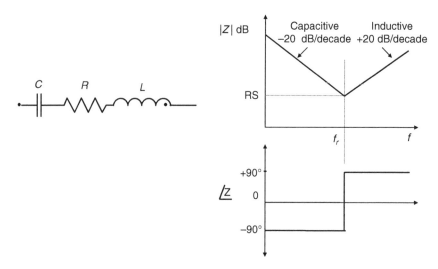

Figure 5.19 Impedance magnitude and phase of a real capacitor

frequency, $1/\omega C > \omega L$ and the capacitance effect is dominant. Above resonance frequency, $\omega L > 1/\omega C$ and the inductance effect is dominant.

The parasitic inductance of a capacitor is mainly determined by the length of the leads and hence does not change. Therefore, for a given type of capacitor, the resonant frequency tends to decrease with an increase in capacitance value. This effect has important practical consequences, which are shown in the following example. Suppose a 470 pF ceramic capacitor is used as a shunt element to divert high-frequency noise currents. The loop formed by the component leads (12 mm length and 6 mm separation) has a parasitic inductance of about 14 nH. The capacitor will resonate at a frequency of 62 MHz and at this frequency it has minimum impedance. As the frequency is increased beyond the resonance frequency, the impedance of the capacitor starts increasing and it behaves more like an inductor. The above capacitor may be suitable for providing a low-impedance path to noise currents at, say, 60 MHz, but may not be suitable for noise currents at, say, 200 MHz. There are many practical considerations in the use of capacitors to divert noise currents.

- The self-resonant frequency of the capacitor should be considered.
- In low-amplitude signal applications, the capacitor should not be placed in such a way that it forms a loop with other components serving as a receiving antenna for radiated electromagnetic interference (EMI).
- A capacitor in parallel with the inductance of the cable it is protecting can form a parallel $L-C$ circuit and can produce 'ringing' due to resonance.
- Shunting capacitors work best in high-impedance circuits (noise current division between the circuit impedance and the capacitor impedance).

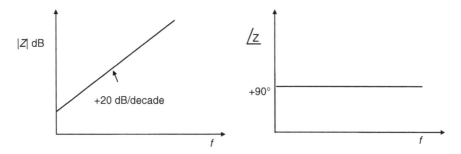

Figure 5.20   Impedance magnitude and phase of an ideal inductor

5.4.2 Inductors

The impedance of an ideal inductor (Figure 5.20) is given by

$$Z_L = j\omega L = \omega L \angle +90°$$
$$20\log|Z_L| = 20\log(2\pi L) + 20\log(f)$$

The magnitude increases at a rate of 20 dB per decade increase in frequency. The general circuit for the inductance is as shown in Figure 5.21, where L is the nominal value of the inductance (much higher than lead inductance), R_s the loss resistance (eddy current losses resistance of the wires), and C_p the effective parasitic capacitance (capacitance between turns, between turns and the core). This is also known as the lead capacitance.

Inductors behave as a parallel-resonant circuit, i.e. $R_s + j\omega L || 1/j\omega C$ and $Z_L(j\omega) = \{(R_s + j\omega L)(1/j\omega C)/R_s + j[\omega L - (1/\omega C)]\}$. At resonance, $\omega L = 1/\omega C$ and the impedance Z_L is a maximum. The resonant frequency f_r is given by $1/2\pi\sqrt{(LC)}$. Below the self-resonant frequency, $\omega L < 1/\omega C$ and the inductor predominantly has an inductive character. Above the self-resonant frequency, $\omega L > 1/\omega C$ and the inductor behaves as a capacitance. At very low frequencies the inductive impedance may be even lower than the series loss resistor, and the inductor behaves more like a resistor.

Inductors are used to block noise current and are used in series, whereas capacitors are used to divert (shunt) noise currents and are used in parallel (Figure 5.22). Inductors behave as a capacitance above the self-resonant frequency, whereas a capacitor behaves as an inductance above the self-resonant frequency. Increasing the value of an inductor does not necessarily give a higher impedance at a given frequency, because the large value of the inductance lowers the self-resonant frequency of the inductor. Practical considerations in the use of an inductor to block noise currents are as follows.

- The self-resonant frequency of the inductor should be considered.
- The impedance of the inductor at noise frequencies should be larger than the impedance of the circuit it is protecting. That is, blocking series inductors are most effective in the low-impedance circuits.

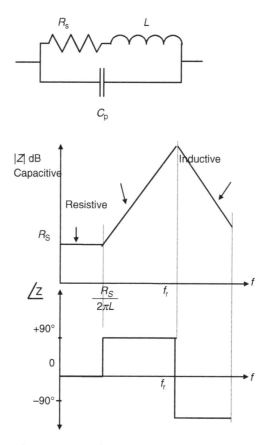

Figure 5.21 Impedance magnitude and phase of a real inductor

- There is the possibility of 'ringing' due to resonance formed by parallel or series (parasitic) capacitance of the circuit.
- In inductors with ferromagnetic cores, the non-linear effect of saturation must be considered.

Figure 5.23 shows the $B-H$ curve of a ferromagnetic core. The magnetic permeability $\mu = \Delta B/\Delta H$ is the slope of the $B-H$ curve and is also a function of $B[\mu(B)]$. At low values of magnetic flux density, the slope of the $B-H$ curve is large, i.e. μ_r is large. As the flux density is increased the core saturates and μ_r is small. Therefore in inductors with a ferromagnetic core the inductance value decreases with increasing current.

5.4.3 Resistors

There are basically three types of resistors: (1) carbon composition, (2) wire wound and (3) thin metal film. Ideally a resistor has impedance equal to the resistance

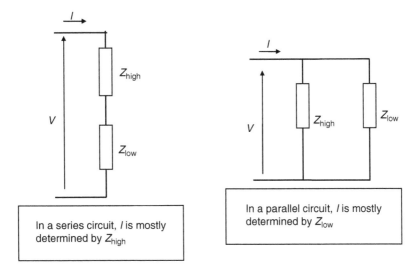

Figure 5.22   Series and parallel circuits

value and a phase angle of zero at all frequencies, i.e. $Z = R\angle 0°$. Therefore the voltage across the resistor is proportional to the current through it. This property is used in the measurement of current in a circuit using shunt, and for deriving control signals proportional to the current. However, the behaviour of a practical resistor is different from ideal and very much depends on the construction method. The equivalent circuit of a real resistor is given in Figure 5.24, where R is the nominal value of the resistor, L_{we} the inductance of the connecting leads (typical value of ~ 15 nH), and L_C is the inductance of the resistance element itself, e.g. the inductance of the wire in a wire-wound resistor. This is usually negligible except for wire-wound resistors. C_p is the parasitic capacitance, which includes the lead capacitance and leakage capacitance of the resistor body. Its typical value is in the range 1–2 pF. Figure 5.25 presents a commonly used model for a resistor.

5.4.3.1   Behaviour of resistors at various frequencies

Under d.c. conditions, an inductance acts as a short circuit and capacitance as an open circuit. As the frequency is increased, the impedance of the capacitor decreases and at a

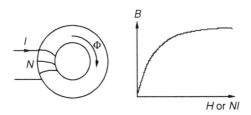

Figure 5.23   B–H *curve of ferromagnetic cores*

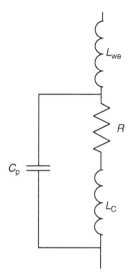

Figure 5.24   Possible equivalent circuit of a resistor

frequency $f = 1/2\pi R C_p$ the impedance of the capacitance and resistance become equal (Figure 5.26). Above this value of frequency the capacitor impedance become dominant (i.e. less than R) and the net impedance decreases by approximately −20 dB per decade and the phase angle approaches $\angle -90°$. The model inductance and capacitance resonate at $f_0 = 1/[2\pi\sqrt{(L_{we}C_p)}]$ and the impedance is a minimum at this frequency. Above the resonant frequency, the inductive impedance become dominant and the magnitude of the impedance increases by +20 dB per decade and the phase angle approaches $\angle +90°$.

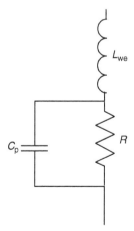

Figure 5.25   Commonly used model for resistor

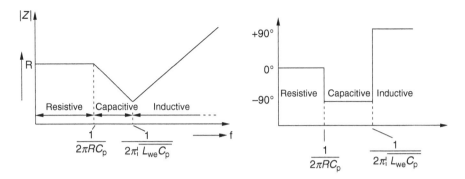

Figure 5.26 Bode plot of impedance magnitude and phase of a real resistor

Starting from d.c., as the frequency is increased the inductive impedance in the resistor model is also increasing, while the capacitive impedance is decreasing. For a given value of the lead inductance and parasitic capacitance, the behaviour of the resistor depends on the nominal value of the resistance. For low-value resistors, the capacitance becomes dominant in the parallel RC circuit only at a very high frequency. In this case, the resistor goes from the resistive phase to the inductive phase directly. A numerical example is given below.

$$C_p = 2 \text{ pF}$$

$$L_{we} = 20 \text{ nH}$$

$$f_0 = 1/[2\pi\sqrt{(L_{we}C_p)}] = 796 \text{ MHz}$$

R	$\dfrac{1}{2\pi RC}$			
1 Ω	79.6 GHz	$>f_0$	resistive	→ inductive
100 Ω	796 MHz	$=f_0$	resistive	→ inductive
1 000 Ω	79.6 MHz	$<f_0$	resistive	→ capacitive → inductive

For low-value resistors the inductive reactance became dominant, even before the resonant frequency $1/2\pi\sqrt{(LC)}$ is reached. For example, at 1 MHz, a 20 nH inductance has an impedance of 0.125 Ω. If the nominal value of the resistor is 0.1 Ω, the resistor is behaving mostly as an inductance.

At high frequencies and below the resonant frequency the following statements apply.

- Capacitance is the main problem for high-value resistors (inductive impedance is only a small fraction of the nominal resistance value).
- Inductance is the main problem for low-value resistors (capacitive impedance is many times the nominal value of the resistor).

302 *Lightning Protection*

To avoid the above two problems, the leads can be shorted as much as possible and connect many smaller value resistors in series to get a high resistance value (parasitic capacitance in series). Also, many higher resistors may be connected in parallel to obtain a low resistance value (lead inductances in parallel).

5.5 Surge protection coordination

Surge protective devices are usually placed at the service entrance of buildings, at some of the branch circuits and at sensitive equipment inside buildings. In the usual concept of coordination of surge protective devices, surge protectors (SPs) at sensitive equipment have lower energy handling capability and lower clamping voltage than the SP at the service entrance. It is expected that the SP at the service entrance (primary SP) will handle the bulk of the surge energy and the SPs at the sensitive equipments (secondary SP) will handle the remaining fraction of the energy. However, in practice, this is not always the case. In some situations the SPs at the sensitive equipment carry the bulk of the surge stress, destroying themselves. The parameters that influence the sharing of surge energy between the primary and secondary SPs comprise the individual characteristics of protectors, the nature of the transient overvoltage, and the distance between the protectors.

For example, consider two varistors separated by a length of conductor as shown in Figure 5.27. If this length is electrically short compared to the rise time of the transient, then the voltage across the primary varistor is the vector sum of the voltage drop in the conductor and the secondary varistor clamping voltage. Because the voltage drop in the conductor is mostly inductive, the dI/dt of the transient is important. The energy handled by the primary varistor is determined by the voltage appearing across it. Therefore we can see that the surge energy division (coordination) between the two varistors is determined by (1) the clamping voltage of the varistors, (2) the voltage–current characteristics of the varistors, (3) the length and impedance of the conductor between, and (4) the characteristics of the transient overvoltage. The influence of transient characteristics on varistor coordination was investigated in Reference 29. Three transient waveforms were used: a $1.2/50$–$8/20$ μs combination

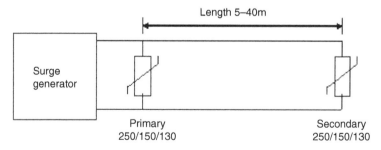

Figure 5.27 *Co-ordination of two varistors separated by some length of conductors in between*

wave, a 0.5 μs–100 kHz ring wave, and 10/1 000 μs long impulse wave. Three varistors, with nominal r.m.s. voltage ratings of 250, 150 and 130 V, were used as primary and secondary in all the nine possible combinations.

During the combination wave and ring wave the primary varistor handled a larger share of the energy when the clamping voltage of the primary was higher than the secondary (high–low combination), even though the energy division was sensitive to the length of conductor between the varistors. However, during the 10/1 000-μs-long wave the high–low combination of varistors resulted in the secondary varistor handling almost all the surge energy. That is, the coordination failed in this case. Usually, the varistors in equipment have a lower energy handling capacity than the one at the service entrance and cannot handle large surge energy. However, the reverse combination, that is low–high, was able to provide good coordination for the 10/1 000-μs voltage wave. The usefulness of numerical simulation in varistor coordination is also shown in Reference 29. A review of the numerical simulation of gas discharge protectors can be found in References 30 to 32.

So far, the discussion on coordination has been based on the assumption that the transient overvoltages are conducted from outside into buildings, i.e. through the 'front door'. Also, it has been assumed that the distances between the surge protectors are small compared to the spatial extent occupied by the rising edge and falling edge of the transients. When transients enter through the 'back door', for example by direct interaction of the electromagnetic fields with the conductors inside the building or by transients generated within the building, the protection coordination is very complex and not well understood. Besides, the equipment impedance characteristics, across which the secondary protector is connected, influence the coordination. The presence of many branch circuits throughout the building and possible resonances complicate the situation. It is not feasible to experimentally investigate all the possible situations for designing optimum protection for an installation. However, if adequate models are developed for various protection components, cabling and loads, the numerical simulation techniques can be very powerful in designing optimum protection for an installation.

References

1. Standler R.B. *Protection of Electronic Circuits for Overvoltages*. New York: John Wiley and Sons, Inc.; 1989.
2. Goedbloed J.J. 'Transients in low-voltage supply networks'. *IEEE Trans. Electromagnetic Compatibility* 1987;**29**:104–15.
3. Rakov V.A. and Uman M.A. *Lighting Physics and Effects*. Cambridge University Press; 2003.
4. Barker P.P., Short T.A., Eybert-Berard A.R. and Berlandis J.P. 'Induced voltage measurements on an experimental distribution line during nearby rocket triggered lightning flashes'. *IEEE Trans. Power Deliv.* 1996;**11**:980–95.
5. Uman M.A., Rakov V.A., Rambo K.I., Vaught T.W., Fernandez M.I., Bernstein R. *et al.* 'Triggered lightning facility for studying lightning effects on power

systems'. *Proceedings of the 23rd International Conference on Lightning Protection*; ICLP, Florence, Italy, 23–27 September 1996.
6. Thottappillil R., Goldberg J., Rakov V.A., Uman M.A., Fisher R.J. and Schnetzer G.H. 'Properties of lightning M component current pulses'. *J. Geophys. Res.* 1995;**100**:25711–20.
7. Morched A., Marti L. and Ottevangers J. 'A high frequency transformer model for EMTP'. *IEEE Trans. Power Deliv.* 1993;**8**(3):1615–26.
8. Manyahi M.J. and Thottappillil R. 'Simplified model for estimation of lightning induced transient transfer through distribution transformer'. *Int. J. Elec. Power Energy Syst.* 2005;**27**(4):241–53.
9. Darveniza M., Roby D. and Tumma L.R. 'Laboratory and analytical studies of the effects of multipulse lightning current on metal oxide arresters'. *IEEE Trans. Power Deliv.* 1994;**9**:764–71.
10. Pérez H., Ye M. and Scuka V. 'Induced overvoltages in low voltage power installations caused by lightning electromagnetic impulses'. *Proceedings of the CIGRE Symposium on Power System Electromagnetic Compatibility; Lausanne Switzerland, 1993*. Paper 500–04.
11. Pérez H., Ye M. and Scuka V. 'Coupling characteristic of lightning electromagnetic fields in a low voltage power installation network'. *Proceedings of the 22nd International Conference on Lightning Protection*; ICLP, Budapest, Hungary, 1994. Paper R3b–07.
12. Silfverskiöld S., Thottappillil R., Ye M., Cooray V. and Scuka V. 'Induced voltages in a low voltage power installation network due to lightning electromagnetic fields'. *IEEE Trans. Electromagnetic Compatibility* 1999;**41**(3):265–71.
13. Thottappillil R. 'Electromagnetic pulse environment of cloud-to-ground lightning for EMC studies'. *IEEE Trans. EMC* 2002;**44**:203–13.
14. Thottappillil R. 'A review of lightning with emphasis on the properties important for the protection of ground based assets'. Report to Swedish Defense Material Administration – FMV, Stockholm, Project No. 272263-LB649823, January 2000.
15. Fisher R.J. and Schnetzer G.H. '1990 Sandia rocket-triggered lightning field tests at Kennedy Space Center, Florida', Sandia Report SAND90-2926, 1990.
16. Fisher F.A. and Plumer J.A. *Lightning Protection of Aircraft*. NASA Reference Publication 1008, 1977.
17. Berger K., Anderson R.B. and Kroninger H. 'Parameters of lightning flashes'. *Electra* 1975;**41**:23–37.
18. Fisher R.J., Schnetzer G.H., Thottappillil R., Rakov V.A., Uman M.A. and Goldberg J. 'Parameters of triggered lightning flashes in Florida and Alabama'. *J. Geophys. Res.* 1993;**98**:22887–902.
19. Cooray V. 'Horizontal fields generated by return strokes'. *Radio Science* 1992;**27**:529–37.
20. Rubinstein M., Rachidi F., Uman M.A., Thottappillil R., Rakov V.A. and Nucci C.A. 'Characterization of vertical electric fields 500 and 30 m from triggered lightning'. *J. Geophys. Res.* 1995;**100**:8863–72.

21. Rakov V.A., Uman M.A., Rambo K.J., Fernandez M.I., Fischer R.J., Schnetzer G.H. et al. 'New insights into lightning processes gained from triggered-lightning experiments in Florida and Alabama'. *J. Geophys. Res.* 1998;**103**:14117–30.
22. Crawford D.E. 'Multiple-station measurements of triggered lightning electric and magnetic fields'. Master's thesis, Department of Electrical and Computer Engineering, University of Florida, Gainesville, USA, 1998.
23. Uman M.A., Rakov V.A., Schnetzer G.H., Rambo K.J., Crawford D.E. and Fisher R.J. 'Time derivative of the electric field 10, 14 and 30 m from triggered lightning strokes'. *J. Geophys. Res.* 2000;**105**:15577–95.
24. Rachidi F., Nucci C.A., Ianoz M. and Mazzetti M. 'Influence of a lossy ground on lightning induced voltages on overhead lines'. *IEEE Trans. Electromagnetic Compatibility* 1996;**38**(3):250–64.
25. Vance E.F. *Coupling to Cable Shields*. New York; Wiley Interscience; 1978.
26. Heinrich C. and Darveniza M. 'The effects of multipulse lightning currents on low-voltage surge arresters'. *10th International Symposium on High Voltage Engineering*; Montreal, 1997.
27. Douglas J. 'Custom power: Optimizing distribution services'. *EPRI J.* 1996;**21**:6–15.
28. 'Superconductivity in electric power'. *IEEE Spectrum* July 1997.
29. Lai J.-S. and Martzloff F.D. 'Coordinating cascaded surge protection devices: high–low versus low–high'. *IEEE Transactions on Industry Applications* 1993;**29**:680–87.
30. Larsson A., Tang H. and Scuka V. 'Mathematical simulation of a gas discharge protector using ATP–EMTP'. *International Symposium on Electromagnetic Compatibility*; 1996.
31. Tang H., Högberg R., Lötberg E. and Scuka V. 'Numerical simulation of a low-voltage protection device using the EMTP program'. *23rd International Conference on Lightning Protection*, Florence, Italy, 1996.
32. Tang H., Larsson A., Högberg R. and Scuka V. 'Simulation of transient protectors co-ordination in a low voltage power installation using ATP–EMTP'. *International Symposium on EMC*; EMC'96, Rome, Italy, 1996.

Chapter 6
External lightning protection system
Christian Bouquegneau

6.1 Introduction

An external lightning protection system (external LPS), is intended to intercept direct lightning flashes to a structure, to conduct the lightning current safely towards earth and to disperse the lightning current into the earth. These three goals can be considered separately by means of three complementary systems: an air-termination system (Section 6.2), a down-conductor system (Section 6.3) and an earth-termination system (Section 6.4).

The internal LPS (see Chapter 7) is supposed to prevent dangerous sparking within the structure to be protected, by using either equipotential bonding or electrical insulation thanks to a separation distance (distance between two conductive parts at which no dangerous sparking can occur) between the external LPS components and other electrically conducting elements internal to the structure.

The complete LPS has to be an effective measure not only for the protection of structures against physical damage, but also for the protection against injury to living beings due to touch and step voltages (Section 6.4). The main purpose is to reduce the dangerous current flowing through bodies by insulating exposed conductive parts and by increasing the surface soil resistivity. Soil conductivity and the nature of the earth are very important.

The type and location of an external LPS should primarily take into account the presence of electrically conductive parts of the structure. The external LPS can be isolated or not from the structure to be protected. In an isolated external LPS, the air-termination system and the down-conductor system are positioned in such a way that the path of the lightning current has no contact with the structure to be protected, and dangerous sparks between the LPS and structure are avoided.

In most cases, the external LPS may be attached to the structure to be protected.

An isolated external LPS is required when the thermal and explosive effects at the point of strike on the conductors carrying the lightning current may cause damage to the structure or to its content. Among the typical cases, let us mention the structures with combustible walls and coverings as well as areas with danger of explosion and fire.

308 *Lightning Protection*

Another important requirement is to avoid dangerous sparking between the lightning protection system and the structure; this is satisfied in an isolated external LPS by sufficient insulation or separation and in a non-isolated external LPS by bonding or by sufficient insulation or separation.

Natural conductive components (i.e. pipeworks, metallic cable elements, metal ducts) installed not specifically for lightning protection can be used to provide the function of one or more parts of the LPS. Particular care is taken with the protection of structures containing solid explosive materials or hazardous zones.

A last section (Section 6.5) is devoted to the selection of materials.

The use of the new international standard IEC 62305 on Protection against Lightning, particularly its part 3 (IEC 62305-3) related to physical damages to structures and life hazards, is greatly recommended. In this chapter, we adopt most of its requirements.

6.2 Air-termination system

The air-termination system is the first part of an external LPS and uses metallic elements such as rods, mesh conductors or catenary wires that are intended to intercept lightning flashes.

6.2.1 Location of air-terminations on the structure

Air-termination systems can be composed of any combination of vertical rods (including free-standing masts), catenary wires, horizontal or meshed conductors. Individual air-termination rods are connected together at roof level to ensure current division.

Air-termination components installed on a structure shall be located preferably at corners, exposed points and edges, especially on the upper level of any facades, in accordance with the rolling sphere method (see Section 6.2.1.1) or more advanced leader inception models (see Chapter 4 on lightning interception).

To take into account the uses in different countries, the international standard IEC 62305 also accepts the mesh method (see Section 6.2.1.2) or the protection angle method (see Section 6.2.1.3). This last method is only suitable for simple-shaped buildings and is subject to the limits of air-termination height indicated in Figure 6.1. The mesh method is suitable when plane surfaces are to be protected. Maximum values of rolling sphere radius R, mesh size W and protection angle α corresponding to four levels of protection are given in Figure 6.1. These values are assessed in order to obtain equivalent protected volumes by using either the protection angle method or the rolling sphere method.

In the international standard IEC 62305, four classes (I, II, III, IV) of lightning protection systems corresponding to a set of construction rules and related to four protection levels (I, II, III, IV) are introduced. At first glance, a respective global protection efficiency of 98 per cent (level I), 95 per cent (level II), 90 per cent (level III) or 80 per cent (level IV) is associated with each of them. To pass over 98 per cent of global protection efficiency (sometimes called level I^+), we need to apply additional protection measures.

External lightning protection system 309

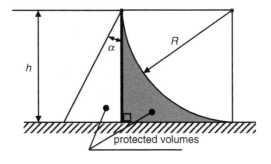

Class of LPS	Protection method		
	Rolling sphere radius R (m)	Mesh size W (m)	Protection angle α
I	20	5 × 5	See figure below
II	30	10 × 10	
III	45	15 × 15	
IV	60	20 × 20	

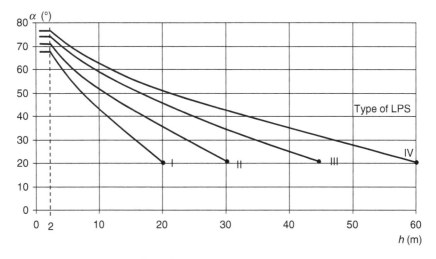

Figure 6.1 Comparison of the three usual lightning protection methods (LPSs). The table gives the maximum values of rolling sphere radius R, *mesh size* W *and protection angle* α *corresponding to the class of lightning protection systems. Note that the protection angle method (bottom panel) is not applicable beyond the values marked with a solid symbol. Only the rolling sphere and mesh methods apply in these cases;* h *is the height of the air-termination above the area to be protected. The angle will not change for values of* h *below 2 m.*

310 Lightning Protection

More precisely, to each protection class (I, II, III, IV) is associated a set of minimum and maximum values of the parameters bound to imposed lightning currents related to a fixed protection level. Maximum values of lightning current amplitudes are respectively fixed to 200 kA (99 per cent of the strokes) at level I, 150 kA (98 per cent of the strokes) at level II and 100 kA (97 per cent of the strokes) at levels III and IV. Minimum values of lightning current amplitudes are related to the application of the rolling sphere method in the design of lightning protection systems: they are fixed to 3 kA (99 per cent of the strokes; $R = 20$ m) at level I, 5 kA (97 per cent of the strokes; $R = 30$ m) at level II, 10 kA (91 per cent of the strokes; $R = 45$ m) at level III and 16 kA (84 per cent of the strokes; $R = 60$ m) at level IV.

The efficiency of such protection measures is supposed to be equal to the probability of finding the lightning current parameters situated between the minimum and maximum limits inside the selected protection level.

6.2.1.1 Positioning of the air-termination system utilizing the rolling sphere method

When applying the rolling sphere method, the positioning of the air-termination system is adequate if no point of the volume to be protected comes into contact with a sphere with radius R depending on the lightning protection level (I–IV, see the table in Figure 6.1), rolling around and on top of the structure in all possible directions. In this way, the sphere only touches the air-termination system (see Figure 6.2, left and right panels, where the height H of the structure is smaller than 60 m).

On tall structures higher than the rolling sphere radius R ($H > R$), side flashes may occur. Each lateral point of the structure touched by the rolling sphere is a possible

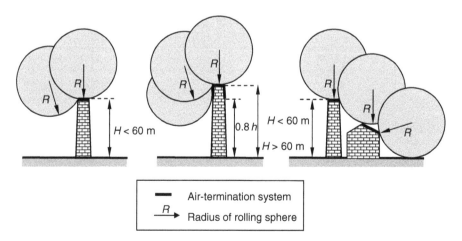

Figure 6.2 Design of a lightning protection system air-termination according to the rolling sphere method. The rolling sphere radius R complies with the selected lightning protection level.

External lightning protection system 311

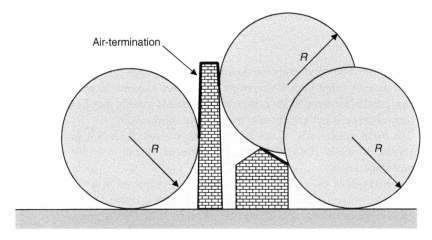

Figure 6.3 Design of an air-termination system according to the rolling sphere method

point of strike. However, the probability of flashes to the sides is generally negligible for structures lower than 60 m.

For taller structures, the majority of flashes will hit the top, horizontal leading edges and corners of the structure. Only a few per cent of all flashes will hit the sides of the structure. Moreover, it has been proven experimentally that the probability of side flashes rapidly decreases as the height of the point of strike on tall structures decreases, when measured from the ground [1]. It appears sufficient to install a lateral air-termination system on the upper part of tall structures, typically the upper 20 per cent of the height of the structure (see middle panel of Figure 6.2).

Returning to the right panel of Figure 6.2, if the height of the structure (chimney) is higher than the rolling sphere radius R selected for a certain level of protection (see the table in Figure 6.1), when rolling the sphere of radius R around and over all the structure until it meets the ground plane or any permanent structure or object in contact with the ground plane capable of acting as a conductor of lightning current, we see (see Figure 6.3) that a lateral air-termination system should be installed on the upper part of the structure to be protected.

We recommend installing an air-termination system on all sides of structures at a height between R and H on structures such that $H > R$, although the international standard is much less stringent, because it only imposes an air-termination system on the topmost 20 per cent of lateral surfaces of structures of height H higher than 60 m, and to protect all parts that may be endangered above 120 m.

6.2.1.2 Positioning of the air-termination system utilizing the mesh method

The mesh method is suitable for protecting buildings and structures having a flat roof (horizontal or inclined roofs with no curvature) or flat surfaces on their top. It is also suitable for flat lateral surfaces to protect the building or structure against side flashes.

312 *Lightning Protection*

The mesh is considered to protect the whole upper surface if the following conditions are fulfilled.

- Air-termination conductors are positioned on roof edge lines, on roof overhangs and on roof ridge lines if the slope of the roof exceeds 10 per cent, in which case parallel air-termination conductors, instead of a mesh, may be used providing their distance is not greater than the required mesh width.
- The mesh dimensions of the air-termination network are not greater than the values of mesh sizes given in the table in Figure 6.1, for the required protection level.
- The network of the air-termination system is constructed in such a way that the lightning current will always encounter at least two distinct metal routes to the earth-termination system. There are no metal installation protrusions outside the volume protected by the air-termination system, and if there are, these protrusions must be electrically connected to the mesh or to the network of the air-termination system.
- The air-termination conductors must follow, as far as possible, the shortest and most direct route.

An example of a flat roof protected according to the mesh method is given in Figure 6.4. Figure 6.5 shows another example of a lightning protection system on

Figure 6.4   Flat roof protected according to the mesh method

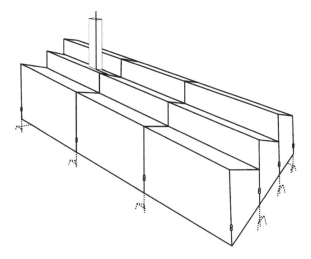

Figure 6.5 Example of a lightning protection system on a shed roof structure

an industrial building. Other interesting examples of protection according to the mesh method can be found in annex E of IEC 62305-3 (see Section E.5.2.2.3) [2].

6.2.1.3 Positioning of the air-termination system utilizing the protection angle method

After Benjamin Franklin, the installation of vertical lightning rods was considered for a long time as 'a salve for every sore' in the lightning protection domain. It was wrongly admitted that the tip had an attractive power in a relatively small volume called the 'protection zone' or 'protection cone' of the lightning rod. This volume is shaped as a right circular cone with the vertex placed on the air-termination axis (see Figure 6.6) with a determined semi-apex angle α of $30°$, $45°$ or $60°$, for example.

These simple empirical models were rapidly denied by experiments on tall structures showing that many points of strike could happen along parts lower than the tips of the air-termination rods, metallic high towers and high-voltage transmission lines.

Only the rolling sphere method can explain such occurrences.

Nevertheless, for lower structures there is a possible equivalence in applying both methods (rolling sphere method and protection angle method). That is why the protection angle method is also presented in the international standard.

The position of the air-termination system (air-termination conductors, rods, masts and wires) is adequate if the structure to be protected is fully situated within the protected volume provided by this metallic air-termination system.

Depending on the class of lightning protection system, the angle α is selected from the table in Figure 6.1. Figures 6.6 and 6.7 show how to proceed to define the volume protected by a vertical air-termination rod. Note that the protective angle α is different

314 *Lightning Protection*

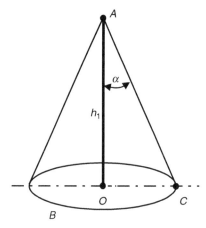

Figure 6.6   *Volume protected by a vertical air-termination rod. A is the tip of the air-termination rod, B is the reference plane, OC is the radius of the protected area, and h_1 is the height of the air-termination rod above the reference plane of protection.*

for different heights of air-terminations above the surface to be protected. If structures on a roof are to be protected with finials and the protection volume of the finials is over the edge of the building, the finials should be placed between the structure and the edge. If this is not possible, the rolling sphere method should be applied.

The volume protected by a horizontal wire is defined by the composition of the volume protected by virtual rods having vertices on the wire. An example of such a protected volume is given in Figure 6.8.

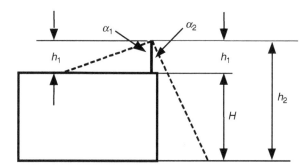

Figure 6.7   *Volume protected by a vertical air-termination rod above a roof surface. h_1 is the physical height of the air-termination rod. The protective angle α_1 corresponds to the height h_1 above the roof surface to be protected, and the protective angle α_2 corresponds to the height $h_2 = h_1 + H$, the ground being the reference plane.*

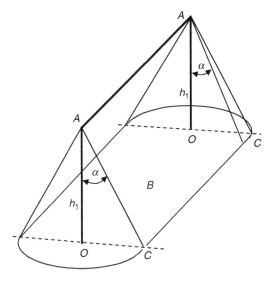

Figure 6.8 Volume protected by a horizontal wire as the air-termination system

Wires can be combined in a mesh. In this case, the volume to be protected is defined by a combination of the volume determined by the single conductors forming the mesh. Figures 6.9 and 6.10 (see Annex A.1.3 in IEC 62305-3 [2]) illustrate examples of a volume protected by wires combined with a mesh, according to either both protection angle method and rolling sphere method (Figure 6.9) or both mesh method and protective angle method (Figure 6.10).

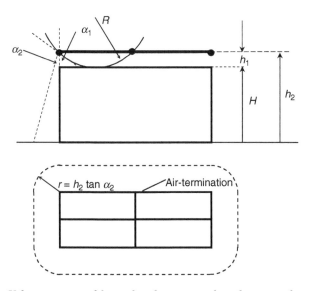

Figure 6.9 Volume protected by isolated wires combined in a mesh

316 Lightning Protection

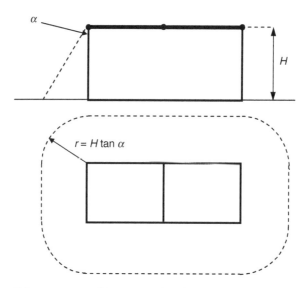

Figure 6.10 Volume protected by non-isolated wires combined in a mesh

6.2.1.4 Comparison of methods for the positioning of the air-termination system

The positioning method may be selected by the lightning protection system designer. Personally we always recommend the use of the rolling sphere method applicable to all cases. However, the following considerations may be valid.

- The protection angle method is only suitable for simple common structures or for small parts of bigger structures. It is not suitable for structures higher than the radius of the rolling sphere relevant to the selected protection level of the lightning protection system.
- The mesh method is suitable for general purposes and particularly for the protection of plane surfaces (flat roofs).
- The rolling sphere method is always suitable, even for complex shaped structures.

In Figures 6.11, 6.12 and 6.13, we give three practical examples (a villa, a building, an industrial plant) with an optimal air-termination system designed according to these methods and completed with the corresponding down-conductor system and earth-termination systems that we shall introduce in Sections 6.3 and 6.4. We shall come back to these three drawings, after studying the down-conductor system and the earth-termination system.

6.2.2 Construction of air-termination systems

Non-isolated air-termination systems can be installed on the surface of the roof when the roof is made of non-combustible material. Otherwise, due care needs to be taken

External lightning protection system 317

Figure 6.11   External lightning protection system of a villa

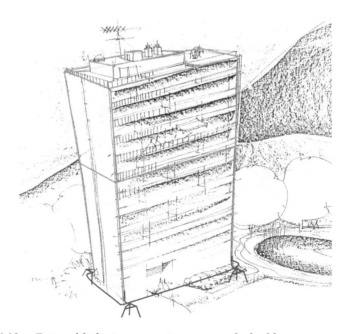

Figure 6.12   External lightning protection system of a building

318 *Lightning Protection*

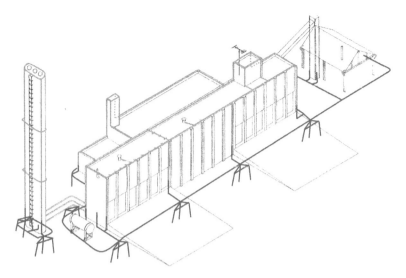

Figure 6.13 External lightning protection system of an industrial plant

when the roof is made of readily combustible material with regard to the distance between the air-termination conductors and the material.

For thatched roofs where no steel bars are used for monitoring of the reed, a distance of at least 15 cm is adequate. For other combustible materials, a distance higher than 10 cm is considered adequate.

Easily combustible parts of the structure to be protected should not remain in direct contact with the components of an external lightning protection system and should not remain directly under any metallic roofing membrane that might be punctured by a lightning flash. On flat roofs where water can accumulate, air-termination conductors should be installed above the maximum probable water level.

As we have already insisted, metallic natural components should be considered as parts of a lightning protection system, if some conditions are satisfied. For example, with metal sheets (not cladded with insulating materials) covering the structure to be protected, electrical continuity between the various parts is made durable by means of careful brazing, welding, crimping, seaming, screwing or bolting. The thickness t of the metal sheet (see Table 6.1) must be sufficient to prevent punctures, hot spots or ignition. If it is not important to prevent puncture of the sheeting or to consider ignition of any readily combustible materials underneath, a thickness t' smaller than t is a minimum. The selection of materials and their dimensions are reviewed in Section 6.5.

Other natural air-termination components include the following:

- metal components of roof construction (trusses, interconnected reinforcing steel, and so on) underneath non-metallic roofing provided that this latter part can be excluded from the structure to be protected;

Table 6.1 Table of minimum thicknesses of metal sheets for air-termination systems (see IEC 62305-3 [2])

Class of LPS	Material	Thickness, t (mm)	Thickness, t' (mm)
I to IV	Lead	–	2.0
	Stainless steel or galvanized steel	4	0.5
	Titanium	4	0.5
	Copper	5	0.5
	Aluminium	7	0.65
	Zinc	–	0.7

- metal parts such as ornamentation, railings, pipes, coverings of parapets, and so on, with cross-sections not less than that specified for standard air-termination components;
- metal pipes and tanks on the roof, provided that they are constructed of material with thicknesses and cross-sections in accordance with Table 6.10 (later);
- metal pipes and tanks carrying readily combustible or explosive mixtures, provided that they are constructed of material with minimum thickness t (see Table 6.1) and that the temperature rise of the inner surface at the point of strike does not constitute a danger (see Annex E in IEC 62305-3 [2]).

If these conditions cannot be fulfilled, the pipes and tanks shall be integrated into the structure to be protected.

Piping carrying readily combustible or explosive mixtures should not be considered as an air-termination natural component if the gasket in the flange couplings is not metallic or if the flange sides are not otherwise properly bonded.

For structures with a risk of explosion and roofs with flammable materials, additional measures have to be taken. Generally the highest level of protection is required in this case. The lightning protection system should then be designed and installed in such a manner that, in the case of a direct lightning flash, there are no melting or spraying effects except at the striking point. For structures containing solid explosive materials sensitive to rapidly changing electric fields or radiated by lightning impulsive electromagnetic fields, an isolated external lightning protection system is imposed. Structures totally contained within a metallic shell thick enough (e.g. 5 mm for steel structures, 7 mm for aluminium structures) may be considered protected by a natural air-termination system.

For structures containing hazardous areas, all parts of the air-termination system (and all parts of the down-conductor system, see Section 6.3) should be at least 1 m away from a hazardous zone. Where this is not possible, conductors passing within 0.5 m of a hazardous zone should be continuous or connections should be made with compression fittings or by welding.

Specific applications (distribution stations, storage tanks, pipelines, and so on) can be found in Annex D of IEC 62305-3 [2].

320 *Lightning Protection*

6.2.3 Non-conventional air-termination systems

Radioactive air-terminals are no longer allowed. In the past, these (supposed) active air-terminals never acted more efficiently than classical lightning rods of the same shape and dimensions.

Very little has been done to improve classical air-termination systems. However, the ambitions and potential earnings involved in the design of more effective lightning receptors have been an obvious motivation for the invention and presentation of a lot of different lightning protection systems and items. The claimed advantages have often been widely advertised, unfortunately without verification of their functions and validation of their effect. So far, parallel tests with simple Franklin rods and various ESE (early streamer emission) devices, lightning repellers, charge transfer systems, ion plasma generators and so on exposed to natural lightning have shown no significant difference in the attraction distance nor in the number of strokes to the different types of air-terminals.

Hopefully, in the future, more effective lightning protection components and systems will be developed. Until such systems are proven in a scientific sense their use should not be allowed for structures to be protected. Let us remain reasonable and careful when issuing new standards and guides. Of course the IEC 62305 international standard, following the opinion of confirmed scientists, does not advertise such devices because the international scientific community disregards them.

Unfortunately, some manufacturers continue to promote and install these fancy devices. Hence the struggle against non-conventional and non-verified systems or models is far from over. These manufacturers will probably continue to produce such devices as long as awful national standards promoting them exist. A new standard could even be published each time a new device appeared on the market.

A main target of the international scientific community should be to succeed in withdrawing misleading national standards and the copies made by others blindly following these nations [3,4].

6.3 Down-conductor system

The down-conductor system is the second part of an external LPS and is intended to conduct the lightning current from the air-termination system to the earth-termination system. In order to reduce the probability of damage due to lightning current flowing in the lightning protection system, several equally spaced down-conductors of minimum length are installed and an equipotential bonding to conducting parts of the structure is performed.

The selection of materials and their dimensions are reviewed in Section 6.5.

6.3.1 Location and positioning of down-conductors on the structure

The choice of the number and position of down-conductors should take into account the fact that if the lightning current is shared in several down-conductors, the risk of side flashes and electromagnetic disturbances inside the structure is reduced.

Table 6.2 Typical values of spacing between down-conductors and between ring conductors according to the class of LPS

Class of LPS	Typical distances (m)
I	10
II	10
III	15
IV	20

It follows that, as far as possible, the down-conductors should be uniformly placed along the perimeter of the structure and with a symmetrical configuration. Current sharing is improved not only by increasing the number of down-conductors but also by equipotential interconnecting rings.

Down-conductors should be placed as far as possible away from internal circuits and metallic parts in order to avoid the need for equipotential bonding with the lightning protection system.

To fulfill the requirements of the electro-geometric model (EGM) and the rolling sphere method, as many down-conductors as possible should be installed at equal spacing around the perimeter, interconnected by ring conductors. In this way, the probability of dangerous sparking is reduced and the protection of internal installations is facilitated (see the IEC 62305-4 standard [2] and Chapter 7).

An equal spacing between down-conductors should never exceed 20 m, or even 10 m at level I (see Table 6.2). Lateral connection of down-conductors is made not only both at the top of the structure and at ground level but also at every 10 to 20 m of height of the structure, according to the Table 6.3. A tolerance of ~ 20 per cent is generally accepted.

To keep inductance as small as possible, the down-conductors should be as short as possible. The geometry of the down-conductors and of the ring conductors affects the separation distance (see Section 6.3.2). The positioning of down-conductors depends upon whether the LPS is isolated from the structure or not. In the isolated case, if the air-termination system consists of rods on separate non-metallic masts, at least one down-conductor is needed for each mast. If it consists of catenary wires, at least

Table 6.3 Level coefficient k_i according to IEC 62305-3

Class of LPS	k_i
I	0.08
II	0.06
III, IV	0.04

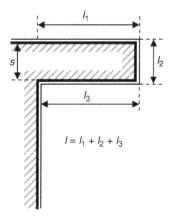

Figure 6.14   Isolation of external LPS, separation distance s on a loop in a down-conductor

one down-conductor is needed at each supporting structure. If it consists of a network of conductors, at least one down-conductor is needed at each supporting wire end. For each non-isolated lightning protection system, the number of down-conductors shall be not less than two and the down-conductors should be distributed around the perimeter of the structure to be protected.

6.3.2   Construction of down-conductor systems

When possible, a down-conductor should be installed at each exposed corner of the structure to be protected. All the down-conductors, straight and vertical, should be installed so that, as far as practicable, they form a direct continuation of the air-termination conductors or rods, such that they provide the shortest and most direct path to earth.

Loops must be avoided, unless it is not possible, in which case the separation distance s, measured across the gap between two points on the conductor and the length l between those points (see Figure 6.14) should be large enough (see Tables 6.3, 6.4 and 6.5).

Table 6.4   Partitioning coefficient k_c (see Section 6.3.6 on partitioning) according to IEC 62305-3

Number n of down-conductors	k_c
1	1
2	1 ... 0.5
4, >4	1 ... 1/n

Table 6.5 Material coefficient k_m according to IEC 62305-3. When there are several materials in series, it is good practice to use the lower value of k_m

Material	k_m
Air	1
Concrete, bricks	0.5

Indeed, as for the electrical insulation between the air-termination or the down-conductor and the metal parts of the structure, the interior metal installations and systems can be achieved by providing a distance d between the parts greater than the separation distance s such that

$$s = k_i (k_c/k_m) l \tag{6.1}$$

where k_i is the level coefficient, depending upon the class of LPS (see Table 6.3), k_c is the partitioning coefficient, depending upon the current flowing on the down-conductors (see Table 6.4), k_m is the material coefficient, depending upon the insulation material used (see Table 6.5), and l is the length along the air-termination of the down-conductor from the point where the separation distance is to be considered to the nearest equipotential bonding point.

In structures with metallic or electrically continuous frameworks, a separation distance is not required. In the case of the lines or external conductive parts connected to the structure, it is necessary to ensure lightning equipotential bonding, by direct connection or connection by surge protective devices, at their point of entry into the structure. Moisture in gutters can lead to intensive corrosion, so down-conductors should not be installed in gutters or down-spouts, even if they are covered by insulating material.

Non-isolated down-conductors may be positioned on the surface of the wall if the latter is made of non-combustible material. Otherwise, they can still be positioned on the surface of the wall, provided that the temperature rise due to the passage of lightning current is not dangerous for the material of the wall. If the wall is made of readily combustible material and the temperature rise of the down-conductors is dangerous, they should be placed in such a way that the distance between them and the wall is always greater than 10 cm. Mounting brackets may be in contact with the wall. If the distance from the down-conductor to a combustible material cannot be assured, the cross-section of the conductor should be raised to become at least 100 mm^2.

As for the air-termination system, natural metallic components can be used as down-conductors, provided the electrical continuity is made durable with secure connections made by means of brazing, welding, clamping, crimping, seaming, screwing or bolting. Their dimensions should be at least equal to those specified in Table 6.10.

324 *Lightning Protection*

In some countries, the metal of the electrically continuous reinforced concrete framework of the structure can be used as a down-conductor. In other countries, this is not the case.

Steelwork within reinforced concrete structures (including pre-cast, pre-stressed reinforced units), with horizontal and vertical bars welded or securely connected (clamped or overlapped a minimum of 20 times their diameters), can be considered as electrically continuous and they can be used as down-conductors if the overall electrical resistance between the uppermost part and ground level is as low as 0.2 Ω. If this value is not achieved, an external down-conductor system is installed. Brazing, welding (over a length of at least 3 cm) or clamping to the steel-reinforcing rods should ensure electrical continuity. Annex E in IEC 62305-3 [2] gives more information with many details of construction (see Figure 6.15 and Section E.4.3 of the standard [2] on reinforced concrete structures).

The interconnected steel framework of the structure, metallic facade elements, profile rails and metallic sub-constructions of facades can also be used as down-conductors, provided they are electrically continuous and made of allowed materials with sufficient thicknesses (see Table 6.10 in Section 6.5).

For structures with a risk of explosion and roofs with flammable materials, additional measures have to be taken. Down-conductors should be installed in such

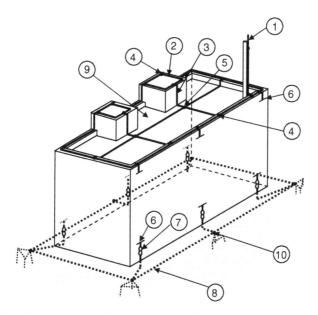

Figure 6.15 *Construction of the external lightning protection system on a low (H < 20 m) structure of steel-reinforced concrete using the reinforcement of the outer walls as natural components. 1, Air-termination rod; 2, horizontal air-termination conductor; 3, down-conductor; 4, T-type joint; 5, cross-type joint; 6, connection to steel reinforcing rods; 7, test joint; 8, ring earth electrode (type B earthing arrangement); 9, flat roof with roof fixtures; 10, T-type joint, corrosion resistant.*

a way that the auto-ignition temperature given by the source of the relative hazardous area will not be exceeded in those applications where it is not possible to install down-conductors outside the hazardous area. Lightning equipotential bonding between the LPS components and other conductive installations, as well as between the components of all conductive installations should be assured inside hazardous areas and locations where solid explosives material may be present, not only at ground level but also where the distance between the conductive parts is less than the separation distance s calculated assuming a partitioning coefficient k_c equal to unity.

For structures containing hazardous areas, all parts of the down-conductor system should be at least 1 m away from a hazardous zone. Connections to piping should be of such a kind (by welding; no screwing nor clapping) that, in the instance of a lightning current passage, there is no sparking.

Specific applications (distribution stations, storage tanks, pipelines, and so on) can be found in Annex D of IEC 62305-3 [2].

6.3.3 Structure with a cantilevered part

When somebody is standing under a cantilevered construction (see Figure 6.16), he can become an alternative path for lightning current flowing in the down-conductor running on the cantilevered wall of length l.

If the height of the person with raised hand is taken to be 2.5 m from the ground level to the tips of his fingers, the separation distance d should satisfy the condition

$$d > 2.5 + s \qquad (6.2)$$

where s is the separation distance calculated according to equation (6.1).

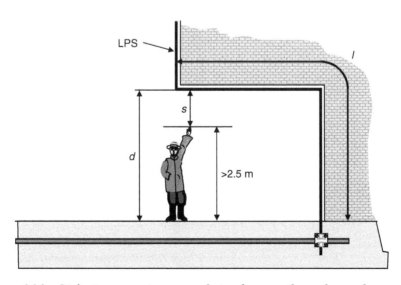

Figure 6.16 Lightning protection system design for a cantilevered part of a structure

326 *Lightning Protection*

Loops as shown in Figure 6.14 can produce high inductive voltage drops, which can cause a lightning discharge to pass through a structure wall, thereby causing damage. If the condition deduced from equations (6.1) and (6.2) are not met, arrangements should be made for direct routing through the structure at the points of re-entrant lightning conductor loops for those conditions shown in Figure 6.14.

6.3.4 *Joints, connections and test joints in down-conductors*

Air-terminations and down-conductors should be firmly fixed so that electrodynamics or accidental mechanical forces (vibrations, slipping of slabs of snow, thermal expansion, and so on) will not cause conductors to break or loosen.

The number of connections along the conductors should be kept to a minimum. Connections shall be made secure by such means as brazing, welding, clamping,

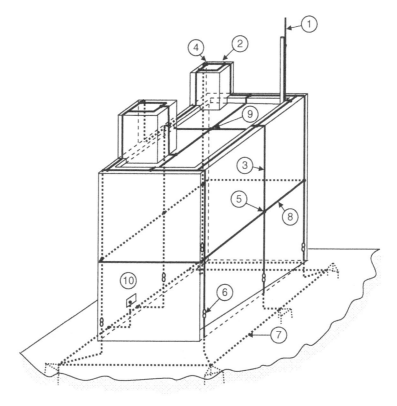

Figure 6.17 *Positioning of the external lightning protection system on a low (H < 60 m) structure made of insulating material (wood, bricks, and so on). 1, Air-termination rod; 2, horizontal air-termination conductor; 3, down-conductor; 4, T-type joint; 5, cross type joint; 6, test joint; 7, ring earth electrode (type B earthing arrangement); 8, equipotentialization ring conductor; 9, flat roof with roof fixtures; 10, terminal for connecting the equipotentialization bar to the internal LPS.*

External lightning protection system   327

Table 6.6   Values of the partitioning coefficient k_c

Type of air-termination system	Number n of down-conductors	k_c for type A earthing arrangement	k_c for type B earthing arrangement
Single rod	1	1	1
Wire	2	0.66*	0.5 … 1[†]
Mesh	≥4	0.44*	0.25 … 0.5[‡]
Mesh	≥4, connected by horizontal ring conductors	0.44*	$1/n$ … 0.5[§]

*Valid for single earthing electrodes with comparable earthing resistances; if earthing resistances of single earthing electrodes are clearly different $k_c = 1$ has to be assumed.
[†]Values range from $k_c = 0.5$ where $w \ll H$ to $k_c = 1$ with $H \ll w$ (see Figure 6.18).
[‡]The relation <q> to calculate k_c in Figure 6.20 is an approximation for cubic structures and for $n \geq 4$; the values of H are assumed to be in the range 5 to 20 m.
[§]If the down-conductors are connected horizontally by ring conductors, the current distribution is more homogeneous in the lower parts of the down-conductor system and k_c is further reduced (especially valid for tall structures, see Figure 6.20 where H, c_s and c_d are assumed to be in the range 5 to 20 m).

crimping, seaming, screwing or bolting. Special care should be taken for connections of steelwork within reinforced concrete structures.

Down-conductors should preferably be connected to junctions of the air-termination system network and routed vertically to the junctions of the earth-termination network. Figure 6.5 shows an example of such a configuration with test joints, which are better and more generally presented on a low building in Figure 6.17.

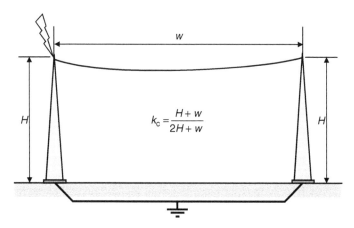

Figure 6.18   Partitioning coefficient k_c for a wire air-termination system and a type B earth-termination system

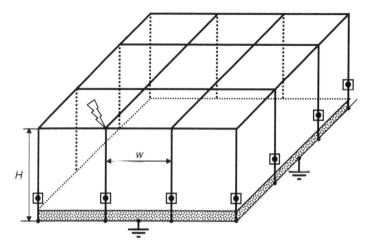

Figure 6.19 Partitioning coefficient k_c in the case of a meshed air-termination system and a type B earth-termination system. $k_c = 1/2n + 0.1 + 0.2 \times \sqrt[3]{(W/H)}$, where n is the total number of down-conductors (add internal down-conductors if they exist), w the spacing between down-conductors and H the height (spacing) between horizontal ring conductors.

Indeed, between the down-conductor system and the earth-termination system a test joint should be fitted on each down-conductor for measuring purposes. The joint should be capable of being opened, although in normal use it should remain closed.

In the case of natural down-conductors combined with foundation earth electrodes, such test joints may not be installed.

6.3.5 Lightning equipotential bonding

Equipotentialization is achieved by interconnecting the lightning protection system with the external conductive parts and lines connected to the structure, but also structural metal parts, metal installations and internal systems (see Chapter 7 on internal lightning protection and IEC 62305-4 [2]). Interconnecting means electrical continuity of bonding conductors (if not provided by natural bonding) or installation of surge protective devices when direct connections with bonding conductors is not feasible.

In the case of an isolated external lightning protection system, lightning equipotential bonding should be established at ground level only.

For a non-isolated external lightning protection system, bonding with connections as direct and straight as possible is achieved not only at ground level but also every 10 m (LPS class I) to 20 m (LPS class IV) from the bottom and along the height of the down-conductors (see, e.g. the building with an intermediate ring conductor in Figure 6.17).

External lightning protection system   329

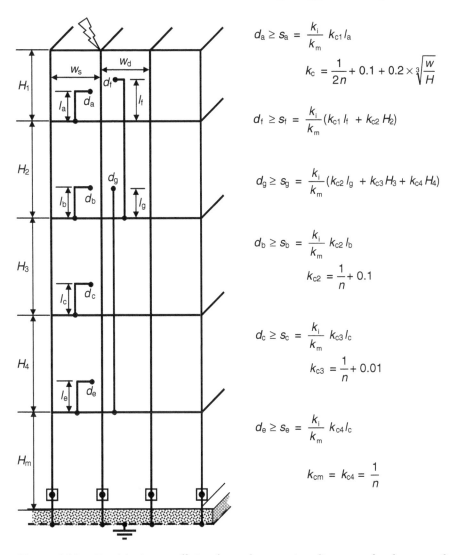

Figure 6.20 Partitioning coefficient k_c and separation distance s for the case of a meshed air-termination system, an interconnecting ring of the down-conductors at each level and a type B earth-termination system, where n is the total number of down-conductors, m the total number of levels, l the height above the bonding point, H the height (spacing) between horizontal ring conductors, w_s the distance from the nearest down-conductor, w_d the distance from the nearest down-conductor on the opposite side and d the distance to the nearest down-conductor

330 Lightning Protection

Table 6.7 Ring earth electrode with vertical rods (small shelter)

ρ ($\Omega \cdot$m) Typical shelter $L = 2$ m	50	100	500
R_4 (Ω)	6.9	13.8	68.8
R_2 (Ω)			
Diagonal	9.0	18.1	90.4
Lateral a	9.1	18.2	91.1
Lateral b	9.6	19.3	96.3
R_1 (Ω)	11.0	22.0	110.1

6.3.6 Current distribution in down-conductors

The partitioning of the lightning current among the down-conductors depends on the overall number n of down-conductors and their position, the presence of interconnecting ring conductors, the type of air-termination system and the type of earth-termination system.

A partitioning coefficient k_c can be evaluated (see Table 6.4), depending on the type (A or B, see Section 6.4) of earthing arrangement (see Table 6.6 associated with the drawings of Figures 6.18, 6.19 and 6.20).

The values collected in Table 6.7 apply for all type B earthing arrangements, but also for type A earthing arrangements, provided that the earthing resistance of each electrode has a similar value.

6.4 Earth-termination system

The earth-termination system is the third part of an external LPS that is intended to conduct and disperse the lightning current into the earth, without causing any danger to people or damage to installations inside the structure to be protected.

In general, a low earthing resistance, if possible lower than 10 Ω when measured at low frequency, is recommended.

6.4.1 General principles

When dealing with dispersion into the earth while minimizing dangerous overvoltages, the transient behaviour of the earth-termination system under high peak lightning currents should be studied. This has been done by many authors both theoretically [5–9] and experimentally [10–12], in frequency and in time domains.

The following are some typical phenomena characterizing the behaviour of earthed electrodes under transient conditions.

- Lightning currents of both polarities propagate into the soil in the 1×10^5 to 1×10^6 Hz frequency ranges and with soil resistivities up to 5 000 Ωm. It is

necessary to measure the soil resistivity and to take into account the possible inhomogeneity of the soil involved in the current discharge.
- Inductive phenomena deeply influence the transient behaviour of earthed electrodes: the rate of rise of current impulses increases the inductive voltage drop with respect to the resistive voltage drop [13,14].
- High lightning current amplitudes associated with very short front durations can result in high current densities near ground electrodes, so that the critical voltage gradients may be exceeded and discharges into soil can occur [15].

The surge impedance is the ratio of the instantaneous value of the earth-termination voltage (potential difference between the earth-termination system and the remote earth) over the instantaneous value of the earth-termination current; in general these do not occur simultaneously. The conventional earth resistance is the ratio of the peak values of the earth-termination voltage and the earth-termination current. The impulse factor is the ratio of the conventional earth resistance over the low-frequency resistance of the earth electrode.

Comparing experimental works and theoretical studies, the following main conclusions may be drawn: the length of an earth electrode contributing to the impulsive current dispersion depends on soil resistivity ρ, on the current risetime T_1 (μs) and on the current amplitude I, and an electrode effective length l_e is defined as the distance between the current injection point and a point at which the value of the conventional earth electrode does not undergo any significant reduction:

$$l_e = K_0 \sqrt{\rho T_1} \qquad (6.3)$$

where K_0 is a factor depending on the geometrical configuration of the earth electrode, with values of $K_0 = 1.40$ for a single conductor energized at one end, 1.55 for a single conductor energized in the middle, and 1.65 for conductors arranged in a star configuration energized at the centre.

Localized earth electrodes experience a conventional earth resistance lower than the power frequency resistance if the value of the lightning impulse current is sufficiently high to initiate soil ionization.

Following Mazzetti [6], Figures 6.21 to 6.24 show the trends of the conventional earth resistance as a function of soil resistivity ρ and current waveshape for different earth-termination arrangements.

The danger to people when dispersing lightning current into earth is reflected by the maximum energy tolerated by a human body in lightning transient conditions, assumed equal to 20 J, and the tolerable risk that this maximum energy value can be exceeded. The corresponding values of the lightning current parameters can be found in IEC 62305-1. The earth-termination system should be designed so that the dispersion does not result in step voltages higher than necessary to dissipate the 20 J energy in the human body resistance of 500 Ω conventionally fixed, so that step voltages stay within the safety limits.

When applying this procedure, we can establish the minimum required dimensions of an earth-termination system according to the IEC 62305 international standard.

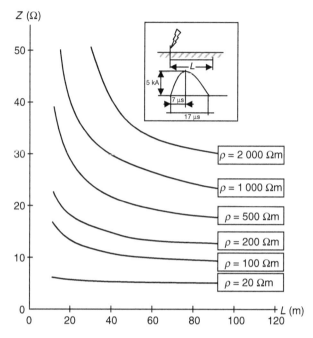

Figure 6.21 Conventional earth resistance of buried horizontal conductors as a function of their length for different values of soil resistivity

6.4.2 Earthing arrangements in general conditions

From the viewpoint of the lightning protection of single structures, a single integrated earth-termination system is preferable and is suitable for all purposes (i.e. lightning protection, power systems and telecommunications systems).

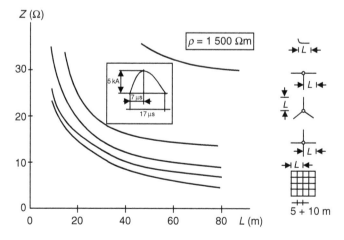

Figure 6.22 Conventional earth resistance of various earth-termination systems

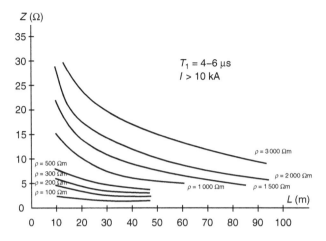

Figure 6.23   *Conventional earth resistance of an earth-termination system (impulse current at the centre of a star arrangement) as a function of length for different values of soil resistivity*

Serious corrosion problems can occur when earth-termination systems made of different materials are connected to each other (see Section 6.5, 'Selection of materials').

According to IEC 62305-3, two basic types of earth electrodes arrangements apply:

- the type A arrangement, comprising horizontal or vertical earth electrodes installed outside the structure to be protected, connected to each down-conductor, with a minimum of two earth electrodes;

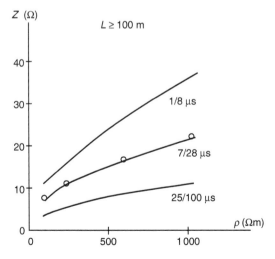

Figure 6.24   *Conventional earth resistance as a function of soil resistivity (open circles are experimental results)*

- the type B arrangement, comprising either a ring conductor external to the structure to be protected, in contact with the soil for at least 80 per cent of its total length, or a foundation earth electrode.

In the type A arrangement, the minimum length of each earth electrode at the base of each down-conductor is l_1 for horizontal electrodes or $0.5l_1$ for vertical (or inclined) electrodes, l_1 being the minimum length of the horizontal electrodes shown in the relevant part of Figure 6.25. For combined (vertical, inclined or horizontal) electrodes, the total length shall be considered. However, this minimum length may be disregarded provided that an earthing resistance of the earth-termination system less than 10 Ω is achieved.

The type A arrangement is suitable for low structures such as family houses, for existing structures or a lightning protection system with rods or stretched wires or for an isolated lightning protection system. Where there is a ring conductor that interconnects the down-conductors, in contact with the soil, the earth-termination system is still classified as type A if the ring conductor is in contact with the soil for less than 80 per cent of its length.

Personally, we never recommend the use of plates as earth electrodes, because of the easy corrosion of the joint.

In a type B arrangement, the ring earth electrode or foundation earth electrode should have a mean radius r_e of the area enclosed that is not less than l_1:

$$r_e \geq l_1 \tag{6.4}$$

where l_1 is represented in Figure 6.25 according to the different LPS classes and then according to the values of lightning current parameters selected for dimensioning.

When the required value of l_1 is larger than the convenient value of r_e, additional horizontal or vertical (or inclined) electrodes should be added with individual lengths l_r (horizontal) and l_v (vertical) according to the following equations:

$$l_r = l_1 - r_e \qquad \text{and} \qquad l_v = 0.5(l_1 - r_e) \tag{6.5}$$

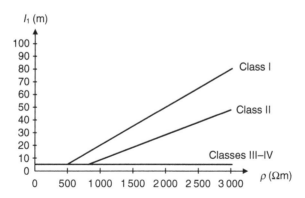

Figure 6.25 Minimum length of an earth electrode according to the class of LPS

The number of additional electrodes connected to the ring earth electrode should not be less than two or less than the number of down-conductors. The additional electrodes should be connected at points where the down-conductors are connected and, for as many as possible, with equal spacing. Examples of efficient earth-termination systems are given in Figures 6.11–6.13.

In order to minimize the effects of corrosion, soil drying and freezing and then to stabilize the value of conventional earth resistance, the ring earth electrode and embedded horizontal electrodes should be buried at a minimum depth of 0.5 m and at a distance of ∼1 m around the external walls. Vertical and inclined earth electrodes should be installed at a depth of the upper end of at least 0.5 m and distributed as uniformly as possible to minimize electrical coupling effects in the earth.

The type B earth-termination system is preferred for meshed air-termination systems, for bare solid rock, for lightning protection systems with several down-conductors and for structures with extensive electronic systems or with a high risk of fire (see IEC 62305-2).

Interconnected reinforcing steel in concrete foundations or other suitable underground metal structures should preferably be used as earth electrodes. When the metallic reinforcement in concrete is so used, special care should be exercised at the interconnections to prevent mechanical splitting of the concrete. In the case of prestressed concrete, consideration should be given to the consequences of the passage of lightning discharge currents, which may produce unacceptable mechanical stresses.

A foundation earth electrode comprises conductors installed in the foundation of the structure below ground. The length of additional earth electrodes should be determined using Figure 6.25. Foundation earth electrodes are installed in concrete and are relatively protected against corrosion.

Materials for earth-termination systems are selected from the Table 6.11 (see Section 6.5) or from other materials with equivalent mechanical, electrical and chemical (corrosion) performance characteristics. A further problem arises from electrochemical corrosion due to galvanic currents. Steel in concrete has approximately the same galvanic potential in the electrochemical series as copper in soil. Therefore, when steel in concrete is connected to steel in soil, a driving galvanic voltage of ∼1 V causes a corrosion current to flow through the soil and wet concrete, dissolving the steel in the soil. Earth electrodes in soil should use copper or stainless steel conductors when these are connected to steel in concrete.

6.4.3 Examples of earthing arrangements in common small structures

In order to achieve both an earth-termination resistance and an earth impedance not exceeding 10 Ω, it is always better [3–5] to design a type B arrangement earth-termination system with short (≥ 2 m length for example) earth electrodes connected to a ring conductor. The selected types of arrangements are defined when considering the various possibilities taking into account the geometric space and the environment.

As a common example, let us protect a small metallic shelter with the dimensions are given in Figure 6.26.

336 Lightning Protection

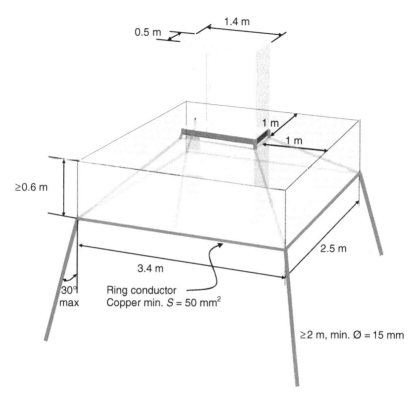

Figure 6.26 Earth-termination system of a metallic shelter

The proposed solution (ring conductor with inclined earth rods) is difficult to build in practice. Inclined rods are generally replaced by vertical rods. Moreover, in Belgium, we accept copper earth conductors of 35 mm^2 cross-section (instead of the 50 mm^2 recommended by the international standard) and, in order to avoid freezing effects, we suggest burying the ring conductor at a depth of 0.8 m. These shelters are installed close to roads so that the 1 m distance from the walls concept is generally impossible to introduce. A practical solution is presented in Figure 6.27.

To calculate the earth resistance, we use Dwight's formulae [17] adapted to a single vertical rod of equivalent physical radius $r_{a,eq} = h + L$ and an equivalent geometric radius $r_{g,eq}$ such that

$$\pi r_{g,eq}{}^2 = a'b'[h/(h+L)] + ab[L/(h+L)] \qquad (6.6)$$

For a set of 2, 3 or 4 parallel vertical rods with respective spacings D_2, D_3 or D_4, the equivalent physical radius is given by

$$r_{eq,2} = \sqrt{(r_c D_2)}; \; r_{eq,3} = \sqrt[3]{(r_c D_3^2)}; \; r_{eq,4} = \sqrt[4]{[\sqrt{2}(r_c D_4^3)]} \qquad (6.7)$$

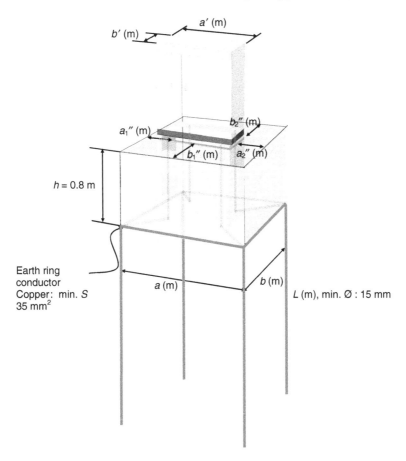

Figure 6.27 Practical earth-termination system of a shelter in reduced space

For structures with a rectangular cross-section ($a'b'$, ab, ...), D_4 differs in the upper $[=\sqrt{(a'b')}]$ and lower $[=\sqrt{(ab)}]$ parts, so the mean equivalent geometric radius is

$$r_{geq,4} = r_{geq,4,up}[h/(h+L)] + r_{eq,4}[L/(h+L)] \qquad (6.8)$$

When only one or two vertical rods (instead of 4) are associated with the rectangular ring, we have

$$r_{geq,1} = r_{geq,4,up}[h/(h+L)] + r_{eq,1}[L/(h+L)] \qquad (6.9)$$

with $r_{eq,1} = r_{cv}$, and

$$r_{geq,2} = r_{geq,4,up}[h/(h+L)] + r_{eq,2}[L/(h+L)] \qquad (6.10)$$

with $r_{eq,2} = \sqrt{(r_{cv} D_2)}$, where D_2 defines the fixed spacing s between two vertical rods.

Pierre Lecomte and the present author have created a convenient and illustrated software in Excel, entitled 'Earth-terminations for common lightning protection systems', to design several kinds of earth-termination systems for common structures. For information contact christian.bouquegneau@fpms.ac.be.

This software provides immediate earth resistance values for the following:

1. *Vertical rods* of length L and radius r_{cv}

$$R = \frac{\rho}{2\pi L}\left(\ln\frac{4L}{r_{cv}} - 1\right) \quad (6.11)$$

where ρ represents the soil resistivity (Ω m). As easily confirmed experimentally, up to a depth h of 1 m, the same formula applies when replacing L by $L+h$ in equation (6.11).

2. *Two parallel vertical rods* of length L with a spacing s,
When $s > L$

$$R = \frac{\rho}{4\pi L}\left(\ln\frac{4L}{r_{cv}} - 1\right) + \frac{\rho}{4\pi s}\left(1 - \frac{L^2}{3s^2} + \frac{2L^4}{5s^4}\right) \quad (6.12)$$

When $s < L$

$$R = \frac{\rho}{4\pi L}\left(\ln\frac{4L}{r_{cv}} + \ln\frac{4L}{s} - 2 + \frac{s}{2L} - \frac{s^2}{16L^2} + \frac{s^4}{512L^4}\right) \quad (6.13)$$

3. *Symmetrical horizontal conductor* of length $2L$, of radius r_{ch}, buried at a depth h

$$R_{h,s} = \frac{\rho}{4\pi L}\left(\ln\frac{4L}{r_{ch}} + \ln\frac{2L}{h} - 2 + \frac{h}{L}\right) \quad (6.14)$$

If this symmetrical horizontal conductor lies at ground level ($h = 0$)

$$R_{h,s,0} = \frac{\rho}{2\pi L}\left(\ln\frac{4L}{r_{ch}} - 1\right) \quad (6.15)$$

4. *Asymmetrical horizontal conductor* of length $2L$, of radius r_{ch}, buried at a depth h

$$R_{h,a} = \frac{\rho}{4\pi L}\left(\ln\frac{8L}{r_{ch}} + \ln\frac{4L}{h} - 2 + \frac{h}{L}\right) \quad (6.16)$$

If this asymmetrical horizontal conductor lies at ground level ($h = 0$)

$$R_{h,a,0} = \frac{\rho}{2\pi L}\left(\ln\frac{8L}{r_{ch}} - 1\right) \quad (6.17)$$

5. *Circular ring* of diameter D, buried at a depth h

$$R_b = \frac{\rho}{2\pi^2 D}\left(\ln\frac{4D}{r_{ch}} + \ln\frac{2D}{h}\right) \quad (6.18)$$

If this ring conductor lies at ground level ($h = 0$)

$$R_{b,0} = \frac{\rho}{\pi^2 D}\left(\ln\frac{4D}{r_{ch}}\right) \quad (6.19)$$

A rectangular ring of length a and width b is equivalent to a circular ring of diameter D such that $D = \sqrt{(4ab/\pi)}$.

These relations are suitable for earth-termination systems of horizontal dimensions greater than the rod length. When the equivalent ring is too small, earth current lines in the ground overlap and the earth resistance value is increased.

6. *Earth-termination arrangement of a small shelter.* This is illustrated in Figure 6.27, with, for example $h = 0.8$ m and $L \geq 2$ m,

$$R_4 = \frac{\rho}{2\pi(L+h)}\left(\ln\frac{4(L+h)}{r_{geq4}} - 1\right) \quad (6.20)$$

$r_{geq,4}$ being defined in equation (6.8).

Our Excel software also allows calculation of the earth resistance values of modular solutions with different numbers of vertical rods associated with the ring conductor. The experimental results were in very good agreement with theoretical results for resistivities up to several hundreds of Ω m.

For higher values of resistivity, we suggest adding to the earth ring conductor a symmetric horizontal earth conductor of length at least equal to $2L = 10$ m (at least 5 m on each opposite side of the shelter). To take into account the effect of common earth current lines in the ground, we multiply the value of the parallel earth resistance by a factor of reduction effect F, estimated to $F_s = 1.18$ for a symmetric horizontal conductor and, if not possible, $F_a = 1.16$ for an asymmetric horizontal conductor of at least $2L = 10$ m.

For example, in the symmetrical situation,

$$R_{//} = F_s \frac{R_4 R_{h,s}}{R_4 + R_{h,s}} \quad (6.21)$$

The software is still more powerful, allowing calculation of the earth resistances when only one or two vertical rods (on line, diagonally fixed and so on) are added to the ring earth electrode. Its application leads to the typical results given in Table 6.7: ring earth electrode with respectively 4 (R_4), 2 (R_2) and only 1 (R_1) vertical rods.

The software also applies to larger earth-termination systems (of decametric dimensions) with ring earth conductor and associated with either vertical rods or

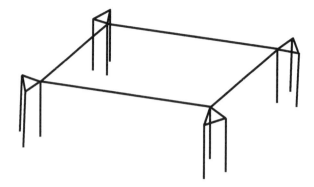

Figure 6.28   Ring conductor with triangular-prismatic vertical arrangements

triangular-prismatic arrangements [5] connected to each corner of the rectangular ring (see Figure 6.28). In this case we adopt and adapt Sunde's formulas [18] and

$$R = \frac{R_w R_4 - R_m^2}{R_w + R_4 - 2R_m}, \qquad (6.22)$$

where R_4 is the equivalent resistance of four vertical rods or four triangular–prismatic vertical arrangements [in which case r_{cv} is replaced by $r_{eq,4}$ as defined in equation (6.7)] connected to the four corners of the rectangular ring conductor buried at a depth h; R_m is the mutual resistance between the ring conductor and the equivalent vertical earth-termination system.

As a first approximation, we consider a rectangular ring conductor of length a and width b, in a soil of resistivity ρ; e is the exponential number (2.71828). Even if the ring is not a pure rectangle, an equivalent rectangular ring $a \times b$, with a mean length a and a mean width b can be used. To calculate R in equation (6.22), the following equations apply:

$$R_4 = \frac{\rho}{8\pi L}\left(\ln\frac{4L}{r_{cv}} - 1 + \frac{L\sqrt{2\pi}}{\sqrt{ab}} + \frac{L\sqrt{\pi}}{2\sqrt{ab}}\right) \qquad (6.23)$$

$$R_w = \frac{\rho}{2\pi(a+b)} \ln\frac{8(a+b)}{\pi\sqrt{2r_{ch}h}} \qquad (6.24)$$

$$R_m = \frac{\rho}{2\pi(a+b)} \ln\frac{8(a+b)\sqrt{e}}{\pi\sqrt{2Lh}} \qquad (6.25)$$

When the ring is trapeziform, we need more parameters: a is the effective length of the basic rectangle, b the effective width of the basic rectangle, a' the corrected length to take beyond part d_a into account, b' the corrected width to take beyond part d_b into account, a_m is the mean length of the equivalent quadrilateral with $a_m = (a+a')/2$, and b_m the mean width of the equivalent quadrilateral with $b_m = (b+b')/2$.

1. With single vertical rods added to the earth ring conductor, the following equations apply:

$$R_4 = \frac{\rho}{8\pi L}\left(\ln\frac{4L}{r_{cv}} - 1 + \frac{L\sqrt{2\pi}}{\sqrt{a_m b_m}} + \frac{L\sqrt{\pi}}{2\sqrt{a_m b_m}}\right) \quad (6.26)$$

$$R_w = \frac{\rho}{2\pi(a_m + b_m)}\ln\frac{8(a_m + b_m)}{\pi\sqrt{2r_{ch}h}} \quad (6.27)$$

$$R_m = \frac{\rho}{2\pi(a_m + b_m)}\ln\frac{8(a_m + b_m)\sqrt{e}}{\pi\sqrt{2Lh}} \quad (6.28)$$

2. With triangular-prismatic vertical rods [with $D_3 = 1.5$ m, $a_{mp} = a_m + 1.2$ m; $b_{mp} = b_m + 1.2$ m and $r_{cv} = r_{eq,3} = \sqrt[3]{(r_{cv}D_3^2)}$] added to the earth ring conductor, the following relations apply:

$$R_4 = \frac{\rho}{8\pi L}\left(\ln\frac{4L}{r_{eq,3}} - 1 + \frac{L\sqrt{2\pi}}{\sqrt{a_{mp} b_{mp}}} + \frac{L\sqrt{\pi}}{2\sqrt{a_{mp} b_{mp}}}\right) \quad (6.29)$$

$$R_w = \frac{\rho}{2\pi(a_{mp} + b_{mp})}\ln\frac{8(a_{mp} + b_{mp})}{\pi\sqrt{2r_{ch}h}} \quad (6.30)$$

$$R_m = \frac{\rho}{2\pi(a_{mp} + b_{mp})}\ln\frac{8(a_{mp} + b_{mp})\sqrt{e}}{\pi\sqrt{2Lh}} \quad (6.31)$$

6.4.4 Special earthing arrangements

6.4.4.1 Earth electrodes in rocky and sandy soils

During construction, a foundation earth electrode should be built into the concrete foundation, because even where a foundation earth electrode has a reduced earthing effect in rocky or sandy soils, it still acts as an equipotential bonding conductor. Additional earth electrodes should be connected to each down-conductor and to the foundation earth electrode.

When a foundation earth electrode is not provided, a ring earth electrode (type B arrangement) should be used instead. If the earth-termination system cannot be buried and has to be mounted on the surface, it should be protected against mechanical damage. Radial earth electrodes lying on or near the earth surface should be covered by stones or embedded in concrete for mechanical protection.

The effect of lightning flashes on rocks, sand and crystals in the ground has been studied by several scientists. Natural fulgurites of different types (sand fulgurites, fulgurites at rock surface, fulgurites in rock crevices and fugurites at crystal surfaces) are the result of long duration strokes that generate remarkable melting effects [19].

6.4.4.2 Earth-termination systems in large areas

Generally, an industrial plant comprises a number of associated structures, between which a large number of power and signal cables are installed. The earth-termination

342 Lightning Protection

systems of such structures are very important for the protection of electrical systems. A low-impedance earth-termination system reduces the potential difference between the structures and so reduces the interference injected into the electrical links. This can be achieved by providing the structure with foundation earth electrodes and additional earth arrangements (type A and type B).

Interconnections between the earth electrodes, the foundation earth electrodes and the down-conductors should be installed below each test joint. Some of the test joints should also be connected to the equipotential bars of the internal lightning protection system.

In order to reduce the probability of direct lightning flashes to cable routes in the ground, an earthing conductor and, in the case of wider cable routes, a number of earthing conductors should be installed above the cable routes.

An example of meshed earth-termination system, including cable trenches, obtained by interconnecting the earth-termination systems of a number of protected structures, is shown in Figure 6.29 (see IEC 62305-3 [2]).

6.4.4.3 Artificial decrease of earth resistance

The earthing resistance of an electrode can be reduced by adding chemical additives. These usually consist of at least two different liquid chemicals, which, when poured together on the electrode, combine in the soil surrounding it to make a jelly-like compound of low resistivity [20–22].

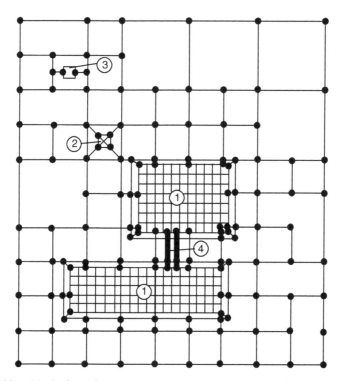

Figure 6.29   Meshed earth-termination system of an industrial plant

Although the earthing resistance can be effectively reduced by this method, its economy is by no means evident. One should always compare the cost of such an operation, which is only useful for very high soil resistivities, with the cost of increasing the effective length of the electrode to obtain the same result. The use of common salt is not recommended.

6.4.5 Effect of soil ionization and breakdown

Lightning earths sometimes have to discharge very high currents, causing high earth-electrode potentials and consequently high voltage gradients at the electrode surface. The soil acts as a dielectric and breaks down at a critical electric field of breakdown strength

$$E_c = \rho \iota \quad (6.32)$$

where ι is the current density.

As the impulse-breakdown voltage of various soils is $\sim$200 to 1 000 kV m^{-1}, corona and streamer formation effectively increases the electrode diameter and thus decreases the effective earthing resistance. This partly counteracts the apparent increase of earthing resistance due to surge phenomena, when a surge impedance stands instead of a conventional earth resistance. The order of magnitude of the diameter of the conducting cylinder around the electrode is $\sim$3 cm for a potential of 100 kV and 60 cm for a potential of 1 MV [23].

The effect of soil ionization is not well understood as yet, and its thermal effect is not easily quantified. Several authors [24,25] have proposed interesting models of non-linear surge characteristics of earthing systems for high currents in order to determine the threshold electric field for ionization.

6.4.6 Touch voltages and step voltages

6.4.6.1 Touch voltages

The vicinity of the down-conductors of a lightning protection system outside the structure may be hazardous to life even if the lightning protection system has been designed and constructed according to the suitable rules of IEC 62305-3 [2] (see Figure 6.30).

The hazard to life is reduced to a tolerable level if one of the following conditions is fulfilled:

- the probability of persons approaching or the duration of their presence outside the structure and close to the down-conductors is very low;
- the natural down-conductor system consists of several columns of the extensive metal framework of the structure or of several pillars of interconnected steel of the structure, with the electrical continuity assured;
- the resistivity of the surface layer of the soil within 3 m of the down-conductor is not less than 5 kΩm; a layer of insulating material, e.g. asphalt of 5 cm thickness or gravel of 15 cm thickness, generally reduces the hazard to a tolerable level.

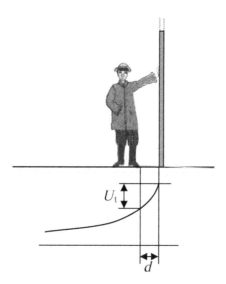

Figure 6.30   Touch voltage

If none of these conditions is fulfilled, protection measures should be adopted against injury to living beings due to touch voltage as follows:

- insulation of the exposed down-conductor, giving a 100 kV, 1.2/50 µs impulse withstand voltage, e.g. at least 3 mm cross-linked polyethylene;
- physical restrictions and/or warning notices to minimize the probability of down-conductors being touched.

6.4.6.2   Step voltages

High current discharges can cause high earth-electrode potentials, and this can cause mortal danger to animals and human beings in the immediate vicinity. The danger increases with increasing potential difference between points of the ground that may be touched simultaneously with different parts of the body, generally the feet are involved and from this comes the notion of step voltage (see Figure 6.31).

Indeed, when lightning strikes open ground the current is discharged into the mass of the earth. On uniform ground, the discharge takes place in a regular pattern, although it might be quite irregular if the ground is non-uniform. A person standing near the striking point is subjected to a potential difference U between the feet, such that

$$U = I \frac{\rho}{2\pi} \frac{s}{d(d+s)} \qquad (6.33)$$

where I denotes the current amplitude, ρ the resistivity, s the step length and d the distance between the striking point and the nearer leg of the person. In human beings this potential difference will cause a current to flow through the legs and lower part of the trunk.

External lightning protection system 345

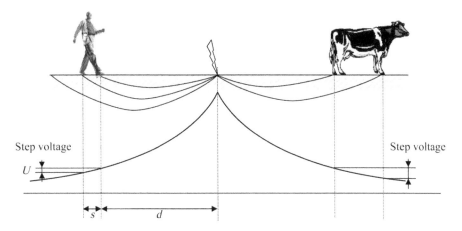

Figure 6.31 Step voltage

In order to keep the step voltage as low as possible the earth electrode should be buried as deeply as is feasible (at least 0.6 m deep).

Close to a down-conductor, the hazard to life is reduced to a tolerable level if one of the following conditions is fulfilled:

- The probability of persons approaching or the duration of their presence in the dangerous area within 3 m from the down-conductor is very low.
- The resistivity of the surface layer of the soil within 3 m of the down-conductor is not less than 5 kΩ m; a layer of insulating material, e.g. asphalt of 5 cm thickness or gravel of 15 cm thickness, generally reduces the hazard to a tolerable level.

If none of these conditions is fulfilled, the following protection measures should be added:

- equipotentialization by means of a meshed earth-termination system;
- physical restrictions and/or warning notices to minimize the probability of access to the dangerous area within 3 m of the down-conductor.

6.4.8 Measurement of soil resistivity and earth resistances

6.4.8.1 Measurement of soil resistivity

A convenient method for measuring the soil resistivity is the four-point method given in Figure 6.32. Four auxiliary electrodes (spikes) are driven into the ground at the same depth d ($d < 1$ m) in a straight line with equal spacings s.

A current I is passed between the outer electrodes, and the voltage drop U is measured between the inner electrodes. The ration U/I gives a resistance value R and the average soil resistivity to a depth of s metres is then

$$\rho = 2\pi s R \qquad (6.34)$$

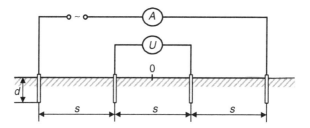

Figure 6.32   Measuring earth resistivity

Indeed the resistance measured is actually a resistance between two hemisphere-like equipotential surfaces surrounding the current electrodes and passing through the voltage electrodes.

For homogeneous soils, the test result is independent of the spacing s used. If it is not, the result will be roughly the average of the resistivity down to the depth s.

By repeating the measurements for different values of separation distance s (required to determine average resistivity to various depths at a given point O: the centre point O of the electrode system is kept fixed and the electrode spacing s increased outward from that point), the average resistivity to various depths can be found and the results will indicate whether an advantage is to be gained by installing deep-driven earth electrodes to reach strata of much lower resistivity than that at the top.

Account should also be taken of seasonal variations in the soil.

Soil resistivity largely depends on the water content of the soil and the resistivity of this water (see Table 6.8).

Table 6.8   Water and soil resistivities

Type of water or soil	Resistivity (Ω m)
Water in oceans	0.1–0.5
Sea water (North Sea coast)	1–5
Groundwater, well, spring water	10–150
Lake and river water	100–400
Rain water	800–1 300
Commercial distilled water	1 000–4 000
Chemically clean water	250 000
Alluvial soils	10–100
Clay	25–100
Sandy clay	40–300
Peat, marshed soil, cultivated soil	50–300
Limestone	100–5 000
Sand, granites	100–10 000
Moraine	1 000–10 000
Calcareous remains	3 000–30 000

6.4.8.2 Measurement of earth resistances

For concentrated earth electrodes, the principle of earthing-resistance measurement is shown in Figure 6.33 where, to avoid errors due to the electrolytic character of most soils, an alternating current I is passed through the electrode T to be tested and an auxiliary electrode A far from the preceding one, so that the distance TA is long compared to the dimensions of the test electrode.

The voltage U between the test electrode and a probe P is also measured. The resistance value measured is then U/I.

When measuring earthing resistances, the probe P is often placed half-way between T and A and the test result is then assumed to be the real earthing resistance of the electrode. In fact, it is slightly too small. Following the work of Curdts [26], correct results are obtained by simply increasing the distance TP to 62 per cent of the distance TA. Convenient lengths in practice are TP = 25 m and TA = 40 m.

In very large earth-termination systems such as power stations with large switching yards, the distance TA must be large compared with the earthing system dimensions, measuring leads of a few kilometres long are necessary, and the 62 per cent TP rule also applies.

If this procedure is not possible for these extended earth electrodes, shorter measuring leads can be used. Figure 6.34 indicates the correct length $b = $ TP for a given length $c = $ TA, for long strip conductors of length L measured either from one end or from the midpoint (upper figure) or semi-spherical electrodes of diameter D (lower figure); although semi-spherical electrodes are not used in practice, they are often a good approximation of an actual extended earthing system, especially when this consists of a large network of vertical electrodes [20–22].

6.5 Selection of materials

The materials used for a lightning protection system should have an excellent electrical conductivity to allow the flowing of the lightning current, sufficient mechanical strength to withstand the electrodynamic stresses caused by the high lightning current peak values, and a suitable resistance to corrosion in aggressive environments.

Components of a lightning protection system should withstand the electromagnetic effects of lightning current and predictable accidental stresses without being damaged.

After IEC 62305-3 [2], Table 6.9 gives a non-exhaustive list (general guidance only) of materials from which components of a lightning protection system should be manufactured.

Stranded conductors are more vulnerable to corrosion than solid conductors. They are also vulnerable where they enter or exit earth/concrete positions, and for this reason stranded galvanized steel is not recommended in earth. Moreover, galvanized steel may corrode in clay soil or moist soil. Galvanized steel in concrete should not extend into the soil due to possible corrosion of the steel just outside the concrete. Gavanized steel in contact with reinforcement steel in concrete may, under certain circumstances, cause damage to the concrete. The use of lead in the earth is often banned or restricted due to environmental concerns.

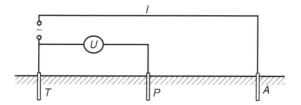

Figure 6.33   *Measuring the resistance of the earthing electrode*

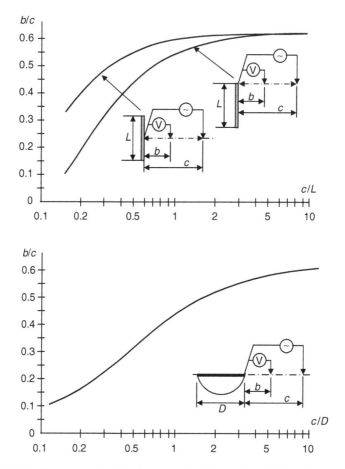

Figure 6.34   *Correct ratio of voltage probe distance b = TP and current electrode distance c = TA as function of the dimensions of a long strip earth-electrode [20–22]*

Table 6.9 Lightning protection system materials and conditions of their use (from IEC 62305-3 [2])

Material	Use in open air	Use in earth	Use in concrete	Corrosion resistance	Corrosion increased by	Destruction by galavanic coupling with
Copper	Solid Stranded	Solid Stranded As coating	Solid Stranded As coating	Good in many environments	Sulphur components Organic materials	–
Hot galvanized steel	Solid Stranded	Solid	Solid Stranded	Acceptable in air, concrete, benign soil	High chlorides content	Copper
Stainless steel	Solid Stranded	Solid Stranded	Solid Stranded	Good in many environments	High chlorides content	–
Aluminium	Solid Stranded	Unsuitable	Unsuitable	When low concentration of sulphur or chloride	Alkaline solutions	Copper
Lead	Solid As coating	Solid As coating	Unsuitable	When high concentration of sulphate	Acid soils	Copper or stainless steel

Other materials of equivalent mechanical, electrical and chemical (corrosion) performance characteristics may be used. Components made of non-metallic materials may also be used for fixing.

Materials and their dimensions should be chosen bearing in mind the possibility of corrosion either of the structure to be protected or of the lightning protection system.

The configurations and minimum cross-sectional areas of air-termination conductors, air-termination rods and down-conductors are given in Table 6.10. For earth electrodes, materials, configuration and minimum dimensions are given in Table 6.11.

The lightning protection system should be constructed of corrosion-resistant materials such as copper, aluminium (never buried directly in the ground!), stainless steel and galvanized steel. Connections between different materials should be avoided, unless they are perfectly protected.

Special attention should be taken in the presence of corrosive and humid atmospheres. Environmental factors such as moisture, dissolved salts (forming an electrolyte), degree of aeration, temperature and extent of movement of electrolyte combine to make this condition a very complex one.

In order to minimize corrosion in a lightning protection system, avoid the use of unsuitable materials in an aggressive environment, avoid contacting metals that are dissimilar from the electrochemical and galvanic points of view, use a sufficient cross-section of conductors and bonding straps, avoid designs where natural corrosion products from a cathodic metal (e.g. copper) could impinge on and plate out the lightning protection system (e.g. metallic copper on an anodic metal like steel or aluminium), and so on.

Stainless steel or copper earthing systems can be connected directly to the reinforcement in concrete. To significantly decrease the risk of corrosion, galvanized steel earth electrodes in soil should be connected to the steel reinforcement in concrete by insulating spark gaps capable of conducting a substantial part of the lightning current. Conductors with lead sheaths should be protected against corrosion; they are not suitable as earth conductors, nor in concrete.

A rule of thumb consists of using, when possible, the same material for all conductive parts of the lightning protection system (air-terminations, down-conductors and earth-terminations) as well as for metal pipes and metallic parts on the structure itself.

Connections by clamping are not generally permissible. Welded joints shall be protected against corrosion as well. Good experience has been gained with crimped joints.

Acknowledgements

The author would like to express his sincere thanks to Pierre Lecomte who designed several figures of this chapter and to Belgacom who authorized him to use his theoretical and experimental results devoted to their installations.

Table 6.10 Materials, configuration and minimum cross-sectional area of air-termination conductors, air-termination rods and down-conductors (from IEC 62305-3 [2])

Material	Configuration	Minimum cross-sectional area (mm^2)	Comments
Copper	Solid tape	50	2 mm min. thickness
	Solid round*	50	8 mm diameter
	Stranded	50	1.7 mm min. diameter for each strand
	Solid round† (only for earth lead-in rods)	200	16 mm diameter
Tin-plated copper (hot dipped or electroplated with coating ≥ 1 μm)	Solid tape	50	2 mm min. thickness
	Solid round*	50	8 mm diameter
	Stranded	50	1.7 mm min. diameter for each strand
Aluminium	Solid tape	70	3 mm min. thickness
	Solid round	50	8 mm diameter
	Stranded	50	1.7 mm min. diameter for each strand
Aluminium alloy	Solid tape	50	2.5 mm min. thickness
	Solid round*	50	8 mm diameter
	Stranded	50	1.7 mm min. diameter for each strand
	Solid round†	200	16 mm diameter
Hot dipped galvanized steel (coating: smooth, continuous, free from flux stains, 50 μm minimum thickness)	Solid tape	50	2.5 mm min. thickness
	Solid round*	50	8 mm diameter
	Stranded	50	1.7 mm min. diameter for each strand
	Solid round† (only for earth lead-in rods‡)	200	16 mm diameter
Stainless steel (chromium ≥ 16% nickel ≥ 8% carbon ≤ 0.07%)	Solid tape	50	2 mm min. thickness
	Solid round*	50	8 mm diameter
	Stranded	70	1.7 mm min. diameter for each strand
	Solid round† (only for earth lead-in rods)	200	16 mm diameter

*When mechanical strength is not an essential requirement, the section can be reduced to 28 mm^2.
†When mechanical stress such as wind loading is not critical, a 10-mm-diameter, 1-m-long maximum air-termination rod with an additional fixing may be used.
‡To avoid melting, the minimum cross-section is 16 mm^2 for copper, 25 mm^2 for aluminium, 50 mm^2 for steel or stainless steel, taking into account a specific energy of 10 MJ Ω^{-1}.

Table 6.11   Materials, configuration and minimum dimensions of earth electrodes (from IEC 62305-3 [2])

Material	Configuration	Earth rod minimum Ø (mm)	Earth conductor minimum values	Earth plate minimum S (mm²)	Comments
Copper	Stranded*		50 mm² S		1.7 mm min. Ø of each strand
	Solid round*		50 mm² S		8 mm min. Ø
	Solid tape*		50 mm² S		2 mm min. thickness
	Solid round	15			
	Pipe	20			
	Solid plate			500 × 500	2 mm min. thickness
	Lattice plate			600 × 600	2 mm min. thickness‡
Steel	G. solid round†	16	10 mm Ø		
	G. pipe†	25			2 mm min. thickness
	G. solid tape†		90 mm² S		3 mm min. thickness
	G. solid plate†			500 × 500	3 mm min. thickness
	G. lattice plate†			600 × 600	30 × 3 mm S
	Copper-coated solid round	14			250 μm min. radial Copper coating 99.9%
	Bare solid round in concrete		10 mm Ø		
	Bare or G. solid tape in concrete		75 mm² S		Copper content
	G. stranded in concrete	50 × 50 × 3	70 mm² S		1.7 mm min. Ø of each strand
	G. cross profile†				3 mm min. thickness
Stainless steel§	Solid round	15	10 mm Ø		
	Solid tape		100 mm² S		2 mm min. thickness

*May also be tin-plated.
†Smooth coating with a minimum thickness of 50 μm for round and 70 μm for flat material.
‡Minimum length of lattice configuration: 4.8 m.
§Chromium ≥16%, nickel ≥5%, molybdenium ≥2%, carbon ≤0.08%. G., galvanized; S, area; Ø, diameter.

References

1. CIGRE SC 33 WG 33.01 (TF03). *Lightning Exposures of Structures and Interception Efficiency of Air-Termination Systems*. Paris: 1989.
2. IEC 62305, *International Standard on Protection Against Lightning*, 2006, particularly Part 3: 62305-1: General principles; 62305-2: Risk assessment; 62305-3: Physical damage to structures and life hazard; 62305-4: Electrical and electronic systems inside structures.
3. Bouquegneau C. *Ready for Lightning to Strike*. February 2006. IEC E-TECH online news. Available from iecetech@iec.ch.
4. Bouquegneau C. *Doit-on Craindre la Foudre?* June 2006. EDP Sciences, les Ulis (F).
5. Bouquegneau C. and Jacquet B. 'How to improve the lightning protection by reducing the ground impedances'. *Proceedings of the 17th ICLP*; The Hague, The Netherlands, 1983. pp. 89–96.
6. Gupta B.R. and Thapar B. 'Impulse impedance of grounding grids'. *IEEE Trans. Power Appar. Syst.* 1980;(6):PAS-99.
7. Liew A.C. and Darveniza M. 'Dynamic model of impulse characteristics of concentrated earths'. *Proc. Inst. Electr. Eng.* 1974;(2):121.
8. Mazzetti C. and Veca G. 'Impulse behaviour of ground electrodes'. *IEEE PES Winter Meeting*; New York, 1983. Paper 83 WM142-7.
9. Verma R. and Mukhedkar D. 'Impulse impedance of buried ground wire'. *IEEE Trans. Power Appar. Syst.* 1980;**5**:PAS-99.
10. Kostaluc R., Loboda M. and Mukhedkar D. 'Experimental study of transient ground impedances'. IEEE PES Summer Meeting; Portland, 1981. Paper 81 SM 399-5.
11. Kouteynikoff P. 'Réponse impulsionnelle des prises de terre aux courants de foudre'. *CIGRE Symposium 22–81*; Stockholm, 1981.
12. Vainer A.L. 'Impulse characteristics of complex earth grids'. *Electrical Technology, USSR* 1966;**1**.
13. Chisholm W.A. and Janischewskyj W. 'Lightning surge response of ground electrodes'. *IEEE Trans. Power Delivery* 1989;**4**(2):1329–37.
14. Oettlé E.E. 'A new general estimation for predicting the impulse impedance of concentrated earth electrodes'. *IEEE Trans. Power Deliv.* 1988;**3**(4):2020–29.
15. Popolansky F. 'Impulse characteristics of soil'. *Proceedings of the 20th ICLP*; Interlaken, Switzerland, 1990. Paper 3.2.
16. CEI Italian Electrotechnical Committee TC 81. 'Lightning protection of structures'. *Parts IV to VI, Energ. Elett.* 1985;**11**:447–75.
17. Dwight H.B. 'Calculation of resistance to ground'. *AIEE Trans.*, Dec 1936;**55**: 1319–28.
18. Sunde E.D. *Earth Conduction Effects of Transmission Lines*. Van Nostrand; 1949. Chapter 3.
19. Schönau J., Brocke R. and Noack F. 'The effect of lightning flashes on rocks, sand and crystals in the ground', *24th ICLP*; Birmingham, UK, 1998. pp. 495–500.

20. Saraoja E. 'Maadoitukseen liittyvien Kasitteiden selvittelya', Voima Valo, Finland, 5/6, cited in Golde R.H. *Lightning*. Academic Press; 1972. Vol. 2, pp. 101–4.
21. Saraoja E. 'Lightning earths' in Golde R.H. *Lightning*. Academic Press; 1972. Vol. 2, Chapter 18.
22. Golde R.H. *Lightning*. Academic Press; 1977. Vol. 2.
23. Norinder H. and Salka O. 'Stosswiderstände der verschiedenen Erdelektroden und Einbettungsmaterialien'. *Bull. Schweiz. Elektrotch. Ver.* 1951;**42**:321–27.
24. Haddad A. *et al.* 'Characterisation of soil ionisation under fast impulse'. *25th ICLP*; Rhodes, Greece, 2000. pp. 417–22.
25. Cooray V. *et al.* 'Physical model of surge-current characteristics of buried vertical rods in the presence of soil ionisation'. *26th ICLP*; Cracow, Poland, 2002. pp. 357–62.
26. Curdts E. 'Some of the fundamental aspects of ground resistance measurements'. *Trans. Am. Inst. Elect. Eng.* 1958;**77**:760–67.
27. Baatz H. 'Protection of structures' in Golde R.H. *Lightning*. Academic Press; 1972. Vol. 2, Chapter 19.
28. Cooray V. *The Lightning Flash*. IEE Power & Energy Series 34. UK; 2003, particularly Chapter 10 (Principles of protection of structures against lightning), by C. Mazzetti.
29. Haddad A. and Warne D. *Advances in High Voltage Engineering*. IEE Power Energy Series 40. UK; 2005, particularly Chapter 3 (Lightning phenomena and protection systems).
30. Rakov V.A. and Uman M.A. *Lightning: Physics and Effects*. Cambridge University Press; 2003.

Chapter 7
Internal lightning protection system
Peter Hasse and Peter Zahlmann

7.1 Damage due to lightning and other surges

In the modern world, industry, the economy and many other public activities are highly dependent on electronic data processing (EDP) systems containing sensitive electronic apparatus. As shown in Figure 7.1, surges caused by lightning and switching operations are responsible for a large number of occurrences of damage to electronic equipment in Germany.

It can be seen in Figure 7.1 that in nearly 24 per cent of cases, lightning and other switching impulses are responsible for defects and breakdowns in electronic circuits. In general, overvoltages first damage the most sensitive section of an electronic system. In the case of a network, the most probable areas to experience damage are the network interface (as shown in Figure 7.2) units of servers, workstations and PC stations. The immediate effect of such damage could be very severe. For example, a stop in the data flow in a network system could paralyse organizations such as banks, halt production lines in industry, or interrupt the customer services of a supermarket.

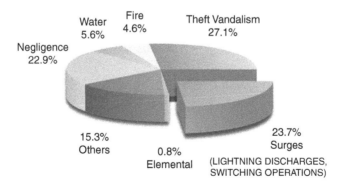

Figure 7.1 Causes of damage to electronic equipment based on the analysis of 7 370 cases in Germany in 2001

356 Lightning Protection

Figure 7.2   Damage caused by a surge

At the beginning of 1990 an analysis was carried out in the United States that highlighted the severity of the problem. When an EDP system breaks down, banks and savings institutes are able to continue operation for only two days, sales-oriented firms for 3.3 days, manufacturing plants for 4.9 days and insurance companies for 5.6 days. Statistics released by IBM Germany show that 4.8 days after an EDP breakdown firms are almost no longer able to operate. One computer safety specialist has stated, 'In real life, nine out of ten firms shut down when computers are not working for 2 weeks'.

In general, voltage and current surges are caused by (1) direct lightning strokes, (2) lightning strokes within 2 km (Figure 7.3) and (3) switching operations in the power

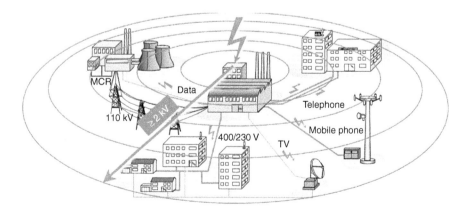

Figure 7.3   Lightning threat within a range of 2 km

7.1.1 Damage in hazardous areas

supply system. Overvoltages that can damage or disturb sensitive equipment can reach these sensitive elements in a number of ways. First, let us consider several examples of damage.

7.1.1 Damage in hazardous areas

The following are examples of lightning damage in hazardous areas.

- In October 1995 lightning struck the Indonesian oil refinery Pertamina in Cilacap on the south coast of Java. The struck tank exploded and the burning oil set fire to six further tanks (Figure 7.4). The reason for this was incomplete equipotential bonding. Thousands of Cilacap inhabitants and 400 Pertamina employees had to be evacuated. The refinery halted operation for about one and a half years. Therefore 34 per cent of Indonesia's domestic requirements during this time in oil, petrol, kerosene and diesel worth ∼€300 000 had to be imported daily to satisfy the demand for supplying Java. New oil tanks were not available until the spring of 1997, when the plant was re-opened.
- In 2004 a lightning strike caused a fire in a wind turbine on a wind power site in Wulfshagen. There was no chance of preventing damage because even with a turntable ladder it would have been impossible to reach the fire at a height of 64 m. As a result, the flames continued for 6 h. The turbine was completely damaged with a loss estimated to be about two million Euros (Figure 7.5).

7.1.2 Lightning damage in city areas

Figures 7.6 and 7.7 show typical damage caused inside structures by lightning strikes. The following are examples of damage to constructions in city areas.

Figure 7.4 Tank explosion caused by lightning in Cilacap, Java in 1995

358   Lightning Protection

Figure 7.5   Lightning damage at a windpower station

Figure 7.6   Punctures to concealed cables due to lightning strike

Internal lightning protection system 359

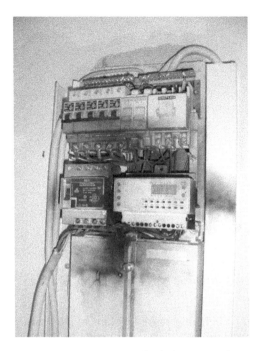

Figure 7.7   A distribution panel damaged by lightning strike

- In 2000, a violent thunderstorm paralysed the telephone network of 2 000 telephone customers in Dueren, Germany, all of whom had a '5' at the beginning of their telephone numbers. The voltage surge in the cables caused by the lightning strike was high enough to cause the communication system to break down (Figure 7.8).
- In 1995, more than 25 000 Telekom customers in the Hamburg suburban area were without cable TV after heavy thunderstorms and lightning strikes. Telekom engineers subsequently repaired about 50 amplifiers that were damaged by the lightning strikes as well as re-installing earth cables that were also destroyed. The extent of the damage was a result of the modern electronics used; as Hans-Joachim Brinckmann, a Telekom spokesman, stated, 'The microchips used for cable transmissions are considerably more sensitive than the old relays'.
- On 22 June 2002, a thunderstorm with heavy rainfall in Krefeld, Germany, caused the railway to come to a standstill after a lightning strike. According to the Deutsche Bahn all train traffic in the Niederrhein area was halted by a lightning strike near the Central Station, Krefeld. Shortly after 6 a.m. a number of trains had to stop in the nearest station because the technical facilities in a switch tower were completely scorched by the lightning strike. Due to the failure of power, some barriers also stopped working. A total of 14 trains were affected in the area of Düsseldorf, Krefeld, Kreis Viersen, and Mönchengladbach, resulting in delays for 41 trains in the suburban and local traffic.

360 *Lightning Protection*

Figure 7.8   Paralysed telephone system caused by lightnings

- On 9 May 2003, a lightning strike in Munich, Germany, on the main control station twice caused the main track to be totally blocked, thus causing complete chaos in tramway traffic. There was total failure at noon for about 2.5 h and another breakdown of the tramway signal system shortly after 3 p.m. Hundreds of thousands of passengers waited for trains in vain, while countless passengers were stuck inside trains for hours as they stopped on the track before they could reach the next station. More than a dozen trains were caught in tunnels for half an hour before an emergency team managed to continue transportation to a station. An examination of the cause of the damage ascertained the following.

    Lightning hit a tree standing about 50 m in front of the five-storey main building of the Federal Railways in Munich (Figure 7.9). Partial lightning currents affected the cable route to a distance of about 5 m away from the tree. Surges affected the electronic control system situated in the basement of the building through cables leading to the electronic control system in the main building. The rectifier control was destroyed, which led to a complete breakdown of the signalling system. There were no surge protectors installed at the time of damage.

7.1.3   Damage to airports

Airports are often subject to lightning damage, and the following are good examples.

- In summer 1995 the famous Changi Airport in Singapore was subject to a direct lightning strike on the radar control station. Equipment was inoperable for 4 h, after which operation resumed with a back-up system.
- In 1992, a lightning strike knocked out air traffic for almost 2 h after hitting the tower at Frankfurt airport, Germany. Seventy flights were affected. About 40 arriving flights had to be rerouted to other airports. The lightning also activated the fire

Figure 7.9 Lightning strikes a tree in front of the main building of the Federal Railways in Munich, 9 May 2003

control system in the tower, releasing halon gas and forcing the evacuation of air traffic controllers from the control tower.

7.1.4 Consequences of lightning damage

Today, many structures contain a spectrum of electrical and electronic systems as well as computer networks or industrial automation systems. Such systems require lightning protection and this has to be provided in addition to the lightning protection systems already installed on structures to prevent material damage and remove risk to lives. Such systems are not only threatened by the immediate effects of direct lightning strikes (fire, explosions, mechanical and chemical damage), but also by the indirect effects of lightning (i.e. surges and magnetic fields).

Beginning in 1980 the International Electrotechnical Commission (IEC) has worked out multiple standards for lightning and surge protection based on practical proven protection methods and protection devices. The most efficient ones will be introduced in the next Section 7.2.

7.2 Protective measures

Lightning protection systems (LPSs) should protect buildings and structures from fire or mechanical destruction, and persons in the buildings from injury or even death. An LPS is composed of external and internal LPSs. The functions of the external LPS are

- to attract direct lightning strikes into an air termination,

- to safely conduct the lightning current to earth by means of a down-conductor system, and
- to distribute the lightning current in the earth via an earth-termination system.

The function of the internal lightning protection is

- to prevent hazardous sparking inside the building or structure.

This is achieved by means of equipotential bonding or a safety distance between the components of the LPS and other conductive elements inside the building or structure. The protection equipotential bonding reduces the potential drops caused by the lightning current. This is achieved by connecting all separate, conductive parts of the installation directly by means of conductors or SPDs (SPDs) (Figure 7.10). In order to ensure the continuous availability of complex information technology installations even in the event of a direct lightning strike, it is necessary to have continuing measures for the surge protection of electronic installations which supplement the lightning protection system.

A detailed description of the external LPS is given in Chapter 6; here we concentrate on the internal LPS.

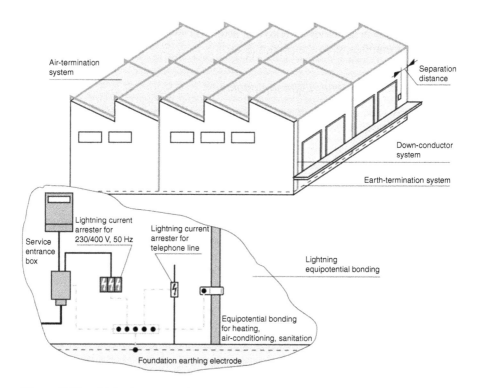

Figure 7.10   Lightning protection system

7.2.1 Internal lightning protection

7.2.1.1 Equipotential bonding in accordance with IEC 60364-4-41 and -5-54

Equipotential bonding is required for all newly installed electrical power installations. Equipotential bonding removes potential differences, i.e. it prevents hazardous contact voltages between the protective conductor of the low-voltage electrical power installation and metal, water, gas and heating pipes. According to IEC 60364-4-41, equipotential bonding consists of the main equipotential bonding and supplementary equipotential bonding.

Every building must be given main equipotential bonding in accordance with the standards stated above (Figure 7.11). The supplementary equipotential bonding is intended for those cases where the conditions for automatic disconnection of supply cannot be met, or for special areas.

The normative definition in IEC 60050-826 of an extraneous conductive component is 'a conductive unit not forming part of the electrical installation and liable to introduce a potential, generally the earth potential'. Note that extraneous conductive components also include conductive floors and walls, if an electrical potential including the earth potential can be introduced via them. The following

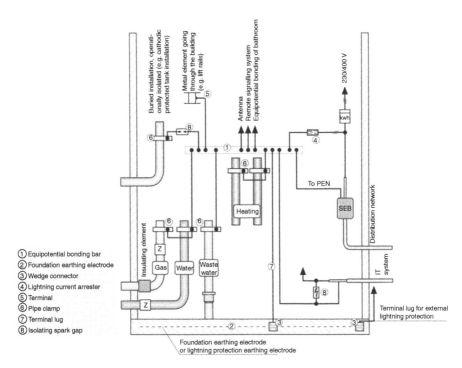

Figure 7.11 *The terminals and respective components of the main equipotential bonding*

364 *Lightning Protection*

installation components have to be integrated indirectly into the main equipotential bonding via isolating spark gaps:

- installations with cathodic corrosion protection and stray current protection measures;
- earth-termination systems of special power installations with nominal voltages above 1 kV, if intolerably high earthing potential can be transferred;
- railway earth for electric a.c. and d.c. systems in accordance with IEC 61133 (railway lines of the German Railway may only be connected with written authorization);
- measuring earth for laboratories, if they are separate from protective conductors.

7.2.1.2   Equipotential bonding for a low-voltage system

Equipotential bonding for low-voltage electrical power installations as part of the internal lightning protection in accordance with IEC 62305-3 represents an extension of the main equipotential bonding bar in accordance with IEC 60364-4-41 (Figure 7.12). In addition to all extraneous conductive parts, this also integrates the low-voltage electrical power installation into the equipotential bonding. The special feature of this equipotential bonding is the fact that a tie-up to the equipotential bonding is only possible via suitable SPDs. Analogous to equipotential bonding with metal installations, equipotential bonding for a low-voltage electrical power installation should also be carried out at the exact location where the installation enters the property.

7.2.1.3   Equipotential bonding for information technology system

Internal lightning protection or lightning equipotential bonding requires that all metal conductive components such as cable lines and shields at the entrance to the building

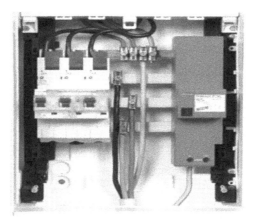

Figure 7.12   *DEHNventil ZP lightning current arrester installed in a busbar terminal field of a meter installation*

shall be incorporated into the equipotential bonding so as to cause as little impedance as possible. Examples of such components include antenna lines, telecommunication lines with metal conductors, and also fibre-optic systems with metal elements. The lines are connected with the help of elements capable of carrying lightning current (arresters and shielding terminals). A convenient installation site is the point where cabling going outside the building transfers to cabling inside the building. Both the arresters and the shielding terminals must be chosen to be appropriate to the lightning current parameters to be expected. In order to minimize induction loops within buildings, the following additional steps are recommended.

- Cables and metal pipes should enter the building at the same location.
- Power lines and data links should be laid spatially close but shielded.
- Unnecessarily long cables should be avoided by laying lines directly.

7.2.2 Lightning protection zones concept

A lightning protection system according to IEC 62305-3 protects persons and valuable material in the buildings, but it does not protect the electrical and electronic systems inside the buildings that are sensitive to transient high-energy surges resulting from the lightning discharge and switching operations. It is precisely such systems – in the form of building management, telecommunication, control and security systems – that are rapidly becoming common in all areas of residential and functional buildings. The owner/operator places very high demands on the permanent availability and reliability of such systems. The protection of electrical and electronic systems in buildings and structures against surges resulting from a lightning electromagnetic pulse (LEMP) is based on the principle of lightning protection zones (LPZ). According to this principle, the building or structure to be protected must be divided into internal lightning protection zones according to the level of threat posed by the LEMP (Figure 7.13). This enables areas with different LEMP risk levels to be adjusted to the immunity of the electronic system. With this flexible

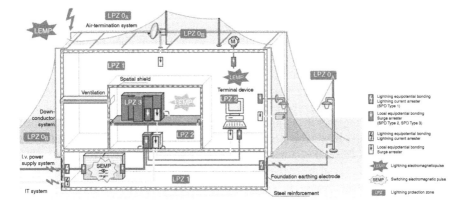

Figure 7.13 Concept of the lightning protection zones

concept, suitable LPZs can be defined according to the number, type and sensitivity of the electronic devices/systems, ranging from small local zones to large integral zones that can encompass the whole building. Depending on the type of threat posed by lightning, the following lightning protection zones are defined.

- **External zones**
    - LPZ 0_A: at risk from direct lightning strikes, from impulse currents up to whole lightning current, and from the whole electromagnetic field of the flash of lightning
    - LPZ 0_B: protected against direct lightning strikes, at risk from impulse currents up to whole lightning current and from the whole electromagnetic field of the flash of lightning
- **Internal zones**
    - LPZ 1: impulse currents limited by the splitting of the current and by surge protective devices (SPDs) at the zone boundaries; the electromagnetic field of the lightning flash can be attenuated by spatial shielding
    - LPZ 2 ... n: impulse currents further limited by the splitting of the current and by
    - SPDs at the zone boundaries; the electromagnetic field of the lightning flash is usually attenuated by spatial shielding

The requirements on the internal zones must be defined according to the immunity of the electrical and electronic systems to be protected. At the boundary of each internal zone, the equipotential bonding must be carried out for all metal components and utility lines entering the building or structure. This is done directly or with suitable SPDs. The zone boundary is formed by the shielding measures.

7.2.3 Basic protection measures: earthing magnetic shielding and bonding

7.2.3.1 Magnetic shielding

Particularly important when shielding against magnetic fields, and hence for the installation of LPZs, are extended metal components such as metal roofs and façades, steel reinforcements in concrete, expanded metals in walls, lattices, metal supporting structures and pipe systems existing in the building. The meshed connection creates an effective electromagnetic shield.

Figure 7.14 shows the principle behind how a steel reinforcement can be developed into an electromagnetic cage (hole shield). In practice, however, it will not be possible to weld or clamp together every junction in large buildings and structures. The usual practice is to install a meshed system of conductors into the reinforcement, said system typically having a size of $a \leq 5$ m (Figure 7.14). This meshed network is connected in an electrically safe way at the crosspoints, e.g. by means of clamps. The reinforcement is 'electrically hitched' onto the meshed network at a typical distance of $b \leq 1$ m (Figure 7.14). This is done on the building side, for example by means of tie connections. Mats made of construction steel in concrete are suitable for shielding purposes.

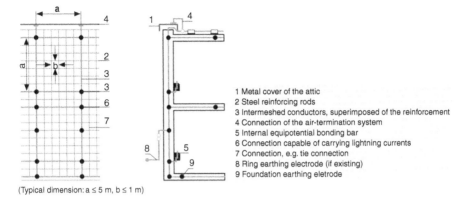

Figure 7.14 Use of reinforcing rods of a building or structure for shielding and equipotential bonding (a = mesh size of the equipotential bonding network; b = mesh size of the reinforcement)

When upgrading existing installations, such steel mats are also laid later. For this type of design, the steel mats must be galvanized to protect them from corrosion. These galvanized steel mats are then laid on roofs, for example, so that they overlap, or are applied externally or internally to the exterior wall to provide shielding for the building.

Figures 7.15a and b show the installation of galvanized steel mats on the roof of a building. To bridge expansion joints, connect the reinforcement of precast concrete components, and for terminals on the external earth-termination system or the internal equipotential bonding system, the building must already be equipped with sufficient fixed earthing points. Figure 7.16 shows an installation of this type, which must be taken into consideration for designing preliminary building works. The magnetic field inside the building or structure is reduced over a wide frequency range by means of reduction loops, which arise as a result of the meshed equipotential

Figure 7.15 (a) Galvanized construction steel shielding roofs. (b) Use of galvanized construction steel mats for shielding, for example in the case of planted roofs.

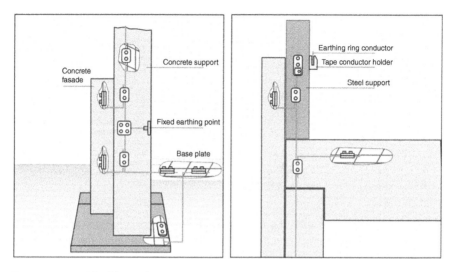

Figure 7.16   Shielding of a structure or building

bonding network. Typical mesh sizes are $a \leq 5$ m. The interconnection of all metal components both insideand on the buildings and structures results in a three-dimensional meshed equipotential bonding network. Figure 7.17 shows a meshed equipotential bonding network with appropriate terminals. If an equipotential bonding network is installed in the LPZs, the magnetic field calculated according to the formulae stated above is typically further reduced by a factor of two (corresponds to 6 dB).

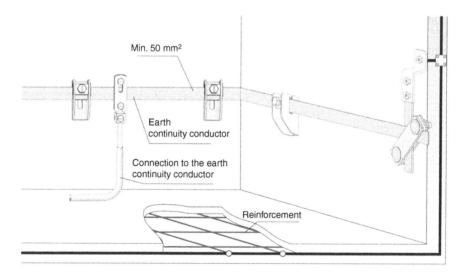

Figure 7.17   Earth continuity conductor

7.2.3.2 Cable shielding

Cable shields are used to reduce the effect of interference on active lines, and the interference emitted from active lines to neighbouring systems. From the point of view of lightning and surge protection, attention must be paid to the following applications of shielded lines.

- *No shield earthing.* Some installation systems recommend a shielded cable, but, at the same time, forbid earthing the shield (e.g. EIB). If there is no shielding terminal, the shield is not effective against interferences and must therefore be considered not to exist (Figure 7.18).
- *Double-ended shield earthing.* A cable shield must be continuously connected along the whole of its length for good conducting performance, and earthed at least at both ends. Only a shield used at both ends can reduce inductive and capacitive inputs. If the shielded cable is laid between two LPSs, the cable shield is capable of carrying lightning currents in accordance with IEC 62305-3 if it has a shield cross-section >10 mm^2 and does not exceed a maximum length of ~ 80 m. The ends of the shield must be earthed (Figure 7.19).
- *Single-ended and indirect shield earthing.* For operational reasons, cable shields are sometimes earthed at only one end. In fact, a certain attenuation of capacitive interference fields is given. Protection against the electromagnetic induction arising with lightning strikes, however, is not provided. The reason for single-ended shield earthing is the fear of low-frequency equalizing currents. In extended installations, a bus cable, for example, can often stretch many hundreds of metres between buildings. In particular with older installations, it can arise that one part of the earth-termination system is no longer in operation, or that no meshed equipotential bonding exists. In such cases, interference can occur as a result of multiple shield earthing. Potential differences of the different building earthing systems can allow low-frequency equalizing currents ($n \times 50$ Hz), and the transients superimposed thereon, to flow. At the same time, currents measuring up to a

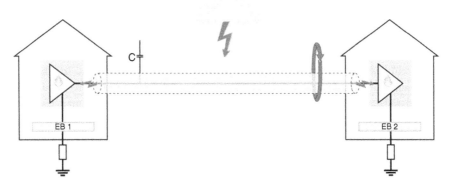

Figure 7.18 No shield connection: no shielding from capacitive/inductive couplings

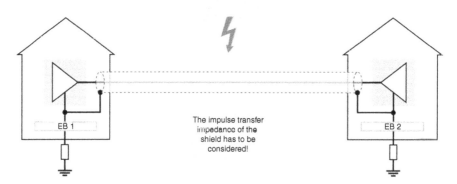

Figure 7.19 Shield connection at both ends: shielding from capacitive/inductive couplings

few amperes are possible, which, in extreme cases, can cause cable fires. In addition, crosstalk can cause signal interference if the signal frequency is in a similar frequency range to the interference signal.

The aim, however, must be to virtually implement the requirements of EMC and prevent equalizing currents. This can be achieved by combining single-ended and indirect shield earthing. All shields are directly connected with the local equipotential bonding at a central point such as the control room. At the far ends of the cable, the shields are indirectly connected to the earth potential via isolating spark gaps. Because the resistance of a spark gap is ~ 10 GΩ, equalizing currents are prevented in surge-free operation. Should EMC interferences such as lightning strikes occur, the spark gap ignites and discharges the interference pulse without consequential damage to equipment. This reduces the residual impulse on the active lines and the terminal devices are subject to even less stress. The BLITZDUCTOR CT arrester is equipped with a patented insert that can take a gas discharge tube, if necessary. This switches between the cable shield and the local earth. The gas discharge tube can be inserted or removed during upgrading or maintenance work in order to change between direct and indirect shield earthing (Figure 7.20).

- *Low-impedance shield earthing.* Cable shields can conduct impulse currents of up to several kA. During the discharge, the impulse currents flow under the shield and through the shield terminals to earth. The impedance of the cable shield and the shielding terminal creates voltage differences between shield potential and earth. In such a case, voltages of up to several kV can develop and destroy the insulation of conductors or connected devices. Coarse-meshed shields and the twisting of the cable shield (pig tail) to the terminal in a rail clamp are particularly critical. The quality of the cable shield used affects the number of shield earthings required. Under certain circumstances, an earthing is required every 20–40 m in order to achieve an efficient shielding effect. Suitable large contacting clamps with slipping spring elements are recommended for the shielding terminal. This is important to compensate for the yield of the synthetic insulation of the conductor (Figure 7.21).

Internal lightning protection system 371

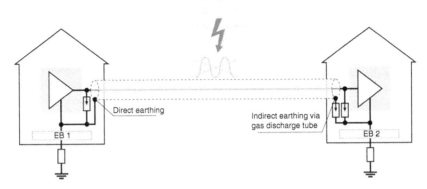

Figure 7.20 Shield connection at both ends preventing equalizing currents: direct and indirect shield earthing

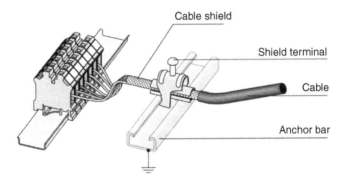

Figure 7.21 Shield connection

7.2.3.3 Equipotential bonding network

The main function of the equipotential bonding network is to prevent hazardous potential drops between all devices/installations in the inner LPZs, and to reduce the magnetic field of the lightning.

The low-inductance equipotential bonding network required is achieved by means of interconnections between all metal components aided by equipotential bonding conductors inside the LPZ of the building or structure. This creates a three-dimensional meshed network (Figure 7.22). Typical components of the network include

- all metal installations (e.g. pipes, boilers)
- reinforcements in the concrete (in floors, walls and ceilings)
- gratings (e.g. intermediate floors)
- metal staircases, metal doors, metal frames
- cable ducts
- ventilation ducts

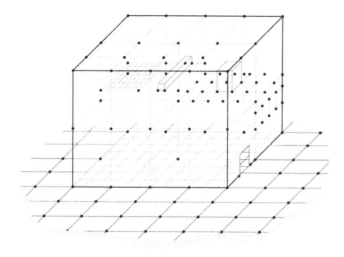

Figure 7.22 Equipotential bonding network in a structure or building

- lift rails
- metal floors
- supply lines

Ideally, a lattice structure of the equipotential bonding network would have an area of around 5×5 m^2. This would typically reduce the electromagnetic lightning field inside an LPZ by a factor of two (corresponding to 6 dB).

Enclosures and racks of electronic devices and systems should be integrated into the equipotential bonding network with short connections. This requires sufficient numbers of equipotential bonding bars and/or ring equipotential bonding bars (Figure 7.23) in the building or structure. The busbars, in turn, must be connected to the equipotential bonding network (Figure 7.24).

Protective conductors (PE) and cable shields of the data links of electronic devices and systems must be integrated into the equipotential bonding network in accordance with the instructions of the system manufacturer. The connections can be made as a mesh or in the shape of a star (Figure 7.25).

When using a star point arrangement S, all metal components of the electronic system must be suitably insulated against the equipotential bonding network. A star-shaped arrangement is therefore usually limited to applications in small, locally confined systems. In such cases, all lines must enter the building or structure, or a room within the building or structure, at a single point. The star point arrangement S must be connected to the equipotential bonding network at one single earthing reference point (ERP) only. This produces the arrangement S_s.

When using the meshed arrangement M, all metal components of the electronic system do not have to be insulated against the equipotential bonding network. All metal components shall be integrated into the equipotential bonding network at as many equipotential bonding points as possible. The resulting arrangement M_m is used for extended and open systems with many lines between the individual

Internal lightning protection system 373

Figure 7.23 Ring equipotential bonding bar in a computer facility

devices. A further advantage of this arrangement is the fact that the lines of the system can enter a building, structure or room at different points.

Complex electronic systems, also allow combinations of star point and meshed arrangements (Figure 7.26) in order to combine the advantages of both arrangements.

7.2.3.4 Equipotential bonding on the boundaries of lightning protection zones

Equipotential bonding for metal installations
At the boundaries of the EMC LPZs, measures to reduce the radiated electromagnetic field must be realized, and all metal and electrical lines/systems passing through the

Figure 7.24 Connection of the ring equipotential bonding bar with the equipotential bonding network via a fixed earthing point

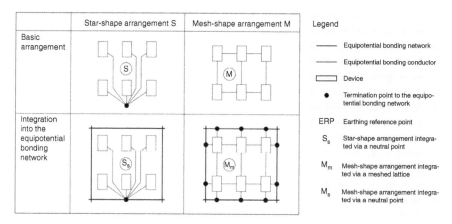

Figure 7.25 Integration of electronic systems into the equipotential bonding network

sectional area must be integrated into the equipotential bonding without exception. This requirement on the equipotential bonding basically corresponds to that on the main equipotential bonding bar in accordance with IEC 60364-4-41 and -5-54. Further towards the main equipotential bonding bar, the lightning equipotential bonding must also be implemented for cables of electrical and electronic systems at this boundary of the zones. This equipotential bonding must be installed as close as possible to the location where the lines and metal installations enter the building or structure. The equipotential bonding conductor should be designed to be as short (low impedance) as possible.

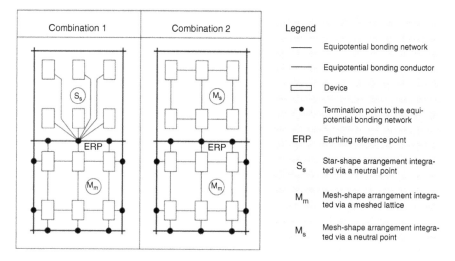

Figure 7.26 Combination of the integration methods according to Figure 7.25

For equipotential bonding, the following minimum cross-sections for tying in the equipotential bonding bar to the earth-termination system, interconnecting the different equipotential bonding bars, and tying in the metal installations to the equipotential bonding bar, must be taken into account:

Material	Cross-section (mm^2)
Cu	16
Al	25
Fe	50

The following metal installations have to be incorporated into the equipotential bonding:

- metal cable ducts
- shielded cables and lines
- building reinforcement
- metal water supply pipes
- metal conduits for lines
- other metal pipe systems or conductive components (e.g. compressed air)

A corrosion-free earth connection can be easily constructed by using fixed earthing points. During this process, the reinforcement can be connected to the equipotential bonding at the same time (Figure 7.27). Figure 7.27 shows the procedure of tying in the equipotential bonding bar to the fixed earthing point, and connecting the conduits to the equipotential bonding.

Equipotential bonding for power supply installations at LPZ 0_A and LPZ 1
In analogy to metal installations, all electrical power lines and data links at the entrance of the building (LPZ boundary 0_A to 1) must be integrated into the equipotential bonding. The following section will look in more detail at the design of equipotential

Figure 7.27   Installation of fixed earthing point

bonding with electrical power lines. The intersections for equipotential bonding at the LPZ boundary 0_A to 1 are defined with the help of the specific design of the property that requires protection. For installations fed by low-voltage systems, the LPZ boundary $0_A/1$ is usually taken to be the boundary of the building (Figure 7.28).

For properties fed directly from the medium-voltage network, LPZ 0_A is extended up to the secondary side of the transformer. The equipotential bonding is carried out on the 230/400 V side of the transformer (Figure 7.29).

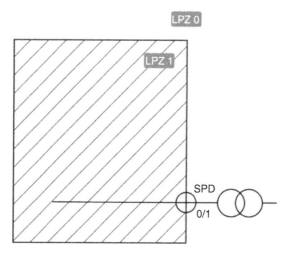

Figure 7.28 *Transformer outside the structure or building. Boundary between the LPZ 0_A and 1 (usually taken to be the boundary of the building)*

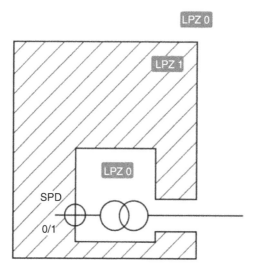

Figure 7.29 *Transformer inside the structure or building (LPZ 0 integrated in LPZ 1)*

To prevent the flow of partial lightning currents in LPZ 0 from affecting parts of the installation/systems in LPZ 1, additional shielding measures are required for the medium-voltage line entering the building or structure. To prevent equalizing currents from occurring between the various equipotential bonding points in an electrical installation, it is recommended to carry out lightning equipotential bonding of all metal lines and electrical power lines and data links entering the building or structure centrally at one point. If local circumstances do not permit this, the use of a ring equipotential bonding bar (Figures 7.30 and 7.31) is recommended.

The ability of the lightning current arrester used (SPD, class I) to discharge the current must correspond to the loads at the location where it is employed, based on the type of LPS used for the property. The type of LPS appropriate for the building or structure under consideration must be chosen on the basis of a risk assessment. If no risk assessment is available, or if it is not possible to make detailed statements about the splitting of the lightning current at the LPZ boundary 0_A to 1, it is recommended to use the type of LPS with the highest requirements (class I) as a basis. The resulting lightning current load of the individual discharge paths is shown in Table 7.1.

When installing lightning current arresters on the LPZ boundary 0_A to 1, it must still be borne in mind that, if the recommended installation site is directly at the service entrance box, this can frequently only be done with the agreement of the power supplier. When choosing lightning current arresters for the LPZ boundary 0_A to 1 then, besides the rating of the discharge capability, the prospective short-circuit current to be expected at the installation site must also be taken into account.

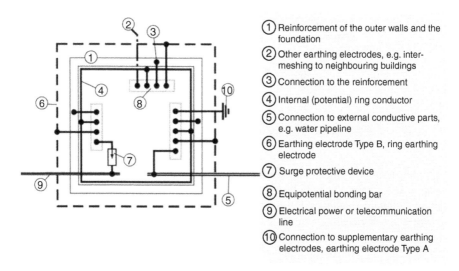

Figure 7.30 Example of equipotential bonding with several entries or the external conductive parts together with an internal ring conductor as a connection between the equipotential bonding bars

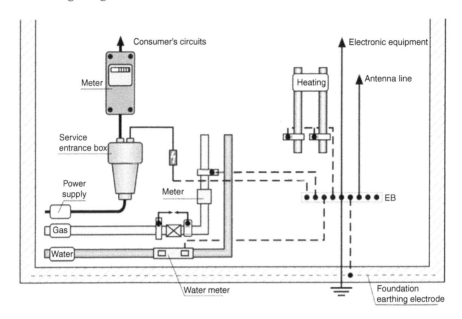

Figure 7.31   Internal lightning protection with a common entry for all supply lines

Table 7.1   Required lightning impulse current capability of SPD class I according to IEC 60364-5-53/A2

Type of LPS	Capability in TN systems (kA/m)	Capability in TT systems (L–N) (kA/m)	Capability in TT systems (N–PE) (kA)
I	≥ 100	≥ 100	≥ 100
II	≥ 75	≥ 75	≥ 75
III	≥ 50	≥ 50	≥ 50

LPS, lightning protection system. m, quantity of conductors, e.g. for L1, L2, L3, N and PE, $m = 5$.

Lightning current arresters based on spark gaps should have a high self-quenshing capacity and a good ability to limit follow currents, in order to ensure that follow currents at the mains frequency are switched off automatically, and to prevent overcurrent on protective devices such as fuses from initiating false tripping (Figures 7.32, 7.33 and 7.34).

Equipotential bonding for information technology installations LPZ 0_A–1
The lightning equipotential bonding from LPZ 0 to 1 must be carried out for all metal systems entering a building. IT lines must be connected as close as possible to the point where they enter the building or structure with lightning current arresters providing a suitable discharge capacity. For telecommunication lines in smaller properties, a

Internal lightning protection system 379

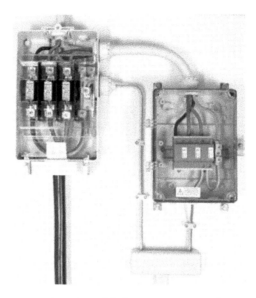

Figure 7.32 DEHNventil combined lightning current and surge arrester

general discharge capability of 5 kA (10/350 μs) is required for the boundary from LPZ 0_A to 1. The generalized approach is not used, however, when designing the discharge capability for installations with a large number of IT lines. After calculating the partial lightning current to be expected for an IT cable (see IEC 62305-3), the lightning current must then be divided by the number of individual cores in the cable actually

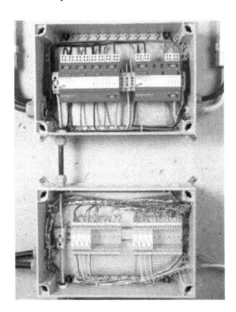

Figure 7.33 Lightning equipotential bonding for power supply and information technology systems situated centrally at one point

Figure 7.34 Lightning current arrester at LPZ boundary LPZ 0_A–1

used, in order to arrive at the impulse current per core. The partial lightning current load is lower for multicore cables than it is for cables with fewer individual cores.

The following surge protective devices can therefore be used:

- arresters designed for a discharge current of (10/350 μs)
- arresters designed for a discharge current of (8/20 μs) if
  - they have no inductance as a decoupling element
  - the specified nominal discharge current is (8/20 μs) >25× the required discharge current (10/350 μs) per core (Figure 7.35)

If the equipotential bonding is carried out for lines on the LPZ boundary 0_B to 1, it is sufficient to use surge protective devices with a discharge capacity of 20 kA (8/20 μs) because no electrically coupled partial lightning currents flow.

7.2.3.5 Equipotential bonding at the boundary of LPZ 0_A and LPZ 2

Equipotential bonding for power supply installations LPZ 0_A–2

Depending on the design of the building or structure, it is often unavoidable to realize a boundary from 0_A to 2, especially with compact installations (Figure 7.36).

Putting such an LPZ transition into practice makes high demands on the SPDs used and the surroundings of the installation. Besides the parameters, a protection level must be achieved that ensures the safe operation of equipment and systems of LPZ 2. A low-voltage protection level and high limiting of the interference energy still transmitted by the arrester form the basis for a safe energy coordination to SPDs in LPZ 2, or to surge-suppressing components in the input circuits of the equipment to be protected. The combined lightning current and surge arresters of the DEHNventil family are designed for such applications and enable the user to

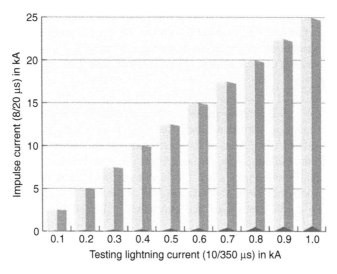

Figure 7.35 Comparison of the amplitudes of test currents waveform $10/350$ μs and $8/20$ μs, each at equal loads

combine lightning equipotential bonding and coordinated terminal device protection in a single device.

Because, for the LPZ boundary from 0 to 2, it is inevitable that the LPZs border each other directly, a high degree of shielding at the zone boundaries is absolutely imperative. As a matter of principle, it is recommended to design the areas of LPZs 0 and 2 that border each other directly to be as small as possible. According to IEC 62305-4 Figure 7.36 shows that the line can enter immediately into LPZ 2. So only

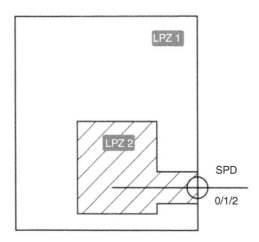

Figure 7.36 Only one SPD (0/1/2) required (LPZ 2 integrated in LPZ 1)

382 *Lightning Protection*

one SPD is required if LPZ 2 is extended into LPZ 1 using shielded cables or shielded cable ducts. In this case the one SPD will reduce the threat immediately to the level of LPZ 2. The attenuation of the electromagnetic field in LPZ 2 obviates the need for systematic shielding of all lines and systems within LPZ 2, which would otherwise be necessary.

Equipotential bonding for information technology installations LPZ 0_A-2
Generally, a lightning current arrester from LPZ 0 to 1 acts like a kind of wave breaker. It conducts a large part of the interference energy away, thus protecting the installation in the building from damage. However, it is frequently the case that the level of residual interference is still too high to protect the terminal devices. In a further step, additional SPDs are then installed at the LPZ boundary from 1 to 2 to make available a low level of residual interference adjusted to the immunity of the terminal device.

When the equipotential bonding from LPZ 0 to 2 is carried out, the first thing is to choose the installation site, and determine the partial lightning current of the individual lines and shields; the requirements on an SPD to be installed changes at the LPZ boundary, as do the requirements on the wiring after this boundary. The protective device must be designed as a combined lightning current and surge arrester and its energy must be coordinated with that of the terminal device (Figure 7.37). On the other hand, combined lightning current and surge arresters have an extremely high discharge capacity and a low level of residual interference to protect the terminal devices. Furthermore, care must be taken to make sure that the outgoing line from the protective device to the terminal device is shielded, and that both ends of the cable shield are integrated into the equipotential bonding.

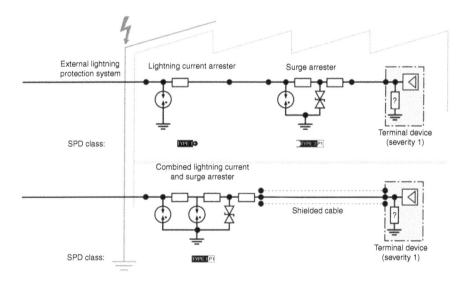

Figure 7.37 SPD combination for the protection of IT equipment

Combined lightning current and surge arresters are recommended if

- the terminal devices are near to the location where the cables enter the building,
- low-impedance equipotential bonding from protective device to terminal device can be created,
- the line from the protective device to the terminal device is continuously shielded and both ends are integrated into the equipotential bonding,
- a particularly cost-effective solution is sought.

The use of lightning current arresters and surge arresters is recommended if

- there are long cable distances from the protective device to the terminal device,
- the SPDs for power systems and IT surge protective devices are earthed via different equipotential bonding bars,
- unshielded lines are used,
- large interferences can occur inside LPZ 1.

7.2.3.6 Equipotential bonding on the boundary of LPZ 1 and LPZ 2 and higher

Equipotential bonding for metal installations
This equipotential bonding must be carried out

- as close as possible to the location where the lines and metal installations enter the zone,
- such that all systems and conductive components must be connected,
- so that the equipotential bonding conductors are designed to be as short (low impedance) as possible,
- so that ring equipotential bonding in these zones facilitates a low-impedance tie-in of the systems into the equipotential bonding.

For the tie-ins of the metal installations to the equipotential bonding, reduced cross-sections can be used for zone boundaries:

Material	Cross-section (mm^2)
Cu	6
Al	10
Fe	16

Equipotential bonding for power supply installations LPZ 1–2 and higher
For LPZ boundaries 1 to 2 and higher, surge limitation and field attenuation are also achieved by systematic integration of the electrical power lines and data links into the equipotential bonding at each LPZ boundary, as is done with all metal systems (see

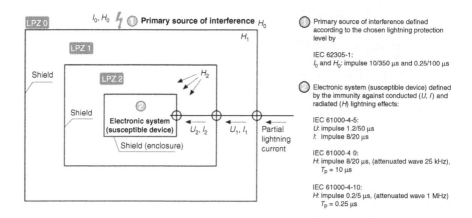

Figure 7.38   Electromagnetic compatibility in the case of a lightning stroke

Figure 7.38). Shielding the rooms and devices leads to the attenuation of the electromagnetic effect.

The function of the SPDs used at LPZ boundaries 1 to 2 or at higher LPZ boundaries is to minimize the residual values of upstream SPDs yet further. They must reduce induced surges affecting the lines laid in the LPZ, and surges generated in the LPZ itself. Depending on the location where the protective measures are taken, they can be either assigned to a device (device protection) (Figure 7.39) or represent the infrastructural basis for the functioning of a device or system in the installation (Figure 7.40). The embodiments of surge protection at the LPZ boundaries 1 to 2 and higher can thus be designed in very different ways.

Equipotential bonding for information technology installations LPZ 1–2 and higher
At LPZ boundaries inside buildings, further measures must be taken to reduce the level of interference. Because, as a rule, terminal devices are installed in LPZ 2 or higher,

Figure 7.39   Device protection: DEHNflex M

Internal lightning protection system

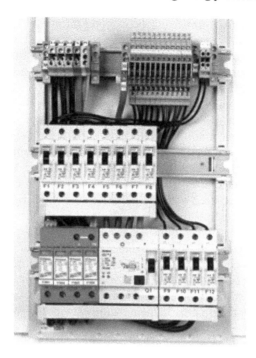

Figure 7.40 System protection: Multipole DEHNguard

the protective measures must ensure that the level of residual interference lies below values the terminal devices can tolerate through

- the use of SPDs in the vicinity of terminal devices,
- the integration of cable shields into the equipotential bonding,
- the use of low-impedance equipotential bonding of the SPD for IT installations to terminal device and SPD for power installations,
- paying attention to the energy coordination of the SPD and terminal device,
- installing telecommunication lines and gas discharge lamps at least 130 mm apart,
- locating the distribution boards of electrical installations and data in different cabinets,
- ensuring that low-voltage lines and telecommunication lines cross at an angle of 90°,
- ensuring that cable intersections are carried out using the shortest route.

7.2.3.7 Coordination of the protective measures at various LPZ boundaries

Power supply installations
Whereas surge protection in the terminal device, or immediately upstream of it, expressly fulfils the function of protecting the device, the function of SPDs in the surrounding installation is twofold. On the one hand, they protect the installation, and, on

the other, they form the protective link between the threat parameters of the complete system and the immunity of the device from the equipment and systems requiring protection. The threat parameters of the system, and the immunity of the device to be protected, are thus dimensioning factors for the protective cascade to be installed. To ensure that this protective cascade, beginning with the lightning current arrester and ending with the terminal device protection, is able to function, one must ensure that individual protective devices are selectively effective; i.e. each protection stage only takes on the amount of interference energy for which it is designed. Synchronization between the protective stages is generally termed coordination. In order to achieve the described selectivity as the protective device operates, the parameters of the individual arrester stages must be coordinated in such a way that, if one protection stage is faced with the threat of an energy overload, the upstream more powerful arrester 'responds' and thus takes over the discharge of the interference energy. When designing the coordination, one must be aware that the pulse waveform with the greatest pulse length must be assumed to be a threat for the complete arrester chain. Even though SPDs, by definition, are only tested with pulse waveforms of 8/20 µs, for coordination between the surge arrester and lightning current arrester, and also for the SPD, it is imperative to determine the ability of the device to carry an impulse current of the partial lightning currents with the waveform 10/350 µs. The SPD family of energy-coordinated products was created to avert the dangers arising from defective coordination and the resulting overloading of low-energy protective stages. These SPDs, which are coordinated both to each other and also to the device to be protected, provide the user with a high level of safety. By designing them as lightning current arresters, surge arresters and combined lightning current and surge arresters, they are ideally matched to the requirements of the corresponding LPZ boundaries (Figures 7.41 and 7.42).

Figure 7.41   *Surge arrester: DEHNguard TT*

Internal lightning protection system 387

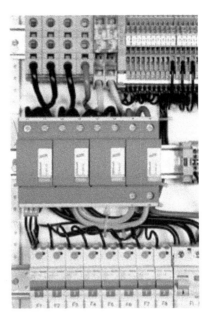

Figure 7.42   Combined lightning current and surge arrester: DEHNventil modular® TNS

Information technology installations
When implementing measures to protect against disturbance variables from nearby, distant and direct lightning strokes within buildings, it is recommended to apply a concept of protective devices with several protective stages. This reduces the high-energy interference (partial lightning current) in stages because an initial energy absorbing stage prevents the main part of the interference from reaching the downstream system. The subsequent stages serve to reduce the interference to values with which the system can cope. Depending on the conditions of the installation, several protective stages can also be integrated into one SPD using a combined protective circuit.

The relevant interfaces where protective devices are used as part of a cascade include the zone boundaries of an LPZ concept that conforms to IEC 62305-4. A cascading of SPDs must be carried out with due regard to the coordination criteria. To determine the coordination conditions, various methods are available (IEC 60364-53/A2), some of which require a certain knowledge about the structure of the protective devices. A 'black box' method is the so-called 'let-through-energy method', which is based on standard pulse parameters and hence can be understood from both mathematical and practical points of view. These methods are, however, difficult for the user to carry out because they are very time-consuming. In order to save time and work, the standard permits the use of information supplied by the manufacturers for coordination (Figure 7.43). All parts of the cascade are considered to be coordinated if the residual values I_p for a short-circuited output and U_p for an open-circuit output are smaller than the input values I_{in}/U_{in}.

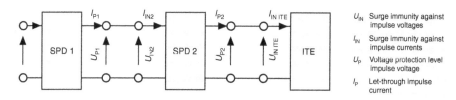

Figure 7.43   Coordination according to let-through method of two SPDs and one terminal device (according to IEC 60364-53/A2)

Lightning current arresters at LPZ 0/1 or higher are, as a rule, specified with a discharge capacity of waveform 10/350 μs. Surge arresters, by contrast, are only specified with a waveform of 8/20 μs. This originates from the fact that surge arresters were developed primarily for interferences of inductive and capacitive inputs. If, however, a line passing out of the building is connected to a cascade comprising a lightning current arrester and a surge arrester, it follows from the coordination conditions that

- the most sensitive element responds first – the surge arrester,
- the surge arrester must also be able to carry part of the partial lightning current with the waveform 10/350 μs, albeit a small one,
- before the surge arrester is overloaded, the lightning current arrester must trip and take over the discharge process.

The SPDs of the yellow/line family are coordinated sequentially and safely with each other and with the terminal devices. Therefore they provide markings indicating their coordination characteristics (Figure 7.44).

7.2.3.8   Inspection and maintenance of LEMP protection

The fundamentals and pre-conditions governing the inspection and maintenance of LEMP protection are the same as those governing the inspection and maintenance of LPSs. The inspections carried out during the construction phase are particularly important for the inspection of LEMP protection, because many components of LEMP protection are no longer accessible when the building work has been completed. The necessary measures (e.g. connecting the reinforcement) must be documented photographically and included with the inspection report.

Inspections must be carried out

- during installation of the LEMP protection,
- after installation of the LEMP protection,
- periodically,
- after each modification to components that are relevant for LEMP protection,
- after a lightning strike to the building or structure, if necessary.

After completion of the inspection, all defects found must be corrected forthwith. The technical documentation must be updated as and when necessary. A

Internal lightning protection system 389

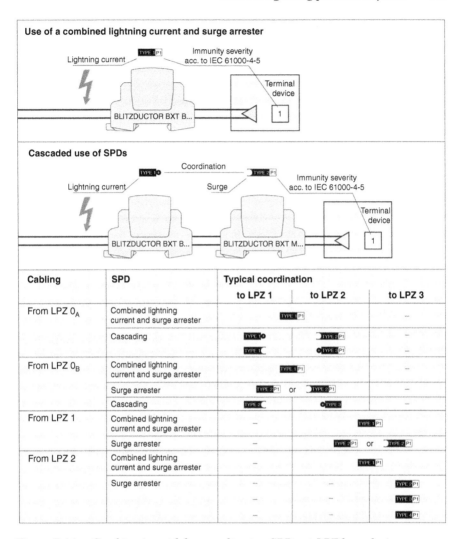

Figure 7.44 *Combination aid for coordinating SPDs at LPZ boundaries*

comprehensive inspection of the LEMP protection must be carried out at least every four years as part of the inspection of the electrical installation in accordance with workplace regulations.

7.3 Surge protection for power systems: power supply systems (within the scope of the lightning protection zones concept according to IEC 62305-4)

The installation of a lightning and surge protection system for electrical installations represents the latest state of the art and is an indispensable infrastructural condition for the trouble-free operation of complex electrical and electronic systems without

Table 7.2   Classification of SPDs according to IEC 61643-12 (Table 5.5.2)

Type Description	IEC 61643-11:1998	EN 61643-11:2001
Lightning current arrester	SPD class I	SPD Typ 1
Combined lightning current and surge arrester		
Surge arrester for distribution boards subdistribution boards, fixed installations	SPD class III	SPD Typ 3
Surge arrester for socket outlets/ terminal units	SPD class III	SPD Typ 3

consequential damage. The requirements on SPDs needed for the installation of this type of lightning and surge protection system as part of the LPZs concept according to IEC 62305-4 for power supply systems are stipulated in IEC 61643-11.

SPDs employed as part of the structure's fixed installation are classified according to the requirements and loads on the chosen installation sites into SPD classes I, II and III (Table 7.2). The highest requirements with respect to the discharge capacity are made in SPD class I. These are employed within the scope of the lightning and surge protection system at the boundary of lightning protection zone 0_A to 1 and higher, as shown in Figure 7.45. These protective devices must be capable of carrying

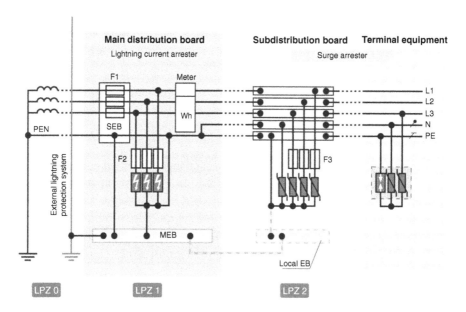

Figure 7.45   Schematic of the use of SPDs in power supply systems

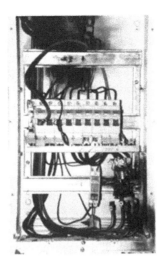

Figure 7.46   Distribution board destroyed by a lightning impulse current

partial lightning currents, waveform 10/350 μs, many times without consequential damage to the equipment. These SPDs class I are called lightning current arresters. The function of these protective devices is to prevent destructive partial lightning currents from penetrating the electrical installation of a structure (Figure 7.46).

At the boundary of lightning protection zone 0_B to 1 and higher, or lightning protection zone 1 to 2 and higher, SPDs class II are employed to protect against surges. Their discharge capacity is ∼10 kA (8/20 μs).

The last link in the lightning and surge protection system for power supply installations is the protection of terminal devices (boundary from LPZ 2 to 3 and higher). The main function of a protective device class III used at this point is to protect against surges arising between L and N in the electrical system. These are particularly switching surges.

7.3.1   Technical characteristics of SPDs

7.3.1.1   Maximum continuous voltage U_C

The maximum continuous voltage (equal to rated voltage) is the root mean square (r.m.s.) value of the maximum voltage, which may be applied to the correspondingly marked terminals of the SPD during operation. It is the maximum voltage on the arrester in the defined non-conductive state, which ensures that this state is regained after it has responded and discharged.

The value of U_C should be selected in accordance with the nominal voltage of the system to be protected and the requirements of the installation provisions (IEC 60364-5-534).

7.3.1.2 Impulse current I_{imp}

This is a standardized impulse current curve with a 10/350 μs waveform. Its parameters (peak value, charge, specific energy) simulate the load caused by natural lightning currents. Lightning impulse currents (10/350 μs) apply to SPDs class I. They must be able to discharge such lightning impulse currents several times without consequential damage to the equipment.

7.3.1.3 Nominal discharge current I_n

The nominal discharge current I_n is the peak value of the current flowing through the SPD. It has an 8/20 μs impulse current waveform and is rated for classifying the test of SPDs class II and also for conditioning the SPDs for class I and II tests.

7.3.1.4 Voltage protection level U_p

The voltage protection level of an SPD denotes the maximum instantaneous value of the voltage on the terminals of an SPD while at the same time characterizing their capacity to limit surges to a residual level. Depending on the type of SPD, the voltage protection level is determined by means of the following individual tests:

- lightning impulse sparkover voltage
- 1.2/50 μs (100 per cent)
- residual voltage for nominal discharge current (in accordance with EN 61643-11: U_{res})

The SPD appropriate to the installation site is chosen in accordance with the overvoltage categories described in IEC 60664-1. It must be noted that the required minimum value of 2.5 kV only applies to equipment belonging to the fixed electrical installation. Equipment in the terminal circuits supplied by the installation requires a voltage protection level much lower than 2.5 kV.

IEC 60364-4-534 also requires a minimum voltage protection level of 2.5 kV for a 230/400 V low-voltage installation. This minimum voltage protection level can be realized by means of a coordinated system of SPDs class I and SPDs class II, or by employing a class I combined lightning current and surge arrester.

7.3.1.5 Short-circuit withstand capability

This is the value of the prospective power-frequency short-circuit current controlled by the SPD in the case where it is furnished with an upstream backup fuse (backup protection).

7.3.1.6 Follow current extinguishing capability at U_C (I_{fi})

The follow current extinguishing capability, also termed breaking capacity, is the unaffected (prospective) r.m.s. value of the mains follow current, which can automatically be extinguished by the SPD when U_C is applied. Either the indicated follow current extinguishing capability of the SPDs corresponds to the maximum prospective short-circuit current at the SPD's installation site, or a corresponding backup fuse shall be chosen for the protective device that interrupts the mains follow current through the protective device.

According to both IEC 60364-5-534 and EN 61643-11, SPDs connected between neutral conductors and PE conductors, where a follow current with mains frequency can arise after the SPD has responded (e.g. spark gaps), must have a follow current extinguishing capability of $I_{fi} \geq 100$ A (r.m.s.).

7.3.1.7 Follow current limitation (for spark-gap based SPDs class I)

Follow current limitation is the capability of a spark-gap based SPD to limit any mains follow currents arising to such a degree that the current actually flowing is noticeably smaller than the possible short-circuit current at the installation site. A high degree of follow current limitation prevents upstream protective elements (e.g. fuses) from tripping because of a too high mains follow current. The follow current limitation is an important parameter for the availability of the electrical installation, particularly for spark-gap based SPDs with a low voltage protection level.

7.3.1.8 Coordination

In order to ensure a selective functioning of the various SPDs, energy coordination among the individual SPDs is absolutely essential. The basic principle of energy coordination is characterized by the fact that each protective stage must only discharge the amount of interference energy for which the SPD is designed. If higher interference energies occur, the protective stage upstream of the SPD, e.g. SPD class I, must take over the discharge of the impulse current and relieve the downstream protective devices. This type of coordination must take into account all possible incidences of interference such as switching surges, partial lightning currents, and so on. According to IEC 62305-4, 'Proof of energy coordination', the manufacturer must prove the energy coordination of its SPDs.

7.3.1.9 Temporary overvoltage

Temporary overvoltage (TOV) is the term used to describe temporary surges that can arise as a result of faults within the medium- and low-voltage networks. To TN systems as well as the L–N path in TT systems and for a measuring time of 5 s, the following applies: $U_{TOV} = 1.45 \times U_0$, where U_0 represents the nominal a.c. voltage of the line to earth.

For TOVs arising in low-voltage systems as a result of earth faults in the high-voltage system, $U_{TOV} = 1\,200$ V for the N–PE path in TT systems has to be taken into consideration for 200 ms.

According to EN 61643-11, the devices of the red/line family of products must be rated and tested in accordance with TOV voltages.

7.3.2 Use of SPDs in various systems

Measures to ensure protection against life hazards always take priority over surge protective measures. Because both measures are directly linked to the type of power supply systems and hence also with the use of SPDs, the following describes TN, TT and IT systems and the variety of ways in which SPDs can be used.

Electric currents flowing through the human body can have hazardous consequences. Every electrical installation is therefore required to incorporate protective measures to prevent hazardous currents from flowing through the human body. Components being energized during normal operation must be insulated, covered, sheathed or arranged to prevent them from being touched if this could result in hazardous currents flowing through the body. This protective measure is termed 'protection against direct shock hazard' (new term: 'protection against electrical shock hazard under normal conditions'). Moreover, it goes without saying, of course, that a hazard must not be caused either by current flowing through the body if, as the result of a fault, e.g. a faulty insulation, the voltage is transferred to the metal enclosure (body of a piece of electrical equipment). This protection against hazards that, in the event of a fault, can result from touching bodies or extraneous conductive components, is termed 'protection against indirect shock hazard' (new term: 'protection against electrical shock hazard under fault conditions').

Generally, the limit of the permanently permissible shock hazard voltage U_L for a.c. voltages is 50 V and for d.c. 120 V. In electrical circuits containing socket outlets and in electrical circuits containing class I mobile equipment normally held permanently in the hand during operation, higher shock hazard voltages, which can arise in the event of a fault, must be disconnected automatically within 0.4 s. In all other electrical circuits, higher shock hazard voltages must be automatically disconnected within 5 s.

IEC 60364-4-41 describes protective measures against indirect shock hazard with protective conductors. These protective measures operate in the event of a fault by means of automatic disconnection or message. When setting up the measures for the 'protection against indirect shock hazard', they must be assigned according to the system configuration and the protective device.

According to IEC 60364-4-41, a low-voltage distribution system in its entirety, from the power source of the electrical installation to the last piece of equipment, is essentially characterized by

- earthing conditions at the power source of the electrical installation (e.g. the low-voltage side of the local network transformer), and
- earthing conditions of the body of the equipment in the electrical consumer's installations.

Hence, essentially, three basic types are defined as distribution systems: the TN, TT and IT systems.

The letters used have the following significance.

- The first letter describes the earthing conditions of the supplying power source of the electrical installation, where T indicates direct earthing of a point of the power source of the electrical installation (generally the neutral point of the transformer), I indicates insulation of all active components from the earth or connection of a point of the power source of the electrical installation earthed via an impedance.

- The second letter describes the earthing conditions of the body of the equipment of the electrical installation, where T indicates the body of the equipment is earthed directly, regardless of any possible existing earthing of a point of the power supply, and N indicates that th body of the electrical equipment is directly connected to the power system earthing (earthing of the power source of the electrical installation).
- Subsequent letters describe the arrangement of the neutral conductor and the protective conductor, where S indicates that the neutral conductor and protective conductor are separate from each other, and C indicates that the neutral conductor and protective conductor are combined (in one conductor).

There are therefore three possible options for the TN system: TN–S, TN–C and TN–C–S systems.

The protective devices that can be installed in the various systems are include

- overcurrent protective devices,
- residual current devices,
- insulation monitoring devices,
- fault-voltage-operated protective devices (special cases).

As previously mentioned, the system configuration must be assigned to the protective device.

This results in the following assignments:

- TN system
  - overcurrent protective device
  - residual current device
- TT system
  - overcurrent protective device
  - residual current device
  - fault-voltage-operated protective device (special cases)
- IT system
  - overcurrent protective device
  - residual current device
  - insulation monitoring device

These measures to protect against life hazards have top priority when installing power supply systems. All other protective measures such as lightning and surge protection of electrical systems and installations are secondary to the protective measures taken against indirect contact with protective conductors under consideration of the system configuration and the protective device. The latter must not be overridden by the use of protective devices for lightning and surge protection. The occurrence of a fault in an SPD, unlikely as it may be, should also be taken into account. This has particular significance because the SPDs are always installed between active conductor (L1, L2, L3, N) and protective conductor (PE, PEN).

In the following sections we therefore describe the use of SPDs in various system configurations. These circuit proposals are taken from IEC 60364-5-534. The concepts

396 Lightning Protection

shown illustrate the use of lightning current arresters mainly in the area of the service entrance box, i.e. upstream of the meter. IEC 60364-5-534 defines the installation site of lightning current arresters as 'close to the origin of the installation'. The preferred design for each kind of supply (system configuration) must be ascertained from the responsible distribution network operator (DNO).

7.3.3 Use of SPDs in TN systems

For 'protection against indirect shock hazard' in TN systems, overcurrent and residual current devices have been approved. For the use of SPDs this means that these protective devices may only be arranged downstream of the devices for 'protection against indirect shock hazard' in order to ensure that the measure to protect against life hazards also operates in the event of a failure of an SPD. If an SPD class I or II is installed downstream of a switching device [e.g. RCD (residual current operate of protective device), fuses], it has to be expected that the discharged impulse current interrupts the circuit due to false tripping of the switching devices or that the distribution board will be overloaded.

Moreover, if an SPD class I is loaded with partial lightning currents it must be assumed that the high dynamics of the lightning current will cause mechanical damage on the residual current device. This would override the protective measure to protect against shock hazards.

Of course, this must be avoided. Therefore both lightning current arresters class I and SPDs class II should be used upstream of the residual current device. Hence, for SPDs class I and II, the only possible measure for 'protection against indirect shock hazard' is using overcurrent protective devices. The use of SPDs must therefore always be considered in conjunction with a fuse as the overcurrent protective device. Whether or not a supplementary separate backup fuse must be designated for the arrester branch depends on the size of the next upstream supply fuse and the backup fuse approved for the SPD. The maximum continuous voltages shown in Figures 7.47 and 7.48 apply to SPDs classes I, II and III when used in TN systems.

Figure 7.49 presents an example of the connections in the use of lightning current arresters and SPDs in TN–C–S systems. It can be seen that SPDs class III are used

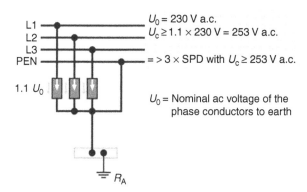

Figure 7.47   '3 + 0' circuit in TN–C systems

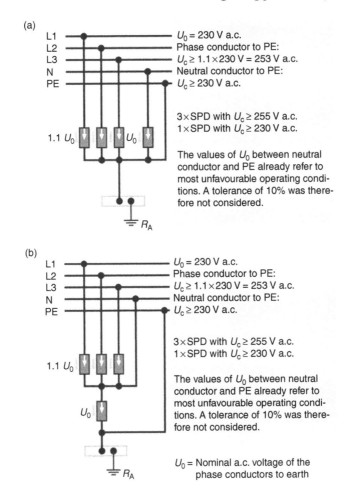

Figure 7.48 (a) '4 + 0' circuit in TN–S systems; (b) '3 + 1' circuit in TN–S systems

downstream of the residual current device (RCD). In this context, please note the following. As a result of the frequency of switching surges in terminal circuits, SPDs class III are primarily employed to protect against differential mode (DM) voltages. These surges generally arise between L and N. A surge limitation between L and N means that no impulse current is discharged to PE. Thus, this process can also not be interpreted as residual current by the RCD. In all other cases, SPDs class III are designed for a nominal discharge capacity of 1.5 kA. These values are sufficient in the sense that upstream protective stages of SPDs class I and II take over the discharge of high-energy impulses. When using an RCD capable of withstanding impulse currents, these impulse currents are not able to trip the RCD or cause mechanical damage. The following diagrams illustrate the use of SPDs as part of the LPZs concept, and the required lightning and surge protective measures for a TN–C–S system. Figure 7.50 illustrates the use of SPDs in a TN–S system.

398 Lightning Protection

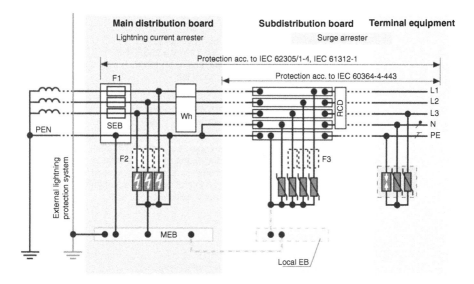

Figure 7.49 Use of SPDs in TN–C–S systems

7.3.4 Use of SPDs in TT systems

For 'protection against indirect shock hazard' in TT systems, the overcurrent protective devices, residual current devices (RCDs) and, in special cases, fault-voltage-operated protective devices have been approved. This means that, in TT systems, lightning current and surge arresters may only be arranged downstream of the above-described

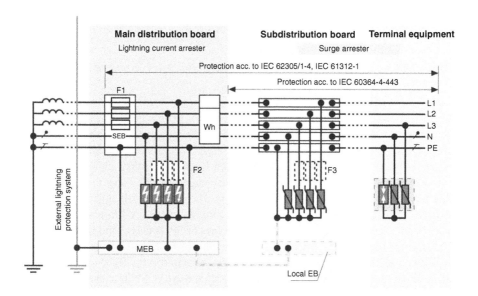

Figure 7.50 Use of SPDs in TN–S systems

protective devices in order to ensure the 'protection against indirect shock hazard' in the event of an SPD failure.

As described in Section 7.3.3, in the case of an arrangement of an SPD class I or II downstream of an RCD, it has to be expected that, because of the impulse current discharged to PE, this discharge process will be interpreted by the RCD as residual current and the circuit is then interrupted by the RCD. If SPDs class I are used, it must further be assumed that the dynamics of the discharged partial lightning current would cause mechanical damage to the RCD as the SPD class I responds as is the case with TN systems. This would damage the protective device for 'protection against shock hazard' and override the protective measure. This type of state, which can result in life hazards, must of course be avoided. Hence, both SPDs class I and SPDs class II must always be installed upstream of the residual current device in TT systems. SPDs class I and II must be arranged in TT systems to meet the conditions for the use of overcurrent protective devices for 'protection against indirect shock hazard'.

In the event of a failure, i.e. a faulty SPD, short-circuit currents must flow to initiate an automatic disconnection of the overcurrent protective devices within 5 s. If the arresters in the TT system were arranged as shown in Figures 7.49 and 7.50 for TN systems then, in the event of a fault, only earth fault currents would arise instead of short-circuit currents. In certain circumstances, however, these earth fault currents do not trip an upstream overcurrent protective device within the time required.

SPDs class I and II in TT systems are therefore arranged between L and N. This arrangement shall ensure that, in the event of a faulty protective device in the TT system, a short circuit current can develop and cause the next upstream overcurrent protective device to respond. However, because lightning currents always occur to earth, i.e. PE, a supplementary discharge path between N and PE must be provided.

These so-called 'N–PE arresters' must meet special requirements because, on the one hand, the sum of the partial discharge currents from L1, L2, L3 and N must be carried but on the other hand, there must be a follow current extinguishing capability of 100 A r.m.s. because of a possible shifting of the neutral point.

The maximum continuous voltages shown in Figure 7.51 apply to the use of SPDs in TT systems between L and N. The lightning current-carrying capability of SPDs class I must be designed to conform to lightning protection levels I, II, III/IV, as per IEC 62305-1. For the lightning current-carrying capability of SPDs between N and PE, the following values must be maintained:

Lightning protection level	I_{imp} (kA) (10/350 µs)
I	≥ 100
II	≥ 75
III/IV	≥ 50

SPDs class II are also connected between L and N and between N and PE. For an SPD between N and PE, in combination with SPDs class II, the discharge capacity must be at least $I_n \geq 20$ kA (8/20 µs) for three-phase systems and $I_n \geq 10$ kA (8/20 µs) for single-phase systems.

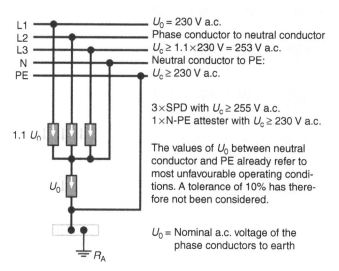

Figure 7.51 TT system (230/400 V): '3 + 1' circuit

Because coordination is always performed on the basis of worst-case conditions (10/350 μs waveform), the N–PE class II arrester from the red/line family is based on a value of 12 kA (10/350 μs).

Figure 7.52 shows an example of the connections for use of SPDs in TT systems. As is the case in TN systems, SPDs class III are installed downstream of the RCD. Generally, the impulse current discharged by this SPD is so low that the RCD does

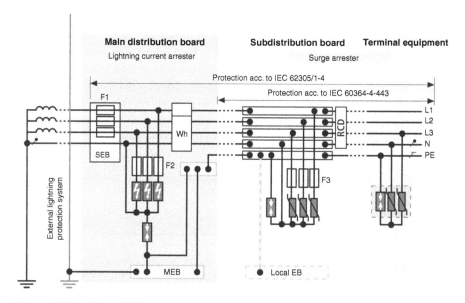

Figure 7.52 Use of SPDs in TT systems

not recognize this process as a residual current. However, it is still important to use an RCD capable of withstanding impulse currents.

7.3.5 Use of SPDs in IT systems

For 'protection against indirect shock hazard' in IT systems, overcurrent protective devices, RCDs and insulation monitoring devices have been approved. Whereas in TN or TT systems, the 'protection against indirect shock hazard' in the event of the first fault is ensured by appropriate automatic disconnection from supply through the overcurrent protective devices or RCDs, the first fault in an IT system only creates an alarm. An excessive shock hazard voltage cannot occur because the first fault in the IT system simply creates an earth connection for the system. The operating state of the IT system then becomes a TN or TT system. Hence, an IT system can be further operated at no risk after the first fault. Thus, work or production processes already begun (e.g. in the chemical industry) can still be completed. For the first fault, the protective conductor adopts the potential of the faulty external conductor, which, however, does not create a risk, because all bodies and metal components with which persons can come into contact adopt this potential via the protective conductor. Hence, no hazardous potential differences can be bridged either. When the first fault occurs, however, it must be noted that the voltage of the IT system of the intact conductors to earth corresponds to the voltage between the external conductors. Hence, in a 230/400 V IT system, in the event of a faulty SPD there is a voltage of 400 V across the non-faulty SPD. This possible operating state must be taken into account when choosing the SPDs with respect to their maximum continuous voltage.

When considering IT systems, a distinction is made between IT systems with neutral conductors entering the building with the others, and IT systems without such neutral conductors. For IT systems with the latter configuration, the SPDs in the so-called '3 + 0' circuit must be installed between each external conductor and the PE conductor. For IT systems with neutral conductors entering the building with the others, both the '4 + 0' and the '3 + 1' circuit can be used. When using the '3 + 1' circuit, it must be noted that, in the N–PE path, an SPD must be employed with a follow current extinguishing capability appropriate to the system conditions.

The maximum continuous operating voltages shown in Figure 7.53 apply to the use of SPDs class I, II and III in IT systems with and without neutral conductors entering the building with the others.

A second fault in an IT system must then cause tripping of a protective device. The statements about TN and TT systems made in Sections 7.3.3 and 7.3.4 apply to the use of SPDs in IT systems in connection with a protective device for 'protection against indirect shock hazard'. The use of SPDs class I and II upstream of the RCD is therefore also recommended for IT systems. A connection example for the use of SPDs in IT systems without neutral conductors entering the building with the others is shown in Figure 7.54.

7.3.6 Rating the lengths of connecting leads for SPDs

Rating the lengths of connecting leads of SPDs is a significant part of the IEC 60364-5-53 installation regulations. The aspects stated below are also frequently the

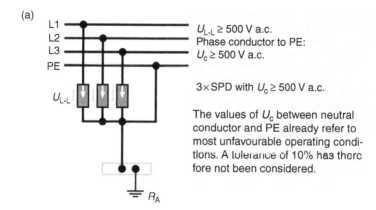

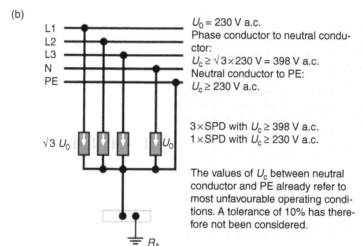

Figure 7.53   (a) IT system without neutral conductor: '3 + 0' circuit. (b) IT system with neutral conductor: '4 + 0' circuit. (c) IT system with neutral conductor: '3 + 1' circuit.

reason for complaints through experts or members of technical inspectorates, for example, inspecting the structure.

7.3.6.1 Series connection (V-shape) in accordance with IEC 60364-5-53

Crucial for the protection of systems, equipment and consumers is the actual level of impulse voltage across the installations to be protected. The optimum protective effect is then achieved when the impulse level across the installation to be protected matches the voltage protection level provided by the SPD.

IEC 60364-5-53 therefore suggests a series connection system (V-shape) as shown in Figure 7.55 to be used for connecting SPDs. This requires no separate conductor branches for connecting the SPDs.

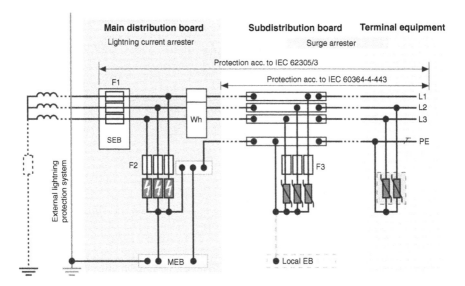

Figure 7.54   Use of SPDs in IT systems without neutral conductor

7.3.6.2   Parallel connection system in accordance with IEC 60364-5-53

The optimum series connection system cannot be used under all system conditions. Nominal currents carried via the double terminals on the SPD as part of the series wiring are limited by the thermal loadability of the double terminals. For this reason, the manufacturer of the SPD prescribes a certain maximum permissible value of the backup fuse, which, in turn, means that series wiring can sometimes not be used for systems with higher nominal operating currents.

Meanwhile, the industry provides so-called 'two-conductor terminals' to solve this problem. Thus, the cable lengths can still be kept short, even if the nominal operating

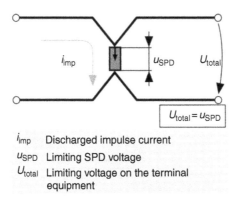

i_{imp} Discharged impulse current
u_{SPD} Limiting SPD voltage
U_{total} Limiting voltage on the terminal equipment

Figure 7.55   Connection of SPDs in serial connection

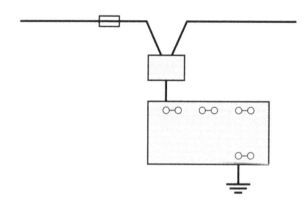

Figure 7.56   Principle of 'two-conductor terminals' (TCT): illustration of a single-pole unit

current is increased. When using the two-conductor terminals, however, it must be ensured that the value of the backup fuse stated by the manufacturer for this particular application is always observed (Figures 7.56 and 7.57).

If series connection is definitely no option, SPDs must be integrated into a separate branch circuit. If the nominal value of the next upstream installation fuse exceeds the nominal current of the maximum permissible backup fuse of the SPD, the branch must be equipped with a backup fuse for the SPD (Figure 7.58).

When the SPD in the conductor branch responds, the discharge current flows through further elements (conductors, fuses), causing additional dynamic voltage drops across these impedances. It can be stated here that the ohmic component is negligible compared to the inductive component.

Figure 7.57   Two-conductor terminals

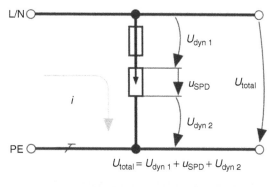

i_{imp}	Discharged impulse current
u_{SPD}	Limiting SPD voltage
U_{total}	Limiting voltage on the terminal equipment
$U_{dyn\,1}$	Dynamic voltage drop on the phase-side connection of the SPD
$U_{dyn\,2}$	Dynamic voltage drop at the earth-side connection of the SPD

Figure 7.58   Connection of SPDs in cable branches

Taking into account the relation

$$U_{dyn} = iR + (di/dt)L$$

and the rate of current change (di/dt) for transient processes of a few 10 kA μs^{-1}, the dynamic voltage drop U_{dyn} is considerably determined by the inductive component. In order to keep this dynamic voltage drop low, the electrician carrying out the work must keep the inductance of the connecting cable and hence its length as low as possible. IEC 60364-5-53 therefore recommends designing the total cable length of SPDs in branch circuits to be no longer than 0.5 m (Figure 7.59).

7.3.6.3   Design of the connecting lead on the earth side

This requirement, which is seemingly difficult to realize, should be explained with the help of the example shown in Figure 7.60. This shows the main equipotential bonding of a low-voltage consumer's installation in accordance with IEC 60364-4-41. Here, the use of SPDs class 1 extends the equipotential bonding to become a lightning equipotential bonding.

In Figure 7.60a, both measures are installed separately. In this case, the protective earth neutral (PEN) was connected to the equipotential bonding bar and the earthing connection of the SPDs was performed via a separate equipotential bonding conductor. Thus, the effective cable length l_a for the SPDs corresponds to the distance

406 Lightning Protection

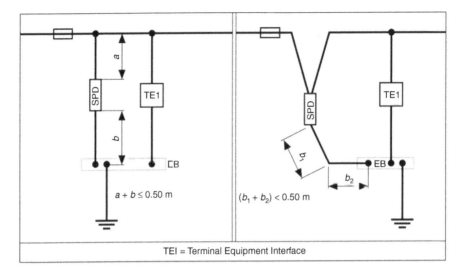

Figure 7.59 Recommended maximum cable lengths of SPDs in branch circuits

between the installation site of the SPDs (e.g. the service entrance box or main distribution board) to the equipotential bonding bar. A connection configuration of this type mostly achieves minimum effective protection of the installation. Without great expense, however, it is possible to use a conductor leading as shown in Figure 7.60b to reduce the effective cable length of the SPDs ($l_b < 0.5$ m). This is achieved by using a 'bypass' conductor (y) from the terminal of the earth side of the arrester to the PEN. The connection from the terminal of the earth side of the arrester to the equipotential bonding bar (x) remains as it was.

When installing the connection y, the distance between the service entrance box or main distribution board and equipotential bonding bar is thus insignificant. The solution for this problem refers only to the design of the connecting cable on the earth side of the SPDs.

7.3.6.4 Design of the phase-side connecting cables

The cable length on the phase side must also be taken into consideration. The following case study illustrates this. In expanded control systems, surge protection must be provided for the busbar system and the circuits attached thereto (A to D) with their consumers (Figure 7.61). For the use of SPDs in this case, installation sites 1 and 2 are taken as alternatives. Installation site 1 is located directly at the supply of the busbar system. This ensures the same level of protection against surges for all consumers. The effective cable length of the SPD at installation site 1 is l_1 for all consumers. If there is not enough space, the installation site of the SPDs is sometimes chosen at a position along the busbar system. In extreme cases, installation site 2 can be chosen for the arrangement shown in Figure 7.61. For circuit A the effective cable length is l_2. Busbar systems in fact have a lower inductance compared to cables and conductors

Internal lightning protection system 407

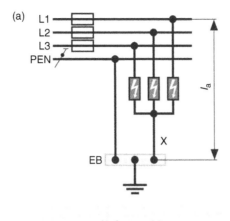

Unfavourable

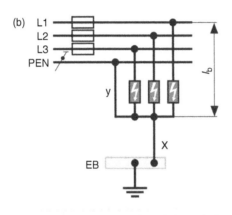

Favourable

Figure 7.60   *(a) Unfavourable connection of SPDs with consideration of the recommended maximum cable lengths. (b) Favourable connection of SPDs with consideration of the recommended maximum cable lengths.*

(approximately 1/4) and hence a lower inductive voltage drop. However, the length of the busbars must not be disregarded.

The design of the connecting cables has considerable influence on the effectiveness of SPDs and must therefore be taken into consideration at the design stage of the installation!

The contents of IEC 60364-5-53 described above were important guidelines for the development of a new combined lightning current and surge arrester (DEHNventil) that was supposed to combine the requirements for lightning current and surge

408 Lightning Protection

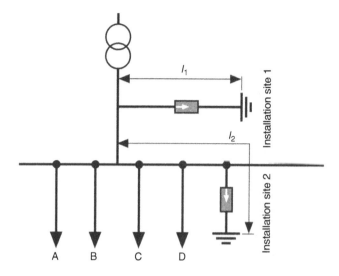

l_1: Total cable length at Installation site 1
l_2: Total cable length at Installation site 2

Figure 7.61 Arrangement of SPDs in a system and the resulting effective cable length

arresters in accordance with the IEC 62305-4 standard series in a single device. This allowed the realization of a series connection directly via the device. Figure 7.62 shows such a series connection in the form of an operating circuit diagram. From Figure 7.63 it can be observed how advantageous it is to implement a series connection with the aid of a busbar.

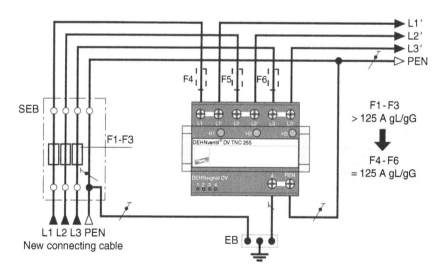

Figure 7.62 Series connection

Internal lightning protection system 409

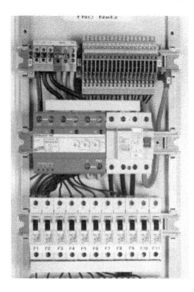

Figure 7.63   Series connection with a busbar

Because of the thermal loading capacity of the double terminals employed, a series connection (also called through-wiring) can be used up to load currents 125 A. For load currents >125 A, the SPDs are connected in the conductor branch (so-called parallel wiring). The maximum cable lengths according to IEC 60364-5-53 must be observed. The parallel wiring can be implemented as shown in Figure 7.64.

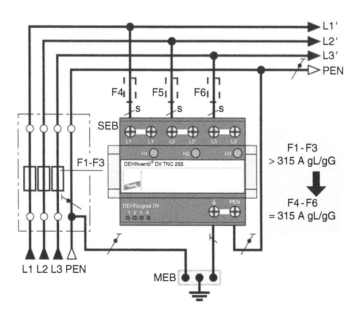

Figure 7.64   Parallel connection

410 *Lightning Protection*

In this context, it should be ensured that the connecting cable on the earth side still benefits from the double terminal for the earth connection. It is often possible to achieve an effective cable length of the order of <0.5 m with a conductor leading from the terminal component 'PE' of the earth-side double terminal to PEN.

7.3.7 Rating of cross-sectional areas and backup protection of SPDs

Connecting leads of arresters can be subjected to loads from impulse currents, operating currents and short-circuit currents. The individual loads depend on various factors:

- *Type of protective circuit*: One-port (Figure 7.65) or two-port (Figure 7.66)
- *Type of arrester*: Lightning current arrester, combined lightning current and surge arrester, SPDs
- *Performance of the arrester on follow currents*: Follow current extinction/follow current limitation

If SPDs are installed as shown in Figure 7.65, the S2 and S3 connecting cables must only be rated on the criteria of short-circuit protection according to IEC 60364-5-53 and the impulse current carrying capability (the data sheet of the protective device provides the maximum permissible overcurrent protection that can be used in this application as backup protection for the arrester). When installing the devices, it must be ensured that the short-circuit current actually flowing is able to trip the

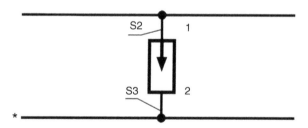

Figure 7.65 One-port protective circuit

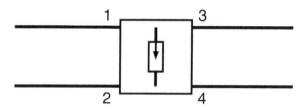

Figure 7.66 Two-port protective circuit

Table 7.3 Material coefficient k for copper and aluminium conductors with different insulation materials according to IEC 60364-4-43 (more information: IEC 60364-4-43:2001; Table 43A)

	Conductor insulation					
	PVC ≤ 300 mm^2	PVC > 300 mm^2	EPR XLPE	Rubber 60 °C	Mineral	
					PVC	Bare
Initial temperature °C	70	70	90	60	70	105
Final temperature °C	160	140	250	200	160	250
Material of conductor						
Copper	115	103	143	141	115	135/115*
Aluminium	76	68	94	93	–	–
Tin-soldered joints in copper conductors	115	–	–	–	–	–

*This value shall be used for bare cables exposed to touch.
NOTE 1 Other values of k are under consideration for:
– small conductors (particularly for cross-sectional areas less than 10 mm^2);
– duration of short-circuit exceeding 5 s;
– other types of joints in conductors;
– bare conductors.
NOTE 2 The nominal current of the short-circuit protective device may be greater than the current-carrying capacity of the cable.
NOTE 3 The above factors are based on IEC 60724.

backup protection. The rating of the cross-sectional area or of the conductor is given by the equation

$$k_2 \cdot s^2 = I^2 \cdot t$$

where t is the permissible time for disconnection in the event of a short circuit in s (duration), s is the conductor cross-section in mm^2, I the current at complete short circuit in A, expressed as an r.m.s. value, and k is the material coefficient in A s mm^{-2} according to Table 7.3.

Furthermore, it must be noted that the information concerning the maximum permissible overcurrent protection circuits in the data sheet of the SPD is only valid up to the value of the stated short-circuit withstand capability of the protective device. If the short-circuit current at the installation site is greater than the stated short-circuit withstand capability of the protective device, a backup fuse must be chosen that is smaller than the maximum backup fuse stated in the data sheet of the arrester by a ratio of 1:1.6.

For SPDs installed as shown in Figure 7.66, the maximum operating current must not exceed the nominal load current stated for the protective device. To protective devices that can be connected in series, the maximum current for through-wiring applies (Figure 7.67).

412   *Lightning Protection*

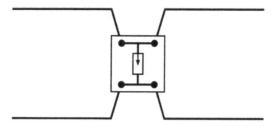

Figure 7.67   SPD with through-wiring

Figure 7.68 shows cross-sectional areas and backup protection for lightning current arresters and combined lightning current and surge arresters class I. Figure 7.69 shows examples of backup protection for SPDs class II. Figure 7.70 shows the same for SPDs class III.

The specific influence of impulse currents must be taken into consideration when rating the backup fuses for SPDs (Figure 7.71). There is a noticeable difference in the way fuses disconnect short-circuit currents compared to the way they disconnect loads with impulse currents, particularly with lightning impulse currents, waveform 10/350 μs. The performance of fuses was determined as a function of the rated current of the lightning impulse current.

Section 1 (Figure 7.71): No melting

The energy brought into a fuse by the lightning impulse current is so too low to cause a melting of the fuse.

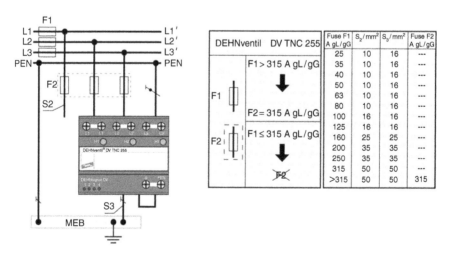

Figure 7.68   Cross-sectional areas and backup protection for SPD class I (DEHNventil, DV TNC 255)

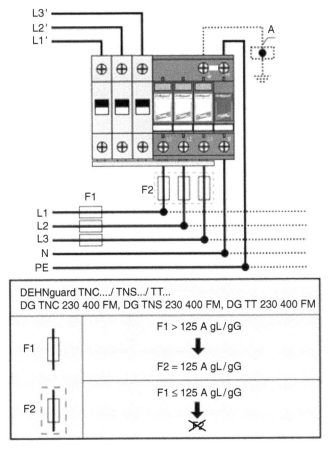

Figure 7.69   Backup protection for SPD class II (DEHNguard TNS)

Section 2 (Figure 7.71): Melting

The energy of the lightning impulse current is sufficient to melt the fuse and hence interrupt the current path through the fuse (Figure 7.72).

It is characteristic for the performance of the fuse that the lightning impulse current, because it is injected, continues to flow, unaffected by the performance of the fuse. The fuse disconnects only after the lightning impulse current has decayed. The fuses are therefore not selective with respect to the disconnection of lightning impulse currents. It must therefore be ensured that, because of the behaviour of the impulse current, the maximum permissible backup fuse according to the data sheet and/or installation instructions of the protective device is always used.

From Figure 7.72 it can also be seen that, during the melting process, a voltage drop builds up across the fuse that in part can be significantly above 1 kV. For applications as illustrated in Figure 7.73 a melting of the fuse can also result in a voltage protection level of the installation being significantly higher than the voltage protection level of the SPD itself.

414 *Lightning Protection*

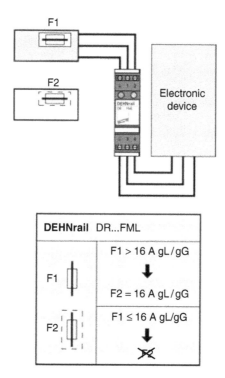

Figure 7.70   Backup protection for SPD class III (DEHNrail)

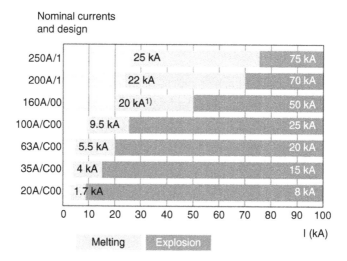

Figure 7.71   Performance of NH fuses bearing impulse current loads

Internal lightning protection system 415

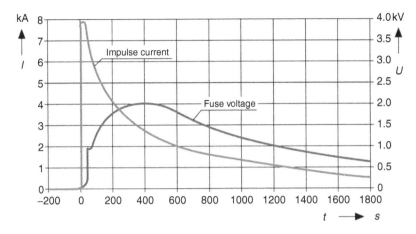

Figure 7.72   Current and voltage of a blowing 25 A NH fuse being charged with lightning impulse currents (10/350 μs)

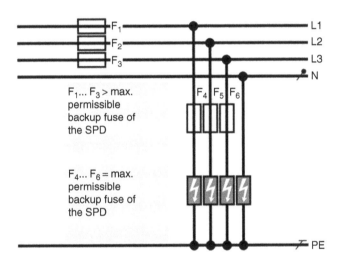

Figure 7.73   Use of a separate backup fuse for SPDs

Section 3 (Figure 7.71): Explosion

If the energy of the lightning impulse current is so high as be much higher than the pre-arcing of the fuse, then the fuse strip can vaporize explosively. This will result in bursting the fuse box. Apart from the mechanical consequences, it must be noted that the lightning impulse current continues to flow through the bursting fuse in the form of an electric arc; the lightning impulse current can thus not be interrupted nor can the required impulse current carrying capability of the employed arrester be reduced.

416   Lightning Protection

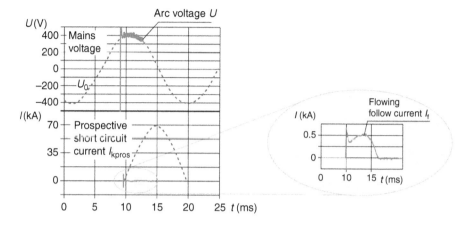

Figure 7.74 Reduction of the follow current with the patented RADAX Flow principle

7.3.7.1 Selectivity to the protection of the installation

When using spark-gap based SPDs, care must be taken that any starting mains follow current is limited to the extent that overcurrent protective devices such as fuses and/or arrester backup fuses cannot trip. This characteristic of the protective devices is called follow current limitation or follow current suppression. Only by using technologies

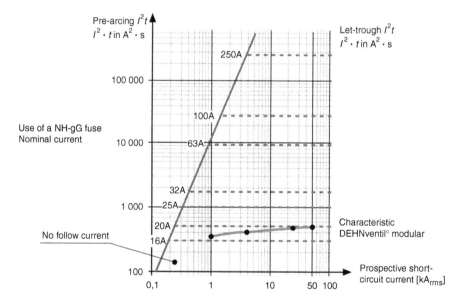

Figure 7.75 Disconnection selectivity of DEHNventil modular to NH fuse holders with different rated currents

such as the RADAX Flow technology allows the development of arresters and combinations of arresters that, even for installations with high short-circuit currents, are able to reduce and extinguish the current to such a degree that upstream fuses for lower rated currents do not trip (Figure 7.74).

The system availability required by IEC 60439-1, even in the event of responding SPDs, can be fulfilled by the aforementioned 'follow current suppression' characteristic of the device. For SPDs with low sparkover voltage, in particular, designed to not only take on the function of the lightning equipotential bonding but also that of surge protection in the installation, the performance of the follow current limitation is more important than ever for the availability of the electrical installation (Figure 7.75).

7.4 Surge protection for telecommunication systems

The primary function of arresters is to protect downstream terminal devices. They also reduce the risk of cables being damaged. The choice of arresters depends, among other things, on the following considerations:

- lightning protection zones of the installation site, if exisiting
- energies to be discharged
- arrangement of the protective devices
- immunity of the terminal devices
- protection against DM and/or CM interferences
- system requirements, e.g. transmission parameters
- compliance with product or user-specific standards, where required
- adaptation to the environmental conditions/installation conditions

Protective devices for antenna cables are classified according to their suitability for coaxial, balanced or hollow conductor systems, depending on the physical design of the antenna cable. In the case of coaxial and hollow conductor systems, the outer conductor can generally be connected directly to the equipotential bonding. Earthing couplings specially adapted to the respective cables are suitable for this purpose.

7.4.1 Procedure for selection and installation of arresters: example BLITZDUCTOR CT

In contrast to choosing SPDs for power supply systems (see Section 7.3), where uniform conditions can be expected with respect to voltage and frequency in 230/400 V systems, the types of signals to be transmitted in automation systems differ with respect to their

- voltage (e.g. 0–10 V)
- current (e.g. 0–20 mA, 4–20 mA)
- signal reference (balanced, unbalanced)
- frequency (DC, LF, HF)
- type of signal (analogue, digital)

418 Lightning Protection

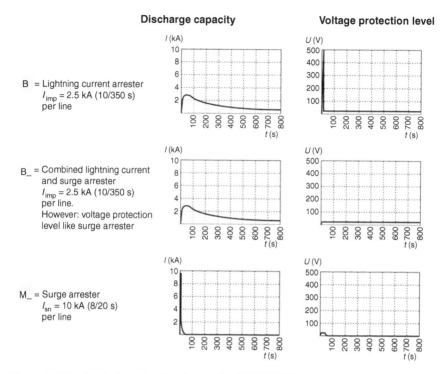

Figure 7.76 SPD classification (example: BLITZDUCTOR CT)

Each of these electrical characterisitcs for the signal to be transmitted can contain the actual information to be transferred. Therefore, the signal must not be influenced intolerably by the use of lightning current and surge arresters in measuring and control installations. Several points must be taken into account when choosing protective devices for measuring and control systems. They are described below for a universal protective devices (BLITZDUCTOR CT) and illustrated by means of application examples (Figures 7.76 and 7.77).

7.4.1.1 Technical data

Voltage protection level U_p

The voltage protection level is a parameter that characterizes the performance of an SPD in limiting the voltage at its terminals. The voltage protection level must be higher than the maximum limiting voltage measured. The measured limiting voltage is the maximum voltage measured at the terminals of the SPD when exposed to a surge current and/or surge voltage of a certain waveform and amplitude.

Measured limiting voltage with a steepness of the applied test voltage waveform of 1 kV μs^{-1}

This test is to determine the response characteristics of gas discharge tubes (GDT). These protective elements have a 'switching characteristic'. The mode of functioning

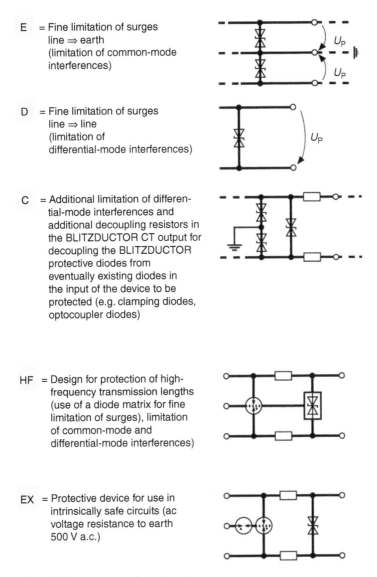

E = Fine limitation of surges line ⇒ earth (limitation of common-mode interferences)

D = Fine limitation of surges line ⇒ line (limitation of differential-mode interferences)

C = Additional limitation of differential-mode interferences and additional decoupling resistors in the BLITZDUCTOR CT output for decoupling the BLITZDUCTOR protective diodes from eventually existing diodes in the input of the device to be protected (e.g. clamping diodes, optocoupler diodes)

HF = Design for protection of high-frequency transmission lengths (use of a diode matrix for fine limitation of surges), limitation of common-mode and differential-mode interferences)

EX = Protective device for use in intrinsically safe circuits (ac voltage resistance to earth 500 V a.c.)

Figure 7.77 SPD circuits and application

of a GTD can be compared to a switch with a resistance that can 'automatically' switch from >10 GW (in non-ignited state) to values <0.1 W (in ignited state) when a certain voltage value is exceeded and the surge applied is nearly short-circuited. The response voltage of the GDT depends on the steepness of the incoming voltage (du/dt).

It generally applies that the higher the steepness du/dt, the higher the response voltage of the GDT. The comparability of different GDTs is made possible by applying

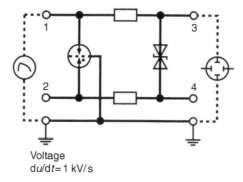

Figure 7.78 Test arrangement for determining the limiting voltage at a rate of voltage rise of $du/dt = 1\ kV\ \mu s^{-1}$

a voltage rise of $1\ kV\ \mu s^{-1}$ at the GDT for determination of the dynamic response voltage (Figures 7.78 and 7.79).

Measured limiting voltage at nominal discharge current
This test is carried out to determine the limiting behaviour of protective elements with constant limiting characteristics (Figures 7.80 and 7.81).

Nominal current I_L
The nominal current of BLITZDUCTOR CT characterizes the permissible continuous operating current. The nominal current of BLITZDUCTOR CT is determined by the

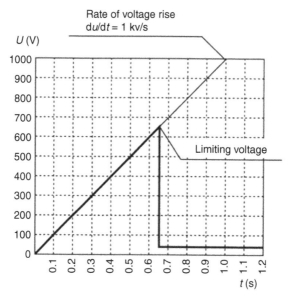

Figure 7.79 Sparkover performance of an SPD at $du/dt = 1\ kV\ \mu s^{-1}$

Internal lightning protection system 421

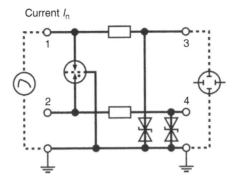

Figure 7.80 Test arrangement for determining the limiting voltage at nominal discharge current

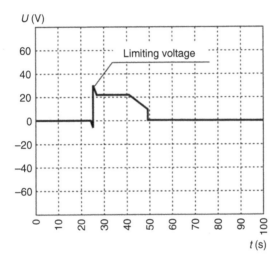

Figure 7.81 Limiting voltage at nominal discharge current

current-carrying capability and the insertion loss of the impedances used for decoupling of GDTs and fine protection elements as well as by the follow current extinguishing capability. The value is stated as a d.c. value (Figure 7.82).

Cut-off frequency fG
The cut-off frequency describes the performance of an SPD depending on the frequency (Figure 7.83). It is that frequency which gives an insertion loss (aE) of 3 dB under certain test conditions (see IEC 61643-21).

7.4.2 Measuring and control systems

The large separations between the measuring sensor and the evaluation unit in measuring and control systems allow a coupling of surges. The consequential destruction of

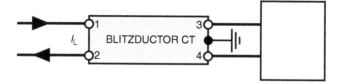

Figure 7.82   Nominal current of BLITZDUCTOR CT

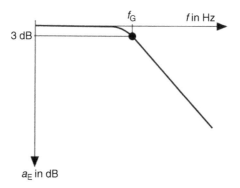

Figure 7.83   Typical frequency response of a BLITZDUCTOR CT

components and the breakdown of complete control units can severely interfere with a process technology procedure. The extent of surge damage caused by a lightning stroke often becomes apparent only some weeks later because more and more electronic components have to be replaced because they no longer operate safely. Such damage can have serious consequences for an operator using a so-called field bus system because all intelligent field bus components together in one segment can break down simultaneously. The situation can be improved by installing lightning and SPDs, which have to be chosen to suit the specific interface. Typical interfaces and the protective devices appropriate to the system can be found, e.g. at www.dehn.de.

7.4.2.1   Electrical isolation using optocouplers

Optoelectronic components (Figure 7.84), which typically produce a dielectric strength between the input and output of a few 100 V to 10 kV, are frequently installed to transmit signals in process technology systems in order to isolate the field side electrically from the process side. Thus their function is similar to that of transformers and they can primarily be installed to block low CM interferences. However, they cannot provide sufficient protection against arising CM and DM interferences when being affected by a lightning stroke (>10 kV) above their transmitter/receiver surge withstand capability.

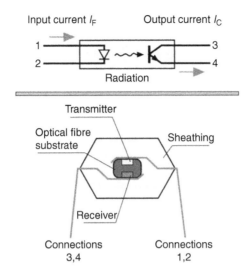

Figure 7.84   Schematic of an optocoupler

Many designers and operators of such installations misleadingly assume that such components provide lightning and surge protection. At this point it should be expressly emphasized that this voltage only provides insulating resistance between input and output (CM interference). This means that, when installed in transmission systems, attention must be paid not only to the limitation of CM interferences but also to sufficient limitation of DM interferences. Furthermore, the integration of supplementary decoupling resistors at the output of the SPD enables energy coordination with the optocoupler diode. Hence, in this case, CM and DM interference limiting SPDs, such as the BLITZDUCTOR XT BXT ML BE C24 (www.dehn.de), must be installed.

7.5   Examples for application

7.5.1   Lightning and surge protection of wind turbines

There is an unabated trend for the utilization of regenerative energies like wind power, solar technology, biomass or geothermics. This comprises an enormous market potential not only for the energy industry but also for suppliers of the energy industry and the electrical trade, worldwide.

In 2005, wind turbines already supplied 5 per cent of gross electric power consumption. Although the installed power capacity was around 7 000 MW in 2000, it was already around 18 500 MW by the end of 2005. With the rapid technological development, German manufacturers and suppliers have become market leaders on the world market. According to calculations of the German Wind Energy Institute, more than 50 per cent of all wind turbines worldwide and their components are

424 Lightning Protection

produced in Germany. The export of wind power technology amounts to ∼60 per cent of the overall turnover of around five billion Euros of German manufacturers and safeguards therefore the most part of the 64 000 or so jobs in this industry.

7.5.1.1 Positive prognoses

The prognoses for the future are positive. The German wind power institute (Deutsches Windenergie-Institut, DEWI) predicts there will be ∼4 000 wind turbines by the year 2030 on the open seas. Thus, a nominal power of ∼20 000 MW could be produced by offshore windfarms. The importance of wind turbines is obvious. Looking at the growth rates of this power market, the reliable availability of energy is an important factor.

7.5.1.2 Danger resulting from lightning effects

An operator of these installations cannot afford downtimes. On the contrary, the high capital investments for a wind turbine must have shown a return over a few years. Wind turbines are comprehensive electric and electronic installations, concentrated in a very small area. Everything that electrical engineering and electronics offer can be found in such installations: switchgear cabinets, motors and drives, frequency converters, bus systems with actuators and sensors. It goes without saying that surges can cause considerable damage there. The exposed position and overall height of wind turbines means that they are greatly vulnerable to direct lightning effects. The risk of being hit by lightning increases quadratically with the height of the structure. MW wind turbines, with their blades, reach a total height of up to 150 m and are therefore particularly exposed to danger. Comprehensive lightning and surge protection is therefore required.

7.5.1.3 Frequency of lightning strokes

The annual number of cloud-to-earth lightning flashes for a certain region is a factor of the well-known isokeraunic level. In Europe, a mean number of one to three cloud-to-earth flashes per km^2 and year applies to coastal areas and low mountain ranges. The mean annual number of lightning strokes to be expected can be determined by

$$n = 2.4 \times 10^{-5} \times N_g \times H^{2.05}$$

where N_g is the number of cloud-to-earth flashes per km^2 and year in $1/(\text{km}^2\, a)$ and H is the height of the object in metres. The maximum and minimum number of lightning strokes to be expected can differ from the mean value by a factor of approximately 3.

An assumed number of annually two cloud-to-earth flashes per km^2 and a height of 75 m results in an expected mean frequency of one lightning stroke in three years.

For dimensioning lightning protection installations, it has to be considered that in the case of objects with a height of >60 m that are exposed to lightning, earth-to-cloud flashes can occur, so-called upward flashes, as well as cloud-to-earth flashes. This results in greater values as given in the above formula. Furthermore, earth-to-cloud flashes originating from high exposed objects carry high charges of lightning current, which are of special importance for the protection measures on rotor blades and for the design of lightning current arresters.

7.5.1.4 Standardization

IEC 61400-24 and the guidelines of Germanischer Lloyd are the basis for the design of the protection concept. The German Insurance Association (GDV) recommends implementation of at least lightning protection systems class II for wind turbines in order to meet the minimum requirements for protection of these installations.

7.5.1.5 Protection measures

The lightning protection zones (LPZs) concept is a structuring measure for creating a defined EMC environment within a structure (Figure 7.85). The defined EMC environment is specified by the electromagnetic immunity of the used electric equipment.

Being a protection measure, the LPZs concept includes therefore a reduction of the conducted and radiated interferences at boundaries down to agreed values. For this reason, the object to be protected is subdivided into protection zones. The protection zones result from the structure of the wind turbine and should consider the architecture of the structure. It is decisive that direct lightning parameters affecting LPZ 0_A from outside are reduced by shielding measures and installation of SPDs to ensure that

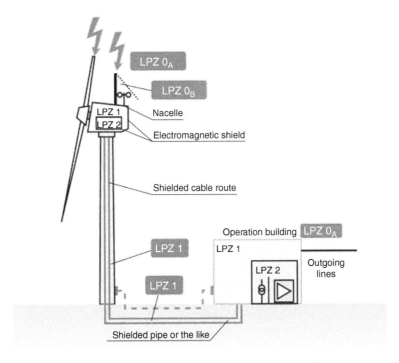

Figure 7.85 Lightning protection zones concept for a wind turbine

426 *Lightning Protection*

the electric and electronic systems and devices situated inside the wind turbine can be operated without interference.

7.5.1.6 Shielding measures

The nacelle should be designed as a metal shield that is closed in itself. Thus a volume can be obtained inside the nacelle with a considerably attenuated, electromagnetic field compared to the outside. The switchgear and control cabinets in the nacelle and, if existing, in the operation building should also be made out of metal. The connecting cables should be provided with an outer, conductive shield. With respect to interference suppression, shielded cables are effective against EMC coupling only if the shields are connected with the equipotential bonding on both sides. The shields must be contacted with encircling contact terminals to avoid long and for EMC improper 'pigtails'.

7.5.1.7 Earth-termination system

In earthing a wind turbine, the reinforcement of the tower should always be integrated. Installation of a foundation earthing electrode in the tower base, and, if existing, in the foundation of an operation building, should also be preferred in view of the corrosion risk of earth conductors.

The earthing of the tower base and the operation building (Figure 7.86) should be connected by an intermeshed earthing in order to get an earthing system with the largest surface possible. The extent to which additional potential controlling ring earthing electrodes must be arranged around the tower base depends on whether too high step and contact voltages must possibly be reduced to protect the operator in the case of a lightning stroke.

For protective circuits for conductors at the boundaries of LPZs 0_A to 1 and higher, besides shielding against radiated sources of interference, protection against

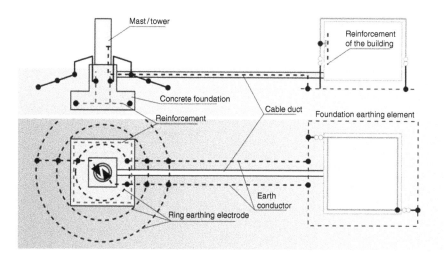

Figure 7.86   *Intermeshed network of earthing electrodes of a wind turbine*

conducted sources of interference at the boundaries of the lightning protection zones must also be provided for reliable operation of the electric and electronic devices.

At the boundaries of LPZs $0_A/1$ (conventionally also called lightning equipotential bonding) SPDs must be used, which are capable of discharging considerable partial lightning currents without damage to the equipment. These SPDs are called lightning current arresters (SPDs class I) and are tested with impulse currents, waveform $10/350$ μs.

At the boundaries of LPZ 0_B to 1 and LPZ 1 to higher, only low-energy impulse currents have to be controlled that result from voltages induced from the outside or from surges generated in the system itself. These protection devices are called SPDs (SPDs class II) and tested with impulse currents, waveform $8/20$ μs.

SPDs should be chosen according to the operating characteristics of the electric and electronic systems. After discharge, the SPDs to be used in the power supply system must be capable of extinguishing safely the follow currents coming afterwards from the mains. Beside the impulse current carrying capability, this is the second important aspect of the design.

This lightning current arrester can be mounted among bare live system parts in the installation to be protected without having to take minimum distances into account. The protective device DEHNbloc is used, for example, for low-voltage lines coming from the wind turbine.

Surge arresters (Figure 7.87) are dimensioned for loads as they occur in the case of inductive couplings and switching operations. Within the scope of energy coordination, they have to be connected downstream of the lightning current arresters. They include a thermally monitored metal oxide varistor.

Figure 7.87   *Application of DEHNguard surge arrester, DG TNC FM, $U_c = 750\ V$*

428 *Lightning Protection*

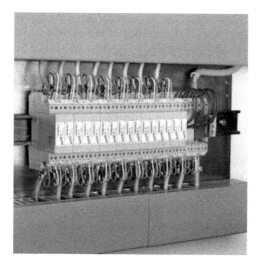

Figure 7.88 Application of BLITZDUCTOR CT lightning current and surge arrester

Contrary to SPDs for power supply systems, special attention has to be paid to system compatibility and the operating characteristics of the measuring and control or data line when installing SPDs in data processing systems. These protective devices are connected in series with the data processing lines and must be able to reduce the interference level below the immunity of the equipment to be protected. Considering a single telephone line within the LPZs concept, the partial lightning current on this conductor can be assumed to be a blanket 5 per cent. For lightning protection systems class III/IV, this would amount to a partial lightning current of 5 kA, waveform 10/350 μs.

Figure 7.88 shows a lightning current arrester suitable for such specific applications. It shows the approved multipurpose device BLITZDUCTOR CT, BCT MOD BE as a lightning current and surge arrester. This protective device can be used for protection of equipment in EMC LPZ I and higher. BLITZDUCTOR CT is designed as a four-terminal network and limits both CM and DM interferences. It can be fixed directly in the course of terminal blocks or, instead of these terminals, on supporting rails, and its special design enables a space-saving arrangement.

7.5.2 Lightning and surge protection for photovoltaic systems and solar power plants

7.5.2.1 Lightning and surge protection for photovoltaic systems

The guaranteed service life of 20 years for photovoltaic generators and their exposed installation sites as well as the sensitive electronics of the inverter really require effective lightning and surge protection. It is not only house owners who install photovoltaic (PV) systems on their rooftops, but also private operating companies, which are investing more and more in shared systems, which are erected on large-surfaced roofs, on traffic structures or on unused open areas.

Because of the large space requirements of a PV generator, PV systems are particularly threatened by lightning discharges during thunderstorms. The causes of surges in PV systems are inductive or capacitive voltages deriving from lightning discharges as well as lightning surges and switching operations in the upstream power supply system. Lightning surges in the PV system can damage PV modules and inverters. This can have serious consequences for the operation of the system. Repair costs for the inverter, for example, are very high, and system failure can result in considerable profit cuts for the operator of the plant.

Necessity of lightning protection
For PV systems, installations on buildings with or without lightning protection must be distinguished. For public buildings such as assembly places, schools, hospitals, and as a result of national building regulations, lightning protection systems are requested for safety reasons. Buildings or structures are differentiated according to their location, construction type, or utilization, allowing for the fact that a lightning stroke could easily have severe consequences. Such buildings or structures in need of protection have to be provided with a permanently effective lightning protection system.

In the case of privately used buildings, lightning protection is often not applied. This happens partly for financial reasons, but also because of a lack of sensibility with regard to the topic. If a building without external lightning protection was selected as location the for a PV system, the question arises if, with the additional installation of the PV generator on the roof, lightning protection should be provided for the entire structure. According to the current scientific state of the art, the installation of PV modules on buildings does not increase the risk of a lightning stroke, so a request for lightning protection cannot be derived directly from the mere existence of a PV system. However, there may be an increased danger for the electric facilities of the building in the event of a lightning stroke. This is based on the fact that, due to the wiring of the PV lines inside the building in existing risers and cable runs, strong conducted and radiated interferences may result from lightning currents. It is therefore necessary to estimate the risk from lightning strokes, and to take the results from this into account in the design. IEC 62305-2 presents the procedures and data for the calculation of the risk resulting from lightning strokes into structures and for the choice of lightning protection systems.

The following valid standards and guidelines must be taken into account for the installation of a PV system:

- IEC 62305
- IEC 60364-7-712

PV systems on buildings without a lightning protection system
Figure 7.89 shows the surge protection concept for a PV system on a building without lightning protection system. Possible installation sites of the SPDs can be

- the generator junction box
- the d.c. input of the inverter
- the 230 V side of the inverter

430 *Lightning Protection*

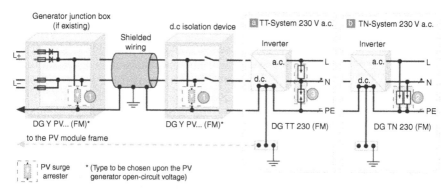

		SPDs DEHNguard Y PV			Special feature
	Type	Mid-central-earthing of the PV generator	Mid-central-earthing of the inverter	Earth-free systems	
SPD Class II	DG Y PV 275	500 V d.c.	350 V d.c.	350 V d.c.	• No leakage current from the varistors to earth
	DG Y PV 320	600 V d.c.	420 V d.c.	420 V d.c.	• No influence on the insulation monitoring system
SPD Class II	DG Y PV 1000	1000 V d.c.	1000 V d.c.	1000 V d.c.	• Fault-resistant circuit to avoid damage to the surge protection due to isolation faults in the generator circuit
					• Despite of high rated voltage control of the air and creepage distances

	Protection for...	SPDs
SPD Class II	a.c. TN system	DEHNguard DG TN 230
SPD Class II	a.c. TT system	DEHNguard DG TT 230

Figure 7.89 Basic circuit diagrams of surge protection for a PV installation on a building without external lightning protection system: (a) TT system 230 V a.c.; (b) TN system 230 V a.c

The maximum continuous operating voltage of the SPDs has to be chosen to be higher than the open-circuit voltage of the PV generator to be expected during maximum insolation on a cold day in winter.

SPDs are available in different designs and for various maximum continuous operating voltages (see the table in Figure 7.89). For generator voltages up to 1 000 V d.c., SPDs can be provided. These SPDs (DEHNguard Y PV) are available with floating contacts for central supervision of the operating state. This avoids expensive inspections of the SPDs after thunderstorms.

If a generator junction box exists, as shown in Figure 7.89, it has to be equipped with SPDs for protection of the PV modules. Protective measures against surges are always only effective locally. This also applies to those installed for protection of the PV modules. Other components of the PV system, especially the PV inverter, which may be mounted some metres away from the generator junction box, have also to be connected to SPDs at the d.c. voltage input of the respective inverter.

For this purpose, SPDs of the same type as in the generator junction box are installed.

The a.c. side of the inverter has also to be equipped with SPDs. The provisions according to IEC 60364-5-54, have to be considered.

The induced DM interference in the main d.c. line can be reduced by arranging the outgoing wire and the return wire of the individual generator cables to be close to one other. The use of a shielded generator main line is also recommended.

PV system on buildings with a lightning protection system
The correct operating condition of the lightning protection system has to be proven by existing test reports or by maintenance tests. If faults are found during the examination of the external lightning protection system (i.e. intense corrosion, loose and free clamping elements), the constructor of the PV system has the duty to inform the owner of the building about these faults in writing. The PV system on the roof surface should be designed under consideration of the existing external lightning protection system. For this purpose, the PV system has to be installed within the protection zone of the external lightning protection system (Figure 7.90) to ensure its protection against a direct lightning stroke. By using suitable air-termination systems, like air-termination rods, for example, direct lightning strokes into the PV modules can be prevented. The necessary air-termination rods to be installed must be arranged to prevent a direct stroke into the PV module within their protection zone and, secondly, any casting of a shadow on the modules. These air-termination

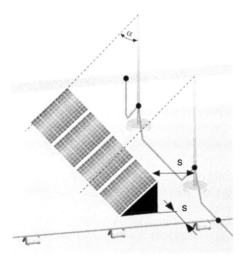

Figure 7.90 *PV modules in the protection area of air-termination rods taking separation distance* s *into consideration*

a) **Separation distance s kept**
The separation distance is calculated acc. to IEC 62305-3.

b) **Separation distance s not kept**
Direct electrically conductive connection between external lightning protection system and PV module frame

Figure 7.91   Separation distance

systems are then connected to the existing ones to create a connection to the down-conductor system and the earth-termination system.

It should be considered that a separation distance s must be kept between the PV components and metal parts such as the lightning protection system, rain gutters, skylights, solar cells or antenna systems in compliance with IEC 62305-3 (Figure 7.91). The separation distance has to be calculated according to IEC 62305-3. If the separation distance cannot be maintained because of unfavourable installation conditions, a direct conductive connection must be provided at these positions between the external lightning protection system and the metal PV components. In all other cases, a direct connection between the external lightning protection system and the metal PV components must be avoided by all means.

Figure 7.92 shows the surge protection concept for a PV system on a building with lightning protection system. Possible application sites for SPDs can be

- a generator junction box
- the d.c. input of the inverter
- the 230 V side of the inverter
- the low-voltage main distribution board

Upstream of the d.c. input of the inverter, the generator main line is furnished with SPDs. The maximum continuous operating voltage of the SPDs has to be chosen to be higher than the open-circuit voltage of the PV generator to be expected during

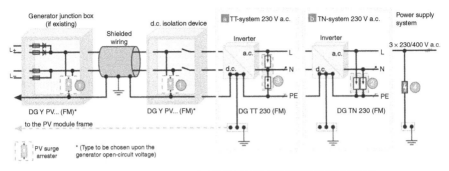

Figure 7.92   Basic circuit diagram: surge protection for a PV installation at a building with an external lightning protection system: (a) TT system 230 V a.c.; (b) TN system 230 V a.c

maximum insolation on a cold day in winter (see the table in Figure 7.92). The SPDs upstream of the d.c. input of the inverter protect the latter against too high peak voltages, which can be coupled in via the unshielded generator on the generator main line.

If the d.c. main line is not situated in the protection zone of the air-termination system, shielding measures must be taken to relieve the d.c. main line and the SPDs used. The shield of this conductor must be capable of carrying impulse currents. The cable shield must be connected to the rack on the generator side, and it must have a flat and well conductive connection to earth upstream of the d.c. input of the inverter.

As well as for the reduction of surges, the use of shielded cable is also advantageous for EMC problems (antenna characteristic of the generator main line) as the emission of electromagnetic interferences from the generator main line is considerably reduced.

434 *Lightning Protection*

The shield diameter should be constructed from at least 16 mm² copper in order to withstand the loads of the partial lightning currents.

Low-voltage power supply
An essential part of a lightning protection system is the lightning equipotential bonding for all conductive systems entering the building from the outside. The requirements of lightning equipotential bonding are met by direct connection of all metal systems and by indirect connection of all live systems via lightning current arresters. The lightning equipotential bonding should be performed preferably near the entrance of the structure in order to prevent a penetration of partial lightning currents into the building. The low-voltage power supply of the building is protected by a multipole combined lightning current and surge arrester with spark gap technology. The SPD has to be chosen according to the type of power supply system. This combined lightning current and surge arrester unites lightning current and surge arrester in one device and is available as a complete prewired unit for every low-voltage system (TN–C, TN–S, TT). There is sufficient protection without additional protective devices between DEHNventil and terminal equipment up to a cable length of 5 m. For greater cable lengths SPDs class III have to be used in addition. If the distance between the 230 V input of the inverter and the application site of DEHNventil is not greater than 5 m, no further protective devices are required for the a.c. side.

The following is a summary of the measures to be taken:

- integration of the photovoltaic generator into the external lightning protection system
- application of shielded generator main lines
- installation of SPDs at the d.c. input of the inverter
- installation of SPDs at the a.c. input of the inverter
- installation of a combined lightning current and surge arrester at the input of the low-voltage power supply for lightning equipotential bonding

7.5.2.2 Lightning and surge protection for solar power plants

The aim is to protect both the operation building and the PV array against damage by fire (direct lightning stroke), and the electric and electronic systems (inverters, remote diagnostics system, generator main line) against the effects of lightning electromagnetic impulses (LEMP).

Air-termination system and down-conductor system
To protect the PV array against direct lightning strokes it is necessary to arrange the solar modules in the protection zone of an isolated air-termination system. Its design is based on lightning protection system class III for PV systems greater 10 kW in compliance with VdS guideline 2010. According to the type of lightning protection system and the height of the air-termination rod, the quantity of air-termination rods required is determined as well as the distance between them by

means of the rolling sphere and/or protective angle method. The air-termination systems must be arranged to cast no shadow on the PV modules, as this would otherwise lead to profit cuts. Furthermore, it has to be ensured that the separation distance s is maintained between the PV supporting frames and the air-termination rods in compliance with IEC 62305-3. Also, the operation building is equipped with an external lightning protection system class III. The down-conductors are connected with the earth-termination system using terminal lugs. Owing to the corrosion risk at the point where the terminal lugs come out of the soil or concrete, they have to be made out of corrosion-resistant material or, in where galvanized steel is used they have to be protected by corresponding measures (applying sealing tape or heat-shrinkable tubes, for example).

Earthing system
The earthing system of a PV system is designed as a ring earthing electrode (surface earthing electrode) with a mesh size of $20 \times 20 \text{ m}^2$ (Figure 7.93). The metal supporting frames to which the PV modules are fixed are connected to the earth-termination system approximately every 10 m. The earthing system of the operation building is designed as a foundation earthing electrode. The earth-termination system of the PV system and the one of the operation building have to be connected to each other via at least one conductor. The interconnection of the individual earthing

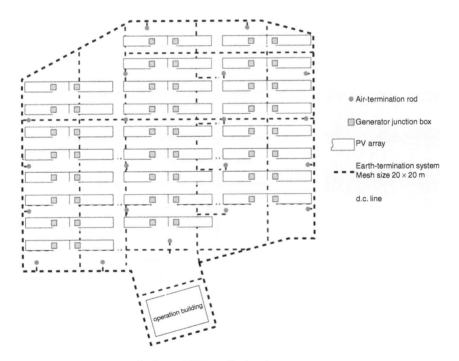

Figure 7.93   Layout of a large PV installation in an open area

systems reduces considerably the total earthing resistance. The intermeshing of the earthing systems creates an 'equipotential surface' that reduces considerably the voltage load of lightning effects on the electric connecting cables between the PV array and the operation building. The surface earthing electrodes are laid at least 0.5 m deep in the soil. The meshes are interconnected with four-wire connectors. The joints in the soil have to be wrapped with an anticorrosion band. This also applies to V4A steel strips laid in the soil.

Lightning equipotential bonding
In principle, all conductive systems entering the operation building from outside have to be generally included into the lightning equipotential bonding. The requirements of lightning equipotential bonding are fulfilled by the direct connection of all metal systems and by the indirect connection of all live systems via lightning current arresters. Lightning equipotential bonding should be performed preferably near the entrance of the structure in order to prevent partial lightning currents from penetrating the building. In the case shown in Figure 7.94, the low-voltage power supply in the operation building is protected by a multipole DEHNventil combined lightning current and surge arrester.

Surge protective measures for the d.c. lines
Being laid in the soil, the d.c. lines of the PV generator must be protected against coupling of partial lightning currents. The d.c. lines are therefore laid in a steel conduit between the generator and the operation building. The steel conduit must be connected with the earth-termination system on the generator side as well as where it enters the building.

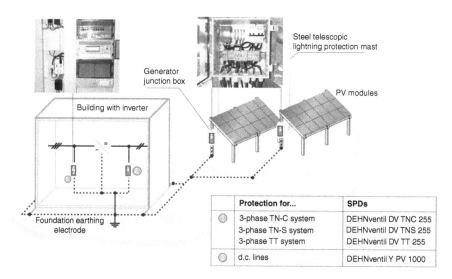

Figure 7.94   SPDs for data acquisition and evaluation

The close mesh design of the earth-termination system and the arrangement of the d.c. lines in the steel conduit provide the advantage that the load on the d.c. cable from partial lightning currents is kept low. The use of SPDs class II is therefore sufficient.

In order to reduce the load on the isolation inside the solar modules following a lightning stroke into the isolated air-termination system, thermally controlled SPDs are installed in a generator junction box as close as possible to the PV generator. For generator voltages up to 1 000 V d.c., SPDs are available.

In practice, it is a proven method to use SPDs with floating contacts to indicate the operating state of the thermal disconnection device. Thus, the intervals between the regular on-site inspections of the protection devices are extended.

The SPDs in generator junction boxes assume the protection for the PV modules locally and ensure that no sparkovers caused by conducted or field-related interferences arise at the PV modules. To protect the inverters in the central operation building, the SPDs are directly installed at the d.c. input terminals of the inverter. Therefore, well known inverter manufacturers ensure that their systems are already equipped with suitable SPDs.

Surge protective measures for data processing systems
The operation building provides a remote diagnostics system that is used for a simple and quick function check of the PV systems. This allows the operator to recognize and remedy malfunctions in good time. The remote supervisory control system constantly provides performance data for the PV generator in order to optimize the output of the PV system.

As shown in Figure 7.95, measurements of wind velocity, module temperature and ambient temperature are performed via external sensors of the PV system. These

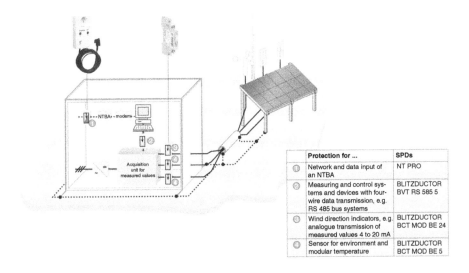

Figure 7.95   Protection concept for data acquisition and evaluation

438 *Lightning Protection*

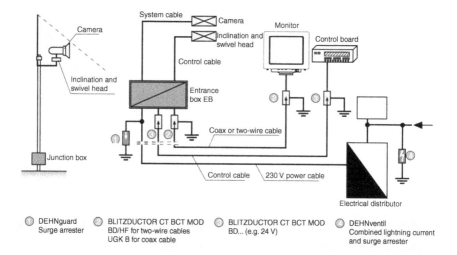

Figure 7.96   Video surveillance system: lightning and surge protection

measurements can be read directly from the acquisition unit. The data acquisition unit provides interfaces such as RS232 or RS485, to which a PC and/or modems are connected for remote enquiry and maintenance. Thus, the service engineers can determine the cause of a malfunction by telediagnosis and then directly eliminate it. The modem in Figure 7.96 is connected to the network termination unit of an ISDN basic access.

The measuring sensors for wind velocity and module temperature shown in Figure 7.95 are also installed in the zone protected against lightning strokes, like the PV modules. Thus, no lightning currents come up in the measuring leads, although there may be conducted transient surges resulting from induction effects in the event of lightning strokes into the isolated air-termination system.

In order to provide a reliable trouble-free and continuous transmission of the measured data to the measuring unit, it is necessary to lead the sensor cables entering the building via SPDs. When choosing the protective devices, it has to be ensured that the measurements cannot be impaired. The forwarding of the measured data via the telecommunication network per ISDN modem must be provided as well, in order to provide continuous control and optimization of the performance of the installation. For this purpose, the Uk0 interface upstream of the network termination unit to which the ISDN modem is connected is protected by a surge protective adapter. This adapter ensures additional protection for the 230 V power supply of the network termination unit.

7.5.3   Surge protection for video surveillance systems

In the industry as well as in the private sector, video surveillance systems are used more and more frequently for entrance monitoring and property supervision. The following sections describe protective measures against surges that meet the requirements of video surveillance systems.

7.5.3.1 Video surveillance systems

A video surveillance system consists at least of one camera, one monitor and one suitable video transmission line. Remotely controllable camera stations are normally equipped with an inclination and swivel support so that the position and viewing angle of the station can be individually adapted by an operator. As shown in Figure 7.96, the video transmission and power supply of the camera are implemented via an interface cable between the terminal box and camera.

The communication line between the terminal box and monitor can be a coaxial cable or a balanced two-wire cable. The transfer of the video signals through coaxial cables is certainly the most common type in video technology. In this case an unbalanced transfer is used, i.e. the video signal is transferred through the core of the coaxial cable (inner conductor). The shielding (earth) is the reference point for the signal transmission. Two-wire transmission is, as well as coaxial cable transmission, a commonly used possibility. If there is already a global telecommunication infrastructure for the object to be monitored, a free twin wire (two-wire cable) in the telecommunication cables is used to transfer the video signal.

Video surveillance systems are partially powered directly from distribution panels, but also via inserted UPS.

7.5.3.2 Choice of SPDs

Building with an external lightning protection system
In Figure 7.96, the camera is installed on a pole. A direct lightning stroke into the camera can be prevented by an air-termination rod mounted at the top end of the pole. With reference to the camera as well as to its connection cable, a sufficient separation distance (IEC 62305-3) must be maintained from parts of the external lightning protection system.

The connecting cable between the terminal box and the camera is usually laid inside the metal pole.

If this is not possible, the camera cable has to be laid in a metal pipe, which must be electrically connected with the pole. For cable lengths of a few metres, a protective circuit in the terminal box is not necessary in these cases.

For the coaxial cable or the two-wire cable as well as for the control cable leading from the terminal box at the pole into a building with an external lightning protection system, lightning equipotential bonding must be implemented. This includes connecting the lightning protection system to pipelines, metal installations within the building and the earth-termination system. Additionally, all earthed parts of the power supply and data processing systems must be integrated into the lightning equipotential bonding. All live wires in the power supply and data processing cables and lines leading in and coming out of the structure are connected indirectly with the lightning equipotential bonding via lightning current arresters. If no lightning current arresters are installed in the low-voltage main distribution board, the operator must be informed that these need to be upgraded.

Figure 7.96 shows the application of a combined lightning current and surge arrester DEHNventil. This combined SPD (lightning current arrester and surge arrester

440 Lightning Protection

Figure 7.97 Camera for video surveillance in the protective area of the air-termination rod

in one device) requires no decoupling coil and is available as a complete prewired unit for each type of low-voltage system (TN–C, TN–S, TT). Up to cable lengths of 5 m between the DEHNventil and the terminal equipment, there is sufficient protection without additional protective devices. For greater cable lengths, additional SPDs are required for the terminal equipment (e.g. DEHNrail).

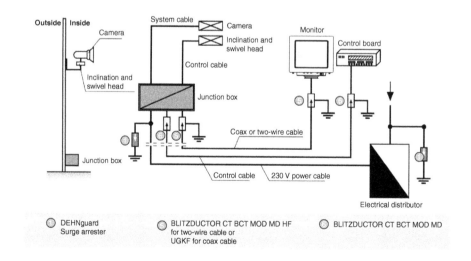

Figure 7.98 Video surveillance system: surge protection

When mounting the camera on an external building façade it should be ensured that the camera is fixed below the outer edge of the roof, in the protection zone. If this is not possible, an area must be created that is protected against lightning strokes by means of additional external lightning protection measures. This can be done with an air-termination system, as shown in Figure 7.97, to safeguard the camera against direct lightning strokes.

Buildings without an external lightning protection system
For buildings without an external lightning protection system, it is assumed that the risk of damage caused by a direct or close lightning stroke into the building is low and therefore accepted. If this risk is also accepted for subsequently mounted video transmission equipment, this can be sufficiently protected by installing SPDs (Figure 7.98).

Bibliography

Lightning Protection Guide (in German and English), Germany: Dehn and Söhne GmbH and Co. KG.; www.dehn.de.

Hasse P., Wiesinger J. and Zischank W. *Handbuch für Blitzschutz und Erdung.* 5th edn. München: Richard Pflaum Verlag; ISBN 3-7905-0931-0.

Hasse P. *Overvoltage Protection of Low Voltage Systems.* 2nd edn. 2000. London: IEEE; ISBN 0-85296-781-0.

Chapter 8
Risk analysis
Z. Flisowski and C. Mazzetti

8.1 General considerations

The risk from lightning to different objects and living beings, as well as the possibility of reducing this risk, are considered in this chapter. To do this in unequivocal way, the terms involved will first be properly defined.

First, the distinction between an object and a structure or a service should be clarified. The term 'object' covers both the structure and the external services [1]. The term 'structure' is reserved for a building and its internal equipment, in particular its electrical and electronic systems, and for those people within it and its external areas. The term 'service' is reserved for the different kinds of external installations or systems (mains, electronics, informatics and telecommunications systems, water and gas piping) connected to a structure for which a lightning hazard is considered.

Similarly we will distinguish between the terms 'lightning flash', 'lightning strike' and 'lightning stroke' [2]:

- A lightning flash to earth means electrical discharge of atmospheric origin between cloud and earth consisting of one or more strokes (on average of 3–4 strokes, with typical time intervals of ~50 ms).
- A lightning stroke means a single electrical discharge in a lightning flash to earth.
- A lightning strike means a lightning flash with defined location on the earth surface itself or on its protruding parts: for example, a structure, lightning protection system (LPS), service, tree.

Depending of their location, the lightning flashes and strikes are divided into direct and nearby ones. Every structure and every external service (incoming lines) may be influenced by direct and by nearby strikes.

Structures, due to a greater use of non inflammable materials, are increasingly resistant to direct lightning flashes. However, the equipment therein, due to an increase in the variety of the equipment and its electromagnetic sensitivity, is at risk to a greater extent from the effects of direct and nearby lightning flashes (the effects of entire and partial lightning currents and their electromagnetic fields).

As a result of direct and nearby lightning flashes a structure, with its equipment and services, is in danger of physical damage or failure to the equipment or injury to living beings. Physical damage is categorized into the mechanical, thermal, chemical or explosive effects of lightning flashes [3]. Failures are mainly considered to be the effects of lightning overvoltages on electrical and electronic equipment and the injury to living beings – from the effects of lightning currents and voltages.

Lightning effects may appear throughout an entire structure or in particular areas having equipment of differing sensitivities to lightning electromagnetic pulses (LEMPs). To consider the danger to different parts of the structure, these are divided into lightning protection zones (LPZs) for which the lightning electromagnetic environment may be defined (for details on the LPZ concept see Reference 4).

The level of hazard for every distinguished zone may be assessed in terms of the risk of damage to the relevant part of the structure and of the failure of its equipment. The risk is formally defined [2] as a probable value of average annual loss due to lightning in relation to the total value of the structure to be protected.

When the level of risk is greater than tolerable, defined as a maximum value of the risk that can be tolerated for the structure to be protected, then this level must be reduced by adequate protection measures. They should be applied in the structure according to the set of lightning current parameters that will result in the relevant lightning protection level (LPL). LPL is a number related to the set of parameters corresponding to the probability that their accepted values will not be exceeded in naturally occurring lightning [1,5]. All protection measures adopted according to the LPZ concept create a complete (external and internal) lightning protection system (LPS) of adequate LPL. An example of a structure influenced by direct and nearby lightning is shown in Figure 8.1. Four different locations D, M, W, Z of lightning strikes have been distinguished. In principle, the lightning currents and electromagnetic fields due to direct strikes and electromagnetic fields of nearby strikes should be taken into account. Successive LPZs – $LPZ0_A$, $LPZ0_B$, LPZ1 and LPZ2 – according to the LPZ concept are visible in Figure 8.1. Every boundary ($S_{0/1}$ and $S_{1/2}$) indicates a place where the location of protection measures [screens, surge protection devices (SPDs)] should be considered. The first is a place for external LPS installation.

The designer, when considering the risk due to lightning and its reduction measures, should be aware that the effects of damages and failures may also be extended to the surroundings of the structure or may involve its environment. Structures with a danger of explosion as well as structures and services with the potential for biological, chemical and radioactive emission as a consequence of lightning discharge should be qualified in an individual way according to the particular requirements.

In order to discuss the lightning hazard (risk), to assess its degree and to control its reduction with respect to the different potential effects of lightning on the structure, its equipment and services, the general concept of risk evaluation as well as different events, phenomena, parameters and conditions should be considered in a particular way. So a description of the general concept is followed by consideration of the relevant number of lightning strikes, damage probabilities, the question of loss assessment, the composition of risk components and the standardized procedure of risk assessment.

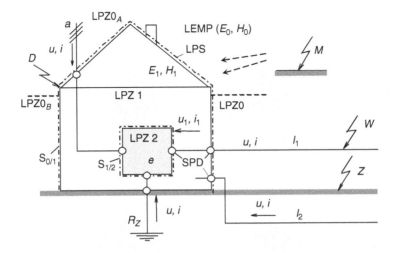

Figure 8.1 An example of the influence of lightning protection on a structure with equipment e and protection measures (screens and SPD on LPZ boundaries). D, M, W, Z, different locations of lightning strikes; SPD, surge protective devices; LPZ0, 1, 2, successive protective zones; $S_{0/1}$, $S_{1/2}$, zone boundaries; u, u_1, i, i_1, voltages and currents in entering services l_1, l_2, a or induced; a, antenna; R, earthing; E, H, electric and magnetic fields.

8.2 General concept of risk due to lightning

Every structure and its equipment is exposed to the natural influences of lightning discharges and their electromagnetic fields, and may be disturbed, damaged or destroyed with a certain risk $R(t)$, which is a result of the correlation of possible stresses x and probable damages y due to these stresses [6–8]. Stresses x are represented by the discrete density function $g_x(k, t)$ of their statistical distribution depending on time t and the number k of dangerous lightning events. Probable damage (or withstand) characteristics may also be represented by a discrete function $P_y(k)$ of the cumulative distribution of damages y depending on the number k of hazardous lightning events. The sum of products $g_x(k, t) P_y(k)$ for successive k, within its entire range $<0, \infty>$, results in the relation

$$R(t) = \sum_{k=0}^{\infty} g_x(k, t) P_y(k) \tag{8.1}$$

Graphical interpretation of this relation is presented in Figure 8.2, where both functions are considered to be discrete.

The process of lightning flashes may be treated as homogeneous and stationary. With the assumption that the entire considered number of lightning flashes is

446 Lightning Protection

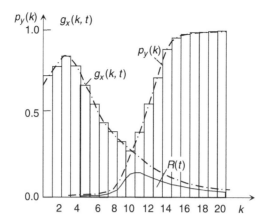

Figure 8.2 Graphical interpretation of the risk R of damages expressed by discrete functions

unlimited, the distribution function of stresses x for a structure may be expressed by a Poissonian expression as follows:

$$g_x(k, t) = \frac{(Nt)^k}{k!} e^{-Nt} \tag{8.2}$$

where N is the frequency of flashes influencing the structure during one year and t the number of years.

The function $g_x(k, t)$ indicates the probability that exactly k lightning strikes will appear within an interval $<0, t>$. The probability that the strike will not appear ($k = 0$) is equal to

$$g_x(0, t) = e^{-Nt} \tag{8.3}$$

The strikes should be treated as separable events, which create an absolute probabilistic system for which the sum of probabilities is equal to unity, so that

$$G_x(k, t) = \sum_{k=0}^{\infty} g_x(k, t) = 1 \tag{8.4}$$

and, according to these expressions the probability that at least one strike will appear is equal to

$$G_x(k \geq 1, t) = 1 - e^{-Nt} \tag{8.5}$$

The cumulative distribution $P_y(k)$ of damages in equation (8.1) represents the different effects of lightning strikes, including fire, explosion, mechanical damages, shocks to people and failure of electrical systems. If the probability of damage y

due to one strike is equal to P, then the probability that with k strikes the damage will not appear is given by

$$(1-P)^k \tag{8.6}$$

and the cumulative distribution of damages may be expressed by

$$P_y(k) = 1 - (1-P)^k \tag{8.7}$$

Substituting equations (8.2) and (8.7) into (8.1), and converting it, the final expression for the risk is given by

$$R(t) = 1 - e^{-NPt} \tag{8.8}$$

which represents simultaneously a cumulative distribution of time of waiting for damage. Usually the risk of damage is related to one year, time $t = 1$, and the product NP is a cumulative level of the risk expressed by

$$R = 1 - e^{-NP} \tag{8.9}$$

When the frequency N of strikes is constant in the period of observation, then the probability P is decisive for the risk level.

The risk cumulative level NP multiplied by the relative value of loss L (human or goods) as a consequence of the damage (or in other words the product NPL) is defined in the IEC Standard [2] as a risk. The probability that a certain amount (NPL) of loss will appear is expressed by

$$R = 1 - e^{-NPL} \tag{8.10}$$

When the product NPL is much smaller than 1, then it is only a measure of the risk, which may be expressed as

$$R = 1 - e^{-NPL} \approx NPL \tag{8.11}$$

According to this formula, for risk evaluation, knowledge of the three elements of the product NPL, i.e. the number N of strikes influencing the structure, the probability P of damage due to one lightning strike and the amount L of relative loss, must be known.

The entire expected risk may be composed of several components depending on the lightning strike location, the features of the structure and its equipment, the type of damage or a group of damages, and the type of loss. Therefore, it is possible to evaluate one component, a selected group of components and the whole combination.

8.3 Number of strikes to a selected location

The location of lightning strikes depends on several features and parameters. Among others, the following should be distinguished and discussed: lightning density in the region of a structure, lightning striking distance, structure environment and its topographic profile, structure dimensions and construction materials (conductive, non-conductive, inflammable, not inflammable), the kind of equipment and its sensibility or resistibility to electromagnetic impulses, and the type and routing of incoming installations.

Lightning density (or the number N_g of lightning strikes per km^2 and per year) is different for different geographic coordinates, in particular for different latitudes, and may be obtained either directly from the data of lightning location and registration systems or indirectly on the basis of isokeraunic level, which is expressed by the number T_D of thunderstorm days in a year, using the following relation valid for temperate regions:

$$N_g = aT_D^b \qquad (8.12)$$

where a and b are coefficients depending on the source of their evaluation. For example, the values $a = 0.04$ and $b = 1.25$ were accepted for previous standards and the values $a = 0.1$ and $b = 1$ are suggested presently for standardized simplifications [2].

Structure location and dimensions are responsible – together with lightning intensity and current parameters – for the number of direct lightning strikes. It is clear that structures located on the top of mountains are exposed to lightning strikes much more often than those located in valleys or between other structures, although it does not mean that a structure located between buildings is less susceptible to overvoltages than those located separately [9].

The number of direct lightning strikes N to an isolated structure on a flat earth surface depends directly on the thunderstorm intensity N_g in the considered region and on the equivalent area A_e from which the structure collects the strikes. The lightning density N_g is available, as mentioned already, either from the data on thunderstorm days or (more properly) directly from the data of lightning location and registration systems. The equivalent area A_{eD} depends on the lightning striking distance R_d, and also indirectly on the lightning current I and structure height h. Different relations for $R_d = f(I)$ and for the average horizontal distance $r_h = f(h)$ from which the lightning strikes the structure are proposed in References 10 and 11. The earth surface occupied by a structure and its surroundings to a distance of r_h is considered as an equivalent area A_{eD}. The general relation for R_d is given by

$$R_d = kI^p \qquad (8.13)$$

The factors k and p in this relation change their values – according to proposals of different researchers – withing broad limits, i.e. from 5.4 to 15.3 for factor k and from 0.65 to 0.84 for factor p. The values $k = 9.4$ and $p = 0.67$ seem to be most reliable for a limited lightning current (not greater than 50 kA) and for structures with limited height h (to $\sim$60 m) [12]. In fact, as observed for a variety of structures, on taller structures (i.e. having heights exceeding $\sim$80–100 m) the flashes of increasing number are initiated by upward progressing leaders, what is in contrast to flashes observed on lower structures, which involve downward leaders [10,12].

Equation (8.13) may also be expressed in another form, one of which is proposed as

$$R_d = 2I + 30[1 - \exp(-I/6.8)] \qquad (8.14)$$

From this relation it results that the average striking distance (corresponding to the average lightning current) is equal to ~90 m. Assuming, for simplification, that the distance R_d to the structure and to the earth surface is the same, it is possible to calculate r_h from the following relation:

$$r_h = \sqrt{[h(2R_d - h)]} \qquad (8.15)$$

For a structure with height $h = 20$ m and $R_d = 90$ m this yields $r_h = 56$ m. In reality the distance R_d to the structure is a little greater than that to the earth surface, so the real average distance $r_h > 56$ m. It may be assumed that in this case $r_h \approx 60$ m, which is also confirmed on basis of different published data [10,13,14]. The interpolation of empirical values leads to the relation for $r_h = f(h)$ or for $m = r_h/h$ as follows:

$$r_h = 13.4\,h^{0.5} \qquad (8.16)$$

$$m = 13.4\,h^{-0.5} \qquad (8.17)$$

Graphical interpretation of these relations is shown in Figure 8.3, where it is seen that for structures with a height of ~20 m, radius $r_h = 60$ m and factor $m = 3$. This value of m has been accepted as a compromised solution for risk assessment in the standards [2]. From Figure 8.3 it is clear that this value is too small for lower structures and too great for higher ones. This means that for structures lower than 20 m the number of strikes may be underestimated and for structures greater than 20 m the factor $m = 3$ leads to a certain margin of safety; when the structure height increases from 60 to ~150 m, this factor decreases further to $m \approx 2$.

The knowledge of factor m allows calculation of the equivalent collection area A_{eD} for structures of different form. Generally it consists of the area covered by the structure itself and the area surrounding it to a distance $r_h = mh$, which may be different for different parts of a structure depending on their heights. For more a complex shape of structure the surrounding area may be evaluated most easily graphically, as shown in Figure 8.4, where area is limited by boundary 4.

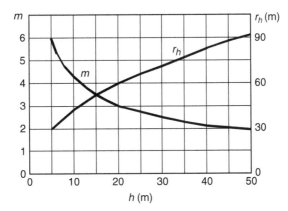

Figure 8.3   Dependence of m and r_h on the structure height h

450 Lightning Protection

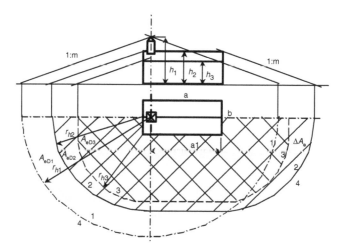

Figure 8.4 *Collection area boundaries for different parts of a structure with complex shape. 1, boundary for part with height h_1; 2, boundary for part with height h_2; 3, boundary for part with height h_3; 4, resulting boundary.*

The main area, limited by boundary 1, is defined by the intersection between the ground surface and a straight line with a slope of 1:m, which passes from the highest part of the structure (touching it there) and rotating around it and, as such, it may be obtained from the formula

$$A_{eD1} = \pi r_{h1}^2 \tag{8.18}$$

This area must be increased by an additional surface ΔA_e connected with the lower part of the structure and located between curves 1 and 2 on the right side of Figure 8.4. Boundary 2 is defined by the intersection between the ground surface and a straight line with 1:m slope, which passes from the roof ridge at height h_2 (touching it), is shifted along it, and rotates around its right end. Boundary 3 is defined by the intersection between the ground surface and a straight line with 1:m slope, which passes from the ridge at height h_3. This boundary is not taken in account because in this case the condition $0.5b + r_{h3} < r_{h2}$ is fulfilled.

Analytical summation of the component areas located along straight sections of the structure and around its corners as shown in Figure 8.4, and particularly the additional area ΔA_e, is very difficult. The following relation confirms the complexity of the analytical evaluation of the latest area

$$\Delta A_e = 0.5 \left[r_{h2} \left(\sqrt{a^2 + 5,3(r_{h2} - a_2)^2} - a \right) + a(r_{h2} - a_2) \right]$$
$$- 0.5 \left[r_{h1} \left(\sqrt{a^2 + 5,3(r_{h1} - a_1 - a_2)^2} - a \right) + a(r_{h1} - a_1 - a_2) \right] \tag{8.19}$$

where

$$a = \sqrt{2(r_{h1}^2 + r_{h2}^2) - a_1^{-2}(r_{h1}^2 - r_{ha}^2)^2 - a_1^2}$$

a_1 – as shown in Figure 8.4

$$a_2 = \frac{r_{h1}^2 - r_{h2}^2 - a_1^2}{2a_1}$$

The entire collection area A_{eD} for this structure is then the sum

$$A_{eD} = A_{eD1} + \Delta A_e \qquad (8.20)$$

To simplify the assessment of equivalent area A_{eD} for a structure with complex shape it is possible to replace (or circumscribe) it by a parallelepiped of the same length a, width b and height h_1. Such a replacement gives, of course, a certain margin of safety (the area is greater than the actual value) and allows us to apply for assessment A_{eD} a very simple expression:

$$A_{eD} = ab + 2r_{h1}(a+b) + \pi r_{h1}^2 \qquad (8.21)$$

Some complications in equivalent area assessment arise when the structure is located between others or is not in a flat area. The graphical method, as presented above (Figure 8.4), also appears to be useful in such cases, but for practical aims the influence of surroundings (structures and topographic profile) may by assessed in a simplified way by introduction of a deterministic location factor C_d. The equivalent area A_{eD} established for an isolated structure on a flat earth surface should be multiplied by this factor. Its value should be greater than 1 for structures located on a hill and somewhat less than 1 for structures located in a valley or between other near objects.

The formulae for the structure collection area calculation are derived with a simplifying but safety-increasing assumption that a structure's structural materials are conductive. It should also be noted that structures are usually equipped from ground to roof with conductive installations and their surfaces are polluted. Furthermore, during thunderstorms they are wet and conducting, so this assumption on their conductivity is close to real conditions.

The structure is also influenced (as mentioned above) by near lightning strikes, by lightning strikes to incoming lines and by lightning strikes near to the lines. The respective equivalent collection areas (A_{eW}, A_{eM} and A_{eZ}) for these cases are shown in Figure 8.5.

The equivalent collection area A_{eW} for lightning striking the incoming services (lines) depends (in a similar way to the case of structures) on the line height h_s, with length l_s taken into account (usually $l_{s\,max} \leq 1$ km) and the distance $r_{hs} = mh_s$ on both sides of the line, which gives

$$A_{eW} = 2l_s r_{hs} \qquad (8.22)$$

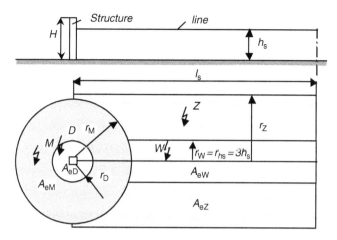

Figure 8.5 Equivalent areas of the structure, its surroundings, its incoming line and its surrounding

Equivalent areas for nearby strikes, i.e. the strikes to the ground near the structure (equivalent area A_{eM}) and the strikes to the ground surface near the incoming lines (equivalent area A_{eZ}) result from equations

$$A_{eM} = \pi r_M^2 - A_{eD} \qquad (8.23)$$
$$A_{eZ} = 2l_s r_Z - A_{eW} \qquad (8.24)$$

where A_{eD}, A_{eW} are the equivalent areas respectively for the structure and the entering line, r_M is the distance from the endangered structure equipment to the place from which the lightning may influence this equipment and r_Z is the distance from the endangered line to the place from which the lightning may influence the line.

Distances r_M and r_Z depend on the electromagnetic sensibility or resistibility of the structure equipment as well as on the kind and routing of incoming and internal installations. In the case of equipment, which is very sensitive to electromagnetic impulses, lightning strikes influence the equipment from a great distance, so the equivalent collection area may be very great and the numbers N_M or N_Z of strikes influencing the equipment during a year may be much greater than 1 [15,16].

8.4 Damage probabilities

The possible simultaneous appearance of different damages during one lightning strike resulting in the probability P in equations (8.6) to (8.10) must be a result of the combination of partial probabilities. Two kinds of probabilities are distinguished:

- The first is the probability p_i of the appearance of a lightning strike in one of four distinguished places, as shown in Figure 8.1, and additionally to different parts of the structure or to the structure and its LPS.
- The second is a probability p_{ij} of the appearance of effects (damages) due to distinguished strikes.

Probabilities p_i create a series composition and may be summarized because it has been assumed that the strikes cannot be intercepted simultaneously by more than one distinguished part. The probabilities p_{ij} appear in parallel combination, which means that their complements to the unity $(1 - p_{ij})$ form a product, which, taken away from unity, gives this combination. In this way it is possible to write [6] that

$$P = \sum_{i=1}^{n} p_i \left[1 - \prod_{j=1}^{m} (1 - p_{ij}) \right] \qquad (8.25)$$

The values p_{ij} are a measure of different damages connected with structure features and with different parameters of lightning strikes and may be determined by the product

$$p_{ij} = p_j p_{sj} \qquad (8.26)$$

where p_{sj} is the probability of the appearance of a j source of damage (e.g. a spark, a shock voltage, and so on) and p_j the probability of the appearance of a medium to be damaged or simply the damage itself due to the j source. The probability p_{sj} may be expressed by the relation

$$p_{sj} = \int_{0}^{\infty} g_j(U) P_j(U) \, dU \qquad (8.27)$$

where $g_j(U)$ is the density function of the distribution of stresses U due to source j and $P_j(U)$ is the cumulated distribution of the effects due to source j. A graphical interpretation of p_{sj} is shown in Figure 8.6.

The possibility of assessment of p_{sj} and p_j values is very limited and usually they must be determined in arbitrary way, but for certain cases their evaluation is possible. For example, this is the case for equipment damage due to overvoltages. Knowing the

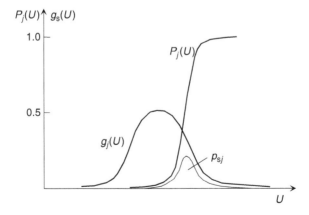

Figure 8.6 Graphical interpretation of interpretation of the damage probability p_{sj} using the continuous function of damage distribution

density function $g_j(U)$ of the overvoltage distribution and the cumulated distribution $P_j(U)$ of damages due to overvoltages it is possible to express the probability p_{sj} by equation (8.27). The function $g_j(U)$ depends on lightning current parameters. Its log-normal form is given by the relation

$$g_j(U) = \frac{1}{\sqrt{2\pi}U\sigma_U} \exp\left[-\frac{(\ln U - \ln U_m)^2}{2\sigma_U^2}\right] \quad (8.28)$$

where U is the lightning overvoltage, U_m its average (expected) value and σ_U the standard deviation of the overvoltage distribution.

The distribution $P_j(U)$ results from the withstand insulation characteristic of the equipment and is approximated by normal one as follows:

$$P_j(U) = \frac{1}{\sqrt{2\pi}\sigma_{Ub}} \int_0^{U_s} \exp\left[-\frac{(U - U_{50\%})^2}{2\sigma_{Ub}^2}\right] dU \quad (8.29)$$

where U is the stressing voltage, $U_{50\%}$ the average value of the breakdown voltage, U_s the voltage corresponding to the selected value of breakdown probability and σ_{Ub} the standard deviation of the breakdown voltage distribution.

For reasons of simplification it may be assumed [in relation to equation (8.29)] that $\sigma_{Ub} = 0$. The considered probability takes the following values:

$$P_j(U > U_{50\%}) = 1 \quad (8.30)$$
$$P_j(U \leq U_{50\%}) = 0 \quad (8.31)$$

This is graphically explained in Figure 8.7.

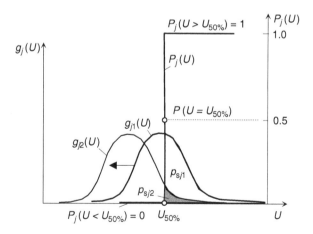

Figure 8.7   Graphical interpretation of overvoltages causing damages with probabilities p_{sj1} and p_{sj2} shown by shaded surfaces in the case of stepped damage distribution function $P_j(U)$

The average value $U_{50\%}$ depends on the features of the individual equipment and is decisive for the damage. In the case of very sensitive equipment it may be damaged at relatively low overvoltages, because the condition $P_j(U > U_{50\%}) = 1$ is usually fulfilled. To avoid the damage, the expected overvoltages must be effectively reduced, which, for example, means that their distribution curve needs to be shifted to the left side as is shown in Figure 8.7. The shaded surface p_{sj1} has been reduced to the dimensions of the surface p_{sj2}.

Thanks to the condition in equation (8.30), the damage probability expressed by equation (8.27) may be significantly simplified as follows:

$$p_{sj} = \int_0^\infty g_j(U) P_j(U > U_{50\%}) \, dU = \int_{U_{50\%}}^\infty g_j(U) \, dU = G(U > U_{50\%}) \quad (8.32)$$

This means that the assessment of damage probability p_{sj} needs only to integrate the distribution density function of stresses in the limits from $U_{50\%}$ of breakdown voltage to infinity. The result is illustrated in Figure 8.7 by shadowed surfaces p_{sj1} and p_{sj2} for two cases – before the reduction of stresses (overvoltages) and after it.

The considerations as given above may be extended to some other kinds of stresses, generally marked by symbol Z instead of symbol U. This extension needs only to insert into equation (8.32) the general symbol Z or other relevant symbol, yielding

$$p_{sj} = \int_{Z_{50\%}}^\infty g_j(Z) \, dZ = G(Z > Z_{50\%}) \quad (8.33)$$

where the average value $Z_{50\%}$ belongs to the cumulative distribution of damages due to relevant stress.

In this way the probabilities of inflammable material ignition p_{s1}, mechanical damage occurrence p_{s2}, metal sheet perforation p_{s3}, melting of metal in a dangerous amount from the conductor p_{s4}, explosive medium ignition p_{s5}, conductor glowing p_{s6}, and so on may be evaluated. Probability p_{s1} depends on the lightning specific energy distribution and on the average of its value causing the ignition. Probability p_{s2} depends on the lightning current distribution and on the average of its value causing the mechanical damage. Probability p_{s3} depends on the lightning charge distribution and on average its value causing sheet perforation. Probability p_{s4} depends on the lightning charge distribution and on average its value causing the dangerous melting of the conductor. Probability p_{s5} depends on the lightning specific energy distribution and on the average of its value causing the explosive medium ignition. Probability p_{s6} depends on the lightning specific energy distribution and on the average of its value causing the glowing of the dangerous conductor [17–20]. Example results of the evaluation are given in Table 8.1.

There are also cases in which the hazard depends not only on one parameter but on more of them. A good example involving two parameter concerns is the shock hazard caused by the effects of lightning current [6] as it is shown in Figure 8.8.

Table 8.1 Examples of damage probability evaluation

Probability symbol	Hazardous parameter	Type of parameter distribution	Average dangerous value	Probability value
p_{s1}	W	log-normal	$W_{50} = 1.3 \times 10^4 \text{ A}^2 \text{ s}$	$p_{s1} = 0.68$
p_{s2}	I	log-normal	$I_{50} = 30 \text{ kA}$	$p_{s2} = 0.26$
p_{s3}	Q	log-normal	$Q_{50} = 32 \text{ A s}$	$p_{s3} = 0.19$
p_{s4}	Q	log-normal	$Q_{50} = 57 \text{ A s}$	$p_{s4} = 0.14$
p_{s5}	W	log-normal	$W_{50} \approx 1 \times 10^{-3} \text{ A}^2 \text{ s}$	$p_{s5} = 1$
p_{s6}	W	log-normal	$W_{50} = 6.2 \times 10^6 \text{ A}^2 \text{ s}$	$p_{s6} = 0.005$

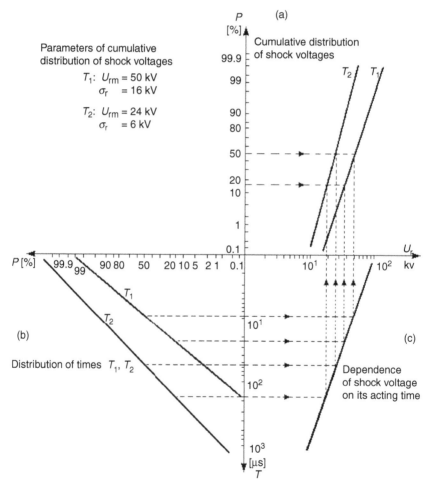

Figure 8.8 Construction of cumulative distribution of shock voltages with lightning current time parameters

The two parameters result from the lightning current wave shape, i.e. from its front and tail times. The induced touch voltages are connected with the current front time, and the step and touch voltages, as voltage drops, are mainly connected with the current tail. The relation between the shock voltage–time characteristics and the lightning current wave shape parameters is given in Figure 8.8. It enables the selection of the average dangerous value of shock voltages U_r both for induced voltages and for voltage drops.

8.5 Simplified practical approach to damage probability

The method of lightning damage probability evaluation as presented in Section 8.4 seems to be in large part sufficiently accurate, but in practice it is much too complicated and even impossible for application, because there is a lack of credible experimental and statistical data for its application to individual cases. In reality, every structure has its own hazard conditions and it would be very difficult to establish average dangerous values for hazardous lightning parameters and to formulate relevant structure conditions and features. This is the main reason why the approach to probabilities p_{ij}, p_j and p_{sj} must be, for practical purposes, significantly simplified. It may be done, for instance, in two steps. The first classifies the damages and the second replaces the probability of damage by factors taking into account the efficiency of protection measures [21].

Damages may be classified into physical damage to the structure, shock to living beings and failure of electrical (electronic) equipment, as well as according to the lightning strike location, which may be assessed by the probability p_i, which is a measure of lightning strike to the structure, near it, to the incoming line, near it and even to different parts of the structure or to its LPS.

The replacement of damage probability p_{ij} by a special parameter representing simultaneously the efficiency of the relevant protection measure is a great simplification, but certainly introduces to the assessment a margin of safety. The concept of such a solution is based on the assumption that every lightning strike may damage an unprotected structure with the natural probability $P_n = 1$, and that the lower value of risk R_X of the damage depends on the annual number N_X of lightning events and on the relative loss value L_X, where X is a general index of the kind of damage or a partial risk. When the resulting value of R_X indicates the need of protection and the relevant protection measure is applied, then the natural probability P_{Xn} is reduced by the factor K_X resulting from the efficiency of the applied protection measures. It allows us to write that

$$P_X = P_{Xn} K_X \qquad (8.34)$$

Factor K_X may reduce the probability P_{Xn} of one kind of damage or may be common for a group of selected damages, and may represent the efficiency of the relevant protection measure or may consist of partial reduction factors K_{si} representing the different protection measures influencing the considered damage. In this case the resulting factor K_s is the product of partial factors and yields

$$K_s = \prod_{i=1}^{k} K_{si} \qquad (8.35)$$

458 Lightning Protection

The number of partial factors K_{si} in their product depends on the number of protection measures involved.

There is no general relation between reduction factor K_X and the factors K_{si} or K_s. In some cases they may be equal but in others K_X is a function of K_s, which may be generally expressed by relation

$$K_X = f(K_s) \tag{8.36}$$

The reason for this is that these factors may be influenced by additional parameters as for instance by lightning current parameters or by their critical values.

8.6 Question of relative loss assessment

The third factor L involved in the product NPL of risk assessment [see equation (8.11)] is a relative value of loss or, in other words, a measurable effect of lightning damages. Together with N and P it creates also the relative value of the risk of damage.

The effects of lightning hazard in a structure may be divided into three basic groups of damages causing different losses:

- shock to living beings (people and animals), which may cause human and economic losses
- physical damage to a structure (fire, explosion, mechanical damage), which may cause human, economic, and cultural losses and loss of services
- failure of electrical and electronic systems, which may cause human and economic loss and loss of services

This means that symbol L may represent not only different types of relative losses, but also different values. However, the approach to their evaluation may be similar or even common, and there are two possibilities:

- to evaluate them on the basis of real relative amounts of probably injured people and damaged goods within a relative time of their exposure to lightning effects
- to evaluate them on the basis of default values of the relative amount of loss (human and physical) corresponding to particular types of damages

According to the first approach the loss evaluation may be performed by the relation

$$L = (n_p/n_t)(t_p/8\,760) \tag{8.37}$$

where parameter n_p is the number of probable injured persons or the mean value of possible loss of goods, n_t is the total number of persons in the structure or total value of the structure itself, and t_p is the time (in hours per year) of a person's presence in a dangerous place (for endangered goods $t_p = 8\,760$). However, for individual cases the evaluation of these parameters, and especially the parameter n_p, may be very difficult and even impossible.

According to the second approach the expected loss for every one of the three basic groups of damages as mentioned above must be evaluated with the assumption that the individual losses in the frame of the damage group have the same value. For instance, the losses resulting from failures of electrical and electronic internal systems due to strikes in different parts of an establishment are the same, but they may be different from those for the damage group. The losses for the damage group may be greater, so special reduction factors must be selected in an arbitrary way. The damage group losses and reduction factors depend primarily on the type of structure and its features.

It should be noted that considerations involving the loss are intended to indicate the way in which they should be involved in the risk assessment. The values of loss may differ significantly depending on the local conditions, so the values given in this chapter should be treated as one of many possible examples.

8.7 Concept of risk components

A proper risk evaluation is possible provided that two kinds of damage are distinguished: damages caused by one lightning strike and damages caused by lightning striking different parts of the structure, its surroundings and connected lines. The probabilities of different damages appearing as a result of one lightning strike create parallel composition and can be combined according to the principle given for probability p_{ij} in the brackets of equation (8.25). It means that the product NPL containing probabilities of different damages should be written as follows:

$$NPL = NL[1 - (1 - P_1)(1 - P_2)(1 - P_3) \cdots] = NL\left[1 - \prod_{X=1}^{n}(1 - P_X)\right] \quad (8.38)$$

where P_1, P_2 and P_3 are the probabilities of parallel damages.

The probabilities of damages appearing due to separate lightning strikes (collected by different parts of a structure, its surroundings and connected lines) create, together with the number of these strikes and the respective losses, the products, which may be combined in series, summarized. In this case the resulting product NPL may be written in the form

$$NPL = N_1 P_1 L_1 + N_2 P_2 L_2 + N_3 P_3 L_3 + \cdots = \sum_{X=1}^{n} N_X P_X L_X \quad (8.39)$$

where X is the number of the successive product $N_X P_X L_X$ taken into account, n is the entire number of products, N_X is the number of strikes collected by a selected part of the structure, its surroundings and the connected lines, P_X is the probability of damage due to one of the strikes N_X, and L_X is the loss connected with this damage.

This shows that in the case of strikes collected by different parts of a structure, its surroundings and connected lines, the risk may be evaluated from the relation

$$R \approx NPL = \sum_{X=1}^{n} N_X P_X L_X \quad (8.40)$$

This means that every distinguished product $N_X P_X L_X$ may contribute to the damage or failure of the same internal piece of structure equipment and may create a separate risk component

$$R_X \approx N_X P_X L_X \tag{8.41}$$

When n components exist, then the risk may be expressed as the following sum of these components:

$$R \approx \sum_{X=1}^{n} R_X \tag{8.42}$$

Quite a different situation is the case when several damages or failures are caused by one strike. Then, according to equation (8.38), it is possible to write

$$R \approx NL \left[1 - \prod_{X=1}^{n} (1 - P_X) \right] \tag{8.43}$$

If equation (8.43) is developed one can obtain the expression

$$R \approx NL \left(\sum_{X=1}^{n} P_X - \Delta P_X \right) \tag{8.44}$$

where ΔP_X is a sum of products of the minimum two probabilities P_X. Because the probabilities $P_X < 1$, their products are much less than the sum of P_X and may be neglected. In this way equation (8.44) may be written as

$$R \approx NL \sum_{X=1}^{n} P_X \tag{8.45}$$

Every product NLP_X in this expression may be treated as the risk component of different damages appearing at the same strike. When n components exist, then the risk in this case may be approximated by the same expression

$$R \approx \sum_{X=1}^{n} R_X \tag{8.46}$$

The difference between equation (8.46) and equation (8.42) lies mainly in the number of strikes. In the first, the number of strikes is the same (i.e. the number collected by the structure) for every risk component; in the second, the number of strikes differs for every component. For example, in the second case they are collected by the structure and the incoming line, as well as by their surroundings.

The variety of possible lightning damages and losses, together with their consideration for different aspects of protection, forces the creation of different risk components and different groups of compositions (sums).

8.8 Standardized procedure of risk assessment

8.8.1 Basic relations

The procedure presented in this section refers to the standard issued in 2006 and it may be modified in the future because the standard is periodically revised. The following considerations intend to indicate the way in which the risk can be evaluated according to this standard. The values of factors involved in the calculation are used as an example and can be changed according to local conditions.

The term 'risk' may be used in the sense of the probability with which the damage should be expected, as discussed in the preceding sections. For practical purposes this term has been replaced in the recent standard [2] by the related value of losses resulting from lightning damage and expressed by the product NPL, so the expression representing the risk can be simplified to the form

$$R \approx NPL = \sum_{X=A}^{Z} N_X P_X L_X = \sum_{X=A}^{Z} R_X \qquad (8.47)$$

where N is the annual number of dangerous events (lightning flashes) influencing a structure or its equipment, P is the probability of damage to the structure or its equipment due to one event, L is the consequent loss due to a damage relative to the total value of humans and goods of the object to be protected, N_X, P_X and L_X are the values of N, P and L selected for the risk distinguished component, R_X is the common symbol for the risk component, and X, A and Z are the symbols for the common respective risk components.

As was shown in the preceding sections, this approach includes significant simplification. The risk and its components are considered as probable annual losses (humans and goods) related to the total value (humans and goods) of the object to be protected.

The number N of flashes influencing a structure contains different kinds of lightning operation or (as was established in the standard [2]) different kinds of damage sources. According to this approach, four types of flashes (as shown in Figure 8.9) should be distinguished depending on the attachment point related to the structure and incoming services:

- flashes to a structure or its LPS (standardized damage source S1)
- flashes near a structure (standardized damage source S2)
- flashes to an incoming service (standardized damage source S3)
- flashes near an incoming service (standardized damage source S4)

Depending on the attachment point of lightning strikes as well as on the lightning current parameters and on the characteristics of the structure and its contents, different kinds of damage may appear, but it is possible to group them into three basic types:

- injury to living beings (standardized as type D1)
- different kinds of physical damages (standardized as type D2)
- failure of electrical and electronic systems (standardized as type D3)

462 Lightning Protection

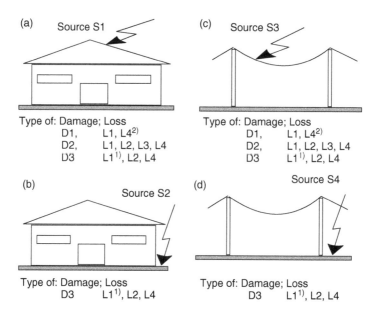

Figure 8.9 Sources of damage (S1, S2, S3, S4), types of damages (D1, D2, D3) and loss (L1, L2, L3, L4) depending on the strike point; 1) for structures where internal system failure immediately endanger human life; 2) for properties where animals may be lost

Selected or combined damages (of a structure or its contents) with their consequences result in specific loss, which may be qualified into four types:

- loss of human life (standardized as type L1)
- loss of service to the public (standardized as type L2)
- loss of cultural heritage (standardized as type L3)
- loss of economic value (standardized as type L4)

Damages or losses may be limited to a part of the structure, may extend to the entire structure or may also involve surrounding structures or its environment. This fact should be taken into account with the consequences weighted for risk limitation.

Lightning flashes may cause different types of damage and losses. Each may be assessed by means of selected risk components R_X (from R_A to R_Z) concerning the successive damage due to one strike. An adequate combination of components gives the whole resulting risk R. Taking into account the simplifying assumption, the combination in the case under consideration has been expressed by the sum shown in equation (8.47).

As mentioned above, the variety of possible lightning damages and losses, together with the tendency to their consideration for different aspects of protection, forces the creation of different risk components and different groups of compositions (sums). Following this tendency, one can describe the group of components depending on the following:

- the location of attachment points (damage sources) about the group: R_{S1}, R_{S2}, R_{S3} and R_{S4}
- the type of damages about the group: R_{D1}, R_{D2} and R_{D3}
- the type of losses about the group: R_{L1}, R_{L2}, R_{L3} and R_{L4}

All of these groups are created by the sum of the relevant components. In order to take into account all possible damages to a structure and its contents, the following risk components should be distinguished [2].

- R_A is the component related to the injury of living beings due to touch and step voltages caused by direct flashes.
- R_B is the component related to physical damage (mechanical, thermal, explosive) due to direct lightning current or sparking.
- R_C is the component related to the failure of electrical or electronic equipment, with all the possible consequences of the failure, caused by overvoltages due to resistive and inductive coupling with lightning current due to direct flashes.
- R_M is the component related to the failure of electrical or electronic equipment, with all the possible consequences of the failure, caused by overvoltages due to resistive and inductive coupling with lightning current due to nearby flashes.
- R_U is the component related to the injury of living beings due to touch voltages caused inside the structure by flashes to the incoming services.
- R_V is the component related to physical damages (mechanical, thermal, explosive) due to partial lightning currents or sparking inside the structure due to flashes to incoming services.
- R_W is the component related to the failure of electrical or electronic equipment, with all the possible consequences of the failure, caused by overvoltages on incoming lines from direct flashes to these lines.
- R_Z is the component related to the failure of electrical or electronic equipment, with all the possible consequences of the failure, caused by overvoltages induced on incoming lines from nearby flashes.

The compositions of the relevant components in all the distinguished groups are presented in Table 8.2. The symbols involved in Table 8.2 allow the selection of the relevant components for assessment of adequate group of components or, in other words, the partial risks of structure damage. They are as follows:

$$R_{S1} = R_A + R_B + R_C \qquad (8.48)$$
$$R_{S2} = R_M \qquad (8.49)$$
$$R_{S3} = R_U + R_V + R_W \qquad (8.50)$$
$$R_{S4} = R_Z \qquad (8.51)$$

464 Lightning Protection

Table 8.2 Composition of the relevant risk components in the distinguished groups

Group of components	Component							
	R_A	R_B	R_C	R_M	R_U	R_V	R_W	R_Z
R_{S1}	×	×	×	0	0	0	0	0
R_{S2}	0	0	0	×	0	0	0	0
R_{S3}	0	0	0	0	×	×	×	0
R_{S4}	0	0	0	0	0	0	0	×
R_{D1}	×	0	0	0	×	0	0	0
R_{D2}	0	×	0	0	0	×	0	0
R_{D3}	0	0	×	×	0	0	×	×
R_{L1}	×	×	*	*	×	×	*	*
R_{L2}	0	×	×	×	0	×	×	×
R_{L3}	0	×	0	0	0	×	0	0
R_{L4}	**	×	×	×	**	×	×	×

Symbols: ×, component contributing to the sum; 0, component not applicable; *, component applicable in the case of life hazard and explosion; **, component applicable in the case of hazard for animals.

$$R_{D1} = R_A + R_U \tag{8.52}$$

$$R_{D2} = R_B + R_V \tag{8.53}$$

$$R_{D3} = R_C + R_M + R_W + R_Z \tag{8.54}$$

$$R_{L1} = R_A + R_B + R_C^* + R_M^* + R_U + R_V + R_W^* + R_Z^* \tag{8.55}$$

$$R_{L2} = R_B + R_C + R_M + R_V + R_W + R_Z \tag{8.56}$$

$$R_{L3} = R_B + R_V \tag{8.57}$$

$$R_{L4} = R_A^{**} + R_B + R_C + R_M + R_U 7^{**} + R_V + R_W + R_Z \tag{8.58}$$

8.8.2 Evaluation of risk components

All the selected components may be assessed according to the principles of entire risk evaluation expressed by equation (8.47). According to this principle, the general relation is applicable to every component as follows

$$R_X = N_X P_X L_X \tag{8.59}$$

where R_X is the distinguished risk component (R_A, R_B, R_C, R_M, R_U, R_V, R_W, R_Z), N_X is the number of flashes corresponding to the selected risk components, P_X is the probability of damage appearing in a structure due to one lightning flash and selected risk component, and L_X is the consequential loss due to the damage corresponding with the selected risk component.

It should be noted that according to the arrangement of Figure 8.9 and equations (8.48) to (8.51) only four different groups of number N_X of flashes should be taken into account. These are

$$N_{S1} = N_A = N_B = N_C \tag{8.60}$$

$$N_{S2} = N_M \tag{8.61}$$
$$N_{S3} = N_U = N_V = N_W \tag{8.62}$$
$$N_{S4} = N_Z \tag{8.63}$$

These numbers depend on local thunderstorm activity, on structure dimensions and on its exposure to flashes relative to the surroundings. This dependence may be expressed by the following formulae [2]:

$$N_{S1} = N_g A_{S1} C_S \times 10^{-6} \tag{8.64}$$
$$N_{S2} = N_g A_{S2} C_S \times 10^{-6} \tag{8.65}$$
$$N_{S3} = N_g A_{S3} C_S \times 10^{-6} \tag{8.66}$$
$$N_{S4} = N_g A_{S4} C_S \times 10^{-6} \tag{8.67}$$

where N_g is the annual local density of flashes (1 km^{-2} yr^{-1}), A_{S1}, A_{S2}, A_{S3} and A_{S4} are the equivalent areas (m^2) for adequate flash interception by the structure, its surroundings, incoming services and their surroundings, and C_S is the coefficient of structure exposure related to the surroundings (topography of the place and influence of other structures and trees).

The annual local density N_g may be established on the basis of data from lightning location system registrations or may be calculated on the basis of isokeraunic level (thunderstorm days T_D). Standardized relations between N_g and T_D for a temperate climate are

$$N_g = 0.04 T_D^{1.25} \tag{8.68}$$

or more simply [2]

$$N_g \approx 0.1 T_D \tag{8.69}$$

The equivalent areas A_{Si} depend, as shown in Figure 8.5, on the structure and the incoming service dimensions. These areas depend on the damage sources (S1 to S4) and may be calculated as follows:

- for a structure with a height H, horizontal rectangular surface S_s and structure perimeter L_s,

$$A_{S1} = S_s + 3HL_s + 9\pi H^2 \tag{8.70}$$

- for structure surroundings with a maximum distance radius r_{S2}

$$A_{S2} = \pi r_{S2}^2 - A_{S1} \tag{8.71}$$

- for incoming overhead services of height h and length l_s (underground services are assumed not to be endangered)

$$A_{S3} = 6hl_s \tag{8.72}$$

- for service surroundings (assuming that neighbouring structures and trees are not present) with a maximum distance radius r_{S4}

$$A_{S4} = 2r_{S4}l_s - A_{S3} \quad (8.73)$$

Coefficient C_S changes from 0.25 for structures located in valleys to 2 for structures elevated on hills. For isolated structures (not shielded by other structures and trees) in flat regions, $C_S = 1$.

Values of the probability P_X related to the individual risk components R_X may be expressed generally by the product

$$P_X = P_{Xn} K_X \quad (8.74)$$

where P_{Xn} is the natural (without protection measures) value of the probability P_X of the damage and K_X is the combined risk reduction factor, which is an effect of the protection measures influencing a selected risk component R_X.

It may be assumed, as shown in Section 8.5, that in every case of an unprotected structure the natural probability $P_{Xn} = 1$, which gives a distinct margin of safety. Every protection measure [1–4] is qualified by its individual factor $K_{si(i=1,2,\ldots,k)}$, which reduces the natural probability P_{Xn} of damage related to a selected risk component R_X. A combined factor K_X of risk reduction is a function of the product K_s of individual factors K_{si}, representing different protection measures, which may be expressed by the relation

$$K_X = f(K_s) = f\left(\prod_{i=1}^{k} K_{si}\right) \quad (8.75)$$

The number of K_{si} factors in the product depends on the number of protection measures involved. In a prevailing number of cases the reduction of the risk component R_X needs only one protection measure, but for the risk component R_M, several measures may be required. Typical values of K_X are specified in Table 8.3.

As can be seen from Table 8.3, the values of factors K_A, K_B, K_C, K_U, K_V, K_W and K_Z are directly related to the efficiency of the relevant protection measures, whereas the value of factor K_M is related to the product K_S of four individual factors as follows

$$K_s = K_{s1} K_{s2} K_{s3} K_{s4} \quad (8.76)$$

The factors K_{s1}, K_{s2}, K_{s3} and K_{s4} of the product represent the efficiency of the external spatial screen of the structure, the efficiency of the internal spatial screens of structure rooms, the characteristics of internal wiring (cable shielding and routing precautions) and the impulse withstand voltage of the protected device, respectively.

The factors K_{s1} and K_{s2} depend on the distance $w \leq 5$ m between the screen conductors and K_{s4} on the impulse withstand voltage U_w [22]. They may be evaluated from the following:

$$K_{s1} = K_{s2} = 0.12\,w \quad (8.77)$$
$$K_{s4} = 1.5/U_w \quad (8.78)$$

Table 8.3 Values of risk reduction factors K_X

Reduction factors of selected risk components	Limit values of factors	
	From	To
K_C	0.03^1	0.001^1
$K_M = f(K_s)$	0.9^2	0.0001^2
$K_W{}^4$	0.95^3	0.02^3
$K_Z{}^6$	0.5^5	0.002^5

[1] Value depends on lightning protection level and characteristics of well coordinated SPDs. [2] Value depends on screening efficiency, wiring characteristics and impulse withstand voltage U_w of protected devices. [3] Value depends on the cable screen resistance R_S and impulse withstand voltage U_w of protected devices. [4] In the case of coordinated SPD application, the lower value of K_V, K_W and K_C values should be chosen. [5] Value depends on the cable screen application, its bonding to the bonding bar, its resistance R_S and impulse withstand voltage U_w of protected devices. [6] In the case of coordinated SPD application, the lower value of K_Z and K_C values should be chosen.

Factor K_{S3} depends on the routing precautions (loop dimensions) and the resistance R_S of cable shields. Its values are placed within the limits from 0.2 to 0.0001 depending on the loop dimensions and cable screen resistances.

The values of reduction factors K_X and K_{Si} for individual cases may be controlled according to standard risk management [2].

The risk component R_X expressed by equation (8.59) depends also on the relative value L_X of losses resulting from lightning damages. The components influencing the four types of distinguished losses (L1 to L4) may be identified by the data of Table 8.4 and equations (8.55) to (8.58) as well as equations (8.80) to (8.82).

The relative values of losses L_X for individual risk components R_X may be evaluated either on the basis of real relative amounts of people and goods within a relative time of their exposure to lightning effects or on the base of default values of relative amounts of losses corresponding with distinguished types of damages (D1, D2, D3) [2]. In the first case the evaluation of losses is performed by means of the relation

$$L_X = (n_p/n_t)(t_p/8\,760) \qquad (8.79)$$

where n_p is the number of endangered persons or mean value of possible loss of goods, n_t is the total number of persons in the structure or total value of the structure itself, t_p is the time (in hours per year) of a person's presence in a dangerous place (for goods $t_p = 8\,760$).

In practice the determination of n_p, n_t and t_p may appear to be uncertain or difficult, so the second assessment process is adopted. In this way the losses L_X related to individual risk components R_X may be established by means of the following:

$$L_A = L_U = r_A\,L_{D1} \qquad (8.80)$$

Table 8.4 Limits of typical values for losses assessment

Loss or factor		Type of structure, soil surface, hazard, precautions and conditions
Symbol	Value	
L_{D1}	1×10^{-4}	All types of structure – persons outside it
	1×10^{-2}	All types of structure – persons outside it
L_{D2}	1×10^{-1}	Hospitals, hotels
	5×10^{-2}	Schools, commercial structures
	2×10^{-2}	Churches, public entertainments
	1×10^{-2}	Other structures
L_{D3}	1×10^{-1}	Structures endangered by explosion, structures like hospitals
	1×10^{-3}	
r_A	1×10^{-2}	Agricultural soil surface, concrete
	1×10^{-3}	Marble, ceramic
	1×10^{-4}	Gravel, carpets
	1×10^{-5}	Asphalt, linoleum, wood
r_B^1 $\quad r_{B1}$	0.5	Manual: extinguishers, alarm install
	0.2	Automatic: extinguishers, alarm install
$\quad r_{B2}$	1	Hazard of explosion
	1×10^{-1} to 1×10^{-3}	Different levels of fire hazard: High, ordinary, low
$\quad r_{B3}$	50	Contamination of surroundings
	20	Hazard for surroundings
	1×10^{-2}	Level of panic: high, average, low

$^1 r_B = r_{B1}\, r_{B2}\, r_{B3}.$

$$L_B = L_V = r_B L_{D2} \qquad (8.81)$$
$$L_C = L_M = L_W = L_Z = L_{D3} \qquad (8.82)$$

where L_{D1} is the loss due to injury by touch and step voltages, L_{D2} is the loss due to physical damage, L_{D3} is the loss due to failure of internal systems, r_A is the reduction factor related to the loss of human life outside or inside the structure, and r_B is the combined reduction/increase factor related to the losses due to physical damages depending on the risk of fire, precautions against its consequences and conditions causing its increase.

Losses L_{D1}, L_{D2} and L_{D3} as well as reduction factors r_A and r_B depend primarily on the type of structure and its features [2]. Their typical mean values are contained in the limits shown in Table 8.4.

8.8.3 Risk reduction criteria

Knowing all the elements of (8.59) it is possible to evaluate the relevant risk components R_X and to combine them (according to (8.47)) into resulting risk R. Its value allows us to make a decision on the requirement for protection and to select adequate protective measures. In both cases the value of the resulting risk should be compared with that of a tolerable one, according to the procedure explained in Figure 8.10.

Two typical values of tolerable risk R_T have been distinguished [2]:

$$R_T = 1 \times 10^{-5} \tag{8.83}$$

$$R_T = 1 \times 10^{-3} \tag{8.84}$$

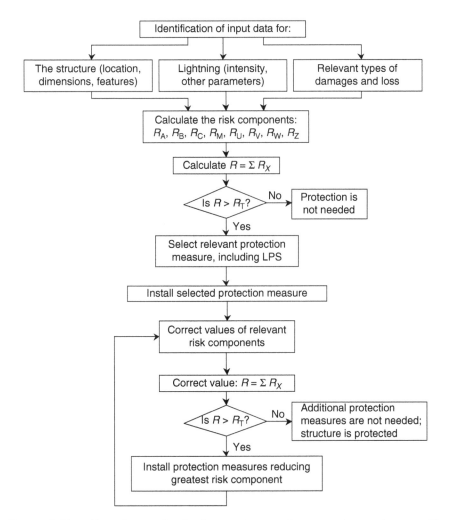

Figure 8.10 Algorithm to decide the requirement for protection and to provide protection measures of required efficiency

470 Lightning Protection

The first value is applicable when the loss of a person's life is considered, whereas the second has to be assumed for every other case.

Any calculated value of the risk R, when greater than tolerable, indicates the need for structure protection or a requirement for an improvement in its efficiency by application of additional protection measures. Such measures should be selected in a certain sequence. The effect of the first and every subsequent measure should be controlled by a risk calculation. New protection measures should reduce the risk component with greatest the value.

To indicate the requirement for the application of a protection measure, the following steps of the procedure should be accepted:

- identification of a structure with its location, dimensions and features
- identification of thunderstorm activity and lightning parameters for the location of the structure
- identification of the relevant types of possible damage and loss as well as the tolerable risk R_T

On the basis of these data the number N_X of flashes influencing the structure with equipment, the probability P_X of damage and the relevant losses L_X should be established and then the risk components R_X and their sum, as the entire risk R, should be calculated.

The risk R should be compared with its tolerable value R_T and, where it is greater than this value, suitable protection measures should be selected and installed. Different types of LPS should be first taken into account.

Proper selection of protection measures should be confirmed by establishing the results of repeating the risk component and their sum calculation. In this case, when the sum is greater than the value of tolerable risk, an additional measure of protection should be selected and installed. The selection should be such that it acts to reduce the greatest risk components [21–24].

In the standards, risk is put into different categories according to the type of loss:

- the risk of loss of human life (standardized as type R1)
- the risk of loss of service to the public (standardized as type R2)
- the risk of loss of cultural heritage (standardized as type R3)
- the risk of loss of economic value (standardized as type R4)

To calculate these the procedure and algorithm shown in Figure 8.10 should be applied and protection is achieved when each category of risk is reduced to a tolerable value.

8.9 Meaning of subsequent strokes in a flash

The present calculations of risk of damage due to lightning overvoltages are based on the assumption that the risk components depend mainly on the current peak value of the first lightning stroke and on the current rate of rise of subsequent strokes [2]. The number of flashes without the number of subsequent strokes in a flash is taken into account. Such simplifications are justifiable for practical reasons, but are not acceptable from the scientific perspective, for which more detailed lightning data should be taken in account. From this angle it would be interesting to know the significance of the discrepancy between lightning strokes and lightning flashes for risk level [25].

In this regard, the cumulative level $N_X P_X$ of risk component or damage frequency F_X [15] should be considered in dependence on the polarity and parameters of the flashes and subsequent strokes. The consideration starts with the formula

$$F_X = N_X P_X \qquad (8.85)$$

The number N_X of strokes is considered assuming that ~ 10 per cent of lightning flashes are positive and ~ 90 per cent are negative [1,26,27,29]. Moreover, the negative flashes may contain subsequent strokes, so that the entire number N_X of strokes (consisting of N_{Xp}, positive, and N_{Xn}, negative first; then N_{Xs}, negative subsequent strokes) is greater than the number N_{Xf} of flashes. Subsequent strokes N_{Xs} are of a random nature and their average number n_m (according to References 27 and 28) may be considered in the range 2.4–4.5. Assuming that $n_m = 4$, the entire number of expected strokes may be evaluated as

$$N_X = N_{Xp} + N_{Xn} + N_{Xs} = 0.1 N_{Xf} + 0.9 N_{Xf} + 0.9 \times 4 N_{Xf} = 4.6 N_{Xf} \qquad (8.86)$$

The share of positive, negative and subsequent strokes in a lightning event may be defined by means of successive coefficients η_p, η_n and η_s [26] with the following condition

$$\eta_p + \eta_n + \eta_s = 1 \qquad (8.87)$$

or in another form,

$$\frac{0.1 N_f}{4/6 N_f} + \frac{0.9 N_f}{4.6 N_f} + \frac{3.6 N_f}{4.6 N_f} = 0.022 + 0.196 + 0.782 = 1 \qquad (8.88)$$

This shows that the subsequent strokes of a negative flash play the most important role in the danger to structure equipment. It is easy to state that the share of subsequent strokes is 3.6 times greater than that of first strokes (positive and negative). This is most important for the D3 type of damage connected with overvoltages.

Probability P_X of the damage due to lightning overvoltages depends on the equipment withstand level and on the lightning current parameters, especially the lightning current amplitude and its front steepness. The current of subsequent strokes, due to the greater front steepness in relation to the first stroke, also causes an increase in the probability. It is therefore clear that the subsequent strokes may be decisive for increment of the risk value.

8.10 Final remarks and conclusions

- The idea for risk calculation is based on the correlation of different kinds of possible lightning stresses with the cumulated distribution of probable damages of the structure and its equipment due to these stresses.

- For risk calculation there are four sources of damages, according to lightning locations, three types of damages and four types of loss.
- Depending on the possible damages, such as injuries to living beings, physical damage and failure of equipment, different risk components have been distinguished and combined into the groups related to individual damage sources, types of damages and losses.
- Every risk component is defined as the product of the number N_X of flashes (strokes), probability P_X of damages and possible loss L_X.
- The damage to an unprotected structure is assumed to be certain ($P_{Xn} = 1$) and protection measures reduce the probability P_{Xn} by a factor K_X resulting from the protection measure's efficiency.
- A requirement for a lightning protection system (LPS) and any other protection measure should be decided on the basis of the special algorithm shown in Figure 8.10.
- The risk evaluated on the basis of lightning flashes and lightning strokes may be different, and this fact should be taken into account.
- The improvement of the accuracy of risk assessment is connected with the progress of results of statistical evaluation of the new data on lightning current parameters and influencing factors.

References

1. International Standard IEC 62305-1. 2006. Protection against lightning. Part 1 – General principles.
2. International Standard IEC 62305-2. 2006. Protection against lightning. Part 2 – Risk management for structures and services.
3. International Standard IEC 62305-3. 2006. Protection against lightning. Part 3 – Physical damage to structures and life hazard.
4. International Standard IEC 62305-4. 2006. Protection against lightning. Part 4 – Electrical and electronic systems within structures.
5. Rakov V.A. and Uman M.A. *Lightning: Physics and Effects*. Cambridge: Cambridge University Press; 2003.
6. Flisowski Z. 'Analysis of lightning hazard to the building structures (in Polish)'. Wydawnictwa Politechniki Warszawskiej. Warszawa 1980, Prace Naukowe Elektryka Z. 63, str. 1–151.
7. Flisowski Z. and Mazzetti C. 'A new approach to the complex assessment of the lightning hazard over buildings'. *Bulletin of the Polish Academy of Sciences*. 1984;**32**(9–10):571–581.
8. Flisowski Z. and Mazzetti C. 'Il rischio di danno per fulminazione delle strutture'. *L'Energia Elettrica*. 1985;**2**:47–58.
9. Technical Report IEC/TR 62066. Surge overvoltages and surge protection in low-voltage a.c. power systems – general basic information, 2002.
10. Darveniza M. *et al.* 'Lightning protection of UHV transmission lines'. *Electra*. 1975;**41**:39–69.

11. IEEE Working Group Report. 'Estimating lightning performance of transmission lines II – updates to analytical models' in W.A. Chisholm (Chairman 1989–1992 of the Working Group) (ed.). *IEEE Transactions on Power Delivery.* July 1993;**8**(3).
12. Eriksson A.J. *A Discussion on Lightning and Tall Structures.* Council for Scientific and Industrial Research (CSIR) Special Report ELEK 152, Pretoria 1978.
13. Cianos N. and Pierce E.T. *A Ground–Lightning Environment for Engineering Usage.* Stanford Research Institute Technical Report 1, Stanford, August 1972.
14. Szpor S. *et al.* 'Lightning current records in industrial chimneys in Poland'. *International Conference on Large High Voltage Electric System (CIGRE)*, Paris 1974; no. 33-10.
15. Amicucci G.L., D'Elia B. and Gentile P. 'A method to assess the frequency of damage due to indirect lightning flashes'. *26th ICLP*; Cracow, 2–6 September 2002, pp. 8b.1.
16. Marzinotto M., Mazzetti C. and Lo Piparo G.B. 'A new model for the frequency of failure evaluation in electrical and electronic systems due to nearby flashes'. *ICLP 2008*; Uppsala, 23–26 September 2008, paper 7.28.
17. Neuhaus H. Blitzgefährdungskennzahlen, 12 Internationale Blitzschutzkonferenz (IBK), Portorož 1972, Special paper.
18. Horvath T. Gleichmässige Sicherheit zur Bemessung von Blitzschutzanlagen. ETZ-b. H. 19, 1975, s. 526–528.
19. Lundquist S. and Hogberg R. Blitzschutz und Zuverlässigkeit. 12 Internationale Blitzschutzkonferenz (IBK), Portorož 1973, paper 4.1.
20. Lundquist S. Über Blitzschutz und Prüfmethoden. 15 Internationale Blitzschutzkonferenz (IBK), Uppsala 1979, paper K4.1.
21. Darveniza M., Flisowski Z., Kern A., Landers E.U., Mazzetti C., Rousseau A. *et al.* 'An approach to problems of risk management for structures and services due to lightning flashes'. *Journal of Electrostatics.* March 2004;**60**:193–202.
22. International Standard IEC 60664-1:2002 Insulation coordination for equipment within low voltage systems. Part 1: Principles, requirements and tests.
23. Łoboda M., Szewczyk M. and Flisowski Z. 'Lightning risk analysis based on draft of new version of IEC 62305-2'. *26-ICLP*; Kraków, 2–6 September 2002, paper 10.10.
24. Surtees A.J., Gillespie A., Kern A. and Rousseau A. 'The risk assessment calculator as a simple tool for the application of the standard IEC 62305-2'. *VIII International Symposium on Lightning Protection (SIPDA)*; São Paulo, Brazil, 21–25 November 2005.
25. Mazzetti C. and Flisowski Z. 'Meaning of the discrepancy between lightning strikes and lightning flashes for risk assessment'. *27th ICLP*; Avignon, France, 13–16 September 2004. pp. 902–907.
26. Amicucci G.L. and Mazzetti C. *Probabilistic Method for Reliability Assessment of Metal Oxide Varistors Under Lightning Stress*, COMPEL, 2004;**23**(1):263–276.
27. Metwally I.A. and Heidler F.H. 'Improvement of the lightning shielding performance of overhead transmission lines by passive shielding wires'. *IEEE Transactions on EMC.* May 2003;**45**(2):378–392.

28. Anderson R.B. and Eriksson A.J. 'Lightning parameter for engineering applications'. *Electra*. 1980;**69**:65–102.
29. Cooray V. *The Lightning Flash*. IEE Power & Energy Series, no. 34; pp. 570, England 2003.

Chapter 9
Low-frequency grounding resistance and lightning protection

Silverio Visacro

9.1 Introduction

Grounding has an important role in lightning protection. This subject is considered in this book in three complementary chapters, with the present chapter providing an introduction. Through a simplified general approach the concept of grounding resistance is explained, together with its role in lightning protection practice. To support this conceptual approach, some experimental results related to the response of grounding electrodes to impulsive currents are presented, and there is a discussion of the relation between grounding resistance and this response. Finally, some relevant conclusive remarks related to lightning protection applications are presented.

Chapter 10 complements this topic, presenting a detailed approach about the high-frequency behaviour of grounding electrodes.

Chapter 11 discusses fundamental aspects regarding the effect of soil ionization resulting from the dispersion of intense leakage currents to the soil through buried electrodes.

9.2 Basic considerations about grounding systems

Cloud-to-ground lightning strikes always involve the flow of intense currents through earth terminations. Sometimes, the current's path to earth passes through natural elements such as trees, or structures of a building, depending on the object that is struck. When lightning strikes protected structures and systems, the current flows to earth through grounding systems.

The grounding system is composed of three components (Figure 9.1): (1) metallic conductors, which drive the current to the electrodes, (2) metallic electrodes buried in the soil, and (3) the earth surrounding the electrodes. This last component is the most relevant.

The function of a grounding system in lightning-related applications is essentially to provide a low-impedance path for the flow of lightning currents towards the soil and

476 Lightning Protection

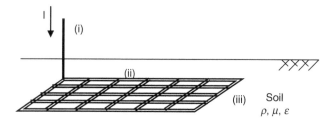

Figure 9.1 Illustration of grounding system components (adapted from Reference 1)

to ensure a smooth distribution of the electrical potentials developed on the surface of the ground in the region where the current enters the soil, due to safety concerns.

In this system, the macroscopic electromagnetic parameters of soil, mainly its resistivity, may vary substantially depending on the local properties of the medium. The electrode consists of any metallic buried body responsible for dispersing the current to the soil. According to the application, it may have different shapes and arrangements, from a single vertical rod or horizontal cylindrical conductor to complex grids. The dimension of the area covered by such systems may vary widely from a few to thousands of square metres, depending on the application and particularly on the value of soil resistivity.

9.3 The concept of grounding resistance

The response of grounding electrodes to lightning currents is expressed by means of the grounding potential rise (GPR) in relation to remote earth, which corresponds to a distant region in the soil with a null electrical potential. Indeed, it is acceptable to consider that this region is not so far away, since the local potential is very reduced, around a few hundredths of the GPR. This usually occurs for distances 10 to 20 times larger than the linear dimension of the area covered by the grounding system.

Any termination to earth presents resistive, capacitive and inductive effects, as is explained in the following for a single 'grounding element' consisting of a short length of electrode and the portion of surrounding earth taken from the grounding system (Figure 9.2a).

The current of the grounding element has two components, as represented in Figure 9.2b: the leakage transversal current I_T spreading into the soil and the longitudinal current I_L transferred to the remaining electrode length. The leakage current flowing into the soil causes a potential rise of the electrode in relation to remote earth and also of the surrounding earth. The flow of the longitudinal current generates a voltage drop along the electrode.

The equivalent circuit that is able to contemplate such effects is shown in Figure 9.3a for a single grounding element [1–5]. In this circuit, the current that enters the electrode is partially dispersed to soil, reaching the remote earth, which is represented in this circuit by the inferior bar, through the transversal parameters

Low-frequency grounding resistance and lightning protection 477

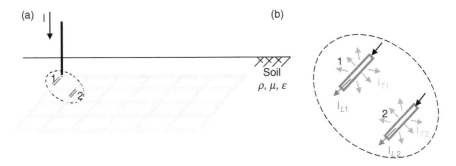

Figure 9.2 Elements of a grounding system: (a) the whole grounding system; (b) two elements taken from the grounding system (adapted from Reference 2)

G and C. The remaining current is transmitted to other grounding elements, where it is finally also dispersed to soil.

The longitudinal current is the source of a magnetic field inside and around the electrode, and causes internal losses along it. In the circuit of Figure 9.3a, the series RL branch (R, resistance; L, inductance) is responsible for these effects and for the voltage drop along the electrode during the flow of current.

On the other hand, the leakage current in the soil has conductive and capacitive components associated respectively with the electric field ($E = \rho J$, where ρ is the electrical resistivity) and to the variation of the electric field ($\partial \varepsilon E/\partial t$, where ε is the electrical permittivity of soil). The ratio between conductive and capacitive currents in the soil does not depend on the electrode geometry, but only on the relation

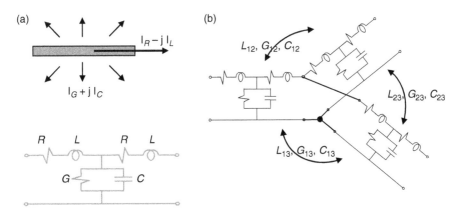

Figure 9.3 Current components in electrode and soil and the equivalent circuit for the grounding system: (a) single grounding element; (b) representation of a set of connected grounding elements with mutual effects among them (inductive, conductive and capacitive) (adapted from Reference 1)

$\sigma/\omega\varepsilon$ ($\sigma = 1/\rho$ and $\omega = 2\pi f$ is the angular frequency). The branch with the parallel conductance and capacitance (G, C) in the equivalent circuit of Figure 9.3a is able to promote the effects associated with the leakage current.

The description above applies only to the single 'grounding element' represented by the simplified circuit of Figure 9.3a. When two elements such as those indicated as elements 1 and 2 in Figure 9.2b are considered together, the circuit becomes a little more complex, because a representation of the mutual effects corresponding to the electromagnetic couplings between the elements is required (capacitive, inductive and resistive couplings), as illustrated in Figure 9.3b.

In the representation of the whole grounding system, all the electromagnetic coupling between any pair of elements has to be computed. Thus, the evaluation of the response of the grounding system to an impressed current requires the solution of a series of circuits similar to the presented one, connected according to the topology of electrodes and taking into account mutual effects among all the grounding elements. The result of this complex circuit is the impedance seen from the point where current is impressed. This impedance is given by the ratio between the potential developed at the electrode V_T in relation to a remote earth and the impressed current I_T. As the solution of this circuit is frequency dependent, this parameter is called the complex grounding impedance $Z(\omega)$:

$$Z(\omega) = V_T(\omega)/I_T(\omega) \tag{9.1}$$

From an electromagnetic perspective, the response of the grounding system is frequently expressed by means of this grounding impedance. Nevertheless, in most applications it is very common to refer to a grounding resistance instead of the grounding impedance. This should be attributed to the fact that reactive effects are negligible for such applications, which usually involve low-frequency phenomena (e.g. short-circuits in power systems). At low frequencies the voltage drop caused by the longitudinal current component along the electrode is negligible and the capacitive current in the soil also. This allows simplification of the equivalent circuit of the grounding system, which is reduced to a set of coupled conductances (or equivalent resistances in parallel), as illustrated in Figure 9.4.

In this condition, the electrical potential remains the same all along the electrodes because there is no voltage drop along them and a constant-potential approach is valid for the electrodes. The ratio between the potential developed in relation to remote earth and the impressed current results in a real number, the grounding resistance R_T, which is defined for a low-current-density condition as

$$R_T = V_T/I_T \tag{9.2}$$

Nevertheless, it is worth keeping in mind that grounding behaves as impedance. The grounding resistance is indeed this complex grounding impedance in the particular low-frequency condition.

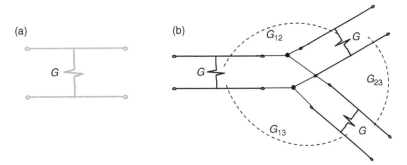

Figure 9.4   *Simplification of the equivalent circuit configuration for low-frequency conditions: (a) single grounding element; (b) a set of grounding elements represented with their conductive mutual effects (adapted from Reference 1)*

9.4 Grounding resistance of some simple electrode arrangements

It can be shown that the grounding resistance of any buried electrode arrangement is proportional to the soil resistivity by a geometric factor K:

$$R_T = K\rho \tag{9.3}$$

In equation (9.3), the resistivity is that of a homogeneous soil or the apparent resistivity found for a stratified medium.

For some simple arrangements, it is relatively easy to determine this geometric factor K, departing from the constant-potential approach for electrodes. Then, the grounding resistance can be expressed by means of approximate expressions that allow expediting the calculation of R_T.

Basically the procedure to determine such a factor follows just a few steps. First, an expression is derived for the electrical potential developed at any point P in the soil in relation to remote earth due to a current impressed on the electrode. This is done through the integral of the electric field from an infinite distance to that point, attributing the null value to the potential there. The electric field is usually calculated from current density J in the earth around the electrode, which is proportional to the injected current I:

$$V_P - V_\infty = -\int_{r_\infty}^{r_P} \vec{E} \cdot d\vec{r} \quad J = \frac{I}{K_I(r)}, \quad \vec{E} = \rho \cdot \vec{J}, \quad V_\infty = 0 \tag{9.4}$$

After solving the integral using analytical or numerical methods, the potential of point P is obtained as the function of its distance r to the electrode:

$$V_P = K(r)\rho I \tag{9.5}$$

When this distance is reduced to zero, meaning that point P is on the electrode, V_P becomes the electrode potential V_T, and the grounding resistance is obtained by dividing it by the impressed current:

$$r = 0$$
$$V_P = V_T$$
$$R_T = V_T/I_T$$
$$R_T = K\rho \tag{9.6}$$

The application of this procedure is very simple for some electrodes that present symmetry, such as hemispheres, rods and so on. Table 9.1 presents some expressions for the grounding resistance obtained from this procedure. It also provides some approaching expressions for the approximate resistance of complex electrode arrangements (first approach formulas).

The accurate calculation of grounding resistances for complex electrode arrangements requires the application of numerical methods. In this respect, the approach suggested by Heppe [6] adopting the constant-potential assumption for electrodes is very robust.

Table 9.1 *Expression for grounding resistance of some common arrangements*

Electrode	Expression for R_T
Hemisphere with radius r	$\dfrac{\rho}{2\pi r}$
Vertical rod Length L, radius a	$\dfrac{\rho}{2\pi L}\left[\left(\ln\dfrac{4L}{a}\right)-1\right]$
Horizontal electrode buried at depth d	$\dfrac{\rho}{2\pi L}\left[\left(\ln\dfrac{2L}{a}\right)+\left(\ln\dfrac{4L}{d}\right)-2+2\dfrac{d}{L}+\cdots\right]$
Disc on the soil surface	$\dfrac{\rho}{4r}$
n aligned rods displaced by distance s	$\dfrac{1}{n}\dfrac{\rho}{2\pi L}\left[\left(\ln\dfrac{4L}{a}\right)-1+\dfrac{L}{s}\left(\ln\dfrac{1.781n}{2.718}\right)\right]$
Three connected vertical rods buried at the vertices of an equilateral triangle with side s	$\dfrac{1}{3}\left\{\dfrac{\rho}{2\pi L}\left[\left(\ln\dfrac{4L}{a}\right)-1+2Ls\right]\right\}$
Circle with n rods displaced by distance s	$\dfrac{1}{n}\dfrac{\rho}{2\pi L}\left[\left(\ln\dfrac{4L}{a}\right)-1+\dfrac{L}{s}\left(\ln\dfrac{2n}{\pi}\right)\right]$
Grid covering area A with a total electrode length L	$0.443\dfrac{\rho}{\sqrt{A}}+\dfrac{\rho}{L}$

In such an approach, the longitudinal current is not considered, because it causes no voltage drop, and the capacitive current is disregarded. The electrodes are partitioned into grounding elements and the variables involved in this problem are only the transversal currents I_T of such elements and their potentials, which have the same value all along the electrodes.

Heppe developed analytical expressions to calculate the resistive coupling between any pair of grounding elements i and j in vertical, horizontal and oblique positions in relation to the soil surface:

$$R_{ij} = \frac{V_i}{I_j}, \qquad R_{ii} = \frac{V_i}{I_i} \qquad (9.7)$$

The self-resistance R_{ii} and the mutual resistance R_{ij} between any pair of grounding elements are then found to compose a resistance matrix $[R]$ that expresses the linear relation between the vector of leakage currents $\underline{I}$ leaving the elements and the vector of their electrical potentials $\underline{V}$ [equation (9.8)]. The method of images is used to take into account the semi-infinite dimension of ground or even the stratification of soil:

$$\underline{V} = [R] \cdot \underline{I} \qquad (9.8)$$

Because this is a low-frequency approach, the electrical potential of all elements in vector $\underline{V}$ has the same value. The solution of this linear system gives the leakage current of all grounding elements. The total current dispersed to soil through grounding electrodes is the summation of such currents and the grounding resistance is determined from the ratio of the constant potential to this total current.

In Reference 6 there is also an analytical expression to determine the contribution of the leakage current I_j of the grounding element j to the potential developed at any point P on the soil surface. Applying this expression repeatedly to all elements allows determination of their contribution to the final potential at this point. The distribution of potential around the grounding electrodes can be determined in this way.

The complexity of the expressions developed by Heppe in his original work [6] made their application very laborious and difficult. However, this approach can be easily improved by using numerical integration to determine the self- and mutual resistances for generally oriented elements. With such improvements, this approach allows implementation of efficient computational codes to determine the grounding resistance and also the potential distribution on the soil surface for complex electrode arrangements. All the grounding resistances calculated and presented throughout the next sections of this chapter were found from such an approach.

It is worth noting that the grounding resistance is directly proportional to soil resistivity. Thus, the resistance $R_T(\rho_0)$ calculated for a given electrode arrangement buried in a soil of resistivity ρ_0 may be used to find directly the resistance $R_T(\rho_i)$ of the same arrangement buried in a soil of any resistivity ρ_i by means of equation (9.9):

$$R_T(\rho_i) = \frac{R_T(\rho_0)}{\rho_0} \cdot \rho_i \qquad (9.9)$$

9.5 Relation of the grounding resistance to the experimental response of electrodes to lightning currents

When subjected to lightning currents, grounding electrodes present a peculiar behaviour that is usually quite different from that of the low-frequency grounding resistance. This behaviour comprises different aspects, the response to impulsive currents and the ionization process being the most relevant ones. This section presents and discusses experimental results related to the first aspect. A detailed theoretical approach about it is developed in Chapter 10. The ionization process is specifically considered in Chapter 11.

When lightning currents are involved, the constant-potential approach is no longer valid for grounding electrodes, because such currents have a frequency content that involves significant high-frequency components. On the other hand, their transient nature makes the complex impedance $Z(\omega)$ inappropriate to represent the grounding response to lightning currents. In this case, the grounding response is frequently represented by the so-called impulsive impedance Z_P given by the ratio of the peak values of developed voltage V_P and current I_P waves:

$$Z_P = V_P/I_P \qquad (9.10)$$

Usually the peaks of voltage and current waves are not simultaneous. Nevertheless, this type of representation is very attractive as it allows prompt determination of the maximum grounding potential rise GPR simply from the product of Z_P by the current peak. This possibility is very appropriate when performing sensitivity analyses in lightning-protection-related evaluations.

On the other hand, in most practical engineering conditions, the measurement of Z_P [or even of $Z(\omega)$ in the high-frequency range] is not a feasible task and the grounding resistance is measured instead. Because it is measured by instruments that usually employ a frequency signal below 1 kHz, it is frequently referred to as the low-frequency resistance R_{LF}.

Because of this, the grounding resistance is still the parameter employed to qualify the lightning performance of grounding systems. However, it is shown clearly in Section 9.6 that the requirements to be fulfilled by the grounding arrangements in lightning-protection-related applications are not directly related to the grounding resistance.

Therefore it is relevant to understand how the low-frequency grounding resistance is related to the impulsive response of grounding electrodes. This is the focus of this section, and an experimental approach is developed to achieve such understanding.

The setup described in Figure 9.5 was implemented to obtain the experimental response of grounding electrodes to impulsive currents [7]. Typically, the front time of natural lightning currents ranges from 10 to 0.4 µs, with median values around 3.5 and 0.85 µs, respectively, for first and subsequent strokes [8,9]. Thus, low-amplitude current waves with front times ranging from 0.4 to 4 µs and time-to-half-peak values around 60 µs were impressed to some simple electrode arrangements, consisting of vertical rods and horizontal electrodes buried 0.5 m

Low-frequency grounding resistance and lightning protection   483

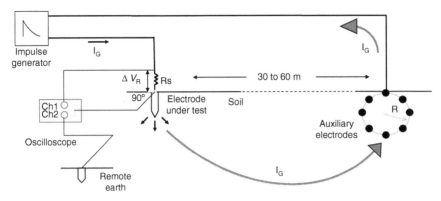

Figure 9.5   *Representation of the experimental setup. Impulsive current waves are impressed by an impulse generator from the electrode under test to an auxiliary grid. Both the current and the grounding potential rise developed in relation to remote earth are measured using a two-channel oscilloscope (adapted from Reference 7).*

deep in the soil. Different electrode lengths were tested in both high- and low-resistivity soils.

Typical results, consisting of the impressed current wave and the developed potential rise measured for each tested electrode and condition, are illustrated in Figure 9.6, which also includes the voltage measured at the generator output.

The curves presented in Figure 9.7 show the results obtained for horizontal electrodes buried in both low- and high-resistivity soils.

In the low-resistivity soil (Figure 9.7a), the current and voltage waveforms are quite similar and have almost simultaneous peaks. This denotes the prevalence of the resistive nature of the grounding impedance, although a short advance of the current wave in relation to the voltage reveals the presence of a small capacitive effect. The

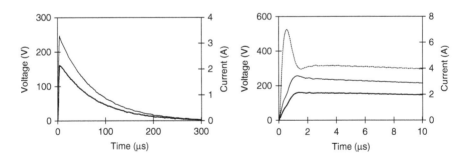

Figure 9.6   *Typical result: measured current (thin line) and grounding potential rise (thick line) waves. Dashed line: voltage at the generator output – current: 1.2/54 µs – soil: 116 Ωm – 3 m vertical rod – rod radius: 0.7 cm (adapted from Reference 7).*

484   Lightning Protection

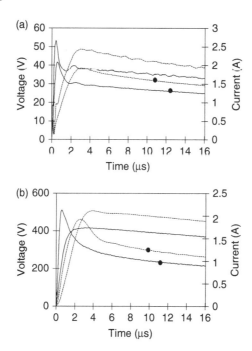

Figure 9.7   Results for a 12-m-long horizontal electrode [0.5 m deep, r = 0.7 cm; approximated value of the current wave (•) front time: 0.5 μs, 3 μs]: (a) 250 Ωm soil, $R_{LF} = 30\ \Omega$; (b) 3.8 kΩ m soil, $R_{LF} = 487\ \Omega$ (adapted from Reference 7)

impulsive impedance is a little lower than the measured low-frequency resistance (Z_P of $\sim 0.75 - 0.9\ R_{LF}$).

There is a significant capacitive effect for the electrode buried in the high-resistivity soil (Figure 9.7b). It is responsible for a clear advance of the current wave in relation to the voltage and also for the value of impulsive impedance significantly lower than the low-frequency grounding resistance ($\sim 0.4 - 0.6\ R_{LF}$).

To complement this analysis, the complex grounding impedance $Z(\omega)$ was determined from the curves of Figure 9.7. The Fast Fourier Transform (FFT) was applied to the current and voltage waves and their frequency components were determined. From the ratio of voltage to current phasors, the complex impedance was found for each frequency. The amplitude and angle of $Z(\omega)$ for the 12-m-long electrode buried in both low- and high-resistivity soils are depicted in the curves of Figure 9.8 in a frequency scale that extends up to 1 MHz.

At low frequencies, the reactive effects are negligible and the complex impedance tends toward the grounding resistance R_{LF}. With increasing frequency, the amplitude of $Z(\omega)$ decreases from R_{LF}.

The impedance angle shows that the capacitive effect is very pronounced in the high-resistivity soil, and this explains the continuous and very significant reduction

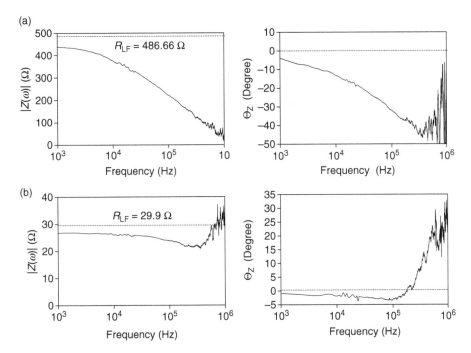

Figure 9.8 *Frequency diagram of the complex grounding impedance Z(ω) for a 12-m-long horizontal electrode. Ratio of the voltage and current phasors [V(ω)/I(ω)]: (a) high-resistivity soil (3.8 kΩm) and (b) low-resistivity soil (250 Ωm) (adapted from Reference 7).*

in impedance of the electrode buried in this soil with increasing frequency. However, this reduction is still relevant in the low-resistivity soil up to a certain frequency. Above that, the impedance amplitude begins to increase due to an inductive effect, determining a minimum value for this impedance around the null impedance angle. Above this threshold the impedance amplitude increases and becomes much larger than the low-frequency resistance. The experimental behaviour described above is finely consistent with the expected response of the equivalent circuit in Figure 9.3 and with the predictions of Reference 2.

The relatively small impedance angle in the low-resistivity soil suggests that the significant reduction of the impedance amplitude with increasing frequency is not due entirely to the capacitive effect. It seems that this reduction results from a combination of the capacitive effect and a decrease of soil resistivity associated with the frequency dependence of this parameter [10–12].

The effect of frequency dependence of soil parameters (resistivity and permittivity) can indeed be significant, as depicted in Figure 9.9. There, experimental voltages developed from impulsive currents impressed on a 12-m-long electrode buried in both low- and high-resistivity soils are displayed along with voltage waves obtained by simulation, assuming that the soil resistivity is constant and equal to its

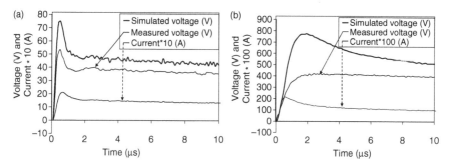

Figure 9.9 Measured voltage and current waves for a 12-m-long horizontal electrode buried 0.5 m deep, with a radius of 0.7 cm: (a) low-resistivity soil (∼250 Ωm) and (b) high-resistivity soil (∼3.8 kΩm) (adapted from Reference 14).

low-frequency value and that the relative soil permittivity is equal to 20. The simulations were implemented using the HEM Model [13], which employs an elaborate electromagnetic approach.

The voltages simulated with the constant-value parameters are significantly higher than the measured ones, leading to impulsive impedances that are ∼30 and 75 per cent higher, respectively, for the low- and high-resistivity soils. This corresponds to values of Z_P larger than the low-frequency resistance, contradicting the experimental results. In this case, in order to make the simulated results match the experimental ones, the resistivity at 1 MHz is required to decrease to one-third of its low-frequency value while the relative permittivity that has a value of ∼20 at 1 MHz is required to increase to a range around 25 to 100 times larger than this value at 100 Hz [14]. As no accurate general formulation is provided in the literature for expressing the frequency dependence of soil parameters, the effect is usually neglected [2].

In general, the impulsive impedance of electrodes that are not too long is expected to be lower than the low-frequency resistance, as has been shown through experimental results. In low-resistivity soils the values of Z_P are expected to range from 0.8 to 0.9R_{LF}, while in high-resistivity soils it is expected to be ∼0.4 to 0.7R_{LF}, depending on the current front time, the electrode length and soil resistivity.

On the other hand, long electrodes might have impulsive impedance values much higher than R_{LF}. The understanding of this question requires introducing the concept of the effective length of electrodes L_{EF} [15]. The high-frequency components of lightning currents are responsible for decreasing the impulsive impedance in relation to the grounding resistance. As is detailed in Chapter 10, the current and voltage waves are attenuated as they propagate along the electrode and the attenuation of the high-frequency components is particularly pronounced. Beyond a determined length, these components are so attenuated that they no longer contribute to the dispersion of current to the soil and, therefore, to the reduction of impulsive impedance. According to Reference 2, the value of L_{EF} is found when the minimum impulsive impedance is achieved while increasing the electrode length. Further increasing the electrode length does not lead to any additional decrease of Z_P, although the electrode grounding resistance R_{LF} still decreases.

Thus, electrodes longer than L_{EF} have values of Z_P larger than R_{LF}. As the electrode length is increased further, the impulsive impedance remains constant while its resistance decreases, leading to values of R_{LF} that decrease continuously in relation to Z_P with increasing electrode length.

9.6 Typical arrangements of grounding electrodes for some relevant applications and their lightning-protection-related requirements

In order to illustrate the role of the grounding resistance in the practices related to lightning protection, some particular applications of major interest are now considered.

9.6.1 Transmission lines

Distributed earth terminations connect shield wires to grounding electrodes all along high-voltage transmission lines. This is done through down-conductors for wooden and concrete structures (poles) or through the tower body in the case of metallic structures (cross-rope, self-sustained and guyed towers).

The arrangement and dimensions of grounding electrodes are mainly determined by the value of soil resistivity. Concentrated electrodes of short length are usual in low-resistivity soils. In this condition, vertical rods are preferentially employed, as their installation is very practical. For wooden or concrete structures, at least one rod is usually installed at each down-conductor termination. For metallic structures, buried conductors are usually derived from the structure base to connect the rods. A typical arrangement used in low-resistivity soils is illustrated in Figure 9.10a.

The use of counterpoise cables is common practice in soils with high and moderate resistivity, composing extensive electrodes [16]. As illustrated in Figure 9.10b, four electrodes buried ~0.5 m deep are derived from the points where the metallic structures reach the soil, in a radial arrangement. Close to the right-of-way border, around 1 m distant, they are bent and laid in parallel to the line route. Cable lengths

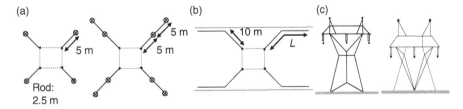

Figure 9.10 Typical grounding electrode arrangements in transmission line tower-footings: (a) concentrated electrodes, radial cables plus rods (4 and 8 rods); (b) extended arrangement, counterpoise wires with length L; (c) types of structure, self-sustained and guyed towers

within a range of 10–90 m are usually employed. For self-sustained towers, the radial conductors are attached to the structure legs. For poles, they are usually bonded to down-conductors. For guyed or cross-rope towers, the counterpoises are derived from the anchors that restrain the steel cables.

Concentrated buried metallic components such as tower grillage or concrete-encased metals of tower foundations also work as electrodes, contributing to the dispersion of current to soil. When counterpoise cables are short, the concentrated components contribute to reduce the grounding impedance.

The fundamental role of grounding in transmission lines (TL) is to influence their lightning performance. Lightning is frequently responsible for line outages, reducing the availability of electrical networks. Direct strokes to phases or to shield wires might cause high overvoltage and electrical discharges across insulator strings, leading to faults. Basically, there are three mechanisms responsible for lightning-related faults in TLs: shielding and midspan failures and the backflashover.

Shielding failure may happen to low-amplitude-current lightning and the direct stroke to the phase might lead to a flashover across insulators from the stricken phase to the grounded structure (tower). Lightning strikes to shield wires at midspan in long spans can develop very high overvoltage there, and a midspan flashover might happen, connecting the shield and phase wires through the air. However, the mechanism that largely prevails as a source of line outages is the backflashover. This happens when a strike to shield wires or to a tower promotes a very high ground potential rise and the overvoltage between grounded structure and one phase exceeds the line insulation withstand, leading to the establishment of an electrical arc across insulator strings [17].

Although grounding has little influence on the occurrence of flashover associated with shielding and midspan failures, it plays a fundamental role in the backflashover mechanism. The overvoltage developed across insulator strings depends very much on the value of tower grounding impedance [18]. Indeed, the amplitude of the ground potential rise and therefore of the voltage developed across insulator strings during the flow of lightning current through the tower increases almost linearly with the grounding impedance.

This is the reason for a common criterion related to the lightning protection of TLs adopted by power utilities – to establish a limit for the acceptable value of tower grounding resistance R_{max}. In countries with typical low-resistivity soils it is common to establish this limit below 10 Ω, whereas this value increases to $\sim$20 Ω in places with high-resistivity soils [18]. This limiting value depends on the nominal voltage level of line, because its insulation withstand increases with this voltage. Reference 18 suggests maximum resistance values of 8, 25, 35, 39 and 50 Ω, respectively, for lines with operational voltages of 69, 138, 230, 345 and 500 kV. It is worth mentioning that this threshold resistance is the acceptable limit for those towers in critical conditions. A lower value of such resistance should be pursued all along the line, with a value of $\sim$0.5 R_{max} considered a reasonable average for the line grounding resistance.

Figure 9.11 presents some curves expressing the grounding resistance value as a function of the grounding system dimension for the typical arrangements presented

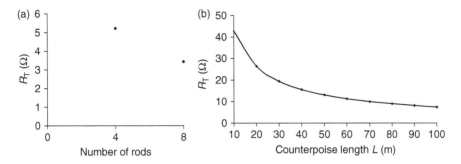

Figure 9.11 Resistance of an electrode arrangement as a function of grounding system dimension: (a) concentrated electrodes, radial cables plus rods (4 and 8 rods), 100 Ωm soil; (b) extended arrangement, counterpoise wires with different length, 1 000 Ωm soil

in Figure 9.10. The results were obtained from simulations using a code developed from an improvement of Heppe's approach [6].

Indeed, the criterion described above aims to limit the value of the grounding impedance to lightning currents assuming it is related to the grounding resistance that is the parameter the measurement of which is feasible. In practice, the length of the grounding electrodes is designed to achieve a determined resistance value and, after installing the designed electrode arrangement, the grounding resistance is measured. If the value exceeds the resistance threshold, then the electrode length is increased to achieve the desired value.

It is relevant to note that the application of this criterion requires the electrode length to be shorter than the effective length. Otherwise, a low value can be found for the grounding resistance without corresponding to a low value of the grounding impedance, which is the parameter that really influences the lightning performance of line.

As explained in Reference 2, the effective length L_{EF} is shorter in low-resistivity soils, because the current attenuation is larger with increasing soil conductivity. Table 9.2, taken from this reference, suggests some values for the effective length

Table 9.2 Estimated effective length of grounding electrodes for a 1.2 μs current front time (horizontal electrodes)

Soil resistivity (Ωm)	L_{EF} (m)
100	10
500	23
1 000	34
2 000	50
5 000	85

490 Lightning Protection

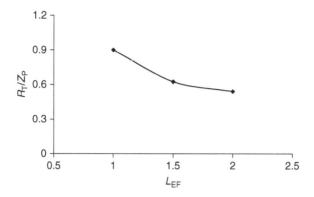

Figure 9.12   Ratio of the grounding resistance to the impulsive impedance as a function of the effective length L_{EF} (1 000 Ωm soil; $\varepsilon_r = 10$; current wave, 1/50 μs) (adapted from Reference 19)

as a function of soil resistivity, considering a 1.2 μs current front time. For longer front times, the values are a little larger, but this reference may still be adopted following a conservative approach.

Ignoring the precaution mentioned above may lead to situations where low values of grounding resistance are achieve but the corresponding values of grounding impedance are much higher. To illustrate this question, the response of a counterpoise arrangement buried in a 1 000 Ω m soil ($\varepsilon_r = 10$) to an impressed current wave 1/50 μs was simulated using the HEM model [13]. The length of the electrode L was varied from L_{EF} to $2L_{EF}$, the value of L_{EF} being $\sim$34 m. The grounding resistance and the impulsive grounding impedance were determined in each case. The curve of Figure 9.12 shows the ratio of the grounding resistance to the impulsive impedance found from simulations.

It is shown that electrodes longer than the effective length have grounding impedance values much larger than the grounding resistance. It is worth mentioning that the simulations have not considered the frequency dependence of soil parameters that would affect this ratio.

9.6.2 Substations

The typical arrangement of buried electrodes in substations consists of grids composed of a large number of meshes, as illustrated in Figure 9.13. The design of this

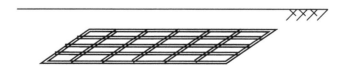

Figure 9.13   Typical electrode arrangement in substations

arrangement is defined only to ensure safety conditions during short-circuit events. The grid design considers the proper position of electrodes to ensure a smooth distribution of the electrical potentials developed at the soil surface during the flow of short-circuit currents, assuming that all the electrodes have the same electrical potential during this event.

Figure 9.14 shows how the grounding resistance is affected by both aspects, increasing the dimension of the area covered by a squared grid and by increasing the number of meshes of such a grid. The thick curve corresponds to a square grid with only one mesh, the linear dimension L (side) of which is increased from 10 to 80 m. It can be seen that increasing length promotes a significant reduction in the grounding resistance. This resistance is reduced from its original value to around one-fifth when the length is multiplied by eight.

On the other hand, it is clear that the inclusion of conductors inside the perimeter (meshes) is not very effective in reducing the grounding resistance. The inclusion of 64 regular meshes in the grid, corresponding to a total length of a conductor almost 20 times longer than in the original configuration (electrodes only at the perimeter), is able to reduce the grounding resistance of the original configuration by no more than 20 per cent. This is explained by the intense mutual resistive effect existing among the elements of the meshes that are too close.

Indeed, it is important to have in mind that the use of meshes is not intended to reduce the resistance, but is basically to control the distribution of the gradient of electrical potential over the soil surface during short-circuit events, due to safety concerns. In this respect, it is a very effective practice.

Figure 9.15 illustrates the potential profile over soil surface for the grids of Figure 9.13 developed when a 1 kA low-frequency current was impressed on the electrodes. The curves denote how the maximum value of difference of potential inside the region covered by the grid is reduced by the presence of meshes. This

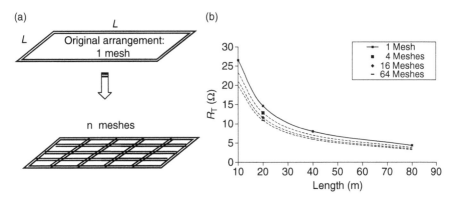

Figure 9.14   *Grounding resistance of grids as a function of covered area and number of meshes: (a) square grid arrangement buried at a depth of 0.5 m; (b) calculated resistance, soil: 500 Ωm*

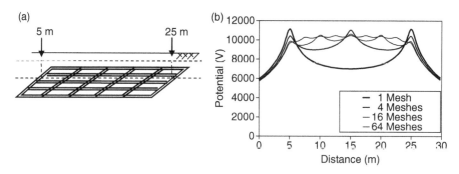

Figure 9.15 The influence of meshes on the distribution of potential on the soil: (a) line on soil surface where the distribution of electrical potential is calculated (square grid 20 × 20 m^2); (b) profiles of developed potential along the indicated line, soil: 500 Ωm

value is decreased to less than one-tenth of the value found for the original arrangement consisting of the electrodes at the perimeter only.

The grid design is not conditioned by lightning concerns. Nevertheless, there are some questions that imply lightning-protection-related requirements. Transmission lines arriving at the station usually have their shield wires connected to the grounding grid at certain points. Also, the lightning protection system LPS of the substation has its down-conductors connected to the grid at some point. Intense lightning currents can be impressed on such points during lightning strike events to shield wires in the substation proximities or to the conductors of the substation LPS. Because the constant-potential approach adopted in the design of the grid is not valid for lightning currents, significant voltages may be introduced along the grid electrodes due to the flow of such currents.

This event may be the source of serious problems. First, intense voltage gradients may be established on the soil close to earth terminations where lightning currents are injected, creating risks to safety [20,21]. It may also cause damage or malfunctions in sensitive devices and pieces of equipment installed at the substation. This last occurrence is quite common, involving systems installed at the substation presenting distributed devices (remote terminals), such as the control system responsible for monitoring and commanding the operation of some pieces of equipment installed in the substation yard, such as circuit-breakers [22].

Terminals connected to different points of the grid may allow circulating intense destructive currents through control cables or shield wires during the mentioned event. Also, control cables connected to the grid only on one side may have this problem if they are very close to metallic parts connected to the grid in another point. In this case, electric discharges might eventually happen to configure connections to different points along the grid.

In this respect, some special practices are usually applied specifically in the area involving the elements under risk. A first practice consists in installing meshes with very small internal areas around the earth terminations to control the voltage gradients.

Owing to the attenuation of lightning currents and associated potentials along the electrodes, the gradient decreases with increasing distance to the earth termination. Therefore, the critical region is limited to the surroundings of this termination and this action should be limited to an area close around it.

Another effective action to reduce the voltage gradients developed at the surroundings of earth terminations due to the flow of lightning currents consists in defining the adequate position(s) to connect the earth termination(s) to the grid. Even for the same electrode configuration, the impedance value may largely vary according to the position of earthing connections to the buried electrode. Because grounding potential rise is basically proportional to such impedance, it also varies with this position. Figure 9.16 illustrates this for two configurations that assume electrodes longer than the effective length (for the particular soil where they are buried).

For the horizontal electrode shown in Figure 9.16a, an earthing connection to point A (electrode extremity) results in a grounding impedance value around twice that obtained for a connection to point B (midpoint). This is explained as, in the second case, the current 'sees' two parallel impedances, the individual values of which are similar to the impedance of the first case.

For the grid shown in Figure 9.16b, the connection to the central point results in an impedance value around one-quarter to one-third of the value found for a connection to a corner. A similar effect can be achieved if the current is distributed to several points along the grid. Figure 9.16c illustrates this case when the current is distributed to the four corners. A behaviour similar to that of the grid fed at the centre is expected in this case, with a grounding impedance around one-quarter of that corresponding to the earth connection only at one corner.

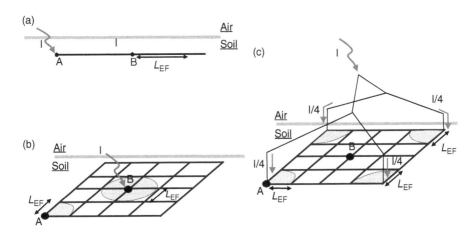

Figure 9.16 The influence of the position of the earth termination connection to buried grid on the grounding impedance: (a) horizontal electrode–earth termination at one extremity and at the midpoint; (b) grid–earth termination centre and at one corner; (c) grid: distribution of the lightning current through the four corners (adapted from Reference 2)

494 *Lightning Protection*

It is worth noting that this effect is very pronounced only for cases in which the electrode exceeds the effective length. In high-resistivity soils, this effect tends to decrease, because the current wave attenuation becomes lower, leading to longer effective lengths of the electrode.

For the actions to reduce risks for the several-terminal sensitive devices and equipment installed at the substation, a two-step practice can be applied. First, it is possible to minimize the difference of the potentials developed at the different points along the grid where the terminals are placed during the lightning current flow. One effective practice to achieve this condition consists of connecting these points with non-buried metallic bars. Because the bar is not buried the current flowing along it is not subject practically to any attenuation and the points of the grid connected to it develop very similar potentials if they are not too distant. In spite of this, the current transit time may be responsible for some difference in potentials. Sometimes, metallic tubes involving control wires are used instead of bars. As a second step, this action has to be complemented by the installation of surge protective devices at the terminal of the sensitive devices and equipment.

9.6.3 Lightning protection systems

Lightning protection systems (LPSs) are designed to avoid direct strikes to structures, such as buildings and houses. In order to ensure no damage to the structure, such systems have to be able to intercept lightning leaders and also to drive the lightning current to the earth through down-conductors connected to grounding electrodes.

The arrangement of the buried electrodes depends on several factors and mainly on the dimension and shape of the protected structure. The two arrangements presented in Figure 9.17 are very common.

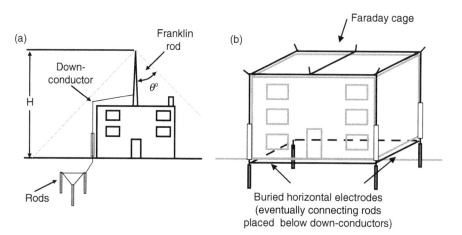

Figure 9.17 *Typical grounding electrode arrangements for an LPS: (a) Franklin-type LPS, with a few connected rods; (b) Faraday-cage LPS, with a buried ring electrode (adapted from Reference 17)*

In this application, the basic role of grounding is to provide a low-impedance path to lightning current while it flows to earth and also to minimize the voltage gradient over the soil surface in the structure surroundings due to safety concerns.

In order to ensure the first objective is achieved, the lightning protection code of different countries usually recommends a maximum acceptable value for the LPS grounding resistance. The value of R_T is usually recommended to be below 10 and 1 Ω, according to the typical level of the local soil resistivity.

Figure 9.18 shows some evaluations related to the grounding resistance of the arrangements mentioned above as a function of their dimension. Figure 9.18a shows an arrangement consisting of three rods at the vertices of an equilateral triangle of side d and connected by horizontal electrodes buried 0.5 m deep in the soil. Departing from a basic case, in which d is 3 m, the variation of the corresponding grounding resistance is indicated for d varying up to 6 m. Figure 9.18b shows the arrangement consisting of a rectangular ring with sides L_1 and L_2 with these parameters equal respectively to 12 and 8 m in the basic case. These lengths are then increased up to four times and the grounding resistance is calculated in each case and the corresponding value indicated in the graph.

Regarding safety concerns, as is shown in References 23 and 24, the buried ring electrode is much more effective. It promotes a lower grounding potential rise when lightning currents flow to the soil through the LPS.

This is a result of two factors. First, this arrangement usually covers a much larger area than that covered by the concentrated configuration. This results in lower values for the grounding resistance and grounding impedance. Also, the several earth

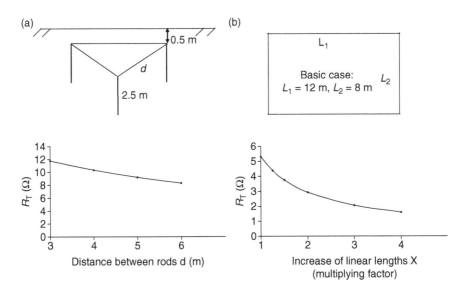

Figure 9.18 Grounding resistance as a function of covered area; 100 Ωm soil: (a) triangular arrangement and calculated grounding resistance; (b) rectangular ring arrangement and grounding resistance variation

connections to the ring, which are typical for this arrangement, contribute to diminishing the grounding impedance in relation to the typical few connections of concentrated electrodes. This is due to an effect that is similar to the one presented in Figure 9.16c, which occurs mainly in soil presenting low and moderate resistivity.

The ring arrangement also tends to promote a smoother potential distribution in the soil surrounding the protected structure. The lower potential gradients developed there diminish the risks to safety during lightning events.

9.6.4 Overhead distribution lines

Lightning is a source of disturbance and damage to overhead distribution lines. Direct strikes to the conductors cause very severe effects, always leading to flashover across insulators and faults. However, nearby strikes are much more frequent and the voltage they induce is a major source of outages and damages to the line. Such lightning induced overvoltage can achieve very high values, in some cases above 200 kV, frequently exceeding the typical insulation withstand of distribution lines.

In spite of the different types of overhead distribution systems, typically their grounding arrangements consist of very simple configurations of concentrated electrodes, most commonly comprising one rod or a few aligned rods. Grounding electrodes are commonly installed along the medium-voltage lines beside those poles supporting pieces of equipment, such as transformers, and are usually connected to the surge arresters responsible for protecting the equipment. Also, grounding electrodes are installed along the low-voltage distribution line at the consumers' service entrance.

Because of safety concerns, some power utilities adopt the criterion to limit the acceptable grounding resistance to a threshold that is $\sim 80 \, \Omega$ in countries with high-resistivity soils. Figure 9.19 shows the grounding resistance calculated for such an arrangement as a function of the number of rods. The curves in Figure 9.19 show that, depending on soil resistivity, it is not so easy to achieve low values of grounding resistance for such a concentrated arrangement.

It is important to observe that, for this type of arrangement, which typically comprises a very short length of electrodes, the grounding impedance is expected to be lower than the grounding resistance. This reduction might be very significant in

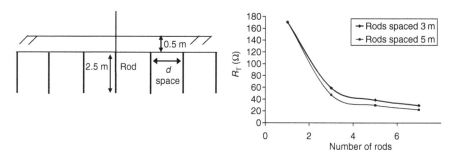

Figure 9.19   *Grounding resistance as a function of the number of rods (spaced 3 and 4 m); soil: 500 Ωm*

high-resistivity soils (to ~50 per cent). Some other questions also deserve to be noted concerning lightning-protection-related requirements.

Direct strikes to medium-voltage lines always cause severe effects to both primary and secondary networks, regardless of the value of the system grounding resistances, because a very high overvoltage is developed in relation to the voltage withstand of distribution lines. Nevertheless, when considering strikes to medium-voltage lines, the value of R_T along the line has a relevant contribution to define the extent of flashover and damages along the line in the surroundings of the striking point. In other words, if the grounding resistances close to this point have low values, a smaller number of poles are affected. On the other hand, the distribution and the value of grounding resistances along the medium-voltage line may be relevant to avoid outages and other effects caused by overvoltages induced by nearby strikes, these being the source of the most frequent lightning stresses.

In general, a very low value of grounding resistance is beneficial. However, in certain situations there are still additional requirements such as in the case of lightning surge transference from medium-voltage distribution networks to consumer facilities through distribution transformers [25]. When a voltage wave (associated with a lightning-induced overvoltage or a distant direct strike) reaches the primary of a protected distribution transformer, surge arresters operate and the associated current is drained to soil. This current is also partially transferred to the low-voltage network through the neutral and phase conductors. If the grounding resistance of a close service entrance is lower than that of the transformer, the surge current tends to be drained to the consumer grounding [25]. The associated overvoltage is driven to that consumer's load. Unless the service entrance is protected, it may be subjected to dangerous voltage levels. Therefore, in this situation there is one more requirement in addition to a low resistance value for the whole set of earth terminations: the transformer grounding resistance should be much lower than that of the consumers.

9.7 Conclusion

In this chapter, some important practical factors related to the grounding resistance and associated lightning-protection-related requirements are remarked on by means of a conceptual approach.

In general, grounding behaves as impedance. Only for very particular imposing phenomena can it be properly represented by a resistance. For lightning currents, the grounding behaviour is usually quite different from that of a resistance, even when non-linearities are disregarded.

In spite of this, the term grounding resistance is the usual reference in problems involving lightning protection, not grounding impedance. This is a reasonable practice that is derived from the practical restrictions on measuring grounding impedance in field conditions. This is a quite complex task for ordinary engineering personnel and, thus, the grounding resistance is measured instead.

The usual assumption of an equalized electrical potential along the electrode length is only a reasonable approach for slow occurrences (typically with low-frequency content) or for very short electrode length. The flow of lightning currents through buried electrodes establishes a significant difference in potential along them due to propagation effects and to the voltage drop associated with inductive and internal resistive effects. Thus, it is prudent to avoid connecting electrically linked pieces of equipment, which configure aerial loops, to electrodes at different points of the grounding grid. This may cause the flow of destructive currents through those loops.

In grounding design, the response of electrodes to lightning currents is evaluated in terms of the grounding potential rise (GPR), which develops a maximum value at the point where the current is impressed. In most situations, this value is calculated from the product of the impulsive grounding impedance Z_P and the peak value of the lightning current.

Several factors contribute to making this impulsive impedance quite different from the low-frequency grounding resistance. The reactive effects (capacitive and inductive currents), the frequency dependence of soil parameters and attenuation effects are the most relevant factors.

In spite of this, in practice the grounding resistance is indeed the parameter used to qualify the lightning performance of grounding electrodes. In some cases, it is possible to develop estimates of the impulsive impedance from this resistance, which is usually the measured parameter. The experimental results presented in Section 9.5 indicate that Z_P is usually lower than R_{LF} unless the electrode is longer than the effective length. According to such results the ratio between such parameters Z_P/R_{LF} ranges from 0.4 to 0.7 in high-resistivity soils, but from 0.75 to 0.9 R_{LF} in low-resistivity soils, for electrodes shorter than L_{EF}.

The effective length is not a constant parameter, its value being influenced mainly by the soil resistivity and the front time of the lightning current impressed on the electrode. The values presented in Table 9.2 give a rough general estimate of this parameter for lightning-protection-related applications.

In terms of lightning protection (and only in this respect), employing electrodes longer than the effective length may be wasteful. Special care is recommended for those applications where the grounding impedance value is estimated from the measured grounding resistance. This is typical for transmission lines, where a limiting value of tower-footing grounding resistance is observed with the expectation that this would limit the value of the grounding impedance as well. Sometimes, a reduced value of grounding resistance does not lead to a reduced value of grounding impedance. When the electrode length is increased, its resistance is decreased continuously. This behaviour holds even after the effective length is exceeded. Therefore, this resistance may reach very low values, while the minimum value of the grounding impedance is limited to that value obtained for the effective length. In this case, a false expectation of reduced value for impedance may be generated.

In applications involving long electrodes, such as grids of substations, it is relevant to observe the proper position to connect the earth termination to the buried electrodes. The grounding resistance is the same, regardless of the number and the position of the connections to the buried electrode. However, the value of the grounding impedance (and therefore of the grounding potential rise) may vary widely according to such a position if the electrodes are longer than the effective length.

Although in a few applications there may be practical interest in increasing the impedance of an earth termination, in most applications low values are recommended for grounding impedance.

The main factor that influences the decrease of grounding impedance and resistance is the area covered by the grounding electrodes. Nevertheless, when lightning currents are involved the attenuation of the current wave limits the efficiency of enlarging the covered area by increasing the electrode length to the effective length. In this case, the action intended to reduce the grounding impedance should be concentrated on a limited region around the current injection point, although mutual effects tend to decrease the effectiveness of this action due to electrode proximity.

When long electrodes are involved, the number of earthing connections may be influential and the use of aerial cables to distribute the current through different earthing connections may play an important role in reducing the grounding impedance value as well. For the specific condition of a layered soil with a second layer that presents a low resistivity value (in comparison to that of the first layer), the use of long rods to reach such a layer can also have a significant influence.

The ionization process [26,27] is not considered in this chapter. However, it is worth mentioning that when this process is very intense, it may affect significantly the grounding impedance, reducing its value. The intensity of this effect depends basically on the density of current on the electrode surface. Short electrodes subjected to intense lightning currents might meet such a condition. Thus, this occurs only for very concentrated arrangements of electrodes and for very high values of lightning current. For arrangements with long electrodes, such as the counterpoise wires in transmission lines, this effect is not able to affect significantly the impedance value [28].

References

1. Visacro S. *Grounding and Earthing: Basic Concepts, Measurements and Instrumentation, Grounding Strategies* (in Portuguese). 2nd edn. São Paulo, Brazil: ArtLiber Edit.; 2002. pp. 1–159.
2. Visacro S. 'A comprehensive approach to the grounding response to lightning currents'. *IEEE Trans. Power Delivery* January 2007;**22**:381–86.
3. Rudenberg R. 'Fundamental considerations on grounding currents'. *Electrical Engineering* January 1945;**64**(1):1–13.
4. Sunde E.D. *Earth Conduction Effects in Transmission Systems*. 2nd edn. New York: Dover Publications; 1968.
5. Meliopoulos A.P. *Power System Grounding and Transients*. New York and Basel: Marcel Dekker Inc.; 1988.
6. Heppe R.J. 'Computation of potential at surface face above an energized grid or other electrode, allowing for non-uniform current distributions'. *IEEE Trans. Power Delivery* 1998;**13**:762–67.
7. Visacro S. and Rosado G. 'Response of grounding electrodes to impulsive currents: an experimental evaluation'. *IEEE Trans. Electromagnetic Compatibility*, February 2009, pp. 161–64.

8. Berger K., Anderson R.B. and Kröninger H. 'Parameters of lightning flashes'. *Electra* 1975;**41**:23–37.
9. Visacro S., Schroeder M.A.O., Soares A.J., Cherchiglia L.C.L. and Sousa V.J. 'Statistical analysis of lightning current parameters: measurements at Morro do Cachimbo Station'. *Journal on Geophysical Research* 2004;**109**:D01105, 1–11.
10. Scott H.S. 'Dielectric constant and electrical conductivity measurements of moist rocks: a new laboratory method'. *Journal of Geophysical Research* 1967;**72**(20):5101–15.
11. Visacro S. and Portela C.M. 'Soil permittivity and conductivity behavior on frequency range of transient phenomena in electric power systems'. *Proceedings of the 1987 International Symposium High Voltage Engineering*; Germany, no.93.06. pp. 1–4.
12. Portela C. 'Measurement and modeling of soil electromagnetic behavior'. *Proceedings of the 1999 IEEE International Symposium Electromagnetic Compatibility*, Vol. 2, pp. 1004–9; USA.
13. Visacro S. and Soares A. Jr. 'HEM: a model for simulation of lightning-related engineering problems'. *IEEE Trans. Power Delivery* April 2005;**20**(2):1026–208.
14. Visacro S., Pinto W.L.F., Almeida F.S., Vale M.H.M. and Rosado G. 'Experimental evaluation of soil parameter behavior in the frequency range associated to lightning currents'. *Proceedings of the 29th International Conference Lightning Protection*; Uppsala, Sweden, 2008, pp. 5c-2.1–5c-2.5.
15. Gupta B.R. and Thapar B. 'Impulsive impedance of grounding grids'. *IEEE PAS-99* November/December 1980; no. 6.
16. Bewley L.V. 'Theory and test of the counterpoise'. *Electrical Engineering* August 1934;1163–72.
17. Visacro S. *Lightning: An Engineering Approach* (in Portuguese). São Paulo, Brazil: ArtLiber Edit.; 2005. p. 1–276.
18. Visacro S. 'Direct strokes to transmission lines: considerations on the mechanism of overvoltage formation and their influence on the lightning performance of lines'. *Journal of Lightning Research* 2007;**1**:60–68.
19. Visacro S. and Pinto W.L.F. 'Is grounding impedance really larger than low frequency resistance?' *Proceedings of the 2006 International Conference Grounding and Earthing – GROUND 2006*; Brazil. pp. 185–88.
20. Grcev L. and Dawalibi F. 'An electromagnetic model for transients in grounding systems'. *IEEE Trans. Power Delivery* 1990;**5**:1773–81.
21. Grcev L. 'Computer analysis of transient voltages in large grounding systems'. *IEEE Trans. Power Delivery* 1996;**11**(2):815–23.
22. Soares A. Jr, Visacro S., Oliveira R.Z., Almeida M.F. and Silva A.P. 'Damages to equipments of a hydroelectric power plant substation due to lightning – case study'. *Proceedings of the 2004 International Conference Lightning Protection*; Avignon, France. pp. 741–46.
23. Visacro S., Soares A. Jr, Schroeder M.A.O. and Murta Vale M.H. 'Evaluation of current and potential distribution for lightning protection system including the

behavior of grounding electrodes'. *Proceedings of the 2000 International Conference Lightning Protection*; Rhodes, Greece. pp. 464–68.

24. Visacro S. 'Performance of typical grounding configuration for residential lightning protection systems'. *Proceedings of the 1998 International Conference Grounding and Earthing – GROUND '98*; Belo Horizonte, Brazil, pp. 116–18.

25. De Conti A. and Visacro S. 'Evaluation of lightning surges transferred from medium voltage to low voltage distribution lines'. *IEE Proceedings* 2005; **152**(3):351–56.

26. Mousa A.M. 'The soil ionization gradient associated with discharge of high currents into concentrated electrodes'. *IEEE Trans. Power Delivery* 1994; **9**(3):1669–77.

27. Liew A.C. and Darveniza M. 'Dynamic model of impulse characteristic of concentrated earths'. *IEE Proceedings* 1974;**121**(2):123–35.

28. Visacro S. and Soares A. Jr. 'Sensitivity analysis for the effect of lightning current intensity on the behavior of earthing systems'. *Proceedings of the 1994 International Conference Lightning Protection*. Hungary. pp. R3a–01(1–5).

Chapter 10
High-frequency grounding
Leonid Grcev

10.1 Introduction

High-frequency and dynamic behaviour of grounding has been the subject of extensive theoretical and experimental research. Here, some representative work is listed containing useful material pertinent to the subject of this chapter. Pioneering but comprehensive work was conducted in the first half of the twentieth century, which is summarized by Sunde and others in well known reference books [1–3]. More recent work is summarized in the books in References 4 to 6. There is a lack of carefully documented experimental works in the literature, but noteworthy examples are found in References 7 to 10. The approaches on which some recent analytical work is based may be classified into the following groups:

- circuit theory [11–15]
- transmission line theory [16–21]
- electromagnetic field theory [22–28]
- hybrid approaches [29–31]

In spite of the large amount of work that has been devoted to this subject, there is still no consensus on how to apply present knowledge to the design of actual grounding for better high-frequency and dynamic performance. As a result, power-frequency ground impedance, which can easily be measured or calculated, is being used in most cases to predict grounding lightning performance [32].

10.2 Basic circuit concepts

High-frequency behaviour of grounding is of interest in lightning protection studies and is dependent on the high-frequency content of the lightning current pulses. Because the highest frequencies in a pulse spectrum are related to the fastest time variation, the lightning pulse usually has the highest frequency content during its rise, that is, during the first moments of the stroke. Grounding is usually designed for safety at industrial frequencies, principally because humans are more sensitive to 50/60 Hz current than to current at high frequencies [33]. However, the performance of grounding systems might be much worse at high frequencies, which reduces the efficiency of the protection during the rise of the lightning current pulse, that is, during the first moments of the stroke. One reason for concern is the occurrence of high-intensity transient voltages that might endanger human safety and cause damage or malfunction in equipment during the first moments of the stroke. The highest voltage appears between the current feed point at the grounding system and a point at remote 'neutral' ground. This voltage, divided by the current, is referred to as ground impedance; lower values indicate better grounding system performance. Usually, the electric potential at the current feed point is used for the definition of the ground impedance instead of voltage. The low-frequency (50/60 Hz) ground resistance, which is practically equivalent to the direct current (d.c.) ground resistance, is a d.c. limit of the ground impedance.

Grounding may be considered by referring to a circuit with an ideal current source I with one terminal connected to the ground electrodes and the other terminal to the remote earth, theoretically at infinite distance (Figure 10.1). The influence of the connecting leads is ignored. The voltage between the current source terminals at d.c. is equivalent to the electric potential of the ground electrodes V with a reference point at remote earth. This enables definition of the ground resistance R as

$$R = \frac{V}{I} \quad (10.1)$$

Neglecting the influence of the current source connecting leads allows for extension of this concept to a general case of arbitrary time-varying excitation, because the path-dependent part of the voltage between current source terminals is ignored. Therefore, the voltage over the current source $i(t)$ is equivalent to the electric scalar potential at the current feed point $v(t)$, which allows for uniquely defined ground transient impedance $z(t)$:

$$z(t) = \frac{v(t)}{i(t)} \quad (10.2)$$

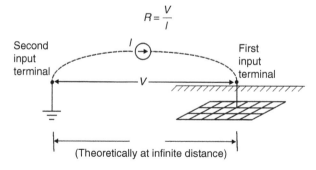

Figure 10.1   Illustration of a theoretical circuit for evaluation of a d.c. ground resistance

High-frequency grounding   505

However, transient impedance is a characteristic that depends on the particular wave shape of the excitation $i(t)$.

Two practically equivalent approaches to excitation-independent ground impedance have been widely used. The first is the time-domain ground surge impedance $z_0(t)$, which is a ratio of the voltage response to a unit step current excitation. The second is the frequency-domain alternative to the surge impedance: ground harmonic impedance. As is well known, the harmonic impedance is a Fourier transform of the unit impulse response [34]. Because the unit impulse function has a constant spectrum, the harmonic impedance $\underline{Z}(\omega)$ may be evaluated simply by determining the voltage phasor $\underline{V}(\omega)$ as a response to a steady-state time harmonic current excitation $\underline{I}(\omega) = 1 - A$ in a frequency range up to the highest frequency of interest for the transient study:

$$\underline{Z}(\omega) = \frac{\underline{V}(\omega)}{\underline{I}(\omega)} \qquad (10.3)$$

Here, $\omega = 2\pi f$, where f is frequency and the underscore denotes a complex variable. It is worth noting that the surge impedance can be determined from the harmonic impedance and vice versa [34]. For example, the surge impedance is given by [34]

$$z_0(t) = R + \frac{2}{\pi} \int_0^\infty \frac{X(\omega)}{\omega} \cos \omega t \, d\omega \qquad (10.4)$$

where R is the d.c. ground resistance [see equation (10.1)] and $X(\omega)$ is the imaginary part of the harmonic impedance $\underline{Z}(\omega)$ [see equation (10.3)] (Figure 10.2).

Both surge and harmonic impedances depend solely on the geometry and the electromagnetic properties of the grounding system and the medium. As is well known, both can be used to determine the time response to an arbitrary excitation [34]. For example, voltage $v(t)$, as a response to an arbitrary current pulse $i(t)$, may be measured or taken from a simulated lightning current pulse, and is given by [25]

$$v(t) = \mathbf{F}^{-1}\{\mathbf{F}[i(t)] \cdot \underline{Z}(\omega)\} \qquad (10.5)$$

where $\mathbf{F}$ and $\mathbf{F}^{-1}$ denote Fourier and inverse Fourier transforms, respectively.

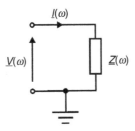

Figure 10.2   Harmonic ground impedance as a circuit element. The d.c. ground resistance is a limit for $\omega \to 0$.

506 Lightning Protection

A basic requirement for a frequency-domain analysis is that the system is linear, which makes this method unsuitable for modelling of non-linear phenomena. However, it is well suited for modelling important frequency-dependent phenomena.

When considering ground electrodes subjected to a lightning-related surge, analyses in the time and frequency domains are mutually related and the following statements apply.

- The initial surge state characterized by time-varying impedance is related to the high-frequency part of the harmonic impedance.
- The latter stationary condition characterized by fixed impedance is related to the low-frequency or d.c. resistance.

10.3 Basic field considerations

Circuit concepts in the previous section are the basis for engineering analysis, but field considerations are necessary for analysis of the limitations of the circuit concepts and for a better physical understanding of the grounding transient and high-frequency behaviour. A field approach is often also necessary to determine more precisely the ground impedances, particularly for cases of more complex grounding electrode arrangements, at high frequencies, and in more conductive earth. The usual simple model of earth is a homogeneous and isotropic half space with a plane interface with air, characterized by frequency-independent constitutive parameters: a fixed electrical conductivity σ on the order of ~ 0.0001 to 0.1 S m^{-1}, permittivity ε (with relative permittivity of ~ 10) and permeability identical to the permeability of air, $\mu = \mu_0$. Of these three constants, the conductivity exhibits by far the largest variations in its magnitude. In this analysis we will disregard the non-linear behaviour of the earth, which may arise as a result of high-intensity currents. However, as discussed in Section 10.8, many practically interesting consequences of frequency-dependent behaviour are not affected by non-linear behaviour.

It is important to distinguish between electromagnetic propagation in air and in the earth for the same frequency. The key quantity that gives a quantitative estimate related to the propagation effects is the TEM wave propagation constant $\underline{\Gamma}$, which in earth is given by [36]

$$\underline{\Gamma} = \alpha + j\beta = \sqrt{j\omega\mu_0(\sigma + j\omega\varepsilon)}$$

$$\alpha = \omega\sqrt{\mu_0\varepsilon}\left\{\frac{1}{2}\left[\sqrt{1+\left(\frac{\sigma}{\omega\varepsilon}\right)^2}-1\right]\right\}^{1/2}, \quad \beta = \omega\sqrt{\mu_0\varepsilon}\left\{\frac{1}{2}\left[\sqrt{1+\left(\frac{\sigma}{\omega\varepsilon}\right)^2}+1\right]\right\}^{1/2}$$

(10.6)

where α and β are attenuation and phase constants, respectively, and $j = \sqrt{-1}$.

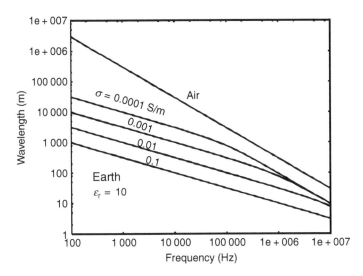

Figure 10.3   Wavelength in air and in earth with different conductivities

The parameters that are important to consider are λ (wavelength), v (velocity of propagation) and δ (skin depth):

$$\lambda = \frac{2\pi}{\beta}, \qquad v = \frac{\omega}{\beta}, \qquad \delta = \frac{1}{\alpha} \qquad (10.7)$$

Figures 10.3 to 10.5 illustrate the strong dependence of λ, v and δ on the conductivity of the earth.

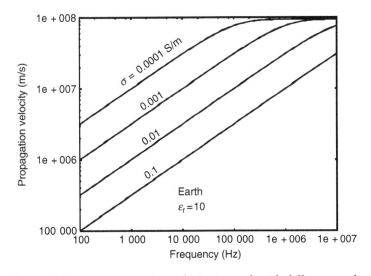

Figure 10.4   TEM wave propagation velocity in earth with different conductivities

508 Lightning Protection

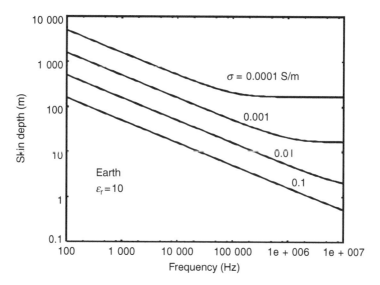

Figure 10.5   TEM wave skin depth in earth with different conductivities

Wavelength λ is smaller in earth than in the air, even for several orders of magnitude in more conductive soil and at lower frequencies (Figure 10.3). An important parameter related to wavelength is the electrical dimension of the system, which is a coefficient equal to the ratio of the physical dimension L and the wavelength λ. If the system's electrical dimensions are much smaller than unity, then the electromagnetic waves do not experience significant change along such distances, and propagation effects can be neglected. In such a case, a quasi-static approximation can be applied for analysis. However, it is important to note that a system might have considerably larger electrical dimensions for the same frequency in earth than in air due to the smaller wavelengths there. As a result, the quasi-static approximation might be applicable in air but not in earth for the same buried system and frequency.

The velocity of propagation v [see equation (10.7)] is much smaller in earth than in air, particularly at low frequencies and for more conductive soil. It approaches the velocity of light at higher frequencies and in less conductive soil (Figure 10.4).

Figure 10.5 illustrates the skin depth δ [see equation (10.7)], that is, the distance the wave must travel in earth to reduce its value to 36.8 per cent of the value at the earth's surface. Figure 10.5 illustrates the field tendency to localize near the source at higher frequencies.

The performance of grounding systems is determined by the rate of energy dissipation and storage in the soil. Based on Poynting's theorem, terms of the mean values of power dissipated and stored in a volume V of soil can be evaluated [35] as follows:

$$\int_V \sigma |\vec{E}|^2 \, dV - \text{dissipated power (related to circuit resistive behaviour)}$$

$\frac{1}{2}\int_V \varepsilon |\vec{E}|^2 dV$ – power stored in the electric field (related to circuit capacitive behaviour)

$\frac{1}{2}\int_V \mu_0 |\vec{H}|^2 dV$ – power stored in the magnetic field (related to circuit inductive behaviour)

Here, $\vec{E}$ and $\vec{H}$ are electric and magnetic field vector phasors, respectively.

The results in Figures 10.4 and 10.5 show that the volumes V available for dissipation and power storage are different at high and low frequencies, and also in the initial surge state and the latter stationary condition. A smaller V at high frequencies and in the surge state results in higher intensities of fields, current and potential near the feed point.

10.4 Frequency-dependent characteristics of the soil

In spite of the wide use of the frequency-independent model, it is well known, however, that in real earth the constitutive parameters σ and ε can be functions of frequency, with rather large variations in their values [36,37]. Tesche [38] recently studied the sensitivity of a calculated transient response to variations in these parameters. As discussed in Reference 36, the complex permittivity $\underline{\varepsilon}$ and the complex conductivity $\underline{\sigma}$ may be assumed as

$$\underline{\sigma} = \sigma' - j\sigma'', \quad \underline{\varepsilon} = \varepsilon' - j\varepsilon'' \tag{10.8}$$

Using real and imaginary parts of the complex permittivity and the complex conductivity (10.8) a real-valued, frequency-dependent effective permittivity and conductivity may be defined [38]:

$$\sigma_{\text{eff}}(\omega) = \omega\varepsilon'' + \sigma', \quad \varepsilon_{\text{eff}}(\omega) = \varepsilon' - \frac{\sigma''}{\omega} \tag{10.9}$$

As an example, Figure 10.6 illustrates the measured frequency dependence of the effective relative permittivity and the effective conductivity of sandy soil with different water contents.

Tesche [38] has shown that Messier's empirical model [40] for fitting the parameters for ε_{eff} and σ_{eff} given by the expressions

$$\varepsilon_{\text{eff}}(\omega) = \varepsilon_\infty + \sqrt{\frac{2\sigma\varepsilon_\infty}{\omega}} \tag{10.10}$$

and

$$\sigma_{\text{eff}}(\omega) = \sigma + \sqrt{2\sigma\varepsilon_\infty\omega} \tag{10.11}$$

can be rather accurate. Moreover, these expressions can be shown to provide causal results for computed responses, and hence they obey the Kramer–Kronig

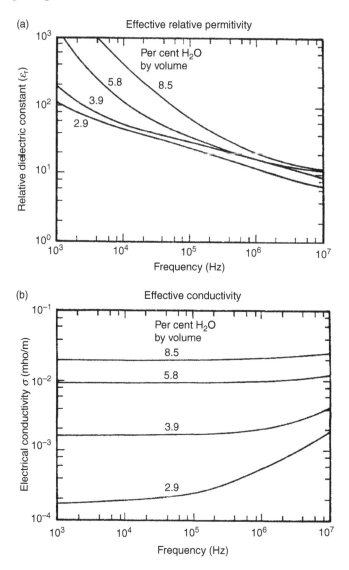

Figure 10.6 Plots of the measured relative effective permittivity (a) and the effective conductivity (b) of sandy soil with different water contents (by volume) (from Reference 39)

relationships [35]. This model for lossy earth depends only on two parameters: high-frequency permittivity ε_∞ and d.c. conductivity σ (together, of course, with the frequency).

For practical application of the above model, the measured data necessary for its implementation are rarely available. However, this phenomenon improves grounding system performance, so ignoring it always gives conservative results.

10.5 Grounding modelling for high frequencies

Vertical ground rods are one of the most simple and most commonly used means for earth termination of electrical and lightning protection systems [6] (Figure 10.7a). Their behaviour at 50 or 60 Hz is well understood using a simplified analysis based on static approximation [41]. By the traditional engineering approach at low frequencies, ground impedance of a single vertical rod is represented by a single resistor (Figure 10.7b), and at high frequencies by a lumped *RLC* circuit [3] (Figure 10.7c).

The classical expressions of lumped ground resistance R, inductance L, and capacitance C for a vertical rod are given by [1]

$$R = \frac{\rho}{2\pi\ell}A_1; \quad L = \frac{\mu_0\ell}{2\pi}A_1; \quad C = \frac{2\pi\varepsilon\ell}{A_1} \quad (10.12)$$

$$A_1 = \ln\frac{4\ell}{a} - 1, \quad (\ell \gg a)$$

where ℓ and a are the length and radius of the rod, respectively (Figure 10.7a).

The parameters of the *RLC* circuit [see equation (10.12)] may be simply used to approximate the per unit length parameters of a distributed-parameter circuit [1] (Figure 10.7d):

$$R' = 1/G' = R\ell\,(\Omega\,\text{m}); \quad L' = L/\ell\,(\text{H}\,\text{m}^{-1}); \quad C' = C/\ell\,(\text{F}\,\text{m}^{-1}) \quad (10.13)$$

The transmission line may be considered to be open at the lower end, and the input impedance (equivalent to the harmonic ground impedance) is given by [1]

$$\underline{Z}(\omega) = \underline{Z}_0 \coth \gamma\ell \quad (10.14)$$

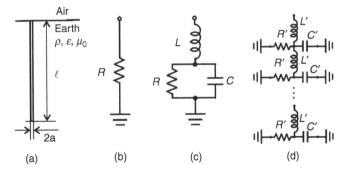

Figure 10.7 Low-current models of a vertical ground rod: (a) physical situation, (b) low-frequency equivalent circuit, (c) high-frequency lumped RLC circuit, (d) high-frequency distributed parameters circuit

Here $\underline{\gamma}$ and $\underline{Z}_0$ are the propagation constant and characteristic impedance, respectively [1]:

$$\underline{\gamma} = \sqrt{j\omega L'(G' + j\omega C')}; \quad \underline{Z}_0 = \sqrt{j\omega L'/(G' + j\omega C')} \tag{10.15}$$

Equations (10.12), (10.13) and (10.15) may be combined for the vertical rod:

$$\underline{\gamma} = \sqrt{j\omega\mu_0(\sigma + j\omega\varepsilon)}; \quad \underline{Z}_0 = \frac{A_1}{2\pi}\sqrt{\frac{j\omega\mu_0\rho}{(1 + j\omega\varepsilon\rho)}} \tag{10.16}$$

The logarithmic term A_1 [see equation (10.12)] involving ℓ cancels in the expression for $\underline{\gamma}$ [see equation (10.16)], but not for the case of $\underline{Z}_0$. However, because the variation with ℓ is logarithmic, the results are not critically dependent on ℓ. The above per unit parameters are based on the TEM mode of propagation approximation. In a theoretical analysis, Roubertou and colleagues [22] have discussed the improper use of transmission line theory for the vertical rod and developed an approach based on electromagnetic theory. Figure 10.8 shows a comparison between computed impedances of two ground rods (the first one 3 m and the second one 30 m long) in earth with resistivity 100 Ω m and relative permittivity 10 by the three modelling approaches, based on lumped RLC circuit, transmission line and electromagnetic theory.

The impedance computed by the RLC model is in agreement with the other methods for rod lengths less than approximately one-tenth of the wavelength, which is in agreement with the conclusions of Reference 27. The model with the distributed-parameter circuit ('2' in Figure 10.8) follows the electromagnetic model ('3' in Figure 10.8), but it still significantly overestimates the values at higher frequencies. Better agreement between results is achieved for small rods in very resistive soil.

Similarly, the approximate modelling procedure for horizontal wires is based on classical lumped ground resistance R, inductance L and capacitance C formulae [1]:

$$R = \frac{\rho}{\pi\ell}A_2; \quad L = \frac{\mu_0\ell}{2\pi}A_2; \quad C = \frac{\pi\varepsilon\ell}{A_2}$$

$$A_2 = \ln\frac{2\ell}{\sqrt{2ad}} - 1, \quad (\ell \gg a, \; d \ll \ell) \tag{10.17}$$

where d is the depth of burial.

Equations (10.12), (10.15) and (10.17) may be combined for the horizontal wire

$$\underline{\gamma} = \sqrt{j\omega\mu_0(1/\rho + j\omega\varepsilon)/2}; \quad \underline{Z}_0 = \frac{A_2}{\pi}\sqrt{\frac{j\omega\mu_0}{2(\sigma + j\omega\varepsilon)}} \tag{10.18}$$

The more accurate transmission line approach is based on Sunde's formulae for per unit length longitudinal impedance $\underline{Z}'$ and transversal admittance $\underline{Y}'$ [1]:

$$\underline{Z}'(\underline{\gamma}) \approx \frac{j\omega\mu_0}{2\pi}\ln\frac{1.85}{a\sqrt{\underline{\gamma}^2 + \underline{\Gamma}^2}}; \quad \underline{Y}'(\underline{\gamma}) \approx \pi(\sigma + j\omega\varepsilon)\left(\ln\frac{1.12}{\underline{\gamma}\sqrt{2ah}}\right)^{-1} \tag{10.19}$$

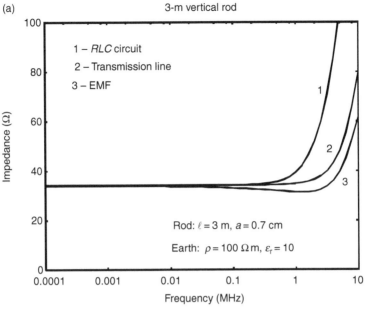
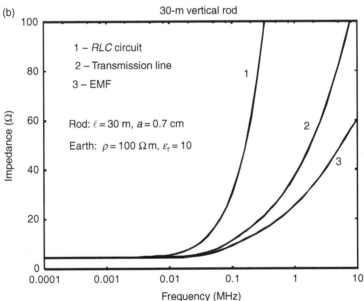

Figure 10.8 Comparison of results for harmonic ground impedance of vertical rods for (1) a lumped RLC equivalent circuit [equation (10.12)], (2) transmission line theory [equation (10.16)], (3) an electromagnetic field approach [22] in (a) a 3-m vertical rod and (b) a 30-m vertical rod

514 *Lightning Protection*

Solution of the transcendent equation

$$\gamma = \sqrt{Z'(\gamma) \cdot Y'(\gamma)} \quad (10.20)$$

yields a value for the propagation constant γ, which with substitution in equation (10.19) evaluates the characteristic impedance as

$$Z_0 = \sqrt{Z'(\gamma)/Y'(\gamma)} \quad (10.21)$$

and subsequently the ground impedance [see equation (10.14)].

The method in [25,26] is developed similarly to the previously developed method in [22]. Figure 10.9 presents a comparison of the results from the three modelling approaches, the transmission line approach based on equation (10.18) and equations (10.19) to (10.21) and the electromagnetic field approach [26]. Results for short wires (10 m) at high frequencies are in good agreement for the case in Figure 10.9a, but the approach based on equations (10.19) to (10.21) is not applicable for frequencies lower than 100 kHz. For longer wires (100 m) the model based on equation (10.18) ('1' in Figure 10.9) overestimates the values at higher frequencies, while the model based on equations (10.19) to (10.21) ('2' in Figure 10.9) follows the electromagnetic model ('3' in Figure 10.9).

Different modelling approaches have been developed for more complex grounding electrodes arrangements (mentioned in Section 10.1). There is no available systematically developed and reliable set of experimental data that would serve as a standard, so the electromagnetic model is used for the results in the rest of this chapter.

10.6 Frequency-dependent grounding behaviour

Figure 10.10 shows typical frequency dependence of the grounding harmonic impedance. The figure shows the ratio of the impedance modulus $|\underline{Z}(\omega)|$ and the low-frequency ground resistance R. Two frequency ranges may be distinguished: the low-frequency range, where the impedance is nearly constant, that is, frequency independent, and the high-frequency range, where impedance is frequency dependent. Speaking in circuit terms, the high-frequency grounding behaviour may be categorized as inductive when $|\underline{Z}(\omega)|/R > 1$, resistive when $|\underline{Z}(\omega)|/R \approx 1$, or capacitive when $|\underline{Z}(\omega)|/R < 1$. The important parameter in the case of inductive behaviour is the limiting frequency between the low-frequency and high-frequency ranges, the characteristic frequency F_c [44]. Figure 10.10 also illustrates the dominant influence of the earth's resistivity on the grounding frequency-dependent behaviour, because the same electrode behaves differently in earths with different resistivity.

Resistive and capacitive behaviour is advantageous because the high-frequency impedance is equal to or smaller than the low-frequency resistance to earth and consequently the grounding high-frequency performance is the same or better than at low frequencies. However, this is typical usually only for electrodes with smaller dimensions and in more resistive soils. More frequently the lengths of the electrodes and the earths' characteristics are such that the grounding exhibits inductive behaviour and

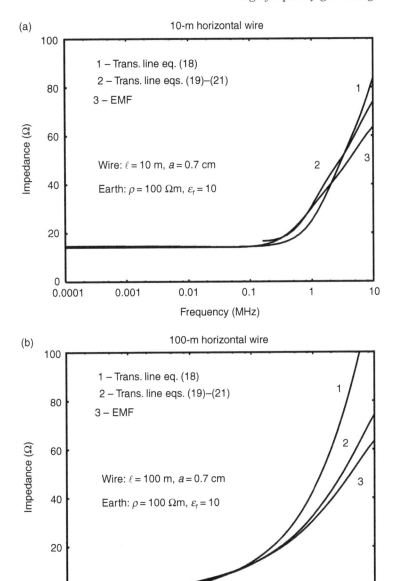

Figure 10.9 Comparison of results for harmonic ground impedance of horizontal wires for (1) transmission line equation (10.18), (2) transmission line equations (10.19) to (10.21), (3) the electromagnetic field approach [26] in (a) a 10-m horizontal wire and (b) a 100-m horizontal wire

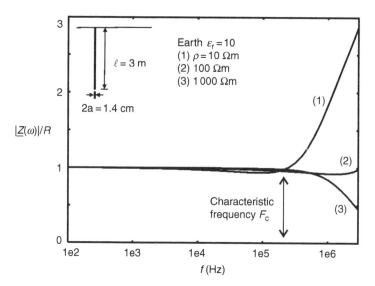

Figure 10.10 Typical frequency-dependent behaviour of the harmonic ground impedance: (1) Inductive behaviour, (2) resistive behaviour, (3) capacitive behaviour

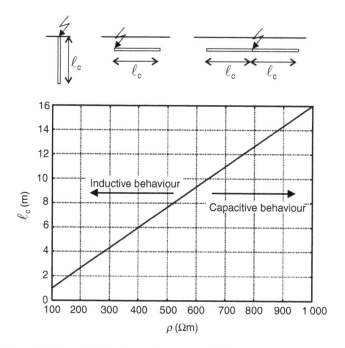

Figure 10.11 Regions of inductive and capacitive behaviour of grounding electrodes

consequently worse high-frequency performance. Figure 10.11 gives the regions of inductive and capacitive behaviour of vertical and horizontal rods depending on the earth's resistivity and the characteristic dimension ℓ_c, which is defined as a distance between the feed point and the most distant point of the grounding electrodes [43,44].

The first possibility for providing good high-frequency performance is to use smaller electrodes with capacitive or resistive behaviour. However, practically, this is seldom possible because longer electrodes are usually required to fulfil safety requirements at 50/60 Hz. Figure 10.12 illustrates typical high-frequency inductive ground impedance dependence on the grounding electrode length. Typically the characteristic frequency F_c is smaller for longer grounding electrodes with larger values of $\underline{Z}(\omega)$ in the high-frequency range and consequently with worse high-frequency performance. In addition F_c is also smaller in more conductive earth. For each characteristic frequency F_c there is a limiting length ℓ_R for low-frequency resistive behaviour, which is also referred to as the harmonic effective length (Figure 10.12).

One relation that determines such harmonic effective length ℓ_R as a function of the earth's resistivity ρ and the characteristic frequency F_c is [43,44]

$$\ell_R = 0.6(\rho/F_c)^{0.43} \tag{10.22}$$

Equation (10.22) is also illustrated in Figure 10.13.

It can be seen from Figure 10.12 that for frequencies higher than F_c, which corresponds to the harmonic effective length ℓ_R, all electrodes longer than ℓ_R exhibit nearly the same behaviour. This means that above F_c only the length ℓ_R of the electrode, measured from the feed point, effectively dissipates current into earth, regardless of how much longer the electrode is.

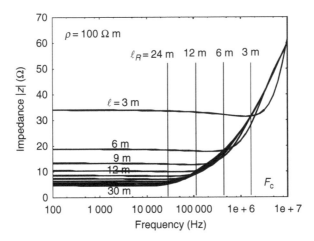

Figure 10.12   Harmonic impedance frequency dependence of horizontal wires in 100 Ω m earth

518 Lightning Protection

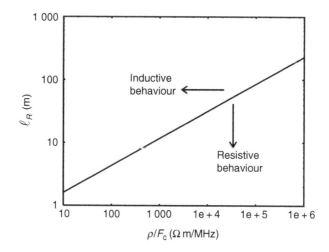

Figure 10.13 *The limiting grounding electrode length for resistive behaviour*

It has to be noted, however, that the practical importance of high-frequency behaviour in the case of pulse excitation depends on the frequency content of the excitation pulses. For example, if the frequency content of the lightning current pulses is lower than the characteristic frequency F_c, then the grounding behaviour will be resistive and not affected by the high-frequency inductive part of the impedance. As a first approximation, Gary [42] has suggested 100 kHz and 1 MHz frequency content for the first and subsequent strokes, respectively. This means that some electrodes that exhibit resistive behaviour for the first stroke might exhibit inductive behaviour for the subsequent stroke. This topic will be further discussed in Section 10.7.

The frequency-dependent analysis of more complex grounding electrode arrangements can also be based on the characteristic dimension, that is, the distance between the feed point and the most distant point of the arrangement. Figure 10.14 illustrates the influence of feed point location [28] on the high-frequency inductive behaviour of grid-like grounding electrode arrangements. It should be noted that for the corner feed point the characteristic dimension of the grounding grid is the length of the diagonal and is twice as large as for the centre feed point. As expected, the central feed point, that is, the smaller characteristic dimension, leads to higher characteristic frequency F_c, i.e. to better high-frequency performance. Having a central feed point rather than a corner one (in other words, twofold smaller characteristic dimension) broadens the low-frequency range with resistive behaviour by a factor of ten, i.e. from 1 kHz to nearly 10 kHz for soil with $\rho = 100\ \Omega$ m and from 10 to 100 kHz for soil with $\rho = 1\ 000\ \Omega$ m. In addition, in the high-frequency range the impedance for the central feed point arrangement is two times smaller than that of the corner feed point.

The influence of grounding grid size on harmonic impedance is illustrated in Figure 10.15. Five square ground grids with 10 m to 10 m square mesh are chosen for computations, with dimensions ranging from 10 m × 10 m to 120 m × 120 m.

High-frequency grounding 519

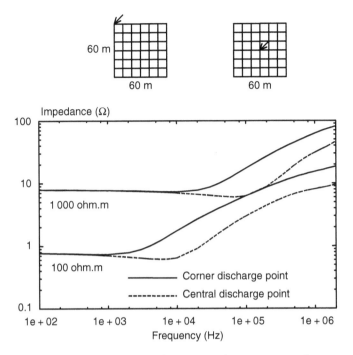

Figure 10.14 Influence of feed point location on harmonic impedance

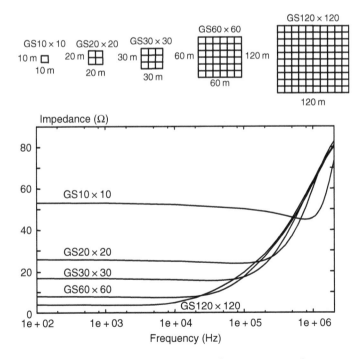

Figure 10.15 Influence of ground grid size on harmonic impedance

All grids are in soil with $\rho = 1\,000\,\Omega$ m, $\varepsilon_r = 9$ and with the feed point at the corner, so the characteristic dimension is always the length of the diagonal. Grounding grid characteristic dimension in Figure 10.15 has a similar influence as the single electrode in Figure 10.12. As for the single grounding wire, grounding grid size has a large influence on the low-frequency value of impedance to ground, but in the high-frequency range the behaviour of the different ground grids above a certain frequency becomes very similar. Clearly, the effective area of the grounding grids becomes smaller for higher frequencies.

The influence of grounding grid conductor separation on harmonic impedance is illustrated in Figure 10.16. Five square grounding grids with the same dimensions, 60 m × 60 m, with the number of meshes ranging from 4 to 124, are chosen for comparison. All grids are in soil with $\rho = 1\,000\,\Omega$ m and $\varepsilon_r = 10$. The feed point is at the corner. As is well known, ground grid conductor separation has only a small influence on the low-frequency ground resistance, for which the dominant influence is the area of the grid. A similar conclusion is valid for the high-frequency behaviour of ground grids when conductor separation is reduced from 30 m to 6 m by introducing denser square meshes. A greater influence on the reduction of impedance in the high-frequency range (in other words, on an improvement in high-frequency performance) is achieved through a further reduction of conductor separation to 3 m near the feed point, that is, in the effective area.

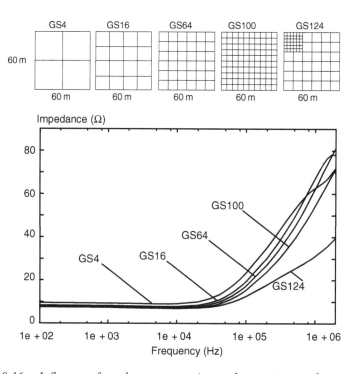

Figure 10.16 Influence of conductor separation on harmonic impedance

10.7 Frequency-dependent dynamic grounding behaviour

Here, we consider only the frequency-dependent aspects of dynamic behaviour. The non-linear behaviour of soil due to high currents is disregarded. However, in Section 10.8 the relation between frequency-dependent and non-linear dynamic behaviour will be discussed.

When harmonic impedance $\underline{Z}(\omega)$ [see equation (10.3)] is determined, then the time function of the electric potential at the feed point $v(t)$ as a response to an injected lightning current pulse $i(t)$ can be evaluated [see equation (10.5)]. Several parameters are used to characterize dynamic behaviour of earth electrodes. One is transient impedance $z(t)$ [see equation (10.2)]. Usually, the impulse impedance Z is used, which is defined as the ratio between the peak values of $v(t)$ and $i(t)$:

$$Z = \frac{\max[v(t)]}{\max[i(t)]} = \frac{V_{\max}}{I_{\max}} \qquad (10.23)$$

Another usual parameter is the impulse coefficient (efficiency), defined as the ratio between the impulse impedance and the resistance at low frequency, Z/R. It is worth noting that values of impulse coefficient larger than one are related to poorer transient performance.

The meaning of these parameters will be illustrated in an example. Consider a 12-m-long vertical rod constructed of copper with radius 0.7 cm in earth with a resistivity of 100 Ω m and relative permittivity of 10. Figure 10.17 shows the modulus of the harmonic impedance.

It can be seen that harmonic impedance is frequency independent and is equal to the low-frequency resistance to ground, $R = 10.3\ \Omega$, up to the characteristic frequency, $F_c \approx 100$ kHz. For higher frequencies, it exhibits inductive behaviour and its value becomes larger than R. The influence of such larger high-frequency values of $\underline{Z}(\omega)$

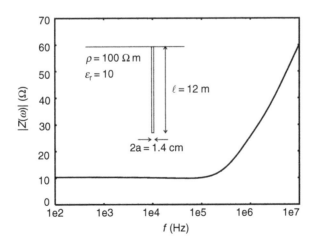

Figure 10.17 Modulus of the harmonic impedance to ground of a 12-m vertical rod

522 Lightning Protection

on the transient response depends on the frequency content of the excitation lightning current impulse.

Figure 10.18a shows the injected lightning current pulse typical for the first stroke $i(t)$, and the response to this excitation, the potential at the feeding point $v(t)$, and the transient impedance $z(t)$ [see equation (10.2)]. The procedure of determining the response $v(t)$ in equation (10.5) may be interpreted as passing the excitation $i(t)$ through a 'filter' with a frequency characteristic given by the harmonic impedance $Z(\omega)$ in Figure 10.17. The first stroke current pulse $i(t)$ (Figure 10.18a) has zero-to-peak time of about 8.4 μs and maximum steepness of ~ 12 kA s^{-1}. Consequently, it does not have significant frequency content above the characteristic frequency of about $F_c \approx 100$ kHz (Figure 10.17), and is not affected by the high-frequency part of the 'filter'. The response is mostly determined by the frequency-independent part, that is, the pure resistive part of the harmonic impedance, below F_c (Figure 10.17). Consequently, the response, that is, the voltage pulse

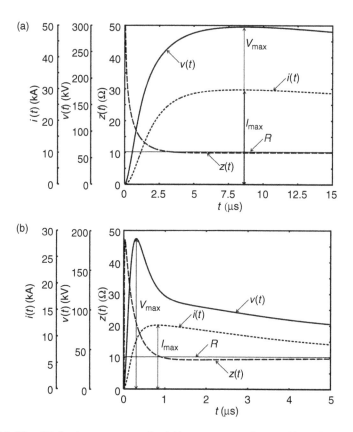

Figure 10.18 Lightning current pulse i(t), transient voltage v(t) and ground impedance of a 12-m ground rod: (a) response to a typical first stroke current pulse; (b) response to a typical subsequent stroke current pulse

waveshape, is not significantly modified in comparison to the current pulse waveshape and their maxima occur at the same time. Although the transient impedance $z(t)$ goes very quickly to some high value (larger than 50 Ω), it quickly settles to the low-frequency resistance value ($R = 10.3\ \Omega$) during the current rise. The transient impedance determines the duration of the transient period; here after about 3 µs the transient period is practically finished and a quasi-static analysis might subsequently be applied. The impulse coefficient is excellent, equal to one, that is, the impulse impedance is equal to the low-frequency resistance $Z = R$.

For the subsequent stroke current pulse injected in the same round rod the situation is different (Figure 10.18b). The subsequent stroke current pulse is more rapidly time-varying than the first stroke. It has a zero-to-peak time of $\sim$0.8 µs and maximum steepness of $\sim$40 kA s^{-1}. Consequently, it has significant frequency content above the characteristic frequency $F_c \approx 100$ kHz of the 'filter' (in Figure 10.17), and so the response is influenced by the inductive part of $Z(\omega)$. The 'filter' amplifies the high-frequency components of the pulse, which results in a large peak V_{max} of the transient voltage $v(t)$ during the rise of the current $i(t)$. Typically for inductive behaviour, the voltage pulse precedes the current pulse. This causes a larger value of the impulse impedance, $Z = 15.7\ \Omega$, than the low-frequency resistance, $R = 10.3\ \Omega$, and the impulse coefficient is equal to 1.5. The transient impedance $z(t)$, similar to the case of the first stroke (Figure 10.18a) rises very rapidly to a high value (of $\sim$47 Ω), but also quickly (in $\sim$1 µs) settles to values near the low-frequency value, $R = 10.3\ \Omega$, shortly after the occurrence of the peak of the current pulse.

The example in Figures 10.17 and 10.18 illustrates the fact that high-frequency inductive behaviour of grounding might result in large peaks of transient potential at the feed point in cases when the current pulses have enough high-frequency content to be influenced by high-frequency inductive behaviour. However, after a few microseconds, transient processes are practically finished and the transient impedance $z(t)$ settles to the value of the power-frequency resistance R.

Figure 10.19 shows the dependence of the impulse coefficient of vertical ground rods on their length and the earth's resistance for the first and subsequent stroke current pulses [45].

It can be concluded that impulse performance is worse for longer rods in better conductive earth and for faster varying pulses, such as subsequent strokes. This effect is less important in less conductive soil and for not so quickly varying pulses, such as first strokes.

The dynamic effective length of a grounding rod is defined as the limiting length above which the impulse coefficient is larger than one. Values from Figure 10.19 are given in Table 10.1.

Figure 10.19 might be used for a first estimate of impulse efficiency and effective lengths of some other ground rod arrangements. Table 10.2 gives the percentage of the reduction of the impulse efficiency of the single vertical rod, given in Figure 10.19, for other multiple or horizontal grounding rod arrangements.

The use of multiple ground rod arrangements improves impulse efficiency. Horizontal rods are slightly less effective at power frequencies in comparison to vertical rods, but have better impulse efficiency.

524 *Lightning Protection*

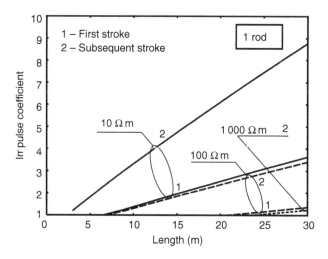

Figure 10.19   Impulse coefficient of a single vertical rod

Table 10.1   Dynamic effective lengths for the grounding rod in Figure 10.19

Earth's resistivity (Ω m)	Lightning stroke [45]	Vertical grounding rod dynamic effective length (m)
10	First	7.5
	Subsequent	3
100	First	22
	Subsequent	7.5
1 000	First	>30
	Subsequent	22

Table 10.2   Impulse coefficient and effective lengths of several ground rod arrangements

Ground rod arrangement						
Impulse coefficient (%)*	100	95	85	85	80	70
Effective length (%)*	100	105	118	118	125	143

*Percentage of values in Figure 10.19.

High-frequency grounding 525

It should be emphasized that the impulse coefficient and the effective length that characterize possible worsening of the grounding performance during the lightning pulse in comparison with low-frequency performance are related only to the transient period, which may be very short (e.g. 3 μs for the first stoke and 1 μs for the subsequent stroke for the case in Figure 10.19).

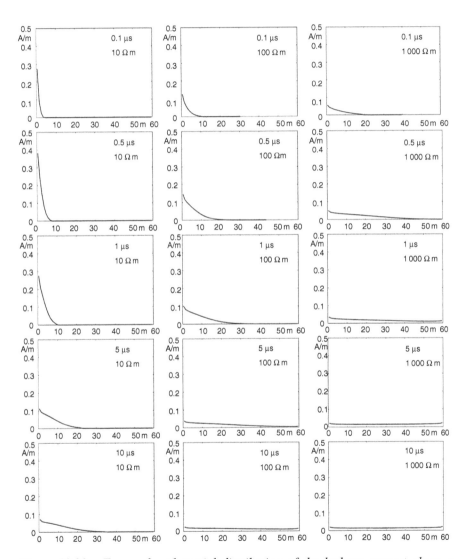

Figure 10.20 Temporal and spatial distribution of the leakage current along a 60-m-long horizontal wire as a response to an injected double-exponential ($T_1/T_2 = 1$ μs/50 μs) current pulse with a peak value 1 kA, at 0 m end in earths with resistivities of 10, 100 and 1 000 Ω m

10.8 Relation between frequency-dependent and non-linear grounding behaviour

When the electric field at the surface of the grounding electrode exceeds a value typically in the range 100 to 500 kV m^{-1}, electrical breakdown occurs in the surrounding earth [11,46]. This practically enlarges the dimensions of the grounding electrode and improves its performance.

The electric field at the surface of the grounding electrode is related to the leakage current:

$$\vec{J} = \sigma \cdot \vec{E} \qquad (10.24)$$

Leakage current distribution characterizes the performance of the grounding electrodes because their main function is to dissipate the current to the earth. Figure 10.20 shows temporal and spatial distribution of the leakage current along a 60-m-long horizontal wire with radius $a = 0.7$ cm as a response to an injected double-exponential current pulse with a peak value 1 kA and zero-to-peak time $T_1 = 1$ μs and time-to-half $T_2 = 50$ μs, at the 0 m end in earths with resistivities of 10 Ω m (first column, Figure 10.20), 100 Ω m (second column, Figure 10.20) and 1 000 Ω m (third column, Figure 10.20). The velocity of electromagnetic field propagation is substantially smaller in less resistive earths, where a practically smaller part of the electrode is effective in dissipating current in the first moments of the stroke, resulting in larger values of the leakage current and electric field near the feed point. However, in more resistive earths, where non-linear phenomena are more likely to occur, due to the higher velocity of the electromagnetic field propagation, the transient period is smaller and a distribution equivalent to a d.c. (or 50/60 Hz) distribution of the leakage current (and electric field at the electrode surface) is spread over the whole length of the electrode very quickly after the beginning of the lightning pulse.

Table 10.3 gives the transient maximal and 50 Hz values of the current density and electric field for injected current with a peak value of 1 kA.

The critical electric field intensity for a soil breakdown (~300 kV m^{-1}) is reached in the case in Figure 10.20 even for 50 Hz in very resistive soil ($\rho = 1\,000$ Ω m).

Table 10.3 Maximal transient and 50 Hz values of the leakage current and electric field at the surface of the 60-m electrode at the feed point

Earth's resistivity (Ω m)		10	100	1 000
Maximal transient values	I' (kA m^{-1})	0.43	0.18	0.08
	E (kV m^{-1})	98	409	1 819
50 Hz values	I' (kA m^{-1})	0.016	0.016	0.016
	E (kV m^{-1})	3.6	36	364

Although the transient maximal values are larger, they last for a very short period of time (less than 1 μs). On the other hand, soil ionization effects may last the large part of the pulse. However, since soil ionization usually improves the performance its neglect may be considered as an assumption on the 'safe' side. More accurate modelling that takes into account both soil ionization and frequency dependent effects is discussed in [48,49].

References

1. Sunde E.D. *Earth Conduction Effects in Transmission Systems*, 2nd edn. New York: Dover Publications, 1968.
2. Bewley L.V. *Traveling Waves on Transmission Systems*, 2nd edn. New York: Wiley, 1951.
3. Rudenberg R. *Electrical Shock Waves in Power Systems*. Cambridge: Harvard University Press, 1968.
4. Ryabkova E.Y. *Grounding in High Voltage Stations* (in Russian). Moscow: Energia, 1978.
5. Meliopoulos A.P. *Power System Grounding and Transients*. New York and Basel: Marcel Dekker, 1988.
6. Hasse P. and Wiesinger J. *Handbook for Lightning and Grounding*, 4th edn (in German). Munich: Pflaum, 1993.
7. Saint-Privat-d'Allier Research Group, 'Eight years of lightning experiments at Saint-Privat-d'Allier'. *Review Generale de l'Electricite (RGE)* 1982;**91**: 561–82.
8. Bourg S., Sacepe B. and Debu T. 'Deep earth electrodes in highly resistive ground: frequency behaviour'. *Proceedings of the 1995 IEEE International Symposium on Electromagnetic Compatibility*, 1995, pp. 584–89.
9. Stojkovic Z., Savic M.S., Nahman J.M., Salamon D. and Bukorovic B. 'Sensitivity analysis of experimentally determined grounding grid impulse characteristics'. *IEEE Transactions on Power Delivery* 1998;**13**(4):1136–42.
10. Sekioka S., Hayashida H., Hara T. and Ametani A. 'Measurements of grounding resistances for high impulse currents'. *IEE Proceedings – Generation, Transmission and Distribution* 1998;**145**(6):693–99.
11. Liew A.C. and Darveniza M. 'Dynamic model of impulse characteristics of concentrated earths'. *Proceedings of IEE* 1974;**121**(2):123–135.
12. Ramamoorty M., Narayanan M.M.B., Parameswaran S. and Mukhedkar D. 'Transient performance of grounding grids'. *IEEE Transactions on Power Delivery* 1989;**PWRD-4**:2053–59.
13. Geri A. 'Behaviour of grounding systems excited by high impulse currents: the model and its validation'. *IEEE Transactions on Power Delivery* 1999; **14**(3):1008–17.
14. Otero A.F., Cidras J. and del Alamo J.L. 'Frequency-dependent grounding system calculation by means of a conventional nodal analysis technique'. *IEEE Transactions on Power Delivery* 1999;**14**(3):873–78.

15. Grcev L. and Popov M. 'On high-frequency circuit equivalents of a vertical ground rod'. *IEEE Transactions on Power Delivery* 2005;**20**(2):1598–1603.
16. Devgan S.S. and Whitehead E.R. 'Analytical models for distributed grounding systems'. *IEEE Transactions on Power Apparatus and Systems* 1973;**PAS-92**: 1763–81.
17. Mazzetti C. and Veca G.M. 'Impulse behavior of grounded electrodes'. *IEEE Transactions on Power Apparatus and Systems* 1983;**PAS-102**:3148–56.
18. Velazquez R. and Mukhedkar D. 'Analytical modeling of grounding electrodes'. *IEEE Transactions on Power Apparatus and Systems* 1984;**103**:1314–22.
19. Menter F. and Grcev L. 'EMTP-based model for grounding system analysis'. *IEEE Transactions on Power Delivery* 1994;**9**:1838–49.
20. Liu Y., Zitnik M. and Thottappillil R. 'An improved transmission-line model of grounding system'. *IEEE Transactions on Electromagnetic Compatibility* 2001;**43**:348–55.
21. Liu Y., Theethayi N. and Thottappillil R. 'An engineering model for transient analysis of grounding system under lightning strikes: nonuniform transmission-line approach'. *IEEE Transactions on Power Delivery* 2005;**20**(2):722–30.
22. Roubertou D., Fontaine J., Plumey J.P. and Zeddam A. 'Harmonic input impedance of earth connections'. *Proceedings of the 1984 IEEE International Symposium on Electromagnetic Compatibility*, 1984, pp. 717–20.
23. Dawalibi F. 'Electromagnetic fields generated by overhead and buried short conductors, part I – single conductor'. *IEEE Transactions on Power Delivery* 1986;**PWRD-1**(4):105–11.
24. Dawalibi F. 'Electromagnetic fields generated by overhead and buried short conductors, part II – ground networks'. *IEEE Transactions on Power Delivery* 1986;**PWRD-1**(4):112–19.
25. Grcev L. and Dawalibi F. 'An electromagnetic model for transients in grounding system'. *IEEE Transactions on Power Delivery* 1990;**5**(4):1773–81.
26. Grcev L. 'Computer analysis of transient voltages in large grounding systems'. *IEEE Transactions on Power Delivery* 1996;**11**(2):815–23.
27. Olsen R. and Willis M.C. 'A comparison of exact and quasi-static methods for evaluating grounding systems at high frequencies'. *IEEE Transactions on Power Delivery* 1996;**11**:1071–81.
28. Grcev L. and Heimbach M. 'Frequency dependent and transient characteristics of substation grounding system'. *IEEE Transactions on Power Delivery* 1997; **12**(1):172–78.
29. Papalexopoulos A.D. and Meliopoulos A.P. 'Frequency dependent characteristics of grounding systems'. *IEEE Transactions on Power Delivery* 1987;**PWRD-2**: 1073–81.
30. Heimbach M. and Grcev L. 'Grounding system analysis in transients programs applying electromagnetic field approach'. *IEEE Transactions on Power Delivery* 1997;**12**(1):186–93.
31. Visacro S. and Soares A. 'HEM: a model for simulation of lightning-related engineering problems'. *IEEE Transactions on Power Delivery* 2005;**20**(2): 1206–208.

32. Chowdhuri P. *Electromagnetic Transients in Power Systems*. New York: Wiley, 1996.
33. IEEE Std. 80-2000, *IEEE Guide for Safety in AC Substation Grounding*. New York: IEEE, 2000.
34. Papoulis A. *The Fourier Integral and its Applications*. New York: McGraw-Hill, 1962.
35. Balanis C.A. *Advanced Engineering Electromagnetics*. New York: Wiley, 1989.
36. Ramo S., Whinnery J.R. and van Duzer T. *Fields and Waves in Communication Electronics*, 3rd edn. New York: Wiley, 1994.
37. Visacro S. and Portela C.M. 'Investigation of earthing system behavior on the incidence of atmospheric discharges at electrical systems'. *Proceedings of ICLP*, Interlaken, Switzerland, 1990, 3.8p.
38. Tesche F.M. 'On the modeling and representation of a lossy earth for transient electromagnetic field calculations'. Theoretical note 367, Kirtland AFB, NM, July 2002.
39. Mallon C. *et al.* 'Low-field electrical characteristics of soil'. Theoretical note 315, Kirtland AFB, NM, January 1981.
40. Messier M.A. *The Propagation of an Electromagnetic Impulse through Soil: Influence of Frequency Dependent Parameters*, MRC-N-415. Santa Barbara, CA: Mission Research Corporation, February 1980.
41. Tagg G.F. *Earth Resistances*. London: George Newnes Ltd, 1964.
42. Gary C. 'L'impédance de terre des conducteurs enterrés horizontalement'. Presented at the *International Conference on Lightning and Mountains*, Chamonix, France, 1994.
43. Grcev L. 'Improved earthing system design practices for reduction of transient voltages'. Presented at *CIGRÉ*, Session, Paris, France, 1998.
44. Grcev L. 'Improved design of transmission line grounding arrangements for better protection against effects of lightning'. Presented at the *International Symposium on Electromagnetic Compatibility (EMC'98 ROMA)*, Roma, Italy, 1998.
45. Rachidi F., Janischewskyj W., Hussein A.M., Nucci C.A., Guerrieri S., Kordi B. and Chang J.-S. 'Current and electromagnetic field associated with lightning – return strokes to tall towers'. *IEEE Transactions on Electromagnetic Compatibility* 2001;**43**:356–67.
46. Wang J., Liew A.C. and Darveniza M. 'Extension of dynamic model of impulse behavior of concentrated grounds at high currents'. *IEEE Transactions on Power Delivery* 2005;**20**(3):2160–65.
47. Grcev L. 'Impulse efficiency of ground electrodes'. *IEEE Transactions on Power Delivery* 2009;**24**(1):441–51.
48. Grcev L. 'Lightning surge characteristics of earthing electrodes'. Presented at the *29th International Conference on Lightning Protection (ICLP)*, Uppsala, Sweden, 2008.
49. Grcev L. 'Time- and frequency-dependent lightning surge characteristics of grounding electrodes'. *IEEE Transactions on Power Delivery*, in press.

Chapter 11
Soil ionization
Vernon Cooray

11.1 Introduction

If a continuously increasing electric field is applied to a medium, a stage will be reached at which electrical breakdown takes place. For example, in air this critical electric field is $\sim 3 \times 10^6$ V m^{-1}. At this electric field the balance between the ionization and deionization processes is destroyed and the ionization processes take over. The same process applies if an electric field is applied to a liquid or solid medium. Again, there is a critical electric field at which the electrical breakdown takes place in the medium. In liquids and solids the critical electric field needed is much higher than in air. For example, the intrinsic breakdown electric fields in highly purified hexane and benzene are $\sim 1.3 \times 10^8$ V m^{-1} and 1.1×10^8 V m^{-1}, respectively. Intrinsic breakdown strength in solids is well in excess of 1×10^8 V m^{-1}. The reason why the breakdown electric field in solids is much larger than that of a gaseous medium is the smaller mean free path available for the electrons to gain energy before collisions in solids. However, practice shows that, unless all extraneous influences are removed and a very small volume of material is tested under best experimental conditions, the electrical breakdown in liquids and solids takes place at electric fields much smaller than the intrinsic breakdown electric fields. One such extraneous parameter that reduces the dielectric strength of liquids and solids is the presence of gaseous bubbles or voids in the medium. Consider the application of an electric field into a solid dielectric. Let E_d be the strength of the electric field in the solid. If air voids are present in the solid, the electric field inside these voids depends on their shape and the dielectric constant of the medium. For example, the electric fields inside an air void having three different shapes, i.e. spherical, elongated in the direction of the electric field and elongated in a direction perpendicular to the electric field (see Figure 11.1) are given by equations (11.1), (11.2) and (11.3), respectively:

$$E = E_d \frac{3\varepsilon}{1 + 2\varepsilon} \qquad (11.1)$$

$$E = E_d \qquad (11.2)$$

$$E = \varepsilon E_d \qquad (11.3)$$

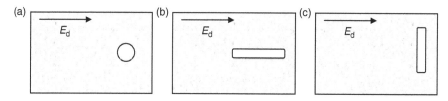

Figure 11.1 The electric field in an air void located in a dielectric medium depends on the shape of the cavity. The approximate electric field in the centre of the cavity for the different configurations shown above is given in the text.

Thus, as the electric field in a solid containing air voids increases, a stage will be reached in which the electric field inside the air void exceeds the breakdown electric field in air and the breakdown process starts inside the void. Because the electric field inside the void, which depends on the dielectric constant of the solid, could be larger than the electric field in the solid, the electric breakdown process may commence in the voids when the electric field in the solid is less than the critical electric field necessary for breakdown in air. Moreover, as can be seen from equations (11.1) and (11.3), the larger the dielectric constant of the solid, the higher the electric field inside the void and, therefore, the electric field in the solid necessary to create electrical breakdown in the void decreases with increasing dielectric constant. Of course, the critical electric field necessary for breakdown in the void depends also on the dimension of the void. The relationship between the dimension of the void and critical electric field necessary for electrical breakdown in it can be estimated from the Paschen curve. Figure 11.2 shows the critical electric field necessary to cause electrical breakdown in a uniform gap at standard atmospheric pressure and temperature as a function of gap length (see Reference 1 for a review of the electrical breakdown mechanism in air). For a given air pressure in a void, there is a minimum size of the void below which the critical electric field necessary for electrical breakdown increases.

As far as electrical breakdown is concerned, soil can be approximated by a solid material containing a large number of air voids of different shapes. Moreover, most of these air voids are connected to each other through air channels of different dimensions. The information presented above shows that these air voids will play a dominant role in causing electrical breakdown in soil. Electrical breakdown in soil is of interest in evaluating the time-varying resistance that will result when a current is injected into a grounding electrode buried in soil. If the current amplitude is not large enough to cause any ionization, the resistance of the buried electrodes is given by

$$R = \frac{\rho}{2\pi r_h} \quad \text{for a buried hemisphere} \quad (11.4)$$

$$R = \frac{\rho}{2\pi l} \ln \frac{r_r + l}{r_r} \quad \text{for a buried rod} \quad (11.5)$$

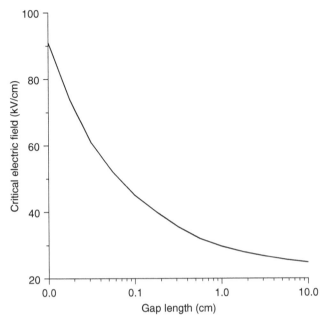

Figure 11.2   Critical electric field necessary to cause electrical breakdown in a uniform gap as a function of the gap length. The data corresponds to standard atmospheric pressure and temperature. The results are based on the Paschen curve.

where r_h is the radius of the hemisphere, r_r the radius of the rod, l its length and ρ the ground resistivity. In deriving equation (11.5) it was assumed that the equipotential lines around the rod are located in a manner as depicted in Figure 11.3.

The electric field that will be generated in soil during the application of a current pulse to the buried electrode is largest at the surface of the electrode and it decreases with increasing distance from the surface. The larger the injected current the larger the electric field in soil. If this electric field exceeds the critical electric field necessary for breakdown in soil, ionization processes take place in the volume of sand in which the electric field is above this critical value. It is reasonable to assume that the breakdown process starts in air voids and spreads through the volume of soil across air channels interconnecting the voids. The ionization causes the air in the voids to change from an insulator to a conductor, leading to a decrease in the resistivity of the soil volume engulfed by ionization. This in turn leads to a decrease in the resistance of the buried conductor. Thus it is important to understand how the resistance of the buried conductors will change during the application of large currents, similar to those that are encountered by buried conductors used to divert lightning currents that are capable of ionization into soil. In this chapter we will elucidate several models used by scientists to simulate soil ionization in evaluating the resistance of buried electrodes as a function of injected current amplitude.

534 *Lightning Protection*

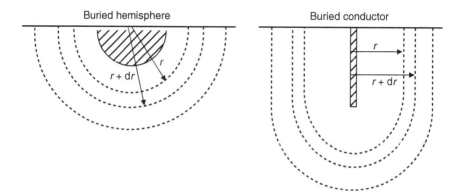

Figure 11.3   The shape of the equipotential surfaces (dotted lines) around a buried hemisphere and a buried conductor assumed in the calculations

11.2 Critical electric field necessary for ionization in soil

As mentioned previously, each medium will have a certain critical electric field, and if the applied electric field exceeds this critical value, ionization processes will start in the medium. The same is true for soil, but a fixed critical electric field cannot be assigned for it because soil in general is an inhomogeneous medium and the constituents of the soil may change from one geographical region to another and also from one season to another. The main constituent of soil is SiO_2 (quartz), and the grain size may change from one type of soil to another. The soils can be separated into different categories, namely very fine (radius 1/16 to 1/8 mm), fine (radius 1/8 to 1/4 mm), medium (radius 1/4 to 1/2 mm), coarse (radius 1/2 to 1 mm), very coarse (radius 1 to 2 mm) and gravel (radius 2 to 64 mm). The grains of the soil can have very different shapes too. Because the soil grains cannot be packed into a given volume to completely fill the space, soil contains air pockets or voids between the grains. Soil can also contain organic matter between the grains and other chemicals deposited on the surface of the grains. In addition, it may also contain a certain amount of moisture. The resistivity of quartz itself is rather high, but the resistivity of soil can be much lower than that of quartz because the soil grains are usually coated with water in which some salts are dissolved. Thus, in general, soil can be treated as a medium having bulk resistivity (and dielectric constant) and consisting of irregularly shaped particles with air voids between.

When a current is injected into soil the flow of current through it will generate an electric field E given by

$$E = \rho_0 J \tag{11.6}$$

where J is the current density at the point of consideration and ρ_0 is the resistivity of soil. With increasing amplitude of the injected current, the electric field in soil increases, and if the strength of this electric field exceeds a critical value, the process of ionization will commence in soil. The mechanism of ionization to a high probability is associated with the electrical breakdown taking place in air voids located between the soil grains. As mentioned previously, the electric field in air voids depends on the shape of the void and on the dielectric constant of soil. As shown in the introduction the electric field in air voids in soil can be larger than the bulk electric field in soil. Because air needs $\sim 3 \times 10^6$ V m^{-1} for electrical breakdown to occur, one can expect the critical electric field necessary to start ionization processes in soil to be smaller than this value. Experiments conducted by Towne [2] using gravel with unknown moisture content gave values in the range of 1.6×10^5 to 5.2×10^5 V m^{-1} for the critical electric field necessary to cause ionization in soil. The tests of Bellaschi [3] generated values in the range of 1.2×10^5 to 4.2×10^5 V m^{-1}. Kosztaluk and colleagues [4] and Laboda and colleagues [5,6] studied different soils and came up with values in the range of 5.6×10^5 to 9.0×10^5 V m^{-1} for the critical electric field. In the experiments carried out by Oettlé [7], which involved several types of soil, critical electric fields of 6.0×10^5 V m^{-1} and 1.85×10^6 V m^{-1} were found for wet and dry soil, respectively. Analysing a large sample of data, Mousa [8] suggested 3×10^5 V m^{-1} as a suitable value for the critical electric field in soil to be used in theoretical evaluations. The value of 4×10^5 V m^{-1} is being used by CIGRE [9]. It is important to stress here that soil is an inhomogeneous medium with a water content that may differ from place to place and from season to season. It may also contain different organic materials that have a lower resistivity than soil grains. For this reason, it is very difficult, if not impossible, to suggest one particular value for the critical electric field necessary for ionization in soil. Moreover, in any experimental attempt to evaluate this critical electric field it is necessary to pinpoint the exact time at which ionization sets in soil. Thus, the derived values of critical electric field may depend on the assumptions made by different workers in deciding the time of onset of ionization, which to some extent may also depend on the electrodes configuration used. Furthermore, in order to generate electrical breakdown in a medium it is not sufficient to increase the strength of the electric field at a given point in the medium to a value larger than the critical one. For electrical breakdown to occur the electric field must exceed the critical value over a critical spatial distance. These facts may cause the critical electric fields obtained in experiments using different electrode arrangements to differ. Indeed, more experimental data to ascertain the critical breakdown fields in soils is a must before setting the limits on this important parameter.

Once the critical electric field necessary to cause ionization in soil is known it can be utilized in soil ionization models to evaluate how the resistance of buried conductors will vary in the presence of soil ionization. In the sections to follow we will illustrate several models, with different grades of physics, that can be utilized to evaluate how the resistance of a buried conductor will be modified by soil ionization. One assumption pertinent to all these models is that ionization takes place more or less uniformly around the buried conductor. In the case of a hemisphere the ionization region forms a concentric spherical region around the conductor, and in the case of cylindrical rods it is assumed that the ionized regions are also cylindrical in shape, with the axis coinciding with the rod (see Figure 11.3). In reality, sparking may take place in soil, and if the ionization process is confined to a few sparks in soil then the symmetry assumed in the models would be completely destroyed.

11.3 Various models used in describing soil ionization

11.3.1 Ionized region as a perfect conductor

This is the simplest soil ionization model available for the engineers and it is frequently utilized by scientists to evaluate the resistance of grounding rods under the application of high currents. In this model, it is assumed that the ionization processes decrease the resistivity of soil to zero and, therefore, the region of ionization can be replaced by a perfect conductor. Because the ionization is assumed to take place uniformly around the buried conductor, then the model basically assumes that the effect of soil ionization is to increase the diameter of the conductor. For example, consider the injection of a current pulse $I(t)$ into a hemispherical conductor buried in soil. Assume that the current pulse is such that at time t the critical electric field necessary for ionization is reached at a radial distance of r_i which, of course, is a function of time. This is the case because the current pulse amplitude changes with time. Thus the resistance of the buried conductor is given by

$$R_{\text{hem}}(t) = \frac{\rho_0}{2\pi r_i(t)} \tag{11.7}$$

with

$$r_i(t) = r_0 \quad \text{if} \quad \frac{I(t)\rho_0}{2\pi r_0^2} < E_c \tag{11.8}$$

$$r_i(t) = \sqrt{\frac{I(t)\rho_0}{2\pi E_c}} \quad \text{if} \quad \frac{I(t)\rho_0}{2\pi r_0^2} > E_c \tag{11.9}$$

where ρ_0 is the resistivity of the soil, r_0 is the radius of the hemispherical electrode and E_c is the critical electric field necessary to produce ionization in soil. In deriving this it is assumed that the ionization takes place uniformly around the conductor. Note that the resistance decreases as the ionization region spreads outwards (i.e. with increasing r_i). Observe that according to this model the resistance will have the same value for a given current amplitude, irrespective of whether the current is increasing or decreasing. In other words the current versus resistance curve does not show any hysteresis.

Recall that this model assumes that the ionized volume of soil can be represented by a medium with zero resistivity. In reality, ionization processes will decrease the soil resistivity to a finite value. In order to take this into account, some researchers have attempted to replace the resistivity of the ionized region by a finite value [10].

According to this procedure, if the ionization spreads at a given time to a radius r_i, then the resistance of the hemisphere is given by

$$R_{\text{hem}}(t) = \frac{\rho_i}{2\pi}\left[\frac{1}{r_0} - \frac{1}{r_i(t)}\right] + \frac{\rho_0}{2\pi r_i(t)} \tag{11.10}$$

where ρ_i is the resistivity of the ionized region. In this case also there is no hysteresis in the current versus resistance curve. The residual ionization in the ionized region is evaluated by different researchers to be ~ 40 to 8 per cent of the resistivity of the undisturbed medium.

These models assume that changes in soil resistivity caused by ionization and deionization processes take place instantaneously at the beginning of the process and remain the same during the process. In the next Section we will consider a model that attempts to improve this assumption.

11.3.2 Model of Liew and Darveniza

Liew and Darveniza [11] rejected the assumption of an instantaneous change in resistivity with ionization and deionization processes. They assumed that there is a relaxation time associated with the changes in resistivity caused by ionization and deionization processes. In constructing the model they also assumed that the soil is homogeneous and that soil resistivity is the same in all directions; i.e. it is isotropic. The model assumes that the ionization in a given volume starts when the electric field increases beyond a critical value. Once the ionization sets in, the resistivity of the soil decays from a higher value to a lower value exponentially as

$$\rho = \rho_0 e^{-t/\tau_1} \tag{11.11}$$

where τ_1 is the ionization time constant and time t is measured from the onset of the ionization. This drop in resistivity continues until the electric field (or the current density) in soil is above the critical value necessary for ionization. When the electric field (or the current density) drops below the critical value during the decaying phase of the current the resistivity will recover to its original value. In the model it is assumed that this recovery, which is current dependent, takes place exponentially. This recovery phase is described mathematically as

$$\rho = \rho_i + (\rho_0 - \rho_i)(1 - e^{-t/\tau_2})\left(1 - \frac{J}{J_c}\right)^B \tag{11.12}$$

where τ_2 is the deionization time constant, ρ_i is the resistivity at the end of the ionization phase, J is the current density, J_c is the critical current density necessary for ionization and B is a constant. In the model the value of B is set to 2. In the recovery phase described by equation (11.12), the last term on the right-hand side makes the deionization process current dependent. This factor enters into the equation in such

538 Lightning Protection

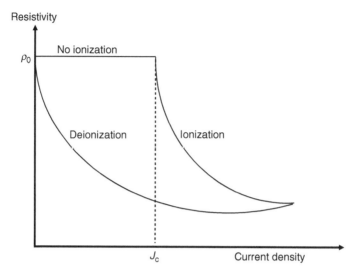

Figure 11.4   General variation of the resistance of a buried conductor as a function of the injected current density as predicted by Liew and Darveniza model [11]. Note that the current versus resistivity curve shows hysteresis. The same general shape is also predicted by the models of Wang et al. [12], Sekioka et al. [13] and Cooray et al. [15].

a way that larger the injected current, the slower the recovery. This is physically reasonable because the more energy is injected into soil, the slower the recover to the non-conducting state. The general variation of the resistance of a buried conductor as a function of the injected current as predicted by this model is depicted in Figure 11.4. Note that the current versus resistivity curve shows hystereses. That is, the recovery of the resistance during the deionization stage does not follow the path taken by decreasing resistance during the ionization state.

11.3.2.1   Application of the model to a single driven rod

In the model, ionization is assumed to take place around the rod symmetrically, as shown in Figure 11.3. Consider an element dr at a radial distance r (also marked in Figure 11.3). The current density flowing across the element is

$$J(t) = \frac{I(t)}{2\pi r^2 + 2\pi r l} \tag{11.13}$$

The voltage, V_r across the element is given by

$$V_r = \frac{I(t)\rho\, dr}{2\pi r^2 + 2\pi r l} \tag{11.14}$$

Assume that the ionization takes place when the electric field exceeds a critical value E_c. Thus breakdown happens when

$$V_r > E_c \, dr \tag{11.15}$$

Then the critical current density necessary for breakdown is given by

$$J_c = \frac{E_c}{\rho_0} \tag{11.16}$$

Now, in any given element as the current density increases the resistivity remains the same until the current density exceeds the critical value. That is,

$$\rho = \rho_0 \quad \text{for } J < J_c \tag{11.17}$$

Once ionization sets in the resistivity changes as

$$\rho = \rho_0 e^{-t/\tau_1} \quad \text{for } J > J_c \tag{11.18}$$

Let us assume that the maximum radius at which ionization takes place is r_{im}. In the decaying phase of the current one has to consider three regions:

1. Region where $r > r_{im}$ and $J < J_c$. In this region no ionization has taken place and therefore the resistivity is equal to the normal value ρ_0.
2. Region where $r < r_{im}$ and $J < J_c$. In this region the ionization has set in but now the current density has decreased below the critical value and the region is in a decaying state. In this region the resistivity is given by

$$\rho = \rho_i + (\rho_0 - \rho_i)(1 - e^{-t/\tau_2})\left(1 - \frac{J}{J_c}\right)^2 \tag{11.19}$$

3. Region where $r < r_{im}$ and $J > J_c$. In this region the current density is still higher than the critical value and therefore the resistivity still continue to decrease. In this region the resistivity is given by

$$\rho = \rho_0 e^{-t/\tau_1} \tag{11.20}$$

Note that in these equations the time t is measured from either the start of ionization [in equation (11.18)] or the start of deionization [in equation (11.19)]. Therefore, at any given time the parameter t is different in different equations. These equations specify the resistivity of any given element as a function of time and the total resistance of the buried conductor can be obtained by summing the contribution from each element.

The model contains three parameters, namely the critical electric field necessary for soil breakdown (i.e. E_c), τ_1 and τ_2. Liew and Darveniza conducted several experiments and evaluated the best values of model parameters that gave the best fit to the experimental data. Based on this comparison they estimated $\tau_1 \approx 1.5-2$ μs and $\tau_2 \approx 0.5-4.5$ μs.

11.3.3 Model of Wang and colleagues [12]

Recall that Liew and Darveniza considered ionization, deionization and no ionization regions in soil in constructing their model. Wang and colleagues [12] modified this model by assuming that at any given time during current injection into ground the physical processes taking place in ground can be separated into four distinct phases. These four phases are sparking, ionization, deionization and no ionization. At any given time these four phases divide the soil volume into four regions, sparking region, ionization region, deionization region and no ionization region. The four regions as visualized in the model are shown in Figure 11.5. As in the Liew and Darveniza model, the ionization region is characterized by a critical current density J_c. The ionization sets in soil whenever the current density increases beyond this value. The sparking region where the ionization takes place through sparks is assumed to be located close to the electrode. In the model the resistivity in the sparking region is assumed to be zero. The sparking region is also characterized by a critical current density J_s. If the current density increases beyond this value the sparking takes place. In the model it is assumed that

$$J_s = \alpha J_c \qquad \alpha > 1 \qquad (11.21)$$

where α is a parameter that varies with current density. Now let us see how the model can be applied to find the current-dependent resistance of buried electrodes. Assume that r_{im} is the maximum radius up to which ionization takes place. One has to separate the analysis into two parts, one during the rising part of the current and the other during

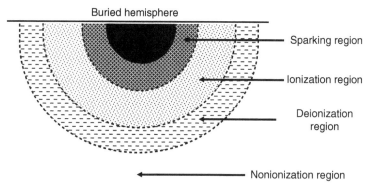

Figure 11.5 Different regions of ionization as visualized in the model of Wang et al. [12]

the decaying part. In the current rising part the soil volume can be divided into three regions:

1. Region where $J < J_c$. In this region the soil resistivity remains at ρ_0, i.e.

$$\rho = \rho_0 \quad \text{for } J < J_c \qquad (11.22)$$

2. Region where $J_c < J < J_s$ and $r < r_{im}$. In this region breakdown takes place without sparks and, as postulated in the original model of Liew and Darveniza, the resistivity decays exponentially, i.e.

$$\rho = \rho_0 e^{-t/\tau_1} \qquad (11.23)$$

3. Region where $r < r_{im}$ and $J > J_s$. This is the sparking region. In this region

$$\rho = 0 \qquad (11.24)$$

When the current decays from the peak value one has to consider four regions:

1. Region where $r > r_{im}$ and $J < J_c$. No ionization has occurred in this region and therefore

$$\rho = \rho_0 \qquad (11.25)$$

2. Region where $r < r_{im}$ and $J > J_c$. In this region soil resistivity recovers in a manner as postulated in the model of Liew and Darveniza and is given by

$$\rho = \rho_i + (\rho_0 - \rho_i)(1 - e^{-t/\tau_2})\left(1 - \frac{J}{J_c}\right)^2 \qquad (11.26)$$

where τ_2 is the deionization coefficient.

3. Region where $r < r_{im}$ and $J_c < J < J_s$. This is the region that was previously ionized by the sparking process. The sparking has stopped in this region but ionization still takes place and the soil resistivity continues to decrease, i.e.

$$\rho = \rho_0 e^{-t/\tau_1} \qquad (11.27)$$

4. Region where $r < r_{im}$ and $J > J_s$. Sparking still occurs in this region and the resistivity is zero, i.e.

$$\rho = 0 \qquad (11.28)$$

In applying the model the volume of soil around the electrode can be divided into elementary volumes symmetrically located around the conductor (see Figure 11.3), and the resistivity appropriate to any particular volume element at a given time can be obtained from the above equations. The total resistance can be obtained by summing the contribution from each element.

Recall that in the model the parameter α in equation (11.21) is assumed to be a function of current. So far we have not discussed how α varies as a function of current. Wang and colleagues assumed a rather complex variation of α with current. First, note that the value of α has to be larger than unity. Initially α is assumed to decrease with increasing current as given by

$$\alpha = \alpha_0[1 - \lambda \exp(I^{\beta_1})] \qquad \alpha > 1 \qquad (11.29)$$

where I is the injected current, α_0 is the initial value of α, and λ and β_1 are to be determined. According to this equation, α continues to decrease with increasing current, but recall that it should not decrease below unity. To realize this, at a certain value of α (equal to α_s) the form of variation is changed from equation (11.29) to the following equation:

$$\alpha = 1 + [\exp(1/I)]^{\beta_2} \qquad (11.30)$$

where the parameter β_2 is given by

$$\beta_2 = I(t - \Delta t) \ln(\alpha_s - 1) \qquad (11.31)$$

where Δt is the time interval in the dynamic time iteration process. In the above equation α_s is the last value of α obtained from equation (11.29). Now, as the current decays from the peak value, α is assumed to recover in the following manner:

$$\alpha = \alpha_p + (\alpha_0 - \alpha_p)\left(1 - \frac{I}{I_p}\right)^{\beta_3} \qquad (11.32)$$

where α_p is the value of α at current peak I_p and β_3 is a value to be determined. The way α varies as a function of current is depicted in Figure 11.6.

The model consists of six parameters including the value of critical gradient necessary for ionization. They are E_c, λ, τ_1, τ_2, β_1 and β_3. Wang and colleagues [12] compared the model predictions with one particular experimental dataset of Liew and Darveniza [10] and came up with the following values for the model parameters: $\tau_1 = 0.5-1.0$ μs, $\tau_2 = 1-1.5$ μs, $E_c = 24$ kV m^{-1}, $\alpha_0 = 70$, $\lambda = 0.00017$, $\beta_1 = 0.8$ and $\beta_3 = 3.0$.

The main problem associated with this model is the large number of parameters, the values of which have to be estimated by comparing model predictions with experiment. Unfortunately, the authors did not study how the model parameters would change from one experiment to another. Moreover, it is difficult to understand from

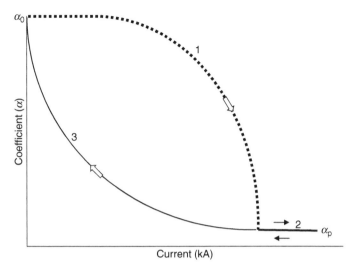

Figure 11.6 Schematic description of the way in which the coefficient α, one of the model parameters in the model of Wang et al. [12], varies as a function of current. The three regions 1, 2 and 3 are described by equations (11.29), (11.30) and (11.32), respectively.

a physical point of view the difference between the sparking region and the ionization region. In the model no physical explanation is given as to the difference between ionization and sparking regions. Of course ionization processes in a medium may take place in different forms. For example, in air, ionization may take place in the form of a spark or as a corona discharge. However, irrespective of whether the ionization takes place in the form of sparking or corona, it is always decided by one critical electric field. Whether the ionization takes place in the form of a spark or silently like a corona discharge is determined by the spatial extension of the region where the electric field is larger than the critical value. In reality, when sparking occurs in air it travels all the way to the edge of the ionization region or sometimes it may even penetrate into the low field region thanks to the electric field enhancement at the tips of the breakdown channels. These observations in air show that there is no physical ground to divide the region of ionization into sparking region and ionization region. Having said that, observe that if α reaches unity then the boundary between the ionization region and the sparking region coincides. According to the model, as the current increases the value of α decreases. This means that the boundaries between the normal ionization region and the sparking region approach each other with increasing current. In this respect the model is on the right line.

11.3.4 Model of Sekioka and colleagues [13]

These authors utilized the energy balance equation corresponding to arcs to study how the conductivity of soil changes with time during ionization. As in other models it is assumed that the ionization takes place when the background electric field or the current density exceeds a critical value. Again, as in other models, it is assumed that the ionization takes place symmetrically around the conductor. Consider a hemispherical element of soil thickness dr at a distance r (see Figure 11.3). The energy balance equation for the discharge taking place across a unit radial length of the element can be written as

$$\frac{dQ}{dt} = ui - P \qquad (11.33)$$

where u is the discharge voltage across the unit radial length, i the discharge current, Q the accumulated energy, P the power loss and t the time. In the model, Sekioka and colleagues assumed that the soil ionization process is similar to that of the switching arc and hence could be described by Mayr's equation [14]:

$$g = Ke^{Q/Q_0} \qquad (11.34)$$

where g is the arc conductance (S/m) and K and Q_0 (J/m) are constants. Combining equations (11.33) and (11.34) one can derive the following differential equation:

$$\frac{1}{g(r,t)} \frac{dg(r,t)}{dt} = \frac{1}{Q_0}(ui - P) \qquad (11.35)$$

In the model it is assumed that the power loss is caused by heat dissipation from the hot region to the cold region. Therefore heat dissipation is a function of temperature of the soil element under consideration. Because the temperature decreases with radial distance the heat is assumed to flow outwards. The power loss is therefore assumed to be proportional to the surface area, S, of the element under consideration. Thus,

$$P = \lambda S \qquad (11.36)$$

where λ (W/m^3) is a constant. Once the conductance is calculated from the above equation, the total resistance can be calculated by

$$R(i,t) = \int_{r_0}^{r_e} \frac{dr}{g(r,t)} + \int_{r_e}^{\infty} \frac{dr}{g(r,t)} \qquad (11.37)$$

where r is the distance (in metres) from the grounding electrode, r_e is the effective radius of the ionization zone and $g(r, t)$ is the conductance of the segment at r. Note that for $r > r_e$ there is no ionization and the conductance is constant. Using the relationship $i = gu$, equation (11.35) was solved by Sekioka and colleagues and the corresponding solutions for conductivity and resistivity were given as

$$g = g_0 e^{-t/\tau} \left[1 + \frac{1}{g_0 Q_0} \int_0^t i^2 e^{s/\tau} \, ds\right] \qquad (11.38)$$

$$\rho = \rho_0 e^{-t/\tau} \left[1 + \frac{1}{g_0 Q_0} \int_0^t i^2 e^{s/\tau} \, ds\right]^{-1} \qquad (11.39)$$

Table 11.1 Values of the model parameters of the Sekioka
et al. model as a function of soil resistivity ρ_0

ρ_0 (Ω m)	160	560	1 070
E_c (kV m^{-1})	330	110	110
λ (MW m^{-3})	680	21.6	11.3
Q_0 (J m^{-1})	1 800	600	150

where t is the time measured from the onset of ionization in a given segment and $\tau = Q_0/P$.

The model contains two constants, Q_0 and λ, in addition to the critical electric field necessary for soil ionization, E_c. Sekioka and colleagues compared the model predictions with experiments and found that the values of model parameters that gave a good fit for experimental data depend on soil resistivity. These values are tabulated in Table 11.1 for different resistivities.

In evaluating the energy balance in the model it is assumed that the energy is dissipated radially. However, in reality most of the energy dissipated in the discharge is being used to heat the sand in which the discharge is in contact. Thus the energy dissipation (at least part of it) would be proportional to the sand volume in which ionization takes place. It is also important to note that the arc equation used in the model is valid for a fully developed arc whereas in soil most of the heating and energy dissipation takes place during the initiation and establishment of the arc. This energy dissipation is not taken into account in the model. It is actually doubtful whether real arc conditions are established in soil, especially when the duration of the current impulse is of the order of hundreds of microseconds. Interestingly, the authors show that from their model they can recover the exponential decrease of resistivity envisaged by Liew and Darveniza during the ionization stage.

11.3.5 Model of Cooray and colleagues [15]

Cooray and colleagues attempt to model the soil ionization by assuming that the breakdown process is taking place completely in air voids in soil. The basic assumptions of the model as described in Reference 15 are given below.

1. In the case of a hemisphere the equipotential surfaces are spherical region concentric with the grounded hemisphere, and in the case of buried rod they can be represented by cylindrical sections with a hemispherical cap (see Figure 11.3). This assumption is similar to the one made in other models. It is assumed that this symmetry is retained even during ionization of the air in the soil.
2. It is assumed that the ionization process in a given elementary volume will commence when the electric field at that element increases beyond a certain critical value. Because each element is assumed to be bounded by equipotential surfaces, the ionization sets in the whole elementary volume simultaneously. This critical

546 Lightning Protection

electric field can be obtained from direct measurements of this parameter in soil under consideration or from the $V - I$ characteristics of the buried conductor.

3. The exact mechanism in which electrical breakdown is initiated in soil is not known. It is possible that the breakdown process in soil consists of a series of electrical discharges taking place in the air gaps between soil grains. Studies of electrical discharges of millimetre to centimetre scale in air show that the breakdown process can be divided into two parts. The first phase consists of the propagation of streamers across the gap and the rearrangement of the charge in the streamer filaments resulting in a glow state, heating and expansion of the filament. During this stage, the channel could be heated to a few thousand degrees. In the second phase the current and the temperature in the channel start to increase, the thermal ionization becomes significant, and the spark forms, leading to thermalization of the discharge. The experimental data show that in small gaps the duration of the first phase is a few hundred nanoseconds [16, and references therein]. In the model the first phase is assumed to take place instantaneously leaving behind a discharge channel, which is raised to a temperature T_0 and with an elevated conductivity. From this point onwards the temporal development of the temperature and the conductivity of the discharge channels are estimated by assuming that the energy dissipated in the channels is used completely in heating the channels. The variation of the conductivity of the discharge channels as a function of time is obtained from the experimental data on the conductivity of air as a function of temperature [17,18].

4. The volume of the heated air is assumed to be a fraction F of the total volume of soil in which ionization takes place. The current flow through this heated volume of air will dissipate more energy, leading to a further increase in the temperature and conductivity of the air. It is also assumed that the heat dissipation from this heated volume to the surrounding soil is proportional to the volume and the temperature of the heated air. S denotes the constant of proportionality. In the model T_0, F and S are the parameters that have to be determined by comparing experimental data with simulations.

11.3.5.1 Mathematical description

Let us consider a volume element of thickness dr located at a radial distance r from the buried rod. The density of the current passing through this element at time t is given by

$$J(t) = \frac{I(t)}{2\pi r l + 2\pi r^2} \qquad (11.40)$$

where $I(t)$ is the current injected into the rod and l is the length of the rod. It is assumed that the ionization of the air in this element will be initiated when the current density passing through this element exceeds a critical value. Let us denote this critical value J_0. This is given by

$$J_0 = \sigma E_c \qquad (11.41)$$

where E_c is the critical electric field necessary to cause breakdown in the soil. For a given peak value of the injected current I_p, the critical radius r_{im}, beyond which no ionization of the soil takes place, can be obtained by solving the equation

$$J_0 = \frac{I_p}{2\pi r_{im} l + 2\pi r_{im}^2} \tag{11.42}$$

The resistance of the volume elements located beyond this critical radius is given by

$$dR_s = \frac{dr}{\sigma_0(2\pi r l + 2\pi r^2)} \quad \text{where } r \geq r_{im} \tag{11.43}$$

where σ_0 is the conductivity of the soil ($\sigma_0 = 1/\rho_0$). Let us consider a volume element located within the critical radius. The ionization of air in this element is initiated at time t_0, at which the current density of the element surpasses the critical value J_0. For values of $t \leq t_0$ the resistance of the volume element is given by

$$dR(t < t_0) = \frac{dr}{\sigma_0(2\pi r l + 2\pi r^2)} \quad \text{where } r \leq r_{im} \tag{11.44}$$

According to assumptions concerning the pre-discharge phase of the ionization, at time $t = t_0$ a volume of air of magnitude equal to $F(2\pi r l\, dr + 2\pi r^2\, dr)$ in this volume element is immediately raised to a temperature T_0. Thus the resistance of the volume element at $t = t_0$ will be changed from that given in equation (11.44) to

$$dR(t = t_0) = \frac{dr}{[\sigma_0 + F(T_0/T_a)\sigma_a(T_0)](2\pi r l + 2\pi r^2)} \tag{11.45}$$

where σ_a is the conductivity of air, which is a function of the temperature, and T_a is the temperature of the volume of air before ionization sets in. The variation of the conductivity of air as a function of temperature is plugged into the model through a polynomial fit constructed from the available experimental data [17,18]. Note that in writing down the above equation, Cooray and colleagues have tacitly assumed that the fraction, F, is much smaller than unity, and that the heated air is at atmospheric pressure. Let us consider what happens during the time interval $t = t_0$ to $t = t_0 + dt$. The joule heat dissipated in the volume during this time interval is given by

$$dH = \frac{dr[I(t_0)^2]dt}{[\sigma_0 + F(T_0/T_a)\sigma_a(T_0)](2\pi r l + 2\pi r^2)} \tag{11.46}$$

and according to the assumption that the energy dissipation is proportional to the volume, the energy dissipated out from the air volume to the surroundings is given by

$$dU = SF(T_0/T_a)(2\pi r l + 2\pi r^2)dr\, T_0\, dt \tag{11.47}$$

Thus the increase in temperature of the air volume during the time interval dt is given by

$$dT = \frac{(dH - dU)}{F(T_0/T_a)(2\pi r l + 2\pi r^2) dr \delta(T) C_p} \quad (11.48)$$

where $\delta(T)$ is the density of air at temperature T and atmospheric pressure and C_p is the heat capacity of air at constant pressure. This procedure can be repeated to evaluate the variation of the temperature of the gas volume under consideration as a function of time and from this the temporal variation of the resistance of the gas volume can be obtained. The same procedure is applied in every element from $r = r_0$ to $r = \infty$. The total resistance of the buried conductor as a function of time, $R(t)$, can thus be obtained from

$$R(t) = \int_{r_0}^{r_{im}} dR(t)\, dr + \int_{r_{im}}^{\infty} dR_s\, dr \quad (11.49)$$

11.3.5.2 Model parameters

The model parameters are T_0, F and S. In testing the model Cooray and colleagues utilized three sets of measurements, two from Bellaschi [3] and one from Liew and Darveniza [11]. In the first setup used by Bellaschi, a rod of length 3.05 m and radius 0.0127 m was buried in soil (clay with 0.002-mm-diameter soil particles) of conductivity 11.5 mS m^{-1} (soil M). The injected current pulse had a rise time of $\sim$20 μs and it decayed almost to zero by 60 μs. In the second setup used by Bellaschi, the length and the radius of the rod were 2.16 m and 0.00794 m, respectively and it was buried in soil of conductivity 6.37 mS m^{-1} (soil F). In the experiment of Liew and Darveniza, a rod of length 0.61 m and radius 0.00635 m was buried in soil of conductivity 20 mS m^{-1}. The critical electric fields used in the calculations were identical to those estimated by Liew and Darveniza. The analysis showed that the values $T_0 = 3\,600$ K, $F = 0.005$ and $S = 7.13 \times 10^{-3}$ J K^{-1}m^{-3} provided a reasonable fit for all experimental data. Figure 11.7 shows the measured results of Bellaschi and colleagues [3] and those predicted by the model.

It is important to note that the three model parameters are physical quantities that might be derived in a self-consistent manner through a detailed analysis of the problem under consideration. The particular values obtained for the model parameters above may differ from the exact values associated with the physical process due to various simplifying assumptions Cooray and colleagues have made in constructing the model. First, it was assumed that the pre-breakdown stage of the discharge process takes place instantaneously at the moment the electric field exceeds a critical value. As mentioned previously, from analysis of experimental data obtained in air this process may take a few hundreds of nanoseconds to be completed. If this is also the case in air voids in sand, then the model may generate unrealistic results for injected currents with very fast rise times. This point needs further experimental investigation. It is interesting to observe, however, that the temperature of the channels after the

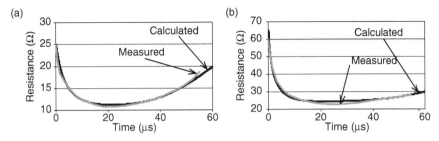

Figure 11.7 Comparison of the measured resistance and the resistance calculated using the model of Cooray et al. *[15]: (a)* Bellaschi et al. *[2], soil M; (b)* Bellaschi et al. *[3], soil F*

pre-breakdown stage, i.e. T_0, estimated in the experiments is in the range of temperatures at which the air changes state from being an insulator to a good conductor. Second, in the analysis they have assumed that the ionization of soil takes place uniformly around the buried conductor. However, in reality, the electric field is enhanced at the buried end of the rod and it is probable that the ionization phenomenon is initiated first at this end. Third, it was assumed that the fractional volume of ionization F remains constant as the discharge phenomenon expands radially away from the conductor. This assumption requires that the number of discharge channels increases as the discharge extends away from the conductor. This is not far from reality, because in general electrical discharges form branches and expand laterally as they proceed away from the high-voltage electrodes. However, the fraction F need not remain constant during discharge development. Fourth, they have assumed that all the energy dissipated in the discharge channels is utilized in heating them. In reality, part of this energy goes into dissociation of nitrogen and oxygen molecules, part goes into ionization, and another part goes into the vibrational excitation of the molecules. However, a significant part of this energy transfer may take place during the pre-breakdown stage; in the model it is conveniently taken care of by the initial temperature T_0. Indeed, the energy stored in the vibrational levels will come back to the channels as heat when the temperature of the channels increases and dissociation sets in. Finally, Cooray and colleagues have assumed that the energy dissipation out of the channels is proportional to the volume of the heated gas and its temperature. If the diameters of the discharge channels are more or less similar in the whole ionization region then the volume of ionized gas is proportional to the surface area of the discharge channels. This validates the assumption made, because the heat dissipation is proportional to the surface area of the discharge channels. On the other hand, the heat dissipation out of the channels is proportional, in reality, to the temperature difference between the discharge channels and the soil particles with which the discharge channel is in contact. This assumption is valid only if the heat absorbed by the soil is transferred quickly away from the contact region so that the soil temperature at the contact points remains smaller than the channel temperature. In reality, this is not true and, therefore, the value of S that Cooray and colleagues have estimated should be treated as an effective value pertinent to the model under consideration.

11.4 Conclusions

In this chapter we have presented several soil ionization models created by scientists to evaluate the voltage–current relationship of buried conductors in the presence of soil ionization. Following through the models one can observe that some models are pure engineering constructions while others attempt to incorporate more physics in model development. The models should be judged according to the validity of the physics underlying them and their ability to make reasonable predictions. In this respect it is desirable to have a model the parameters of which do not change from one experimental configuration to another. This is so because a model with parameters that change from one configuration to another could not be used in making predictions. Because all the discussed models involved assumptions that may not be valid in real situations, caution should be applied in making use of the results of the models. Moreover, all the models assume that discharges takes place uniformly around the electrodes. In practice, the grounding may mediate through a few sparks, and in such cases all the models that assume uniform ionization break down. One parameter that all models require is the critical electric field necessary for ionization of the soil. More experimental data gathered under realistic conditions are necessary to evaluate this critical electric field.

References

1. Cooray V. 'Mechanism of electrical discharges', in V. Cooray (ed.). *The Lightning Flash*. London: IEE; 2003. IEE Power & Energy Series 34.
2. Towne H.M. 'Impulse characteristics of driven grounds'. *Gen. Elect. Rev.* 1928; **31**(11):605–609.
3. Bellaschi P.L., Armington R.E. and Snowden A.E. 'Impulse and 60-cycle characteristics of driven grounds'. *Part II, AIEE Trans.* 1942;**61**:349–63.
4. Kosztaluk R., Laboda M. and Mukhedkar D. 'Experimental study of transient ground impedances'. *IEEE Trans. Power Appar. Syst.* 1981;**100**:4653–60.
5. Laboda M. and Pochanke Z. 'Experimental study of electrical properties of soil with impulse current injections'. *Proceedings of the 18th International Conference on Lightning Protection*; Munich, Germany, 1985.
6. Laboda M. and Scuka V. 'On the transient characteristics of electrical discharges and ionization processes in soil, experimental study of electrical properties of soil with impulse current injections'. *Proceedings of the 23rd International Conference on Lightning Protection*; Florence, Italy, 1996.
7. Oettlé E.E. 'The characteristics of electrical breakdown and ionisation processes in soil'. *Trans. SA Inst. Elec. Eng.* 1988;**12**:63–70.
8. Mousa A.M. 'The soil ionization gradient associated with discharge of high currents into concentrated electrodes'. *IEEE Trans. Power Deliv.* 1994;**9**:1669–77.
9. CIGRE working group on lightning. *Guide to Procedures for Estimating the Lightning Performance of Transmission Lines*. Paris: CIGRE; October 1991.

10. Liu Y., Theethayi N., Gonzalez R.M. and Thottappillil R. 'The residual resistivity in soil ionization region around grounding system for different experimental results'. *Proceedings of IEEE International Symposium on EMC*; Boston, USA, 2003, pp. 794–99.
11. Liew A.C. and Darveniza M. 'Dynamic model of impulse characteristics of concentrated earths'. *Proc. IEE* February 1974;**121**(2):123–35.
12. Wang J., Liew A.C. and Darveniza M. 'Extension of dynamic model of impulse behaviour of concentrated grounds at high currents'. *IEEE Trans. Power Deliv.* 2005;**20**:2160–65.
13. Sekioka S., Lorentzou M., Philippakou M.P. and Prousalidis J.M. 'Current-dependent grounding resistance model based on energy balance of soil ionization'. *IEEE Trans. Power Deliv.* 2006;**21**:194–201.
14. Mayr O. 'Beitrage zur theorie des statschen und des dynamischen lichtbogens'. *Arch. fur Elektrtechnik* 1943;**37**:588–608.
15. Cooray V., Zitnik M., Manyahi M., Montano R., Rahman M. and Liu Y. 'Physical model of surge-current characteristics buried vertical rods in the presence of soil ionization'. *J. Electrostat.* 2004;**60**:193–202.
16. Larsson A. *Inhibited Electrical Discharges in Air*. PhD Thesis, University of Uppsala, Sweden, 1997.
17. Yos J.M. 'Transport properties of nitrogen, hydrogen and air to 30 000 K'. *Tech. Mem.* AVCO-RAD-TM-63-1, AVCO Corp., March 1963.
18. Borovsky J.E. 'An electrodynamic description of lightning return strokes and dart leaders: guided wave propagation along conducting cylindrical channels'. *J. Geophys. Res.* 1995;**100**:2697–726.

Chapter 12
Lightning protection of low-voltage networks
Alexandre Piantini

12.1 Introduction

In recent years, the growing use of sensitive electronic devices and components, as well as the increasing demand of utility customers for stability of the power supply, have stressed the importance of improving the reliability and power quality levels of electric systems. As lightning is a major source of faults on overhead lines and damages to or malfunction of sensitive electronic equipment, it is essential to evaluate the lightning electromagnetic environment in order to mitigate its effects and improve power system quality. Many studies have been carried out, especially on medium-voltage (MV) lines, aiming at obtaining a better understanding of the characteristics of the lightning overvoltages.

More recently, special attention has been drawn to the transients on low-voltage (LV) systems. As the surge withstand capabilities of LV networks are much lower than those of MV lines, they are more susceptible to lightning-caused disturbances. There are various ways by which lightning can disturb low-voltage lines. Transients may originate from direct strokes (to the MV or LV networks or to structures) or indirect ones (either intracloud or cloud-to-ground flashes).

The magnitudes and waveforms of these transients depend on many lightning parameters and are substantially affected by the LV network configuration, which is usually complex and may vary widely. The evaluation of the overvoltages associated with indirect strokes entails the calculation of lightning fields, which are defined by the spatial and temporal distribution of the stroke current along the channel, as well as by the earth electrical parameters. A suitable coupling model is required for the analysis of the electromagnetic interaction between the field and the line conductors. Additionally, the frequency responses of distribution transformers and LV power installations have a great influence on lightning surges. This scenario demonstrates that the evaluation of lightning transients on LV networks is an intricate matter.

Nevertheless, it is possible to evaluate the general characteristics of the overvoltages and assess their dependence upon the network configuration and stroke parameters. This is the scope of this chapter, which aims also at appraising the effectiveness of the installation of secondary arresters at strategic points of the network on the mitigation of lightning overvoltages. Emphasis is given to the voltages induced by indirect strokes and to those transferred from the MV system, which are the most important ones on account of their magnitudes and frequencies of occurrence.

This chapter describes, initially, some typical LV network configurations and earthing practices, as well as models that can be used to represent the high-frequency behaviour of distribution transformers and LV power installations. As underground networks are much less prone to lightning disturbances than overhead ones, focus is given to the latter. The major mechanisms by which overvoltages originate from lightning are presented and the voltage characteristics evaluated. The last part of the chapter is dedicated to the lightning protection of secondary networks.

12.2 Low-voltage networks

The LV network comprises that part of the electric distribution system in which the voltage levels are up to 1 000 V. This includes the LV side of distribution transformers, the secondary circuit and consumers' installations.

There are various possible configurations for the LV grids. The most common layouts are radial, mesh, open ring and link arrangements. The first has just one infeed point, whereas in a meshed network there are at least two possible electrical paths through which consumers can be supplied. The open ring arrangement provides at least two alternative paths to each consumer. In normal operation, each section of the ring can be treated as a radial feeder. In the event of a fault, after its isolation a normally open switch is closed and supply can be restored to the other parts of the ring. In the link arrangement two secondary substations are interconnected. However, because of there being a normally open switch the system operates as two radial feeders.

The optimum design depends strongly on the load density, because each arrangement has its own advantages and disadvantages in terms of cost, simplicity, reliability, flexibility, voltage drop, short-circuit power and degree of protection sophistication. Link, open ring and meshed networks are employed in densely populated urban areas, where the consumers are normally fed through underground cables. Radial overhead networks are used in both rural and urban regions.

There is a large diversity of combinations of grid structure, operating criteria and load types. Distribution transformer rating and connections, voltage level, number and type of conductors, total line length, as well as earthing practices may vary from country to country and are much dependent on the characteristics of the particular district concerned [1,2]. Nevertheless, it is possible to identify the most important network parameters required for the assessment of the basic features of lightning surges on LV networks.

12.2.1 Typical configurations and earthing practices

Overhead networks are much more vulnerable to lightning than underground ones, and therefore the configurations and parameters presented hereafter will, unless otherwise specified, refer to this type of network.

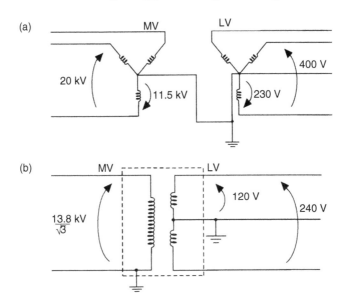

Figure 12.1 *Typical examples of LV systems: (a) three-phase, four conductors; (b) single-phase, three conductors*

Three-phase systems may have either three or four conductors, depending on whether the neutral is present or not, whereas single-phase systems are in general composed of three conductors, two phases plus neutral. Figure 12.1 presents two typical examples of transformer connections used in secondary lines.

Both bare and separately covered conductors are commonly used at low-voltage. Also common is the bunched cable, illustrated in Figure 12.2, in which the phase conductors are isolated and twisted around the bare neutral, which also has the mechanical function of supporting the line.

In urban areas the LV network is commonly installed below the MV line. A typical vertical clearance between the lines is 3 m. The minimum height of the conductors above ground level is usually between ∼4 and 6 m, depending on locality. Aluminium conductors with cross-sections in the range 35–70 mm^2 are normally used.

The total length of a LV network may vary from 100 to ∼2 000 m. In rural areas a typical length is 1 000 m, while 300 m is more representative of urban networks [2]. Average distances between loads are typically 300 and 50 m on rural and urban networks, respectively. Distances between neutral earthing points are usually in the range of 150 to 300 m; a common spacing is 200 m [3]. The maximum length of the service drop from the pole to the customer's premises depends on the load, but usually does not exceed 30 m.

Different types of system earthing arrangements are used in LV distribution networks. The IEC Publication 60364 series classifies these practices according to the relationships of the power system conductors and of the exposed conductive parts

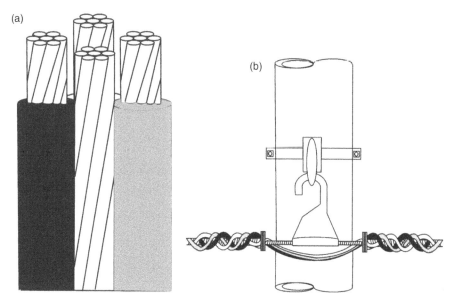

Figure 12.2   Bunched cable with twisted conductors: (a) conductors; (b) detail of the cable suspended from a pole

of components and equipment in the electrical installation with respect to the earthing system [4], as illustrated in Figure 12.3.

The IEC nomenclature essentially adopts two letters to designate the systems. The first, which may be either 'I' or 'T', is related to the system earthing; 'I' indicates that all live parts of the system are isolated from earth or that points of the network are connected to earth through impedances. On the other hand, 'T' signifies a direct connection of at least one point in the network to earth. The second letter refers to the equipment earthing, and may be 'T' or 'N'. 'T' signifies that accessible conductive parts of the equipment in the installation are directly connected to earth, independently of the earthing of any point of the power system, whereas 'N' means a direct connection of accessible conductive parts of the equipment to the earth points of the power system. These connections may be made by means of a protected earth neutral (PEN) or protected earth (PE) conductor.

There are basically three types of system earthing, namely IT, TT and TN, although the latter can be further subdivided into TN–C, TN–S and TN–C–S depending on the way the neutral and protective functions are provided. In the TN–C system, a single conductor (PEN) performs both functions, whereas in the TN–S system separate conductors (neutral and PE) are used. The TN–C–S system is a combination of the TN–C and TN–S systems, as the neutral and protective functions are combined in one conductor from the distribution transformer to the service entrance equipment and provided by separate conductors beyond this point. The TN–C–S system is the most commonly used in public LV networks [1,5].

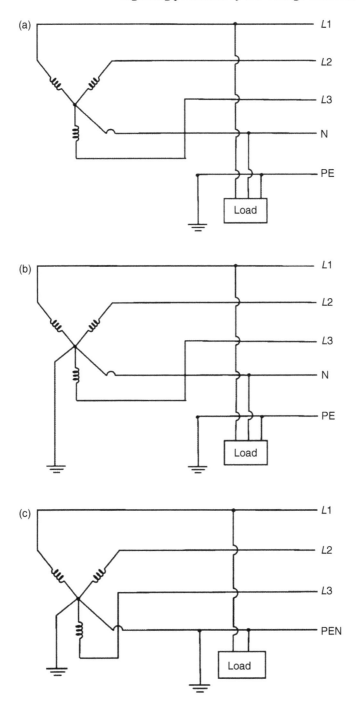

Figure 12.3 Types of system earthing arrangements used in LV distribution networks: (a) IT system; (b) TT system; (c) TN–C system;

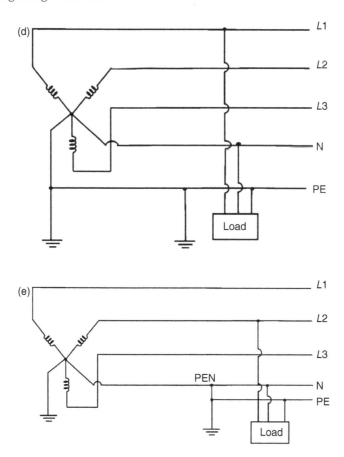

Figure 12.3 (Continued) (d) TN–S system; (e) TN–C–S system

12.2.2 Distribution transformers

Distribution transformers can be either three-phase or single-phase and range in size from 5 to over 2 000 kVA. There is a wide variety of possibilities concerning type, connections and winding arrangement. In rural areas the majority are pole-mounted and rated up to 300 kVA, whereas in highly concentrated urban areas, where power ratings can exceed 2 000 kVA, transformers rated above 250 kVA are usually of the pad-mounted type.

There are various types of transformer connections that can be applied in three-phase systems. Both three-phase transformers and banks of single-phase units are used, the winding arrangements being related to the type of primary supply (effectively earthed, impedance-earthed or unearthed), to the load characteristics (rated voltage, degree of unbalance), and to considerations such as primary and secondary earth faults and susceptibility to ferroresonance. The delta–delta, delta–wye, wye–delta, wye–wye and open–delta configurations are among the most common ones. Guidance for application of transformer connections in three-phase distribution

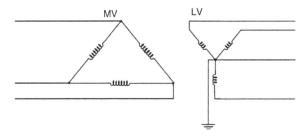

Figure 12.4   Delta–earthed wye connection

systems is given in Reference 6, where all combinations of delta and wye, earthed and unearthed, as well as other special configurations, are considered. The earthed wye–earthed wye connection is shown in Figure 12.1a. The delta–earthed wye configuration, which is also very common [7], is presented in Figure 12.4.

The evaluation of lightning-induced surges on a LV network requires the use of reliable models to represent all the elements involved in the phenomenon. The distribution transformer high-frequency behaviour plays a major role regarding both the voltages induced by indirect strokes and those transferred from the primary to the secondary side. In the first case the most important information is the impedance seen from the LV line, while the analysis of transferred surges requires a more detailed representation.

In order to investigate the input impedances seen from the LV side, frequency-domain measurements were performed in Reference 8 on 15 distribution transformers with rated power from 50 to 1 250 kVA. The results showed that, in the range 10–500 kHz, the average impedance can be represented by means of either an inductance or a capacitance, depending on the condition of the transformer neutral. If the neutral is connected to earth (TN or TT systems), the impedance can be approximated by an inductance between 4 and 40 µH, strongly correlated with the transformer rated power and voltage. On the other hand, if the neutral is isolated (IT system), a satisfactory approximation is obtained through a capacitance between 2 and 20 nF which, however, is poorly correlated with the transformer ratings [9]. Although lightning-induced voltages may have important higher-frequency components, e.g. 1 MHz, such simplified representations are usually adequate for practical purposes.

A comparison between measured and calculated input impedances seen from the LV terminals of two 13.8 kV–220/127 V delta–earthed wye transformers with power ratings of 30 and 112.5 kVA is presented in Figure 12.5. The differences among the impedances of the terminals were not significant and therefore only one curve is shown for each transformer. It can be seen that up to ∼1 MHz the impedances of both transformers can be reasonably approximated by a simple *RLC* parallel circuit. The values of the parameters to represent the impedance of each terminal are $R = 1\,100\,\Omega$, $L = 48\,\mu H$ and $C = 0.76\,nF$ for the 30 kVA transformer and $R = 760\,\Omega$, $L = 12.4\,\mu H$ and $C = 1.0\,nF$ for the 112.5 kVA transformer. The influence of the terminations at the high-voltage (HV) side on the measured input

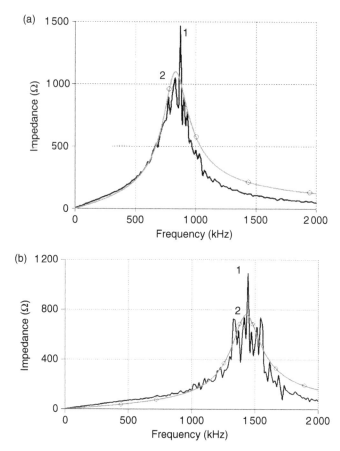

Figure 12.5 Comparison of the magnitudes of the transformer input impedance seen from each LV terminal (1) and the impedance of an RLC parallel circuit (2): (a) 30 kVA transformer (R = 1 100 Ω, L = 48 μH and C = 0.76 nF); (b) 112.5 kVA transformer (R = 760 Ω, L = 12.4 μH and C = 1.0 nF)

impedance was found to be negligible. In the tests, voltages were applied between the secondary terminals short-circuited and the transformer tank. This procedure is valid as long as the voltages induced on the phase conductors have approximately the same amplitude and waveform. This is indeed the normal situation as far as nearby strokes are concerned, especially in the case of a bunched cable, where the twisted conductors are very close to each other and can be assumed at the same height.

The evaluation of the voltage surges transferred from the primary to the secondary windings requires the knowledge of the voltages induced at the HV side as well as of the transformer behaviour with respect to high-frequency signals. Several models for calculating transferred voltages in the case of fast transients have been proposed in the literature.

The model presented by Morched and colleagues in Reference 10 can be used to simulate the high-frequency behaviour of any type of multi-phase, multi-winding transformer by means of combinations of RLC networks that match the frequency response of the transformer at its terminals. The frequency response can be obtained either from measurements or from models based on the physical layout of the transformer. The accuracy and numerical stability of the model were verified from tests and its validation was confirmed through comparisons between measured and simulated – using the Electromagnetic Transients Program (EMTP) [11] – step responses of a 125 MVA, 215/44 kV, wye–wye connected transformer.

The voltages transferred to the LV transformer terminals when lightning strikes close to a distribution line were calculated by Borghetti and colleagues [12] for some simple configurations considering two different transformer models: the well-known capacitive PI-circuit and the model proposed by Vaessen [13], which is valid for unloaded transformers. The results showed that the voltage waveforms can be completely different depending on the model used, and that in some cases the ratio between the amplitudes of the transferred voltages is about 20:1.

The model developed by Piantini and Malagodi [14] for unloaded transformers was shown to satisfy the requirements concerning accuracy and simplicity and was applied in References 15 and 16 to analyse voltages transferred to the secondary in the case of nearby lightning. In Reference 17 it was modified to also enable the representation of a specific transformer under the loaded condition. Based on the transfer characteristics of nine typical distribution transformers, delta–earthed wye connected, and with power ratings ranging from 15 to 225 kVA, a further improvement was proposed by Piantini and Kanashiro [18]. The model was used in Reference 19 to study the effect of the secondary loads on the transferred voltages. The transformer was treated as a quadripole and its transfer function and input, output and transfer impedances were obtained through the application of impulse voltages to the short-circuited primary terminals with the secondary ones open, and vice versa [18]. Details of the modelling procedure, which does not account for saturation effects of the transformer core, are described in Reference 20. The circuit parameters corresponding to four typical distribution transformers, together with the test results that validated the model, are presented in Reference 18.

The influence of different loads on the secondary side of a MV distribution transformer was analysed by Richter and Zeller [21]. Three cases were considered for the transformer termination: open, open cable, and cable with a resistive load. A simple single-phase model was presented and a relatively good agreement was obtained between measurements and computations for the situations considered.

A simple linear distribution transformer model, capable of predicting transient voltages transferred from primary to secondary and vice versa, was presented by Manyahi and Thottappillil [22]. The model was based on the admittance network representation proposed in Reference 10, with the assumption that the transient voltages at the transformer terminals are the same in all phases. Tests were performed in the open-circuit condition and a good agreement was found between the measured and simulated transferred voltages.

The voltages transferred from the MV to the LV distribution system were studied by Montaño and Cooray [23] using the Agrawal *et al.* coupling model [24] for the calculation of voltages induced on the primary line and the transformer models presented in Reference 18 and in Reference 25. From the simulations results, the authors concluded that the transformer has a significant impact on the signature of the overvoltages transferred to the secondary network.

The studies conducted in References 26 to 29 dealt with the voltages transferred to the LV network in the case of direct strikes to the MV line. The investigations were performed using the model described in Reference 18, applied however to different transformers. In the calculations done in Reference 30 for the case of indirect strokes, the transformer was simulated by the circuit adopted in Reference 26.

The investigation carried out by Borghetti and colleagues [31] emphasized the importance of an adequate representation of the distribution transformer in the simulation of transferred voltages. The calculations were done with the LIOV-EMTP code [32–34] considering two transformer models: the pure capacitive PI-circuit and the more complex and accurate model presented in Reference 10. The parameters of both models were obtained experimentally and the results showed that the capacitive PI-circuit model overestimates the transferred voltages by about one order of magnitude in comparison with the high-frequency model proposed in Reference 10. The main conclusion of the paper describes the crucial role played by the transformer in the simulation of transferred overvoltages.

A model of core-type distribution transformers that can be used to evaluate transferred surges was presented by Noda and colleagues [35]. It consists of the equivalent circuit of the transformer with circuit blocks representing winding-to-winding and winding-to-enclosure capacitance, skin effects of winding conductors and iron core, and multiple resonances due to the combination of winding inductance and turn-to-turn capacitance. The model parameters can be determined by measurements using an impedance analyser. Transient simulations using the EMTP [11], considering a 10 kVA unloaded transformer modelled with the proposed method, agreed well with laboratory test results. However, it was later observed that the model gives inaccurate results for a specific type of transformer because the skin effect of the secondary windings was neglected. Therefore, an improved model taking the skin effect into account was proposed by Honda and colleagues [36], and field tests were performed using an actual-scale distribution line. The primary line was three-phase, three-wire, 430 m long and matched at both ends, while the LV line was one-phase, three-wire, 164 m long and open-ended. Through an impulse generator, currents were injected at different points of the system, simulating direct and indirect lightning hits. Comparisons of measured and calculated voltages at the secondary side of a 10 kVA transformer confirmed the accuracy of the model.

An adequate representation of the high-frequency behaviour of the power transformers is essential for the appraisal of the surges transferred to the LV network. Models differ remarkably and the desired compromise between accuracy and simplicity depends on the specific application. Figure 12.6 shows the simplified model obtained in References 18 and 19 for various three-phase, 13.8 kV–220/127 V, delta–earthed wye connected distribution transformers. The parameters to represent each phase of

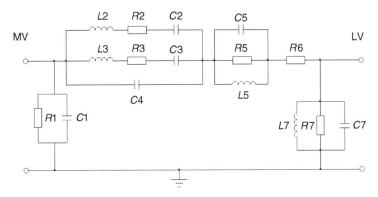

Figure 12.6 Simplified distribution transformer model for evaluation of transferred voltages (adapted from References 18 and 19)

two transformers, rated 30 and 112.5 kVA, are given in Table 12.1. Figure 12.7 presents comparisons between measured and calculated transferred voltages for the secondary both in open circuit and connected to a balanced load, for impulses with different waveforms applied to the short-circuited primary terminals of the 30 kVA transformer. The load was simulated by three resistors of the same value connected in an earthed wye configuration, and the voltages were measured across one of the

Table 12.1 Parameters of the model shown in Figure 12.6 for the transformers rated 30 and 112.5 kVA [18,19]

Parameters	Transformer	
	30 kVA	112.5 kVA
$R1$ (kΩ)	–	–
$R2$ (kΩ)	14	3
$R3$ (kΩ)	0.8	5
$R5$ (kΩ)	–	–
$R6$ (kΩ)	1.1	0.35
$R7$ (kΩ)	1.615	1.5
$C1$ (pF)	493	600
$C2$ (pF)	94.8	1126
$C3$ (pF)	21.5	146
$C4$ (pF)	50	600
$C5$ (pF)	–	400
$C7$ (pF)	760	850
$L2$ (mH)	16	35
$L3$ (mH)	1.84	15
$L5$ (mH)	–	–
$L7$ (mH)	0.05	0.0124

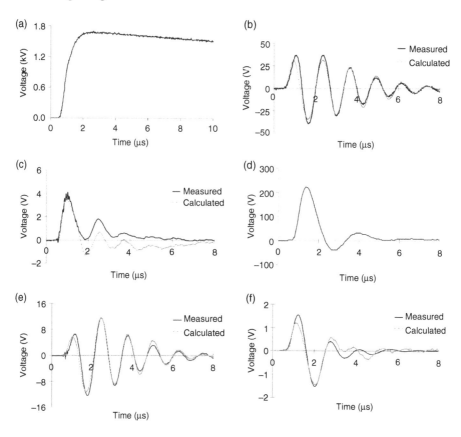

Figure 12.7 Measured and calculated voltages transferred to the secondary of the 30 kVA transformer under different load conditions: (a) voltage applied to the HV terminals (standard 1.2/50 μs waveform, with amplitude of 1.70 kV for panel b and 0.95 kV for panel c; (b) transferred voltages (no-load condition) corresponding to panel a; (c) transferred voltages (balanced load of 50 Ω) corresponding to panel a with amplitude of 0.95 kV; (d) voltage applied to the HV terminals (panels e and f); (e) transferred voltages (no-load condition) corresponding to panel d; (f) transferred voltages (balanced load of 50 Ω) corresponding to panel d

resistors. Similar comparisons are presented in Figure 12.8 for the transformer rated 112.5 kVA. The relatively good agreement observed also in comparisons involving other transformers and load conditions [18,19] indicates the suitability of the model for the evaluation of transferred lightning surges.

12.2.3 Low-voltage power installations

Lightning transients on secondary networks are greatly affected by the characteristics of the loads. The input impedance of a LV power installation, i.e. the impedance seen

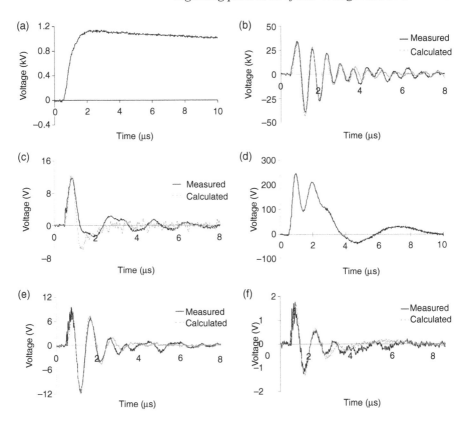

Figure 12.8   Measured and calculated voltages transferred to the secondary of the 112.5 kVA transformer under different load conditions: (a) voltage applied to the HV terminals (standard 1.2/50 µs waveform, panels b and c; (b) transferred voltages (no-load condition) corresponding to panel a; (c) transferred voltages (balanced load of 50 Ω) corresponding to panel a; (d) voltage applied to the HV terminals (panels e and f); (e) transferred voltages (balanced load of 100 Ω) corresponding to panel d; (f) transferred voltages (balanced load of 10 Ω) corresponding to panel d

by the power line at the service entrance, varies according to the earthing system arrangement, number and configuration of the circuits, type, size and length of the conductors, and frequency response of the connected electric appliances. As a consequence, remarkable differences may be observed between the impedances of distinct installations. This justifies, to a certain extent, the use of simple models in the analysis of lightning transients, where loads are often simulated as lumped resistors [37–39], inductors [27] or capacitors [40]. A more accurate representation of the load frequency-dependent behaviour, however, may allow for a more precise evaluation of the main features of the overvoltages.

566 Lightning Protection

Measurements in the range of 5 kHz to 2 MHz, performed on several residential units, indicate that the connected electric appliances have a major influence on the input impedance of TN systems, although for frequencies above 100 kHz the impedance can be approximated by an inductance in the range 2–20 μH [8,9]. The inductance decreases with the number of circuits and branches, so that lower values correspond to larger installations. On the other hand, the impedance of an IT system is affected to a lesser degree by the connected consumer loads. Its behaviour is typically capacitive up to ∼100 kHz, becoming inductive for higher frequencies. The equivalent circuit can be modelled by a capacitance between 2 and 200 nF in series with an inductance in the range 2–20 μH. Larger capacitance values are associated with larger installations. Figure 12.9 presents the frequency responses of the equivalent circuits of the input impedances of a 127 m^2 residential apartment [9]. The measurements were performed at the meter cabinet between the phase conductors and the PE conductor, with the installation disconnected from the distribution

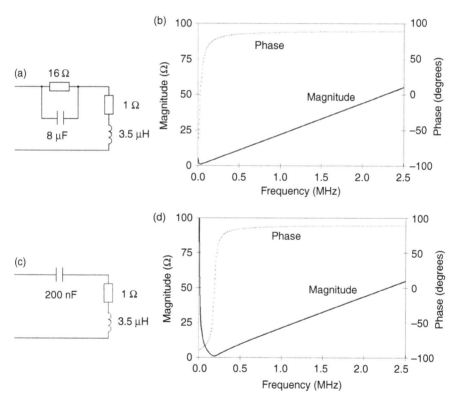

Figure 12.9 Equivalent circuits and corresponding frequency responses of the input impedances of typical LV power installations: (a) equivalent circuit (TN system); (b) frequency response (TN system); (c) equivalent circuit (IT system); (d) frequency response (IT system) (adapted from Reference 9)

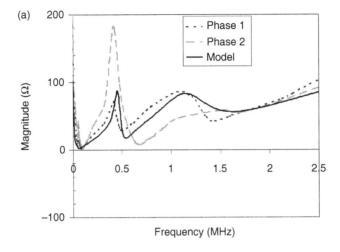
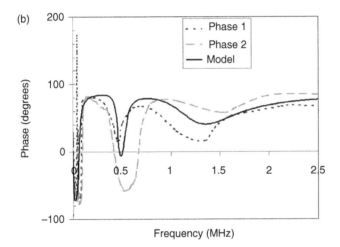
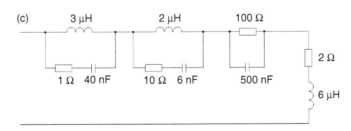

Figure 12.10 Measured and calculated input impedances of each phase of a 90 m² residential apartment (TN system) as function of frequency and corresponding equivalent circuit: (a) input impedance (magnitude); (b) input impedance (phase); (c) equivalent circuit (adapted from Reference 41)

network. Data relative to the IT system were obtained by connecting the neutral to the phase wires, while the tests related to the TN system were carried out with the neutral connected to the PE conductor.

To illustrate the diversity of responses of LV power installations, Figure 12.10 presents the measured input impedance of a residential apartment (TN system) of approximately 90 m^2, as well as the results relative to the corresponding equivalent circuit [41]. Unlike in References 8 and 9, the phase wires were not connected with each other, and the measurements were performed between each phase and the neutral. However, notwithstanding the differences between the input impedances, mainly for lower frequencies, the same general behaviour, predominantly inductive, is observed both in Figures 12.9b and 12.10a, b. Input impedances of various electric appliances, as well as their respective approximations by means of simple circuits, are presented in Reference 42.

12.3 Lightning surges on low-voltage systems

Lightning transients on LV networks can be produced by several mechanisms, which can be classified into the following categories:

- direct strikes to the LV system (either to the line conductors or to end-user installations);
- intracloud or cloud-to-cloud lightning;
- indirect strikes (cloud-to-ground lightning);
- transference from the MV system.

In this section the characteristics of the surges associated with each of these types are evaluated. Emphasis is given to the overvoltages caused by indirect strikes and to those transferred from the MV system, which are among the most important ones on account of their magnitudes and frequencies of occurrence.

12.3.1 Direct strikes

If a flash hits an overhead line, the current injected into the conductor is divided at the strike point, giving rise to two voltage waves that propagate in opposite directions. The prospective magnitude of these voltages can be estimated by multiplying the current that flows in each direction (half of the stroke current) by the characteristic impedance of the line, which is normally in the range 400–550 Ω. Therefore, for a line characteristic impedance of 400 Ω and a stroke current of 10 kA, whose probability of being exceeded is larger than 90 per cent, the corresponding overvoltage is 2 000 kV, which is far beyond the line insulation level. As a consequence, multiple flashovers occur between all the conductors and also to earth in various points of the line, causing many current and voltage reflections as well as reduction of the effective earth impedance.

After the occurrence of the disruptive discharges a rough estimation of the overvoltage can be obtained, if the propagation effects are disregarded, by multiplying the stroke current by the equivalent earth impedance. However, even in the case of a low value for the effective impedance (10 Ω, for example), voltages much larger than the line lightning impulse withstand level would result (100 kV in the above example), which would lead to further flashovers.

In general, LV networks are not that prone to direct strikes due to their relatively short lengths and to the shielding provided by the MV line, trees and nearby structures. However, in some rural and semi-urban areas, exposed LV lines longer than 1 000 m do exist, and in case of direct lightning hits, the resulting overvoltages can damage unprotected connected equipment.

A direct strike to the lightning protection system or to other parts of an end-user building causes an earth potential rise that may lead to the operation of surge protective devices or to flashovers between the structure and the line conductors. In both situations a portion of the stroke current is injected into the power line, producing overvoltages that propagate along the network. This portion depends mainly on the relative impedance of the line with respect to the impedances of all the other possible current paths (local earth, metallic pipes and other services such as telecommunications lines). Figure 12.11 illustrates the situation, showing a case in which 50 per cent of the stroke current enters the earth termination system and 50 per cent is distributed evenly among the services entering the structure.

The division of the lightning current between the earthing system of a test house and the neutral of the power supply was investigated at the International Center for Lightning Research and Testing (ICLRT) at Camp Blanding, Florida, by means of the rocket-triggered lightning technique [43,44]. Small rockets trailing thin wires were used to trigger lightning and inject its current into the lightning protection system of the test house, whose electrical circuit was connected to the secondary of a pad-mounted distribution transformer [45,46]. The distance between the house and the transformer was ~50 m, and the primary was connected to a 650-m-long

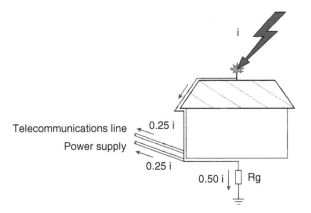

Figure 12.11 Injection of surges into the LV power line due to a direct strike to an end-user installation. The current division depends on the relative impedances of all the possible current paths.

unenergized underground power cable. Test configurations varied with respect to the lightning current injection point, number of down-conductors, earthing system of the test house, and use of surge protective devices. The current waveforms observed in the earth rods differed substantially from those recorded in other parts of the system. The ratio between the peak values of the currents entering the power supply neutral to the injected lightning current varied from ~ 22 to over 80 per cent, depending on the test configuration.

In the event of a direct strike to the MV line, part of the stroke current is injected into the neutral conductor, causing overvoltages on the LV network. This mechanism will be discussed in Section 12.3.4.1.

12.3.2 Cloud discharges

Cloud discharges, which include intracloud, cloud-to-cloud and cloud-to-air flashes, last typically between 200 and 500 ms [47] and are the most frequent type of discharges, representing ~ 75 per cent of the global lightning activity [48,49]. Nevertheless, the number of studies conducted about this phenomenon is relatively small in comparison with that relating to cloud-to-ground flashes, which have a much greater impact in terms of deleterious effects. The main practical interest in cloud flashes lies in the protection of aircraft and space craft, although the short interval between the associated induced voltage pulses may cause degradation, damage and failure of electronic components of sensitive apparatus connected to the LV power supply.

The voltages induced on complex LV power installations by cloud (IC) and cloud-to-ground (CG) discharges have been studied in References 50 to 53. The investigation carried out by Galván and colleagues [50] refers to two small networks isolated from the power supply. Simultaneous measurements of the incident vertical lightning electromagnetic fields and the corresponding induced voltages across a 50 Ω resistor connected between one of the phase conductors and earth were performed. The technique proposed by the authors uses these measurements to extract the transient response of the power installation, irrespective of its complexity. The peak-to-peak values of the four voltages induced by cloud flashes shown in the paper are below 2 V, and in all cases a relatively good agreement was found between measured and calculated results.

The experimental study conducted by Silfverskiöld and colleagues [51] compares the amplitudes of the common-mode voltages induced on a residential installation in the complete duration of typical CG (negative and positive) and IC flashes. The installation was disconnected from the power distribution line. The measurements of the vertical component of the electric field and the corresponding induced voltages on the power installation showed that the discharge events that take place inside the cloud, preceding CG and IC flashes, give rise to bipolar pulses with very fast rise times. The pulse trains associated with such processes may induce voltages with magnitudes of the same order of (and even higher than) those induced by the return stroke itself. These events are therefore important and should be taken into account in the evaluation of the interference problems caused by lightning electromagnetic pulses

(LEMP). From the analysis of the obtained results the authors estimate, from a typical lightning within a distance of a few kilometres from the LV power installation network, a few tens of induced voltage pulses exceeding 400 V peak-to-peak.

A further investigation of the transient response of LV power installations to lightning electric fields was performed by Galván and Cooray in Reference 52, where comparisons between measured and calculated induced voltages using the measured lightning electric field (both inside and outside the installation) as the driving source are also presented. The amplitudes and waveforms of the induced voltages were found to be highly dependent on the soil resistivity and on the loads connected to the LV power installation. In Reference 53, Galván and colleagues apply the technique proposed in Reference 50 to a simple circuit and to a complex wiring system. Comparisons between measured and simulated induced voltages, presented for both systems, are in good agreement. Discussions are provided on the advantages and limitations of the method, which represents a useful tool for evaluating induced voltages in electrical installations with linear behaviour.

Voltages induced by cloud discharges at both open-circuited terminations of an unenergized 460-m-long distribution line, together with the corresponding electric fields, are reported by Rubinstein and Uman [54]. The line consisted of two conductors arranged in a vertical configuration, and the peak-to-peak voltages induced at the top conductor by a flash at an altitude greater than $\sim$5 km were around 140 V.

Even though further investigations are necessary to better characterize the voltages induced by cloud discharges as well as the significance of their effects on sensitive loads, protection measures against the more severe types of lightning surges are also likely to be effective against such transients.

12.3.3 Indirect strikes

When lightning strikes the earth or an object in the vicinity of a distribution network, the voltages that arise on the LV conductors may be subdivided as follows:

- voltages induced 'directly', due to the electromagnetic coupling between the line and the stroke channel;
- voltages associated with the part of the stroke current that is intercepted by the earthing points of the neutral conductor;
- voltages transferred from the MV line, which will be dealt with in Section 12.3.4.2.

From experiments performed at Camp Blanding using the rocket-and-wire technique to trigger lightning, Rakov and Uman [43] and Fernandez and colleagues [55] showed that when the strike point is at tens of metres from the line, an appreciable fraction of the total current enters the system from the neutral earth connections. In three cases reported in Reference 43, in which the distances between the line and the strike point were 60, 40 and 19 m, the observed peak values of the currents entering the system from its earth connections were, respectively, 10, 5 and 18 per cent of the stroke current peak.

The voltages 'directly' induced are in general the most important on account of their severity and frequency of occurrence. However, although the number of studies conducted in this field has been increasing consistently [8,9,30,40,56–60], there is still a stark contrast to that of MV lines, to which much more attention has been given in the past.

The analysis carried out by Hoidalen in Reference 9 made use of the Agrawal and colleagues coupling model [24] for the calculation of lightning-induced voltages on LV systems. From frequency response measurements, simple models were proposed for the input impedances of typical distribution transformers and LV power installations, and their influences upon the lightning-induced voltages on simple TN and IT systems were investigated. The line considered in the simulations was 500 m long and the phase conductors were simulated by a single wire with characteristic impedance of 300 Ω. In one end there was a transformer, modelled as an inductance of 10 μH, while an impedance representing the power installation was connected at the other termination. The voltage magnitudes were found to have a high dependence on the load, the lowest values being associated with larger installations. A comprehensive investigation was conducted by Hoidalen in Reference 8, where the effect of the finite earth conductivity on the induced voltages on TN and IT systems is thoroughly discussed.

In References 40 and 57 the authors concluded that the induced voltages are characterized by a high-frequency damped oscillation with a period equal to twice the travel time of a span (portion of the line between two adjacent neutral earthings). The simulations were done with the LIOV-EMTP code [32–34], which is based upon the Agrawal *et al.* coupling model, and considered overhead cables with two or four twisted conductors, with neutral earthing spacing in the range of 250 to 400 m. Owing to the high transient electromagnetic coupling between the conductors, for the configuration examined the wire-to-wire voltages were disregarded and line-to-earth voltages, assumed to be the same on the different conductors, were presented.

The influences of various parameters on the lightning-induced voltages were evaluated by Piantini and Janiszewski [58] for the case of a 300-m-long, single-phase line. In Reference 59, a line with twisted conductors was considered. In both cases the distribution transformer and the LV power installations were represented according to the models shown in Figure 12.6 (30 kVA transformer) and in Figure 12.9a, respectively. The main difference between the bunched cable and the open-wire configuration is that the former is characterized by a stronger coupling between the wires, which are much closer. The greater the mutual surge impedance between the neutral and phase conductors, the smaller the induced voltage magnitudes. If the conductors are twisted, this impedance varies along the line. However, as the distance between the wires is much smaller than their heights above ground, the variation is small and for practical cases it can be neglected. The simulations, performed by means of the 'Extended Rusck Model' (ERM) [61–63], showed that the induced voltages have a great impact on the lightning performance of LV distribution lines.

Measurements performed by Hoidalen [60] in Norway, where the ground flash density is mostly below 1 km^{-2} yr^{-1}, show that more than 1 000 voltages above

500 V should be expected per year in a typical rural LV overhead line with isolated neutral. Voltages up to 5 kV were recorded, and according to the study, overvoltages can be induced by strokes more than 20 km away from the line.

Rocket-triggered lightning experiments with simultaneous measurements of induced voltages on a 210-m-long overhead LV line with twisted conductors are reported by Clement and Michaud [3]. The line was connected to a transformer at one of the ends, and to a 60-m-long underground cable, terminated by LV arresters, at the other. The stroke location was either on a tower close to the underground cable termination or on the firing area, 50 m away from this point. The induced voltages were measured at the LV transformer terminals for a total of 12 launchings. The stroke currents varied in amplitude from 4 to 50 kA, and the corresponding phase-to-earth and neutral-to-earth voltages reached maximum values in the range 2–12 kV.

The analysis of the characteristics of such surges is of great importance, because they have a high frequency of occurrence and can often reach large magnitudes. The severity of the induced voltages depends on many lightning parameters and is also substantially affected by the network configuration. The understanding of the way the various parameters involved in the induction mechanism affect their amplitudes and waveforms is, therefore, of vital importance.

12.3.3.1 Calculation of lightning-induced voltages

Owing to the impact of lightning-induced overvoltages on the performance and power quality of distribution systems, several theoretical and experimental studies have been conducted in order to better understand their characteristics or to assess the effectiveness of the methods that can be used for their mitigation [64–96]. With regard to experimental investigations, Yokoyama and colleagues presented in References 97–99 simultaneous measurements of stroke currents and the corresponding induced voltages, thus allowing direct comparisons between measured and calculated results.

The appraisal of three different theories for computation of lightning-induced voltages on overhead lines presented in Reference 100 concludes that the Rusck model [81] leads to consistent results. However, in its original form the electric field is assumed to be constant in the region between the line and the ground, the lengths of the line and of the stroke channel are assumed to be infinite, only straight lines can be considered, and thus realistic configurations cannot be taken into account. These restrictions limit the application of the model and an extension was proposed by Piantini and Janiszewski [61]. The so-called Extended Rusck Model (ERM) overcomes these limitations and enables us also to take into account the incidence of lightning flashes to nearby elevated objects, the occurrence of upward leaders, and the presence of a periodically earthed shield wire (or neutral conductor) and equipment such as transformers and surge arresters. Lines with various sections of different directions can be considered through the evaluation of the correct propagation time delays for the elementary voltage components that determine the induced voltage at a given point of the line.

The validity of the ERM has been demonstrated from various comparisons of theoretical and experimental results [61–63] for the case of an electromagnetic field radiated by a lightning channel perpendicular to the earth plane. This is indeed the hypothesis under which the Rusck model was developed and Cooray [101] and Michishita and Ishii [102] have shown that, for this condition, it leads to results identical to those obtained from the more general coupling model proposed by Agrawal and colleagues [24], the adequacy of which has been confirmed in References 69, 75, 88 and 103.

The calculation of lightning-induced voltages on overhead lines through the ERM is based on electric and magnetic potentials due to the charges and currents in the channel. The inducing scalar potential associated with the charges acts as a distributed source and is responsible for the generation of waves that propagate along the conductors. On the other hand, the magnetic potential associated with the currents contributes with its time derivative to the total induced voltage in each point of the line.

In the case of direct strikes to a metallic elevated object, the return stroke starts at the top of the structure. The currents in the object and in the stroke channel are assumed to have equal magnitudes and polarities, but different speeds and directions of propagation, as illustrated in Figure 12.12 for the case of a downward negative flash. The current through the strike object propagates at a speed very close to that of light in free space (c), whereas in the channel the speed is a fraction of this value.

The voltages $U(x, t)$ induced on an overhead line located in the vicinity of the strike object are obtained by adding the component associated with the charges in the stroke channel (electrostatic component) to those associated with the currents that propagate in the lightning channel and in the strike object (magnetic components). Thus,

$$U(x, t) = V(x, t) + \int_0^h \frac{\partial Ai(x, t)}{\partial t} dz + \int_0^h \frac{\partial Ai_t(x, t)}{\partial t} dz \qquad (12.1)$$

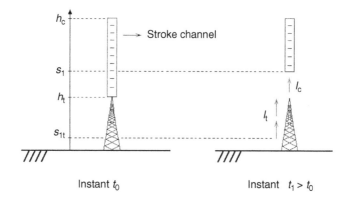

Figure 12.12 Lightning strike to an elevated object (h_t, height of the strike object; h_c, height of the upper extremity of the stroke channel; I_c and I_t, currents in the stroke channel and in the strike object, respectively; s_l and s_{lt}, heights, at instant t_l, of the current front in the channel and in the strike object, respectively)

where $V(x, t)$ is the induced scalar potential, h is the height of the line, and $Ai(x, t)$ and $Ai_t(x, t)$ are the vector potentials associated with the currents that propagate through the stroke channel and through the strike object, respectively. The procedure for calculating the voltage induced on an overhead line in the presence of a tall strike object, disregarding the reflections at the bottom and top of the structure, is described in Reference 61.

Baba and Rakov, using the finite-difference time-domain (FDTD) method, verified that the ratio between the magnitudes of lightning-induced voltages for strikes to a tall object and to flat ground increases with increasing distance from the lightning channel (ranging from 40 to 200 m), decreasing the current reflection coefficients at the top and at the bottom of the strike object (ρ_{top} and ρ_{bot}, respectively), and decreasing return stroke speed [83,104]. Also, the ratio increases with decreasing lightning current rise time. Under realistic conditions such as $\rho_{top} = -0.5$ and $\rho_{bot} = 1$, the ratio is larger than unity (the tall strike object enhances the induced voltages), but it becomes smaller than unity under some special conditions, such as $\rho_{top} = 0$ and $\rho_{bot} = 1$.

The lightning-induced voltage waveforms presented by Yokoyama and colleagues in References 97–99 were measured on an experimental line with two sections, as shown in Figure 12.13. They were obtained simultaneously with the stroke currents that hit a 200-m-high metallic tower situated at a distance of 200 m from the experimental line. Electrical–optical converters were used for transmission of the obtained waveforms by optical cables and, after optical–electrical conversion, the data were stored in magnetic tapes. Owing to the characteristics of the converters, the recorded waveforms present a faster decay than the real ones [98]. The length of the lightning channel was assumed to be equal to 3 km and the stroke current was assumed to propagate through the lightning channel at a constant speed of 30 per cent of that of light in free space, as these parameters were not measured.

Figure 12.14 presents measured and calculated voltages induced on the experimental line by downward negative flashes that struck the tower. The corresponding stroke current waveforms are also shown. As shown in Reference 63, the calculations presented in Figure 12.14, which take into account the effects of the tower, line topology

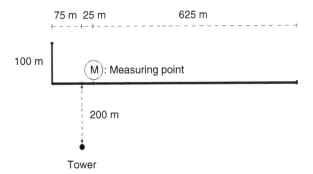

Figure 12.13   *Top view of the experimental line for the measurements carried out in References 97–99*

576 Lightning Protection

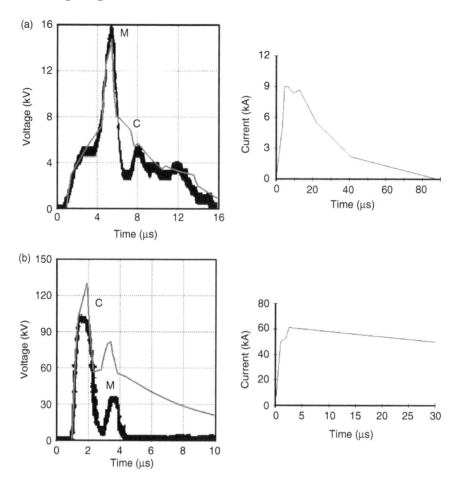

Figure 12.14 *Measured (M) and calculated (C) lightning induced voltages and corresponding stroke current waveforms: (a) case 81-02 [97]; (b) case 86-03 [99] (adapted from Reference 56)*

and finite length of the lightning channel, are in a much better agreement with the measured voltage waveforms than those performed with the original Rusck model. In spite of the differences on the wavetails, both the voltage magnitudes and wavefronts are reasonably well reproduced by the ERM.

The observed discrepancies can be attributed partially to the representation of the stroke current waveform, which has a significant influence on the induced voltages, especially in the case of strikes to elevated metallic objects. Other reasons for the differences are the features of the electrical–optical converters, as already mentioned, the Transmission Line (TL) model [105] adopted for the determination of the current distribution along the lightning channel, and the fact that no reflections were considered at the top and bottom of the structure. Greater discrepancies would probably be observed in the case of stroke currents with steeper fronts, because in this case the current waveforms along the tower would be affected more significantly by the reflections at the tower extremities. Finally, the assumptions of a constant current propagation velocity of 30 per cent that of light in free space and of a lightning channel perpendicular to a perfectly conducting ground plane also contribute to the deviations. For these reasons the comparisons can only be made under a qualitative perspective.

Even so, the overall agreement between theoretical and experimental results is very reasonable and indicates the adequacy of the ERM, for which the validity has also been confirmed from many other comparisons of measured and calculated voltages, mostly using data obtained from scale model experiments performed under different network configurations.

A comparison using data obtained from the 1:50 scale model described in References 75, 80 and 103 is presented in Figure 12.15, where the voltage and time scales are referred to the real system by applying the corresponding scale factors (1:18 000 and 1:50, respectively). In this simple configuration, the line was matched at both terminations. A good agreement is observed between the measured and calculated voltages.

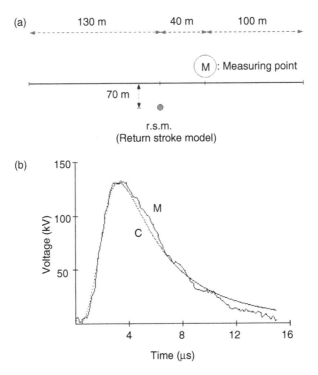

Figure 12.15   Calculated (C) and measured (M) induced voltages obtained from a 1:50 scale model (all the parameters are referred to the full-scale system). Stroke current with peak value of 34 kA, front time of 2 μs, and propagation velocity of 11 per cent c: (a) line topology (top view); (b) induced voltages.

578 Lightning Protection

In References 75 and 103 the experimental facility was used to validate the Agrawal *et al.* coupling model [24] through comparisons with simulations performed using the LIOV–EMTP code [69,75,88] considering much more complex network configurations. This program allows for the calculation of lightning-induced voltages on homogeneous, multiconductor, lossy overhead lines, also taking into account the effects of downward leader electric fields and corona [64].

12.3.3.2 Sensitivity analysis

Let us consider, initially, the situation illustrated in Figure 12.16, in which lightning strikes a point 50 m from a LV line, midway between its terminations. The line is single-phase, 1 km long, matched at both ends and lossless. In Reference 106, Rachidi and colleagues demonstrate that when the line length does not exceed a certain critical value (typically 2 km), the surge propagation along it is not appreciably affected by the finite earth conductivity as long as this conductivity is not lower than ~ 0.001 S m^{-1}.

The heights of the phase and neutral conductors are, respectively, 6.5 and 6.48 m, and both wires have the same diameter, namely 1 cm. The neutral is earthed at a single

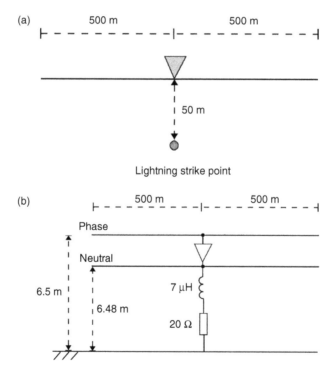

Figure 12.16   LV line matched at both ends, without any loads connected. The triangle denotes the input impedance of the LV side of the distribution transformer: (a) top view; (b) side view.

point, very close to the transformer. The value of the earth resistance (Rg) is 20 Ω and the earth lead inductance is 7 μH.

The transformer is located at the middle of the line. For the estimation of the lightning-induced voltages on a secondary network, the most important information regarding the transformer is the high-frequency behaviour of its impedance seen from the LV side, which is represented by a simple RLC parallel circuit ($R = 367\ \Omega$, $L = 16\ \mu$H, $C = 2.28$ nF). All the LV power installations are disconnected from the network and earth is assumed to be a perfectly conducting plane.

Two stroke current waveforms are considered in order to simulate typical negative downward flashes. The first stroke current has a peak value of 30 kA and its waveform is mathematically described by the Heidler function [107]

$$i(t) = \frac{I_0}{\eta} \frac{(t/\tau_1)^n}{[(t/\tau_1)^n + 1]} e^{-(t/\tau_2)}; \quad \eta = e^{-\left(\frac{\tau_1}{\tau_2}\right)\left(n\frac{\tau_2}{\tau_1}\right)^{1/n}} \quad (12.2)$$

where $I_0 = 28.3$ kA, $\tau_1 = 1.75\ \mu$s, $\tau_2 = 130\ \mu$s and $n = 2$.

The subsequent stroke current has a peak value of 12 kA and its waveform is simulated by the sum of two Heidler functions with the following parameters: $I_{01} = 10.7$ kA, $\tau_{11} = 0.25\ \mu$s, $\tau_{12} = 2.5\ \mu$s, $I_{02} = 6.5$ kA, $\tau_{12} = 2.1\ \mu$s, $\tau_{22} = 230\ \mu$s and $n_1 = n_2 = 2$. These current peak values have a probability of $\sim$50 per cent of being exceeded [108]. The propagation velocity is assumed to be 60 per cent that of light in free space for both currents. The lightning channel is vertical, 4 km long, has no branches, and is modelled according to the TL model [105]. All the induced voltage calculations presented hereafter have been performed using the ERM.

The waveforms of the two stroke currents are presented in Figure 12.17, together with the corresponding voltages induced at the middle of the line in the absence of the transformer and of the neutral conductor. The first stroke induces a voltage with slightly larger magnitude, but as the subsequent stroke current has a higher time-variation rate, the ratio between the voltage peak values is much smaller than that between the current peaks (approximately 1.02 versus 2.5).

Figure 12.18 presents the phase-to-earth and phase-to-neutral voltages induced by the first and subsequent strokes at the middle of the line shown in Figure 12.16, i.e. at the LV transformer terminals. All the voltage magnitudes are much lower than those presented in Figure 12.17, as the connection to earth give rise to currents on the neutral that, by coupling, reduce the voltage on the phase conductor. This voltage is further reduced by the connection that exists between the wires through the transformer. As the transformer input impedance is predominantly inductive in the frequency range of interest, the phase-to-neutral voltage peak value is associated with the maximum time derivative of the current that flows through the inductance, which is lower in the case of the first stroke current. As a consequence, unlike the previous case, shown in Figure 12.17b, the larger voltages are induced by the subsequent stroke.

Phase-to-neutral voltages affect directly the majority of the equipment connected to the LV network, and stress the insulation between the phase and the earthed conductive parts of apparatus in TN systems. On the other hand, phase-to-earth and

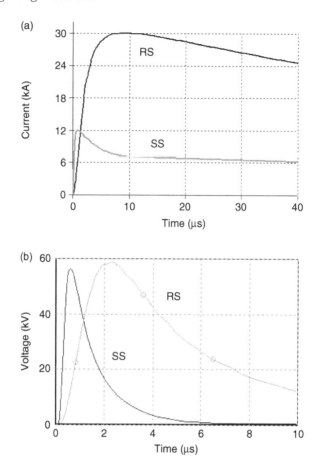

Figure 12.17 *Stroke currents and corresponding induced voltages in the middle of the line shown in Figure 12.16 in the absence of the transformer and of the neutral conductor: (a) first (RS) and subsequent stroke (SS) currents waveforms; (b) induced voltages*

neutral-to-earth voltages are important in the TT arrangement, as well as in the case of equipment connected to an independent earthing system.

These simple simulations indicate that the analysis of the lightning-induced voltages is not so straightforward. Many factors must be taken into account, and it is convenient to define a base case in order to evaluate how the voltages are affected by the various parameters involved in the induction mechanism. Therefore, unless otherwise indicated, all the induced voltage calculations presented henceforth are made under the following assumptions:

- the lightning channel is vertical, 4 km long, has no branches and is modelled according to the TL model [105];

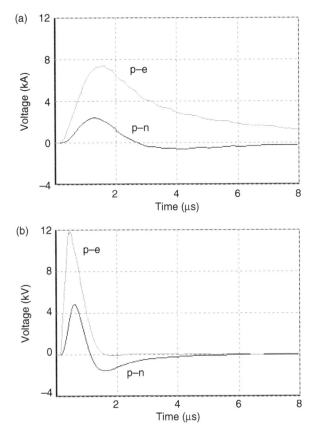

Figure 12.18 Phase-to-earth (p–e) and phase-to-neutral (p–n) induced voltages in the middle of the line for the situation indicated in Figure 12.16 and stroke currents shown in Figure 12.17a: (a) first stroke; (b) subsequent stroke

- the stroke current is typical of a subsequent stroke; the waveform is that shown in Figure 12.17a, with amplitude of 12 kA; its propagation velocity is 60 per cent of that of light in free space;
- the LV line is three-phase, lossless and 1 km long. All the wires have the same diameter, namely 1 cm, and the heights of the phase and neutral conductors are, respectively, 6.5 and 6.48 m;
- the self and mutual characteristic impedances of the conductors are 472 and 389 Ω, respectively. As the phase conductors are at the same height and at approximately the same distance to the lightning channel, the induced voltages on the three wires are assumed to be equal and therefore the three phases are simulated by an equivalent conductor with characteristic impedance of 417 Ω;
- the distance between adjacent neutral earthings is 200 m;

582 *Lightning Protection*

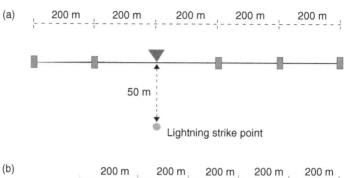

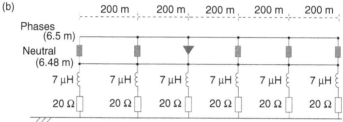

Figure 12.19   *Low-voltage line configuration (base case). The triangles and the rectangles denote, respectively, the distribution transformer and the LV power installations: (a) top view; (b) side view; (c) relative position between MV and LV lines (example of an actual system).*

- the transformer input impedance seen from each LV terminal is that shown in Figure 12.5a, so that the parameters of the equivalent RLC parallel circuit in the equivalent single-phase representation are $R = 367\,\Omega$, $L = 16\,\mu H$ and $C = 2.28\,nF$;
- the LV power installations are simulated according to the equivalent circuit indicated in Figure 12.9a (TN system) and are distributed along the line as shown in Figure 12.19;
- the LV power installations are close to the line and, thus, in Figure 12.19b each impedance formed by an inductance of $7\,\mu H$ in series with a resistance of $20\,\Omega$ corresponds to the equivalent impedance of the earth connections of the neutral and of the closest consumer's installation. In this way, each equivalent impedance takes into account the inductances of the earth lead and of the service drop, as well as the earth resistances at that neutral earthing point and at the closest service entrance;

- the lightning strike point is 50 m from the line and in front of the distribution transformer, as shown in Figure 12.19;
- earth is assumed as a perfectly conducting plane;
- voltages are calculated at the point closest to the stroke location, i.e. at the transformer LV terminals;
- neither the line nor the LV power installations have secondary arresters.

The values adopted for the lightning parameters are typical of downward negative flashes (subsequent strokes). Although the load model has been derived from measurements performed on a residential apartment, the configuration depicted in Figure 12.19 is more representative of a rural network.

The presence of the MV conductors causes a slight reduction on the induced voltages on the LV line. However, unlike the case of urban regions, where the LV network is usually below the MV conductors, in rural areas the lines frequently form angles with each other, as illustrated in Figure 12.19c, which shows an example of an actual system. This results in a smaller coupling, which can usually be disregarded.

Lightning channel
The length of the lightning channel may vary widely, but is generally in the range 1–6 km [109]. Its influence on the induced voltage magnitude is negligible, as for cases of practical interest the peak value is reached well before the stroke current wavefront has reached the top of the channel. Very short channels may have a minor effect on the wavetail, causing a slightly faster decay time [62], which can usually be disregarded.

The induced voltages depend on the temporal and spatial distribution of the stroke current along the channel, and thus they are affected by the return stroke model used to calculate this distribution. However, despite the differences that may occur on the amplitudes and maximum steepness, the voltage waveforms are similar, as shown by Nucci and colleagues in Reference 66. The voltages computed in Reference 8 on a 1-km-long overhead line matched at both ends using the TL model and the Modified Transmission Line Model with Exponential Current Decay (MTLE) [110] have approximately the same amplitude, front time and time to half-value.

Stroke current magnitude and waveform
Because, in practice, the induced currents that flow through the neutral earthings are not high enough to cause soil ionisation and the line is supposed without secondary arresters, the system is linear and, therefore, the induced voltages are directly proportional to the stroke current magnitudes as long as the current waveform is kept unaltered.

Figure 12.20 presents the phase-to-earth and phase-to-neutral induced voltages at the transformer LV terminals for the typical current waveforms shown in Figure 12.17. The equivalent front times (tf_{30}) are about 4.9 and 0.5 μs for the first and subsequent stroke currents, respectively. This parameter is defined as the time to peak of a current with linearly rising front, which has the same time interval between the points corresponding to 30 and 90 per cent of the maximum value. As far as induced voltages are concerned, subsequent stroke currents described by the

584   Lightning Protection

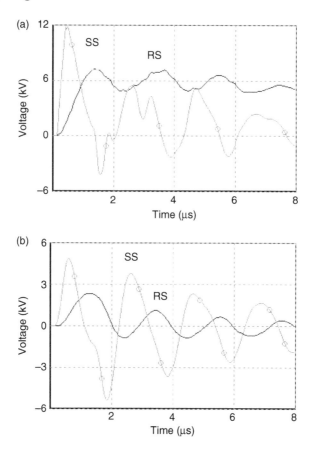

Figure 12.20   Induced voltages at the transformer LV terminals for the first (RS) and subsequent (SS) stroke currents shown in Figure 12.17. Line configuration indicated in Figure 12.19 (a) phase-to-earth (b) phase-to-neutral.

Heidler function can be reasonably approximated by either triangular or trapezoidal waveforms with the same equivalent front time (tf_{30}). In the case of first strokes, however, the correspondence is not so direct.

Figure 12.21 compares phase-to-earth and phase-to-neutral voltages at the transformer LV terminals for stroke currents with peak value of 12 kA and triangular waveforms. Front times (tf, which for a triangular waveform correspond to the time to peak) of 0.5, 1 and 2 µs are considered, while the time to half-value is in all cases equal to 50 µs. The current front time influences significantly both the voltage magnitude and waveform. Induced voltages associated with faster currents are characterized by larger amplitudes, shorter front times and more pronounced oscillations.

The phase-to-earth and neutral-to-earth (not shown in the figure) voltages tend to reach their maximum values at an instant that is closely related to the stroke current front time. However, the reflections that occur at the transformer and at the entrances

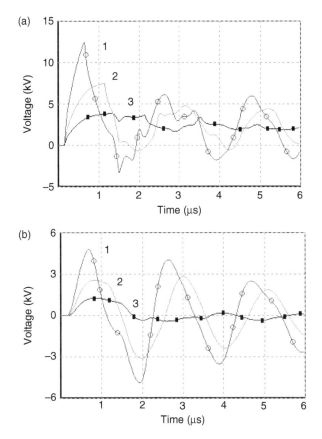

Figure 12.21 Induced voltages at the transformer LV terminals corresponding to stroke currents with triangular waveforms and various front times (tf): (a) phase-to-earth (b) phase-to-neutral. Line configuration indicated in Figure 12.19. Curve 1, tf = 0.5 μs; curve 2, tf = 1 μs; curve 3, tf = 2 μs.

of the LV power installations may also have an appreciable effect on the time to peak of the phase-to-earth voltages, especially when the stroke current front time is longer than the propagation time of the reflected waves. For this reason, the voltage time to peak corresponding to the case of tf = 2 μs (curve 3 of Figure 12.21a) is shorter than the current front time. The influence of the current time to half-value is very small and can usually be neglected as far as the voltage peak value is concerned.

Stroke current propagation velocity
Measurements of return stroke propagation velocities of natural and triggered negative lightning using optical techniques indicate that this parameter varies from ∼7 to over 90 per cent of that of light in free space c. This range includes both first and subsequent strokes and measurements made at different portions of the lightning channel. The

velocity tends to be higher at the bottom part of the channel, where it is typically one-third to two-thirds that of light. As pointed out by Rakov [48], the existence of a relationship between the velocity and the peak current is not generally supported by experimental data.

The influence of the propagation velocity on the induced voltages depends on the current front time; for stroke currents with steep fronts (time to peak shorter than ~1 μs), an increase in the propagation velocity leads to an increase in the induced voltages. This is usually the case of subsequent strokes. On the other hand, for currents with longer front times, such as typical first strokes, the induced voltages diminish as the propagation velocity increases. Both situations are illustrated in Figure 12.22, which presents the phase-to-neutral induced voltages at the transformer LV terminals considering current propagation velocities of 60 per cent c (base case) and 20 per cent c.

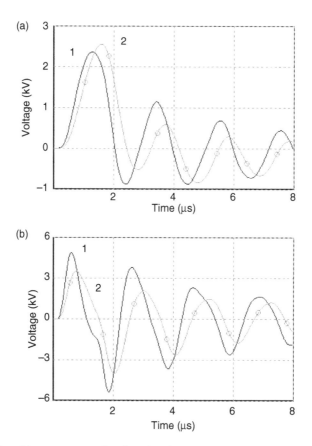

Figure 12.22   Phase-to-neutral induced voltages corresponding to the stroke currents of Figure 12.17a with different propagation velocities v: (a) first stroke; (b) subsequent stroke. Line configuration indicated in Figure 12.19. Curve 1, v = 60 per cent c; curve 2, v = 20 per cent c.

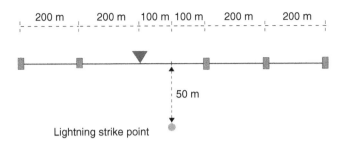

Figure 12.23 Lightning strike point midway between the transformer and a LV power installation (top view)

Relative position between the line and the stroke location
The situation considered in Figure 12.19a, where the lightning strike point is in front of the transformer, corresponds to the most critical one in terms of the amplitudes of the overvoltages induced at the transformer LV terminals. The voltages at this point tend to decrease as the flash moves in direction to one of the line ends. In the situation indicated in Figure 12.23, the strike point is equidistant from the transformer and one of the LV power installations, while its distance to the line is the same as in the base case (50 m). The phase-to-earth and phase-to-neutral voltages induced at the transformer terminals in the two situations are compared in Figure 12.24. The closer the distance between the transformer and the stroke location, the larger the voltage magnitude across its terminals.

The phase-to-earth voltage at the point of the line closest to the stroke location in Figure 12.23 is much larger than that induced at the transformer LV terminals in the situation shown in Figure 12.19a (strike point in front of the transformer). The reason for this is that in the former case the effect of the reflections that occur at the transformer and at the power installations are delayed due to the time required for their propagation to the observation point. This can be readily seen in Figure 12.25a. For the stroke current waveform considered, the voltage peak value is reached before the arrival of the reflections and therefore the magnitude of the first positive peak is exactly the same as that shown in Figure 12.17b, relative to a line without transformer, power installations and neutral conductor. The arrival of the reflections (at ~ 0.54 µs) causes the voltage to oscillate with a frequency f determined by the distance between the transformer and the closest power installation xg. In this case $xg = 200$ m, and thus the frequency is approximately $f = c/(2xg) = 750$ kHz. As shown in Figure 12.25b, the difference between the phase-to-neutral voltages is also important, although less significant.

Distance between the line and the lightning strike point
The distance between the line and the lightning strike point has a considerable influence on the induced voltages, particularly on their amplitudes. This is illustrated in Figure 12.26, which presents the phase-to-earth and phase-to-neutral voltages at the transformer LV terminals for distances d of 25, 50 (base case) and 100 m. As

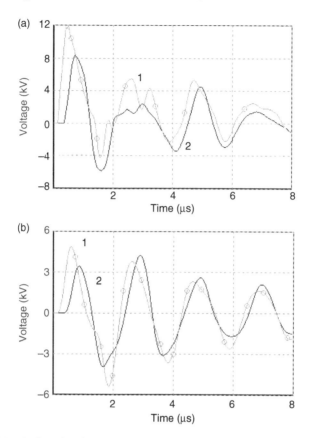

Figure 12.24 Induced voltages at the transformer LV terminals for two relative positions of the stroke location with respect to the line: (a) phase-to-earth; (b) phase-to-neutral. Curve 1, stroke location in front of the transformer (Figure 12.19, base case); curve 2, stroke location midway between the transformer and a LV power installation (Figure 12.23).

expected, the shorter the distance, the larger the induced voltage. For the situations considered, the ratios between the magnitudes of the phase-to-earth voltages corresponding to the distances of 50 and 25 m with respect to that relative to $d = 100$ m were approximately 2.0 and 3.5. Regarding the phase-to-neutral voltages, the variation was smaller, the corresponding numbers being about 1.7 (for $d = 50$ m) and 2.7 (for $d = 25$ m).

Line length and topology
The consideration of a finite line length does not alter the magnetic component of the induced voltage in equation (12.1), because this component does not cause charge flow along the line conductor axis. On the other hand, the electrostatic component

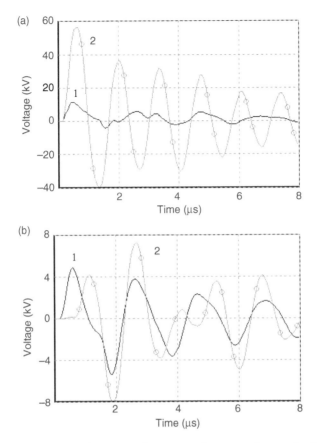

Figure 12.25 Induced voltages at the point closest to the stroke location: (a) phase-to-earth; (b) phase-to-neutral. Curve 1, situation indicated in Figure 12.19a (base case); curve 2, situation indicated in Figure 12.23.

depends upon the line length, because the number of elements that contribute to the voltage increases with this parameter. For this reason, a line matched at both terminations does not represent an infinite line.

Induced voltage calculations on lines with lengths varying from 500 m to 5 km were performed by Piantini and Janiszewski [62] for a stroke 60 m from the line and equidistant from its terminations. The current front time was equal to 3 μs and earth was assumed as a perfectly conducting plane. The voltages, computed in the middle of the line, reached the same peak value in all cases except for the shorter line, where the amplitude was ~3 per cent lower. The effect of the line length tends to be even less significant for the case of stroke currents with shorter front times.

Lightning-induced voltages on H-shaped networks were studied by Nucci and colleagues [86] and by Hoidalen [111]. The influence of the line topology on the

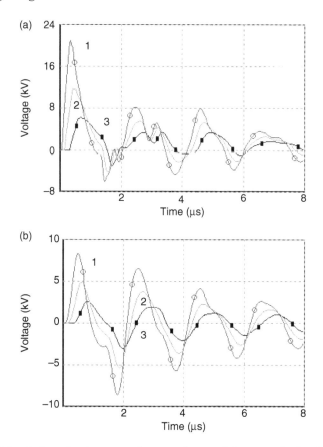

Figure 12.26 Induced voltages at the transformer LV terminals for different distances d between the line and the lightning strike point: (a) phase-to-earth; (b) phase-to-neutral. Curve 1, d = 25 m; curve 2, d = 50 m (base case); curve 3, d = 100 m.

induced voltages is closely related to the stroke location, observation point, and earth resistivity. However, for the situation considered in the base case shown in Figure 12.19, in which the observation point is close to the stroke location, the low impedance of the power installations lessens the effect of line sections that are more distant from the region of interest. Thus, an increase in the length of the line by adding other sections at its ends, independently of the direction of these sections, has a negligible impact on the induced voltage at the transformer LV terminals. Additionally, the possible presence of secondary arresters along the network would also contribute to decrease the influence of more distant line sections. Further discussion about the effect of the line length and topology on the induced voltages will be presented later in this section, on the item concerning the influence of the number of LV power installations connected to the LV network.

Line height

The vertical component of the lightning electric field below a single conductor line above a perfectly conducting earth plane is practically uniform if the conductor height is much smaller than the length of the stroke channel, which is always the case for distribution networks. Hence, in this situation the induced voltage increases linearly with the line height [81,87], a conclusion that is supported by data obtained from experiments conducted in a reduced scale model [80].

On the other hand, this is not the case when the neutral is present. Let us consider the case of a line with two wires. All the power installations are disconnected and the neutral is earthed at a single point (x_1), in which the value of the earth resistance is Rg. Neglecting the earth lead inductance, the current Ig that flows to earth is

$$Ig(x_1, t) = \frac{Un'(x_1, t)}{(Rg + Zn/2)} \qquad (12.3)$$

where $Un'(x_1, t)$ is the voltage that would be induced at x_1 in the absence of the earth connection and Zn is the characteristic impedance of the neutral conductor. The voltage $Up(x_2, t)$ at point x_2 on the phase conductor is given by

$$Up(x_2, t) = Up'(x_2, t) - \frac{Zm}{2} \cdot Ig\left(x_1, t - \frac{|x_2 - x_1|}{c}\right) \qquad (12.4)$$

where $Up'(x_2, t)$ is the voltage that would be induced at point x_2 of the phase conductor in the absence of the neutral earthing and Zm is the mutual impedance. The voltages Un' and Up' increase linearly with height, but the self and mutual characteristic impedances vary with the position of the wires, and therefore the voltage Up is no longer directly proportional to the conductor height. The variation of the characteristic impedances will also affect the reflections when the presence of LV power installations and multiple earth connections are considered. However, the deviation is not significant for the usual line heights, as illustrated in Figure 12.27, which presents the ratio between the voltage peak values as a function of the ratio between the heights of the phase conductor, for both the first and subsequent stroke currents shown in Figure 12.17a. The base case ($h = 6.5$ m) is taken as reference, and the distance between the phase and neutral conductors is in all cases equal to 2 cm.

When the earth conductivity cannot be assumed as infinite, the voltage increase with height is not linear and varies with the stroke location, the observation point and the soil resistivity, as shown by Nucci and Rachidi in Reference 64.

Variations of the conductors' cross-sections affect the characteristic impedances, but the influence on the induced voltages is not important for the range of diameters commonly used.

Distribution transformer

In order to assess the impact of the transformer input impedance on the phase-to-earth and phase-to-neutral induced voltages, three cases are considered in Figure 12.28. Two of them refer to the 30 and 112.5 kVA transformers for which input impedances are

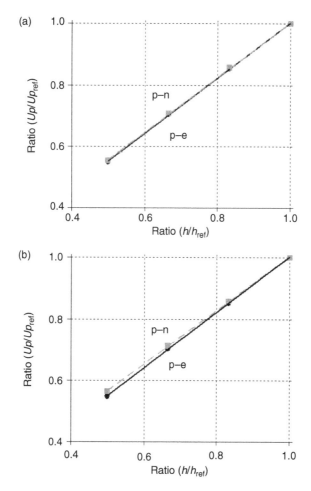

Figure 12.27 Ratio between the induced voltage peak values (Up/Up$_{ref}$) at the transformer LV terminals as function of the ratio between the height of the phase conductor h and h$_{ref}$ = 6.5 m for (a) first stroke and (b) subsequent stroke. Up$_{ref}$ refers to the absolute values of the peak voltages corresponding to the first and subsequent strokes shown in Figure 12.20. The distance between the phase and neutral conductors is 2 cm. p–e, phase-to-earth voltages; p–n, phase-to-neutral voltages.

shown in Figure 12.5a and b, respectively. In the third case the transformer is simulated by an equivalent single inductance of 40 µH. The parameters of the equivalent *RLC* parallel circuits, in the equivalent single-phase representation, are as follows:

- 30 kVA transformer, $R = 367 \, \Omega$, $L = 16 \, \mu H$ and $C = 2.28 \, nF$ (base case);
- 112.5 kVA transformer, $R = 253 \, \Omega$, $L = 4.1 \, \mu H$ and $C = 3 \, nF$.

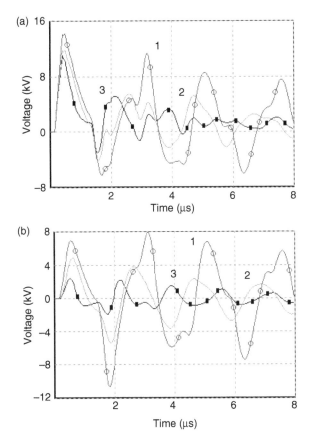

Figure 12.28 Induced voltages at the transformer LV terminals for different transformer models: (a) phase-to-earth; (b) phase-to-neutral. Curve 1, L = 40 µH; curve 2, parallel RLC circuit (R = 367 Ω, L = 16 µH, C = 2.28 nF) – base case; curve 3, parallel RLC circuit (R = 253 Ω, L = 4.1 µH, C = 3 nF).

As expected, the phase-to-neutral voltages are highly dependent on the transformer characteristics. The greater the input impedance seen from the LV terminals in the frequency range of ∼10 kHz to 1 MHz, the larger the magnitude of the induced voltages. The phase-to-earth voltages are less affected, while the variation of the neutral-to-earth voltages (not shown in the figure) is usually negligible.

It is relevant to mention that, although in Figure 12.28 the 30 and 112.5 kVA transformers were simulated by their corresponding *RLC* circuits, the representation of transformers by only their inductances in general does not introduce appreciable differences as far as lightning-induced voltages are concerned. This applies to TN and TT systems, as for the IT earthing arrangement the input impedance is better approximated by a capacitance in the range 2–20 nF [9].

594   *Lightning Protection*

Low-voltage power installations
As discussed in Section 12.2.3, an inductance between 2 and 20 µH is in general a good approximation for the input impedance of a typical LV power installation in the TN system [8,9]. The influence of this impedance on the induced voltage at the transformer terminals is not significant for the basic configuration considered. The major reason for this is that the effect of the installations is felt only after the voltages at the transformer have reached their peak values. Additionally, considering the frequency spectra of the induced voltages and the above range for the inductances, the impedance of the power installations is low in comparison with the characteristic impedance of the line, and therefore the amplitudes of the reflected waves are not different enough to dramatically change the induced voltages.

A larger influence is observed on the voltages at the service entrances, although for the line configuration considered in the base case the variation of the phase-to-earth voltages is not significant. The phase-to-neutral voltages are more affected. Figure 12.29a presents the voltages computed for loads represented as inductances of 2, 10 and 20 µH in the situation shown in Figure 12.23, where the lightning strike point is midway between the transformer and a consumer installation.

The phase-to-neutral voltage at the installation entrance depends on the input impedance of the installation and on the current that flows through it. The current diminishes as the impedance increases, but to a lesser degree, so that the overall effect is an increase in the voltage. The degree of current reduction depends on the relative position between the phase and neutral conductors. The stronger the coupling between the wires, the lower the current that enters the installation. Consequently, the influence of the load impedance tends to be more important for greater distances between the phase and neutral conductors, as illustrated in Figure 12.29b for the case of neutral at the height of 7 m (instead of 6.48 m as in Figure 12.29a). In any case, larger voltages are induced in the case of higher impedances, which are generally associated with smaller installations [9]. Induced voltages on predominantly resistive or capacitive loads are discussed in Reference 58.

Earth resistance and type of earthing system
It has been assumed so far that the LV power installations are close to the line and thus each impedance to earth corresponds to the equivalent impedance of the earth connections of the neutral and of the closest consumer's installation. The value of 20 Ω, adopted for the earth resistance in the base case, can be visualized as equivalent, for instance, to $Rg = 25$ Ω at the neutral and $Rg = 100$ Ω at the service entrances of the LV power installations.

The earth resistance may appreciably affect the phase-to-earth and neutral-to-earth voltages. Both voltages increase with Rg, but as the latter has a greater dependence, the difference between them diminishes as Rg increases. Therefore, the influence on the phase-to-neutral voltages tends to be less significant. This is illustrated in Figure 12.30, which presents the results corresponding to the equivalent resistances Rg of 5 and 100 Ω.

In the situations considered, the peak value of the phase-to-earth voltage increased 120 per cent (from 10 to 22 kV), whereas the corresponding value for the

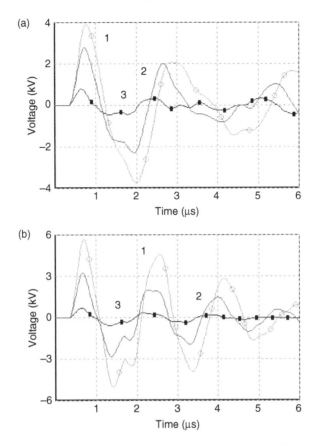

Figure 12.29 Phase-to-neutral induced voltages at the entrance of the closest consumer to the stroke location for the configuration shown in Figure 12.23: (a) height of neutral conductor: 6.48 m; (b) height of neutral conductor: 7.0 m. Low-voltage power installations represented as inductances (L) of different values. Curve 1, L = 20 µH; curve 2, L = 10 µH; curve 3, L = 2 µH.

neutral-to-earth voltage was ∼170 per cent (from 6.9 to 18.5 kV). For the same variation of R_g, the absolute value of the phase-to-neutral voltage decreased ∼14 per cent (from 5.5 to 4.7 kV).

The effect of the earth lead inductance may be important in the case of low earth resistance values, especially on the neutral-to-ground voltages. However, even in this condition the influence on the phase-to-neutral voltages is negligible.

Concerning the earthing arrangement, voltages of much higher magnitudes are induced on IT systems, as pointed out by Hoidalen [8,9]. IT systems are also more sensitive to the soil resistivity, which can bring about substantially larger voltages. A more elaborate representation of the load characteristics, such as the typical resonance peaks that appear on the impedance curves, has only a minor effect on the voltage features.

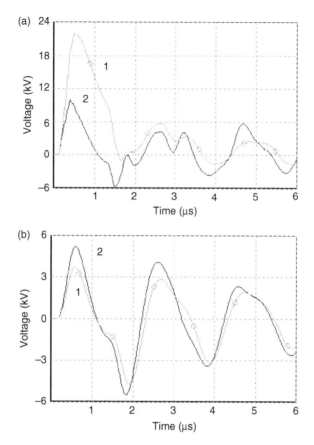

Figure 12.30 *Induced voltages at the transformer LV terminals for different values of the earth resistance (R_g): (a) phase-to-earth; (b) phase-to-neutral. Curve 1, $R_g = 100\ \Omega$; curve 2, $R_g = 5\ \Omega$.*

Number of services

In order to assess the effect on the induced voltages of the number of LV power installations connected to the line, three situations are considered. In the first, all consumers are disconnected. In the second configuration only one of the closest installations is connected, while in the third case two of them, one at each transformer side, are considered. The distance between the transformer and the closest power installations is 200 m, as shown in Figure 12.31. In all situations the line is assumed to be matched at both terminations and the neutral is earthed every 200 m. The value of the earth resistance is 25 Ω in all points except at the transformer, in which the value of 20 Ω is considered. These values were chosen to match the assumptions adopted for the base case, as discussed in the item relative to the influence of the earth resistance.

The results of the calculations are presented in Figure 12.32. The reflections associated with the loads are felt at the transformer terminals after the voltages have reached

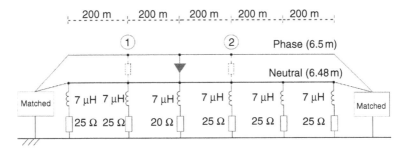

Figure 12.31 Low-voltage line configuration for the induced voltages shown in Figure 12.32 (side view). The triangle and the dotted rectangles denote, respectively, the secondary of the distribution transformer and LV power installations.

their first peak values, and in the first and second cases only the waveforms are affected. The presence of the loads intensifies the oscillations, and this is particularly evident on the phase-to-neutral voltages. However, in the third case, in which power installations are connected at both sides of the transformer, the reflected waves provoke a negative peak whose magnitude is slightly larger than the first one.

The voltages corresponding to the base case (e.g. Figure 12.24) and to the third configuration are virtually the same, and this holds for loads represented by inductances in the whole range that characterize typical installations in TN systems (from 2 to 20 µH [9]). This result indicates that the contributions to the induced voltages at the transformer LV terminals are limited basically to the portions of the line located between the transformer and the closest customers' installations. In other words, the conditions of the line terminations and the connection of other loads do not practically affect the induced voltages at the transformer terminals.

The influence of the distance between the power installations will be discussed later in this section, in the item regarding the characteristics of the lightning-induced voltages on rural versus urban lines.

Bunched cables versus open wire lines
The bunched cable has some advantages over the open wire line in terms of, for example, aesthetics, safety, clearance requirements, erection costs and frequency of tree cutting. It has also a superior lightning performance, as the stronger coupling between phase and neutral conductors results in lower overvoltages.

Considerable differences are observed between the voltage magnitudes in the two line configurations when only the conductors are considered. However, when the presences of the transformer and of the power installations are taken into account, the voltages induced on the open wire line are drastically reduced, as in this situation the phase and neutral conductors are connected through relatively low impedances. The major consequence is that the difference between the induced voltages relative to

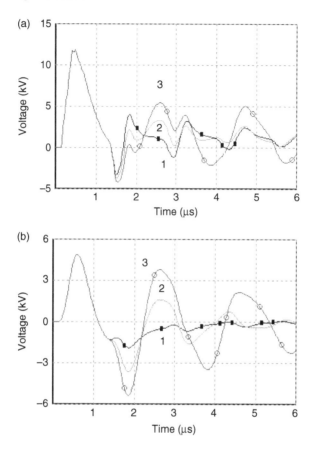

Figure 12.32 Induced voltages at the transformer LV terminals for a different number of LV power installations connected to the line, according to the configuration shown in Figure 12.31: (a) phase-to-earth; (b) phase-to-neutral. Curve 1, all power installations disconnected from the line; curve 2, just one power installation connected to the line, at point 1; curve 3, two power installations connected to the line, at points 1 and 2.

the two line configurations is not so large as it would be if neither the transformer nor the loads were connected.

The voltages induced at the LV transformer terminals for both line configurations are compared in Figure 12.33, with the open wire line simulated as two-conductor, single-phase and neutral. Both wires have the same diameter, namely 1 cm, and their heights are 6.5 m (phase) and 7.0 m (neutral). According to Dugan and Smith [112], the cable capacitance does not influence significantly the lightning surges on the LV side of distribution transformers, and therefore this parameter was neglected in the simulations.

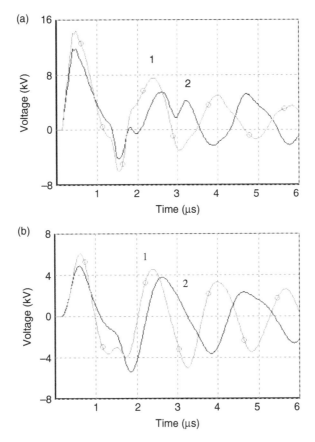

Figure 12.33   Induced voltages at the transformer LV terminals for the bunched cable versus open wire line configuration: (a) phase-to-earth; (b) phase-to-neutral. Curve 1, open wire line; curve 2, bunched cable.

Rural versus urban lines
The configuration adopted in the base case can be considered representative of a rural line. However, urban LV lines usually have smaller lengths and are characterized by shorter distances between consumers' installations.

In order to evaluate how these features affect the induced voltages, let us compare the results of the base case with those corresponding to the situation depicted in Figure 12.34. The geometries are similar, but the lines differ in terms of length (250 m against 1 km) and distance between power installations (50 m against 200 m). Besides, unlike in the base case, in which the distance between adjacent neutral earthing points coincides with the load spacing, in the urban line configuration the neutral is earthed only at the transformer and at the line terminations. Therefore, in order to match the assumptions adopted for the base case regarding the earth resistances, two values are considered for Rg in the urban line: 20 Ω at the line ends and

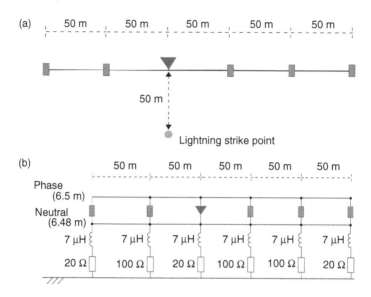

Figure 12.34 *Configuration representing an urban LV line: (a) top view; (b) side view. The triangle and the rectangles denote, respectively, the secondary of the distribution transformer and the LV power installations. The neutral is grounded at the transformer and at the line ends.*

at the transformer, and 100 Ω at the service entrances of the LV power installations. The value of 20 Ω at the line ends corresponds to the equivalent resistance to earth of the neutral (25 Ω) and of the closest power installation (100 Ω), as discussed in the item relative to the influence of the earth resistance. However, the induced voltages suffer only minor changes if the value of the earth resistance is kept the same (20 Ω) at all points.

Regarding the value of the equivalent inductance that is in series with the earth resistance, it should be rigorously smaller at the line ends, where the consumers' installations are close to the neutral earthing points. As, however, the voltages are calculated at the transformer LV terminals, in the situation considered only the earth lead inductance at that point has a visible – although not significant – influence on the induced voltages, and hence the value of the inductance to earth seen from the neutral was assumed to be the same (7 μH) at every earthing point.

The phase-to-earth and phase-to-neutral voltages corresponding to the two line configurations are presented in Figure 12.35. As both voltages tend to increase with the distance between adjacent earthing points (xg), rural lines are in principle more prone to experience induced voltages with larger magnitudes. It is important to bear in mind, however, that lightning may strike very close to an urban line due to the presence of tall objects in its vicinity and this may result in severe overvoltages. The frequency of the oscillations increases as xg decreases, and therefore lightning transients on urban lines tend to have a broader frequency spectrum.

The influence of the distance between adjacent earthing points depends to a great extent on the stroke current front time. This is illustrated in Figure 12.36, in which the

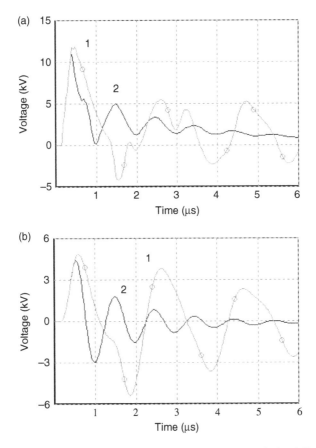

Figure 12.35 *Induced voltages at the transformer LV terminals for different line configurations: (a) phase-to-earth; (b) phase-to-neutral. Curve 1, rural line (base case, Figure 12.19); curve 2, urban line (Figure 12.34).*

same comparisons of Figure 12.35 are presented for the case of the typical first stroke current waveform depicted in Figure 12.17a. The peak values of the phase-to-earth voltages are about the same in the two line configurations, whereas the magnitude of the phase-to-neutral voltage induced on the urban line is considerably lower than that of the rural line.

Earth electrical parameters
The voltages induced on a given line depend on the lightning electromagnetic fields, which in turn are affected by the soil electrical parameters. The results obtained by Rachidi and colleagues [106] show that the approximation of a perfectly conducting earth is in general reasonable for the calculation of both the azimuthal magnetic field and the vertical component of the electric field for distances not exceeding ~1 km, although the time derivatives of both fields may be significantly affected by the propagation effects, as pointed out by Cooray [113,114]. On the other hand, the

602 Lightning Protection

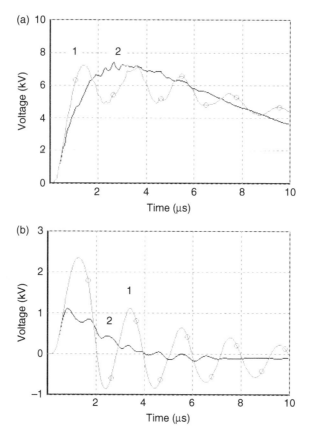

Figure 12.36 *Induced voltages at the transformer LV terminals for different line configurations, considering the typical first stroke current depicted in Figure 12.17a: (a) phase-to-earth; (b) phase-to-neutral. Curve 1, rural line (Figure 12.19); curve 2, urban line (Figure 12.34).*

earth resistivity has a remarkable effect on the horizontal electric field [82,93, 106,115–118] and, by extension, on the lightning-induced voltages.

Many investigations have been carried out about this topic [64,66,82,92,93, 106,119–125]. According to Nucci [82], the assumption of a perfectly conducting earth is reasonable for distribution systems located above a soil with resistivity lower than $\sim 100\ \Omega$ m. In general, the induced voltages tend to increase with earth resistivity, although they may also decrease and/or change polarity depending on the stroke location and the observation point. The induced voltages are particularly affected when the stroke location is close to one of the line ends. In the case of side strokes (such as in the situations shown in Figures 12.16, 12.19, 12.23 and 12.34), the influence of the horizontal component of the electric field is much smaller and therefore the induced voltages are far less sensitive to the propagation effects.

The specific case of LV networks is addressed by Hoidalen in Reference 8, where it is shown that, depending on the situation, the magnitudes of the induced voltages on IT systems may increase more than ten times when the earth resistivity increases from zero (case of perfectly conducting earth) to 1 000 Ω m. The effect is much less pronounced in TN systems, although also important. In either case the front time of the induced voltage increases when the earth losses are taken into account.

12.3.4 Transference from the medium-voltage line

Lightning overvoltages on the LV network can originate from the primary circuit either in the case of a direct hit or a nearby stroke. In both situations the transformer plays an important role in the transference mechanism.

12.3.4.1 Direct strikes

Direct strikes to the primary circuit produce overvoltages on the LV network due to transference from the distribution transformer and to injection of current into the neutral conductor. The latter is a consequence of the earth potential rise caused by the flow of current through the earth resistance following the operation of MV surge arresters and the occurrence of flashovers across the MV insulators. The transferred voltages may vary widely depending on the strike and observation points, stroke current magnitude and waveform, and line configuration.

The overvoltages that result on a typical LV distribution network in the case of direct lightning hits on the primary line were studied in References 26–30 and 40. A typical distribution network was considered in Reference 26 and the voltages at different points of the LV line were calculated in order to study the basic characteristics of the surges transferred to a typical secondary network for the case of direct strokes to the MV line. The simulations were performed with the Alternative Transients Program (ATP) [126], with the transformer represented by the model shown in Figure 12.6. Flashovers across the MV and LV insulators were taken into account according to the Disruptive Index Model [127]. It was shown that a correct representation of the distribution transformer is essential and that the well-known purely capacitive PI-circuit is generally not adequate for the evaluation of transferred surges. The simulations showed also that voltage magnitudes of some tens of kilovolts may occur at the transformer LV terminals and at the consumers' entrances in the absence of surge protective devices. Even the installation of surge arresters very close to the primary transformer terminals does not prevent its failure from lightning transients, because severe stresses may be caused by surges coming from the LV side. Although the application of secondary arresters to the LV transformer terminals is highly recommended, it does not prevent large overvoltages at the customers' entrances.

The influences of the stroke current magnitude and front time, earth resistance and number of LV power installations on the voltages transferred to the LV network were discussed in Reference 27. The results showed that the flashovers that take place across MV and LV insulators affect significantly the magnitudes and waveforms of the transferred voltages and thus should always be taken into account. In most of the cases the larger phase-to-neutral voltages occur at the transformer terminals. The voltages are characterized by oscillations originating from the various reflections throughout the secondary network and therefore are strongly affected by the spacing between adjacent earthing points. The effect of the earth resistance of the neutral conductor on the phase-to-neutral voltages is greater than that of the earth resistance at the consumers' entrances. For a given line, the general trend of the phase-to-neutral voltage magnitudes is to decrease with the number of consumers.

In general, the shorter the distances between the transformer and the lightning strike point, the higher the transferred voltages. However, the insulator flashovers tend to diminish this effect. In Reference 28, the LV power installations were represented as pure inductances, resistances or capacitances, and were connected to the LV network through 20-m-long service drops. The reflections that occur at both terminations of the service drops led, in the situations considered, to a decrease of the phase-to-neutral voltages at the transformer terminals when the values of the load impedance increased. At intermediate points and at the line terminations the voltage amplitudes may increase or decrease with load impedance, depending on the magnitudes of the reflected waves at the various discontinuity points. Concerning the line configuration, the results obtained in Reference 128 showed that the overvoltages are usually much larger in the open-wire line than in the bunched cable and that the energy dissipated in the power installations may be, in some cases, ten times higher.

Figure 12.37 shows, as an example, typical overvoltages transferred to the secondary side of the transformer due to a direct lightning strike at the primary circuit, at a distance of 800 m from the observation point. The simulations, performed with the ATP, considered a 10-km-long primary line, with four conductors (three phases plus neutral), wood poles, and horizontal configuration, being the distance between adjacent phases equal to 0.75 m. The heights of the phase and neutral conductors were 10 and 6.48 m, respectively, and all the wires had the same diameter, namely 1 cm. The neutral, which was shared between the primary and secondary circuits, was below the middle phase. The MV line configuration, as well as its position with respect to the secondary circuit, is presented in Figure 12.38. The LV line is the same as that depicted in Figure 12.19, but a section of 400 m is below the primary circuit, as indicated in Figure 12.38b. The transformer, located at the end of the MV line, is represented by the model shown in Figure 12.6, with the parameters given in Table 12.1 for the 30 kVA transformer.

The stroke current waveforms are those depicted in Figure 12.17a for the first and subsequent strokes, with amplitudes, respectively, of 30 and 12 kA. A perfectly conducting earth plane is assumed, and the gapless ZnO surge arresters connected at the HV transformer terminals are simulated by non-linear resistors with the V/I characteristic shown in Figure 12.39a. The possibility of the occurrence of flashovers across the MV and LV insulators, as well as from the neutral conductor to earth, was taken into account by means of switches placed at every pole between all the conductors and earth. The V/t characteristic curves corresponding to the standard lightning impulse waveform (1.2/50 μs), obtained with the integration method presented in Reference 127, are shown in Figure 12.39b.

The application of surge protective devices in the LV network will be considered in Section 12.4.

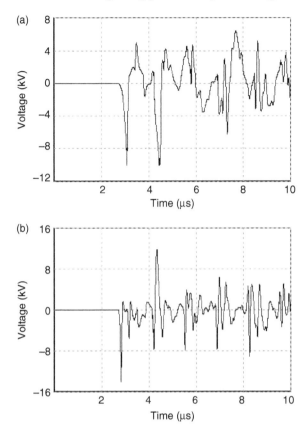

Figure 12.37 Phase-to-neutral transferred voltages at the transformer LV terminals: (a) first stroke; (b) subsequent stroke. The lightning strike point is 800 m from the transformer. Stroke current waveforms are depicted in Figure 12.17a.

12.3.4.2 Indirect strikes

The assessment of the voltages transferred from the MV line due to lightning strikes in the vicinity of the distribution network requires the knowledge of the transformer high-frequency behaviour as well as the voltages induced at the primary side. This subject is discussed in References 12, 15 and 16 for the case of unloaded transformers, while the presence of the LV line is considered in References 31 and 56.

A suitable transformer model is essential for the evaluation of transferred surges, and this issue has been examined somewhat in Section 12.2.2. In the well-known purely capacitive PI-circuit, the transformer is represented by the capacitances C_1 (between primary and earth), C_2 (between secondary and earth) and C_{12} (between primary and secondary). The measured values corresponding to the 30 kVA transformer discussed in Section 12.2.2 are $C_1 = 0.138$ nF, $C_2 = 0.423$ nF and

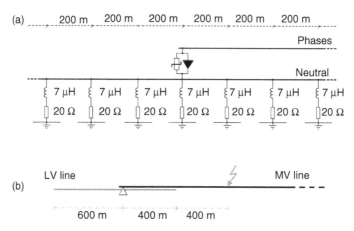

Figure 12.38 MV line configuration: (a) side view (LV line not shown); (b) relative position between the MV and LV lines. The triangle and the rectangle denote, respectively, the distribution transformer and the set of surge arresters at the MV side. The configuration of the LV line is depicted in Figure 12.19.

$C_{12} = 0.305$ nF, and its equivalent PI-circuit is depicted in Figure 12.40a. This circuit is, however, not adequate for evaluating transferred surges, as it greatly overestimates the overvoltages. This is illustrated in Figure 12.40b, which presents the measured and calculated voltages transferred to the secondary (in open circuit) when a standard 1.2/50 μs impulse voltage of 1.7 kV (Figure 12.7a) is applied to the primary terminals short-circuited. For comparison purposes, the voltage calculated using the model of Figure 12.6 is also shown. The absolute value of the ratio between the peak values of the calculated – using the PI-circuit – and measured voltages is 17.5 (700 V against 40 V). It can be readily seen that not only the magnitudes, but also the voltage waveforms, differ considerably.

Let us now consider the situation indicated in Figure 12.41. The stroke location is in front of the transformer, at a distance of 50 m from the MV network. The primary has the same configuration as shown in Figure 12.38a, except that the transformer is not at its end, but midway between the line terminations. The LV line configuration is that depicted in Figure 12.19, and no coupling is considered with the MV conductors. The other conditions are the same as those indicated in Section 12.3.2. Although important voltages would be induced 'directly' on the secondary line, let us focus only on the voltages transferred from the primary side through the transformer.

Figure 12.42 presents the phase-to-neutral voltages at the transformer terminals, both with and without surge arresters at the MV side, considering the typical subsequent stroke current depicted in Figure 12.17a. The calculations corresponding to the first stroke current are presented in Figure 12.43. The voltages oscillate with a frequency governed by the transformer transient response and by the configuration of the LV network, with the distance between adjacent loads playing a major role.

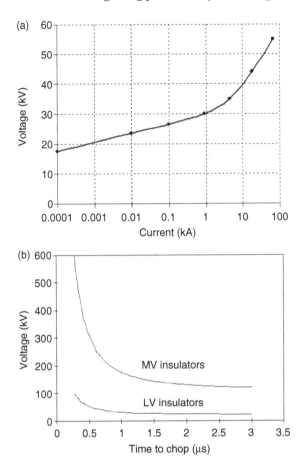

Figure 12.39 *Characteristics of the surge arresters and insulators relevant to the simulations presented in Figure 12.37: (a) non-linear V/I characteristic of the ZnO MV surge arresters; (b) V/t characteristic curve for MV and LV insulators*

As can be readily seen by comparing the two figures, even though on the primary side the voltages of higher magnitudes are induced by the first stroke, the larger transferred voltages are associated with the subsequent stroke. The reason for this is that the voltages induced by the subsequent stroke have steeper fronts and, thus, broader frequency spectra.

It can also be observed both in Figure 12.42 and in Figure 12.43 that the difference between the magnitudes of the voltages at the LV transformer terminals is much less than that at the primary side. The same outcome was obtained by Borghetti and colleagues for the system studied in Reference 31. The influence of the MV arresters on the transferred surges depends, however, on the configuration of the secondary line. The presence of arresters close to the transformer terminals causes a larger

608 Lightning Protection

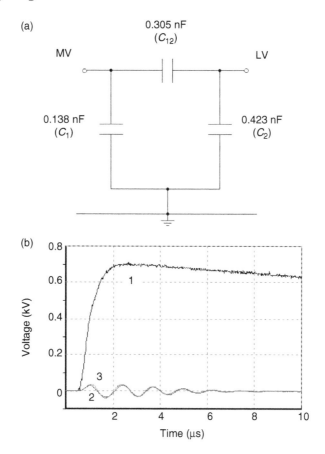

Figure 12.40 Equivalent capacitive PI-circuit and comparison of measured and calculated phase-to-neutral voltages transferred to the secondary of the 30 kVA transformer, in the no-load condition, for a standard 1.2/50 μs impulse voltage of 1.7 kV applied to the primary terminals short-circuited: (a) equivalent capacitive PI-circuit; (b) transferred voltages (curve 1, capacitive PI-circuit; curve 2, model indicated in Figure 12.6; curve 3, measured)

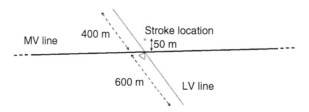

Figure 12.41 Relative position between the MV and LV lines

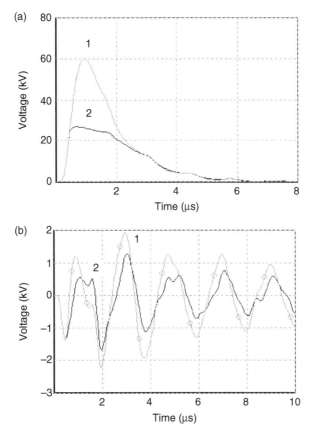

Figure 12.42 Phase-to-neutral voltages at the transformer terminals considering the typical subsequent stroke current depicted in Figure 12.17a: (a) induced voltages at the MV side; (b) transferred voltages to the LV side. Curve 1, without MV surge arresters; curve 2, with MV surge arresters.

current to flow through the earth lead, which may bring about a larger neutral-to-earth voltage. In the configuration considered in Reference 56, this resulted in a greater influence of the MV arresters.

The influence of the line configuration can be assessed by comparing the transferred voltages shown in Figures 12.42b and 12.43b with those relating to the case of LV power installations disconnected from the line. These comparisons are presented in Figure 12.44 for the case of surge arresters installed at the transformer primary side. The curves labelled '2' and '1' correspond to lines with and without power installations connected, respectively. As in the former situation the distance between the transformer and the closest power installations is 200 m, the voltage waveforms start to deviate at ~ 1.5 μs. This corresponds to the sum of the times required for the initiation of the transient at the transformer terminals (0.17 μs) and for the arrival of the reflected waves at the observation point (1.33 μs).

610 Lightning Protection

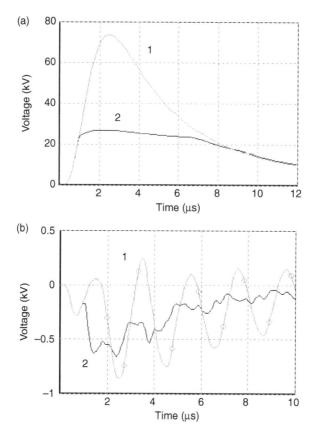

Figure 12.43 Phase-to-neutral voltages at the transformer terminals, considering the typical first stroke current depicted in Figure 12.17a: (a) induced voltages at the MV side; (b) transferred voltages to the LV side. Curve 1, without MV surge arresters; curve 2, with MV surge arresters.

It can be observed in Figure 12.44a, relevant to the subsequent stroke, that even when the power installations are not connected, the transferred voltage has an important frequency component of 750 kHz (curve 1), which is associated with the distance between adjacent neutral earthings. These oscillations are intensified with the connection of the power installations to the secondary line (curve 2).

The transferred voltages relevant to the first stroke are presented in Figure 12.44b. In the absence of the power installations, the frequency component of 150 kHz, which is determined by the reflections that occur at the open line ends, is shown to be of greater importance than in the previous case. The voltage waveform has a superimposed oscillation of 750 kHz, which is related to the neutral earthings. As the current wavefront is much slower than that of the subsequent stroke, the oscillations present on the voltage calculated with the loads connected to the line (curve 2) are far less pronounced than those observed in curve 2 of Figure 12.44a. On the other

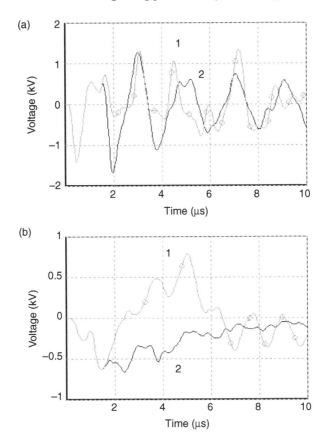

Figure 12.44 Phase-to-neutral voltages at the transformer LV terminals, with and without the LV power installations: (a) subsequent stroke; (b) first stroke. Surge arresters at the MV side. Curve 1, LV power installations disconnected from the line; curve 2, LV power installations connected.

hand, the lower current steepness featured by the first return stroke results in a greater divergence between curves 1 and 2, i.e. in a larger effect of the power installations.

12.4 Lightning protection of LV networks

The overvoltages originating from lightning often have high magnitudes and are usually the major cause of failures or damage to transformers and consumers' electric appliances, especially for lines characterized by poor pole earthing conditions and located in regions with high lightning incidence.

A fault on a transformer is always sustained and the corresponding costs are related to the repair or replacement of the equipment and to the service interruption. Protection measures such as the application of secondary arresters and the reduction of earth resistance at the transformer pole can improve the lightning performance of LV

612 Lightning Protection

networks. However, although there has been an increasing awareness of the effectiveness of the application of LV arresters, this practice has not been widespread among electric utilities. With a few exceptions, e.g. Reference 129, this measure is usually taken only to meet cases where the action – mainly for improving the earth connections – did not give satisfactory results [3] or to solve recurrent problems. This is in fact an economic issue, a trade-off between the lightning costs and the investment in the protection scheme. The failure rate of the surge protective devices has also to be considered in the cost – benefit analysis. However, even without taking into account the costs of damages to consumers' equipment, their application to transformer secondaries can be justified in areas with high lightning damage rates, as pointed out by Darveniza [130] and by Dugan and colleagues [131].

Gapless secondary arresters of the metal oxide varistor (MOV) type are the most appropriate to protect the LV network. The impact on lightning overvoltages of the application of such devices to the transformer and service entrances will be discussed in this section.

12.4.1 Distribution transformers

Although several factors can cause distribution transformers to fail, in lightning-prone regions failure rates can be more than twice the norm, which is typically between 0.8 to 1.5 per cent for non-interlaced and 0.4 to 0.7 per cent for interlaced transformers [132]. Most of the additional failures may be due to current surges in the LV windings [133]. These surges can be created whenever a significant portion of the stroke current is injected into the neutral between the transformer and the power installations and, with the exception of cloud discharges, this situation can occur for all the mechanisms discussed in Section 12.3. The problems related to the so called 'LV side surges' or 'secondary side surges' have been discussed in References 38, 39, 112, and 131 to 133.

Let us consider the case of a direct strike to the primary line, as illustrated in Figure 12.45. The MV arrester discharge current splits so that one portion flows

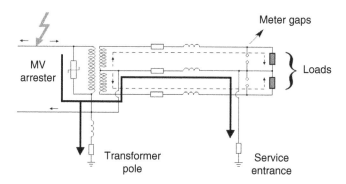

Figure 12.45 Injection of current into the neutral due to a direct strike to the MV line

through the pole earth lead and another is injected into the neutral conductor. The division of the bulk of the stroke current is determined by the earth resistances, but in the beginning of the transient it is highly dependent on the ratio between the inductances of the two paths. Therefore, as the path to the pole earth is shorter and the corresponding inductance is lower, initially a greater portion of the current flows through the transformer earth lead. After a few microseconds, when the current time derivative becomes smaller, the influence of the inductances decreases and the division is controlled by the resistances. The lower the earth resistance of the service entrance in comparison with that of the transformer pole, the larger the magnitude of the surge current that enters the neutral and the worse the problem.

The voltage drop across the neutral, produced by the surge current, gives rise to equal currents in the two phase conductors – if the configuration is symmetric – which can damage loads and cause meter gaps to flash over. These currents flow through the transformer secondary, as indicated in Figure 12.45, and induce a surge voltage in the primary that can cause part of the winding to short out. This would change the transformer ratio and subject the load to sustained overvoltages, resulting in damage to consumers' equipment. The surge voltage can also cause a layer-to-layer insulation breakdown and a subsequent transformer failure and power outage. As pointed out by Dugan and colleagues [131], the surge voltage distribution inside the transformer is such that the primary arrester has little effect on the prevention of this failure.

Although for the symmetric configuration the surge current divides equally into the secondary windings, there is a significant, equal and opposite voltage induced in each half of the MV winding so that the net voltage across the primary terminals is nearly zero. The magnitude of this induced voltage is approximately proportional to the voltage across the secondary windings, and it can be estimated reasonably well by considering only the inductances [131]. Hence, transformers with lower secondary impedances have a better performance against this type of surge. This is the case when the secondary windings are interlaced, as in this condition the impedance at surge frequencies is about a tenth that of non-interlaced transformers [132].

On the other hand, interlacing the secondary windings is not effective to solve the problem in the case of unbalanced surges, i.e. when the currents through the secondary windings are not equal. This situation happens, for example, if flashover or arrester operation occurs on only one side of a service. According to Marz and Mendis [132], up to half of all interlaced transformer failures may be due to secondary side surges. A better solution involves the application of LV arresters to the transformers, as in this case protection is provided against both balanced and unbalanced surges, regardless of winding connection.

The situation illustrated in Figure 12.45 refers to just one power installation connected to the LV line. In the case of multiple services from the transformer, a lower voltage drop will develop across the neutral and therefore less current will be forced into the secondary transformer terminals. The stresses on both the transformer and the consumers' loads will then be reduced in comparison with the single service case. On the other hand, multiple services mean higher line lightning exposure and a possible increase in the number of surges may counteract this effect.

Longer lengths of the LV circuit lead to greater voltage drops and consequently to current surges of higher magnitudes impressed on the transformer secondaries. This voltage increase is, however, not linear, as the ratio of resistance to inductance of the entire circuit generally increases, changing the dynamic response of the circuit and reducing the rate of rise of the surge currents [133].

Comparisons between surge currents in secondary windings for three types of cable, namely the open wire, triplex, and shielded, are presented in Reference 112. The best results, i.e. lower currents, are obtained with the shielded cable. Owing to the greater spacing between conductors and lower mutual coupling between wires, an open wire line has higher inductance than a triplex cable of the same length, and therefore a larger net voltage develops across the neutral, causing a surge current of higher magnitude to flow in the transformer and consumers' loads.

Long-term studies performed by Darveniza and Mercer [129] led to an improved lightning protection scheme for exposed pole-mounted transformers in Australia. The protection measures consisted in the relocation of primary surge arresters close to the terminals and in the fitting of secondary arresters. The lightning damage rates decreased from ~ 2 to 0.3 per 100 transformers per year after the implementation of the protection system [130], and the authors attribute this reduction mainly to the installation of secondary arresters.

It is important to note that protecting the transformer by means of interlaced secondaries or LV arresters results in larger surge currents, as both measures provide a low impedance for the surge. As a consequence, customers' devices may be subjected to higher voltage stresses.

Concerning the transformer LV arrester, discharge levels of 2–5 kV are considered adequate. Although the lightning impulse withstand level of a transformer secondary is typically in the range of 20 to 30 kV, insulation degradation may be caused by overloading, so that lower protective levels may be beneficial. Secondary arrester classes between 175 and 650 V are in principle suitable, but 440 or 480 V arresters have some advantages over both limits. They have better coordination with the primary arrester than 175 V arresters, which are susceptible to thermal failures caused by switching events [39,132,133], and have a lower discharge voltage than a 650 V arrester, thus reducing the risk of damage to sensitive consumers' devices.

As pointed up by Dugan and colleagues [131], LV transformer arresters must not substitute one type of failure for another and should be designed so that the possibility of failure is remote. Although the magnitude of a typical secondary current surge may be less than 1 kA, the arresters should have a current discharge capability of at least half that for a standard distribution class arrester, i.e. in the range 20–40 kA.

In all the simulations performed in Section 12.4.2 the transformer is protected with 440 V secondary gapless MOV arresters, for which the equivalent circuit is a capacitance of 780 pF in parallel with a non-linear resistor with the V/I characteristic depicted in Figure 12.46. The inductance of the connecting leads is disregarded.

12.4.2 Low-voltage power installations

The characteristics of the lightning overvoltages on the secondary network depend on a number of parameters, but in general those induced by nearby strokes or transferred

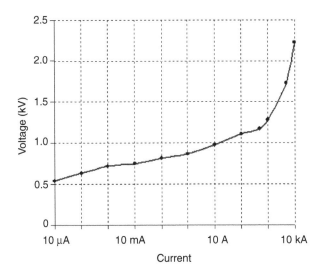

Figure 12.46   V/I characteristic of the secondary MOV arresters

from the MV line due to direct strikes to the primary conductors play a more significant role in network performance. These overvoltages may have a damaging effect on the customers' loads, and the use of properly coordinated surge protective devices at the service entrance and at susceptible equipment is recommended, especially in regions with high lightning incidence.

The arresters applied at the service entrance should be similar in rating to the transformer arrester and have a discharge voltage of less than 2 kV. This level, which is already too high for sensitive electronic equipment, may be much higher within the installation due to voltage oscillations caused by reflections at various points. Therefore, local protection for such loads is always required.

The arresters at the service entrance should have higher energy handling capability than those placed at internal parts of the premises – sometimes referred to as 'suppressors' – and divert the bulk of the surge current. None of the protective devices should be overloaded, and this is the concept of arrester coordination. Besides the energy absorption capability, the clamping voltages and the distances between the secondary arresters and the suppressors should also be considered for achieving a successful coordination. This topic is specifically addressed in References 134 to 138, and guidelines for installing surge protective devices are given in Reference 139.

In order to evaluate how the overvoltages induced at the service entrances by indirect strokes are affected by the application of secondary arresters, the following line configurations, indicated in Figure 12.47, will be considered:

- arresters not installed (configuration 1);
- arresters only at the transformer terminals (configuration 2);
- arresters at the transformer terminals and at all service entrances, except at point 4 (configuration 3).

616 *Lightning Protection*

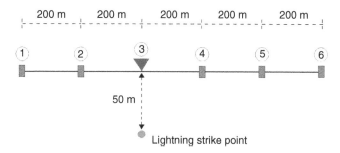

Figure 12.47   Low-voltage line configurations. The triangles and rectangles denote, respectively, the distribution transformer and the LV power installations. Configuration 1, line without arresters; configuration 2, arresters only at point 3; configuration 3, arresters at points 1, 2, 3, 5 and 6.

The arresters at the service entrances and at the transformer are assumed to have the same characteristics. As in Section 12.3.3, the three-phase line is represented by an equivalent single conductor, so that the corresponding arrester capacitance to be considered in the simulations is 2.34 nF. Likewise, the values of currents in the horizontal axis of Figure 12.46 should be multiplied by three. The transformer input impedance is simulated by a simple inductance of 16 μH, which corresponds to the inductance of the equivalent single-phase model of the 30 kVA transformer considered in Section 12.3.3. Owing to the presence of the secondary arresters at its LV terminals, the transformer model does not drastically influence the results.

The circuit adopted in the sensitivity analysis of Section 12.3.3 to represent the impedance seen by the power line at the service entrances is typical of a relatively large power installation. The equivalent load impedance has been shown to significantly affect the induced voltages, larger impedances leading to higher voltages. A more conservative condition is considered henceforth, and the loads, in the TN system, are simulated by just an inductance of 10 μH, which is representative of smaller installations [8]. The other parameters remain the same as in the base case of Section 12.3.3.

The most critical situation, from the point of view of a power installation, is that where the stroke location is in front of it. This case is illustrated in Figure 12.48, where the phase-to-neutral induced voltages at point 4, relative to the line configurations 1 and 2, are compared. The results are presented for both the first and the subsequent stroke, and it can be seen that indeed the latter induces considerably higher voltages for typical secondary line configurations. The voltage induced by the first stroke at the transformer is relatively low, so that the effect of the arresters is not significant. As shown in Figure 12.48a, the voltage induced by the first stroke has already reached its peak value and is still positive when the effect of the arresters is felt, so that its maximum value does not vary. On the other hand, in Figure 12.48b a slight increase is observed in the amplitude of the voltage induced by the subsequent stroke. This is due to the higher magnitude of the currents reflected at the arresters, which arrive at the observation point at an instant in which the voltage is negative, thus contributing to increase the absolute value of the negative peak.

Lightning protection of low-voltage networks 617

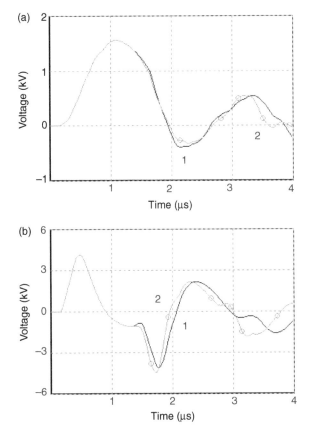

Figure 12.48   *Phase-to-neutral induced voltages at point 4 of Figure 12.47 when the stroke location is in front of it, at a distance of 50 m: (a) first stroke; (b) subsequent stroke. Stroke currents depicted in Figure 12.17a. Curve 1, configuration 1; curve 2, configuration 2.*

Therefore, if the stroke location is in front of an unprotected service entrance, the presence of arresters at other points of the line will not be effective in reducing the induced voltage magnitude at that point. For the situation considered, the voltage peak value induced by the subsequent stroke is about twice the recommended protective level of 2 kV.

A more favourable situation occurs when the lightning strike point is in front of the transformer, as indicated in Figure 12.47. The phase-to-neutral voltages induced by the subsequent stroke at points 2, 3, 4 and 5 are presented in Figure 12.49 for the line configuration 2, which corresponds to arresters installed at the transformer terminals. In this case all the voltage magnitudes are lower than 2 kV.

Let us now consider a more severe condition. The first and subsequent stroke currents have the same waveforms depicted in Figure 12.17a, but their magnitudes are, respectively, 88.5 and 29.2 kA. These values have a probability of only 5 per cent of being exceeded [108], and the corresponding maximum time derivatives are ~36

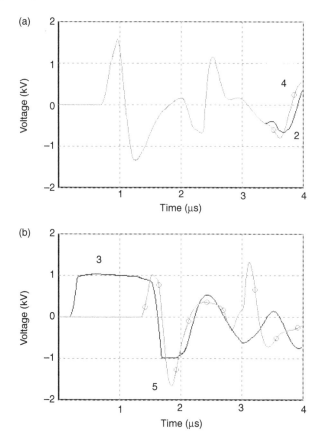

Figure 12.49   *Phase-to-neutral induced voltages at points 2, 3, 4 and 5 of Figure 12.47 considering the typical subsequent stroke current depicted in Figure 12.17a: (a) points 2 (curve 2) and 4 (curve 4); (b) points 3 (curve 3) and 5 (curve 5). Configuration 2 (arresters only at the transformer terminals).*

and 120 kA μs^{-1}. Figure 12.50 shows that, in the case of the first stroke, the presence of arresters at the transformer terminals prevents the induced voltages at the monitored points reaching levels above the recommended limit. The problem is, however, much more acute for the subsequent stroke, for which the level of 2 kV is exceeded at all service entrances. The voltages induced at points 2, 3, 4 and 5 are presented in Figure 12.51.

The application of secondary arresters to a power installation can effectively reduce the local overvoltages to acceptable limits. However, in some circumstances this may result in higher voltage stresses at unprotected premises. This situation is illustrated in Figure 12.52, which depicts the phase-to-neutral induced voltages corresponding to the line configuration 3. At all service entrances, with the exception of the unprotected one (point 4), the voltages are kept below ∼1.1 kV. A comparison of Figures 12.52a

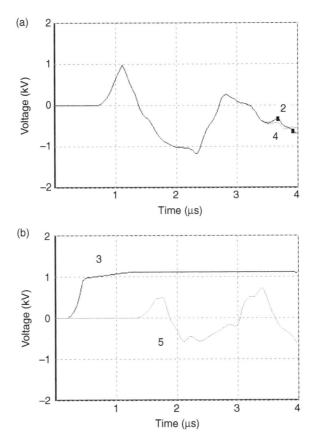

Figure 12.50 *Phase-to-neutral induced voltages at points 2, 3, 4 and 5 of Figure 12.47 considering a first stroke with the current waveform depicted in Figure 12.17a but with amplitude of 88.5 kA: (a) points 2 (curve 2) and 4 (curve 4); (b) points 3 (curve 3) and 5 (curve 5). Configuration 2 (arresters only at the transformer terminals).*

and 12.51a shows that the voltage magnitude at point 4 is indeed larger than that relative to configuration 2 (5 kV against 3.5 kV), in which the arresters were placed just at the transformer terminals. For network configuration 3 and the 'severe' subsequent stroke current considered, only when the distance between the line and the stroke location is greater than 200 m will the peak voltage at point 4 be lower than 2 kV. Thus, in order to protect LV power installations against lightning overvoltages, properly rated and coordinated surge protective devices should be installed at all service entrances.

In comparison with TN systems, IT systems are in general subject to much larger induced voltages and are also far more affected by the finite earth conductivity. A meticulous analysis of this topic is presented by Hoidalen in Reference 8. Higher values of the soil resistivity usually lead to larger voltage magnitudes, particularly

620   Lightning Protection

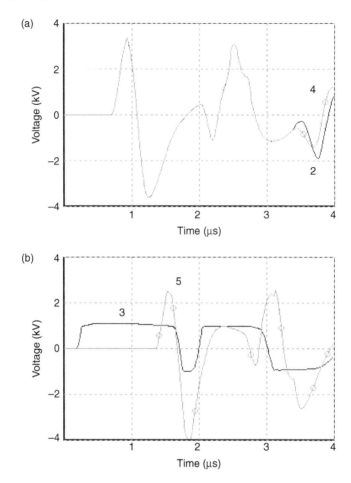

Figure 12.51   Phase-to-neutral induced voltages at points 2, 3, 4 and 5 of Figure 12.47 considering a subsequent stroke with the current waveform depicted in Figure 12.17a but with amplitude of 29.2 kA: (a) points 2 (curve 2) and 4 (curve 4); (b) points 3 (curve 3) and 5 (curve 5). Configuration 2 (arresters only at the transformer terminals).

in the case of IT systems. On the other hand, as the voltage front time, which has a remarkable influence on the effectiveness of the surge protective devices, also increases, the net result may be, in some cases, a decrease of the voltage magnitude.

When lightning strikes the MV network, short duration pulses of several tens of kilovolts may be transferred to the secondary circuit either by the first or subsequent strokes. Some examples of typical transferred voltages are presented in Figures 12.53 and 12.54 for the LV line depicted in Figure 12.47. The primary line, for which the characteristics are described in Section 12.3.4.1, forms a T-configuration with the secondary circuit, similar to that shown in Figure 12.19c. The strike point is 800 m from

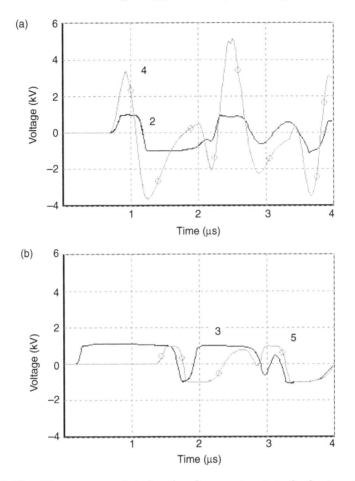

Figure 12.52 Phase-to-neutral induced voltages at points 2, 3, 4 and 5 of Figure 12.47 considering a subsequent stroke with the current waveform depicted in Figure 12.17a but with an amplitude of 29.2 kA: (a) points 2 (curve 2) and 4 (curve 4); (b) points 3 (curve 3) and 5 (curve 5). Configuration 3 (arresters at points 1, 2, 3, 5 and 6).

the transformer, which is represented by the simplified model of Figure 12.6 (30 kVA transformer).

The voltages at Point 4 and at the transformer windings for configuration 2 and the subsequent stroke current of Figure 12.17a are shown in Figure 12.53. Multiple insulator flashovers on the primary side bring about heavy voltage oscillations at the transformer terminals, which can also be observed on the secondary. The voltages at the service entrances have similar waveforms and at all points the peak values exceed 40 kV.

As in the case of nearby lightning, higher voltage amplitudes are usually related to the subsequent stroke. This is illustrated in Figure 12.54, in which the phase-to-neutral voltages at point 4, considering line configuration 3 and the stroke currents depicted in Figure 12.17a, are put side by side.

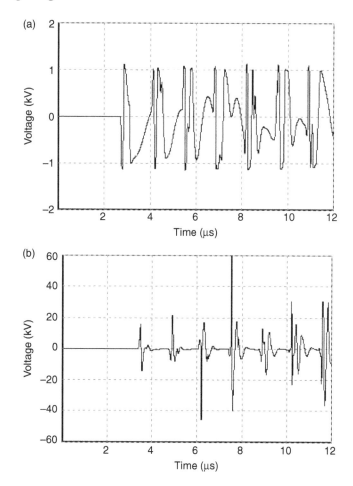

Figure 12.53   Phase-to-neutral transferred voltages at points 3 and 4 of Figure 12.47, considering the typical subsequent stroke current depicted in Figure 12.17a: (a) point 3 (transformer); (b) point 4. The strike point is 800 m from the transformer. Configuration 2 (arresters only at the transformer terminals).

A comparison between Figure 12.53b (configuration 2) and Figure 12.54b (configuration 3) shows that the presence of arresters at various places in the LV line does not prevent high voltages from arising at unprotected points. In fact, the stronger reflection that occurs at point 5 when it has an arrester (configuration 3) causes the voltage at point 4 to increase significantly in magnitude after ∼4.67 µs, when the first reflection arrives. Because the transformer arrester, the contribution of point 2 to the voltage at point 4 is much smaller. Reflections originating at all the other service entrances arrive at 6 µs, and after that the analysis is more complex. Eventually, the voltage peak value is reduced in comparison with the case of configuration 2.

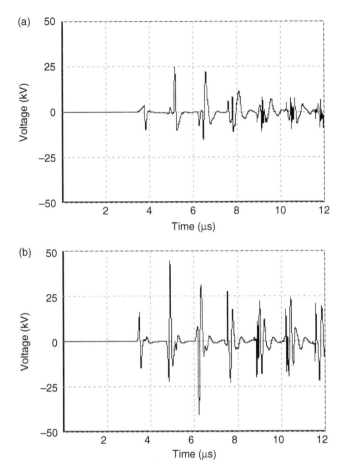

Figure 12.54 *Phase-to-neutral transferred voltages at point 4 of Figure 12.47, considering the stroke currents depicted in Figure 12.17a: (a) first stroke; (b) subsequent stroke. The strike point is 800 m from the transformer. Configuration 3 (arresters at points 1, 2, 3, 5 and 6).*

12.5 Concluding remarks

Lightning causes various power quality problems and usually has a considerable impact on the number of equipment damages and failures, voltage sags, and unscheduled power supply interruptions experienced by LV customers. Owing to the widespread use and growing dependency on the continuous operation of sensitive electronic equipment, there has been an increasing awareness of the importance of mitigating such effects.

In this chapter, the major mechanisms by which overvoltages stem from lightning were discussed. Particular emphasis was given to the voltages induced on overhead LV networks by nearby strokes and to those transferred from the MV system, which are the most important on account of their magnitudes and frequencies of occurrence. Simple but effective models were used to represent the high-frequency behaviour of typical distribution transformers and LV power installations.

Surge magnitudes and waveforms depend considerably on many line and lightning parameters, which may combine in an infinite variety of ways. Therefore, a sensitivity analysis was carried out and a typical LV distribution network, which may be considered representative of rural lines, was taken as reference. The basic characteristics of the overvoltages, as well as their dependence upon the network configuration and the most important stroke parameters, were assessed. The analysis revealed that secondary systems are in general more susceptible to subsequent strokes, although severe surges can also be produced by the first.

Phase-to-ground voltages induced by nearby strokes can reach some tens of kilovolts at various points along the network, especially if the stroke location is not in front of a neutral earthing point. Lower magnitudes are observed at the transformer and customers' entrances, but the value of 10 kV may often be exceeded in the case of strikes closer than $\sim$50 m. Phase-to-neutral voltages of some kilovolts are common if surge protective devices are not applied. In the case of direct strikes to the MV line, short duration pulses of several tens of kilovolts are transferred to the secondary circuit.

In regions of high lightning activity, surges originating in the LV side can be responsible for a great number of transformer failures or damage, even if arresters are placed close to the primary terminals. The application of arresters on transformer secondaries can significantly reduce lightning damage rates of exposed transformers, but it does not prevent overvoltages from arising at the service entrances.

Similarly, the application of secondary arresters to a power installation can effectively reduce the local overvoltages to acceptable limits, but in some circumstances this may result in higher voltage stresses at unprotected premises. Therefore, unless they are applied at every service entrance, exposed sensitive electronic equipment can be damaged. In fact, voltage oscillations caused by reflections at various points within the installation can give rise to internal overvoltages with higher magnitudes than that limited by the arresters placed at the service entrance. Therefore, local protection is required for such susceptible loads.

References

1. Lakervi E. and Holmes E.J. *Electricity Distribution Network Design*, 2nd edn (IEE Power Engineering Series, London, **21**, 1996).
2. Silvestro F., Massucco S., Fornari F., Burt G. and Foote C. 'Classification of low-voltage grids based on energy flows and grid structures'. Project DISPOWER, Contract no. ENK5-CT-2001-00522, 2004.
3. Clement M. and Michaud J. 'Overvoltages on the low voltage distribution networks. Origins and characteristics. Consequences upon the construction of Electricité de France networks'. *Proceedings of the 12th International Conference on Electricity Distribution (CIRED)*, Birmingham, 1993, pp. 2.16.1–2.16.6.

4. IEC 60364-4-41. 'Low-voltage electrical installations – Part 4–41: Protection for safety – Protection against electric shock'. 2005.
5. IEEE Std. C62.41.1. 'IEEE Guide on the surge environment in low-voltage (1 000 V and less) a.c. power circuits'. 2002.
6. IEEE C57.105. 'IEEE Guide for application of transformer connections in three-phase distribution systems'. 1978.
7. Massucco S. and Silvestro F. 'Appendix – Structure and data concerning electrical grids for Italy, Germany, Spain, UK, Poland'. Project DISPOWER, Appendix_D2.5 G, Contract no. ENK5-CT-2001-00522, December 2003.
8. Hoidalen H.K. 'Lightning-induced overvoltages in low-voltage systems'. PhD Thesis, Norwegian University of Science and Technology, 1997.
9. Hoidalen H.K. 'Lightning-induced voltages in low-voltage systems and its dependency on overhead line terminations'. *Proceedings of the 24th International Conference on Lightning Protection (ICLP)*, Birmingham, September 1998, pp. 287–92.
10. Morched A., Martí L. and Ottevangers J. 'A high frequency transformer model for the EMTP'. *IEEE Transactions on Power Delivery*, 1993;**8**(3):1615–26.
11. Dommel H.W. *Electromagnetic Transients Program Reference Manual (EMTP Theory Book)* (The University of British Columbia, Vancouver, 1986).
12. Borghetti A., Iorio R., Nucci C.A. and Pelacchi P. 'Calculation of voltages induced by nearby lightning on overhead lines terminated on distribution transformers'. *Proceedings of the International Conference on Power Systems Transients (IPST)*, Lisbon, September 1995, pp. 311–16.
13. Vaessen P.T.M. 'Transformer model for high frequencies'. *IEEE Transactions on Power Delivery*, 1988;**3**(4):1761–68.
14. Piantini A. and Malagodi C.V.S. 'Modelling of three-phase distribution transformers for calculating lightning induced voltages transferred to the secondary'. *Proceedings of the 5th International Symposium on Lightning Protection (SIPDA)*, São Paulo, May 1999, pp. 59–64.
15. Piantini A. and Malagodi C.V.S. 'Voltage surges transferred to the secondary of distribution transformers'. *Proceedings of the 11th International Symposium on High Voltage Engineering (ISH)*, London, **1**, August 1999, pp. 1365–68.
16. Piantini A. and Malagodi C.V.S. 'Voltages transferred to the low-voltage side of distribution transformers due to lightning discharges close to overhead lines'. *Proceedings of the 5th International Symposium on Lightning Protection (SIPDA)*, São Paulo, May 1999, pp. 201–205.
17. Piantini A., Bassi W., Janiszewski J.M. and Matsuo N.M. 'A simple transformer model for analysis of transferred lightning surges from MV to LV lines'. *Proceedings of the 15th International Conference on Electricity Distribution (CIRED)*, Nice, 1999, paper 2–18/ 1–6.
18. Piantini A. and Kanashiro A.G. 'A distribution transformer model for calculating transferred voltages'. *Proceedings of the 26th International Conference on Lightning Protection (ICLP)*, Cracow, **2**, September 2002, pp. 429–34.
19. Kanashiro A.G. and Piantini A. 'The effect of the secondary loads on the voltage surges transferred through distribution transformers'. *Proceedings of the 13th International Symposium on High Voltage Engineering (ISH)*, Rotterdam, 2003.

20. Kanashiro A.G., Piantini A. and Burani G.F. 'A methodology for transformer modelling concerning high frequency surges'. *Proceedings of the 6th International Symposium on Lightning Protection (SIPDA)*, São Paulo, November 2001, pp. 275–80.
21. Richter B. and Zeller P. 'Lightning and leakage currents through MO-surge arresters in distribution systems'. *Proceedings of the 26th International Conference on Lightning Protection (ICLP)*, Cracow, **2**, September 2002, pp. 527–30.
22. Manyahi M.J. and Thottappillil R. 'Transfer of lightning transients through distribution transformer circuits'. *Proceedings of the 26th International Conference on Lightning Protection (ICLP)*, Cracow, **2**, September 2002, pp. 435–40.
23. Montaño R. and Cooray V. 'Penetration of lightning induced transient from high-voltage to low-voltage power system across distribution transformers'. *Proceedings of the 27th International Conference on Lightning Protection (ICLP)*, Avignon, **2**, September 2004, pp. 303–308.
24. Agrawal A.K., Price H.J. and Gurbaxani S.H. 'Transient response of a multiconductor transmission line excited by a nonuniform electromagnetic field'. *IEEE Transactions on Electromagnetic Compatibility*, 1980;**EMC-22**(2):119–29.
25. Bachega R., De Araujo J., De Souza A. and Martinez M.L.B. 'Digital models of transformer for application on electromagnetic transient studies'. *Proceedings of the International Conference on Grounding and Earthing and 3rd Brazilian Workshop on Atmospheric Electricity*, Rio de Janeiro, November 2002, pp. 279–84.
26. Piantini A., Kanashiro A.G. and Obase P.F. 'Lightning surges transferred to the low-voltage network'. *Proceedings of the 7th International Symposium on Lightning Protection (SIPDA)*, Curitiba, November 2003, pp. 216–21.
27. Obase P.F., Piantini A. and Kanashiro A.G. 'Overvoltages transferred to low-voltage networks due to direct strokes on the primary'. *Proceedings of the 8th International Symposium on Lightning Protection (SIPDA)*, São Paulo, November 2005, pp. 489–94.
28. Obase P.F., Piantini A. and Kanashiro A.G. 'Overvoltages on LV networks associated with direct strokes on the primary line'. *Proceedings of the 28th International Conference on Lightning Protection (ICLP)*, Kanazawa, **1**, September 2006, pp. 479–84.
29. Obase P.F., Piantini A. and Kanashiro A.G. 'Overvoltages on distribution networks caused by direct lightning hits on the primary line'. *Proceedings of the International Conference on Grounding and Earthing and International Conference on Lightning Physics and Effects*, Belo Horizonte, November 2004, pp. 198–203.
30. De Conti A. and Visacro S. 'Calculation of lightning-induced voltages on low-voltage distribution networks'. *Proceedings of the 7th International Symposium on Lightning Protection (SIPDA)*, São Paulo, November 2005, pp. 483–88.
31. Borghetti A., Morched A.S., Napolitano F., Nucci C.A. and Paolone M. 'Lightning-induced overvoltages transferred from medium-voltage to

low-voltage networks'. *Proceedings of the IEEE St. Petersburg Power Tech.*, St. Petersburg, June 2005.
32. Nucci C.A., Bardazzi V., Iorio R., Mansoldo A. and Porrino A. 'A code for the calculation of lightning-induced overvoltages and its interface with the Electromagnetic Transient program'. *Proceedings of the 22nd International Conference on Lightning Protection (ICLP)*, Budapest, September 1994, pp. 19–23.
33. Paolone M., Nucci C.A. and Rachidi F. 'A new finite difference time domain scheme for the evaluation of lightning induced overvoltages on multiconductor overhead lines'. *Proceedings of the 5th International Conference on Power System Transients (IPST)*, Rio de Janeiro, June 2001.
34. Borghetti A., Gutierrez J.A., Nucci C.A., Paolone M., Petrache E. and Rachidi F. 'Lightning-induced voltages on complex distribution systems: models, advanced software tools and experimental validation'. *Journal of Electrostatics*, 2004;**60**(2–4):163–74.
35. Noda T., Nakamoto H. and Yokoyama S. 'Accurate modeling of core-type distribution transformers for electromagnetic transient studies'. *IEEE Transactions on Power Delivery*, 2002;**17**(4):969–76.
36. Honda H., Noda T., Asakawa A. and Yokoyama S. 'Improvements to a pole-mounted distribution transformer model for electromagnetic transient studies and validation by field tests'. *Proceedings of the 28th International Conference on Lightning Protection (ICLP)*, Kanazawa, **2**, September 2006, pp. 789–94.
37. Smith D.R. and Puri J.L. 'A simplified lumped parameter model for finding distribution transformer and secondary system responses to lightning'. *IEEE Transactions on Power Delivery*, 1988;**4**(3):1921–36.
38. Puri J.L., Abi-Samra N.C., Dionise T.J. and Smith D.R. 'Lightning induced failures in distribution transformers'. *IEEE Transactions on Power Delivery*, 1988;**3**(4):1784–801.
39. Goedde G.L., Dugan R.C. and Rowe L.D. 'Full scale lightning surge tests of distribution transformers and secondary systems'. *IEEE Transactions on Power Delivery*, 1992;**7**(3):1592–600.
40. Mirra C., Porrino A., Ardito A. and Nucci C.A. 'Lightning overvoltages in low-voltage networks'. *Proceedings of the International Conference on Electricity Distribution (CIRED)*, Birmingham, Conference Publication No. 438, June 1997, pp. 2.19.1–2.19.6.
41. Bassi W. 'Input impedance characteristics and modeling of low voltage residential installations for lightning studies'. *Proceedings of the 29th International Conference on Lightning Protection (ICLP)*, Uppsala, June 2008.
42. Bassi W., Burani G.F. and Janiszewski J.M. 'Impedance frequency characteristics and modelling of residential appliances for lightning transient analysis'. *Proceedings of the 8th International Symposium on Lightning Protection (SIPDA)*, São Paulo, November 2005, pp. 495–98.
43. Rakov V.A. and Uman M.A. 'Artificial initiation (triggering) of lightning by ground-based activity'. *Lightning: Physics and Effects* (Cambridge University Press, Cambridge, 2003), Chapter 7, pp. 265–307.

44. Rakov V.A. 'Rocket-triggered lightning experiments at Camp Blanding, Florida'. *Proceedings of the 5th International Symposium on Lightning Protection (SIPDA)*, São Paulo, May 1999, pp. 375–88.
45. Decarlo B.A., Rakov V.A., Jerauld J., Schnetzer G.H., Schoene J., Uman M.A. et al. 'Triggered-lightning testing of the protective system of a residential building: 2004 and 2005 results'. *Proceedings of the 28th International Conference on Lightning Protection (ICLP)*, Kanazawa, September 2006, pp. 628–33.
46. Rakov V.A., Uman M.A., Fernandez M.I., Mata C.T., Rambo K.J., Stapleton M.V. et al. 'Direct lightning strikes to the lightning protective system of a residential building: triggered-lightning experiments'. *IEEE Transactions on Power Delivery*, 2002;**17**(2): 575–86.
47. Cooray V. 'The mechanism of the lightning flash'. *The Lightning Flash* (IEE Power Engineering Series, London, **34**, 2003), Chapter 4, pp. 127–239.
48. Rakov V.A. 'Lightning phenomenology and parameters important for lightning protection'. *Proceedings of the 9th International Symposium on Lightning Protection (SIPDA)*, Foz do Iguaçu, November 2007, pp. 541–64.
49. Rakov V.A. and Uman M.A. 'Cloud discharges'. *Lightning: Physics and Effects* (Cambridge University Press, Cambridge, 2003), Chapter 9, pp. 321–45.
50. Galván A., Cooray V. and Thottappillil R. 'Effects of lightning electromagnetic field pulses (LEMP) in low-voltage power installations'. *Proceedings of the 24th International Conference on Lightning Protection (ICLP)*, Birmingham, September 1998, pp. 729–34.
51. Silfverskiöld S., Thottappillil R., Cooray V. and Scuka V. 'Induced voltages in a low-voltage power installation network due to lightning electromagnetic fields: an experimental study'. *IEEE Transactions on Electromagnetic Compatibility*, 1999;**41**(3):265–71.
52. Galván A. and Cooray V. 'Analytical simulation of lightning induced voltage in low-voltage power installations'. *Proceedings of the 25th International Conference on Lightning Protection (ICLP)*, Rhodes, September 2000, pp. 290–95.
53. Galván A., Cooray V. and Thottappillil R. 'A technique for the evaluation of lightning-induced voltages in complex low-voltage power-installation networks'. *IEEE Transactions on Electromagnetic Compatibility*, 2001;**43**(3):402–409.
54. Rubinstein M. and Uman M.A. 'Review of the University of Florida research on lightning induced voltages on power distribution lines'. *Proceedings of the 21st International Conference on Lightning Protection (ICLP)*, Berlin, September 1992, pp. 189–93.
55. Fernandez M.I., Rakov V.A. and Uman M.A. 'Transient currents and voltages in a power distribution system due to natural lightning'. *Proceedings of the 24th International Conference on Lightning Protection (ICLP)*, Birmingham, September 1998, pp. 622–29.
56. Piantini A. 'Lightning transients on LV networks caused by indirect strokes'. *Journal of Lightning Research*, 2007;**1**:111–31.
57. Joint Cired/Cigré Working Group 05. 'Protection of MV and LV networks against lightning. Part I: basic information'. *Proceedings of the International*

Conference on Electricity Distribution (CIRED), Birmingham, Conference Publication No. 438, June 1997, pp. 2.21.1–2.21.7.
58. Piantini A. and Janiszewski J.M. 'Lightning induced overvoltages on low-voltage lines'. *Proceedings of the 5th International Symposium on Lightning Protection (SIPDA)*, São Paulo, May 1999, pp. 234–39.
59. Silva Neto A. and Piantini A. 'Induced overvoltages on LV lines with twisted conductors due to indirect strokes'. *Proceedings of the 8th International Symposium on Lightning Protection (SIPDA)*, São Paulo, November 2005, pp. 234–39.
60. Hoidalen H.K. 'Lightning-induced overvoltages in rural isolated neutral low-voltage systems'. *Proceedings of the 28th International Conference on Lightning Protection (ICLP)*, Kanazawa, September 2006, pp. 485–90.
61. Piantini A. and Janiszewski J.M. 'Induced voltages on distribution lines due to lightning discharges on nearby metallic structures'. *IEEE Transactions on Magnetics*, 1998;**34**(5):2799–802.
62. Piantini A. and Janiszewski J.M. 'The Extended Rusck Model for calculating lightning induced voltages on overhead lines'. *Proceedings of the 7th International Symposium on Lightning Protection (SIPDA)*, Curitiba, November 2003, pp. 151–55.
63. Piantini A. and Janiszewski J.M. 'An improved model for lightning induced voltages calculations'. *Proceedings of the IEEE/PES Transmission & Distribution Conference and Exposition: Latin America*, São Paulo, November 2004, pp. 554–59.
64. Nucci C.A. and Rachidi F. 'Interaction of electromagnetic fields with electrical networks generated by lightning', in V. Cooray (ed.). *The Lightning Flash* (IEE Power Engineering Series, London, **34**, 2003), Chapter 8, pp. 425–78.
65. Rakov V.A. and Uman M.A. 'Deleterious effects of lightning and protective techniques'. *Lightning: Physics and Effects* (Cambridge University Press, Cambridge, 2003), Chapter 18, pp. 588–641.
66. Nucci C.A. *et al*. 'Lightning-induced voltages on overhead power lines. Part III: Sensitivity analysis'. *Elektra*, October 2005, pp. 27–30.
67. Borghetti A., Henrikssen T., Rojas P.M., Nucci C.A., Paolone M., Rachidi F. *et al.* 'Lightning protection of medium voltage networks'. *Proceedings of the 8th International Symposium on Lightning Protection (SIPDA)*, São Paulo, November 2005, pp. 17–23.
68. Piantini A. and Janiszewski J.M. 'The effectiveness of surge arresters on the mitigation of lightning induced voltages on distribution lines'. *Proceedings of the 8th International Symposium on Lightning Protection (SIPDA)*, São Paulo, November 2005, pp. 777–98.
69. Paolone M., Nucci C.A., Petrache E. and Rachidi F. 'Mitigation of lightning-induced overvoltages in medium voltage distribution lines by means of periodical grounding of shielding wires and of surge arresters: modeling and experimental validation'. *IEEE Transactions on Power Delivery*, 2004;**19**(1):423–31.
70. Piantini A., De Carvalho T.O., Silva Neto A., Janiszewski J.M., Altafim R.A.C. and Nogueeira A.L.T. 'A system for simultaneous measurements of lightning

induced voltages on lines with and without arresters'. *Proceedings of the 27th International Conference on Lightning Protection (ICLP)*, Avignon, **1**, September 2004, pp. 297–302.
71. Shostak V., Janischewskyj W., Rachidi F., Hussein A.M., Bermudez J.L., Chang J.S. et al. 'Estimation of lightning-caused stresses in a MV distribution line'. *Proceedings of the 27th International Conference on Lightning Protection (ICLP)*, Avignon, **2**, September 2004, pp. 678–83.
72. Yokoyama S. 'Lightning protection of MV overhead distribution lines'. *Proceedings of the 7th International Symposium on Lightning Protection (SIPDA)*, São Paulo, November 2003, pp. 485–507.
73. Piantini A. and Janiszewski J.M. 'Lightning induced voltages on distribution transformers: the effects of line laterals and nearby buildings'. *Proceedings of the 6th International Symposium on Lightning Protection (SIPDA)*, Santos, November 2001, pp. 77–82.
74. Borghetti A., Nucci C.A., Paolone M. and Bernardi M. 'Effect of the lateral distance expression and of the presence of shielding wires on the evaluation of the number of lightning induced voltages'. *Proceedings of the 25th International Conference on Lightning Protection (ICLP)*, Rhodes, **A**, September 2000, pp. 340–45.
75. Nucci C.A., Borghetti A., Piantini A. and Janiszewski J.M. 'Lightning-induced voltages on distribution overhead lines: comparison between experimental results from a reduced-scale model and most recent approaches'. *Proceedings of the 24th International Conference on Lightning Protection (ICLP)*, Birmingham, **1**, September 1998, pp. 314–20.
76. Darveniza M. 'A practical approach to lightning protection of sub-transmission and distribution systems'. *Proceedings of the 5th International Symposium on Lightning Protection (SIPDA)*, São Paulo, May 1999, pp. 365–72.
77. Joint Cired/Cigré Working Group 05. 'Protection of MV and LV networks against lightning: Part II: application to MV networks'. *Proceedings of the International Conference on Electricity Distribution (CIRED)*, Birmingham, Conference Publication No. 438, 1997, pp. 2.28.1–2.28.8.
78. Barker P.P., Short T.A., Eybert-Berard A.R. and Berlandis J.P. 'Induced voltage measurements on an experimental distribution line during nearby rocket triggered lightning flashes'. *IEEE Transactions on Power Delivery*, 1996;**11**(2):980–95.
79. Piantini A. and Janiszewski J.M. 'Lightning induced voltages on overhead lines: the effect of ground wires'. *Proceedings of the 22nd International Conference on Lightning Protection (ICLP)*, Budapest, September 1994, pp. R3b/1–R3b/5.
80. Piantini A. and Janiszewski J.M. 'An experimental study of lightning induced voltages by means of a scale model'. *Proceedings of the 21st International Conference on Lightning Protection (ICLP)*, Berlin, September 1992, pp. 195–99.
81. Rusck S. 'Induced lightning over-voltages on power-transmission lines with special reference to the over-voltage protection of low-voltage networks'. *Transactions of the Royal Institute of Technology*, 1958, 120.

82. Nucci C.A. 'Lightning-induced voltages on distribution systems: influence of ground resistivity and system topology'. *Proceedings of the 8th International Symposium on Lightning Protection (SIPDA)*, São Paulo, November 2005, pp. 761–73.
83. Baba Y. and Rakov V.A. 'On calculating lightning-induced overvoltages in the presence of a *tall strike object*'. *Proceedings of the 8th International Symposium on Lightning Protection (SIPDA)*, São Paulo, November 2005, pp. 11–16.
84. Piantini A. and Janiszewski J.M. 'Protection of distribution lines against indirect lightning strokes through the use of surge arresters'. *CIGRE Surge Arrester Tutorial*, Rio de Janeiro, April 2005.
85. Piantini A., De Carvalho T.O., Silva Neto A., Janiszewski J.M., Altafim R.A.C. and Nogueira A.L.T. 'A system for lightning induced voltages data acquisition – preliminary results'. *Proceedings of the 7th International Symposium on Lightning Protection (SIPDA)*, São Paulo, November 2003, pp. 156–61.
86. Nucci C.A., Guerrieri S., Correia de Barros M.T. and Rachidi F. 'Influence of corona on the voltages induced by nearby lightning on overhead distribution lines'. *IEEE Transactions on Power Delivery*, 2000;**15**(4):1265–73.
87. Borghetti A., Nucci C.A., Paolone M. and Rachidi F. 'Characterization of the response of an overhead line to lightning electromagnetic fields'. *Proceedings of the 25th International Conference on Lightning Protection (ICLP)*, Rhodes, **A**, September 2000, pp. 223–28.
88. Nucci C.A. and Rachidi F. 'Lightning induced overvoltages'. *IEEE Transmission and Distribution Conference, Panel Session 'Distribution Line Protection'*, New Orleans, April 1999.
89. Piantini A. and Janiszewski J.M. 'Use of surge arresters for protection of overhead lines against nearby lightning'. *Proceedings of the 10th International Symposium on High Voltage Engineering (ISH)*, Montreal, **5**, August 1997, pp. 213–16.
90. Piantini A. and Janiszewski J.M. 'The influence of the upward leader on lightning induced voltages'. *Proceedings of the 23rd International Conference on Lightning Protection (ICLP)*, Florence, **1**, September 1996, pp. 352–57.
91. Nucci C.A., Rachidi F., Ianoz M. and Mazzetti C. 'Comparison of two coupling models for lightning-induced overvoltage calculations'. *IEEE Transactions on Power Delivery*, 1995;**10**(1):330–39.
92. Ishii M., Michishita K., Hongo Y. and Ogume S. 'Lightning-induced voltage on an overhead wire dependent on ground conductivity'. *IEEE Transactions on Power Delivery*, 1994;**9**(1):109–18.
93. Nucci C.A., Rachidi F., Ianoz M. and Mazzetti C. 'Lightning-induced overvoltages on overhead lines'. *IEEE Transactions on Electromagnetic Compatibility*, 1993;**35**(1):75–86.
94. Chowdhuri P. 'Response of overhead lines of finite length to nearby lightning strokes'. *IEEE Transactions on Power Delivery*, 1991;**6**(1):343–51.
95. Chowdhuri P. 'Estimation of flashover rates of overhead power distribution lines by lightning strokes to nearby ground'. *IEEE Transactions on Power Delivery*, 1989;**4**(3):1982–89.

96. Liew A.C. and Mar S.C. 'Extension of the Chowdhuri–Gross model for lightning induced voltage on overhead lines'. *IEEE Transactions on Power Systems*, 1986;**1**(2):240–47.
97. Yokoyama S., Miyake K., Mitani H. and Takanishi A. 'Simultaneous measurement of lightning induced voltages with associated stroke currents'. *IEEE Transactions on Power Apparatus and Systems*, 1983;**102**(8):2420–27.
98. Yokoyama S., Miyake K., Mitani H. and Yamazaki N. 'Advanced observations of lightning induced voltage on power distribution lines'. *IEEE Transactions on Power Delivery*, 1986;**1**(2):129–39.
99. Yokoyama S., Miyake K. and Fukui S. 'Advanced observations of lightning induced voltage on power distribution lines (II)'. *IEEE Transactions on Power Delivery*, 1989;**4**(4):2196–203.
100. Piantini A. and Janiszewski J.M. 'Analysis of three different theories for computation of induced voltages on distribution lines due to nearby lightning'. *Proceedings of the International Conference on Electricity Distribution (CIRED)*, Buenos Aires, December 1996, Session 1/127–132.
101. Cooray V. 'Calculating lightning-induced overvoltages in power lines: a comparison of two coupling models'. *IEEE Transactions on Electromagnetic Compatibility*, 1994;**36**(3):179–82.
102. Michishita K. and Ishii M. 'Theoretical comparison of Agrawal's and Rusck's field-to-line coupling models for calculation of lightning-induced voltage on an overhead wire'. *IEE Transactions of Japan*, 1997;**117**(9):1315–16.
103. Piantini A., Janiszewski J.M., Borghetti A., Nucci C.A. and Paolone M. 'A scale model for the study of the LEMP response of complex power distribution networks'. *IEEE Transactions on Power Delivery*, 2007;**22**(1):710–20.
104. Baba Y. and Rakov V.A. 'Voltages induced on an overhead wire by lightning strikes to a nearby tall grounded object'. *IEEE Transactions on Electromagnetic Compatibility*, 2006;**48**(1):212–24.
105. Uman M.A. and Mclain D.K. 'Magnetic field of the lightning return stroke'. *Journal of Geophysical Research*, 1969;**74**:6899–910.
106. Rachidi F., Nucci C.A., Ianoz M. and Mazzetti C. 'Influence of a lossy ground on lightning-induced voltages on overhead lines'. *IEEE Transactions on Electromagnetic Compatibility*, 1996;**38**(3):250–64.
107. Heidler F. 'Analytische Blitzstromfunktion zur LEMP- Berechnung'. *Proceedings of the 18th International Conference on Lightning Protection (ICLP)*, Munich, September 1985, pp. 63–66.
108. Anderson R.B. and Eriksson A.J. 'Lightning parameters for engineering application'. *Elektra*, 1979;**69**:65–102.
109. Uman M.A. *The Lightning Discharge* (Academic Press, San Diego, 1987).
110. Nucci C.A., Mazzetti C., Rachidi F. and Ianoz M. 'On lightning return stroke models for LEMP calculations'. *Proceedings of the 19th International Conference on Lightning Protection (ICLP)*, Graz, April 1988, pp. 463–69.
111. Hoidalen H.K. 'Calculation of lightning-induced overvoltages using MODELS'. *Proceedings of the 4th International Conference on Power Systems Transients (IPST'99)*, Budapest, June 1999, pp. 359–64.

112. Dugan R.C. and Smith S.D. 'Low-voltage-side current-surge phenomena in single-phase distribution transformer systems'. *IEEE Transactions on Power Delivery*, 1988;**3**(2):637–47.
113. Cooray V. 'On the validity of several approximate theories used in quantifying the propagation effects on lightning generated electromagnetic fields'. *Proceedings of the 8th International Symposium on Lightning Protection (SIPDA)*, São Paulo, November 2005, pp. 112–19.
114. Cooray V. 'Propagation effects due to finitely conducting ground on lightning generated magnetic fields evaluated using Sommerfeld's integrals'. *Proceedings of the 9th International Symposium on Lightning Protection (SIPDA)*, Foz do Iguaçu, November 2007, pp. 151–56.
115. Zeddam A. and Degauque P. 'Current and voltage induced on a telecommunication cable by a lightning stroke'. *Electromagnetics*, 1987;**7**(3&4):541–64.
116. Romero F. and Piantini A. 'Evaluation of lightning horizontal electric fields over a finitely conducting ground'. *Proceedings of the 9th International Symposium on Lightning Protection (SIPDA)*, Foz do Iguaçu, November 2007, pp. 145–50.
117. Barbosa C.F. and Paulino J.O.S. 'An approximate time-domain formula for the calculation of the horizontal electric field from lightning'. *IEEE Transactions on Electromagnetic Compatibility*, 2007;**49**(3):593–601.
118. Shoory A., Moini R., Sadeghi S.H.H. and Rakov V.A. 'Analysis of lightning-radiated electromagnetic fields in the vicinity of lossy ground'. *IEEE Transactions on Electromagnetic Compatibility*, 2005;**47**(1):131–45.
119. Rachidi F., Nucci C.A. and Ianoz M. 'Transient analysis of multiconductor lines above a lossy ground'. *IEEE Transactions on Power Delivery*, 1999;**14**(1):294–302.
120. Borghetti A. and Nucci C.A. 'Frequency distribution of lightning-induced voltages on an overhead line above a lossy ground'. *Proceedings of the 5th International Symposium on Lightning Protection (SIPDA)*, São Paulo, May 1999, pp. 229–33.
121. Ishii M., Michishita K. and Hongo Y. 'Experimental study of lightning-induced voltage on an overhead wire over lossy ground'. *IEEE Transactions on Electromagnetic Compatibility*, 1999;**41**(1):39–45.
122. Guerrieri S., Nucci C.A. and Rachidi F. 'Influence of the ground resistivity on the polarity and intensity of lightning induced voltages'. *Proceedings of the 10th International Symposium on High Voltage Engineering (ISH)*, Montreal, August 1997, pp. 24–30.
123. Hoidalen H.K., Sletbak J. and Henriksen T. 'Ground effects on induced voltages from nearby lightning'. *IEEE Transactions on Electromagnetic Compatibility*, 1997;**39**(4):269–78.
124. Hermosillo V.F. and Cooray V. 'Calculation of fault rates of overhead power distribution lines due to lightning induced voltages including the effect of ground conductivity'. *IEEE Transactions on Electromagnetic Compatibility*, 1995;**37**(3):392–99.
125. Cooray V. and De La Rosa F. 'Shapes and amplitudes of the initial peaks of lightning-induced voltage in power lines over finitely conducting earth: theory

and comparison with experiment'. *IEEE Transactions on Antennas and Propagation*, 1986;**34**(1):88–92.
126. ATP. Alternative transients program rule book. Leuven EMTP Center, 1987.
127. Darveniza M. and Vlastos A.E. 'The generalized integration method for predicting impulses volt-time characteristics for non-standard wave shapes – a theoretical basis'. *IEEE Transactions on Electrical Insulation*, 1988;**23**(3):373–81.
128. De Conti A. and Visacro S. 'Evaluation of lightning surges transferred from medium voltage to low-voltage networks'. *IEE Proceedings Generation Transmission Distribution*, 2005;**152**(3):351–56.
129. Darveniza M. and Mercer D.R. 'Lightning protection of pole mounted transformers'. *IEEE Transactions on Power Delivery*, 1989;**4**(2):1087–95.
130. Darveniza M. 'Lightning arrester protection of distribution transformers – revisited'. *Proceedings of the 6th International Symposium on Lightning Protection (SIPDA)*, São Paulo, November 2001, pp. 385–89.
131. Dugan R.C., Kershaw S.S. Jr and Smith S.D. 'Protecting distribution transformers from low-side current surges'. *IEEE Transactions on Power Delivery*, 1990;**5**(4): 1892–901.
132. Marz M.B. and Mendis S.R. 'Protecting load devices from the effects of low-side surges'. *IEEE Transactions on Industry Applications*, 1993;**29**(6):1196–203.
133. Task force on low-side surge requirements for distribution transformers. 'Secondary (low-side) surges in distribution transformers'. *IEEE Transactions on Power Delivery*, 1992;**7**(2):746–755.
134. Hasse P. 'Surge protective devices (SPDs) – Part I: Requirements of energy coordinated SPDs according to IEC 61312-3'. *Proceedings of the 5th International Symposium on Lightning Protection (SIPDA)*, São Paulo, May 1999, pp. 130–35.
135. Hasse P. 'Surge protective devices (SPDs) – Part II: Practical application of energy coordinated SPDs according to IEC 61312-3'. *Proceedings of the 5th International Symposium on Lightning Protection (SIPDA)*, São Paulo, May 1999, pp. 136–41.
136. Martzloff F.D. and Lai J.-S. 'Coordinating current surge protection devices: high-low versus low-high'. *IEEE Transactions on Industry Applications*, 1993;**29**(4):680–87.
137. Standler R.B. 'Coordination of surge arresters and suppressors for use on low-voltage mains'. *Proceedings of the 9th EMC Zurich*, Zurich, 1991, pp. 517–24.
138. Martzloff F.D. 'Coordination of surge protectors in low-voltage a.c. power circuits'. *IEEE Transactions on Power Apparatus and Systems*, 1980;**99**(1): 129–33.
139. Hasse P. *Overvoltage Protection of Low Voltage Systems* (IEE Power and Energy Series, no. 33, 2nd edn, 2000).

Chapter 13
Lightning protection of medium voltage lines
C.A. Nucci and F. Rachidi

13.1 Introduction

The problem of lightning protection of medium-voltage (MV) networks[1] has been seriously reconsidered in recent years due to the proliferation of sensitive loads and the increasing demand by customers for good quality in the power supply. Overvoltages originated by lightning are indeed a major cause of flashovers on overhead power lines. These flashovers may cause permanent or short interruptions, as well as voltage dips, on the above-mentioned distribution networks. Additionally, lightning-originated surges, depending on their amplitude and energy content, can also damage the power components connected to these networks as well as electronic devices.

The objective of this chapter is to provide a survey of the basic concepts and general principles applicable to the protection of MV networks against lightning-originated overvoltages, taking into account the two aspects of the problem:

- protection of the components connected to the line (e.g. distribution transformers) against the disruptive effect of lightning-caused surges;
- insulation/protection coordination in order to minimize the number of flashovers (and also voltage interruptions or voltage dips) along the distribution lines.

In general, in books or standards dealing with the topic [1–4] the following approach is adopted. Analytical expressions are used for the calculation of directly or indirectly caused lightning overvoltages and the lightning performance of distribution lines is evaluated by means of these simple expressions and some statistical procedures aimed at representing the random nature of lightning. When presented, analysis of surge propagation along the line is carried out making reference to simple/effective reasoning and equations – generally without the aid of computer simulations – often disregarding the presence of surge protective devices along the line.

---

[1] By medium-voltage networks we here mean networks operating at a voltage level from 1 kV up to 69 kV.

In this chapter we attempt to present a more modern approach to the problem, which makes use not only of simple analytical formulae, but also of advanced models and computer codes. Owing to the complexity of the problem, it is indeed almost impossible to achieve sound results without the use of adequate modelling and relevant software. In general, the community dealing with lightning protection is prepared to use Electromagnetic Transient Program (EMTP)-like computer codes for the calculation of the lightning response of high-voltage (HV) transmission lines, but for distribution networks, due to the inherently larger complexity of both topology and configuration, the use of advanced computer codes has not been very popular so far. This is probably due to the fact that, besides the mentioned complexity of the system to be studied, the phenomena involved are in general more complex for distribution lines than for transmission lines. Think, for instance, of the electromagnetic coupling between the lightning electromagnetic pulse (LEMP) radiated by a nearby return stroke and the conductors of an overhead distribution network. It is reasonable to imagine that in the future the use of 'coupling-transient' codes will progressively enhance its importance with larger benefits for the power quality and reliability of power distribution networks. After all, the use of sophisticated computer codes for the calculation of power flow and the stability of the power network now represents a common practice.

13.2 Lightning strike incidence to distribution lines

Lightning overvoltages can be classified into two categories.

- *Overvoltages due to direct lightning.* Owing to the relatively low levels of insulation in MV networks, a lightning return stroke to the phase conductor, to the neutral conductor or to the support structure in general causes insulation flashover on the line. However, the presence of surge arresters, if adequately combined with the use of shield wires, may prevent surge flashover.
- *Overvoltages due to indirect lightning* (*induced overvoltages*). Indirect lightning, that is, lightning hitting the ground or structures in the vicinity of a line, induces on the line overvoltages that can well exceed the insulation levels. As for direct strokes, the presence of neutral conductors, shield wires and/or surge arresters can have a great influence on the resulting induced-lightning overvoltages, and their mitigation effect can be more effective even than that for direct strokes.

Direct and induced lightning overvoltages differ one from another from the point of view of the parameters that are important for lightning protection (e.g. amplitude, steepness, energy, etc.). For example (see Section 13.3.2) induced surges are in general characterized by lower peaks, faster rise times and faster duration (lower time to half-width). It is thus important to evaluate the expected number of events of each type affecting a distribution line. The expected number of direct/indirect events depends strongly on the exposure of the line and on its screening by nearby objects.

When the lightning leader approaches the ground in a downward cloud-to-ground lightning flash, it continues its downward motion unperturbed unless critical field conditions develop to initiate an attachment with a nearby grounded conductor (final jump). By assuming the leader channel is perpendicular to the ground plane, it is generally accepted that the flash will strike the object if its prospective ground termination point, i.e. its strike location in absence of the object, lies within the attractive radius r. The attractive radius depends on several factors, such as the charge of the leader, its distance from the structure, the type of structure (vertical mast or horizontal conductor), its height, the nature of the terrain (flat or hilly) and the ambient ground field due to cloud charges. Several expressions have been proposed to evaluate such a radius. Some of them are based on the electrogeometric model [5–7] (see also Chapter 4 of this book, where the concepts of final jump, attractive radius and electrogeometric models are introduced):

$$r = \sqrt{[r_s^2 - (r_g - h)^2]} \quad \text{for } h < r_g \quad (13.1a)$$

$$r = r_s \quad \text{for } h \geq r_g \quad (13.1b)$$

where r_s and r_g are the so-called striking distances to the structure and to the ground, respectively, and h is the height of the structure, all expressed in metres. The striking distance is related to lightning current peak I_p, expressed in kA, through the following expressions

$$r_s = \alpha I_p^\beta \qquad r_g = k r_s \quad (13.2)$$

where the values of α, β and k are assumed to be independent of I_p. Other more physically oriented models than the electrogeometric model have been proposed [8–11]. From such models, simple expressions of the following type have been inferred, relating the attractive radius r and the lightning current peak I_p:

$$r = c + a I_p^b \quad (13.3)$$

where a, b and c are again constants and independent of I_p.

Tables 13.1 and 13.2 give some of the values of the parameters of the above-mentioned models. Model 2 in Table 13.1 is an approximation of the formula

Table 13.1 Constants of the striking distance equation (13.2): values proposed by different authors

Striking distance model	α	β	k
1. Armstrong and Whitehead [6]	6.7	0.80	0.9
2a. Adopted by IEEE Std. 1243 [12]	10	0.65	0.55*
2b. Adopted by IEEE Std. 1410 [4]			0.9†
3. Adapted from Golde [5,13]	3.3	0.78	1

*For an average conductor height larger than 40 m.
†For distribution lines.

Table 13.2 Constants of attractive radius equation (13.3): values inferred from the models by different authors

Attractive radius model	c	a	b
4a. Eriksson [8]	0	$0.84\,h^{0.6}$*	$0.7\,h^{0.02}$
4b. Eriksson [8]		$0.67\,h^{0.6\dagger}$	
5a. From Rizk [14,15]	0	$4.27\,h^{0.41}$*	0.55*
5b. From Rizk [9]		$1.57\,h^{0.45\dagger}$	$0.69^{\dagger}$
6. From Dellera and Garbagnati [10]	$3\,h^{0.6}$	$0.028\,h$	1

*For towers.
†For horizontal conductors.

by Love [7] using the exponential format [16] adopted by IEEE Std. 1243 [12] and Std. 1410 [4]. Model 3 in Table 13.1 is an approximation of the formula by Golde [5], proposed in Reference 13.

Concerning model 4 (Table 13.2, derived from the Eriksson model), in Reference 8 two set of values are proposed, one for towers with heights up to 100 m, and another for horizontal conductors, with an 80 per cent reduction of parameter a. The values of model 5b are those proposed in Reference 9 for horizontal conductors with a height range of 10–50 m and for lightning currents with I_p in the range 5–31 kA. As described in Reference 14, the values relevant to model 5a for free-standing structures have been inferred, using non-linear least-squares curve-fitting, from the results given by Rizk in Reference 15, which refers to the case of towers on flat terrain disregarding the ambient ground field. For model 6, the values reported in Table 13.2 were derived by Bernardi [17], by interpolation of plots of the lateral distance of a slim structure against its height (in the range 5–100 m), calculated using the leader progression model of Dellera and Garbagnati [10,11].

The assessment of the above models is clearly beyond the scopes of this chapter. For the problem of interest, in order to distinguish between direct and indirect strokes around a MV overhead line, we shall use model 2b, as adopted within the IEEE framework.

13.2.1 Expected number of direct lightning strikes

The expected number N_d of direct strikes per year, per 100 km, to a distribution line on flat ground can be evaluated by means of the following formula [18], which is based on the leader progression model by Dellera and Garbagnati [10]:

$$N_d = K_o N_g (d_h + 10.5 h^{0.75})/10 \qquad (13.4)$$

where N_g is the ground flash density (number of flashes per square km per year), h is the average height of the line in metres, d_h the horizontal distance between the outer conductors in metres, and K_o the orographic coefficient.

Figure 13.1 gives the coefficient K_o as a function of the orographic parameters defined in Figure 13.2. If the orographic condition is not known, a value $K_o = 1.8$

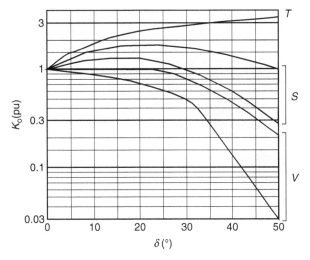

Figure 13.1 Direct strikes to an MV line. Orography correction coefficient K_o as a function of the orography parameters defined in Figure 13.2 (adapted from Reference 19).

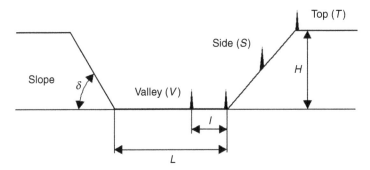

Figure 13.2 Parameters used for the evaluation of the influence of orography on the expected number of direct strokes to a MV line (adapted from Reference 19).

is suggested. This gives results in line with Eriksson's formula developed for transmission lines [22]. [$N_d = N_g(b + 28h^{0.6})/10$.] As an example for a line with $h = 10$ m and $d_h = 2$ m in an area with $N_g = 1$ km^{-2} yr^{-1} and $K_o = 1.8$, the expected number of direct strokes is $N_d = 11$ per year, per 100 km.

13.2.2 Shielding by nearby objects

Usually MV lines share, either totally or partially, right of way with LV lines and even with telecommunication lines. The latter are provided with conductors that are frequently grounded (neutral conductor, suspension conductors, etc.). Coupling with these circuits, frequently not anticipated at the design stage, certainly has an influence on the overvoltages induced on the line.

Shielding by nearby objects having a similar height as that of the line (trees, buildings, other power lines, etc.) may be very effective in reducing the incidence of direct strikes; however, it may increase the number of induced overvoltages. Shielding by nearby objects is more usual in LV overhead lines, which generally cross areas with buildings and are often strung between buildings or even suspended along the walls of buildings. Nonetheless, shielding by nearby objects can also have significant effects for distribution lines.

Hereafter, the shielding effect of nearby objects is considered only from the point of view of the expected number of direct strikes.

Quantitative evaluations of the shielding effect by structures parallel to a line can be found in References 21 and 22. These evaluations assume that a uniform row of objects of the same height is parallel to the line. The shielding effect is taken into consideration by means of a shielding factor S_f that allows the evaluation of the reduced number of direct strikes to a shielded line, N_d^s, by means of the formula

$$N_d^s = N_d(1 - S_f) \qquad (13.5)$$

where N_d is evaluated using equation (13.4) and S_f varies between 0 and 1 (where 1 corresponds to a perfect shielding). IEC [23] gives the environmental coefficient C_e through which S_f can be evaluated ($C_e = 1 - S_f$).

Generally, the shielding effect of a single object on the lightning performance of the line may be disregarded, but it may become important for lines in forested areas or parallel to other power lines. As an example, for a 10-m-high line at a constant distance of 20 m from a forest where the trees are 20 m high on average, the shielding factor is estimated to be 0.75 and 0.5 according, respectively, to References 21 and 22. These values are in agreement with the environmental coefficient C_e given in Reference 22.

13.3 Typical overvoltages generated by direct and indirect lightning strikes

13.3.1 Direct overvoltages

Lightning striking a phase conductor injects current waves in both directions. If we assume for simplicity a single-conductor line, the two voltage waves reach an amplitude that equals the corresponding current (half of the injected current) multiplied by the line characteristic impedance, which is $\sim$400 Ω. Because more than 90 per cent of the strokes have a peak current of at least 10 kA, the overvoltage will exceed 2 000 kV for 90 per cent of the strokes. (We are assuming here that the channel impedance is much greater than 400 Ω.) These voltages are far above the lightning impulse withstand voltage of distribution lines. For this reason flashover(s) between phase and earth (and between phases) will normally occur.

The resulting short circuit and the subsequent fault clearance are associated with (short) voltage interruptions at the network bus connected to the line where the flashover occurs, and with voltage dips at the network buses nearby. Note that this occurs independently from the original source for the short circuit, whether it is direct or indirect lightning.

Lines with covered conductors represent a special problem with respect to direct lightning. A flashover between phases for such lines may cause a mechanical breakdown of the conductors. This is due to the coating, which prevents the footpoint of the power frequency arcing current from moving along the line. Special precautions should therefore be taken to limit the risk of flashover between phases for lines with covered conductors exposed to lightning. The application of arresters or spark gaps along the line is analysed in Reference 24. Efficient protection requires a short distance between the protective devices (e.g. applying the devices at every pole along the line).

Components like transformers and cables are permanently damaged if they are exposed to overvoltages that cause insulation breakdown, unless protected by surge arresters. It is therefore important to analyse the lightning overvoltages at the location where these apparatuses are connected.

The surge propagating from the point of strike along the line is altered by flashovers, if any, occurring between the strike location and the point of interest. Practically, all flashovers to ground occur at the poles, as on overhead distribution lines the weakest insulation is generally at a pole structure rather than between conductors through air. A flashover at a pole reduces the amplitude of the voltage wave, which propagates further. If the voltage wave still exceeds the flashover voltage, new flashovers will occur at subsequent poles, further reducing the overvoltage.

Figure 13.3 shows a typical overvoltage (evaluated by calculations) due to a direct lightning strike of 30 kA current amplitude. The calculations have been performed using the Electromagnetic Transient Program (EMTP) [25] for a single-wire line with no ground wires. The line is composed of eight spans (9 poles) of 200 m length, each with a characteristic impedance of 440 Ω. Each pole, 8 m high, is modelled as a transmission line with a characteristic impedance of 300 Ω. The footing d.c. resistance was assumed to be nonlinear (current-dependent), with 30 Ω at zero current. The insulator flashover voltage was fixed at 150 kV. The voltage is calculated 600 m from the stroke location.

The example shows the general characteristics of a direct lightning overvoltage, which presents a few very short spikes, followed by an impulse voltage with a smoother shape.

For the case of Figure 13.3, the voltage at the flashover location equals the voltage between the cross arms and remote earth. This voltage may be higher than the flashover voltage and this causes a flashover at the next pole as the voltage wave propagates along the line. The first spikes are very steep and are chopped due to line insulation breakdown at the first pole the current wave meets, as well as at the following poles. The peak values of these spikes may exceed the lightning impulse withstand voltage of the line insulation due to the behaviour of air insulation at very steep front overvoltages.

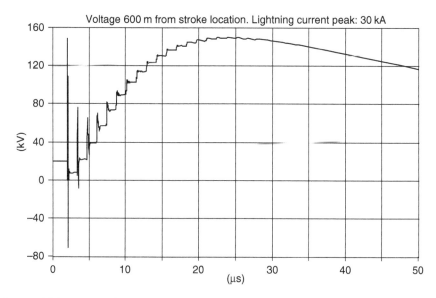

Figure 13.3   Example of a typical lightning overvoltage due to a direct strike to the MV line (adapted from Reference 19)

The smoother impulse voltage has a shape that is similar to that of the lightning current. Its amplitude is approximately equal to the lightning impulse withstand voltage of the line insulation. For this reason it is not chopped.

Figure 13.3 applies to a line with earthed cross arms. The flashover voltage to ground is much larger (of the order of 2 000 kV) for lines with wooden poles and no earthing of the cross arms. This implies that the initial spikes of the voltage in Figure 13.3 can become much larger for such lines.

As may be seen from this example, the lightning overvoltage occurring at some distance from the point of strike is in general strongly dependent on the flashover voltage to earth at the poles. Other important parameters are the number of poles between the strike location and the point of interest and the grounding impedance of the poles where flashover occurs.

13.3.2   Induced overvoltages

A cloud-to-ground lightning flash generates a transient electromagnetic field that can induce overvoltages of significant magnitude on overhead power lines situated in its vicinity. The return stroke phase of the lightning discharge is considered to be the major factor responsible for the induced voltages, because the most intense electromagnetic radiation occurs during this phase.

Although in studies analysing direct lightning strikes to a line, the lightning channel is generally represented by an equivalent current source injecting the return stroke current to the line at the attachment point, as we have seen in Section 13.3.1, for the case of an indirect lightning strike, the analysis is more complex and the calculation of lightning-induced voltages requires the following stages:

- adoption of a return-stroke model that specifies the spatial and temporal distribution of the lightning current along the channel during the return-stroke phase;
- calculation of the electromagnetic field change produced by such a current distribution along the line, including propagation effects on the field;
- using the field-to-transmission line coupling model to obtain the induced voltages resulting from the electromagnetic interaction between the field and the line conductors.

As opposed to simple analytical formulae [such as the popular simplified formula by Rusck [26], see Appendix A13, equation (A13.2)], which are restricted to unrealistically simple configurations, more elaborate models based on 'field-to-transmission line coupling equations' [27–32] allow for an accurate treatment of realistic line configurations.

Moreover, the presence of distribution transformers, of surge protection devices at the line terminations, as well as the presence of surge arresters and shielding wire groundings along the line, should be taken into account. In fact, these more complex models allow for the appropriate treatment of these non-linear components and/or line discontinuities in a convenient way.

The complexity of these models, which have been described thoroughly in Reference 31, calls for their implementation into computer codes because, in general, they require a numerical integration of the relevant coupling equations. The computed results presented in this chapter are indeed obtained by using two computer codes: the LIOV computer code [29–33], developed in the framework of an international collaboration involving the University of Bologna (Department of Electrical Engineering), the École Polytechnique Fédérale de Lausanne (EMC Group), and the University of Rome 'La Sapienza' (Department of Electrical Engineering), and the LIOV-EMTP code, developed also in collaboration with CESI [34–36]. A description of the field-to-transmission line coupling equations implemented in the LIOV code and the main characteristics of both LIOV and LIOV-EMTP codes is given in Appendix C13, which includes also the link address to a free version of the LIOV code.

Detailed analysis of the amplitude and waveshape of lightning-induced voltages as a function of various parameters (lightning current amplitude, front steepness, return stroke speed, line geometry, ground resistivity, etc.) can be found in Reference 31, and the interested reader is encouraged to read that book chapter.

For the purposes of the present chapter what is important to keep in mind is that the relative position between the line and the lightning strike position, combined with the effects of the ground resistivity have a significant influence on the amplitude and waveshape of lightning-induced voltages [31]. For an illustrative purpose, Figure 13.4 shows the voltage induced at both ends of a 1-km-long, 10-m-high single-wire overhead line (the conductor diameter is 1 cm) matched at the line terminations, for various strike positions and for three values of the ground conductivity, namely

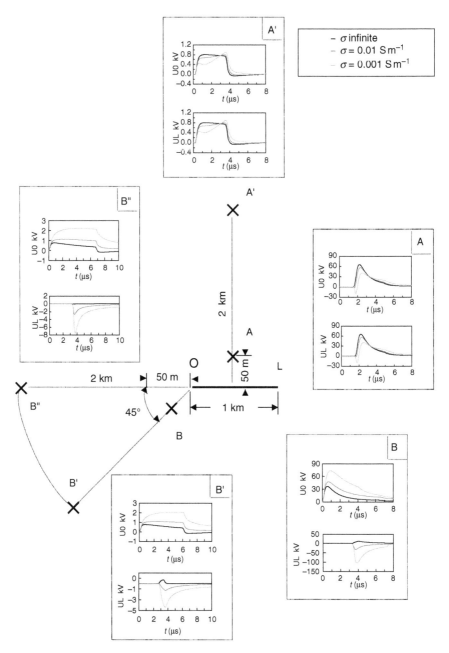

Figure 13.4 Effects of ground resistivity and position of the stroke location with respect to the line on the amplitude and waveshape of lightning-induced voltages. Solid line, perfect ground; dashed line, ground conductivity $0.01\ S\ m^{-1}$; dotted line, ground conductivity $0.001\ S\ m^{-1}$ (adapted from Reference 43).

infinite, 0.01 and 0.001 S m^{-1}. The current parameters (peak value I_p and maximum front time-derivative) are $I_p = 12$ kA and $(di/dt)\max = 40$ kA μs^{-1}, which correspond to a typical subsequent return stroke current according to Berger and colleagues (50 per cent of probability of occurrence of maximum di/dt [39]). It can be seen that the amplitude and waveshape of the induced voltages are indeed strongly dependent upon the relative distance between the lightning strike and the line, as well as on the ground resistivity. In Reference 31 it is also shown that – as expected – the larger is the current amplitude, the larger the induced overvoltage.

It can also be shown that, in general, for the same lightning current amplitude, the larger the ground resistivity, the larger the amplitude of the induced overvoltages. Theoretical and experimental evidence of this conclusion – somewhat unexpected – is presented in several papers [38–43] (see Reference 31 for a comprehensive review).

The points summarized above are fundamental to understanding the results that we shall present in Section 13.5.3.

Note, also, that buildings may reduce the lightning electromagnetic field around the line and, consequently, the induced voltage magnitudes. On the other hand, tall structures may attract lightning discharges very close to the line and therefore large overvoltages may be induced even in the case of stroke currents of moderate intensity.

As a final comment, it is worth mentioning that the topological complexity of real distribution networks requires that the influence of the network topology on the lightning-induced voltages be taken into account. An analysis of the influence of the presence of branches on the lightning-induced voltages has been carried out [45] for different cases. It has been shown that the magnitude and waveshapes of the induced voltages are strongly dependent on the topology. Therefore, in principle, the reduction of a network to a single line is not appropriate and it is important to take into account the actual configuration of the network in the evaluation of lightning-induced voltages [45]. The above is corroborated by experimental investigations conducted by means of a reduced scale model [47]. The induced voltages may be substantially affected by the line branches and by the presence of nearby buildings. In this chapter, however, we will limit the discussion to assuming straight overhead lines, as classically done in the literature on the subject.

13.4 Main principles in lightning protection of distribution lines

An effective reduction of the lightning fault rate of a MV line can be achieved with special measures, such as converting the overhead line into an underground cable or installing surge arresters at every pole and every phase. Other measures, such as the upgrading of the line insulation or the addition of a shield wire, may reduce the number of faults due to induced overvoltages. On the other hand, in order to mitigate the effects of direct strikes, the shield wire should be grounded at every pole, the ground resistances should be low and the critical impulse flashover voltage – to be discussed next – of the line structures should be greater than ∼250 kV. Clearly, the cost-effectiveness of these solutions should always be evaluated. In this respect, the availability of adequate modelling and computer codes is crucial.

13.4.1 Basic impulse insulation level and critical impulse flashover voltage

The basic lightning impulse insulation level (BIL) is the crest value of withstand voltage when insulation is subject to a standard lightning impulse, for dry conditions. It is also known as the lightning impulse withstand voltage (IEC).

For self-restoring insulations, the BIL is statistical, namely the crest value of standard lightning impulse for which the insulation exhibits a 90 per cent probability of withstand.

For non-self-restoring insulations, the BIL is conventional, namely the crest value of standard lightning impulse for which the insulation withstands for a specific number of applications of the impulse.

The standard lightning impulse waveshape, shown in Figure 13.5, has a time to crest equal to 1.2 μs and a time to half value equal to 50 μs. These times are evaluated by constructing the linear characteristic passing through the times corresponding to 30 and 90 per cent of the crest value; the time corresponding to zero voltage on this characteristic is the virtual origin. The time to crest is the time interval between the virtual origin and the time corresponding to the crest voltage on the linear characteristic. The time to half value is the time interval between the virtual origin and the time at which the voltage decreases to 50 per cent of the crest value. As known, the standard lightning impulse waveshape has been chosen in order to reproduce the short fronts and the relatively long tails of lightning surges, but above all because it may be easily produced in all laboratories.

The critical impulse flashover voltage (CFO) is the crest value of the standard lightning impulse wave that causes flashover through the surrounding medium on 50 per cent of the applications. If a Gaussian distribution of flashover data is assumed, then any specific probability of withstand may be calculated from the CFO value and the standard deviation.

Figure 13.6 shows the BIL and CFO for an insulating system having the probability of flashover described by the solid curve.

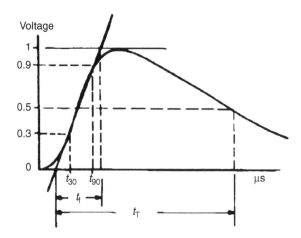

Figure 13.5 Standard lightning impulse waveshape (adapted from Reference 3)

Lightning protection of medium voltage lines 647

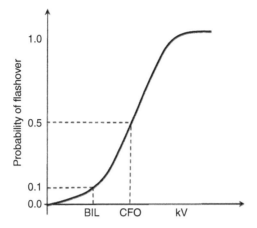

Figure 13.6 Basic impulse insulation level and critical flashover voltage (adapted from Reference 3)

It is important to recall that because the construction of the MV line, including its insulation, makes use of different materials, such as wood, concrete, polymer, porcelain, fibre glass, air, and so on, the insulation level of a given line will be a function of the levels associated with the different components. A typical example of a pole of a distribution line is shown in Figure 13.7.

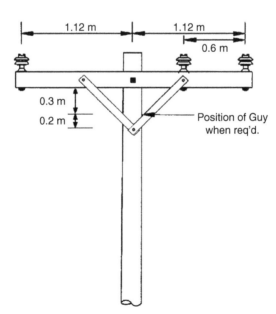

Figure 13.7 Pole of a MV overhead line with wooden cross arm design (adapted from Reference 4)

When insulating materials are used in series, the resulting insulation level will be lower than the summation of the single levels. The estimation of such an insulation level is beyond the scope of this chapter. It is worth mentioning, however, that the lightning flashover levels of distribution lines are affected by several factors, namely atmospheric conditions (air density, humidity, rainfall, etc.) and contamination, the polarity and steepness of the voltage, the line/insulator configuration (mounted vertically, horizontally or at some angle). Also, if wood is present in the lightning-originated discharge path, the insulation may respond in a quite variable way, depending basically on the moisture on the wood surface.

Historically, electrical engineers have been constructing distribution lines using wooden cross arms and poles in series with basic insulators to increase the BIL/CFO of the distribution line insulation. A number of important studies have been carried out to investigate how much lightning-voltage insulation the wood adds to the primary insulation (the insulators). Most of these papers are referred to in Reference 4 in the section devoted to distribution-line insulation level. Also, in Reference 4, a complete section on how to determine the CFO voltage of structures with series insulation, a subject that – as previously mentioned – is beyond the scope of this chapter.

As a final important remark concerning this matter, we feel it worth mentioning that, in general, equipment and support hardware on distribution structures may severely reduce the CFO. As mentioned in Reference 4, these 'weak link' structures may greatly increase flashovers from induced voltages. Some elements are given in the following.

Guy wires (stranded cables used for a semiflexible tension support between a pole or structure and the anchor rod, or between structures), for example, may play a major role in reducing a structure's CFO, because they provide a path to the ground. On the other hand, a fibre-glass strain insulator may provide an increase of the CFO (a 50-cm fibre-glass strain insulator has a CFO of $\sim$250 kV [4]).

Mounting of fuse cutouts may lower a pole's CFO and reduce – depending on how it is mounted – the CFO of the entire structure for a 15 kV class system to 95 kV BIL [4]. A good practice on wooden poles is to arrange the attachment brackets of the cutouts on a pole away from any grounded conductors (guy wires, neutral wires, etc.).

Neutral wires can reduce the CFO too. On wooden poles, the closer the neutral wire to the phase wires, the lower the CFO.

The use of concrete for poles, which is a common practice in some European countries, greatly reduces the CFO. For these structures, practically the whole insulation is supplied by the insulators, which are therefore to be selected with high CFO.

13.4.2 Shield wires

Shield wires are grounded conductors placed close to the phase conductors with the purpose of (i) intercepting lightning return strokes that would otherwise directly strike the phases (protection against direct return strokes) or (ii) reducing, or at least exercising some control on, the electric and magnetic fields that affect the voltage between the phase conductors and the local ground (protection against indirect strokes).

Concerning protection against direct strokes, shielding is used successfully for transmission networks, but it is usually accepted that it is not effective in the same way when applied to distribution lines. A shield wire may reduce the number of flashovers caused by nearby lightning, but in order to be effective against direct strikes it should be grounded at every pole and the ground resistances should be very low. The ground resistance has to be low in order to avoid the back-flashover phenomenon. As a matter of fact, when lightning surge current flows through the pole grounding resistance/impedance, it causes a potential rise that results in a large voltage difference between the ground lead and the phase conductor; such a voltage difference may cause a flashover (the back-flashover) across the insulation from the ground lead to the phase conductor, which represents a major constraint to the effectiveness of shield wires against direct strokes. The lower the lightning impulse strength of the line, the lower the ground resistance value should be.

In summary, the application of shield wires may provide effective protection only if the following apply:

1. Good insulation design practices are used to provide sufficient withstand voltage between the local ground and the phase conductors.
2. Good design practices ensure that most lightning strikes would terminate on the shield wire rather than on the phase conductors.
3. Sufficiently low pole ground resistances are obtained.

It is worth mentioning that in some cases the application of protective devices, to be discussed next, is needed in order to make the use of shield wire effective against direct strokes.

Concerning the protection against lightning-induced overvoltages, the purpose of the shield wire is essentially that of an electromagnetic shield: it affects the induced voltage at the phase conductors through capacitive and inductive coupling. The closer the phase conductors are to the shield wire, the better the coupling and the smaller the induced voltages tend to be, independently of the shield wire being above or below the phase conductors. An example mentioned in Reference 20 shows a reduction of the peak values of the induced voltages due to the presence of shield wires in the range of 25–35 per cent. Similar values are reported in Reference 49.

Note, further, that the final structure height – larger, as the pole height has to be greater to support the shield wire in order to guarantee a sufficient shielding angle (see Chapter 4 of this book) between the shield wire and the other conductor – results in a larger attraction of direct strokes, which may be able to offset the consequent flashover rate reduction.

Although expensive and requiring major design efforts, shield wires have been used by quite a few utilities with some success.

13.4.3 Protective devices

Protective devices are intended to limit transient overvoltages and divert surge currents. They contain at least one nonlinear element. Depending on their principal function they are divided into voltage-switching types (typical example being a spark gap), and voltage-limiting types [typical examples being varistors or metal oxide (MO) arresters].

Capacitors reduce the steepness of incoming overvoltages, and short circuit overvoltages of very high frequencies. In most of the today's applications, gapless MO arresters are used exclusively. In some special cases in MV networks, nevertheless, combinations of gaps and MO arresters are used.

13.4.3.1 General considerations using protective devices [48]

The protective device limits the voltage at its terminals to the residual voltage U_{res}. Voltage reflections along the connections a, between line and apparatus, and b, between line and protective device, cause an apparatus voltage U_a higher than U_{res} (Figure 13.8). The voltage difference $\Delta U = U_a - U_{res}$ increases with the lengths a and b and with the steepness S of the incoming overvoltage. As S (overvoltage steepness) can reach very high values in MV lines, the overvoltage at the apparatus ($U_a = U_{res} + \Delta U$) can also reach large values. In order to reduce the voltage U_a at the apparatus, ΔU must be at a minimum by selecting a and b to be as short as possible. The following formula [48] can be used as a reasonable approximation for U_a:

$$U_a = U_{res} + \frac{2 \times S \times (a+b)}{v} \quad (13.6)$$

where v is the speed of light (300 m μs^{-1}).

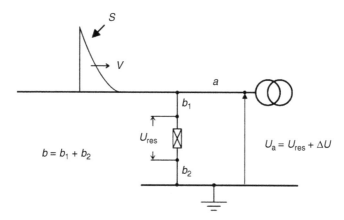

Figure 13.8 Sketch of the protection scheme of a MV transformer directly connected to an overhead line. The protective device limits the voltage at its terminals to the protective level U_{res} (adapted from Reference 49).

As guidance, values of b lower than 1 or 5 m are recommended for the case of wooden pole lines and earthed cross-arm lines, respectively.

A rough estimate of the acceptable maximum distance between the terminals of the protective device and the apparatus to be protected can be obtained based on equation (13.6).

Experience has shown that a safety factor of 1.2 is sufficient between the BIL of the electric apparatus and the maximum occurring voltage U_a at the apparatus:

$$\frac{\text{BIL}}{1.2} \geq U_a = U_{\text{res}} + \frac{2 \times S \times (a+b)}{v} \tag{13.7}$$

If the limiting value is set at $L = a + b$, then the maximum distance can be calculated from

$$L = \frac{v}{2 \times S}\left(\frac{\text{BIL}}{1.2} - U_{\text{res}}\right) \tag{13.8}$$

All the above considerations are based on the obvious assumption that the apparatus and the protective device are connected to the same grounding system. If they were connected to separate grounding systems, the protection will in most cases become useless due to the ground potential rise at the protective device.

Based on these considerations two basic rules have to be followed [48]:

1. The apparatus and the protective device shall be connected to the same grounding system. The galvanic connection between the earth side of the protective device and of the apparatus should be as short as possible.
2. The total length along the connection a between line and apparatus and b between line and protective device should be as short as possible.

13.4.3.2 Spark gaps

A spark gap is an intentional gap or gaps between spaced electrodes. Two different designs have to be considered: the so-called arcing horn, where the insulating gap between the electrodes is in open air, and plate spark gaps, where the spark gap elements are moulded in an insulating and gas-tight material, providing a controlled gas atmosphere between the electrodes.

The protection against lightning overvoltages is given by an intentional disruptive discharge between the electrodes, leading to a collapse of the voltage and passage of current.

In Figure 13.9, a typical application of an arcing horn is shown. The shape of the electrodes is designed to elongate and cool the arc, in order to facilitate its extinction. A third electrode at floating potential is sometimes installed in the middle between the two main electrodes. The main purpose of this electrode is to prevent birds from causing a short circuit of the gap when there are no overvoltages.

Spark gaps with controlled gas atmosphere have one or more plate spark gaps in series, moulded in gas-tight and insulating material. Thus, the sparkover voltage is not influenced by external factors like humidity, pressure and pollution.

652 Lightning Protection

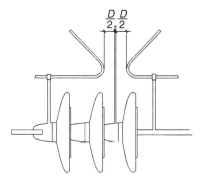

Figure 13.9 *Typical shape of an arcing horn for protection of a MV apparatus, mounted on a rigid string of three cap and pin insulator units (adapted from Reference 48)*

13.4.3.3 Surge arresters

Two principally different designs of arresters are installed in MV networks. The so-called 'conventional' surge arresters were exclusively used in MV networks up to the middle of the 1980s. They consist of a series connection of SiC resistors and plate spark gaps. In the case of an overvoltage, the spark gap would flash over and the follow current from the system would be limited by the SiC resistors, and extinguished in the first natural current zero. The disadvantages of this technology are the unfavourable voltage–time characteristic of the spark gaps and the limited

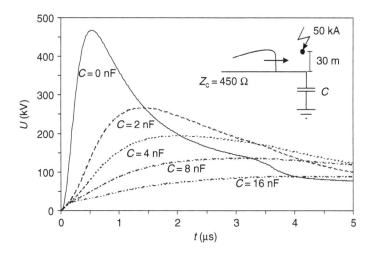

Figure 13.10 *Overvoltage induced by a lightning stroke to the ground on the termination of an overhead line, as a function of the value C of a lumped capacitance connected to the line termination (adapted from Reference 50)*

energy capability of SiC resistors. This type of arrester is no longer manufactured, but is still installed in MV systems in a large numbers.

In the 1990s there were two fundamental improvements of surge arrester technology. On the one hand, the series connection of spark gaps and SiC resistors were replaced by metal oxide (MO) resistors without gaps, and, on the other hand, the porcelain housings were replaced by polymer housings.

A fundamental advantage of MO arresters is the fact that, because of the extreme nonlinear voltage–current characteristic of the MO material, the leakage current is negligibly low, so they do not need any spark gap.

13.4.3.4 Capacitors

Capacitors have a frequency-dependent impedance and can be used to short circuit high-frequency voltages. A capacitance to earth of an apparatus has, in general, a positive effect in reducing the steepness of the voltage in the same time as the amplitude of induced voltages, as shown in Figure 13.10.

13.5 Lightning protection of distribution systems

As mentioned earlier, lightning may cause flashovers on distribution lines from both direct strikes and induced voltages from nearby strikes. However, experience and observations show that many of the lightning-related outages of low-insulation lines are due to lightning that hits the ground in the vicinity of the line [4]. Moreover, due to the limited height of distribution lines of MV and LV distribution networks compared to that of the structures in their vicinity, indirect lightning strokes are more frequent events than direct ones. For this reason, the literature on this subject (see bibliography of Reference 4) focuses mostly on this type of lightning event. In the following sections we will thus limit our discussion to deal with this type of disturbance.

13.5.1 Effect of the shield wire

To assess the effect of the shielding wire on the induced overvoltages, let us consider the line geometry presented in Figure 13.11 [32]. Note that, as already demonstrated in References 31 and 33 the shield wire cannot be viewed as a zero-potential conductor and has to be considered as one of the conductors of the multiconductor overhead line. Therefore, the distance between two consecutive groundings and the value of the grounding resistance will play an important role in the performance of the shield wire.

The computed peak amplitudes of the induced voltages along the line are presented in Figure 13.12 for the case of a perfectly conducting ground. In the same figure, we also present the results obtained using the Rusck expression given by Reference 26:

$$\eta = \frac{U'_i}{U_i} = 1 - \frac{h_{sw}}{h_i} \cdot \frac{Z_{sw-i}}{Z_{sw} + 2R_g} \quad (13.9)$$

where h_i is the height of conductor i, h_{sw} the height of the shielding wire, R_g the grounding resistance of the shielding wire footing, U_i the lightning-induced voltage

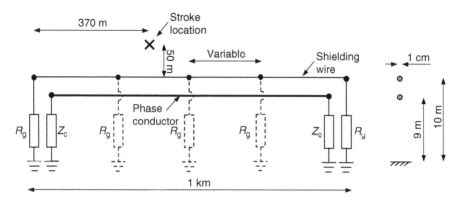

Figure 13.11 Configuration adopted to evaluate the mitigation effect of shielding wires on lightning-induced voltages. Lightning current: maximum amplitude 30 kA maximum time derivative 100 kA μs^{-1} (adapted from Reference 32).

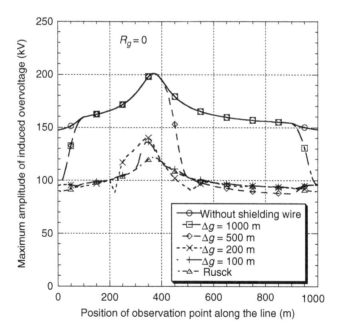

Figure 13.12 Maximum amplitude of the induced voltage along the line as a function of the spacing (Δg) between two adjacent groundings of the shielding wire. Line configuration of Figure 13.11, stroke location 50 m from the line and 370 m from left line termination. Perfectly conducting ground. Lightning current: maximum amplitude 30 kA maximum time derivative 100 kA μs^{-1} (adapted from Reference 32).

amplitude on the ith conductor in the absence of the shielding wire, U'_i lightning-induced voltage amplitude on ith conductor in the presence of the shielding wire, Z_{sw} the surge impedance of the shielding wire and Z_{sw-i} the mutual surge impedance between the shielding wire and the ith conductor.

It is worth recalling that the Rusck expression does not cover the case of multiple groundings of the shielding wire (it assumes that the shielding wire is at ground potential) and, furthermore, it assumes the ground to be perfectly conducting. [As shown in References 53 and 54, however, when used for line configurations and stroke locations consistent with the assumptions with which it was derived, the Rusck expression, equation (13.9), provides good results.]

In Figure 13.12, we can see that an overall effective protection of the line can be achieved only if the spacing between two consecutive groundings is less than ~200 m. This value approximately corresponds to the rise time of the lightning electromagnetic field illuminating the line, for the assumed lightning current waveshape. For larger values of spacing (namely 500 and 1 000 m), only the portion of the line in the immediate vicinity of the grounding points appears to be protected.

From Figure 13.12, one can also see that the Rusck expression gives quite accurate results only for short spacings between two adjacent groundings (less than 200 m) and when assuming a perfectly conducting ground.

Figure 13.13a, b presents similar computed results as those presented in Figure 13.12, but considering a finite value for the ground conductivity equal to 0.001 S m^{-1}. In Figure 13.13a we adopt the value of 10 Ω for the grounding resistance, while in Figure 13.13b the value of 300 Ω is adopted. These values correspond to two different lengths of the grounding electrodes. The results presented in Figure 13.13a, b show clearly that the mitigation effect of the shielding wire depends, in general, more on the spacing between two consecutive groundings rather than on the value of the grounding resistance. This differs from the case of direct stroke for which the effectiveness of the shielding wire depends strongly on the grounding resistance. Additional calculation results show that it is only when the grounding resistance becomes poor (100 Ω or larger) that it starts to affect in a more significant way the distribution of the induced voltage along the line.

Note, additionally, that it is mainly when the stroke location is located in front of a grounding point that the attenuation of the induced voltage on the phase conductors is very dependent on the value of the grounding resistance.

13.5.2 Effect of surge arresters

A typical surge arrester $V-I$ characteristic is shown in Figure 13.14.

To compare the mitigation effect of surge arresters with that of the shielding wire, we consider the same single-conductor line configuration of Figure 13.11 (see Figure 13.15), as done in Reference 33. This is equivalent to assuming that the differential-mode coupling among the line phases is negligible and that the common-mode voltage induced on the three phases is the same. This is certainly a more reasonable assumption for the case of a line illuminated by a lightning electromagnetic field (the illumination is basically the same for the three conductors), rather than for a line

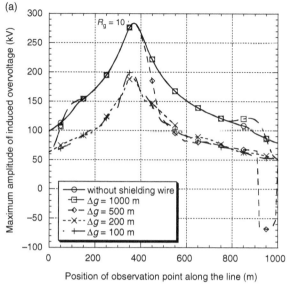

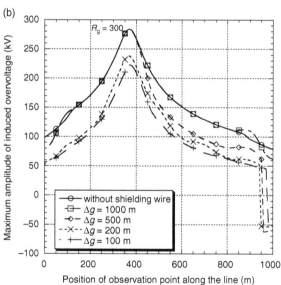

Figure 13.13 Maximum amplitude of the induced voltage along the line as a function of the spacing (Δg) between two adjacent groundings of the shielding wire. Line configuration of Figure 13.11, stroke location 50 m from the line and 370 m from left line termination. Lossy ground $\sigma = 0.001\ S\ m^{-1}$. Lightning current: maximum amplitude 30 kA maximum time derivative 100 kA μs^{-1}. (a) Grounding resistance equal to 10 Ω (b) Grounding resistance equal to 300 Ω (adapted from Reference 32).

Lightning protection of medium voltage lines 657

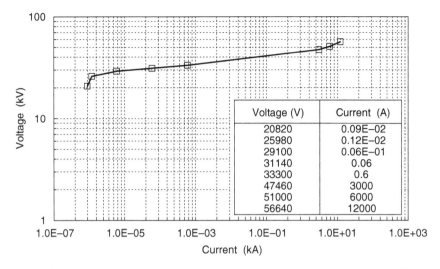

Figure 13.14 Surge arrester V–I characteristic (adapted from Reference 32)

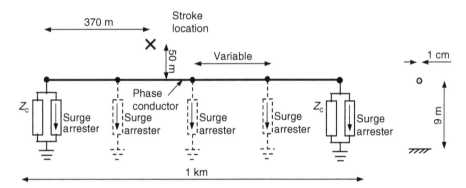

Figure 13.15 Configuration adopted to evaluate the mitigation effect of surge arresters on lightning-induced voltages. Lightning current: maximum amplitude, 30 kA maximum time derivative 100 kA μs^{-1} (adapted from Reference 32).

directly struck by lightning (the lightning current in this case can be injected on one conductor of the transmission line).

A variable number of surge arresters placed along the line is assumed: 2 (at the line terminal only); 3 (every 1 000 m); 5 (every 500 m) and 11 (every 200 m). The amplitudes of the induced overvoltages along the line are shown in Figure 13.16 (perfectly conducting ground) and Figure 13.17 (lossy ground).

The computed results show that an important reduction of the induced overvoltages' amplitude can be achieved only with a large number of surge arresters,

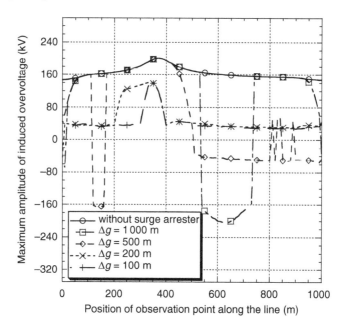

Figure 13.16 Maximum amplitude of the induced voltage along the line as a function of the spacing between two adjacent surge arresters (Δsa). Line configuration of Figure 13.15, stroke location 50 m from the line and 370 m from left line termination. Perfectly conducting ground. Lightning current: maximum amplitude 30 kA, maximum time derivative 100 kA μs^{-1} (adapted from Reference 32).

namely one surge arrester every 200 m. It can also be seen that for some configurations with a low number of surge arresters (e.g. one every 1 000 m), their presence could result in important negative peaks of the induced voltage, which are due to surge reflections occurring as a result of surge arrester operation. Indeed, depending on the line configuration, on the stroke location and on the distance between two consecutive surge arresters, the negative voltage wave due to the arrester non-linear characteristic makes it possible for the largest amplitude of the induced overvoltage to occur at a point on the line different from that closest to the stroke location. In addition, this overvoltage can be more severe than the maximum voltage amplitude induced in the absence of surge arresters (see Figure 13.17).

By increasing the number of surge arresters, the maximum amplitude of the induced overvoltage tends to be confined within the range defined by the positive and negative values of the threshold voltage of the surge arrester's non-linear $V-I$ characteristic (see Figure 13.14).

The influence of some line and lightning parameters on the effectiveness of surge arresters in terms of the reduction of the induced voltage magnitudes is analysed by means of scale model experiments in Reference 55.

Lightning protection of medium voltage lines 659

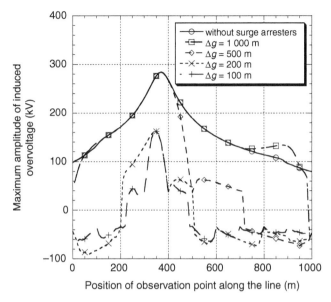

Figure 13.17 *Maximum amplitude of the induced voltage along the line as a function of the spacing between two adjacent surge arresters (Δsa). Line configuration of Figure 13.15, stroke location 50 m from the line and 370 m from left line termination. Lossy ground $\sigma = 0.001$ S m^{-1}. Lightning current: maximum amplitude 30 kA, maximum time derivative 100 kA μs^{-1} (adapted from Reference 32).*

Figure 13.18 *Lightning performance of a distribution line*

In summary, we can conclude the following:

- The effectiveness of shielding wires and surge arresters depends mostly on the spacing between two adjacent grounding points or surge arresters.
- The Rusck expression allows for an accurate prediction of the mitigation effect of the shielding wire only when the number of groundings is large.

13.5.3 Lightning performance of MV distribution lines

The results presented in the two previous subsections, along with the main conclusions of Section 13.3, represent the basis for understanding the matter discussed in the present section.

What power engineers need in order to improve the lightning performance of a given distribution line is the possibility to compute the annual number of flashovers that the line will experience as a function of the line design/configuration, line CFO and annual ground flash density. This is generally expressed by means of a plot in an x–y system having in the horizontal x-axis the CFO of the line and in the vertical axis the number of flashovers, as shown in Figure 13.18. Such a computation can be rather complex and some procedures have been proposed in the literature [4,54].

These procedures start from the statistical distribution of lightning current parameters, assume a lightning incidence model to distinguish between direct and indirect strokes (e.g. model 2b of Section 13.2), make use of a model for the calculation of the lightning-induced voltages and, by means of a iterative/statistical process in which the stroke location and lightning current parameters are randomly varied, infer the number of lightning-induced voltages capable of resulting in line flashover.

In Appendices A13 and B13 we report two of these computation procedures, namely the one presently adopted by the IEEE in Reference 4, and the one, more general, proposed in Reference 54, which we will denote for convenience LIOV–MC.

Although all the details are reported in the mentioned Appendices, for sake of clarity it is worth summarizing the main differences between the two procedures:

- the different number of lightning current parameters taken into account in the statistical process (peak value I_p only for the IEEE one, I_p and time to peak value t_f – including their correlation coefficient – for the LIOV–MC one);
- the different model for the calculation of the lightning-induced voltages [simplified Rusck formula equation (A13.3) of Appendix A13 for the IEEE procedure, versus the coupling model by Agrawal and colleagues described in Appendix C13 for the LIOV–MC procedure];
- the taking into account of the ground resistivity in the calculation of the exciting electromagnetic field in the LIOV–MC procedure, inherently disregarded in the IEEE one as it makes use of the above-mentioned simplified Rusck formula, equation (A13.3), which applies to an infinitely long line above a perfectly conducting soil and for a step-like current waveshape for the lightning current.

On the other hand, the incidence model is the same used for both procedure (model 2b of Section 13.2) and the statistical approaches, as shown in Reference 56, are also equivalent.

Figure 13.19 shows a comparison between the line flashover rate presented in the IEEE Std. (curve A) and that obtained by using the LIOV–MC procedure (curve B), for the case of a 10-m-high, infinitely long line consisting of a single conductor above an ideal ground. (Note that the LIOV–MC procedure is applied to a 2-km-long line matched at both terminations, with stroke locations equidistant from the line ends; this configuration is assumed to be equivalent to an infinitely long line [31] – considered in Reference 4 – and is adopted throughout this section.) Figure 13.19 includes additional curves with a lossy ground, which will be discussed later in this section.

Note that when the line is lossless and situated above a perfectly conducting ground, the two methods used to compute the lightning-induced voltages (Rusck simplified formula and Agrawal and colleagues model) are equivalent [55]. The differences between the two curves (A) and (B) are ascribed to the fact that while the Rusck simplified formula assumes a step-like current waveshape for the lightning current, in the LIOV–MC procedure both the peak value of the lightning current and its front time are varied. In fact, by forcing t_f to be constant and equal to a small value, namely 1 µs, the LIOV–MC computation, curve (B), moves upwards and is nearly superimposed on curve (A), as shown in Reference 54.

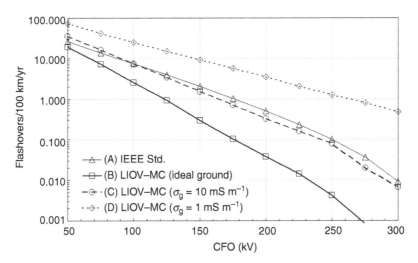

Figure 13.19   Comparison between the line flashover rate of the IEEE Std. (A) and those obtained by using the proposed procedure (LIOV–MC), for the case of a single-conductor 10-m-high infinitely long line, above an ideal ground (B) and above a lossy ground: (C) for ground conductivity $\sigma_g = 10\ mS\ m^{-1}$; (D) for $\sigma_g = 1\ mS\ m^{-1}$ (adapted from Reference 54).

13.5.3.1 Effect of soil resistivity

The additional curves reported in Figure 13.19 refer to the results obtained by applying the proposed procedure when taking into account the more realistic case of a lossy ground with conductivity σ_g equal to 10 mS m^{-1} (curve C) and equal to 1 mS m^{-1} (curve D). As known [56], when evaluating lightning-induced voltages, the finite value of the ground conductivity on the one hand increases the transient propagation losses in the line but, on the other hand, also has an influence on LEMP propagation. Although the former effect tends to decrease the surges propagating along the line, the latter tends to enhance the amplitude of the induced voltages. It is this second effect that, overall, results in induced voltage amplitudes higher than those calculated for the case of an ideal ground [31,39,43,56], as mentioned earlier in Section 13.3.

13.5.3.2 Effect of the presence of shield wires

For the calculations of Figure 13.20, a single-conductor line has been considered. Now, a grounded conductor (shielding or neutral conductor) is added. The IEEE method takes into account the presence of a shielding or neutral conductor by using a shielding factor η expressed by equation (13.9). In the LIOV–MC method the grounded conductor is dealt with by considering it in the same way as the other conductors of the multi-wire lines [30], and by taking into account the various groundings by means of a lumped resistance [32], assumptions in line with the current practice for

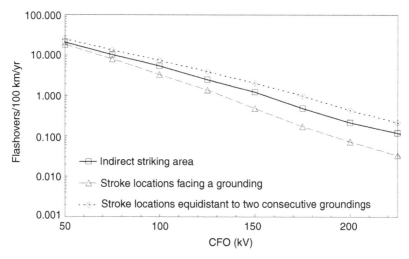

Figure 13.20   *Line flashover rates obtained by using LIOV–MC in the presence of a shielding wire grounded every 500 m for three cases: (i) for stroke locations spread over an indirect striking area; (ii) for stroke locations all facing a grounding; and (iii) for stroke locations equidistant to two consecutive groundings (adapted from Reference 54).*

engineering calculations of this type [32,57]. For illustrative purposes, the line configuration with the same shielding factor value used for the relevant results shown by the IEEE Std, namely 0.75, is assumed. Note that such a value can be obtained by several combinations of line configurations and grounding resistances (R_g). In the presented calculations, the line is assumed to be composed of a 10-m-high conductor, with diameter equal to 1 cm, with a shielding wire placed 8.37 m above a perfectly conducting ground, having the same diameter, and $R_g = 0$, in accordance to the ideal ground assumption.

The shielding or neutral wire is then a LEMP-coupled conductor grounded at regular intervals, whereas in the IEEE Std. such a wire is assumed to be a non-illuminated conductor with continuous grounding connections. It is therefore important to take into account not only the distance (y) of the stroke location from the line but also the relative position of the stroke location with respect to the wire groundings, with particular reference to the closest ones. For that purpose, in the LIOV–MC procedure, the random stroke locations are, for this case, uniformly spread within a given surface that surrounds the line (henceforth called indirect striking area). Each lightning stroke location is therefore characterized by two coordinates, x and y, which completely define its relative position with respect to the line.

Figure 13.20 shows the results obtained by using LIOV–MC for a line having a wire grounded every 500 m for three different cases: (i) for stroke locations spread over an indirect striking area including all strokes at a distance equal to or less than 1 km from the line; (ii) for stroke locations all facing a grounding; and (iii) for stroke locations all aligned at the same distance from two consecutive groundings. As expected, the curve of case (ii) tends to overestimate the shielding effect of the ground wire and, opposite to that, the curve of case (iii) tends to underestimate such an effect.

The influence of the spacing between adjacent groundings on the lightning performance of the line is shown in Figure 13.21, where the flashover rate curve of the IEEE method (curve A) and those obtained by using LIOV–MC, with the shielding wire grounded every 30 m (curve B), and grounded every 500 m (curve C) are presented (the same procedure adopted to infer the curve of case (i) of Figure 13.20 has been followed). Both curves B and C are obtained by forcing t_f to be equal to 1 µs, in order to make the comparison consistent and to emphasize the impact of the grounding spacing on the results. Figure 13.21 shows that the η equation (13.9) gives quite accurate results only for short spacing values between two adjacent groundings, as expected.

Note that the results of Figures 13.20 and 13.21 have been obtained by assuming that the flashover occurs only from the phase conductor to ground. In principle, however, the line could experience flashovers between the phase conductor and the grounded conductor too. Figure 13.22 shows the flashover rates calculated by considering the two different flashover paths, namely the phase-to-ground path and the phase-to-grounded wire paths. Note that the results of Figure 13.22 must be interpreted by keeping in mind that the two different flashover paths are characterized by different CFOs, especially for wooden poles and cross arms. If we assume, for instance, a CFO

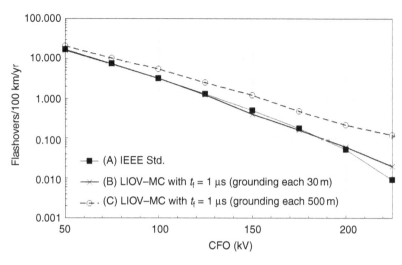

Figure 13.21 Comparison between the line flashover rate curve of IEEE Std. (A) and those obtained by using LIOV–MC, enforcing $t_f = 1$ μs for each event, for two different shielding wire grounding spacings, 30 m (B) and 500 m (C) (adapted from Reference 54).

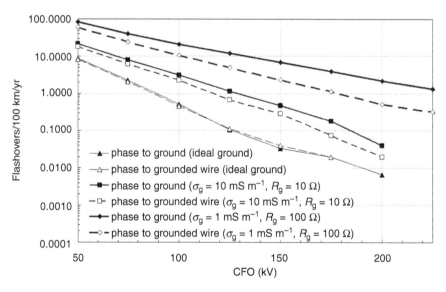

Figure 13.22 Comparison between phase-to-ground and phase-to-grounded-wire flashover rate curves calculated for different ground conductivity σ_g and grounding resistance R_g. Results were obtained by using LIOV–MC for the case of a line with a shielding wire grounded every 200 m and stroke locations spread over the indirect striking area (adapted from Reference 54).

of 200 kV for a phase-to-ground path and a CFO of 130 kV for a phase-to-grounded wire path, we obtain 2.2 flashovers/100 km/year and 4.8 flashovers/100 km/year (and not 0.52 flashovers/100 km/year as would the case by improperly assuming the same CFO for the two cases).

The results in Figure 13.22 are also shown for various values of ground conductivity and grounding resistance. The simulations are carried out assuming a linear model for the grounding impedance of the neutral or shielding wire.

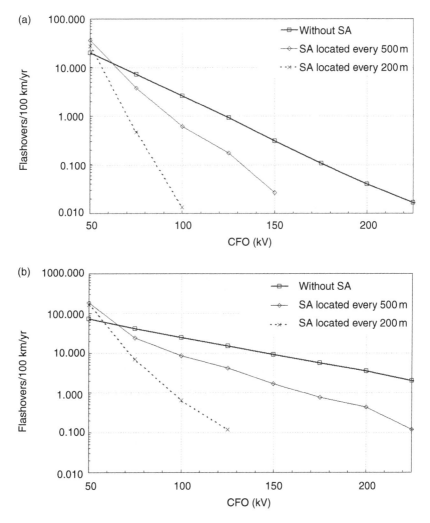

Figure 13.23 Line flashover rate curves obtained by using LIOV–MC for the case of a line with and without surge arresters located every 200 and 500 m: (a) ideal ground and (b) lossy ground ($\sigma_{g\;0.01} = 1\; mS\; m^{-1}$) (adapted from Reference 54).

13.5.3.3 Effect of the presence of surge arresters

With the LIOV–MC procedure, more realistic line configurations than those considered in the IEEE method of Reference 4, for instance including the presence of active protection devices such as surge arresters can be analysed. This subsection is devoted to illustrate this point.

Again the 10-m-high, single-conductor line, with surge arresters placed at regular intervals along the line, is considered. According to the indications reported in Reference 57, the surge arresters are modelled using a V–I non-linear characteristic, which was obtained by the standard 1.2/50 µs pulse test on a typical 20 kV surge arrester [58].

Figure 13.23 shows the influence of the presence of the surge arresters and of the spacing between two consecutive surge arrester stations, for ideal ground and lossy ground. The results show that a significant improvement in the lightning performance of the considered distribution line can be obtained by reducing the spacing between the surge arresters below 300 m, as expected from that illustrated in previous sections.

Appendix A13 Procedure to calculate the lightning performance of distribution lines according to IEEE Std. 1410-2004 (from Reference 4)

For the case of a line without a shielding conductor, the procedure adopted in IEEE Std 1410 [4] for the calculation of indirect lightning performance of a line is defined by the following steps.

1. The assumed range of peak values of lightning current I_p at the channel base from 1 to 200 kA is divided into 200 intervals of 1 kA. Note that such a current is assumed to have a step-like waveshape, as use is made of the simplified Rusck formula (see below).
2. For each interval i, the probability p_i of current peak value I_i to be within interval i is found as the difference between the probability for the current to be equal to or larger than the lower limit and the probability for current to reach or exceed the higher limit. These probabilities are obtained by using the formula

$$P(I_p \geq I_p^*) = \frac{1}{1 + (I_p^*/31)^{2.6}} \qquad \text{(A13.1)}$$

   which gives the cumulative probability of lightning current peak I_p to be equal or greater than value I_p^*.
3. For each interval i, two distances from the line (in metres) are also calculated: (i) the minimum distance $y_{\min,i}$ for which lightning of peak current I_i (in kiloamperes) will not divert to the line and (ii) the maximum distance $y_{\max,i}$ at which lightning may produce an insulation flashover, i.e. an induced voltage equal to the line critical flashover voltage CFO (in kilovolts), multiplied by a factor equal to 1.5 in order to take into account the turn-up in the insulation volt–time curve for short front-time surges.

- $y_{\min,i}$ is calculated by adopting the electrogeometric lightning incidence model 2b of Table 13.1.
- $y_{\max,i}$ is calculated as mentioned above by using the so-called Rusck simplified formula [59], which has been inferred by the same Rusck from the more general model proposed in Reference 26:

$$1.5 \cdot \text{CFO} = Z_0 \frac{I_i \cdot h}{y_{\max,i}} \left(1 + \frac{v}{\sqrt{2}v_0} \frac{1}{\sqrt{[1 - 0.5 \cdot (v/v_0)^2]}} \right) \quad \text{(A13.2)}$$

where $Z_0 = 1/4\pi\sqrt{(\mu_0/\varepsilon_0)} = 30\ \Omega$, h is the line height, v is the return stroke velocity and v_0 the speed of light (the ratio v/v_0 is assumed equal to 0.4). Note that such a formula applies for a single-wire, infinitely long, overhead line above a perfectly conducting ground plane. (It is worth mentioning that in Reference 60, starting from results obtained in References 43 and 31, an empirical extension of the above-mentioned Rusck formula, aimed at taking into account the effect of finite conductive ground on the amplitude of lightning-induced voltages, has been proposed.)

4. Finally, the number of annual insulation flashovers per 100 km of distribution line F_p is obtained as the summation of the contributions from all the current intervals

$$F_p = 0,2 \cdot \sum_{i=1}^{200} (y_{\max}^i - y_{\min}^i) N_g p_i \quad \text{(A13.3)}$$

where N_g is the annual lightning ground flash density (in $\text{km}^{-2}\ \text{yr}^{-1}$).

The results relevant to a line with grounded neutral or shielding wire are obtained from the preceding ones by using the following equation [same as (13.9)] (again proposed in Reference 26) which provides the ratio between the lightning-induced voltage on the line conductor U' and the value of the induced voltage on the conductor without the shielding wire U:

$$\eta = \frac{U'}{U} = 1 - \frac{h_{\text{sw}}}{h} \cdot \frac{Z_{\text{sw}-c}}{Z_{\text{sw}} + 2R_g} \quad \text{(A13.4)}$$

where h_{sw} is the height of the shielding wire, h is the height of the line conductor, Z_{sw} the surge impedance of the shielding wire, $Z_{\text{sw}-c}$ the mutual surge impedance between the shielding wire and the line conductor and R_g the d.c. grounding resistance. This formula has been obtained by assuming the grounded neutral or shielding wire to be a non-illuminated conductor with continuous grounding connections [26]. The results shown in the IEEE Std. refers to the case of $\eta = 0.75$ [61].

Appendix B13 The LIOV-Monte Carlo (LIOV–MC) procedure to calculate the lightning performance of distribution lines (from Reference 54)

This procedure [54] is based on the application of the Monte Carlo method and on the calculation of the induced voltages by using the LIOV code. This procedure will be referred to as the LIOV–MC. It is defined by the following steps.

1. A large number of lightning events n_{tot} is randomly generated. Each event is characterized by three parameters: I_p, t_f and the distance y from the line. The first two values, I_p and t_f, characterize the chosen lightning current waveform and are assumed to follow the log-normal probability distributions adopted by Cigré [20,62]. (Note that these statistical distributions have been inferred mostly from measurements obtained by using instrumented towers. The measurements at the towers are affected by reflections [63]. Moreover, the current amplitude distributions of the lightning events collected by towers are biased toward values higher than those of the distributions of the flashes to ground [14]. These aspects are deliberately disregarded in this chapter.) In particular, the distributions for negative first stroke parameters of Table B13.1 are adopted, with a correlation coefficient between t_f and I_p equal to 0.47 [62]. The distance of the perspective stroke location from the line y is assumed to be uniformly distributed as far as a value y_{max} (in km) – beyond which it is assumed that none of the lightning events could cause a flash on the line – is reached. The numerical procedure used for the random generation of I_p, t_f and y is described below, after point 3.

   From the total set of events, those relevant to indirect lightning are selected by adopting a lightning incidence model for the line. For the sake of the comparison

Table B13.1 Parameters of log-normal distribution for negative downward first strokes (from Reference 20)

Parameter	Median parameter value		Standard deviation value of the parameter logarithm (base e)	
	($I_p \leq 20$ kA)	($I_p > 20$ kA)	($I_p \leq 20$ kA)	($I_p > 20$ kA)
I_p (kA)	61 kA	33.3 kA	1.33	0.605
t_f (μs)	3.83 μs		0.553	

Only first return strokes are taken into account by the analysis. Further investigations are needed to include the effects of subsequent return strokes, solving several open issues, such as (i) the relationship between the subsequent stroke's path and that of the first stroke, (ii) the correlation between first and subsequent stroke current parameters, (iii) the number of subsequent strokes.

with the IEEE Std. results, the same electrogeometric lightning incidence model 2b of Table 13.1 is here adopted.

2. For each indirect lightning event, the maximum induced voltage value on the line is calculated – as earlier mentioned – by means of the LIOV code (see Appendix C13).

3. With n the number of events generating induced voltages larger than the insulation level – here assumed equal to the line critical flashover voltage (CFO), multiplied by a factor equal to 1.5, as in the IEEE Std. (see Appendix A13) – the number of annual insulation flashovers per 100 km of distribution line F_p is obtained as

$$F_p = 200 \cdot \frac{n}{n_{tot}} \cdot N_g \cdot y_{max} \quad (B13.1)$$

where N_g is the annual lightning ground flash density (in km^{-2} yr^{-1}).

As mentioned at point 1 above, for the application of such a procedure, a large number of lightning events n_{tot} is randomly generated. Each nth event is characterized by three values: I_{pn}, t_{fn} and y_n. The joint probability density function of random variables I_p and t_f are assumed to be of log-normal type [20,62]:

$$f(I_p, t_f) = \frac{1}{2\pi \cdot I_p \cdot t_f \cdot \sigma_{\ln I_p} \cdot \sigma_{\ln t_f} \cdot \sqrt{1-\rho^2}} \exp(-A) \quad (B13.2)$$

with

$$A = \frac{1}{2(1-\rho^2)} \left[\left(\frac{\ln I_p - \mu_{\ln I_p}}{\sigma_{\ln I_p}}\right)^2 + \left(\frac{\ln t_f - \mu_{\ln t_f}}{\sigma_{\ln t_f}}\right)^2 - 2\rho \left(\frac{\ln I_p - \mu_{\ln I_p}}{\sigma_{\ln I_p}}\right)\left(\frac{\ln t_f - \mu_{\ln t_f}}{\sigma_{\ln t_f}}\right) \right]$$

and where ρ is the correlation coefficient, $\mu_{\ln I_p}$ the mean value of $\ln I_p$ ($\mu_{\ln I_p} = \ln \bar{I}_p$, where $\bar{I}_p$ is the median value of I_p), $\sigma_{\ln I_p}$ the standard deviation of $\ln I_p$, $\mu_{\ln t_f}$ the mean value of $\ln t_f$ ($\mu_{\ln t_f} = \ln \bar{t}_f$, where $\bar{t}_f$ is the median value of t_f) and $\sigma_{\ln t_f}$ the standard deviation of $\ln t_f$.

The joint probability density function of the three random variables (I_p, t_f and y) can be expressed as the product of the density probability function of I_p, henceforth indicated by $f_1(I_p)$, times the conditional probability density function of t_f given I_p [$f_2(t_f \mid I_p)$], times the probability density function of y coordinate $f_3(y)$:

$$f(I_p, t_f, y) = f_1(I_p) \cdot f_2(t_f \mid I_p) \cdot f_3(y) \quad (B13.3)$$

with

$$f_1(I_p) = \frac{1}{\sqrt{2\pi} \cdot I_p \cdot \sigma_{\ln I_p}} \exp\left[-\frac{(\ln I_p - \mu_{\ln I_p})^2}{2\sigma_{\ln I_p}^2}\right] \quad (B13.4)$$

$$f_2(t_f \mid I_p) = \frac{1}{\sqrt{2\pi} \cdot t_f \cdot \sigma^*_{\ln t_f}} \exp\left[-\frac{(\ln t_f - \mu^*_{\ln t_f})^2}{2\sigma^{*2}_{\ln t_f}}\right] \quad \text{(B13.5)}$$

$$f_3(y) = \begin{cases} 1/y_{\max} & 0 < y < y_{\max} \\ 0 & \text{otherwise} \end{cases} \quad \text{(B13.6)}$$

where

$$\mu^*_{\ln t_f} = \mu_{\ln t_f} + \rho \frac{\sigma_{\ln t_f}}{\sigma_{\ln I_p}} (\ln I_p - \mu_{\ln I_p}) \quad \text{(B13.7)}$$

$$\sigma^*_{\ln t_f} = \sigma_{\ln t_f} \sqrt{1 - \rho^2} \quad \text{(B13.8)}$$

By applying the inverse transform method [56], if U_n, U_{n+1}, U_{n+2} are three uniformly distributed random variates over the interval (0,1), three values of I_p, t_f, and y, say I_{pn}, t_{fn}, and y_n, are obtained from

$$\begin{cases} F_1(I_p) = U_n \\ F_2(t_f \mid I_p) = U_{n+1} \\ F_3(y) = U_{n+2} \end{cases} \quad \text{(B13.9)}$$

where F_i (with $i = 1$ or 3) is the distribution function relevant to f_i and $F_2(t_f \mid I_p)$ is the conditional distribution function of t_f given I_p. However, because f_1 and f_2 are log-normal, it is convenient to use directly standard normal variates Z_n and Z_{n+1}, instead of U_n, U_{n+1}. Therefore, equations (B13.9) are solved by the following procedure:

1. For the calculation of I_{pn}:
   1.1 A standard normal variate Z_n over the interval (0,1) is generated, and then
   1.2 $I_{pn} = e^{a_n}$, where $a_n = \mu_{\ln I_p} + \sigma_{\ln I_p} \cdot Z_n$.
2. For the calculation of t_{fn}:
   2.1 $\mu^*_{\ln t_{fn}}$ and $\sigma^*_{\ln t_{fn}}$ are calculated by using equations (B13.7) and (B13.8),
   2.2 A standard normal variate Z_{n+1} over the interval (0,1) is generated, and then
   2.3 $t_{fn} = e^{b_n}$, where $b_n = \mu^*_{\ln t_{fn}} + \sigma^*_{\ln t_{fn}} \cdot Z_{n+1}$.
3. Finally, for the calculation of y_n:
   3.1 An independent uniformly random variate U_n over the interval (0,1) is generated, and then
   3.2 $y_n = U_n y_{\max}$.

Appendix C13 The LIOV code: models and equations

The LIOV (lightning-induced overvoltage) code is a computer program developed in the framework of an international collaboration involving the University of

Bologna (Department of Electrical Engineering), the Swiss Federal Institute of Technology (Power Systems Laboratory), and the University of Rome 'La Sapienza' (Department of Electrical Engineering). It is based on the Agrawal and colleagues [65] formulation of the field-to-transmission line coupling equations, suitably adapted for the calculation of induced overvoltages when lightning strikes near a horizontal overhead transmission line [33,66]. In the LIOV code, the electromagnetic field radiated by the lightning channel is calculated using the field equations in the form given by Uman and colleagues [67] with the extension to the case of lossy ground introduced by Cooray and Rubinstein [55,68] and assuming the modified transmission line return-stroke current model with exponential decay (MTLE) [69,71] for the description of the spatial-temporal distribution of the lightning current along the return-stroke channel.

The equations are solved in the time domain using the finite-difference time-domain (FDTD) technique. The LIOV code allows for taking into account the finite ground conductivity [66], the presence of multiconductor lines [47], shielding wires and surge arresters [32], the contribution of the downward leader phase of the lightning discharge that precedes the return stroke phase [71] and the corona effect [72].

The coupling with the distribution network is also dealt with through an interface realized with EMTPrv [73].

C13.1 Agrawal and colleagues field-to-transmission line coupling equations extended to the case of multiconductor lines above a lossy earth

Making reference to the geometry of Figure C13.1, the field-to-transmission line coupling equations for the case of a multi-wire system along the x-axis above an imperfectly conducting ground and in the presence of an external electromagnetic excitation are given by [30]

$$\frac{d}{dx}[V_i^s(x)] + j\omega[L'_{ij}][I_i(x)] + [Z'_{g_{ij}}][I_i(x)] = [E_x^e(x, h_i)] \quad \text{(C13.1)}$$

$$\frac{d}{dx}[I_i(x)] + [G'_{ij}][V_i^s(x)] + j\omega[C'_{ij}][V_i^s(x)] = [0] \quad \text{(C13.2)}$$

where

$[V_i^s(x)]$ and $[I_i(x)]$ are frequency-domain vectors of the scattered voltage and the current along line conductor i;

$[E_x^e(x, h_i)]$ is the vector of the exciting electric field tangential to the line conductor located at height h_i above ground;

[0] is the zero matrix (all elements equal to zero);

$[L'_{ij}]$ is the matrix of the per-unit-length line inductance. Assuming that distance r_{ij} between conductors i and j is much larger than their radii, the general expression

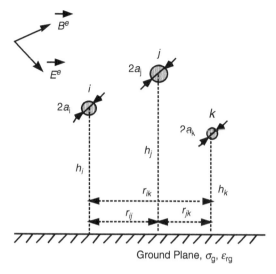

Figure C13.1 Cross-sectional geometry of a multiconductor line in the presence of an external electromagnetic field

for the mutual inductance between the two conductors is given by [74]

$$L'_{ij} = \frac{\mu_0}{2\pi} \ln\left(\frac{d^*}{d}\right) = \frac{\mu_0}{4\pi} \ln\left(\frac{r_{ij}^2 + (h_i + h_j)^2}{r_{ij}^2 + (h_i - h_j)^2}\right) \quad (C13.3)$$

The self-inductance for the conductor i is given by

$$L'_{ii} = \frac{\mu_0}{2\pi} \ln\left(\frac{2h_i}{r_{ii}}\right) \quad (C13.4)$$

$[C'_{ij}]$ is the matrix of the per-unit-length line capacitance. It can be evaluated directly from the inductance matrix using the expression [74]

$$\left[C'_{ij}\right] = \varepsilon_0 \mu_0 \left[L'_{ij}\right]^{-1} \quad (C13.5)$$

$[G'_{ij}]$ is the matrix of per-unit-length transverse conductance. The transverse conductance matrix elements can be evaluated starting either from the capacitance matrix or the inductance matrix using the relations

$$\left[G'_{ij}\right] = \frac{\sigma_{\text{air}}}{\varepsilon_0} \left[C'_{ij}\right] = \sigma_{\text{air}} \mu_0 \left[L'_{ij}\right]^{-1} \quad (C13.6)$$

However, for most practical cases, the transverse conductance matrix elements G'_{ij} are negligible in comparison with $j\omega C'_{ij}$ and can therefore be neglected in the computation.

Finally, $[Z'_{gij}]$ is the matrix of the ground impedance. The general expression for mutual ground impedance between two conductors i and j derived by Sunde is given by [75]

$$Z'_{gij} = \frac{j\omega\mu_0}{\pi} \int_0^\infty \frac{e^{-(h_i+h_j)x}}{\sqrt{x^2 + \gamma_g^2} + x} \cos(r_{ij}x)\, dx \qquad (C13.7)$$

In a similar way as for the case of a single-wire line, an accurate logarithmic approximation has been proposed by Rachidi and colleagues [30], and is given by

$$Z'_{gij} \cong \frac{j\omega\mu_0}{4\pi} \ln \left[\frac{\left(1 + \gamma_g \left(\frac{h_i+h_j}{2}\right)\right)^2 + \left(\gamma_g \frac{r_{ij}}{2}\right)^2}{\left(\gamma_g \frac{h_i+h_j}{2}\right)^2 + \left(\gamma_g \frac{r_{ij}}{2}\right)^2} \right] \qquad (C13.8)$$

Note that in equations (C13.1) and (C13.2) we have neglected the terms corresponding to wire impedance and the so-called ground admittance. Indeed, for typical overhead lines and for the typical frequency range of interest (below 10 MHz), these parameters can be disregarded with reasonable approximation [29,66].

The boundary conditions for the two line terminations are given by

$$[V_i^s(0)] = -[Z_A][I_i(0)] + \left[\int_0^{h_i} E_z^e(0, z)\, dz \right] \qquad (C13.9)$$

$$[V_i^s(L)] = [Z_B][I_i(L)] + \left[\int_0^{h_i} E_z^e(L, z)\, dz \right] \qquad (C13.10)$$

in which $[Z_A]$ and $[Z_B]$ are the impedance matrices at the two line terminations.

A time-domain representation of field-to-transmission line coupling equations is sometimes preferable because it allows us to handle in a straightforward manner non-linear phenomena such as corona, the presence of non-linear protective devices at the line terminals, and also variation in the line topology (opening and reclosure of switches). On the other hand, frequency-dependent parameters, such as the ground impedance, need to be represented using convolution integrals, which require significant computation time and memory storage.

The field-to-transmission line coupling equations (C13.1) and (C13.2) can be converted into the time domain to obtain the following expressions:

$$\frac{\partial}{\partial x}[v_i^s(x, t)] + [L'_{ij}]\frac{\partial}{\partial t}[i_i(x, t)] + [\xi'_{gij}] \otimes \frac{\partial}{\partial t}[i_i(x, t)] = [E_x^e(x, h_i, t)] \quad \text{(C13.11)}$$

$$\frac{\partial}{\partial x}[i_i(x, t)] + [G'_{ij}][v_i^s(x, t)] + [C'_{ij}]\frac{\partial}{\partial t}[v_i^s(x, t)] = 0 \quad \text{(C13.12)}$$

in which $\otimes$ denotes the convolution product and the matrix $[\xi'_{gij}]$ is called the transient ground resistance matrix. The elements of this matrix are defined as

$$[\xi'_{gij}] \cong F^{-1}\left\{\frac{Z'_{gij}}{j\omega}\right\} \quad \text{(C13.13)}$$

The inverse Fourier transforms of the boundary conditions written, for simplicity, for resistive terminal loads are given by

$$[v_i(0, t)] = -[R_A][i_i(0, t)] + \left[\int_0^{h_i} E_z^e(0, z, t)\,dz\right] \quad \text{(C13.14)}$$

$$[v_i(L)] = [R_B][i_i(0)] + \left[\int_0^{h_i} E_z^e(L, z, t)\,dz\right] \quad \text{(C13.15)}$$

where $[R_A]$ and $[R_B]$ are the matrices of the resistive loads at the two line terminals.

The general expression for the ground impedance matrix terms in the frequency domain (C13.7) does not have an analytical inverse Fourier transform. Thus, the elements of the transient ground resistance matrix in the time domain are to be, in general, determined using a numerical inverse Fourier transform algorithm. However, analytical expressions have been proposed that have been shown to be reasonable approximations to the numerical values obtained using an inverse fast Fourier transform (FFT) [76].

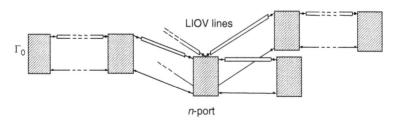

Figure C13.2 Electrical distribution system illuminated by LEMP

C13.2 Lightning-induced voltages on distribution networks: LIOV code interfaced with EMTPrv

In order to analyse the response of realistic configurations such as an electrical MV and LV distribution network to the excitation of a lightning electromagnetic field, the original LIOV code was first interfaced with the Electromagnetic Transient Program (EMTPM39 [34] and EMTP96 [35,36,44]).

The concept at the basis of the first interface is the following. A distribution line can be viewed as a group of lines, electrically connected by means of equivalent n-port circuits, as shown in Figure C13.2. Each of these n-port circuits represents a power component located along the line (such as surge arresters, or distribution transformers), or the periodical grounding of neutral conductor for LV lines, of shielding wires for MV lines, and so on. The LIOV code computes the voltages induced along the various lines that form the overall network (which we shall therefore call 'LIOV lines'), while the EMTP solves the boundary conditions equations relevant to the various n-ports currents of the network. In the first version of this code the physical/data link between the various n-port circuits and the LIOV lines was realized by means of a short non-illuminated line [34–36], which introduced a short time shift in the computed results.

A new, more efficient interface between the LIOV Code and EMTPrv has recently been proposed [73] that does not require any time shift introduced between each illuminated LIOV line and the boundary solution provided by the EMTPrv.

The LIOV Code and its interface with EMTP have been thoroughly tested on different sets of experimental data obtained using a reduced-scale model, EMP simulators and artificially initiated lightning [36].

A free version of the LIOV Code can be downloaded from http://www.liov.ing.unibo.it/

Acknowledgements

Special thanks are due to A. Borghetti and M. Paolone for their helpful comments in the preparation of this chapter. Some of the material of this chapter has been adapted from the publications by CIGRE Working Group C4.401 'Lightning' and Joint CIGRE/CIRED Working Group C4.402 'Protection of MV and LV Networks Against Lightning', whose members are gratefully acknowledged.

References

1. Golde R.H. *Lightning*, vol. 1. London: Academic Press, 1977.
2. Greenwood A. *Electrical Transients in Power Systems*. J. Wiley and Sons, New York, 1991.
3. Hileman R. *Insulation Coordination for Power Systems*. CRC Press, New York, 1999.
4. 'IEEE Guide for Improving the Lightning Performance of Electric Power Overhead Distribution Lines (IEEE Std. 1410)', IEEE, New York, 2004.

5. Golde R.H. 'The frequency of occurrence and the distribution of lightning flashes to transmission lines'. *AIEE Transactions* 1945;**64**:902–10.
6. Armstrong H.R., Whitehead E.R. 'Field and analytical studies of transmission lines shielding'. *IEEE Transactions on Power Apparatus and Systems* 1968; **PAS-87**:270–81.
7. Love E.R. 'Improvements on lightning stroke modeling and applications to the design of EHV and UHV transmission lines'. University of Colorado, Master's thesis, 1973.
8. Eriksson A.J. 'An improved electrogeometric model for transmission line shielding analysis'. *IEEE Transactions on Power Delivery* 1987;**3**:871–86.
9. Rizk F.A.M. 'Modelling of transmission line exposure to direct lightning strokes'. *IEEE Transactions on Power Delivery* 1990;**5**:1983–97.
10. Dellera L., Garbagnati G. 'Lightning stroke simulation by means of the leader progression model, Parts I and II'. *IEEE Transactions on Power Delivery* 1990;**5**:2009–29.
11. Bernardi M., Dellera L., Garbagnati E., Sartorio G. 'Leader progression model of lightning: updating of the model on the basis of recent test results'. *23rd International Conference on Lightning Protection*, Florence, Italy 1996, pp. 399–407.
12. 'IEEE Guide for Improving the Lightning Performance of Transmission Lines'. IEEE Standard, 1997; 1243–1997.
13. Suzuki T., Miyake K., Shindo T. 'Discharge path model in model test of lightning strokes to tall mast'. *IEEE Transactions on Power Apparatus and Systems* 1981; **PAS-100**:3553–62.
14. Borghetti A., Nucci C., Paolone M. 'Estimation of the statistical distributions of lightning current parameters at ground level from the data recorded by instrumented towers'. *IEEE Transactions on Power Delivery* 2004;**19**:1400–409.
15. Rizk F.A.M. 'Modelling of lightning incidence to tall structures, part I: theory and part II: application'. *IEEE Transactions on Power Delivery* 1994;**9**:162–93.
16. Anderson J.G. 'Lightning performance of EHV–UHV lines', in *Transmission Line Reference Book, 345 kV and Above*. Palo Alto: EPRI, 1982.
17. Bernardi M. Private communication.
18. Porrino A., Alexandri I., Clement M., Finlay G.S., Henriksen T., Hoeffelman J., Ishii M., Martinez Cid P., Nucci C.A., Popolansky F., Rachidi F., Roelofs G.P.T., Rudolph R. 'Protection of MV and LV networks against lightning. I. Basic information'. *CIRED. 14th International Conference and Exhibition on Electricity Distribution. Part I: Contributions* (Conf. Publ. No. 438), Vol. 2, 1997, p. 21.
19. 'CIGRE–CIRED JWG C4.4.02: Protection of MV and LV networks against lightning. Part I: Common topics'. *CIGRE Technical Brochure No 287*, December 2005.
20. 'Guide to procedures for estimating the lightning performance of transmission lines'. CIGRE WG 01 of SC 33, CIGRE Publication 63, 1991.
21. McDermott T.E., Short T.A., Anderson J.G. 'Lightning protection of distribution lines'. *IEEE Transactions on Power Delivery* 1994;**9**:138–52.

22. 'Working group report: calculating the lightning performance of distribution lines'. *IEEE Transactions on Power Delivery* 1990;**5**:1408–17.
23. 'IEC protection of structures against lightning. Part 1: General principles'. IEC, New York, 1990.
24. Alstad K. *et al.* 'Lightning protection of overhead lines with covered conductors'. CIRED Conference, Brussels, Belgium, 1995, Paper 3.16.
25. Dommel H.W. *Electromagnetic Transient Program Reference Manual (EMTP Theory Book)*. Bonneville Power Administration, Portland, OR, August 1986.
26. Rusck S. 'Induced lightning overvoltages on power transmission lines with special reference to the overvoltage protection of low voltage networks'. *Transactions of the Royal Institute of Technology, Stockholm* 1958;**120**:47.
27. Nucci C. 'Lightning-induced voltages on overhead power lines. Part I: Return stroke current models with specified channel-base current for the evaluation of the return stroke electromagnetic fields'. *Electra* 1995;**161**:74–102.
28. Nucci C.A. 'Lightning-induced voltages on overhead power lines. Part II. Coupling models for the evaluation of the induced voltages'. *Electra* 1995;**162**:120–45.
29. Rachidi F., Nucci C.A., Ianoz M., Mazzetti C. 'Importance of losses in the determination of lightning-induced voltages on overhead lines'. *EMC '96 ROMA. International Symposium on Electromagnetic Compatibility*. Univ. Rome 'La Sapienza', Rome, Italy, vol. 2, 1996.
30. Rachidi F., Nucci C.A., Ianoz M. 'Transient analysis of multiconductor lines above a lossy ground'. *IEEE Transactions on PWDR* 1999;**14**:294–302.
31. Nucci C., Rachidi F. 'Interaction of electromagnetic fields generated by lightning with overhead electrical networks', in V. Cooray (ed.) *The Lightning Flash*. London: IEE, 2003, 425–78.
32. Paolone M., Nucci C.A., Petrache E., Rachidi F. 'Mitigation of lightning-induced overvoltages in medium voltage distribution lines by means of periodical grounding of shielding wires and of surge arresters: Modelling and experimental validation'. *IEEE Transactions on Power Delivery* 2004;**19**:423–31.
33. Nucci C.A., Rachidi F., Ianoz M., Mazzetti C. 'Lightning-induced voltages on overhead power lines'. *IEEE Transactions on EMC* 1993;**35**:75–86.
34. Nucci C.A., Bardazzi V., Iorio R., Mansoldo A., Porrino A. 'A code for the calculation of lightning-induced overvoltages and its interface with the Electromagnetic Transient program'. *22nd International Conference on Lightning Protection (ICLP)*, Budapest, Hungary, 1994.
35. Paolone M., Nucci C.A., Rachidi F. 'A new finite-difference time-domain scheme for the evaluation of lightning induced overvoltages on multiconductor overhead lines'. *5th International Conference on Power System Transients*, Rio de Janeiro, 2001.
36. Borghetti A., Gutierrez A., Nucci C.A., Paolone M., Petrache E., Rachidi F. 'Lightning-induced voltages on complex distribution systems: models, advanced software tools and experimental validation'. *Journal of Electrostatics* 2004;**60**:163–74.
37. Berger K., Anderson R.B., Kroninger H. 'Parameters of lightning flashes'. *Electra* 1975;**41**:23–37.

38. De la Rosa F., Valdivia R., Pérez H., Loza J. 'Discussion about the inducing effects of lightning in an experimental power distribution line in Mexico'. *IEEE Transactions on Power Delivery* 1988;**3**.
39. Ishii M., Michishita K., Hongo Y., Oguma S. 'Lightning-induced voltage on an overhead wire dependent on ground conductivity'. *IEEE Transactions on Power Delivery* 1994;**9**:109–18.
40. Hermosillo V.F., Cooray V. 'Calculation of fault rates of overhead power distribution lines due to lightning induced voltages including the effect of ground conductivity' *IEEE Transactions on Electromagnetic Compatibility*, August 1995;**37**:392–99.
41. Michishita K., Ishii M., Imai Y. 'Lightning-induced voltages on multiconductor distribution line influenced by ground conductivity'. *23rd International Conference on Lightning Protection (ICLP)*, Florence, Italy, 1996, vol. 1, pp. 30–35.
42. Hoidalen H.K., Sletbak J., Henriksen T. 'Ground effects on induced voltages from nearby lightning'. *IEEE Transactions on Electromagnetic Compatibility* 1997;**39**:269–78.
43. Guerrieri S., Nucci C.A., Rachidi F. 'Influence of the ground resistivity on the polarity and intensity of lightning induced voltages'. *10th International Symposium on High Voltage Engineering*, Montreal, Canada, 1997.
44. Borghetti A., Nucci C.A., Paolone M., Rachidi F. 'Characterization of the response of an overhead line to lightning electromagnetic fields'. *International Conference on Lightning Protection (ICLP)*, Rhodos, Greece, 2000.
45. Nucci C.A. 'Lightning performance of distribution systems: Influence of system configuration and topology'. *International Symposium on Lightning Protection (SIPDA)*, São Paulo, Brazil, 2005.
46. Piantini A., Janiszewski J.M. 'Lightning induced voltages on distribution transformers: the effects of line laterals and nearby buildings'. *VI International Symposium on Lightning Protection (SIPDA)*, Brazil, 2001, 77–82.
47. Rachidi F., Nucci C.A., Ianoz M., Mazzetti C. 'Response of multiconductor power lines to nearby lightning return stroke electromagnetic fields'. *IEEE Transactions on Power Delivery* 1997;**12**:1404–11.
48. CIGRE/CIRED. 'Protection of medium voltage and low voltage networks against lightning. Part 2: Lightning protection of medium voltage networks (draft)', CIGRE–CIRED Joint Working Group C4.402 2008.
49. CIGRE/CIRED. 'Joint CIRED/CIGRE Working Group 05: Lightning protection of distribution networks. Part II: Application to MV networks'. *14th International Conference on Electricity Distribution*, Birmingham, UK, 1997.
50. CIGRE/CIRED. 'Joint CIRED/CIGRE Working Group 05: Lightning protection of distribution networks. Part I: Basic Information'. *14th International Conference on Electricity Distribution*, Birmingham, UK, 1997.
51. Yokoyama S. 'Calculation of lightning-induced voltages on overhead multiconductor systems'. *IEEE Transactions on Power Apparatus and Systems* 1984;**103**:100–108.
52. Piantini A. 'Lightning protection of overhead power distribution lines'. *29th International Conference on Lightning Protection (ICLP)*, Uppsala, Sweden, 2008.

53. Piantini A., Janiszewski J.M. 'The effectiveness of surge arresters on the mitigation of lightning induced voltages on distribution lines'. *VIII International Symposium on Lightning Protection (SIPDA)*, 2005, 777–98.
54. Borghetti A., Nucci C.A., Paolone M. 'An improved procedure for the assessment of overhead line indirect lightning performance and its comparison with the IEEE Std. 1410 Method'. *IEEE Transactions on Power Delivery* 2007; **22**:684–92.
55. Cooray V. 'Lightning-induced overvoltages in power lines: validity of various approximations made in overvoltage calculations'. *22nd International Conference on Lightning Protection*, Budapest, Hungary, 1994.
56. Borghetti A., Nucci C.A. 'Estimation of the frequency distribution of lightning induced voltages on an overhead line above a lossy ground: a sensitivity analysis'. *International Conference on Lightning Protection*, Birmingham, UK, 1998, 306–13.
57. 'IEEE fast front transients task force: Modeling guidelines for fast front transients'. *IEEE Transactions on Power Delivery* 1996;**11**:493–506.
58. Borghetti A., Nucci C.A., Paolone M., Bernardi M., Malgarotti S., Mastandrea I. 'Influence of surge arresters on the statistical evaluation of lightning performance of distribution lines'. *Proceedings of 8th International Conference on Probabilistic Methods Applied to Power Systems*, AMES, Iowa, USA, 2004, pp. 776–81.
59. Rusck S. 'Protection of distribution systems', in R.H. Golde (ed.) *Lightning*, vol. 2. New York: Academic Press, 1977.
60. Darveniza M. 'A practical extension of Rusck's formula for maximum lightning induced voltage that accounts for ground resistivity'. *IEEE Transactions on Power Delivery* 2007;**22**:605–12.
61. Short T.A. 'Communication to the IEEE WG on the lightning performance of distribution lines', 2001.
62. Andersson R.B., Eriksson A.J. 'Lightning parameters for engineering application'. *Electra* 1980;**69**:65–102.
63. Guerrieri S., Nucci C.A., Rachidi F., Rubinstein M. 'On the influence of elevated strike objects on directly measured and indirectly estimated lightning currents'. *IEEE Transactions on Power Delivery* 1998;**13**:1543–55.
64. Rubinstein R.Y. *Simulation and the Monte Carlo Method*. New York: Wiley, 1981.
65. Agrawal A.K., Price H.J., Gurbaxani S.H. 'Transient response of multiconductor transmission lines excited by a nonuniform electromagnetic field'. *IEEE Transactions on Electromagnetic Compatibility* 1980;**EMC22**:119–29.
66. Rachidi F., Nucci C.A., Ianoz M., Mazzetti C. 'Influence of a lossy ground on lightning-induced voltages on overhead lines'. *IEEE Transactions on Electromagnetic Compatibility* 1996;**38**:250–63.
67. Uman M.A., McLain D.K., Krider E.P. 'The electromagnetic radiation from a finite antenna'. *American Journal of Physics* 1975;**43**:33–38.
68. Rubinstein M. 'An approximate formula for the calculation of the horizontal electric field from lightning at close, intermediate and long range'. *IEEE Transactions on Electromagnetic Compatibility* 1996;**38**:531–35.

69. Nucci C.A., Mazzetti C., Rachidi F., Ianoz M. 'On lightning return stroke models for LEMP calculations'. *19th International Conference on Lightning Protection (ICLP)*, Graz, Austria, 1988.
70. Rachidi F., Nucci C.A. 'On the Master, Uman, Lin, Standler and the Modified Transmission Line lightning return stroke current models'. *Journal of Geophysical Research* 1990;**95**:20389–94.
71. Rachidi F., Rubinstein M., Guerrieri S., Nucci C.A. 'Voltages induced on overhead lines by dart leaders and subsequent return strokes in natural and rocket-triggered lightning'. *IEEE Transactions on Electromagnetic Compatibility* 1997;**39**:160–66.
72. Nucci C.A., Guerrieri S., Correia de Barros M.T., Rachidi F. 'Influence of corona on the voltages induced by nearby lightning on overhead distribution lines'. *IEEE Transactions on Power Delivery* 2000;**15**:1265–73.
73. Napolitano F., Borghetti A., Nucci C.A., Paolone M., Rachidi F., Mahserejian J. 'A link between the LIOV code and the EMTP-rv for the calculation of lightning-induced voltages on distributions networks'. *29th International Conference on Lightning Protection (ICLP)*, Uppsala, Sweden, 2008.
74. Tesche F.M., Ianoz M., Karlsson T. *EMC Analysis Methods and Computational Models*. New York: Wiley Interscience, 1997.
75. Sunde E.D. *Earth Conduction Effects in Transmission Systems*. New York: Dover Publications, 1968.
76. Theethayi N., Thottappillil R. 'Surge propagation and crosstalk in multiconductor transmission lines above ground', in F. Rachidi, S. Tkachenko (eds) *Electromagnetic Field Interaction with Transmission Lines. From Classical Theory to HF Radiation Effects*. WIT Press, 2008.

Chapter 14
Lightning protection of wind turbines
Troels Soerensen

14.1 Introduction

Windmills have been used for grinding grain and pumping water for centuries, and lightning striking windmills must have been a well known and feared aspect of the milling business for just as long. However, the risk of lightning striking traditional windmills is not high, as the height of such windmills (Figure 14.1), say 10 or 15 m, is comparable to surrounding structures and trees, and although traditional windmills were placed at windy locations, they were not necessarily placed at the

Figure 14.1   Windmill in the open air museum in Odense, Denmark. It is a wood construction thatched with straw (photo by the author).

highest points in the terrain, but preferably at less exposed locations, as structural damage and brakes failing during storms were always a threat. Considering the materials used in their construction (wood, straw, canvas, etc.), lightning and fires have always been a serious threat, and obviously lightning protection was of interest to owners as soon as such techniques were available. One example of early lightning protection is shown in Figure 14.2, which shows a Franklin rod placed on a windmill cap. However, many traditional windmills remaining today are not protected against lightning.

Modern windmills for the generation of electrical energy are usually called wind turbines, and wind turbines have been used for generating electricity since the early days of electrical power. However, wind turbines were unsuccessful in the competition with diesel-fuelled generator sets and centralized power generation, and were almost completely abandoned after World War II when the expanding electricity networks reached most consumers. The renaissance for wind turbines came following the oil crises in 1973 and 1979, which spurred renewed interest in alternative energy

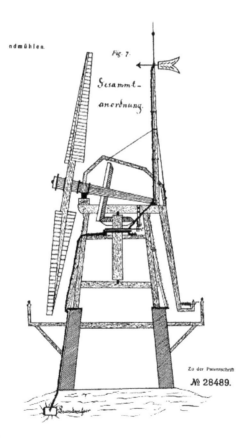

Figure 14.2   Lightning protection of a traditional windmill as depicted in a German patent (from Reference 1)

Lightning protection of wind turbines 683

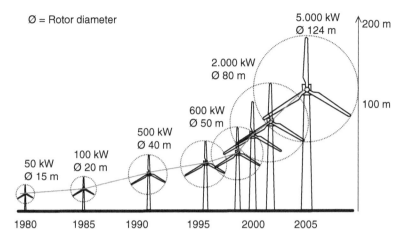

Figure 14.3   Indicative size and rated power development of commercial wind turbines (from Reference 2)

sources. Since then, the development of commercial wind turbines has been impressive, with a 100-fold increase in power, and total height now exceeding 180 m, as indicated in Figure 14.3.

The growth in installed wind turbine power capacity since 1983 is shown in Figure 14.4, with annual growth in later years reaching ∼25 per cent. With these growth rates wind power has already become a significant source of electrical energy in some countries. In 2004 wind turbines produced almost 18.5 per cent of the electric energy consumed Denmark and 5 to 6 per cent in Germany and Spain. Most wind turbines have been installed on land, but large offshore wind projects

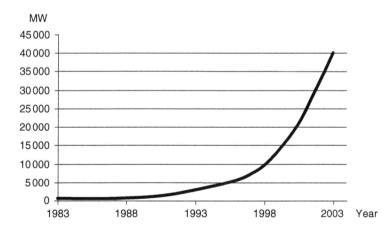

Figure 14.4   Installed wind power capacity in the world (cumulative rated power) (from Reference 2)

have been installed in recent years. Examples in Denmark are the 40 MW (20 × 2 MW turbines) wind farm Middelgrunden outside Copenhagen harbour (2000), the 160 MW (80 × 2 MW) wind farm Horns Reef in the North Sea (2002), and the 165 MW (72 × 2.3 MW) wind farm Nysted in the Baltic Sea (2003). Several large offshore wind projects have been developed in the United Kingdom, the first being the two 60 MW (30 × 2 MW) wind farms North Hoyle and Scroby Sands built in 2003 and 2004, respectively, followed by the 90 MW (30 × 3 MW) wind farm Kentish Flats built in 2005 [3]. Many offshore wind farm projects are currently being developed in Denmark, Sweden, the United Kingdom and Germany.

Lightning protection does not appear to have been of much concern in the production of the 25–50 kW wind turbines in the early 1980s. One explanation sometimes offered for the lack of lightning protection is that the development took place in northern Europe and California, both regions where lightning occurrence is low. With their relatively low total height of 15–20 m typical for wind turbines of that time, lightning strikes were rare and probably simply not very important compared to more immediate problems such as preventing wind turbines from breaking down in high winds. However, serious efforts were put into lightning protection, particularly of blades, in electric utility and government-sponsored projects developing large wind turbines [4–8]. The lightning protection techniques used in those projects originated in the aircraft industry and tended to be complicated and expensive [8,9]. However, in the wind turbine industry lightning protection did not receive much attention until the mid-1990s, when series produced wind turbines with total heights of $\sim$50 m started to appear. At that time it also became clear that lightning strikes each year caused damage to 4–5 per cent of wind turbines in Denmark (see Figure 14.7) [11], which, considering the 20 year technical lifetime normally expected of wind turbines, still corresponded to no more than one lightning damage in a wind turbine lifetime, on average. With larger wind turbines the situation is different. As can be seen from Figure 14.3 wind turbines of 2 MW rated power or more exceed 100 m in height, and because of their height such wind turbines will statistically all be hit by lightning several times in their 20 year technical lifetime as is usually required of wind turbines, even in relatively low lightning intensity regions in northern Europe. Hence, lightning has become a condition of operation for large wind turbines rather than a rare 'Act of God' event, and large wind turbines must therefore have effective lightning protection, particularly when located offshore where access is difficult and maintenance costs are very high.

The all-dominating wind turbine design today is the three-bladed horizontal axis up wind type with tubular tower, as shown in Figure 14.5, and the lightning protection discussed in the following will deal with this type of wind turbine, although the methods will be adaptable to other types as well. The main components in the wind turbine nacelle are shown in Figure 14.6 in a generalized form. Furthermore, a large number of smaller components such as sensors, motors, yaw drives, pumps, hydraulics and so on are also placed in and on the nacelle. In wind turbines as large as 2–3 MW the generator is predominantly low voltage (LV) (i.e. stator voltage less than 1 kV) and a machine transformer is used for stepping up the voltage to that of

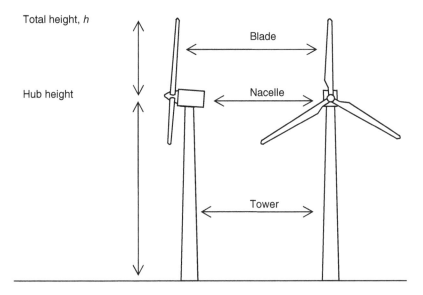

Figure 14.5 Three-bladed horizontal axis wind turbine with tubular tower

the power grid to which the wind turbine is connected. Most large wind turbines have the machine transformer placed in the nacelle or in the tower, and connect to the power grid via switch gear usually placed in the bottom of the tower. Alternatively, the machine transformer may be placed in a cubicle next to the wind turbine (see Figure 14.10).

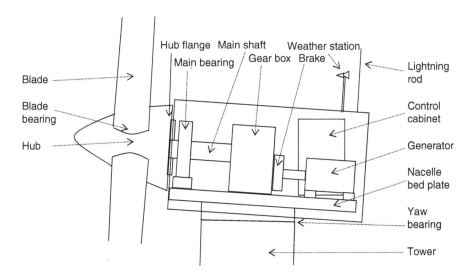

Figure 14.6 Generalized wind turbine nacelle design

14.2 Nature of the lightning threat to wind turbines

For reasons of maximizing earnings, modern wind turbines are preferably placed at high wind sites such as in coastal regions or at elevated positions in the landscape such as on hills or on mountain ridges. Such locations tend to have relatively high lightning occurrence, and furthermore tend to be remote in terms of access to power grid, telecommunication network and access road. The remote locations enhance the impact of lightning-caused operational disturbances, as the number of disturbances to power and telecommunication lines is proportional to the length of the service lines, and the length of the access road influences the repair time.

The lightning threat to wind turbines can be split into two sources of interference: lightning striking the wind turbine and lightning affecting the power and telecommunication lines to which the wind turbine is connected. Lightning affecting the lines may either cause interruption of the power or telecommunication services or may generate transients on the services, which may cause damage when reaching the wind turbine.

The nature of lightning striking wind turbines is influenced by the height and location of the wind turbine. Although wind turbines with total heights up to ~60 m (~500 kW) in relatively flat terrain are predominantly affected by lightning developing from thunderclouds towards earth (i.e. downward lightning), the same size of wind turbine placed exposed at elevated locations may also initiate lightning developing from the wind turbine towards the thundercloud (i.e. upward lightning). Hence the distinction often made in lightning protection between structures lower and higher than 60 m does not apply very well to wind turbines, as the location of the wind turbine has to be considered. The proportion of upward lightning relative to the total number of lightning events affecting a structure becomes significant for structures higher than 60 m and upward lightning dominates for structures exceeding 100 m [12,13]. Furthermore, in areas where lightning appears during winter when thunderclouds are at a low height above ground, say a few hundred

Table 14.1 *Lightning current parameters for dimensioning lightning protection systems according to IEC 62305-1*

Current parameters	Symbol	Unit	Lightning protection level		
			I	II	III–IV
Peak current*	I	kA	200	150	100
Impulse current time parameters*	$T1/T2$	µs/µs		10/350	
Charge[†]	Q	C	300	225	150
Specific energy[†]	W/R	kJ/Ω	10.000	5.625	2.500
Current impulse steepness[‡]	di/dt	kA/µs	200	150	100
Impulse current time parameters[‡]	$T1/T2$	µs/µs		0.25/100	

*First stroke; [†]complete lightning; [‡]subsequent stroke.

metres as opposed to several thousand metres during summer, conditions evidently exist for upward lightning developing from wind turbines. In fact, it has been shown that the occurrence of lightning striking wind turbines during winter is considerably higher than what should be expected from thunderstorm day (keraunic level) and lightning density statistics [10,14–17].

The electrical parameters of both downward and upward lightning are documented in the literature [10,12,13,16,18,19], and the particular parameters relevant to the design and dimensioning of lightning protection systems for wind turbines are compiled in IEC 62305-1 and IEC 61400-24. The lightning current parameters needed for dimensioning lightning protection systems are included in Table 14.1. Lightning protection levels I, II and III–IV are defined in IEC 62305-1 and correspond to the currents appearing in 98, 95, and 90–80 per cent of lightning striking ground, respectively (downward as well as upward lightning).

14.3 Statistics of lightning damage to wind turbines

Obviously there is a relationship between the risk of lightning damages and the number of thunderstorms in the area where wind turbines are situated. The relationship can be seen in Figure 14.7, where the number of lightning-caused faults per 100 wind turbines is shown together with the number of thunderstorm days registered in Denmark in the years 1985–1999. The annual average of wind turbines damaged by lightning in the period was 4 per cent. For Germany in the years 1991–1998 the annual average was 14 per cent in the mountain areas in the south, while in the low lands in the north it was 7.4 per cent and in the coastal areas in the north 5.6 per cent. Lightning damages were reported for a total of 900 wind turbines in Denmark in the period 1990–1999, with the distribution of damaged components shown in Figure 14.8. It can be seen that most damages, 51 per cent, affected the control system, 12 per cent the power system, 7 per cent the generator and 11 per cent the

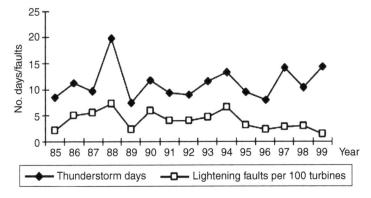

Figure 14.7 Registered lightning faults per 100 turbines in Denmark in the period 1985–1999 shown together with the registered thunderstorm days (keraunic level) (from Reference 11)

688 *Lightning Protection*

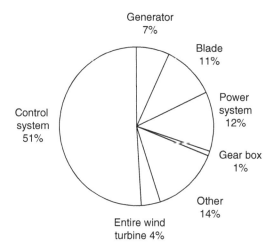

Figure 14.8 *Distribution of damaged components in a total of 900 wind turbines affected by lightning in Denmark in the period 1990–1999 (from Reference 11)*

blades. A significant part of the damages to the power and control systems are probably due to lightning affecting the power and communication lines connected to the wind turbines, but the Danish database does not distinguish between lightning damages caused by lightning striking the wind turbine and lightning affecting the power and communication lines. However, the German statistics indicate that 70 per cent of the lightning damages to wind turbines are caused by lightning affecting the power and telecommunication lines connected to the wind turbines. The German statistics also show that the repair costs increase with size of wind turbine, and show that damages to the blades are the most expensive followed by damages to the generator, due to the relatively high costs of these components and also due to the repair costs, crane costs and loss of production. In general some consideration of the quality of the lightning damage statistics is necessary as such statistics are influenced by the population of wind turbines for which damages are reported. In the case of the Danish database reporting was voluntary, and it was dominated by relatively small wind turbines in the range 100–300 kW. Reporting to the German database was a condition for obtaining subsidies, which clearly makes a difference as to how reliable and complete the statistics should be considered [11,20].

14.4 Risk assessment and cost–benefit evaluation

Assessing the risk of lightning striking a wind turbine or affecting it via the connections to the power and telecommunication lines can be made according to IEC 62305-2 Annex A.

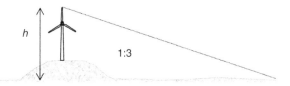

Figure 14.9   *Evaluation of collection area by including local terrain variations in the total height* h

The lightning ground flash (i.e. lightning striking ground) collection area A_g (m^2) for a wind turbine can be evaluated by drawing a line with a 1:3 slope from the tip of the wind turbine blade as shown in Figure 14.9:

$$A_g = \pi(3h)^2 \tag{14.1}$$

Multiplied by the annual ground flash density N_g per km^2, this gives the average number of lightning striking the wind turbine, N_d:

$$N_d = C_d N_g A_g \times 10^{-6} \tag{14.2}$$

Annual ground flash density statistics (N_g) are available from meteorological institutions and operators of lightning location systems. As wind turbines are usually placed at exposed positions a location factor $C_d = 2$ could be used as specified in IEC 62305-2 Annex A. Or, if relevant, an alternative assessment can be made by including local terrain variations in the total height as shown in Figure 14.9.

It can be argued that because the blades rotate the average height is lower than h, and that an elliptical collection area would be more relevant, but it makes very little difference and whatever error introduced by approximating a wind turbine with a simple mast is certain to be made irrelevant by the uncertainty and natural variations of the ground flash density data [21–23].

The number of lightning strikes to the typically buried communication lines connecting the wind turbine can be assessed according to IEC 62305-2 Annex A as

$$N_l = C_d C_t N_g (L_c - 3h) \sqrt{(\rho)} \times 10^{-6} \tag{14.3}$$

And the number of lightning strikes close enough to communication lines to affect the lines can be assessed as

$$N_i = C_d C_t N_g \times 25 \, L_c \sqrt{(\rho)} \times 10^{-6} \tag{14.4}$$

where L_c (m) is the length of the service line from the wind turbine to the next structure on the line. A maximum value of $L_c = 1\,000$ m should be assumed. ρ (Ω m) is the resistivity of the soil where the service is buried. A maximum value of

$\rho = 500\ \Omega$ m should be assumed. C_d is a location factor, which can be applied if relevant. C_t is a transformer factor.

The transformer factor C_t is equal to 1 if there is no transformer between the point of lightning strike and the wind turbine, and $C_t = 0.2$ if there is. As there is usually a medium voltage transformer in large wind turbines $C_t = 0.2$ can be assumed for the medium-voltage (MV) cables connecting the wind turbine to the grid (see IEC 62305-2 Annex A).

The annual number of lightning strikes to the wind turbine N_d, the annual number of lightning strikes affecting the wind turbine via the power and communication lines N_l and N_i, and the corresponding repair costs and lost production in repair time needs to be assessed in order to make a cost–benefit based decision on the economically optimal level of lightning protection. It is, however, quite difficult to find reliable information on repair costs and lost production due to lightning. Some information regarding relatively small wind turbines can be found in IEC 61400-24, but for new and large wind turbines it is advisable to make an assessment based on the actual price of the wind turbine in question. For this purpose Table 14.2 includes a breakdown of the price of a 2 MW wind turbine on individual components, and an assessment of the level of risk of damages to an unprotected component, which will allow a rough assessment of the potential costs caused by lightning. The cost of the wind turbine control system is not included in Table 14.2, possibly because the cost of the control system hardware is less than 1 per cent of the wind turbine.

The methods briefly outlined above make it possible to make simple cost–benefit assessments that will make it possible to decide on the optimal lightning protection level, and which could be useful when comparing the risks of lightning damages for instance in different wind projects.

However, in the author's experience such assessments are of little relevance when considering large wind turbines – say 1 MW or more – as such wind turbines are in

Table 14.2 The relative prices of individual components in per cent of the total wind turbine price [24] and an assessment of the risk of lightning damages for an unprotected component

Component	Relative cost (%)	Risk	Component	Relative cost (%)	Risk
Steel tower	33	None	Yaw system	2	Medium
Rotor blades (3)	18	High	Blade pitch system	5	Medium
Rotor hub	2	None	Power converter	6	Medium
Rotor (main) bearing	1	Medium	Transformer	3	High
Rotor shaft	2	None	Brake system	1	Low
Main carrier (nacelle bed plate)	3	None	Nacelle cover	2	High
Generator	4	Medium	Cables	2	High
Gears	14	Medium	Screws	1	None

practice certain to be hit by lightning one or more times even in a low lightning intensity area such as northern Europe. Consider, for instance, that the cost of a wind turbine blade is 6 per cent of the wind turbine price, say 100.000 € for a blade for a 1 MW wind turbine, and the fact that an unprotected blade will almost certainly be completely destroyed when hit by lightning, then it is immediately clear that lightning protection for the blades is indispensable and that quite substantial sums could be spent on it. Similar considerations with regard to other individual components such as machine transformers, generators, bearings and gears lead to the same general conclusion, that the costs in the case of lightning damages are so high that lightning protection is always economically justified. Fortunately, lightning protection is quite straightforward and low in cost, as will be discussed in the following, so in practice there is really no reason not to have an effective lightning protection system in large wind turbines.

14.5 Lightning protection zoning concept

Lightning protection zoning is a convenient systematic method to evaluate the levels of currents, voltages and electromagnetic fields caused by lightning that may influence different parts of a wind turbine, and to make sure that all parts have adequate protection. The method is thoroughly described in the IEC 62305 series of standards dealing with lightning protection. The same methodology forms the basis of the IEC 61000 series of standards dealing with electromagnetic compatibility (EMC) and the standards and legislation concerning electric power installations (i.e. IEC 60364/ Cenelec HD 384 and Cenelec HD 637 S1). The method is simple. First, identify the levels of the influencing factors (i.e. currents, voltages and electromagnetic fields appearing in the structure when lightning strikes), and then make sure that the components exposed to those influences have sufficient immunity or adequate protection. Applying the method to a wind turbine is not quite as simple. The challenge is to conduct lightning current safely through a complicated structure with large rotating blades of composite materials, mechanical systems with bearings and gears, electrical systems with generators, cables, power converters and transformers, electronic systems with wiring, actuators and sensors, and internal and external systems for control and communication. As an additional challenge lightning protection solutions should be low cost and robust. Fortunately, lightning protection based on protection zoning fits very well with EMC requirements and electrical installation practices of today, in fact so well that lightning protection of electrical and electronic systems can be achieved at little or no extra cost, as it is mainly a matter of securing adequate immunity and insulation levels when specifying the components to be used and using EMC correct installation practices.

Lightning protection zones are defined and ranged according to the level of lightning influences (i.e. currents, voltages and electromagnetic fields) appearing in the zones when lightning strikes. The surface area LPZ 0_A in which lightning may attach is identified with the rolling sphere method. Components in LPZ 0_A may be subjected to attachment of the lightning arc and to conducting the full lightning current, and may be exposed to the unattenuated lightning electromagnetic field. LPZ 0_B is the surface area not exposed to direct attachment, and the lightning protection zones of higher orders LPZ1, LPZ2 and so on represent parts of the wind turbine with higher protection and corresponding lower levels of lightning currents, voltages and electromagnetic fields, as the levels in each zone are controlled by applying shielding, bonding and transient protection of conductors at the zone boundaries. The rolling sphere method is defined in IEC 62305-1, according to which LPZ 0_A is the surface area touched by a sphere rolled back and forth over the structure as shown in Figure 14.10. The sphere radius is selected according to the lightning protection level defined in Table 14.1. It is evident that due to the geometry of wind turbines the rolling sphere will be able to touch most of the surface of the blades, and therefore indicate the possibility of lightning attaching almost anywhere on the blades. This may be true in theory, but it should be realized that practical experience is such that most lightning attaches at or within a few tens of centimetres from the blade tip [25]. In practice it may be convenient to define LPZ1 as the nacelle and tower interiors, and LPZ2 as inside control cabinets, making the maximum use of the nacelle cover, tower and metal cabinets for electromagnetic shielding, and systematically consider the need for transient protection of each conductor crossing into LPZ1 from the outside and into each cabinet (i.e. LPZ2).

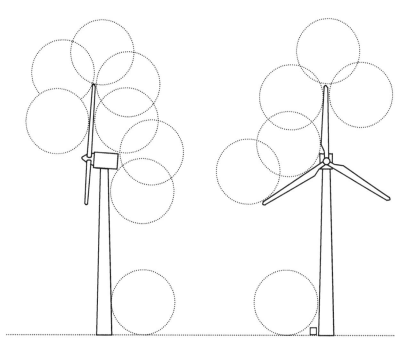

Figure 14.10   *Application of the rolling sphere method for identifying lightning protection zones 0_A in which lightning can attach and 0_B which lightning cannot reach. A transformer cubicle next to the tower can be placed within lightning protection zone 0_B.*

14.6 Earthing and equipotential bonding

Wind turbine construction usually consists of three large metal constructions: the hub, nacelle bed plate and tower. Some nacelle constructions have solid metal plate nacelle cover, and others have a glass fibre cover mounted on a steel frame construction. These large 'natural' metal components should be used as much as possible for shielding, earthing and equipotential bonding purposes, thereby providing the best possible backbone for an effective lightning protection system at close to no extra cost. This is also in accordance with the lightning protection standard IEC 62305, which recommends the use of natural metallic components for shielding, earthing and equipotential bonding purposes, and with electrical codes corresponding to IEC 60364/ Cenelec HD 384, part 6, and Cenelec HD 637 S1, part 2, which allow the use of large metal components as part of the earthing system and that require equipotential bonding of large metallic components to an earthing system. Hence, the main earthing connection of the wind turbine should be via the hub, nacelle construction and the tower (i.e. these large metal constructions should be used both as Faraday cages and as earthing connections through which the lightning current is conducted to earth). All major metal objects should be equipotential bonded to this earthing connection system (e.g. ladders, hoist cables, platforms, etc.) with equipotential bonding connections or preferably via the means mounting the objects to the tower (clamps, brackets, bolts etc.). Dedicated earthing terminals should be provided at each assembly and at each major component of the generator circuit and high-voltage (HV) system. Earthing terminals should be without paint, and should be designed to ensure direct and permanent metal-to-metal contact to equipotential bonding and earthing connections with sufficient contact area. For earthing of cable shields earthing terminals should be positioned close to where the cables enter the wind turbine.

In general it should be realized that a modern wind turbine with a tubular steel tower is an almost ideal Faraday cage, and therefore it is impossible to conduct more than a fraction of a lightning current or any transient current from the tower via a looped earthing conductor from inside the tower through the foundation to an outside earthing system [26]. The majority of current will pass directly from the tower to the foundation reinforcing steel, if not via metallic connections then via sparks between the long bolts connecting the tower to the foundation.

The wind turbine earthing system should be a foundation earthing system complying with the requirements in the electrical code (Cenelec HD 637 S1) and the lightning protection standard IEC 62305. Some foundation contractors and even some wind turbine manufacturers shy away from using foundation earthing systems for reasons of fear of corrosion and fear of causing damage to the foundation when lightning strikes. However, in reality it is impossible to avoid lightning current passing from the say 140 long foundation bolts holding the tower to the reinforcing steel bars. Trying to separate the foundation bolts from the foundation reinforcing steel bars will probably just make certain that potentially harmful sparking takes place when lightning strikes. The tower bottom flange or foundation bolts should ideally be in metal-to-metal contact with the foundation earthing system, or alternatively a minimum of four connection points to the foundation earthing system should be

694 Lightning Protection

Figure 14.11 Steel reinforcement of a foundation for a 1 MW wind turbine. The dimensions of the horizontal plate $L \times W \times H$ are $\sim 10\,m \times 10\,m \times 1\,m$ (photo by the author).

positioned on the outside of the tower, at or as close as possible to the tower bottom flange separated by 90°. The abovementioned standards Cenelec HD 637 S1 and IEC 62305 give detailed instructions on the construction of foundation earthing systems and choice of materials. The standards recommend welded and clamped connections of reinforcing steel bars, but as foundations for large wind turbines have horizontal dimensions of say 15 m × 15 m and are 1–2 m in depth or bigger, and include many tons of reinforcing steel bars, it should be realized that in practice it is not possible to weld or clamp secure connections of more than a fraction of the crossings of the reinforcing steel bars (see Figure 14.11). The foundation contractor will in many cases not allow welding of the reinforcing steel bars out of fear of reducing their strength, and he will probably just use tie wire for fixing and holding the reinforcing steel bars in place while pouring the concrete. Hence, it is advisable to construct an overall bolted or welded foundation earthing network embracing the entire body of reinforcing steel bars, which ensures equipotential bonding and distribution of lightning and electrical fault current through the foundation earthing system without causing potential differences in the foundation hazardous to the concrete (see Figure 14.12). The overall foundation earthing network should have at least four conductors (displaced 90°) arranged radially along the surface of the foundation, connecting (bolted or welded) to circumferential conductors at the interface to the

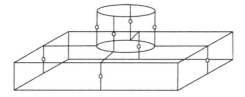

Figure 14.12 Overall foundation earthing network. Circles indicate connection points to which additional external earthing systems may be connected.

wind turbine tower and along all of the foundation edges as needed to achieve a foundation earthing network with opening size of say 3 m × 3 m². The overall foundation earthing network can be made of reinforcing steel bars or extra iron pieces of the same type of steel. An iron profile different from the armouring is preferred as it facilitates effective visual inspection. Copper wire could be used for the overall foundation earthing network, but galvanized iron should not be used as that may lead to corrosion problems in some environments. Welded connections of the overall foundation earthing network should be made according to IEC 62305-3. Clamps type tested according to EN 50164-1 should be used for the bolted connections. All other iron armouring crossing the overall earthing network should be bonded to it with ordinary tie wire.

Connection points for earthing and equipotential bonding should preferably be standard earthing connection point components. These are usually a stainless steel plate flush with the concrete surface, ~ 80 mm diameter, with a treaded hole at the centre for a M10 or M12 bolt. The plate is preassembled with a length of 10–12 mm ø steel, long enough to reach into the concrete and be clamped or welded onto the overall foundation earthing network. If needed, connection points to the overall foundation network can be positioned so that the earthing system can be extended with external earthing systems, for example with concentric earthing conductors for lowering the earthing resistance controlling touch- and step-voltages on the surface near the wind turbine.

14.7 Protection of wind turbine components

14.7.1 Blades

Historically, wind turbine blades have been manufactured from a range of materials such as wood, wood canvas, wood laminate and steel, but modern wind turbines appearing since about 1980 have predominantly been equipped with blades made of composite materials such as glass fibre reinforced plastics (GFRP, polyester or epoxy), sometimes in combination with wood, wood laminates and even carbon fibre. Most composite material blades are made as two shells produced separately in a manual dry or wet lay up process very similar to the way glass fibre boats are made. The two shells are subsequently glued together along the leading and trailing edges and glued to an internal beam structure. In recent years more advanced methods have been developed where the blade is infusion cast in one piece. Such full cast blades are probably stronger generally with respect to lightning as the composite can be made more homogeneous and there are no glued interfaces. In any case, the resulting blade structure has large air-filled compartments stretching the length of the blade separated in the lengthwise direction by an internal beam structure. Glued interfaces are clearly relatively weak parts of the blade structure as a characteristic feature of lightning damages to blades is that such glued interfaces are ripped apart by the excessive pressure waves developed by lightning arcs inside the blades.

There are two types of blades used in large wind turbines: the stall regulated blade with tip air brake used for wind turbines as large as 2 MW (see Figure 14.13a), and the pitch type blade (see Figure 14.13b–d), which is used on large wind turbines

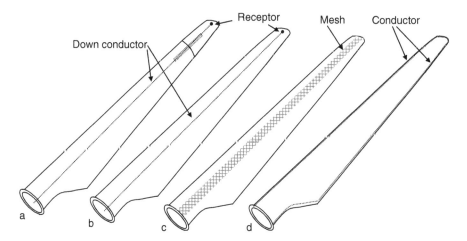

Figure 14.13   Main types of lightning protection for wind turbine blades

including all turbines larger than 2 MW where the whole blade is pitched (pivoted around the lengthwise axis) to regulate production and to brake the wind turbine. Large wind turbines such as 1, 2, 3.6 and 5 MW have blade lengths of about 27, 38, 52 and 61 m, respectively, and the blades represent as much as 15–20 per cent of the costs of the wind turbine (see Table 14.2).

The tip air brake on stall regulated blades (Figure 14.13a) is controlled with a steel wire retained by hydraulics (see Figure 14.19, later) placed in the blade root. The tip is mounted on a shaft, which, forced by the centrifugal force and a spring at one end of the shaft, provides a 90° turning of the tip when the hydraulics is released. Blades with tip air brake not protected against lightning are usually hit by lightning at the tip or within few tens of centimetres from the tip [25]. Lightning strikes usually penetrate the composite tip at the edges or through the sides and connect to the outer end of the tip shaft inside the blade. The pressure wave from this lightning arc usually rips the tip sides apart along the glued interfaces and away from the shaft. An example is shown in Figure 14.14. The damage is limited to the tip if the shaft and spring construction and the steel wire to the hydraulics at the blade root are able to conduct the lightning current. Whenever a lightning arc has appeared inside the main part of the blade the pressure wave typically rips the blade open along the trailing edge, and if the structural strength of the blade is weakened too much the blade may even be completely destroyed (Figure 14.15). Lightning arcing inside the main part of the blade typically appears in cases where the steel wire has melted or broken because of heating by the lightning current. Hence careful construction and dimensioning is necessary. An effective lightning protection system was developed in the mid-1990s for GFRP blades with tip air brakes. The system consists of a lightning receptor (air terminal) placed on each side of the blade tip flush with the surface to avoid noise and connected with a down-conductor through the tip shaft to the control wire. The tip shafts in large blades are wound with carbon fibre reinforced plastic (CFRP),

Lightning protection of wind turbines 697

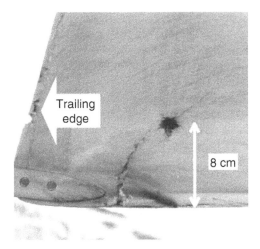

Figure 14.14   Blade tip with tip air brake ripped apart by a lightning attaching 8 cm from the tip (photo by the author)

Figure 14.15   Blade ripped apart in the glued interface along the trailing edge by a lightning arc inside the blade (photo by the author)

which is electrically conducting (see Section 14.7.1.1 for further discussion of CFRP) and which therefore needs to be bonded to the down-conductor to avoid arcing. There have been cases of damages to the CFRP shaft, but in most cases it seems that the wall thickness of the shaft is sufficient to give the necessary conductivity. The cross-section of the steel wire retaining the tip has to be dimensioned to sustain conducting lightning current. Stainless steel wire has to be a minimum of 10 mm in diameter and preferably 12 mm. At the root end of the blade the hydraulic cylinder retaining the wire to the tip also must be protected (see Section 14.7.6 for further discussion of lightning protection of hydraulics).

The pitch type blade (Figure 14.13b–d) can be made completely of non-conducting elements except the root end flange and bolts, and therefore it has been speculated in the past whether such blades would attract lightning at all and whether lightning protection is needed for this type of blade. However, a GFRP wind turbine blade cannot be considered as an insulator in the electro-technical meaning of the word, as both the materials and the structures are very inhomogeneous not least because of the manual production processes and issues of quality of work. In reality 'non-conducting' blades do attract lightning, because the blades influence the electrical field as a result of the permittivity of GFRP being higher than air, because streamer discharges develop more easily along composite surfaces as compared to air, and probably not least because both the outside and inside surfaces over time become increasingly semi-conducting because of ageing, pollution and moisture. In any case it has been proven in practice by numerous damaged blades that lightning protection is absolutely necessary even for 'non-conducting' blades.

There are three different main types of lightning protection for the pitch type blade. Type b in Figure 14.13 which is used by Danish manufactures of blades, has a pair of lightning receptors at the tip, and on large blades additional receptor pairs on the sides of the blade interspaced by 5–8 m. All receptor pairs are connected to an internal down-conductor to the root end (see Figure 14.16). This type of lightning protection has proven quite effective. A typical example of a 40-m blade struck by lightning at a tip receptor is shown in Figure 14.17. It can be seen that the lightning has attached at the edge of the receptor, and that the lightning arc has been drawn towards the trailing edge of the blade by the wind, as the blade was rotating at the time of lightning strike. The heat from the arc has superficially scorched the orange aircraft warning paint, and the receptor is slightly eroded at the edge at the arc attachment point. The amount of surface scorching and receptor erosion depends on the lightning current parameter values (peak current and charge). At locations with high lightning intensity or much winter-time lightning activity it may be necessary to enhance the durability of the lightning protection, particularly at the tip. Methods of enhancing the durability of lightning protection sometimes used by blade manufacturers include receptor

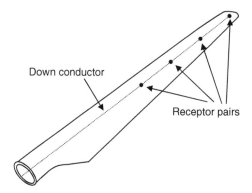

Figure 14.16   *Lightning protection for large blades with receptor pairs at the tip and on the sides of the blade*

Figure 14.17 Typical scorching of the painted surface of a blade tip on a 40-m blade where lightning has attached to the receptor (photo by the author)

materials with higher melting temperature [21], all-metal tip ends, or metal plate caps covering the tips.

Type c in Figure 14.13 has a metal mesh in the blade surface placed under the surface coating. An example of this solution was the Messerschmidt–Bölkow–Blom (MBB) blades for the Näsudden wind turbine in Sweden [5]. A metal mesh solution may be most relevant where carbon fibres are used in blade construction (see Section 14.7.1.1 for further discussion of CFRP). Type d in Figure 14.13, which is used by the German manufacturer Enercon, has a metal conductor placed in the surface all along the leading and trailing edges [27]. The large bearing between the blade root flange and the hub for the pitching motion is usually protected either with sliding contacts directly across the bearing or with sliding contact or spark gaps from the root end of the blade to the nacelle structure [27,28]. The pitching motion is controlled by hydraulics or electrically with actuator systems, which are usually placed so as to be protected inside the hub.

14.7.1.1 Blades with carbon fibre

Carbon fibres are used in some very large wind turbine blade constructions to improve the strength to weight ratio. Hitherto, the use of CFRP has been limited for reasons of cost and because blade constructors have been able to design still longer blades in glass fibre without having to use carbon fibres. CFRP blade constructions have large amounts of unidirectional carbon fibres, which in some constructions are used for a spar or several beams stretching the length of the blade while others have carbon fibres in the blade sides. Examples of blades with CFRP are the MBB blades for the Näsudden wind turbine in Sweden [5], the blades for the German Growian wind turbine [8] and the LM Glasfiber 61.5-m blade first built in 2004 for the Repower 5 MW prototype.

CFRP constructions are problematic with regards to lightning protection because carbon fibres are electrically conducting, but as the conductivity of carbon fibre is 1 000–2 000 times less than that of aluminium, a very large cross-section is needed to conduct lightning current safely. Furthermore, CFRP structures in a wind turbine blade influence the electrical field around the blade, and thereby influence where lightning attaches to the blade. Depending on the position and shape of CFRP structures in blades it must therefore be expected that the efficiency of lightning protection such as conductors and receptors is influenced by the presence of CFRP. High-voltage laboratory tests were made on the CFRP blade spar (beam) for the German Growian wind turbine, and showed that a large part of the discharges went directly into the CFRP and not to metal conductors placed on top of the spar. The conclusion was that lightning protection along the edges of the blades as in Figure 14.13d could not be expected to be efficient and that conductors along the edges of the CFRP spar would be necessary [8]. Another problem with CFRP is that the individual carbon fibres are insulated from each other by the polymer matrix, and the electrical resistance of CFRP structures with unidirectional fibres is therefore much higher in directions perpendicularly to the fibres than in the direction along the fibres. When a lightning current is injected into a carbon fibre structure the current is therefore distributed into the structure via electrical breakdown of the insulating polymer matrix between the fibres. Close to the injection point where the current density is high, heating and evaporation of the polymer matrix cause the top 1–5 layers of carbon fibre to be ripped apart within a radius of $\sim$10 cm, and the CFRP material can be heated and weakened in an area 2–10 times larger than the visibly damaged area [29–33]. Because of the length of the CFRP structures it is necessary to provide equipotential bonding to avoid potential differences and sparking between CFRP components and other conducting elements such as lightning protection down-conductors. Providing effective equipotential bonding to carbon fibre components is difficult as the task is to establish electrical connection to a large number of fine fibres embedded in insulating polymer and thereby avoid the breakdown and evaporation of the matrix that otherwise would weaken and even rip the material apart. Possible methods of improving electrical contact to the CFRP known from the aircraft industry are metal-coated fibre mats, metallic net, thin metal wires woven into the carbon fibre mats and metallic paint [9,29,31,33]. The most practical method may be to cover the whole of the CFRP component with a metallic net, which provides both equipotential bonding of surfaces and a reduction in the amount of current going into the CFRP by making use of the skin effect.

14.7.1.2 Guidelines, quality assurance and test methods

Practical experience accumulated over the period 1995–2005 clearly shows that apart from the design of the lightning protection system components, the quality of workmanship and the presence of other conduction components and even water in the blades influence the efficiency of the lightning protection system [25,34].

Electrical discharges from metallic components inside the blades is a concern, as electrical discharges (streamers) progress very easily and uncontrollable along

GFRP surfaces and may cause lightning to attach to the blade away from the lightning protection system. It is recommended, therefore, that conducting components such as down-conductor systems, balancing weights, inside the blades are encapsulated in insulating material corresponding to 50 kV lightning impulse breakdown voltage or say 3-mm glass fibre. The purpose is to impede the formation of discharges inside the blade. In general it is recommended that no components in the blade construction should be made of electrically conductive or semiconductive materials that are not absolutely necessary. Examples of components that should be avoided or only used with great care are metallic weights, wires, screws, nails and clamps, but also mouldings and other components of foam materials, which are semiconductive either in themselves or when moist. Typical examples of the latter are components coloured with carbon black. Even drawing lines with conducting colour should be avoided as discharges have been seen to track lines drawn with an ordinary graphite pencil. Accumulation of water in the blades must not be possible, and no metals should be used that may corrode and thereby pollute the inside of the blade with semi-conductive water and corrosion products.

Quality assurance procedures in the production of blades should be established to ensure the quality of build in order to reduce the number of damages due to failing lightning protection components and errors made in the mounting of lightning protection components in the blades. Tests of the correct connection of the components of the lightning protection system should be made after completion of the blade as part of quality control, for example, an electrical connection test such as measurement of resistance through the system.

The lightning protection system components used in the blades (primarily receptors and connection components) should be type tested. Tests should be made in mock ups representing the application in or on the blades and tests should be made with impulse current impulses with current stress parameters according to IEC 62305-1, and test procedures such as described in EN 50164 should be followed. The tests should prove that the surface erosion of receptors is within acceptable limits with a view to service intervals. The tests should also prove that the lightning protection components, assembly methods and mounting in the blade are all sufficiently sturdy for withstanding electrodynamic forces and for conducting the current without any damage. A full-scale structural test of the blade (see IEC 61400-23) should be made on blades including the lightning protection system in order to demonstrate that the lightning protection system is intact after testing.

The lightning attachment efficiency of the air terminals of the lightning protection system should be documented, but as of 2006 there is no standardized procedure available, and it is proposed that such documentation for blades could include the following*:

- statistically well documented operational experience
- qualifying HV laboratory test (e.g. in accordance with SAE ARP 5416 ED105, aircraft lightning test methods [35,36]), and

---

*High voltage and high current test methods for wind turbine blades will be described in edition 2 of IEC 61400-24: Wind Turbine Generator Systems—Part 24: Lightning Protection.

- HV test of the blade material as it is designed in the areas around the capturing system (air terminals/receptors) when subjected to repeated HV stress corresponding to the electric field appearing at the blade during thunderstorms.

Statistics regarding the efficiency of lightning protection systems are not well documented in the open literature, and the few investigations that have been published tend to document problems with lightning protection systems [25,34,37]. Some of the reported problems can reasonably be explained as limitations in early lightning protection designs and issues of inadequate quality of implementation. In any case it is recommended that thorough statistical documentation of operation experiences should be required of blade manufacturers.

High-voltage testing of blades with lightning protection has been performed by several researchers [8,21,35,38,39]. In general, the earlier tests have been made with a HV rod electrode positioned within a distance of ~ 1 m from the blade surface and with the blade lightning protection system connected to earth (i.e. tests performed according to MIL STD 1557A [40]). In such tests discharges tend to go through the blade surface whenever the HV rod is pointed at positions on the blade more than ~ 1 m from the air terminals (receptors), and as this is not what is seen in the field it is reasonable to conclude that this test procedure tends to stress the blade in an unrealistic way. Another test procedure for aircraft described in the SAE ARP 5416 ED105 standard has recently been proposed for wind turbine blades and reproduces lightning attachment to the blade more realistically [35]. Clearly, lightning protection of blades will be improved by applying HV laboratory tests, as insufficient solutions will be identified. It should, however, be realized that such tests are to some extent demonstrations of lightning protection concepts, and do not provide statistical documentation of the attachment efficiency unless very many tests are made. Many of the failures seen in practice are due to the inhomogeneous GFRP and the glue used for mounting the receptors and down-conductors in the blades simply not being very good HV insulating materials, which leads to breakdown through the blade materials before safe lightning attachment to an air terminal receptor has been established. It is, therefore, recommended that documentation should be required of the HV breakdown level of the blade construction materials, or at least of the materials used in areas of the blade within some metres from lightning capturing systems. Test methods for GFRP blade materials have been described in References 41–43. In other words the air terminal receptors are HV electrodes and should be considered and documented as such in terms of wear of electrode material over time and stress of the surrounding insulation systems.

14.7.2 Hub

The hub for large wind turbines is a hollow cast iron sphere of 2 or 3 m in diameter. It is usually an almost perfect Faraday cage as the openings towards the blades and the nacelle are usually electromagnetically well blocked by the blade flange plates and shaft flange and therefore the contents of the hub require no particular lightning protection. The task of lightning protection of the hub is therefore limited to equipotential bonding and transient protection of blade actuator systems, and electrical and control circuits extending to the outside of the hub.

14.7.3 Nacelle

The nacelle structure should be part of the lightning protection, so that it is ensured that lightning striking the nacelle will either strike to natural metal parts able to withstand the stress or strike to a lightning capturing system designed for the purpose. Nacelles with GFRP cover or similar should be provided with a lightning capturing system (air terminal system) and down-conductors forming a Faraday cage around the nacelle. The lightning capturing system including the exposed conductors in this cage should be able to withstand lightning strikes corresponding to the chosen lightning protection level. Other conductors in the Faraday cage should be dimensioned to withstand the share of lightning currents to which they may be exposed. Lightning air terminal systems for the protection of instruments etc. on the outside of the nacelle should be designed according to the general rules in IEC 62305-3, and down-conductors should be connected to the above-mentioned Faraday cage. A metal net (e.g. mesh width 10 cm × 10 cm) could be applied to nacelles with GFRP cover to provide magnetic shielding. Alternatively, all circuits inside the nacelle could be placed in closed metal conduits. An equipotential bonding system must be established in which the major metal structures in and on the nacelle are included, as is required in the electrical codes, and so as to provide an efficient equipotential plane to which all earthing connections should be made.

Lightning current from lightning striking the blades should preferably be conducted directly to the above-mentioned Faraday cage, thereby completely avoiding lightning current passing through the blade pitch bearings and drive train bearings. (See Sections 14.7.1 and 14.7.5 for discussion of protection of blades and bearings.) Several manufacturers use different kinds of brushes for diverting lightning currents away from bearings. However, the efficiency of such discrete brushes is low, as it is very difficult to construct the brush and earth lead systems with impedance low enough to significantly reduce the current going through the low impedance of the electrically parallel shaft and bearing systems to the nacelle bed plate.

14.7.4 Tower

A tubular steel tower, as used for large wind turbines, can also be considered an almost perfect Faraday cage, as it is electromagnetically almost closed both at the interface to the nacelle and at ground level. In order to keep the tower as closed as possible there should be direct electrical contact all the way along the joint between tower sections. The tower and all major metal parts in it should be integrated into the protection earth conductor (PE) and equipotential bonding systems to make the best of the protection offered by the Faraday cage. Ladder systems should be bonded to the tower at the top and bottom of each ladder section, and at each platform. Through-going ladder systems should be bonded to the tower every 20 m. Wires, rails and guides for hoists should, as a minimum, be bonded to the tower at the top and bottom and if possible every 20 m.

704 *Lightning Protection*

The interface towards the nacelle is usually closed with metal platforms and hatches, which can also serve as an electromagnetic shield closing the tower Faraday cage (see Section 14.7.5 for discussion of lightning protection of the yaw bearing).

The tower interface to the earthing system is discussed in Section 14.6. If the tower is constructed as a Faraday cage, as described above, then the contents of the tower require no particular lightning protection. The task of lightning protection of the tower is therefore limited to equipotential bonding and transient protection of electrical and control circuits extending to the nacelle and to the outside of the tower.

14.7.5 Bearings and gears

Few studies have been made on the effects of lightning current on bearings, and very few certain reports of damage are available [21,38,44–46]. The general picture is that lightning current passes through bearings via multiple plasma arcs through the thin oil or grease layer between the bearing rolling elements and raceways. Such arcs cause damage ranging from the size of pin holes to abrasions several millimetres in diameter depending on the current passing through the individual arcs.

The absence of reported damages may indicate that the large and slow-moving bearings for pitching blades are to some extent unaffected by lightning currents, most likely because the rolling elements are in sufficient contact with the raceways to prevent damages that can be detected without disassembling the bearings. There have been reports of damaged main bearings on older wind turbines, but not on the main bearing of large wind turbines. It is possible that lightning currents have a limited effect on large main bearings due to the size and the relatively slow rotation. A main bearing for a large wind turbine is ~ 1 m in diameter, it holds more than 100 rolling elements arranged in two rows, and usually rotates at 15–20 r.p.m.

Bearing protection is usually done by diverting the lightning current away from the bearing via brushes or spark gaps. Diverting the current from the root end of the blade via brushes or spark gaps directly to the nacelle is obviously the most effective method as current through the blade bearings and main bearings can be completely avoided [27,28]. Protection systems where brushes are applied electrically in parallel with the blade bearings and main bearings should obviously be expected to be less efficient as the fraction of current diverted through such brushes will be decided by the impedances of the brush system including connection leads and the low impedance current path through the large bearing constructions. Insertion of resistive or insulating layers in bearings to increase the impedance through the bearing and make a parallel brush system more effective is a possibility that has not been used until recently, and then only to protect against the well-known problems with generator bearing currents caused by frequency converters, which has forced the industry to give more attention to the protection of bearings.

The yaw bearing at the tower interface to the nacelle should be constructed and bridged with sliding contacts as needed to avoid too large potential differences in the case of lightning strikes and in the case of faults in electrical systems. Potential differences across the yaw bearing should be kept well below the withstand levels of the circuits crossing the bearing. Sliding contacts on the inside of the yaw bearing should be considered as low-frequency equipotential bonds only, as it is virtually impossible to divert higher-frequency currents away from the large bearings and inwards, and hence the optimal position of sliding contacts would be on the outside of the bearing. It is, however, quite unlikely that yaw bearings are damaged by lightning, as the yaw bearings are very large and slow moving.

14.7.6 Hydraulic systems

Hydraulic systems are used in wind turbines for pitching the blades, in brake systems for the blades, in the mechanical drive train, and for other purposes such as cranes. Most of these systems are situated inside the wind turbine, and are therefore not directly exposed to lightning currents. One important exception is fixed-speed wind turbines up to 2 MW, where the blade tip air brake is controlled with a wire to a hydraulics piston placed in the blade root end or in the hub as shown in Figure 14.18. When the wire controlling the system conducts lightning current this particular hydraulic system is placed directly in the down-conductor system exposed to the full lightning current, and consequently damages to the piston rods and housing can occur as seen in Figure 14.19. Different systems are used for protecting the hydraulic system holding

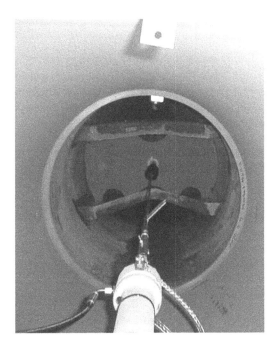

Figure 14.18   A hydraulic piston controlling a steel wire extending into the blade towards the tip air brake. A black hydraulic hose can be seen on the left side of the piston housing and on the right a flexible copper conductor connecting the steel wire and the piston housing (photo by the author).

706 *Lightning Protection*

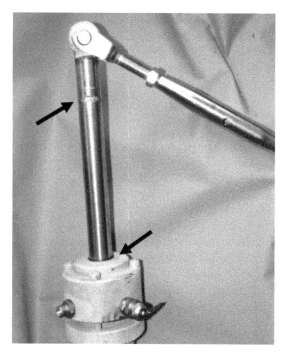

Figure 14.19 Hydraulic piston damaged by lightning. The rod was pulled into the housing when the lightning hit the blade, and the abrasions at the upper arrow were caused by a lightning arc between the piston rod and the housing at the position indicated by the lower arrow (photo by the author).

the wire to the tip brake: one is simply to bypass the hydraulic system with a flexible copper cable from the wire to the blade root as in Figure 14.18, other systems use sliding contacts on the wire or on the piston rod to avoid arcing damaging the piston rod. As hydraulic oil is flammable there is a risk of arcing causing a fire, and hydraulic pipes and hoses that may be exposed to lightning currents should therefore be protected against lightning currents. Pipes and hoses with metallic reinforcement must therefore have cross-sections able to sustain conducting the lightning currents to which they may be subjected and must be securely connected to earth at both ends.

14.7.7 *Electrical systems, control and communication systems*

Wind turbines and particularly large wind farms are being equipped with increasingly advanced electrical systems and control and communication systems in order to meet the grid connection requirements formulated by transmission system operators, and also in order to the meet the wind farm operators' general need for remote control and surveillance of the wind turbines. These systems are of course critical for the operation and control of the wind turbine and must be protected against lightning.

However, in general, the lightning protection needed for systems inside wind turbines can be done by careful selection of insulation levels and transient voltage immunity levels for the components in the systems, careful mounting of cables on ladders and in trays and conduits providing shielding against magnetic fields, and a limited number of lightning protection components (e.g. surge arresters). The circuits entering and leaving the wind turbine tower at the bottom and the circuits entering and leaving the nacelle and hub are the ones that need the most careful lightning protection consideration.

The backbone of lightning protection of the electrical system is the rated insulation level and rated impulse voltage category of all the components. This information is available from the manufacturers for all components in the LV system as it is required in the electrical codes (IEC 60364), and it is also available for MV components (Cenelec HD 637 S1). Likewise for control and communication systems, detailed information about the withstand capabilities of each component is available in the EMC documentation, as at least within the European Union all electronic equipment must be CE-marked and therefore tested according to EN61000-6-2. Insulation coordination for components in a lightning protection zone is ensured by choosing LV components and equipment of the correct impulse voltage withstand category and control and communication system components with sufficiently high EMC test levels (i.e. higher than the stress level evaluated for the zone after installation of surge protection etc. at the zone boundaries).

14.7.7.1 Electrical systems

The electrical systems can conveniently be divided into

- the generator circuit
- the MV circuit
- the auxiliary power circuit(s)

14.7.7.2 Generator circuit

The generator circuit connects the generator to the MV transformer. The auxiliary power circuit usually branches off as a TN–C–S system on the LV side of the MV transformer. The MV system (e.g. 10 kV) connects the MV transformer to a MV cable system, which connects a row of wind turbines either directly to the grid or to a transformer station stepping up the voltage to that of the sub-transmission system at for example 132 kV. Several types of generators are used in wind turbines. The two most common configurations are shown in Figure 14.20, where configuration a dominates fixed-speed wind turbines up to $\sim$2 MW, and configuration b, with a frequency converter in the rotor circuit, is used in variable-speed wind turbines from $\sim$1.6 MW and up to the largest 5 MW machines. Other configurations are also used, particularly noteworthy being the gearless synchronous generators connected via a frequency converter in the generator circuit to the MV transformer and also squirrel-cage induction generators connected via frequency converters.

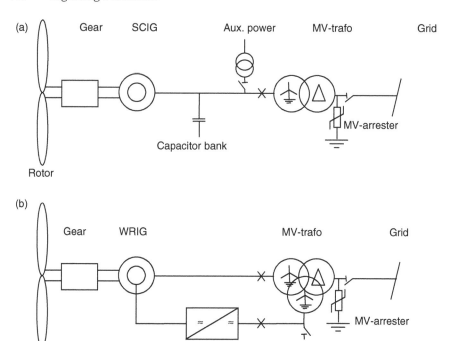

Figure 14.20 Two typical electrical system configurations in wind turbines. (a) An asynchronous squirrel cage induction generator (SCIG) with capacitor bank for phase compensation and two winding machine transformer. (b) A wound rotor induction generator (WRIG) with a frequency converter in the rotor circuit and three-winding machine transformer.

With regard to lightning protection the type of generator is of little importance. The generator, as it is situated at the back of the nacelle, is usually well protected against lightning currents. There is a possibility that small parts of the lightning current (a few per cent) could reach the generator via the mechanical drive train (i.e. hub, main shaft, gear and generator shaft) [45], but usually there is an insulating coupling in the high-speed shaft connection between the gearbox and the generator. Therefore, lightning protection as such is usually not needed for the generator, but there may be a need for equipotential bonding with arresters from the generator phase terminals to local earth in the case where local earth potential rise may exceed the insulation level of the generator or generator circuit. It is well known that frequency converters will cause destructive currents circulating between stator and rotor and though the generator bearings, if appropriate protection is not provided. Such protection includes insulation of the generator bearings, careful earthing connection of the generator housing and sliding contacts for equipotential bonding connection of the generator shaft, and although this is not exactly a lightning protection issue it will also provide protection of the bearings if some fraction of a lightning current should reach the generator.

The need for surge arresters for protection of other components in the generator circuit such as capacitor banks or frequency converters should be decided upon by comparing the stress levels evaluated for the lightning protection zone to the information available from the manufacturers about the insulation levels and transient voltage withstand levels of the components. However, for components placed in well protected nacelles and towers it may be adequate to assess whether transient protection is needed for each of the circuits connecting the components. In such an assessment the possibility of transient voltage reflections (doubling) should be considered if distances between surge arresters (e.g. at the LV terminals of the machine transformer) and, for example, a frequency converter are larger than say 10 m.

14.7.7.3 Medium-voltage system

The machine transformer (MV-transformer) may be placed in the back of the nacelle, in the bottom of the tower or next to the wind turbine tower. Surge arresters on the MV side of the transformer is probably always needed for wind turbines on land, as protection against earth potential rise when lightning strikes the wind turbines and as protection against transients entering the wind turbines from the MV collection system outside the wind turbine. MV surge arresters should preferably be placed at the transformer terminals as shown in Figure 14.20, thereby providing maximum protection for the transformer, but if the transformer is placed in the nacelle and the MV switchgear in the bottom of the tower, it may be convenient to place surge arresters at the switchgear, thereby avoiding transients from the outside being conducted up inside the tower. A closer study will be necessary to decide if arresters at the bottom of the tower can provide the needed protection of the transformer. If the machine transformer is placed outside the tower it is important that the transformer earthing system is connected to the wind turbine earthing system, and preferably there should be one earthing system.

Surge arresters on the LV side of the machine transformer are probably an appropriate general precaution, particularly if significant transients may pass through the transformer from the HV side, in which case a type of arrester suitable for a transformer application should be chosen (i.e. arresters with high energy absorption capability). The capacitive coupling between MV and LV sides of a transformer, and therefore, also the transient levels transferred to the LV side, depend very much on the design of the transformer, and particularly on the design and earthing connection of the LV winding. It is therefore advisable to obtain a sufficiently detailed transformer model from the manufacturer for transient studies in order to decide if arresters are required on the LV side of the transformer. The recommendations regarding installation practices, cable routing etc. in IEC 61000-5-2 should always be observed, and the requirements to cable insulation and transient voltage withstand levels, installation, earthing and bonding etc. in IEC 60204-1 and IEC 60204-11 should be observed.

14.7.7.4 Auxiliary power circuit(s)

As mentioned, all component manufacturers are required to provide documentation for the insulation level (rated insulation voltage category and rated impulse voltage category, IEC 60364) for any of the components that are used in the LV systems in wind turbines. Therefore, insulation coordination for components in a lightning protection zone can be ensured by choosing LV equipment of impulse voltage withstand category higher than the stress level evaluated for the zone. Alternatively, additional lightning protection zones could be defined (e.g. one or more cabinets) for which additional protection means can be applied to reduce the stress level regarding current, voltage and/or electromagnetic field. In general, such additional protection could be circuit routing in metallic cable ducts and closed cable ladders, magnetic shielding and surge arresters, or any combination thereof. The recommendations regarding installation practices, cable routing and so on in IEC 61000-5-2 should always be observed, and the requirements for cable insulation and transient voltage withstand levels, installation, earthing and bonding in IEC 60204-1 should be adhered to. A schematic of a cabinet arranged as lightning protection zone LPZ2 is shown in Figure 14.21. The metal cabinet serves as a magnetic shield – an equipotential bonding bar connected to the cabinet is placed in the cabinet to which all earthing leads are connected. In a practical application, cables would usually enter through the bottom of the cabinet, and it would be proper EMC practice to keep the power system in one side of the cabinet and C&I systems in the other. Cable shields should be earthed in EMC-type glands directly to the cabinet or to an earthing bar immediately inside the cabinet, avoiding long 'pig-tail' earthing connections. Surge protection devices (SPDs) should be placed as close as possible to the entrance of the cables, and earthing leads should be kept as short as possible. The standards provide very detailed instructions about methods both for deciding the need for arresters and practical solutions. The reader is referred to, for example, to IEC 61000-5-2, IEC 61643 or the specialized literature for further guidance [47].

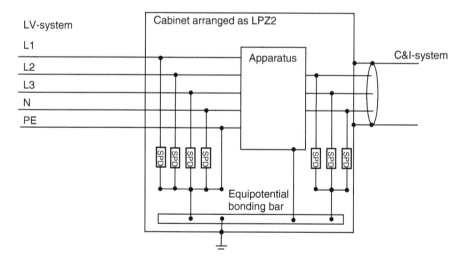

Figure 14.21   Arrangement of a cabinet arranged as a lightning protection zone LPZ2 holding apparatus too sensitive for the transient level in the outside LPZ1

14.7.7.5 Control and communication systems

C&I components to be used in a wind turbine should be tested in accordance with the requirements in EN61000-6-2, which deals with components for industrial environments. In fact this is required according to the European Union EMC directive, and hence it should be possible to obtain information about EMC immunity levels of all C&I-system components including sensors and actuators. With this information in hand it should be quite straightforward to choose components with sufficient immunity levels, and to identify components that require additional protection. In the case where additional protection is required for several components it will probably in general be convenient to place such components in one or more protected cabinets (see Figure 14.21), for which additional protection means can be applied to reduce the stress level regarding current, voltage and electromagnetic field. As with LV circuits such additional protection could be circuit routing in metallic cable ducts/closed cable ladders, magnetic shielding and arresters, or any combination thereof. The recommendations regarding installation practices, cable routing and so on in IEC 61000-5-2 should always be observed, and the requirements for cable insulation and transient voltage withstand levels, installation, earthing and bonding in IEC 60204-1 should be observed. Table 14.3 summarizes a range of test parameter values used for emc testing of components, which can be compared to Table 14.4, which shows typical levels of lightning surges appearing on conductors in lightning protection zones in buildings, and Table 14.5, which shows typical protection levels that can be achieved by applying arresters on lines crossing interfaces of lightning protection zones. As compared to ordinary buildings it should always be possible to achieve lower values of surges inside wind turbines by making the maximum use of the Faraday cage effect of the hub, nacelle and tower.

Assemblies (switchboards, cabinets, junction boxes, etc.) should be made of well connected metal with an earthing bar connected to the local earthing system in the wind turbine with short low impedance cables. Attention is drawn to the requirements

Table 14.3 Ranges of test values for tests on components with magnetic field, electrostatic discharge and surges

Test parameter	Test value	Test standard
Magnetic field	100 to 1 000 A m^{-1}	IEC 61000-6-9
Electrostatic discharge		IEC 61000-4-2
• Contact discharge	± 2 to 8 kV	
• Air discharge	± 2 to 15 kV	
Surges 1.2/50 (8/20)		IEC 61000-4-5
• Line to line power ports	± 0.5 to 2 kV (0.25 to 1.0 kA)	
• Line to earth power ports	± 0.5 to 2 kV (0.25 to 1.0 kA)	
• Line to line signal ports	± 0.5 to 2 kV (0.25 to 1.0 kA)	
• Line to earth signal ports	± 0.5 to 1 kV (0.25 to 0.5 kA)	

Table 14.4 *Typical values of surges appearing on conductors in lightning protection zones LPZ 0 and LPZ 1 in buildings (from Reference 47)*

Test parameter	LPZ 0_A	LPZ 0_B	LPZ 1
Surge amplitude and wave shape	100 kA, 10/350 μs* 25 kA, 0.25/100 μs* 200 A, 0.5 s*	10 kV, 1.2/50 μs 5 kA, 8/20 μs	6 kV, 1.2/50 μs 3 kA, 8/20 μs

*To be divided between the number of cables n and number of conductors per cable m.

Table 14.5 *Typical protection levels provided by arresters on lines crossing interfaces of lightning protection zone LPZ 0/1 and LPZ 1/2 (from Reference 47)*

Test parameter	LPZ 0/1	LPZ 1/2
Surges 1.2/50 (8/20) • Power lines • Signal lines	 4 kV (2 kA) 2 kV (1 kA)	 1.5 kV (0.75 kA) 1.0 kV (0.5 kA)

in IEC 60204-1 §14.1.3, stipulating that where conductors belonging to two different circuits are placed together, for example in cable ducts, such conductors shall be insulated for the highest voltage to which any of the conductors in the same duct may be subjected. This means that where, for example, signal conductors are routed in a duct together with 690 V conductors, the signal conductors are also to be insulated as 690 V conductors.

Insulation coordination for C&I system equipment in a lightning protection zone can be ensured by documenting that the impulse voltage testing levels for C&I system equipment are higher than the stress level evaluated for the zone where the equipment is placed. Cables for C&I systems in wind turbines should have a metal braid screen of sufficient cross-section for conducting the lightning current or induced current that may appear at the place in question. The screen should be earthed (bonded) at both ends. Screening for cables and sensors may consist wholly or partially of screening cable routing such as closed cable trays or metal conduits earthed at both ends. Cables for the C&I system connecting equipment in different lightning protection zones should preferably be optical cables. Connections to the external communication should always be optical cables, as it is extremely difficult to protect metallic wire communication circuits leaving the wind turbines, mainly because of the earth potential rise of the wind turbine earthing system when lightning strikes, which will force significant parts of the lightning current to flow into the communication cable screen and signal conductors.

14.8 Wind farm considerations

Large wind turbine projects where many wind turbines are erected and operated together are usually called wind farms. In flat areas the wind turbines are usually placed in rows, forming a simple geometric pattern, but in hilly or mountainous areas it is the high points and ridges that decide where the individual wind turbines are placed. In some wind farms it has been experienced that certain wind turbines are more exposed than others. Typically, these are the wind turbines that are placed at the wind farm boundaries facing the typical direction of approaching weather systems.

The risk of lightning striking the wind turbines in a wind farm can be evaluated with the method for individual wind turbines as outlined in Section 14.4, and, if necessary, correcting for overlapping collection areas. The wind turbines in a wind farm are connected in rows to a collection system of MV cables feeding the power to a substation, stepping up the voltage to that of the power grid connection. In modern wind farms there is also a communication system connecting all the wind turbines either to the public telephone and data network or in most cases to a wind farm computer system that handles communications and control for all the wind turbines for the wind farm. In wind farms it is therefore obviously important that lightning striking one wind turbine cannot affect the other wind turbines in the wind farm via the MV power cables and the wind farm communications and control system. The MV power cable system should be protected with arresters in each wind turbine as described in Section 14.7.7.3 and at the substation. The earthing systems of the individual wind turbines and the wind farm substation should be interconnected with horizontal earthing conductors running in parallel to and in the same trench as the MV power cables. Such an interconnected earthing system will mainly serve to reduce the earthing resistance in the wind farm in general, reduce potential differences in the wind farm, and reduce lightning currents affecting communication cables and the power cable shields. Previously, there have been cases where during one storm many wind turbines in a wind farm have been affected by lightning because transient overvoltages could spread through the metallic wire communication systems. As examples of widespread lightning damages to control systems in wind farms, three cases in Denmark are mentioned. In a thunderstorm in June 1988 lightning damaged the control systems in all 36 wind turbines in a wind farm. In a thunderstorm in September 1988 lightning damaged the control systems in 17 wind turbines in another wind farm. This same wind farm was hit again by a thunderstorm in May 1998 when lightning damaged control systems in 22 wind turbines. Obviously, if metallic wire communication systems are used for wind farm communication systems it is necessary to provide transient protection as described in Section 14.7.7.5 in order to protect the systems in the wind turbine. Even with transient protection of the communication systems it is very difficult to avoid lightning current from spreading as transient overvoltages into metallic wire communication cables outside the wind turbines unless large-cross-section cable shields are used, or alternatively the cables are placed in shielded cable ducts. In general, communication and control cables in wind farms should therefore always be optical fibre cables, as such cables are not affected by

714 *Lightning Protection*

lightning. Fortunately, in modern wind farms optical cables may also be the natural choice anyway, as modern communications and control systems used in wind farms are high-capacity 10–100 Mbit computer networks.

14.9 Off-shore wind turbines

Lightning protection of offshore wind turbines does not differ much from wind turbines on land. The main difference is that accessing wind turbines offshore is not as easy as on land. First, winds and waves must be below certain limits and then it usually involves a two-person service crew and a service boat or even a helicopter with another two-person crew. Hence, just the costs of transportation and man hours is 2 000–3 000 € per trip, or 5 to 10 times the costs of sending a service crew to a wind turbine on land. This example is included to illustrate the importance of systematic lightning protection of all circuits and systems in an offshore wind turbine. The trial and error approach sometimes applied on wind turbines on land would certainly result in very high maintenance costs with offshore wind turbines, which obviously puts an end to all discussions about whether or not a 50 € transient protection should be spent on for example the circuits for a wind turbine weather station.

Offshore wind farms are so far not located more than ∼15 km from the coast, and therefore lightning ground flash densities are not much different from what is seen at the coast. In the future, wind farms will probably be placed farther from the coast, where lightning occurrence generally is lower. Risk assessments can be made as described in IEC 62305 part 2, where it would probably be reasonable to use a location factor $C_d = 2$ or maybe even higher, as the offshore wind turbines are the only tall objects in the area, with the exception of possible meteorological masts and offshore transformer platforms. However, to date (2006) there is no published information available about practical experience regarding lightning striking frequency for off-shore wind turbines.

Offshore wind turbines are often equipped with relatively advanced equipment such as communication antennas and transponders, GPS receivers, sea marking lights, air traffic warning lights, visibility detectors, fog-horns, meteorological instruments, all of which are usually connected with cables to electronic systems placed inside the wind turbine. Each and every one of such circuits must be considered carefully and must be lightning protected with air terminals (see Figure 14.22), surge protection and metallic conduits for outside cabling, as is appropriate.

There are two main types of foundations used for offshore wind turbines: concrete foundations (see Figure 14.23) and mono-pile foundations. The first type is basically a very heavy steel reinforced concrete plate structure placed on the sea bed with a tower extending to well over the water level of highest tide, whereas the mono-pile foundation is a steel construction that is driven or drilled say 20 m into the sea bed. The earthing systems for offshore concrete foundations are in principle made the same way as described in Section 14.6 for foundations on land. Steel mono-pile foundations should be considered as an extension of the tower, so electrical continuity must be

Lightning protection of wind turbines 715

Figure 14.22 Meteorological instruments on an offshore wind turbine protected with lightning air terminals (photo by the author)

Figure 14.23 Concrete offshore foundations under construction on a barge for later deployment at offshore (photo by the author)

ensured between the tower bottom flange and the foundation top flange and any other interface between metallic parts of the foundation structure. In both cases care must be taken to ensure good connections from the overall foundation earthing network in concrete foundations and from the mono-pile steel structure, respectively, to all metal objects in contact with the sea water, for example boat landings, ladders, sacrificial

716 *Lightning Protection*

anodes for corrosion protection and marine cable armouring. Under no circumstances should copper be used as earthing electrodes in sea water, as that would increase the corrosion of the foundation steel parts.

14.10 Lightning sensors and registration methods

As wind turbines are placed at remote locations and are remotely operated it is difficult to know when a turbine has been hit by lightning. In the past damages may often have led to an immediate stop of the turbine or to loss of communication connections, which would cause the operator to send someone to inspect the turbine. However, wind turbines have in some cases continued operating after being hit by lightning, which involves risks of small damages developing into unnecessarily large damages. To address such concerns, lightning sensors have been developed for wind turbines, which register the magnetic field from the lightning current and either send an alarm signal to the wind turbine controller or send an alarm and register information about the lightning current pulses. An example of the first type is the Jomitek sensor, which picks up the lightning current magnetic field with a pair of loop antennas placed on opposite sides of the wind turbine tower (see Figure 14.24) [48,49]. Wind turbine blade manufacturers such as Vestas and LM Glasfiber have developed lightning current sensors that are placed on down-conductors in the blades. A more simple magnetic-link registration can be made with magnetic strip cards (similar to a credit card), which can be placed on lightning conductors for registering the peak lightning current amplitude by simple erasure of a signal pre-recorded on the magnetic strip (see Figure 14.25).

Figure 14.24 *Lightning sensor antenna on wind turbine tower (photo by the author)*

Lightning protection of wind turbines 717

Figure 14.25   Magnetic strip card for registering peak lightning current (photo by the author)

14.11   Construction phase and personnel safety

Erection of large wind turbines on land takes several days when including the time it takes to assemble and disassemble the very large cranes that are used. Offshore wind turbines on the other hand may be erected in less than a day by the use of specially designed vessels or jack ups. In any case there is usually up to a few weeks of post-erection completion work before the wind turbine is commissioned. During this time many people are at work in, on and around the wind turbine, and they are at considerable risk of being affected if lightning strikes the wind turbine. Therefore, it is highly recommended that the construction site manager organizes proper safety procedures with regard to lightning, which should include the following:

- checking local weather forecasts regularly (e.g. every morning)
- applying intermediate earthing system connections as soon as possible
- keeping a look out for developing thunderclouds, audible thunder and visible lightning
- identifying safe locations
- Providing an acoustic warning signal

Weather offices will usually provide reasonably accurate forecasts with regard to the possibility of thunderstorms, but may have difficulties in saying when and how active it will be. Some weather offices provide warning services by telephone, fax or internet, which should definitely be considered, but it should not replace instructing people on site keeping lookout for developing thunderclouds, audible thunder (within 10–15 km) and visible lightning (within ~30 km). Local area and even portable lightning detection and thunderstorm warning devices are available from different manufacturers. High-end and recommended examples are the ESID or TSS924 devices from Vaisala (previously Global Atmospherics), which have been successfully used in connection with wind turbines [50]. During construction work connections of cranes, generators and so on to the foundation earthing system should be made as soon as possible. Safe locations in a wind turbine are on platforms inside the tower, as the tower is a near-to-perfect Faraday cage. People should be instructed to stop work and go sit down on the middle of the closest platform inside the tower until the thunderstorm has passed. Being on the outside of the nacelle is definitely not safe. People who, while standing on a nacelle, experience that their hair stands on its ends and who hear crackling noise coming from the weather station and other extremities, should know that they are being given the very last warning. People stepping out of the wind turbine, standing next to the tower, climbing ladders or entering boats etc. will be at risk if lightning strikes the wind turbine. They should therefore be instructed to stay in the wind turbine tower until the danger is over. Other safe places are inside metal roof vehicles, containers, and so on. As it may be difficult to communicate effectively in a construction area, some kind of acoustic warning signal should be agreed. This could just be a repeated honking of a car horn or a compressed air horn.

References

1. Wachtmann H. Patentschrift no. 28489. Kaiserliche Patentamt, Germany, 1883.
2. BTM Consult, World Market Update 2003, March 2004.
3. BWEA, Briefing Sheet Offshore Wind, at www.bwea.com
4. Bankaitis H. Evaluation of Lightning Accommodation Systems for Wind Driven Turbine Rotors. DOE/NASA/20320-37, NASA TM-82784, USA, 1982.
5. Dalén G. 'Lightning protection of large rotor blades. Design and experiences'. *IEA 26th Meeting of Experts*, Milan, Italy, 1994.
6. Dodd C.W., McCalla T., Smith J.G. How to Protect a Wind Turbine from Lightning. DOE/NASA/0007-1, NASA CR-168229, USA, 1983.
7. Jaeger D. 'Design provisions for wind energy systems lightning protection and EMC'. *5th Meeting of Experts*. Environmental and Safety Aspects of the Present Large Scale WECS, IEA, 1980.
8. Molly J.P. Blitzschutzversuche an einem CFK-Roterblatt, Deutsche Forshungs – und Versuchsanstalt für Luft – und Raumfahrt e.V. Insitut für Bauweisen- und Konstrucktionsforshung, Stuttgart.
9. Fisher F.A., Plumer J.A. *Lightning Protection of Aircraft*, Lightning Technologies Inc., Pittsfields MA, USA, 1990.
10. Goto Y., Narita K., Komuro H., Honma N. 'Current waveform measurement of winter lightning struck on isolated tower'. *20th ICLP*, Interlaken, Switzerland, 1990.

11. Soerensen T., Jensen F.V., Raben N., Lykkegaard J., Saxov J. 'Lightning protection for offshore wind turbines'. *CIRED 2001*, Amsterdam, The Netherlands, June 2001; 18–21.
12. Berger K. 'Blitzstrom parameter von Aufwärts Blitzen'. *Bull. SEV/VSE, Bd.* 69, 1978;**8**: 353–360.
13. Eriksson A.J., Meal D.V. 'The incidence of direct lightning strikes to structures and overhead lines'. *IEE Conference on Lightning and Power Systems*, IEE Conference Publications 1984; 236.
14. Schei A. 'Lightning protection of wind generator systems. Service experience from norway and proposed solutions to reduce lightning damage'. *IEA 26th Meeting of Experts*, Milan, Italy, 1994.
15. Soerensen T., Pedersen Aa., Jeppesen R.T. 'Lightning parameters contra climatic conditions in Denmark'. *21st ICLP*, Berlin, Germany, 1992.
16. Wada A., Asakawa A., Shindo T. 'Characteristics of lightning flash initiated by an upward leader in winter'. *23rd ICLP*, Florence, Italy, 1996.
17. Zundl T., Fuchs F., Heidler F., Hopf Ch., Steinbigler H., Wiesinger J. 'Statistics of current and field measurements at the Peissenberg Tower'. *23rd ICLP*, Florence, Italy, 1996.
18. Anderson R.B., Eriksson A.J. 'Lightning parameters for engineering application'. *Cigré Electra*, 1980;**69**:65–102.
19. Berger K., Anderson R.B. 'Parameters of lightning flashes'. *Cigré Electra*, 1975;**41**:23–37.
20. Cotton I., McNiff B., Soerensen T., Zischank W., Christiansen P., Hoppe-Kilpper M. et al. 'Lightning protection for wind turbines'. *25th ICLP*, Rhodes, Greece, 2000.
21. Cotton I., Jenkins N., Hatziargyriou N., Loretzou M., Haigh S., Hancock A. Lightning Protection of Wind Turbines, a Designers Guide to Best Practice. University of Manchester, UMIST, 1999.
22. DEFU Recommendation 25, Lightning Protection for Wind Turbines, DEFU, 1999.
23. Hermoso B. et al. 'Risk assessment on a windmill higher than 90 meters'. *26th ICLP*, Cracow, Poland, 2002.
24. Weinhold N. 'Inconspicuous world champions'. *New Energy* **5**, 2005.
25. Soerensen T., Brask M.H., Olsen K., Olsen M.L., Grabau P. 'Lightning damages to power generating wind turbines'. *24th ICLP*, Birmingham, UK, September 1998, 14–18.
26. Shiranishi Y., Otsuka T., Matsuura H. 'The observation of direct lightning stroke current to the wind turbine system'. *28th ICLP*, Avignon, France, 2004.
27. Wobben A. Windenergieanlage mit Blitzschutzeinrichtung. German patent DE 44 36 197 A1, 1994.
28. Scheibe K. Blitzschutzmassnahmen für eine Windkraftanlage. VDE/ABB-Blitzschutztagung, Neu Ulm, Germany, 1999.
29. Brocke R. et al. 'The stress caused to mesh protected composite materials at the attachment point of lightning arcs'. *25th ICLP*, Rhodes, Greece, 2000.

30. Clarke B., Reid G.W. 'Further investigations into the damage of various types of unprotected carbon fibre composites with a variety of lightning arc attachments'. *16th International Aerospace and Ground Conference on Lightning and Static Electricity*, Mannheim, Germany, 1994.
31. Evans R.E. et al. 'Nickel coated graphite fibre conductive composites'. *SAPE Quarterly*, July 1986;**17**:(4).
32. Gondot P. et al. 'Ligtning protection of aeronautical structural materials'. *22nd ICLP*, Berlin, Germany, 1996.
33. Jones C.C.R., Burrows B.J.C. A Designers Guide to the Correct use of Carbon Fibre Composite Materials and Structures Used in Aircraft Construction, to Protect Against Lightning Strike Hazards. Culham Laboratory, UK, 1990.
34. McNiff B. Wind Turbine Lightning Protection Project 1999–2001, National Renewable Energy Laboratory, USA, Report NREL/SR-500-31115, 2002.
35. Larsen F.M., Soerensen T. New Lightning Qualification Test Procedure for Large Wind Turbine Blades. International Conference on Lightning and Static Electricity, Blackpool, UK, 2003.
36. SAE ARP 5416, Ed 105, Aircraft Lightning Test Methods.
37. Yoh Y., Toshihisa F. 'Lightning analysis on wind farm – sensitivity analysis on earthing'. *28th ICLP*, Avignon, France, 2004.
38. Cotton I., Jenkins N., Pandiaraj K. 'Lightning protection for wind turbine blades and bearings'. *Wind Energy*, January/March 2001;**4**(1):23–37.
39. Schmid B.R. 'Investigations on GRP-rotor blade samples of wind power plants regarding lightning protection'. *24th ICLP*, Birmingham, UK, September 1998, 14–18.
40. MIL-STD-1757A: Military Standard: Lightning Qualification Test Techniques for Aerospace Vehicles and Hardware, Department of Defence, USA, 1983.
41. Madsen S.F., Holboell J., Henriksen M., Bjaert N. 'Tracking tests of glass fiber reinforced polymers (GRP) as part of improved lightning protection of wind turbine blades'. *28th ICLP*, Avignon, France, 2004.
42. Madsen S.F., Holboell J., Henriksen M., Krog-Pedersen S. 'Breakdown tests of composite materials, and the importance of the volume effect'. *Nordic Insulation Symposium*, NORDIS, 2005.
43. Madsen S.F., Holboell J., Henriksen M., Assentoft J. 'Experimental investigation of the relationship between breakdown strength and tracking characteristics of composites'. *Nordic Insulation Symposium*, NORDIS, 2005.
44. Celi O., Pigini A. 'Evaluation of damage caused by lightning current flowing through bearings'. *IEA 26th Meeting of Experts*, Milan, Italy, 1994.
45. Garbagnati E., Pandini L. 'Current distribution and indirect effects on a wind power generator following a lightning stroke'. *IEA 26th Meeting of Experts*, Milan, Italy, 1994.
46. Soerensen T., Brask M.H., Jensen F.V., Raben N., Saxov J., Nielsen L. et al. 'Lightning protection of wind turbines'. *European Wind Energy Conference and Exhibition*, Nice, France, 1999.
47. Hasse/Wiesinger, EMV Blitzshutzzonen Koncept, VDE-Verlag 1993, ISBN 3-8007-1982-7.

48. Soerensen T., Johansen P., Jensen F.V. 'Lightning sensor'. *European Wind Energy Conference and Exhibition*, Copenhagen, Denmark, 2001.
49. Soerensen T., Brask M.H., Johansen P., Jensen F.V., Raben N., Soerensen J.T. et al. 'Lightning strike sensor for power producing wind turbines'. *European Wind Energy Conference and Exhibition*, Nice, France, 1999.
50. McNiff B. Wind Turbine Lightning Protection Project: Final Report, National Renewable Energy Laboratory, USA, MLI Report No. 0901-nrel, 2001.
51. Avrootskij V.A., Sergievsaya I.M., Sobolevskaya E.G. 'Electrical characteristics of graphite epoxy and their lightning protection'. *20th ICLP*, 1990, Interlaken, Switzerland.
52. Wada A., Yokoyama S. 'Lightning damages of wind turbine blades in winter in Japan – lightning observation on the Nikaho-Kogen wind farm'. *28th ICLP*, Avignon, France, 2004.
53. IEA: Recommended Practices for Wind Turbine Testing and Evaluation–9. Lightning Protection for Wind Turbine Installations, IEA, 1997.
54. Ukar O., Zamora I., Idiondo R., Mugica A. 'Analysis for high frequencies of grounding systems for wind turbines'. *28th ICLP*, Avignon, France, 2004.

Referenced standards

IEC 62305-1: Protection against lightning – Part 1: General principles.
IEC 62305-2: Protection against lightning – Part 2: Risk management.
IEC 62305-3: Protection against lightning – Part 3: Physical damage to structures and life hazard.
IEC 62305-4: Protection against lightning – Part 4: Electrical and electronic systems within structures.
IEC 62305-5: Protection against lightning – Part 5: Services (draft).
IEC 61400-23: Wind turbine generator systems – Part 23: Full-scale structural testing of rotor blades.
IEC 61400-24: Wind turbine generator systems – Part 24: Lightning protection.
IEC 61000-4-2: Electromagnetic compatibility (EMC) – Part 4-2: Testing and measurement techniques – Electrostatic discharge immunity test.
IEC 61000-4-5: Electromagnetic compatibility (EMC) – Part 4-5: Testing and measurement techniques – Surge immunity test.
IEC 61000-4-9: Electromagnetic compatibility (EMC) – Part 4-9: Testing and measurement techniques – Pulse magnetic field immunity test.
IEC 61000-6-2: Electromagnetic compatibility (EMC) – Part 6-2: Generic standards – Immunity for industrial environments.
IEC 61000-6-4: Electromagnetic compatibility (EMC) – Part 6-4: Generic standards – Emission standard for industrial environments.
IEC 61000-5-2: Electromagnetic compatibility (EMC) – Part 5: Installation and mitigation guidelines – Section 2: Earthing and cabling.
EN 50164: Lightning Protection Components (LPC).
IEC 60364/Cenelec HD 384: Electrical installations of buildings.
Cenelec HD 637 S1: Power installations exceeding 1 kV a.c.

IEC 61643-1: Surge protective devices connected to low-voltage power distribution systems – Part 1: Performance requirements and testing methods.

IEC 61643-12: Surge protective devices connected to low-voltage power distribution systems – Part 12: Selection and application principles.

IEC 60204-1: Safety of machinery, Electrical Equipment of machines – Part 1: General requirements.

IEC 60204-11: Safety of machinery, Electrical Equipment of machines – Part 11: Requirements for HV equipment for voltages above 1 000 V a.c. or 1 500 V d.c. and not exceeding 36 kV.

Chapter 15
Lightning protection of telecommunication towers
G.B. Lo Piparo

15.1 Lightning as a source of damage to broadcasting stations

15.1.1 General

Regarding the protection of broadcasting stations, the effects of lightning flashes in general create problems that are difficult to solve. Lightning phenomena are extremely random in nature and, because of the wide use of solid-state techniques, it is necessary to ensure that the overvoltages affecting such equipment do not exceed very low values. In unattended stations, which are commonly very exposed to lightning and difficult to access, damage from lightning flashes, whether or not the station is actually struck, can be particularly serious because of the period of time during which the station is out of service, resulting in unacceptable loss of public service.

Figure 15.1 gives a typical distribution of the number of breakdowns per year and the mean lost transmission time in hours per year caused by lightning flashes to unattended unprotected broadcasting stations in Europe [1].

The lightning current is the source of damage. The type of damage depends both on the position of the point of strike relative to the station and on the characteristics of the station involved:

1. *Injury to people.* This arises primarily due to touch and step voltages and may result in attended stations as a consequence of flashes to the station or to the lines connected to the station.
2. *Physical damage*
    - There may be immediate mechanical damage due to thermal and/or electrodynamic stresses.
    - There may be fire due to the hot lightning plasma arc itself or triggered by sparks caused by overvoltages resulting from resistive and inductive coupling and to the passage of part of the lightning currents, or due to the current resulting in ohmic heating of conductors (over-heated conductors), or due to the charge resulting in arc erosion (melted metal).

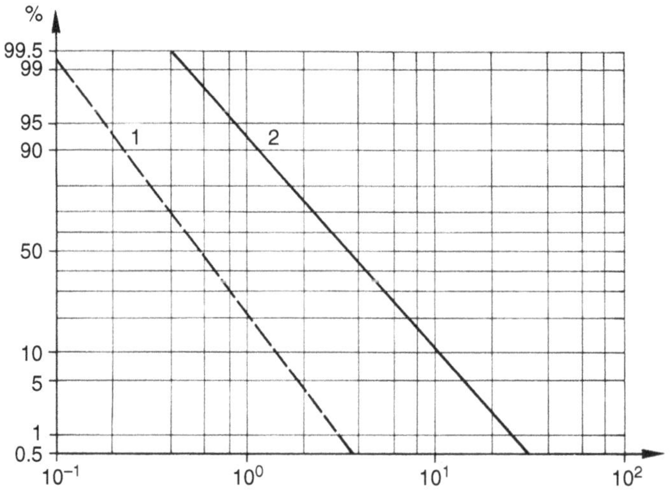

Figure 15.1 *Typical distribution of (1) the number of breakdowns per year and (2) the mean lost transmission time in hours per year caused by lightning flashes to unattended unprotected broadcasting stations in Europe*

3. *Failure of internal electrical and electronic systems due to electromagnetic effects of lightning current (LEMP)*. This results from surges (overvoltages and overcurrents) arising from resistive and inductive coupling the lightning currents flowing across the struck station or appearing on connected lines and transmitted to the station.

Depending on the point of strike, the following may occur.

- Flashes to the station and/or to lines connected to the station may result in physical damage, in failures of internal systems, and even in injury to people if the station is attended.
- Flashes to the ground near to the station and/or near to lines connected to the station may cause failures of internal systems only.

In practically every case of lightning damage to a station, it is the power supply system or the antenna support structure that is struck. Lightning flashes to or near the supply line give rise to overvoltage surges, which may reach the radio and telecommunication equipment by way of the step-down or isolating transformer, with risk of failure of insulation, whereas flashes to the antenna support structure itself tend to engender heavy currents in the down-conductors and earth-termination system, causing large differences of potential between different parts of the earthing system, as well as thermal and mechanical stresses.

15.1.2 Injury to people

In principle, the effect on a person inside or near a broadcasting station when it is struck by lightning is the same as if he received an electric shock from any apparatus under tension. In brief, there will be temporary or permanent, possibly fatal, paralysis of one or more organs, usually with burning at the points of contact.

Based on the flow of lightning current through the down-conductor systems into the earth-termination system and its dispersal in the soil, two specific concepts have been adopted, namely touch-voltage and step-voltage.

Touch-voltage may be defined as the potential difference caused by a lightning current between any conducting object that a person can touch and the floor or ground on which he stands. Moreover, the lightning current in the down-conductors may induce dangerous voltages to earth in nearly all the metallic loops in the station.

The step-voltage is related to the differences of potential in the Earth's surface while the energy of the lightning current dissipates in the soil; the voltage between the two points touched by a person's feet may be injurious or even fatal. This risk will usually exist up to ~ 50 m from the point where lightning strikes the ground or, in the case of a flash to the station, in the vicinity of the earth-termination system, up to distances depending upon the magnitude of the lightning current, the configuration of the earth-termination system and the soil resistivity.

15.1.3 Physical damage

Physical damage in a broadcasting station mainly arises from the thermal and electrodynamic effects of the lightning current.

15.1.3.1 Thermal effects

Typically, heat is generated at the point where the lightning current enters a good electric conductor, and the quantity of heat may be sufficient to cause fusing of the conducting material and/or ignition of adjacent non-conductors. Methods have been developed for deducing the charge in lightning discharges from the amount of metal melted. For example, at the point where a mast structure is struck, 4.4 mm^3 As^{-1} for steel, 5.4 mm^3 As^{-1} for copper and 12 mm^3 As^{-1} for aluminium may be melted, and the piercing of metal sheet of up to 2 mm in thickness has been reported.

Because the duration of a pulse of lightning current is very short, the heating effect within good conductors is not usually enough to cause fusing; calculation and experience agree that copper of cross-section 16 mm^2 and steel of 25 mm^2 can carry typical lightning currents without damage from thermal effects, a temperature rise of 100 K being generally tolerable. However, lead sheaths of cables can be damaged because of the low melting point of lead.

The case of non-conductors is quite different. Very large quantities of heat are generated when the lightning current passes through a non-conductor, such as wood or brickwork. The moisture occluded in the material is instantaneously evaporated and the resulting very high pressure causes an explosion. This effect is most likely to occur where humidity can collect – in cracks, cavities, joints, etc. – and where the

726 *Lightning Protection*

lightning current is concentrated on entering or leaving a poor conductor, at its junction with a good conductor, such as a water pipe, reinforcement bar or electric cable.

15.1.3.2 Electrodynamic effects

Where two more-or-less parallel conductors both carry lightning current, a mechanical force acts between them, its magnitude depending upon the intensity of the current and the distance between the conductors. If a fixed conducting rod is struck perpendicularly to its longitudinal axis, it undergoes a mechanical stress in the direction of the lightning channel. When carrying the lightning current, bends or loops in conductors tend to straighten.

In addition to those disruptive forces, the attraction between close, parallel conductors carrying the lightning current can cause tubular elements to collapse or the insulation between conductors to be squeezed out of place.

15.1.4 Failure of internal electrical and electronic systems

Breakdowns resulting from lightning flashes occur most commonly in the power supply installations of broadcasting stations, most often in the sections directly connected to the feeding line. A complete failure will occur only if the operational, the reserve and/or the control equipment are directly exposed to overvoltage surges on the supply line. Failures of step-down or isolating transformers can occur if the surge protective devices (SPDs) provided for their protection are not properly selected and/or installed. The high magnetizing current taken by a transformer subjected to a voltage surge may cause the fuses to burn.

Two categories of breakdowns of radio and telecommunication equipment, including the low-voltage (LV) sections of the electricity supply system, may be distinguished, namely those of equipment directly subjected to the lightning current and those subjected to transferred overvoltages.

The costs of lightning damage to broadcasting stations are predominantly those incurred for the attendance of maintenance personnel. The cost of equipment replacement is usually relatively small.

15.2 Effects of lightning flashes to the broadcasting station

15.2.1 Effects of lightning flashes to the antenna support structure

These are relatively frequent in the case of broadcasting stations, because, for technical reasons, the antennae must usually be erected at a considerable height above the ground, and often on high-altitude sites in regions of high keraunic activity. The metallic supporting structure thus tends to favour the origination of upward leaders and thereby flashes to the structure itself.

During the dispersion of the lightning current into the earth, considerable potential differences exist in the ground in the vicinity of the station, so it is necessary to define 'zero potential' rather vaguely as being that of the earth far enough away from the station as to be unaffected by the lightning flash in question.

Because the lightning current wave has in general a very steep front and a less abrupt decay, not only the resistance of the earthing system, but more importantly its inductance and to a lesser extent its capacitance determine the resultant potential differences in the station. The significant parameter is therefore surge impedance; unfortunately, no rigorous definition applicable to lightning protection has yet been adopted internationally. However, the notion of surge impedance is sufficiently well known to enable its effects to be gauged, at least qualitatively. A convenient definition in this context is the ratio between the maximum voltage and the peak value of the current at the input of, for example, a buried earth termination. This makes it possible to estimate the maximum voltage in a particular case, because the peak current can be assumed. Such a surge impedance is defined in the standard IEC 62305 as 'conventional earth impedance'.

Figure 15.2a, b depicts plots of the conventional earth impedance and the surge coefficient of a wire buried in ground having soil resistivities of 100 and 1 000 Ω m. The surge coefficient is the ratio between the conventional earth impedance and the resistance of the wire at 50 Hz. In the case of an earth termination consisting of a long buried wire, the potential difference between the wire and the surrounding soil diminishes with distance from the point of entry of the lightning current, the more rapidly the lower the soil resistivity. Figure 15.3 shows typical curves of the ratio of the potential difference between the end where the lightning current enters and points along the wire, from which it can be seen that no advantage is gained with long buried earth terminations in soil of low resistivity.

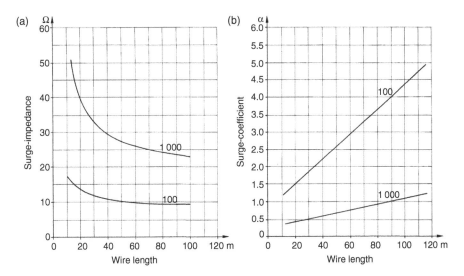

Figure 15.2   *(a) Conventional earth impedance of a buried wire as a function of its length. Impulse current: 5 kA, 7/17 μs. Parameters soil resistivity (Ω m). (b) Surge coefficient of a buried wire as a function of its length. Parameters soil resistivity (Ω m).*

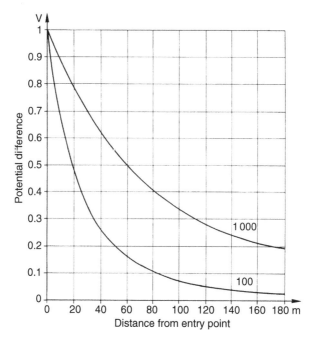

Figure 15.3 Decrease of potential along a buried wire. Parameters soil resistivity ($\Omega\,m$).

Flashes to the antenna support structure can have three types of effects.

1. The lightning current I flowing through the impedance Z of the earth-termination system raises it up to the potential $U = ZI$ referred to the zero potential of the remote earth. The difference of potential between these two points is usually called the 'earth-termination voltage'. The equipment of the station connected to the electrical or telecommunication lines entering the station are stressed by the earth-termination voltage, being the conductors of the lines at the potential of the remote earth at the far end. The earth-termination voltage may easily rise up to several hundreds of kilovolts even for values of impedance Z of some ohms; as a consequence, failure of the equipment of the station connected to lines is practically guaranteed. Following the breakdown of the insulation of the equipment, live conductors of the lines – as well as the screen of the cables and external conductive parts such as pipe works, metal ducts, etc. bonded to the earth-termination system of the station – carry a part of the lightning current towards the remote earth.
2. The lightning current I, on its way to earth and due to the voltage drop on the inductance of the conductors of the earthing system, may cause large potential differences between different parts of it, with resultant damage to sensitive equipment connected to them.
3. The passage of the typically very heavy lightning current can, in addition to thermal and electrodynamic effects, cause, by induction, surges (overvoltages

and overcurrents) in nearby conducting loops such as electrical circuits that may cause damage to the connected equipment.

15.2.1.1 Lightning current flowing through external conductive parts and lines connected to the station

The evaluation of the part of lightning current flowing through external conductive parts and lines connected to the station is needed to select SPD for protection against surges of equipment connected to lines. Namely, the threat due to these surges must be lower than the withstand of the SPD used (defined by adequate tests).

When conducted to earth, the lightning current I (the value of I depends on the lightning protection level (LPL) needed for protection of the station according to a risk assessment, see Section 15.4.4) is divided between the earth-termination system, the external conductive parts and the lines, directly or via the SPDs connected to them. If

$$I_f = k_e I$$

is the part of the lightning current relevant to each external conductive part or line, then k_e depends on

- the number of parallel paths
- their conventional earth impedance for underground parts of line, or their earth resistance, where overhead parts connect to underground parts, for overhead parts of line
- the conventional earth impedance of the earth-termination system

Assuming as a first approximation that one-half of the lightning current flows in the earth-termination system, the value of k_e may be evaluated for an external conductive part or line by

$$k_e = 0.5/n$$

where n is the overall number of external parts or lines connected to the station.

If entering lines (e.g. electrical and telecommunication lines) are unshielded or not routed in a metal conduit, each of the n' conductors of the line carries an equal part of the lightning current

$$k'_e = k_e/n'$$

For shielded lines bonded at the entrance, the values of current k'_e for each of the n' conductors of a shielded service are given by

$$k'_e = k_e R_s/(n' R_s + R_c)$$

where R_s is the ohmic resistance per unit length of shield and R_c the ohmic resistance per unit length of inner conductor.

In an LV power supply line, detailed calculations should take into account several factors that can influence the sharing of current I among line conductors:

- the cable length;
- the different impedances of neutral and phase conductors (e.g. if the neutral (N) conductor has multiple grounds, the lower impedance of N compared with

730 Lightning Protection

L1, L2 and L3 could result in 50 per cent of the current flowing through the N conductor with the remaining 50 per cent being shared by the other three lines (17 per cent each), and if N, L1, L2 and L3 have the same impedance, each conductor will carry ~25 per cent of the current)
- the different transformer impedances
- the relation between the conventional earth impedances of the transformer and the items on the load side
- the parallel consumers, which may increase the partial lightning current flowing into the line due to the reduction of its effective impedance.

15.2.1.2 Potential differences between different parts of the earth-termination system of the station

For any but the smallest of installations it is not usually practicable to adopt one single earth-termination system. Typically, there will be three: for the antenna support structure, the apparatus building and the transformer. The effect on the station of a lightning

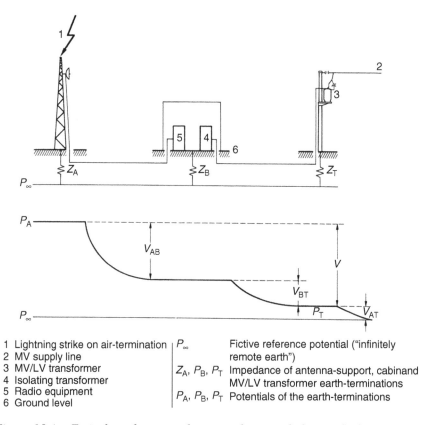

1 Lightning strike on air-termination
2 MV supply line
3 MV/LV transformer
4 Isolating transformer
5 Radio equipment
6 Ground level

P_∞ — Fictive reference potential ("infinitely remote earth")
Z_A, P_B, P_T — Impedance of antenna-support, cabin and MV/LV transformer earth-terminations
P_A, P_B, P_T — Potentials of the earth-terminations

Figure 15.4 *Typical earth potential pattern during a lightning flash to a station having three separate earth-terminations*

flash to the antenna support structure will of course depend on the nature and relative positions of those three earth-termination system, that is to say, on their conventional earth impedances.

In the case of a lightning flash to the antenna support structure of a station with three separate earth-termination systems, the potential situation would be as indicated in the lower part of Figure 15.4. It is apparent that the potential differences, with lightning currents of typically several tens of kiloamperes, between the two ends of the antenna feeders and between the two ends of the power supply cable are likely to exceed the withstand voltages of their insulation.

If the three earth-termination systems are connected together, the situation is rather different. The composite earthing system is practically equipotential at 50 Hz, provided that the inductance and the capacitance of earthing conductors are negligible at low frequencies, but is very unlikely to be so for the impulse currents engendered by a lightning flash to the antenna support structure. As a consequence, the earthing system of the station is no longer equipotential, in spite of the conductors bonding its different parts, and it does not matter how large their cross-sections are.

The large number of variables renders it difficult to indicate typical effects, but measurements made on a typical station are described and discussed in the Annex to this chapter. A survey of typical damages due to lightning flashes to the broadcasting station is shown in Figure 15.5.

15.3 Lightning flashes affecting the power supply system

In order to investigate the effects of lightning flashes on the power supply system, it is essential to estimate the cause and magnitude of surges (overvoltages and overcurrents) that may propagate along the line. The cases of overhead lines and buried cables must be considered separately as well as the cases of flashes to the line and flashes to ground near the line.

15.3.1 Power supply by overhead lines

A surge on an overhead line may be caused either by a lightning flash to some component of the line (conductor, shield-wire, support, etc.) or by a lightning flash to some object or to ground near the line (a distance up to 1 km should be considered). Lightning flashes to line tend to cause more damage, but they are relatively rare; lightning flashes to ground near the line, although more frequent, usually have less serious consequences.

15.3.1.1 Lightning flashes to an overhead line

In general, only the case of flash to the line conductors need be considered, because shield-wires are very rarely provided except for extra high voltage (EHV) lines, and because the supports are less exposed than the conductors. The overvoltage caused by the lightning between the conductors of a three-phase system and between conductors and earth depends upon the surge impedance of the individual conductors

Figure 15.5 *(a) Lightning damage to a 220 V isolating transformer (photo courtesy of Radiotelevisione italiana). (b) Vestige of an air-termination rod fused by lightning current (photo courtesy of Radiotelevisione italiana). (c) Lightning damage to an 11 kV oil-filled circuit breaker (photo courtesy of Independent Broadcasting Authority).*

and its relation to the surge impedance between conductors, as well as, of course, the magnitude of the lightning current. Typical values of these parameters indicate that even moderate lightning currents can produce overvoltages of several thousands of kilovolts (and associated overcurrents of several kiloamperes) and thus cause the failure of the insulation of the line. From the point of strike, the surge propagates in both directions and its behaviour differs depending upon the arrangement of the line insulators. If the studs or brackets of the insulators are connected to earth, the insulation level of the line is equal to the withstand voltage of an insulator. If the surge voltage at the first insulator encountered is greater than that withstand voltage, the insulator will flash over, part of the surge current will pass to earth, and the surge will propagate further along the line with a reduced voltage. This process will repeat at each insulator until the ongoing surge voltage is lower than the insulator withstand voltage. For points of strike several spans distant from the station, it may be assumed that the overvoltage arriving at the station does not exceed the withstand voltage of the line insulators. In cases of extremely high front steepness voltage waves, a flash-over occurs between conductors and the surge would thereafter propagate along all the conductors.

In installations where the insulator studs are not earthed (e.g. wooden-pole lines), the effective withstand voltage is that of the insulator plus that of the pole, so the reduction of the surge voltage at each insulator along the line is negligible. The station equipment may therefore be subjected to practically the same overvoltage as at the point of strike, and that on all phases.

15.3.1.2 Lightning flash to ground near an overhead line

In these cases, the induced surge is the same on all the phase conductors and depends, of course, on the position of the object struck, relative to the line. Very commonly, the object struck is the station's antenna support structure.

It can be shown that LV overhead lines with an impulse voltage withstand level in the range 6–15 kV are very susceptible to insulation failure resulting from nearby lightning flashes, whereas that risk is very slight in the case of high-voltage (HV) lines without shield-wires and with earthed insulator studs, for which the withstand level of the insulation is usually $\sim$100 kV. Where the insulator studs are insulated from earth (e.g. by wooden poles), the insulation level will in all cases be adequate to eliminate that risk.

The large number of variables makes it difficult to indicate typical characteristics of the waveform of induced overvoltages, but the literature suggests that the maximum amplitude rarely exceeds 100 kV, and that 20 kV μs^{-1} is a typical value of the slope of the wave front, with maximum values of $\sim$60 kV μs^{-1}.

15.3.1.3 Surges at the point of entry of an overhead line in the station

Although a surge undergoes a gradual reduction of amplitude and also a decrease in the steepness of its front, as it propagates along a line, flashes very close to the station must also be anticipated and precautions taken against overvoltages at the transformer primary winding that are considerably in excess of the basic insulation level of the line. Methods of estimating the conditions at the transformer input are to be found in the literature, but the unavoidable assumptions and simplifications seriously compromise their validity in practice. Measurements on typical installations indicate that overvoltages of three or four times the line insulation level may be expected. Moreover, the overvoltage effectively arriving at the HV/LV transformer primary winding depends also on whether the overhead line is connected directly to the transformer, or through a length of buried cable, as well as upon the SPD, if any, provided for the protection of the transformer.

15.3.1.4 Overhead line connected directly to the transformer

The maximum amplitude of the surge voltage is in this case twice the peak amplitude of the incident voltage wave and varies from twice the withstand level of the line insulation in the case of lightning flashes remote from the transformer, to six to eight times for flashes near the transformer.

15.3.1.5 Buried cable connection between the overhead line and the transformer

In this case, the waveform of the incident surge and the type and length of the cable are determinant factors. Thus, the following apply.

- The overvoltage increases with an increased ratio between the half-amplitude decay time of the incident wave and the propagation time along the cable, as well as with an increased ratio between the surge impedance of the cable and that of the overhead line.
- For each cable there is a critical length, depending on the incident waveform, for which the maximum overvoltage is equal to the peak incident amplitude.
- With cables shorter than the critical length, the overvoltages are greater, whereas with cables longer than that length, the overvoltages are lower than the incident amplitude.
- For every type of cable, there is a minimum length for auto-protection. This is a function of the critical length and it defines the length for which overvoltages are reduced to the cable insulation level. Cables longer than that minimum length need no further protection against surges, those very much shorter require an SPD at the interface between the line and the cable, and intermediate cases require an SPD at each extremity of the cable.

The exponential attenuation of surges along the cable is reduced in cases of surges having longer rise and decay times (typically those due to flashes to the overhead line very remotely from the cable). It can be neglected for cables shorter than 300 m. For flashes near the line/cable interface, however, overvoltages of between twice and 2.4 times the incident peak amplitude may occur, because of successive reflections between the point of strike and the interface. It follows that the maximum surge voltage can vary, in the case of distant flashes, from a value less than the withstand level of the insulation for cables longer than the critical length, up to twice that level for cables shorter than the critical length, and from seven to ten times that level for nearby flashes. The provision of adequate SPDs between the cable and the transformer of course restricts the voltage effectively applied to the transformer to the protection level (i.e. spark-over voltage) of the SPD.

15.3.2 Power supply by underground cable

Underground cables are not immune to damage by lightning, because lightning to ground can give rise, near the point of strike, to potential gradients that are greater where the soil resistivity is higher. When the lightning current entering the ground and the soil resistivity are both high, a disruptive discharge can occur between the point of strike and a buried cable in the vicinity, depending upon the dielectric strength of the soil. The distance over which such a discharge can occur rarely exceeds 30 m, common values being between 5 and 10 m. When such a discharge occurs, the full lightning current propagates along the cable in both directions from the point where the current enters the sheath. When there is no such discharge, the particular effect that occurs will depend upon the construction of the cable, as described below.

If the cable has a metallic bare sheath in contact with the soil, it will at first equipotential with the soil; close to the point of strike, the sheath current is very low, but rapidly increases in both directions and finally decays. For typical cables, a flash to ground ~ 5 m from the cable would result in a maximum sheath current of between 10 and 15 per cent of the lightning current with a soil resistivity not exceeding 300 Ω m, and between 20 and 30 per cent if it were about 2 000 Ω m.

If the cable has a metallic sheath with an insulating covering, there is a considerable risk, in soil of high resistivity, that the covering will be perforated and the sheath itself damaged, perhaps even followed by complete breakdown of the cable insulation.

The case of cables having non-metallic sheaths is, with appropriate changes in the details, the same as that of cables with insulated metallic sheaths. In this case, the cable conductors will be at risk, and the insulation level corresponds to the withstand voltage of the cable insulation.

The current wave propagating in the metallic sheath of a cable gives rise to a potential difference stressing the insulation between the cores and the sheath, and having its maximum at the point of entry of the lightning current. It falls off much more rapidly with distance from that point when soil resistivity is low. If, however, a discharge occurs at the point of entry, the voltage between the cores and the sheath increases with distance from that point, and more rapidly when soil resistivity is high.

Because the magnitude of the potential difference between the metallic sheath and the cores of a cable is directly proportional to the coupling impedance between the sheath and cores, it is clearly important to reduce that impedance. The commonly adopted solutions include the provision of shield conductors laid parallel to the cables, laying the cables in a steel tube or duct, and the use of cables of layered construction or of cables having specially increased insulation.

15.3.3 Transfer of overvoltages across the transformer

Voltage surges, caused by lightning flashes to or near the supply line and arriving at the primary windings of the transformer, give rise to spurious voltage waves in the secondary windings that depend on the characteristics of the incident surge, on those of the transformer and on the load connected to its secondary windings.

Depending upon those conditions, the transfer may be either capacitative or inductive. In the case of capacitative transfer, the transferred voltage would under typical conditions be about 6–10 per cent of an overvoltage occurring between all the phases and earth, although the presence of an electrostatic screen between the primary and secondary windings, which cancels the capacitance between those windings, almost completely eliminates the capacitative transfer of the overvoltages. In the case of inductive transfer, the voltage transferred to the secondary winding depends upon the impedances connected across the primary and secondary windings of the transformer, upon its turns ratio and the waveform of the applied overvoltage; it is therefore desirable that those impedances should be kept as low as practicable.

It can be shown, on the basis of typical values of the various parameters, that an HV supply with stepdown transformer is likely to introduce less potentially damaging overvoltages into the station than an LV supply with isolating transformer.

15.4 The basic principles of lightning protection

The purpose of this section is to outline the basic principles underlying the precautions to be taken against lightning damage when stations are being planned, with the object of arriving at an optimum compromise, for each particular case, between the maximum of reliability and personal safety on the one hand and economic considerations on the other.

15.4.1 The protection level to be provided

The statistical nature of the phenomenon of lightning implies that, within the limits of a reasonable outlay, and taking account of all practical constraints, no system or means of protection can guarantee absolute security. This can be approached only with a more or less high probability, dependent on the nature and the scale of the precautions adopted.

It is neither necessary, economic nor advantageous for every station to be given the maximum protection level that is technically possible. Less complete protection can usually be accepted for stations on sites that are clearly less exposed to lightning, for those that are at all seasons readily accessible for rapid repair, and even for those serving only very small communities.

This practical rule is theoretically standardized in IEC 62305: the objective of protection against lightning is reduction to or below the tolerable level R_T of the risk R of damage to the stations, including injury to persons in the case of attended stations. Injury to visiting maintenance personnel may usually be disregarded provided that the time of exposure to danger in a year is very low.

If the risk R is higher than the tolerable level R_T, suitable protection measures should be adopted in order to reduce the risk R to the tolerable level R_T. Protection measures will be more and more effective as the ratio R/R_T is greater and greater: the higher the ratio R/R_T, the higher will be the protection level of measures to be provided.

The risk assessment should be performed according the procedure standardized in IEC 62305-2. According to this standard, the following types of loss and relevant risks should be considered in a station:

- risk R_1 of loss of human life
- risk R_2 of loss of service to the public
- risk R_4 of loss of economic value

The values of the tolerable risk suggested by standard IEC 62305-2 are as follows:

- $R_T = 1 \times 10^{-5}$ for risk R_1 of loss of human life
- $R_T = 1 \times 10^{-3}$ for risk R_2 of loss of service to the public

To evaluate the cost-effectiveness of providing protection measures in order to reduce economic loss, the risk R_4 of loss of economic values should be assessed.

When determining the risk R_4 to the station in question, the following must be taken into account:

- *The cost of repairing the breakdowns.* This cost might be very high, because of the remoteness of the station from maintenance centres and, if the station is located in a mountainous region, because of difficulty of access, especially in winter.
- *The importance of the service provided by the station.* The cost of service lost may be very high due to time needed to repair and to the number of users served directly by the station and, indirectly, by the radio-linked stations and the rebroadcast installations.

15.4.2 Basic criteria for protection of stations

An ideal protection for a broadcasting transmitting station would be to enclose the station to be protected within an earthed perfectly conducting continuous shield of adequate thickness, and by providing adequate bonding at the entrance point into the shield, of the services connected to the structure. This would prevent the penetration of lightning current and the related electromagnetic field into the station to be protected and prevent any dangerous thermal and electrodynamic effects of the current, as well as dangerous sparkings and overvoltages for internal systems. In practice, it is often not possible nor cost-effective to go to such lengths to provide such optimum protection.

Lack of continuity of the shield and/or its inadequate thickness allow the lightning current to penetrate the shield, causing the following:

- physical damage and life hazard
- failure of internal systems
- failure of the connected services

This general protection principle means that in order to enclose the station to be protected – in agreement with IEC 62305-1 – within a lightning protection zone (LPZ), this zone is such that the lightning electromagnetic environment is compatible with the capability of the station to withstand stress, causing the damage to be reduced (physical damage, failure of electrical and electronic systems due to overvoltages).

An LPZ is implemented by protection measures such as an LPS (lightning protection system), shielded circuits, magnetic spatial shields and SPDs.

15.4.3 Protection measures

The usually adopted protection measures in a station are as follows.

1. Protection measures to reduce injury of living beings due to touch- and step-voltages
   - by increasing the surface resistivity of the soil inside and outside the structure
   - through equipotentialization by means of a meshed earthing system
   - through physical restrictions and warning notices
2. Protection measures to reduce physical damage by means of
   - a lightning protection system

A LPS for the building of the station is usually not required provided that the building lies inside the volume protected by the antenna structure.

3. Protection measures to reduce failure of the electrical and electronic systems:
   - earthing and bonding measures
   - magnetic shielding of cables, apparatus and/or of the building
   - line routing
   - 'coordinated SPD protection'
   - increased withstand voltage of the insulation of equipment and cables
   - an isolating transformer at the entrance point of the LV power lines in the station

As a general rule, according to their type, the protection measures to reduce the failure of electrical and electronic systems work as

- preventive measures, which avoid the rise of surges, or
- suppressive measures, which limit the surge to a predefinite value, once it is build up

Protection measures listed under measure (3), above, are of a preventive type, except the 'coordinated SPD protection' and the isolating transformer.

Moreover, as a general rule, route redundancy, redundant equipment, autonomous power generating sets, uninterruptible power systems, fluid storage systems and automatic failure detection systems are effective protection measures to reduce the loss of activity of the service.

15.4.4 Procedure for selection of protection measures

Protection measures, adopted to reduce such damages and relevant consequential loss, should be designed for the defined set of lightning current parameters against which protection is required. In standard IEC 62305 this set of lightning current parameters is defined as the lightning protection level (LPL). The required protection measures and the relevant LPL are determined by performing a risk assessment according to the procedure reported in IEC 62305-2.

Planning and coordination of protection measures begins with an initial risk assessment to determine the required protection measures needed to reduce the risk to, or below, the tolerable level. When possible, protection measures should be selected taking into account both technical and economical aspects; this is not the case for a LPS, which is the only measure to reduce the physical damage.

As far as the protection measures to reduce failure of electrical and electronic systems are concerned, those of a preventive type are structural protection measures that it is generally possible to implement only at the stage of erection of the station and at the installation of electrical and electronic systems.

Even the design and installation of such measures is a heavy task and their cost may be high; however, it is recommended when high reliability and availability of protection is required, as is the case for large unattended stations, remote from maintenance centres and located in mountainous regions, with difficulty of access, especially in winter.

Therefore, the following steps should be carried out in implementing protection measures to reduce the failure of electrical and electronic systems.

1. An earthing system, comprising a bonding network and an earth termination system, should be provided.
2. External metal parts and incoming services should be bonded directly or via suitable SPDs.
3. The internal system should be integrated into the bonding network.
4. Spatial shielding in combination with line routing and line shielding may be implemented.
5. Requirements for a coordinated SPD protection system should be determined.

Following this, the cost–benefit ratio of the selected protection measures should again be evaluated and optimized using the risk assessment method.

15.4.5 Implementation of protection measures

The lightning protection of transmitting stations requires a large number of precautions against the effects of lightning flashes, and those precautions must be effectively coordinated. In the case of lightning flashes to the station, the lightning energy must be dispersed harmlessly, whereas in the case of lightning flashes affecting the energy supply system, telecommunication connections, cable ways, and the like, dangerous voltage surges must be limited to tolerable levels.

740 *Lightning Protection*

The problems that arise in the planning and implementation of lightning protection installations result mainly from the high peak amplitudes and steep wavefronts of the current surges and the resultant potential differences of up to several thousands of kilovolts, and high electromagnetic field strengths.

The essential precautions must be primarily directed towards the avoidance of potential differences dangerous for the personnel, building and equipment, as well as towards the limitation within predetermined limits of the voltages induced in electrical and electronic systems.

For this practical purpose the abovementioned protection measures should be integrated with each other. In the following sections, as an aid for explanation, two examples are taken as typical in order to facilitate the task of lightning protection engineers in the design and installation of the overall LPS.

1. A low-power television and VHF/FM rebroadcasting installation, as depicted in Figure 15.6, its essential components being shown in Figure 15.7.
2. A large-scale television and VHF/FM station with radio relay facilities, as depicted in Figure 15.8.

Figure 15.6 A typical low-power television and VHF/FM rebroadcasting station

Lightning protection of telecommunication towers 741

1. LV open-wire supply line;
2. Surge diverter;
3. Cable-termination pylon;
4. Prefabricated sheet-steel transposer housing;
5. Supply distribution board;
6. Isolating transformer;
7. Cable termination and meter cabinet;
8. LV supply cable;
9. Station earth termination;
10. Potential-equalization busbar;
11. Earthing of RF cable sheath;
12. Transposer;
13. RF cables;
14. Earth conductors;
15. Earthing system of transmission-line pylon.

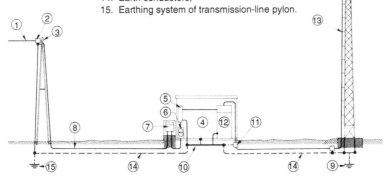

Figure 15.7   Elements of a low-power television and VHF/FM rebroadcasting station, such as that depicted in Figure 15.6

Figure 15.8   A typical large-scale television and VHF/FM broadcasting and radio relay station

15.5 Erection of protection measures to reduce injury of living beings

The protection measures outlined in the following are aimed at providing protection against touch- and step-voltages due to lightning and are additional to those needed to provide protection against touch- and step-voltages due to fault to earth of the power system feeding the station.

Moreover, this hazard is to be disregarded if the probability of persons approaching is low, or the duration of their presence in the dangerous area is very low. This is usually the case for unattended stations.

15.5.1 Protection measures against step voltages

The outside area within 3 m from the down-conductors of a LPS (e.g. the legs of the metallic lattice tower supporting the antenna systems, or the installed down-conductors in the case of non-metallic towers or buildings) may be hazardous to life even if the LPS has been designed and constructed according to IEC 62305-3. In the case of attended stations, protection measures must be taken.

One or more of the following protection measures is suitable for this purpose.

- The resistivity of the surface layer of the soil, within 3 m of the down-conductor, should be increased to 5 kΩ m or more. This can be achieved by covering the soil with a layer of insulating material, e.g. asphalt of 5 cm thickness (or a layer of gravel 15 cm thick).
- Equipotentialization should be achieved by means of a meshed earthing system, as better specified in Sections 15.6.3.1 and 15.6.3.2.
- Physical restrictions and/or warning notices should be provided to minimize the probability of access to the dangerous area, within 3 m of the down-conductor.

All protection measures should obviously conform to the relevant standards (see ISO-3864).

15.5.2 Protection measures against touch-voltages

This hazard is in general tolerable if the metallic tower supporting the antenna acts as a natural air-termination system and down-conductor system. If this is not the case, one of the following protection measures should be provided.

- The resistivity of the surface layer of the soil, within 3 m of the down-conductor, should be increased to 5 kΩ m or more. This can be achieved by covering the soil with a layer of insulating material, e.g. asphalt of 5 cm thickness (or a layer of gravel 15 cm thick).
- Physical restrictions and/or warning notices should be provided to minimize the probability of access to the dangerous area, within 3 m of the down-conductor.

All protection measures should obviously conform to the relevant standards (see ISO-3864).

15.6 Erection of the LPS to reduce physical damage

15.6.1 Air-termination system

The air-termination system has to be positioned according to Annex A of IEC 62305-3. A typical point of strike for downward flashes (or point of departure for upward flashes, which are much more frequent in the case of transmitting stations) is the top of the antenna structure; both the support structure and the antennas themselves may be affected.

Although metal lattice masts are self-protected by their nature, the antennas can sometimes suffer damage when the charge is high enough to perforate the antenna by fusion at the point of strike, with consequent infiltration of water. Usually, however, the antennas are already protected, either because of their design (e.g. folded dipoles), or because they incorporate protective devices such as, for example, star gaps or shorting stubs that act as short circuits for the lightning current.

Microwave antennas (paraboloid and horn types), when they are entirely of metal, are in general sufficiently robust to withstand without damage even the strongest lightning flashes; they can, however, suffer considerable damage if they are constructed of plastic material with an internal conducting coating.

Whenever it is not certain that the antenna is capable of withstanding without damage the stresses of a lightning stroke, it must be located within the protected volume provided by the support structure itself or by a suitably placed air-termination. The uppermost antennas (typically, those for Band IV/V) must be surmounted either by the support structure itself or by air-terminations (Figure 15.9a). However, for heights of more than 20 m, there is a considerable risk that the uppermost section of the support structure will be struck directly by lightning. In that case, where it is necessary to protect also the highest parts of the structure, recourse must be had to air-terminations having a more complex geometry, with, for example, rods extending horizontally from the top of the structure (Figure 15.9b).

Where the antenna arrays are enclosed in plastic housing as protection against icing, it is necessary to make the uppermost part of such a housing of metal and to connect it, at low impedance, with the down-conductors.

Any high antenna structure with a platform at the top must, for the protection of persons working there, be provided with air-termination rods extending at least 3 m above the platform (Figure 15.10).

15.6.2 Down-conductors system

Stations having lattice towers or masts very commonly have metal topmasts for supporting the antennas. In such cases, it is not as a rule necessary to provide down-conductors from the air-termination system to the earth-termination system, because the cross-sectional area of the support structure is usually adequate, provided that there are ample contact areas between the various sections of the lattice tower or mast. This is valid also where coaxial cables are attached to the support structure, provided that their sheaths are of adequate cross-section and solidly earthed at

744 Lightning Protection

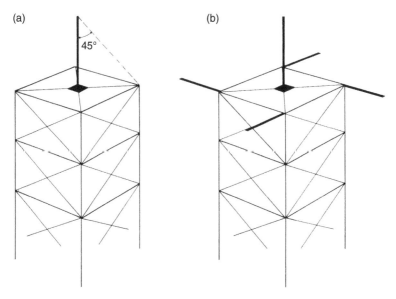

Figure 15.9 Air terminations: (a) single vertical air termination rod for low structures; (b) additional horizontal rods for structures of height greater than 20 m

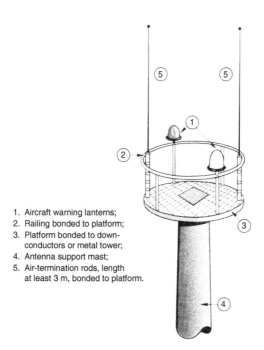

1. Aircraft warning lanterns;
2. Railing bonded to platform;
3. Platform bonded to down-conductors or metal tower;
4. Antenna support mast;
5. Air-termination rods, length at least 3 m, bonded to platform.

Figure 15.10 Air terminations rods on a tower platform

both extremities. If, instead, the coaxial or supply cables have only thin sheaths, it is advisable to install a down-conductor along the support to which the cables are attached.

In other cases, the antennas, typically those of radio-relay links, are installed on the tops of buildings. Where the buildings are of steel construction, the discharge of the lightning energy is effected by the steel framework itself, and no separate down-conductors are required. It is necessary merely to ensure that the connection with the earth-termination system has the lowest possible impedance. The steel framework provides a path of relatively low impedance, the resistance of such structures being low. This is important because the magnitude of that impedance determines the potential differences produced along the path of the discharge, as well as those between the apparatus and the building at different vertical levels.

In the case of ferro-concrete towers, it is useful to utilize the steel reinforcement as down-conductors to the greatest practicable extent, provided that the reinforcement is effectively bonded. This is best done by bonding the outermost reinforcement bars at the top and foot of the tower to form horizontal loops, with additional loops at vertical intervals of ~ 3 m (approximately corresponding to the heights of the storeys). In addition to the reinforcement bars, however, further conductors, the number of which will depend on the scale of the structure, must be arranged at regular intervals over its surface and bonded to the reinforcement.

For towers consisting of separate sections, the reinforcements at the junctions must be connected together conductively, or external conductors must be provided and, if possible, connected to the reinforcements of the sections. The air-terminations and down-conductors must also be bonded to all metal structures on the top (platforms, etc.).

In those rare cases where the transmitter building is not located within the protected volume provided by the antenna structure, appropriate precautions must be taken, depending upon the nature of the building. In addition, any national regulations specifying the protection of buildings against lightning must be complied with.

15.6.3 Earth-termination system

The function of the earth-termination system is to provide a low-impedance path to earth for the lightning currents resulting from lightning flashes to the station, in order to minimize the potential rise of the internal systems. It is, in effect, the interface between the discharge path and geological earth (the more-or-less conducting substance constituting our planet).

Being sited on mountains, most transmitting stations are located on terrain with poor soil conductivity, and the first requirement, once the site has been fixed, is to measure the soil resistivity.

For designing the earth-termination system and correctly dimensioning the earth electrodes, it is essential to know the soil resistivity exactly, as it is involved in all calculations concerning conventional earth impedance of the earth-termination system. This is a ratio of the peak values of the earth-termination voltage and the earth-termination current which, in general, do not occur simultaneously. In the published literature, reference is often made to tables in which the resistivity of different types of soil is indicated, but these can be used only as a guide and must not be applied in design work. In particular, in the case of broadcasting stations for which reliability is important and of sites where there is a high probability of lightning flashes, measurements of the soil resistivity must be carried out in every individual case, to provide a reliable basis for the design of the earth-termination system. In order to design an earth-termination system suitable for obtaining predetermined values of conventional earth impedance, it is necessary to determine the average resistivity of that part of the ground that will be directly involved in dispersing the current. For this, it is necessary to assume a homogeneous terrain upon which the principal methods of calculation for earth-electrodes are based.

Moreover, the choice of the type of electrodes to be used should be based on the resistivities and positions of the soil strata. In principle, electrodes of the vertical type (rods) may be used whenever strata of low resistivity can be reached, otherwise electrodes of the horizontal type (radial, ring, etc.) are more suitable. For example, with low-resistivity topsoil above high-resistivity subsoil of considerable depth, horizontal electrodes would be the most suitable (Figure 15.11a). With high-resistivity topsoil above low-resistivity subsoil (Figure 15.11b), horizontal electrodes would be appropriate where the uppermost stratum is of considerable thickness, whereas it would be better to use rod electrodes where it is easy to reach the better conducting underlying stratum. In the case of Figure 15.11c, the electrodes could be either horizontal or vertical, depending upon the relative thickness of the strata.

As regards the dimensioning criteria for the earth-termination system, it must be borne in mind that the attainment of a low conventional earth impedance is not strictly

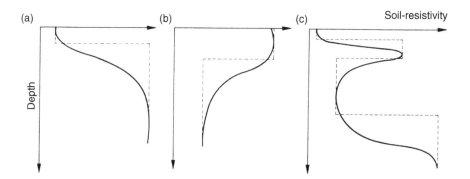

Figure 15.11   *Adaptation of earth electrode design to the soil resistivity configuration: (a) low-resistivity topsoil over high-resistivity subsoil; (b) high-resistivity topsoil over low-resistivity subsoil; (c) topsoil and subsoil strata alternately of low and high resistivity. Continuous line, measured resistance; dashed line, deduced stratification.*

necessary. The protection of the connections and the equipment of the station, as well as that of the personnel, is in practice obtained by potential equalization (Section 15.7), screening (Section 15.8) and by a coordinated SPD system (Section 15.9). Nevertheless, a low conventional earth impedance of the earth-termination system is desirable, in that it enables the lightning current to pass to the ground as directly as possible, minimizing any risk from current surges (currents and overvoltages) to the lines and other services that enter the station, and preventing propagation to a distance of a significant part of the lightning energy.

The optimum value of the conventional earth impedance depends frequently on technico-economic considerations. From engineering considerations, it is more appropriate, when fixing the dimensioning criteria, to base them on the form and dimensions of the earth electrodes and the geometry of the earth-termination system, rather than to fix only a low value for the resistance of the termination itself.

In many stations, the earth-termination system has to provide protection both against lightning and against power-supply system faults; in order to deal adequately with this second function, the earth-termination system must make it possible to restrict, within pre-determined limits, the 'step-voltages' and 'touch-voltages' resulting from fault currents to earth in the power supply system of the station. The earth-termination system must also conform with any relevant safety regulations in force in the country in question.

The station, of whatever type it may be, must have a single integrated earth-termination system for the antenna support structure, the equipment building and, if provided, the transformer HV/LV substation. The configuration of that system will nevertheless differ, depending upon whether the station has an autonomous electricity supply (aeolian generator, thermogenerator, diesel–electric set, etc.) or whether it is supplied by line from an external source (public power supply).

15.6.3.1 Earth-termination system for stations with an autonomous power supply

In stations with an autonomous supply, it is important only to equalize the potential between the various parts of the earth-termination system, no importance being attached either to the area covered by the electrodes or to the value of their conventional earth impedance. When lightning strikes the antenna support, the whole of the earth-termination system assumes, relative to a remote reference earth, a very high potential, which is higher the greater the conventional earth impedance. The difference in potential has, however, no effect, because no equipment is held (e.g. over an external line) to the potential of the remote earth. The functioning of the equipment is therefore not affected, provided that there are no appreciable differences of potential between the various parts of the earthing system. These can be avoided by inserting, in the floor of the building, a meshed metallic network of regular size, connecting it by way of the cable ducts to the antenna support, and making use of a perimeter ring of interconnected small rod electrodes, surrounding the building and the base of the antenna support (Figure 15.12).

748 Lightning Protection

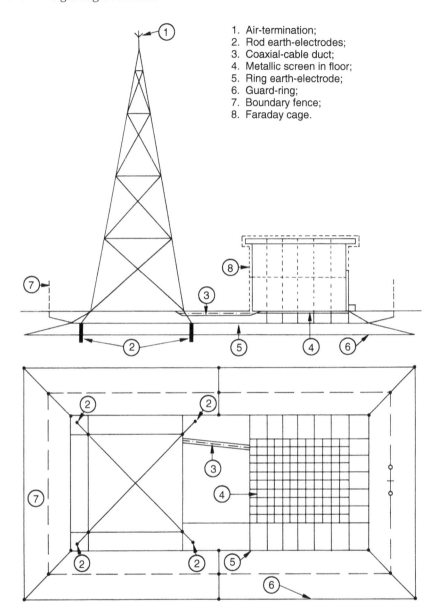

1. Air-termination;
2. Rod earth-electrodes;
3. Coaxial-cable duct;
4. Metallic screen in floor;
5. Ring earth-electrode;
6. Guard-ring;
7. Boundary fence;
8. Faraday cage.

Figure 15.12   Typical earthing for a station having an autonomous power supply

In installations of this type, the area affected is almost always quite small, so the metal boundary fence will be within the field of influence of the station earth-termination system. In order to avoid any dangerous touch-voltages, it is opportune to connect the fence itself to earth, providing it on the outside with a perimetric loop buried at a depth of ~ 1 m and connected to the earth-termination system.

15.6.3.2 Earth-termination system for stations powered by an external source

In the case of stations supplied from an external source, the lightning protection measures must ensure potential equalization between the various parts of the installation itself, and also a predetermined low conventional earth impedance, in order to maintain the total earth-termination system voltage (voltage between supply-line conductors and earthing system) within acceptable limits. To perform this task the geometry and the dimensions of the earth-termination system are the main tools.

The earth-termination system should be installed in direct connection with the down-conductors system; in most cases this implies an installation around the base of the tower supporting the antennas. The earth-termination system will typically have a configuration similar to that depicted in Figure 15.13.

Depending upon the geoelectrical characteristics of the soil, electrodes can have various geometrical forms, but in each case their dimensions must be such as to

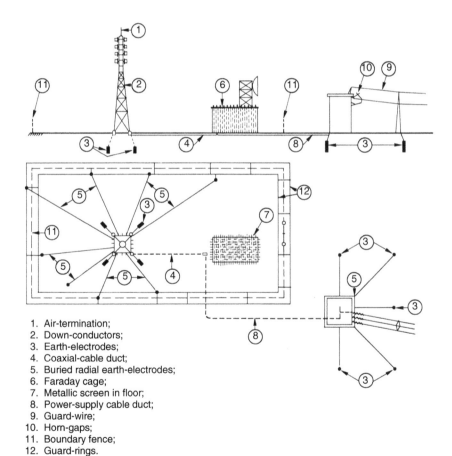

1. Air-termination;
2. Down-conductors;
3. Earth-electrodes;
4. Coaxial-cable duct;
5. Buried radial earth-electrodes;
6. Faraday cage;
7. Metallic screen in floor;
8. Power-supply cable duct;
9. Guard-wire;
10. Horn-gaps;
11. Boundary fence;
12. Guard-rings.

Figure 15.13 Typical earthing for a station connected to an external power supply

attain a sufficiently low conventional earth impedance. On terrain of low resistivity and where subsoil strata of low resistivity are easily reached, the electrodes may consist of small elements (localized electrodes: small rods, plates, very short conductors). On terrain of high resistivity, it is better to use radial electrodes having at least five or six 'spokes', the length of which can be chosen once the limiting values of the soil resistivity and of the conventional earth impedance have been determined. Around the base of the antenna support, one or more concentric ring conductors should be buried and well bonded to the support, the down-conductors (if any) from the air-termination system and to the centre point of the radial earth terminations. This precaution aimed at equipotenzializing the soil may be necessary to limit touch- and step-voltages in the vicinity of the support. With concentric rings spaced at 5 or 10 m, the maximum step-voltage within the area thus delimited is respectively about 5 or 10 per cent of the earth-termination voltage; the limitation of the step-voltage is thus a matter of limiting the earth-termination voltage, the configuration of the earth-termination system being assumed fixed. The dimensioning of the earth-termination system can, therefore, be effected with the help of Figures 15.14 and 15.15 once the required protection level has been specified.

Where the concentric rings are not provided and soil equipotenzialization is not achieved, other protection measures to be adopted against touch- and step-voltages may consist in restrictions impeaching people to reach the dangerous zone or in

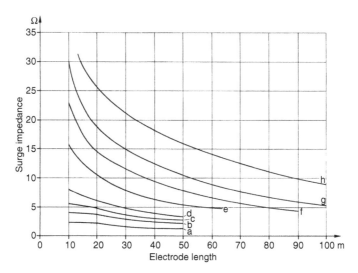

Surge-impedance of symmetrical radial earth terminations
Parameter: soil-resistivity ρ (Ω m)
a: 100 Ω m   e: 1 000 Ω m
b: 200 Ω m   f: 1 500 Ω m
c: 300 Ω m   g: 2 000 Ω m
d: 500 Ω m   h: 3 000 Ω m

Figure 15.14   *Conventional earth impedance of symmetrical radial earth termination. Parameter: soil resitivity ρ (Ω m).*

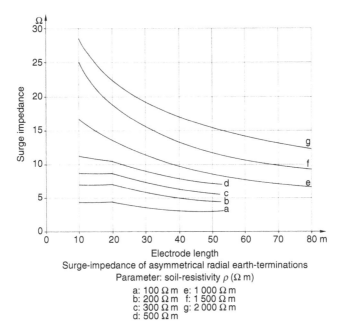

Figure 15.15 *Conventional earth impedance of asymmetrical radial earth termination. Parameter: soil resitivity* ρ *(Ω m).*

covering the soil with a layer of insulating material having an adequate thickness (5 cm for asphalt or 15 cm for gravel).

In practice the dimensions of an electrode earth-termination system depends upon whether the average soil resistivity is less or greater than 1 000 Ω m. For a resistivity of less than 100 Ω m the electrode may consist simply of a buried ring conductor around the base of the aerial support, possibly connected to rods and radial conductors several metres in length, in such a manner that its earth resistance (measured at 50 Hz) does not exceed 10 Ω. For such electrodes, the surge efficiency (i.e. the ratio between the conventional earth impedance and the earth resistance) is always less than unity, so the conventional earth impedance cannot then be more than 10 Ω.

For a resistivity greater than 1 000 Ω m, a radial electrode system is required, its dimensions being such as to ensure a conventional earth impedance not greater than 10 Ω. The length of the spokes of such a system, as a function of the mean soil resistivity, is indicated in Figure 15.16, which depicts two cases: curve (a) corresponds to the optimum effect/cost ratio and curve (b) leads to the minimum conventional earth impedance.

In Figure 15.17 it is shown that the minimum length of each earth electrode according to the class (lightning protection level) of LPS as reported in IEC 62305-3; curve (b) is in agreement with such requirements.

It is important to note that the choice of 10 Ω as the maximum value of the conventional earth impedance is purely indicative; it serves simply for determining the

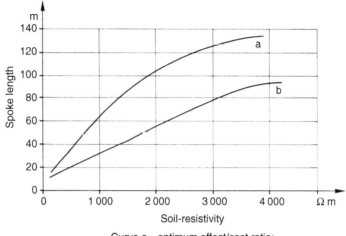

Curve a = optimum effect/cost ratio;
Curve b = lowest surge-impedance.

Figure 15.16 Minimum lengths of radial earth-electrodes for soil of high resistivity

dimensions that an earth-termination system should have, in order to ensure effective dispersal to earth of the lightning current, while respecting economic constraints.

It may be appropriate to assign lower values to the conventional earth impedance and, therefore, greater dimensions to the electrodes, if it is desired to maintain the total earth-termination voltage within particular limits, for example, in order to limit overvoltages at the point of entry in the station of the power supply line.

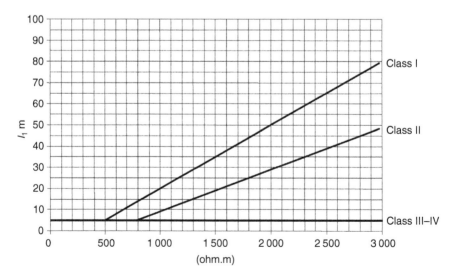

Figure 15.17 Minimum lengths l_1 of each earth-electrode according to the class of LPS

15.6.4 Protection against corrosion

Metals in direct contact with damp or acidic soil or with water can be corroded by parasitic currents and the formation of galvanic cells. Corrosion, therefore, constitutes a serious factor affecting the long-term effectiveness of earth electrodes.

It is rarely possible to prevent the formation of galvanic cells (the commonest cause of corrosion) by electrically separating earth electrodes acting as anodes from other objects acting as cathodes, because all the earth electrodes have to be connected to each other and to all other metal objects in the soil in order to ensure potential equalization and thereby a maximum degree of safety from overvoltages caused by electricity supply faults or lightning discharges. Accordingly, the only remaining option is that of minimizing the risk of corrosion by using suitable materials for the electrodes.

For direct burial in the soil, a solid conductor is preferable to a stranded conductor of similar gauge, being less subject to corrosion and more resistant to mechanical stresses.

Earth electrodes of copper or steel with copper sheathing may not be bonded to earth electrodes of more electronegative materials (e.g. galvanized steel) or to other buried steel objects (e.g. pipework and tanks). Relative to copper, lead has a considerably more negative polarity, not very different from that of steel. Conductors with lead sheathings may therefore be connected with earth electrodes of galvanized steel and other buried steel objects.

The steel reinforcement of concrete foundations usually has a potential similar to that of copper, so earth electrodes directly connected with foundation reinforcement should be lead-sheathed; this applies in particular to short connecting conductors and bonding in the immediate vicinity of the concrete. Lead sheathing embedded in concrete must be protected against corrosion by a moisture-proof cover (e.g. butyl-rubber tape).

Except in the case of the electricity supply system, for which the earth electrodes must always be connected to the operational equipment, it is advisable to interrupt the connection between buried objects having very different potentials by inserting isolating spark-gaps. In normal circumstances, no more corrosion current can then flow. When a surge occurs, the spark-gap breaks down and interconnects the objects for the duration of the surge.

Because of the increased risk of corrosion, earth-terminations of galvanized steel must be protected for at least 300 mm above and below ground level. Connection points under ground must be so constructed that their resistance to corrosion is the same as that of the protective coating of the electrode. All connection points and connecting devices that contain cavities and are not otherwise protected against corrosion must be enclosed in a corrosion-resisting covering after assembly.

It is in some circumstances opportune to provide a massive additional electrode to act as an anode relative to the whole earthing system and thereby relieve that system of the effects of corrosion.

15.6.5 Earthing improvement

The resistance of a mountain cannot be modified by technical means. It is determined by the natural properties of the substratum and by the weather. Contact and bedding resistance, however, can be reduced by appropriate measures, termed earthing improvement, such as increasing the soil conductivity by injecting highly conducting solutions and thereby reducing the contact and bedding resistance. Formerly, use was made of saline solutions, although they were very conducive to corrosion. More recently, hygroscopic emulsions have been developed for this purpose. In order to facilitate the penetration of the emulsion into the rock, blasting is sometimes necessary.

In the case of high-altitude transmitting stations, where rock of very high resistivity is often found, earthing improvement is in most cases essential. By emulsion injection, lower earthing resistances are obtained, independent of fluctuations in air temperature and humidity, even where the soil conductivity is exceptionally low. In very acidic soil, the application of this method will, moreover, provide adequate corrosion protection.

Care must also be taken that the earth electrodes are buried below the frost line because freezing of the soil considerably increases its resistivity. An earthing installation for a power supply system can have an entirely satisfactory resistance during the warm season, but an unacceptably high one during the winter. For lightning-protection earthing systems this is, however, not so important, because the lightning current is impulsive, and in icy ground the conventional earth impedance of an earth electrode is less than its resistance.

15.6.6 Foundation earth electrode

In recent years, special attention has been paid to the possibility of using the steel reinforcement of the building foundations as the earth electrodes and, in particular, that of the plinths of the antenna support structure. Experience and laboratory tests have confirmed that, if they are electrically well bonded, the reinforcement bars of the foundations are useful as earth electrodes and they can make a decisive contribution to the reduction of the conventional earth impedance of the earth-termination system, particularly in soil of high resistivity.

Generally the foundation earth electrode alone does not constitute an adequate earth-termination system for lightning protection and should be integrated with additional earth electrodes. The foundation electrode, embedded in the foundation concrete, can be merely considered as an interface with the geological earth, that is to say, the rock or soil, whose resistance depends to a large extent on the conductivity of the concrete. Geologically, concrete may be compared to a conglomerate, but it differs from geological conglomerates in that its conductivity does not remain constant, but changes with time as a result of chemical processes that occur in the concrete during hydration. Initially, the wet, unset concrete, as well as the water contained in the interstices, has a very low resistivity ($2-5$ Ω m). In due course, however, the unset concrete changes into set concrete and colloids, and the free water into gel water and chemically bound water, resulting in a considerable increase in its resistivity. Measurements carried out on numerous recent and older concrete structures have shown that usually, after only about five years, resistivities of ~ 500 Ω m are reached, which is three to five times the normal resistivity of humus. After ten years, values in excess of 1 000 Ω m can occur, and very old concrete has a resistivity similar to that of hard rock. This increase of the resistivity must be taken into account when designing foundation electrodes.

Furthermore, the resistivity of concrete depends on the composition of the concrete mixture; the higher the proportion of ballast, the lower the resistivity. Thus, the mechanical and geoelectrical requirements conflict. In addition, mechanical damage and sudden rises of contact resistance have been observed at the passage of heavy lightning currents.

For all these reasons, the foundation electrode alone cannot be regarded as adequate for lightning protection; it must be supplemented by specifically designed lightning-protection electrodes. The best solution is to drive vertical electrodes from the base of the foundation into the ground, with their upper extremities welded to the foundation reinforcement. This solution ensures good conduction of the lightning currents, and the vertical electrodes also improve the effectiveness of the foundation electrode.

The earth electrodes embedded in the foundation function by virtue of the interaction of the cement and water, which has a decisive influence on electrical conductivity. The complicated hydration processes of the cement give rise to shrinkage, with the formation of additional capillary space ($\sim$75 ml kg^{-1} of concrete), which retains soil humidity and thus leads to extensive integration of the foundation electrode in the surrounding soil or rock.

In the design of foundation electrodes, the configuration of the foundation has, of course, to be taken into account. It is necessary also to make sure that there is no insulation below the foundation electrode, because the foundation concrete must be in equilibrium with the moisture of the underlying and surrounding soil. For earth electrodes, use may be made of a round steel bar of diameter of at least 10 mm, or a steel strip with a minimum cross-section of 90 mm^2 and a minimum thickness of 3 mm (at least 30 $\times$ 3 mm^2). The use of ungalvanized steel is recommended, because zinc is attacked by the concrete, causing an increase in the contact resistance between the earth electrode and the concrete.

In most cases, however, for mechanical reasons, the foundations will contain steel reinforcement bars that, when welded into a mesh, can partially or wholly render the insertion of specific electrodes superfluous, but where there are points that are more than 5 m distant horizontally from a foundation electrode, additional electrodes must be provided. As far as possible from the constructional point of view, the foundation electrodes should form closed loops in the foundation of the building. Foundation electrodes in several sections should be reliably bonded by means of clamped, welded or bolted connections.

Where the reinforcement steel acts as the foundation electrode, there is a large number of connection points, so the usual clamps generally suffice. With foundations with unreinforced piles, electrodes must be inserted in the piles, the extremities being bonded together. Foundation electrodes of strip steel must be laid on edge, so as to prevent the formation of water pockets beneath them. In order to maintain the electrodes in position while the concrete is poured, suitable supports should be used.

Figure 15.18 shows a typical foundation electrode and its connection with the reinforcement of a plinth for a 40-m steel lattice mast.

The foundation electrodes are protected against corrosion only when they are surrounded on all sides by at least 50 mm of concrete; this is particularly important when the foundation reinforcement is used as the earth electrode. Electrodes embedded in foundations that are exposed to water or soil that readily attacks concrete must be made of rust-resisting steel.

The foundation earth electrode must be provided with two or more connection lugs protruding from the concrete and connected to the potential-equalization busbar inside the building (see Section 15.7.3). Other connection lugs must be led outside the building at maximum spacings of 5 m. Connection lugs and bonding between foundation earth electrodes protruding from the foundation concrete, as well as bridges across expansion joints, must be protected against corrosion by means of self-adhesive plastics tape, wound overlapping. Corrosion protection must begin at least 50 mm below the surface of the concrete and extend for at least 300 mm above the ground. Where connection lugs are enclosed in masonry of natural or artificial stone, they must be

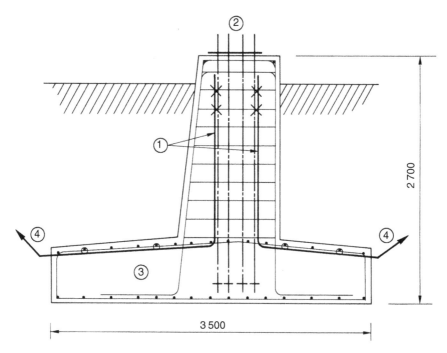

Figure 15.18 Bonding of the tower feet to reinforcement bars in the concrete foundations and to the soil or rock beneath: galvanized-steel strips (about 30×3 mm^2, 1), welded to the feet of two diagonally opposed tower corner elements (2) and welded or clamped to the reinforcement bars in the concrete foundation (3) and extended (4) to connect to the buried earth system. Typical dimensions are shone in mm.

protected against corrosion over their entire length. Only standard cement mixes may be used for foundation electrode concrete. Ballast with a screen of more than 32 mm may not be used, and the use of additives containing chlorides is also to be avoided.

15.7 Potential equalization to reduce failures of electrical and electronic systems

15.7.1 Potential equalization for the earth-termination system of the station

In order to avoid differences of potential, due to lightning effects, between the individual earth-termination systems, such as the earth-termination of the LPS, usually installed at the base of the tower supporting the antennas, the earth-termination of the equipment building and the earth-termination of the HV/LV substation, if any, of the power supply system, all these must be effectively bonded and integrated in a single earth-termination system.

As mentioned in Section 15.2.1.2, the lightning current that flows on the bonding conductors on its way to earth leads to a voltage drop on the inductance of the conductors of the earthing system, and may cause large potential differences between different parts of it, with resultant damage to equipment connected to them. From the physical view point the voltage drop on the inductance of the earthing conductors is related to the voltage induced by the lightning current flowing on it, which appears on the loop made up by the earthing and live conductors connecting the equipment. Reducing the induced voltage – and thus achieve equipotentialization of equipment connected to different parts of an earthing system – is possible by

- reducing the induction current, by sharing it among different parallel paths
- reducing the area of the induced loop, by use of cables with integrated earthing and live conductors and/or laying live and earthing conductors in the same cableway
- reducing the inducing electromagnetic field, by use of screened cables and/or running live and earthing conductors in a metallic close duct

The main way to equipotenzialize the earth-termination system, avoiding potential differences between equipment connected to different parts of the earthing system, is to use screened cables for connection of equipment to each other, with the screen having an adequate cross-section bonded to the earth-termination system at both extremities.

With both extremities of the screen connected to earth, the screen is electrically in parallel with the earth conductors of the station and will carry part of the lightning current that the earth-termination system disperses in the ground. The overvoltages that build up on the cables (common-mode overvoltages) due to the flowing of the lightning current on the screen can be calculated, and precautions must be taken to reduce them to safe values.

The overvoltages are reduced by increasing the cross-section and the screening factor of the sheath. Continuous metallic screening is the best, while plaited screening

758   Lightning Protection

is better than helicoidal tape screening. Moreover, the magnetic permeability of the material should be as high as possible and it is advantageous to increase the number, the cross-section and the symmetry of the shield-wires connected to the screen at both extremities.

A useful rough way to estimate the overvoltage U affecting a screened cable takes into account its geometrical and physical surroundings and uses the equation

$$U = \frac{K_p \rho \cdot l}{S} \cdot I_F$$

where K_p is the installation coefficient, S the screen cross-section, ρ the resistivity of the cable, l the length of the cable and I_F the current flowing inside the cable. Values of K_p for different arrangement of conductors parallel wired to the cable's screen are reported in Figure 15.19.

The shielding effect of a metallic open duct increases with the ratio h/w between the height h and the width w of the open duct because of the reduced penetration of the magnetic field. This reduces the installation coefficient K_p. The reduction factor M is reported in Figure 15.20.

The effectiveness of a cable screen in reducing the overvoltages results directly from the cross-section of the screen. If this is not adequate and cannot easily be changed (i.e. because the cable is already installed) it is possible to increase the shielding effectiveness of the cable screen by decreasing the current flowing on it. This is achieved by having one or more conductor parallel wired to the cable screen. If the

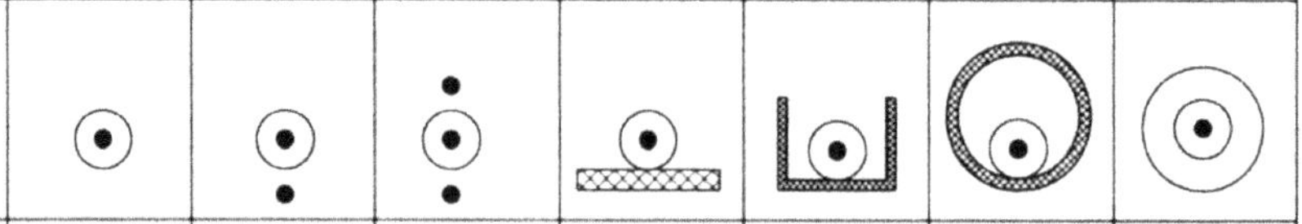

	1	2	3	4	5	6	7
K_p	1	$\frac{S}{S+S_c}$	$0.5\frac{S}{S+2S_c}$	$0.3\frac{S}{S+S_c}$	$0.3M\frac{S}{S+S_c}$	$0.05\frac{S}{S+S_c}$	$0.005\frac{S}{S_e}$

Solutions
1. Screened cable
2. Screened cable with one additional conductors
3. Screened cable with two additional conductors
4. Screened cable with plaited conductor
5. Screened cable in open metalic duct (height equal to width)
6. Screened cable metalic tube
7. Double screened cable ($S_e = 2S$)

Legenda
M = Mutual inductive coefficient
S = Screen cross section
S_c = Cross section of additional conductor
S_e = Cross section of external screen

Figure 15.19   Values of K_p for different arrangements of conductors parallel wired to the cable's screen

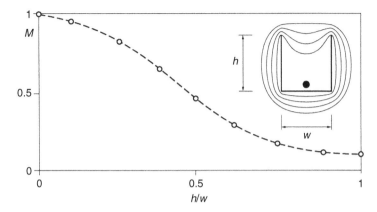

Figure 15.20 Reduction factor M of a metallic open duct of width w and height h compared to a solid plate conductor whose width is 2h + w

conductors are adequately selected and installed, the final result can be better than increasing the cross-section of the screen itself.

An improved performance can be achieved by using solid plate conductors symmetrically installed in relation to the cable. For the high-frequency domain, solid plate conductors have a lower impedance compared to round ones, and current is forced to flow on the external symmetrically installed conductors.

The best performance is achieved by installing the cable in a metallic close duct, preferably of the ferromagnetic type. This practice, although expensive, has the effect of reducing in a very effective manner the value of current on the cable screen, quite avoiding resistive and inductive coupling phenomena.

Implementation of the potential equalization of the earth-termination system may be performed as follows.

1. For effective functioning of equipment in a building and safety of the personnel, it is necessary to prevent the formation of dangerous potential gradients on the floor of the building and large voltage drops in the down-conductors inside the building itself. This can be achieved by inserting in the floor a regularly spaced metallic mesh and by preventing lightning currents from passing through that mesh. For small buildings of up to 10 m^2, it is sufficient to bury a ring electrode around the building. The cable sheaths and earth conductors entering and leaving the building must do so at a single point and must be effectively connected to a potential-equalization busbar (see Section 15.7.3). The earth connections of equipment having no external connections may be directly connected to the floor mesh.
2. It is usually advantageous to arrange a single point of entry into the building for all the cables and pipework of every kind. In this case, the busbar can consist of a metal bar of appropriate dimensions at the point of entry, to which are connected the outer conductors of the coaxial cables and waveguides, the pipework and the

760 Lightning Protection

screens of the cables of other types, as well as the shield-wires in the ducts between the building and the antenna-support structure, and between the building and the HV/LV transformer.

3. The connection between the ring electrodes around the base of the antenna support and the earth-termination system of the building should be effected by means of two earth conductors ('shield wires') laid symmetrically in the antenna-feeder duct (Figure 15.21a). Where only coaxial cables of large diameter are laid in the duct, it is sufficient to provide one shield wire (Figure 15.21b); if supply cables also pass through the duct, particular arrangements must be adopted for the shield wires (Figure 15.21c).

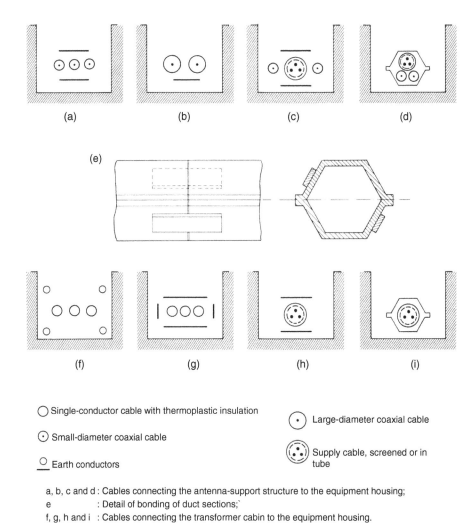

○ Single-conductor cable with thermoplastic insulation

⊙ Small-diameter coaxial cable

○ Earth conductors

(•) Large-diameter coaxial cable

(:·:) Supply cable, screened or in tube

a, b, c and d : Cables connecting the antenna-support structure to the equipment housing;
e   : Detail of bonding of duct sections;
f, g, h and i : Cables connecting the transformer cabin to the equipment housing.

Figure 15.21   Arrangements of cables in the ducts

In particularly critical circumstances (very long duct, very high soil resistivity, high probability of lightning), it may be necessary to provide a completely enclosed metallic conduit accommodating all the cables (Figure 15.21d). Where it is not possible to install a conduit in a single section, its several sections must be bonded by means of suitably welded straps (Figure 15.21e).

4. The transformer substation (HV/LV supply) should be provided with a ring earth electrode with interconnected rods or horizontal radial electrodes, depending on the soil stratification. The connection between the ring earth electrode of the transformer housing and the earth-termination system of the building should be effected by means of earth conductors placed in the power-cable duct, in accordance with the principles described for the cables between the antenna and the building. Typical arrangements of the cables between the transformer and the equipment housing are indicated in Figure 15.21f–i.

5. A metal perimeter fence needs to be connected to the general earth-termination system of the station only where it is not entirely outside the earth-termination system and at a distance at least 5 m from the nearest electrode of the earth-termination system. In that case, the power-cable duct, in the area where it is crossed by the fence, must consist of non-conducting tubes for at least 5 m on either side of the fence.

15.7.2 Potential equalization for the antenna support structure

All the waveguides, the sheaths of the radio-frequency (RF) and supply cables and their metal protective conduits must be connected to the antenna support structure at the top and at the base. Bonding, even if it is not strictly essential, is desirable also at all the points where the ducts or cables are attached to the structure. If the electrical conductivity of the structure is not excellent over its entire height, it must be improved by installing one or more substantial down-conductors connecting the air-terminations at the top of the structure to the buried earth-electrodes at its base. The several sections of the structure itself, as well as the cable sheaths, waveguides and ducts, must be connected to the down-conductors, at least at the top and base. Flexible sections of waveguides must be bridged by conducting straps connecting the rigid sections.

A tubular steel support-mast constitutes an ideal low impedance bonding network, although steel-lattice construction is also adequate. All the cabling, such as the antenna feeders and the cables for the aircraft-warning lanterns, must be completely enclosed within the metallic structure.

Where the antenna support consists of a cylinder of, for example, epoxy-resin reinforced with glass-fibre, which protects the antenna array from the weather, but not from the electromagnetic fields such as occur with lightning flashes, other precautions must be taken to protect the cables, as described in Section 15.6.

15.7.3 Potential equalization for the equipment within the building

A low impedance bonding network is needed to avoid dangerous potential differences between all equipment inside the building. This can be realized by a meshed bonding network integrating the conductive parts of the building, the parts of the internal electrical and electronic system to be earthed (i.e. metal frames, cabinets, screen of cables, etc.), and by bonding it to the earth-termination system directly or by using a suitable SPD.

The bonding network can be arranged as a three-dimensional meshed structure with a typical mesh width of 5 m. This requires multiple interconnections of metal components in and on the building (such as the concrete reinforcement, elevator rails, cranes, metal roofs, metal facades, metal frames of windows and doors, metal floor frames, service pipes and cable trays). In the same way, the bonding bars (e.g. ring bonding bars, several bonding bars at different levels of the structure) should be integrated. Conductive parts (e.g. cabinets, enclosures, racks) and the protective earth conductor (PE) of the electrical and electronic system should be connected to the bonding network.

In a building of poorly conducting material such as concrete, masonry or the like, it is necessary to bond electrically the metal framework or reinforcement bars at roof and floor levels, and thus to bond them with multiple connections to the earthing system.

In the case of transmitting and radio-relay towers, which are often constructed as in Figure 15.8, the power supply and radio equipment being housed in a ferro-concrete tower, with the antennas mounted on a steel support structure (of tabular or lattice construction), the following solution should be adopted.

In ferro-concrete towers containing transmitting equipment, the outermost reinforcement bars should be connected together by encircling conductors clamped or, preferably, welded at the level of each storey to the down-conductors between the air- and earth-termination system.

In order to avoid dangerous potential differences within the tower, it is necessary, also at the level of each storey, to install a conductor connecting the encircling conductor to all the metallic objects in the building, such as lift-cages, cable shafts and apparatus racks.

In order to facilitate earthing of the frameworks of the machines, the apparatus racks and the other metallic objects inside the building, a potential-equalization busbar should be provided to ensure their direct and effective connection to the earth electrodes of the building. A good solution is to install a conductor, preferably of copper of an appropriate cross-section, mounted directly along the internal surface of the outside walls, to form a ring, which is then bonded to the earth-electrodes at intervals of not more than 10–15 m.

A potential-equalization busbar is preferably installed in the basement of the building, in order to obtain the shortest connections to the earth-electrodes and to the buried pipework. In the case of buildings without basements, the potential-equalization busbar should be installed as low as possible, not more than 300 mm above ground level.

In single-storey buildings with a mesh electrode in the floor, the potential-equalization ring can be omitted and the earthing of the metal objects can be more easily effected by connecting them directly to the mesh. For small buildings, a ring busbar is not usually essential; a short busbar on one wall is often sufficient.

For buildings of several storeys, such a ring must be installed at each level, the rings at the several levels being connected together by vertical conductors spaced ~10–15 m from one another. For buildings equipped with LPS and thus having down-conductors from an air-termination system to ground, the vertical conductors should be connected to all the potential-equalization busbars.

Miscellaneous metallic objects inside the building, such as ventilation grills, filter frames and metal doors, must also be bonded to busbars. The connectors to the busbars should have minimum cross-sections in accordance with the requirements of IEC 62305-3.

15.7.4 Potential equalization for metallic objects outside the building

Because the potential rise at the instant of a lightning flash can amount to several hundreds of kilovolts, it is essential to connect all nearby conducting objects external to the building, such as structures, cables, pipes and fencing, to the earth-termination system, if they are located within ~ 5 m from earth electrodes; this is to avoid flash-overs and insulation failures, and the consequent danger. These connections are of great importance, in particular where there are installations belonging to third parties, such as cableways, lifts, hotels and electricity-supply or telecommunication equipment.

Parts of buildings and equipment that, for operational or corrosion reasons, cannot be permanently earthed, must be connected, by way of isolating spark-gaps, with the earth-termination system or a potential-equalization busbar.

In the case of a supply from an HV line, the screen of the HV and LV cables, the frame of the HV/LV transformer, the star point of its LV winding, the earth of the HV protection devices and the guardwire (if any) of the incoming line, must be connected to the station earth-termination system. Where this is not possible or opportune (e.g. to avoid affecting the spreading of the lightning current dispersed by the earth-termination system of the antenna support structure, pipework or cable sheath) it may be necessary to insert insulating joints (at least for some 10 m in the case of pipework) and to protect the cable cores against overvoltages. At a site shared with other services, it is in many cases necessary to make a simulation study of the potential distribution in the area concerned to determine the potential situation to be expected in the case of lightning and to provide a basis for appropriate precautions.

15.8 Screening to reduce failures of electrical and electronic systems

Because of the high amplitudes of lightning currents, strong magnetic fields occur in the vicinity of the conductors through which they flow. Furthermore, as a consequence of their very steep wavefronts, the magnetic field strength in their vicinity also changes very rapidly, causing induction effects in the circuits inside the building. To minimize these effects the equipment and cabling should be screened. The connections entering the building must also be screened or protected by the precautions indicated in Section 15.9.

15.8.1 Screening of circuits within the building

An effective screening of circuits wthin the building can be achieved by

- a spatial shield enclosing all circuits and equipment of the building (to be effective the shield should be continuous or meshed with a mesh width not greater than some tens of centimetres)
- use of screened cables and metallic enclosures for equipments

Table 15.1 Magnetic screening attenuation (K_s) as a function of the mesh width or the thickness of the shield

Type of screen	Screen characteristics	K_s
Mesh	w = 5 m	0.6
	w = 4 m	0.5
	w = 3 m	0.35
	w = 2 m	0.24
	w = 1 m	0.12
	w = 0.5 m	0.06
	w = 0.2 m	0.024
	w = 0.1 m	0.012
Sheet	s = 0.1 mm	10–4
	s = 0.5 mm	10–6

w = mesh width; s = sheet thickness.

Although this latter method is easier, cheaper and more flexible to use, the former is sometime used in special cases of low-power stations for small zones of large stations.

The simplest way to screen a low-power station is to use a transmitter housing of sheet-steel, closed on all sides, as depicted in Figure 15.6. This is also the optimum solution for the dispersion of lightning currents, as well as for screening from electromagnetic fields. The screening against the electric component is effective because no lines of force can penetrate the housing, whereas attenuation of the magnetic component increases with rising frequency, because of the Kelvin effect within and near the outer surface of the screen.

In a large station, in special cases, an effective screening may be required for a particular zone of the station; it can be achieved by constructing a room, enclosed on all sides with metal sheet or mesh, for housing particularly sensitive apparatus. Table 15.1 indicates the magnetic screening attenuation (K_s) of such a room as a function of the mesh width or the thickness of the shield.

Several internationally standardized methods of measuring screen attenuation are given in the literature. The attenuation (K_s) of either the electric or the magnetic field may be expressed as the ratio of their strengths with and without screening.

15.8.2 Circuits of the station entering the building

It is equally necessary to protect external cabling and equipment connected inside the station, but that extend beyond the screen of the building and which, therefore, must all be connected to the single earthing system. They, too, must be enclosed in a screen, ideally an extension of that of the building, which can in practice be done by using suitably screened cables or by laying the cables in metallic tubes or ducts with, if necessary, further measures (see Section 15.7.1) for limiting the voltages induced in the cables. These circuits can be screened effectively by connecting to the earth-termination system the screens and the metal sheaths of all the cables, the chassis of the electrical and radio apparatus and the frames of the machines and the isolating transformer. The metal sheaths and the screens of the cables must be connected to earth at both extremities; all the earth conductors must have the least possible impedance, and therefore short and straight routes, frequent bonding, and so on. The RF cables do not pose any particular problems, because they are always screened. On the other hand, the power supply circuits must also be carried in screened cables, and in some cases this gives rise to technico-economic problems.

The practical solutions to be adopted depend on the required protection level for the station. In the case of stations for which only a lesser protection level is necessary, partial screening for external cables laid in ducts can be used. The power cables should be arranged symmetrically in the ducts with respect to the shield-wires electrically connecting earth-electrodes located at the two extremities of the duct itself. The number and nature of the shield-wires depend upon the required protection level, bearing in mind that the best protection can be obtained with completely closed metal ducts (Figure 15.21). The shield-wires must have the fewest possible joints and all the joints must be welded.

The induced overvoltages between the cable cores are generally termed differential-mode voltages and can be a danger to electronic equipment. It is possible to reduce such differential-mode voltages by, for example, transposing the cores at the inputs and outputs, or by screening.

For stations requiring a particularly high protection level, screened cables are essential in every case. For stations requiring the highest protection level, only screened cables laid in steel tubes or cables of layered construction may be used; compared with a normal screened cable, correct installation in a steel tube of suitable properties reduces the common-mode voltage by ~ 25 V, whereas the use of an appropriate cable of layered construction reduces it by 500 V.

Supply cables of aircraft warning lanterns and other possible auxiliary services attached to the metal antenna support structure can be screened by enclosing the cables in a continuous metallic tube, the cables preferably having improved insulation. The tube must be metallically continuous along its entire path (from the top of the antenna support to the earth-termination system of the building), any joints being welded. Naturally, the screen thus produced must also be effectively connected to the support structure. In practice, good electrical bonds must be made at least at the two extremities of the cable and at the base of the antenna support structure.

Aircraft warning lanterns must be protected by metal grids, electrically well connected to the upper end of the tube. A more convenient solution is to use cables incorporating a continuous metal sheath of adequate thickness, bonded at both extremities; this ensures greater reliability, by avoiding discontinuities along the surface of the screen. It is advisable for greater mechanical strength that the sheath be made of a lead–antimony alloy. In particularly difficult cases (very high antennas on very exposed sites), the sheath may be too thin to prevent excessive voltages between the cores and sheath of the cable. It is then opportune to insert, at the lower end of the cable, an isolating transformer with an electrostatic screen and with an SPD on the cable.

15.9 Coordinated SPD protection system to reduce failures of electrical and electronic systems

A main protection measure for the electrical and electronic systems of the station against surges is a 'coordinated SPD protection system' for both power and signal circuits.

The coordinated SPD protection is a set of SPDs properly selected, energy coordinated and installed.

15.9.1 Selection of SPDs with regard to voltage protection level

The impulse withstand voltage U_w of the equipment to be protected should be greater than or equal to the voltage protective level U_P of the SPD plus a margin necessary to take into account the voltage drop of the connecting conductors.

The inductive voltage drop ΔU due to the flowing of partial current I_f (see Section 15.2.1.1) on the connecting conductors, will add to the protective level U_P of the SPD. The resulting effective protective level, $U_{P/f}$, defined as the voltage at the output of the SPD resulting from the protection level and the wiring voltage drop in the leads/connections, can be assumed to be

$$U_{P/f} = U_P + \Delta U \quad \text{for voltage-limiting type SPD}$$
$$U_{P/f} = \max(U_P, \Delta U) \quad \text{for voltage-switching type SPD}$$

When the SPD is carrying the partial lightning current, $\Delta U = 1$ kV m^{-1} length, or at least a safety margin of 20 per cent, should be assumed when the length of the connection conductors is ≤ 0.5 m. This case is usually relevant to class I SPDs installed at the point of entry of the line in the station (main distribution board, MB, or primary windings of insulating transformer, if any). [Note that in particular conditions (following strokes of lightning flashes, high values of the partial current I_f) the voltage drop ΔU can be higher than 1 kV m^{-1}; the value of $\Delta U = 1$ kV m^{-1} is the one suggestd by IEC 62305-4.]

When the SPD is carrying induced surges only, ΔU can be neglected. This case is usually relevant to class II SPDs installed at the secondary distribution board, SB, or at the equipment's terminals.

15.9.2 Selection of SPD with regard to location and to discharge current

At the line entrance into the station, namely at the MB or at the primary windings of the insulating transformer, if any, class I SPDs should be used to discharge a partial lightning current, namely current with a high energy content.

The required impulse current I_{imp} of the SPD should provide for the (partial) lightning current to be expected at this installation point based on the chosen LPL according to IEC 62305-1:

$$I_{imp} \geq I_f$$

and

$$I_{imp} \geq 10\,kA \quad \text{for LPL I or LPL II}$$
$$I_{imp} \geq 5\,kA \quad \text{for LPL III or LPL IV}$$

Close to the apparatus to be protected (downstream of the insulating transformer, if any, and in any case downstream of a class I SPD), for example, at the SB, or at the equipment's terminals, class II SPDs may be used for the current to be discharged arising from induction effects in installation loops or as a remaining threat downstream of class I SPD.

The required nominal discharge current I_n of the SPD should provide for the surge level to be expected at the installation point based on the chosen LPL according to IEC 62305-1:

$$I_n \geq I_f$$

and

$$I_n \geq 5\,kA \quad \text{for LPL I or LPL II}$$
$$I_n \geq 2.5\,kA \quad \text{for LPL III or LPL IV}$$

15.9.3 Installation of SPDs in a coordinated SPD protection system

The efficiency of a coordinated SPD protection system depends not only on the proper selection of SPDs, but also on the proper installation of these. An over-long distance between the SPD and the equipment to be protected may make the protection ineffective due to the propagation of surges in the circuit connecting the SPD and the equipment, and to induced additional surges in the circuit itself.

The maximum length of the circuit between the SPD and the equipment, for which the SPD protection is still adequate, taking into account propagation and/or induction phenomena, is called the protective distance l_P.

15.9.3.1 Protective distance l_P

During the operating state of an SPD, the voltage between the SPD terminals is limited to $U_{P/f}$ at the location of the SPD. If the length of the circuit between the SPD and the equipment is too long, propagation of surges can lead to an oscillation phenomenon that can increase the overvoltage up to $2U_{P/f}$ and failure of equipment may result even if $U_{P/f} \leq U_w$.

Moreover, lightning flashes to the station can induce an overvoltage U_i in the circuit loop between the SPD and the equipment, which adds to $U_{P/f}$ and thereby reduces the protection efficiency of the SPD. Induced overvoltages increase with the dimensions of the loop (line routing: length of circuit, distance between PE and active conductors, loop area between power and signal lines) and decrease with attenuation of the magnetic field strength (spatial shielding and/or line shielding).

On the safety side, the equipment can be considered protected if $U_w \leq 2U_{P/f} + U_i$.

The induced overvoltage is reduced by

- minimizing the loop between the SPD and the equipment
- spatial shielding of the building of the station
- circuit shielding by use of the shielded cables or cable ducts

In general, the overvoltage U_i can be disregarded if one of these provisions is taken.

15.9.3.2 Induction protective distance l_{Pi}

Lightning flashes to the station can induce an overvoltage in the circuit loop between the SPD and the equipment, which adds to $U_{P/f}$ and thereby reduces the protection efficiency of the SPD. Induced overvoltages increase with dimensions of the loop (line routing: length of circuit, distance between PE and active conductors, loop area between power and signal lines) and decrease with attenuation of the magnetic field strength (spatial shielding and/or line shielding).

The induction protective distance l_{Pi} is the maximum length of the circuit between the SPD and the equipment for which the protection of the SPD is still adequate (taking into account the induction phenomena). In general, the induced overvoltage is reduced by

- minimizing the loop between the SPD and the equipment
- spatial shielding of the building of the station
- circuit shielding by the use of shielded cables or cable ducts

According to IEC 62305-4, if one of these provisions is taken, the equipment can be considered protected. If this is not the case, the protective distance l_{Pi} may be checked by the method reported in IEC 62305-4, Annex D.

15.10 Protection of lines and services entering the station

All the lines connected to the station must be protected, because they are also connected to earth at points more or less distant from the station and are therefore under stress because of the earth-potential differences.

The proper choice of individual precautions is dependent upon the magnitude of the risk and, therefore, upon the protection level required for the station. For stations requiring low protection levels, often no more than the usual precautions need to be taken. In that case, use can be made, for the lines connected to the station, of overhead open-wire lines or overhead or buried cables. The potential equalization described in Section 15.7 should, nevertheless, be adopted.

When high protection levels are required, screening is indispensable and the other precautions indicated in Section 15.8.2 may be required. For the most important stations, the highest protection level is likely needed. In this case it is essential to specify cables of layer construction and/or to install buried sheet-metal ducts, with low-impedance bonding, in which the cables and connectors are laid. Thus, the choice of a suitable cable is very important.

15.10.1 Overhead lines

The criteria for protecting open-wire lines or overhead cables are the same, in principle, for telecommunication and for energy transmission. There is, nevertheless, a difference in the protective devices, depending upon the energy involved in the circuit to be protected, upon its nominal voltage and upon its withstand insulation level. It should be noted that, in the case of open-wire lines, the line insulation is self-restoring, whereas that is not the case for cables.

The principal criteria of protection are as follows.

1. All the live conductors, at their points of entry in the building, must be protected by SPDs.
2. Any spare conductors in telecommunication cables and the neutral conductor of LV supply lines must be connected to the earthing system of the station.
3. The catenary wires of overhead cables, their screens, if any, and the guard-wires of open-wire lines must be connected to earth.
   Precautions (2) and (3) contribute to an efficient screening of the live conductors of the lines.
4. The earth connections mentioned in (1)–(3) must be as direct as possible:
   - to the ring earth-electrode around the HV/LV transformer in the case of the connections to its primary and
   - to the ring earth-electrode around the building or to the potential-equalization busbar for the lines entering directly into the building.
5. Use must be made of cables having thermoplastic insulation, which withstands surges much better than impregnated paper.
6. To avoid the upstream propagation of surges along the lines connected to the station, and to avoid damage to the lines or to the onnected equipment, it is necessary to install SPDs
   - at each point of discontinuity of the insulation characteristics of the line (change from open-wire line to cable, change from cable having thermoplastic insulation to cable having impregnated-paper insulation or from screened cable to unscreened cable);
   - at the point where the line branches off the supply network (in the case of telecommunication cables, all the conductors must be protected, whether they be connected or spare, in the principal circuit or in the branch circuit);
   - at a point 300–500 m from the station.

In all these cases it is necessary to provide a local earth-termination system having a conventional earth impedance that is as low as possible, and to connect to it the earthy terminals of the SPD, the screen, if any, of the cable and the guard-wire or the cable catenary, if any. The SPD used must be capable of withstanding the high energy associated with the lightning current; for this reason, it is, in general, recommended to use class I SPD. Use may be made for

- telecommunication circuits, of gas-filled spark-gap SPD;
- HV power supply circuits, of horn gaps directly mounted on the insulators of open-wire lines;
- LV supply circuits, of open or gas-filled spark-gap SPDs (in the latter case, it is usually necessary to ensure correct coordination of the insulation, to improve that of the line and of the connected equipment).

15.10.2 Screened cables

Unscreened cables, even if buried, require the same protection as that indicated above for overhead lines. For protecting screened cables, the fundamental criterion is a reduction of the overvoltages between the core and sheath, caused by the passage of the lightning current along the screen of the cable, to a level compatible with the insulation of the cable, which implies reducing to predetermined values the core-to-screen coupling impedance. For this purpose, the cables may be laid in steel tubes or they are of layered construction.

If the cable has been buried in ground having uniform soil conductivity, the current carried by the cable sheath will disperse progressively into the ground, thus attenuating the lightning current along the cable sheath. However, with overhead insulated cables and open-wire lines, the lightning current is the same over the entire length of the cable sheath and there is no progressive attenuation of the impulse current. For that reason, for stations requiring very high protection levels, buried cables are preferred, and the cables should be laid in ground having the best possible soil conductivity and with the best possible bedding.

To reduce the stress on the cable, a supplementary earth tape should be laid in parallel; this should be bonded to the cable sheath at all the sleeves. Where cables are close to other earth electrodes, they should, if possible, be bonded (potential equalization). If that is not possible, appropriate insulation must be provided.

Regarding the laying of the cables in steel ducts, profiles of the so-called Zorès protective ducts for this purpose are sketched in Figure 15.21e. For the longitudinal electrical bonding of these ducts, use should be made of 5-mm copper wires. The half-shells must be welded together at each end.

The necessary length of the protective duct is essentially dependent on the soil conductivity of the surrounding ground. For stations sited in mountainous areas having poorly conducting ground, such ducts may be necessary for up to 2 km. In most cases, however, the duct will extend as far as the next part of the station, where again potential equalization, as explained in Section 15.7, will be required.

For exposed sites on low ground, too, where direct lightning flashes must be expected, a duct constitutes the optimum protection, particularly if it can be buried in well-conducting ground with good contact with the surrounding soil, to facilitate the rapid dissipation of the lightning current. If several cables or conductors have to follow the same route, they must be laid in the same duct.

15.11 Arrangement of power supply circuits

The most important part of the equipment of a transmitting station that must be protected by SPDs is the power supply system. Apart from telecommunication towers within urban areas, the power supply for most transmitting stations is provided by way of HV or LV open-wire lines terminated at a pole or pylon located at between fifty and several hundreds of metres from the station.

For protection against lightning, a station may be categorized as one of three types, depending upon whether the power supply is at HV or LV, or from a self-contained generator.

15.11.1 Stations supplied at high voltage

The earthing system of the HV/LV transformer is a part of the single integrated earth-termination system of the station. The arrangements described in the following should be applied when the transformer is the property of the broadcasting organization; when it belongs to the supply authority, other arrangements may be specified.

The points where the appropriate SPD must be inserted in stations supplied at HV are indicated in Figure 15.22. Regarding the overhead HV line, the SPD (1) needs to be installed only where it is desirable to prevent the propagation of surges along the HV line as a result of lightning flashes to the antenna support structure; for that purpose, moreover, connection to earth of the studs of the line insulators is also very useful.

In all cases of supply by overhead line, the SPD (2) must be provided for protecting the HV/LV transformer. If the supply is entirely by cables protected according to the criteria indicated in Section 15.10, SPD (2) could be the limiting type.

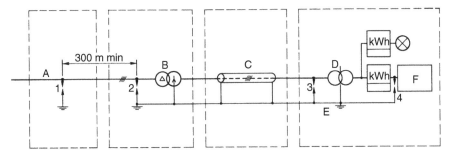

A: Open-wire overhead line (MV);   D: Isolating transformer (LV);
B: Step-down transformer (MV/LV); E: Station earthing system;
C: Buried cables (LV);   F: Radio equipment;
1,2,3,4: Arresters.

Figure 15.22   *Placing of arresters in a station supply at HV by an overhead line*

772 Lightning Protection

Where a short section of protected cable is inserted between the overhead line and the HV/LV transformer, it is necessary to install an SPD of the switching type (spark-gaps) at the interface. SPDs need to be inserted at the secondary terminals of the HV/LV transformer only when it is not possible to connect any point of the secondary winding to earth-termination. It is always necessary to connect to the single earthing system of the station, the metal casings of the HV and LV units, the frame of the transformer, the star point of the LV winding, the earth terminals of the HV protection devices, the shield-wire (if any) at the arrival end of the line and the metal sheaths (if any) of the LV cables between the HV/LV transformer and the station building.

The supply meters are installed downstream of the isolating transformer. The SPD (3) can, in general, be omitted if the arrangements of Figure 15.21h, i are adopted for the cable duct between the transformer and the building.

The isolating transformer is not indispensable when the HV/LV transformer is located inside the station building; it can also be omitted where the cable duct is constructed in accordance with the criteria of Figure 15.21h, i, and is not more than 20–30 m in length; in that case, however, it is essential to protect the equipment with an SPD (4). In any case, for stations of considerable complexity and importance, SPDs must also be installed on the main distribution board to protect the main and service equipment.

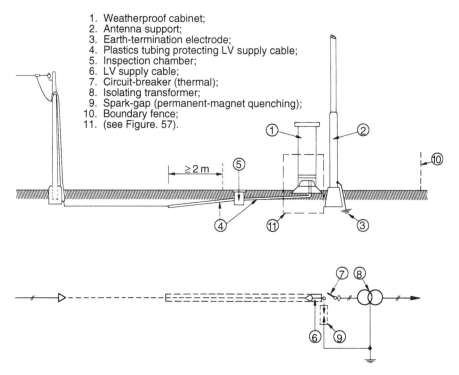

Figure 15.23 *Typical arrangement of a low-power station fed at LV and having the equipment housed in a weatherproof cabinet*

Lightning protection of telecommunication towers 773

An area particularly at danger from the effects of lightning is the aircraft-warning lighting, especially on high antennas and telecommunication towers. In addition to using screened cables, it is necessary to earth the cable screens at the point of entry and to install SPD at the same place, which may be class II tested SPD.

15.11.2 Stations supplied at low voltage

Supply is by overhead lines or by cable, and the earthing system of the substation from which the supply comes is not connected in any way to the earthing system of the broadcasting station (TT system). Arrangements suitable for stations supplied at LV are indicated in Figures 15.23 and 15.24.

For small installations using a modest amount of power (<2 kVA) and with the equipment in weatherproof casings, meters can often advantageously be dispensed with (subject to agreement with the supply authority to charge a fixed sum for the energy consumed), thus eliminating a potential source of breakdown.

For larger stations with equipment housed in a conventional building a possible arrangement is indicated in Figure 15.25.

In each case, to protect the equipment it is necessary that a switching-type SPD (an open or gas-filled spark-gap) be installed at the arrival end of the line, that an isolating transformer be provided, and that the impulse withstand voltage between the primary

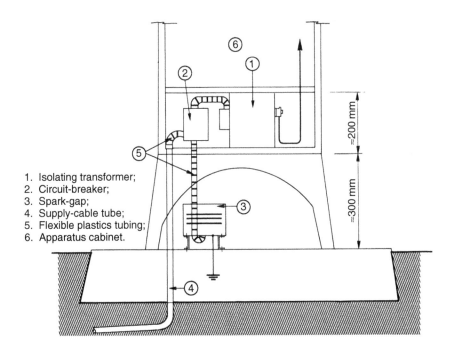

1. Isolating transformer;
2. Circuit-breaker;
3. Spark-gap;
4. Supply-cable tube;
5. Flexible plastics tubing;
6. Apparatus cabinet.

Figure 15.24   Variant of the installation depicted in Figure 15.23, having a vitreous-resin apparatus cabinet and internal isolating transformer

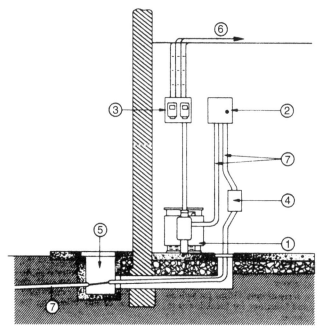

1. Isolating transformer;
2. Circuit-breaker;
3. Consumption meters for radio and domestic loads;
4. Spark-gap;
5. Inspection chamber;
6. Feeds to LV switchboard;
7. Plastics tubing for power-supply line.

Figure 15.25   Power supply protection arrangements for a station comprising a conventional equipment building

winding of the transformer and earth and the secondary, and that of the circuit upstream of the primary of the transformer itself, be high enough to facilitate coordination of the insulation levels.

Moreover, for correct functioning of the station, the main circuit breaker at the arrival end of the supply line must be provided with a thermal trip only (magnetic trips are unsatisfactory), in order to avoid untimely, sudden actuation due to the surge current, when the isolating transformer is saturated because of an overvoltage across its primary. Where appropriate, and subject to agreement with the supply authority, protection in the HV/LV substation that feeds the station should be coordinated, in order to avoid tripping of the circuit breaker by the surge current or by the sparking-over of the spark-gaps of the station. For that purpose, the three arrangements described below and depicted in Figure 15.26 are suitable.

1. *Arrangement with delayed fuses (Figure 15.26a)*. Protection coordination obviates interruptions of the supply to the station when there is a surge current on the line due to the causes described earlier. Fuses F1 must have a fusion current greater than the net three-phase short-circuit current on the LV side of the transformer T. These fuses function for breakdowns on the primary of the transformer, whereas for breakdowns on the secondary, upstream of the F2

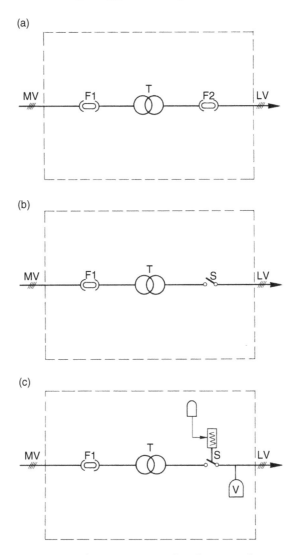

Figure 15.26 Three typical arrangements for the overvoltage protection in the step-down transformer housing

fuses, the overvoltage protection in the substation will function. Fuses F2 must have a delay of 300 ms for the transient current and function also for phase-to-earth short-circuits.
2. *Arrangement with delay switches (Figure 15.26b).* Fuse F1 is as for (1) above. Switch S has a delay of only 150 ms for sudden magnetic changes, and functions for breakdowns to earth or between phases.
3. *Arrangement with reclosing switches (Figure 15.26c).* Fuse F1 is as for (1) above. Switch S is of the motor-driven type with instantaneous action.

776 *Lightning Protection*

For a transient flash-over on the overhead line, in the cable, on the isolating transformer or in the associated circuits, switch S opens and, after a predetermined time, closes again automatically, controlled by a simple device, which locks the switch in the open position if the flash-over recurs upon closure.

Regarding the propagation upstream along the LV line of surges due to lightning flashes to the antenna support structure, it should be noted that the surges affect lines that are directly connected to the equipment of other users served by the same supply authority, so the means for blocking their propagation are less effective and more difficult to carry out (in view of the complexity of LV distribution networks) than in HV cases.

In such a case, a more drastic but effective solution would be to feed the station directly at HV or by an autonomous source. If, for technical or economic reasons, those solutions are not adopted, the shared use of a single HV/LV transformer may be envisaged provided that an isolating transformer, equipped with an earthed electrostatic screen and having a phase-to-earth impulse withstand level of the winding on the line side of at least 45 kV, is inserted between the HV/LV transformer and the line. The isolating transformer must be as close as possible to the HV/LV transformer (e.g. mounted on the same pole as the HV/LV transformer), but nevertheless downstream of any branching for other users.

15.11.3 *Stations with self-contained power supplies only*

Stations powered exclusively by diesel-electric sets, photovoltaic generators, aeolian generators, thermogenerators, batteries, and so on, do not need any particular protection for the supply circuits.

Annex A15: Surge testing of installations

A15.1 Simulation of surge phenomena

A15.1.1 General

The need for evaluating, by means of pulse tests, the parameters of the overvoltages of atmospheric origin that manifest themselves at transmitting stations, derives mainly from the following considerations.

1. For stations already in existence, pulse tests are practically the only method by which useful knowledge of the characteristics of the overvoltages can be gained, and by which the points where potentials are reached that are dangerous for the station itself can be identified rapidly. By that method, moreover, it is possible to discover any places where the potential equalization of the installation is defective, as well as to determine the voltage to which the insulation of the input transformer of the supply system is actually subjected. This enables suitable precautions to be taken for reducing the overvoltages to levels that are harmless to the equipment and for evaluating the relative effectiveness of the various solutions.

2. For stations to be erected in the future, and for which the appropriate protection level has already been determined, in accordance with risk assessment, pulse tests will be necessary for checking the effectiveness of the precautions adopted for reducing the overvoltages, as well as for checking the adequacy of the design of the station in these respects. However, the determination of the overvoltages transferred across the transformers, voltage regulators and the like is feasible only in the case of stations functioning under actual service conditions, because all those transfers depend on the impedances upstream and downstream of the equipment.

The tests to be carried out must evidently reproduce with good approximation the stresses to which the equipment of the station is subjected by atmospheric discharges.

A15.1.2 Tests for determining overvoltages due to lightning flashes to the antenna-support structure

Lightning phenomena affecting the antenna-support structure are simulated by injecting high impulse currents at the base of the structure itself and by measuring the resulting overvoltages on the various parts of the installation. In particular, it is possible to measure the overvoltages between separate points on the earthing system (verification of its potential equalization), as well as the total voltage to earth to which the insulation at the input of the power supply system is subjected.

For carrying out the tests, a high-voltage pulse generator is necessary that is capable of causing the circulation, through the earthing system of the station, of current impulses or oscillations of a highly damped form, having peak amplitudes of a few thousand amperes. One terminal of the generator must, therefore, be connected to the earth-termination of the station, close to the base of the mast or tower, the other terminal being connected by means of insulated connections to an auxiliary earth electrode located at a distance of at least five times the maximum dimension of the earth-termination of the station and consisting of numerous small stakes, spaced not too closely and all connected in parallel (Figure A15.1).

In order to isolate the power supply circuits of the pulse generator from the overvoltages, it should be energized by an independent diesel-electric set located alongside the pulse generator. The currents and the overvoltages can be measured on an oscilloscope in association with a highly insulated probe and a shunt having a short response time.

A15.1.3 Determination of the transferred overvoltages

These measurements are made with lower voltages, using pulses repeating at a certain frequency, obtained from a recurrent-pulse generator. Measurements corresponding to the service conditions of the station can be made by applying trains of low-voltage pulses at the desired points and by noting the voltages transferred to other points of the station itself. The tests must make it possible to determine the amplitude and the waveform of the voltage surges to which the equipment of the station is subjected, as well as the effect of their transfer across the components in question; the absolute values of such magnitudes can be determined only if the overvoltages at the points

778 Lightning Protection

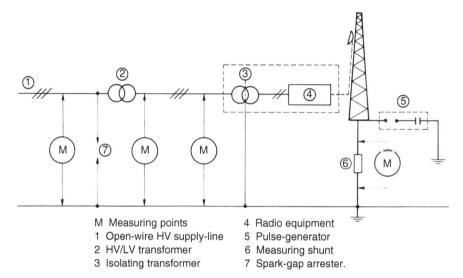

M Measuring points
1 Open-wire HV supply-line
2 HV/LV transformer
3 Isolating transformer
4 Radio equipment
5 Pulse-generator
6 Measuring shunt
7 Spark-gap arrester.

Figure A15.1   Measurement of overvoltages caused by simulated lightning flash to the antenna support

where they appear (earthing systems, if not perfectly equipotential, power supply system, etc.) are known. These measurements, therefore, make it possible to solve the problems of the qualitative choice and the quantitative dimensioning of the protective measures to be adopted.

For the tests, a low-voltage pulse-train generator ($V_x < 1\,000\ V_{peak}$) is required, capable of providing, under the various test conditions, an approximately step-function waveform that is almost constant in amplitude, as well as a single-trace oscilloscope, with a camera, having the facility of displaying the difference between two applied signals. This permits recording of voltages relative to earth, and those between points at different potentials.

The measurements must be made with the equipment energized and under normal operating conditions, the test-pulses being superposed on the supply voltage. Where this is impossible (e.g. pulses applied to the primary of an HV/LV transformer), it is sufficient to carry out the measurements with the equipment connected, but not energized.

The measuring circuit is illustrated in Figure A15.2.

A15.1.4 Results

Regarding the validity of the tests and the possibility of extrapolating the results to correspond with real lightning conditions, it should be noted that with the values of the current and voltage that can be achieved in practice under test conditions, it is not possible to take into account any nonlinear phenomena such as corona effect, saturation of the transformers or dispersion of the current in the ground; the values of overvoltage measured are, therefore, generally to be considered as being on the side of safety and

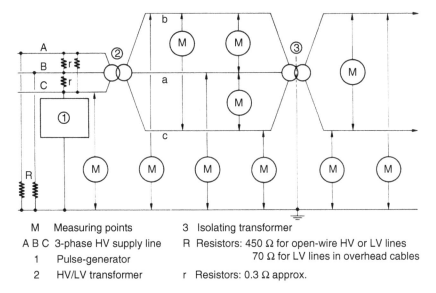

Figure A15.2 Measurement of overvoltages transferred from the power supply line

the results should be applied with prudence. A station can be considered to be effectively protected when in no circumstances are the measured overvoltages dangerous for the equipment.

These same tests, by means of which the equipment described above was tested during construction, can evidently be used also for assessing the quality of the lightning protection systems of stations already in service. They make it possible to identify the principal characteristics of the overvoltages of atmospheric origin that affect the station and the places where these assume dangerous values. It is then possible to determine where it is necessary to make modifications and to what extent the protective system of the station can be improved.

A15.2 Description of a typical test programme [2]

A15.2.1 Introduction

A typical programme of tests has been carried out at RAI's television and VHF/FM rebroadcasting station at Foligno, Italy, which has on numerous occasions in the past been damaged by lightning flash. The station is at an altitude of 900 m above sea level, in a very exposed position on the summit of a mountain. Measurements of the soil resistivity, using the 'four-stake' method at two different points sufficiently distant from the earthing system of the station, and for each of those points in two directions at right angles (Figure A15.3), indicated the existence of a low-resistivity topsoil and a subsoil of fairly high resistivity (Figure A15.4). The earth resistance of the station's earth termination system was 7 Ω, measured at 50 Hz.

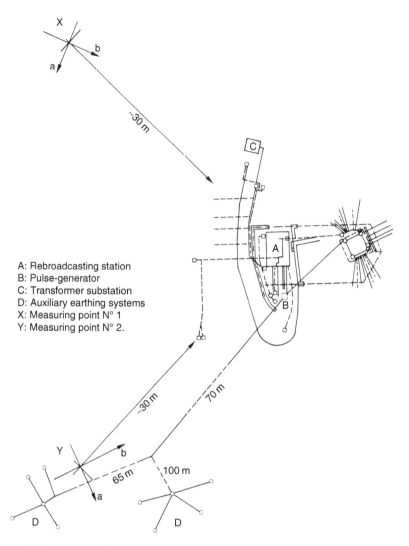

Figure A15.3   Plan of RAI's rebroadcasting station at Foligno, showing the arrangements for the soil resistivity method

A15.2.2   The power supply arrangements of the station

The general characteristics of the power supply installation are those shown in Figure A15.5 and listed in Table A15.1. For the voltage-stabilization system there are three possibilities:

1. Automatic voltage regulator consisting of an autotransformer continuously variable by a slider moved by an electronically controlled motor;
2. Static voltage-regulator with magnetic amplifiers;

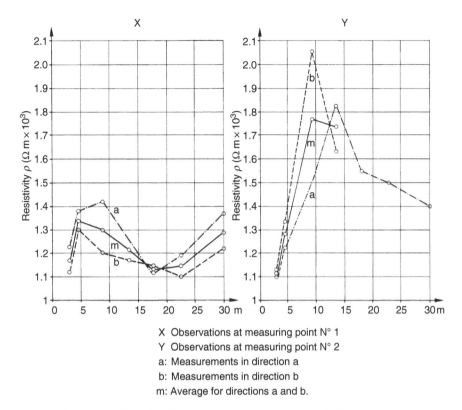

Figure A15.4   *Results of soil resistivity measurements at Foligno, as functions of the depth*

3. 'No-break' supply system, consisting of an inverter with regulated output voltage and a rectifier for charging an accumulator (1 500 A h – 24 V); by this means the station can function in spite of failure of the public supply.

A15.2.3   The test equipment

The tests were carried out by means of a transportable pulse generator (7.5 kJ, 6 stages, 600 kV maximum) and a pulse-train generator. The pulse-train generator was capable of supplying 1/500 to 5/500 µs pulses at a rate of $25\,\text{s}^{-1}$ and with a maximum peak amplitude of 700 V. The source impedance was a few tens of ohms.

A15.2.4   The test procedure and results

The measurements were made on a wideband oscilloscope (15 MHz) with a self-contained supply and with probes of high input impedance (10–100 MΩ) suitable for peak voltages of up to 40 kV and with a shunt having a response time of 8 ns.

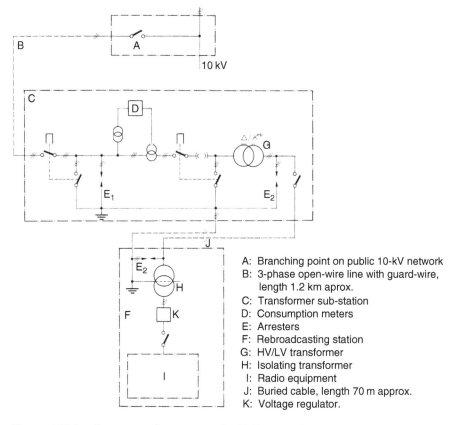

Figure A15.5 Power supply system at the Foligno station

Table A15.1 Characteristics of electricity supply equipment of the Foligno rebroadcasting station

		Characteristics
HV/LV transformer	Rating	50 kVA
	Nominal voltages	10 000–380 V
Isolating transformer	Rating	15 kVA
	Nominal voltages	220–220 V
	Impulse-withstand voltage (1/50 μs pulse).	28 kV
Arrester E_1	Nominal voltage	10 kV
	100% spark-over voltage	44 kV
	Rated discharge capability (10/20 μs pulse).	10 kA
Arrester E_1	Nominal voltage	500 kV
	100% spark-over voltage (10 kV/μs)	2 500 V
	Rated discharge capability	5 kA

A15.2.4.1 Measurement of the overvoltages due to lightning flashes to the antenna-support structure

The tests were made with a peak current of 2.1 kA applied to the earth termination system of the station (in operation). The readings were made

- at the HV/LV transformer substation, recording the overvoltages between the terminals of the HV and LV windings and earth;
- at the station, measuring the overvoltages between one of the terminals of the primary winding of the isolating transformer and earth, and between the same primary terminal and one terminal of the secondary winding (some results are shown in Figure A15.6).

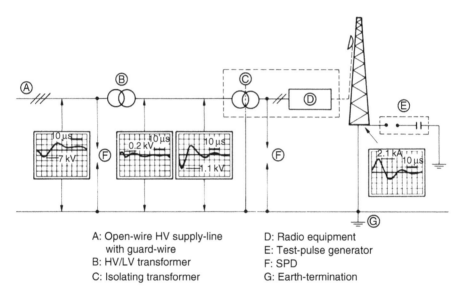

A: Open-wire HV supply-line with guard-wire
B: HV/LV transformer
C: Isolating transformer
D: Radio equipment
E: Test-pulse generator
F: SPD
G: Earth-termination

Figure A15.6 Observation of a simulated lightning strike on an antenna support

A15.2.4.2 Measurement of the transferred overvoltages

- The pulses, of positive polarity and waveform 0.5/600 μs, were applied between terminal C (Figure A15.7) of the HV winding of the HV/LV transformer isolated from the line, and earth; terminals A and B were individually connected to earth through 500 Ω. The pulse generator was connected to the terminals corresponding to the highest voltages transferred to the secondary of the isolating transformer of the station (for an equal peak amplitude of the pulses applied). Under those conditions, observations were made of the voltages transferred to
    - the terminals of the secondary winding of the HV/LV transformer when on no load and when terminated with an operational load;
    - the secondary terminals of the isolating transformer of the station, with loads of various impedances.

784　*Lightning Protection*

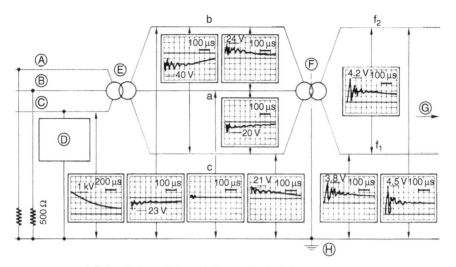

A B C : 3-phase HV supply-line　　F : Isolating transformer
D : Test-pulse generator　　　　　G : To radio equipment
E : HV/LV transformer　　　　　　 H : Earth-termination.

Figure A15.7　Observation of simulated overvoltages surges transferred from the supply line

- The tests were partly repeated under more severe conditions, with all the equipment of the station functioning under normal service conditions (HV/LV transformer fed by the HV line), by connecting the recurrent-pulse generator between the terminals A and C of the primary winding of the isolating transformer of the station. The waveform of the pulses, of positive polarity and superposed on the supply voltage corresponding to the negative maximum, was approximately 0.5/50. The tests were made with various values of the operational load.
- Finally, tests with common-mode transfer, representing the case when the lightning voltage wave flows along the three phase conductors, were carried out by applying the pulses between earth and the three HV terminals of the HV/LV transformer, isolated from the supply line and connected together. The transferred voltages at the terminals of the secondary winding were less than 1 per cent of the peak voltage of the pulses applied.

The results of the tests described above are given in Figure A15.7 and in Table A15.2.

A15.3　Discussion of the results

From the examination of the results obtained and taking account of the effect of the arresters at the entry of the HV line for phase-to-earth voltages exceeding a peak of 44 kV, the following conclusions may be drawn.

Lightning protection of telecommunication towers 785

Table A15.2 Measurements, made at the Foligno station, of transferred overvoltages expressed as percentages of the overvoltage on the supply line

Test conditions			Measuring points								
HV line circuit breaker	HV/LV transformer secondary	Isolating transformer secondary	HV/LV transformer						Isolating transformer		
			a to earth	b to earth	c to earth	a to b	a to c	b to c	f_1 to earth	f_2 to earth	f_1 to f_2
Open	No load			5	4						
	Connected to service load	(1)		2.3	2	2.3	2	4	2	2.5	4
		(2)							0.5	0.5	1
Closed	Connected to service load plus load	(1)							2	2	4
		(2)									1.3
		(3)									0.7
		(4)									1
Open			1	1	1				Negligible		

a, b and c: LV terminals of the HV/LV transformer.
f_1 and f_2: secondary terminals of the isolating transformer.

(1) lighting load disconnected.
(2) lighting load connected.
(3) air-conditioning load connected: 4 kW unity power-factor.
(4) capacitance of 8 μF connected between each phase and earth.

A15.3.1 Lightning flashes to the antenna-support structure

The primary winding of the HV/LV transformer is subjected, either to earth or to the secondary winding, to the voltage drop across the conventional earth impedance of the earth-termination system, namely the total earth voltage.

An effective voltage to earth of 3.35 kV kA^{-1} will cause the flash-over of the lightning arresters located at the HV terminals of the transformer, for peak lightning currents in excess of 13–15 kA and, therefore, for ~50 per cent of the lightning strikes on the mast.

The three phases of the star-connected secondary winding of the HV/LV transformer are, because their centre-point is connected to the earthing system, affected by common-mode voltages induced on the cable connecting to the transmitter, so that the lightning current flows to earth along the earth connections laid alongside it. Those voltages, which are ~100 V kA^{-1} at the HV/LV transformer, become ~500 V kA^{-1} at the primary of the isolating transformer. It is, therefore, desirable to provide LV lightning arresters at the terminals of the primary winding of the isolating transformer.

The common-mode voltages transferred to the secondary of the isolating transformer and, therefore, to the input of the radio equipment, are negligible, because of the electrostatic screen.

A15.3.2 Overvoltages transferred by the power supply line

Two types of transfer should be distinguished in this case:

1. *Common-mode transfer.* In this way, only modest phase-to-earth overvoltages (1–1.5 kV peak) can be transferred across the HV/LV transformer. These, therefore, do not constitute any danger for the insulation. In view of their common-mode character, there is no overvoltage between the LV phases. In addition, any possible remaining components that are not attenuated by the capacitance of the cable between the HV/LV transformer and the isolating transformer would be blocked by the electrostatic screen of the latter.
2. *Differential-mode transfer.* Under normal service conditions, the maximum amplitudes of the overvoltages on the secondary winding of the HV/LV transformer, mainly transferred by electromagnetic effects, are ~1 kV between phase and earth and ~2 kV between the phases. Such values are doubled when the transformer is not loaded.

These overvoltages suffer no appreciable attenuation during their transfer across the isolating transformer, so under these conditions overvoltages of ~2 kV would reach the equipment directly connected to its secondary terminals. The overvoltage can be reduced by augmenting the load. It would decrease from 2 kV with only the radio equipment in operation, to 600 V by connecting the lighting circuits (load: 135 Ω) and to 300 V by connecting also the air-conditioning equipment (load: 10 Ω).

Alternatively, the overvoltage can be reduced by connecting capacitors between the terminals of the secondary winding and earth (450 V with 8 μF). Regarding the voltage-stabilization system, the following may be observed.

- The voltage-regulator (1) mentioned in Section A15.2.2 does not attenuate the overvoltages; these, therefore, arrive at full amplitude at the primary windings of the supply transformers of the individual items of equipment.
- The static voltage-stabilizer (2) completely isolates the equipment supplied by it from overvoltages.
- The no-break system (3) is an effective obstacle to the transfer of overvoltages. In no case were overvoltages exceeding 1–2 per cent of the working voltage recorded downstream of it. Nevertheless, in view of the magnitude of the voltages that can exist at the secondary of the isolating transformer, the no-break unit itself could be damaged by disruptive discharges.

In view of the importance for the station of continuity of service, it was necessary to adopt solution (3). In addition, reliability was augmented by inserting the voltage stabilizer described under (2) between the isolating transformer and the no-break unit.

A15.4 Conclusions

A campaign of measurements on some of the RAI's rebroadcasting stations of typical design has made it possible to determine the principal characteristics of the overvoltages of atmospheric origin that affect stations, as well as the places where they assume dangerous values. It thus became practicable to determine which, in principle, are the optimum precautions to be taken at stations of that type, to reduce the overvoltages to tolerable values.

Individual tests are, however, necessary for those stations that are particularly endangered by lightning flashes, or that are exceptional on account of the importance of the service or of the complexity of the circuitry.

Acknowledgements

The author is grateful to Dr Eng. Fabio Fiamingo for his precious contribution in preparing the manuscript. Without his help this chapter could not have been completed.

References

1. Lo Piparo G.B., Belcher J., Graf W. Kikinger H., 'The protection of broadcasting installations against damage by lightning'. UER Technical Monograph No. 3117 (1986).
2. Garbagnati E, Lo Piparo G.B., 'Protezione degli impianti di telecomunicazione contro le scariche atmosferiche' (Protection of telecommunication installations against atmospheric discharges), *Elettronica e Telecomunicazioni*, No. 4 (1970).

Chapter 16
Lightning protection of satellite launch pads
Udaya Kumar

16.1 Introduction

Satellites are indispensable in the modern world. There are several types operating at different orbits that are useful for communication, earth observation, weather prediction, astronomy, scientific exploration, navigation, military reconnaissance, and so on. Among them, communication and weather satellites are most directly seen to affect day-to-day life. In fact, communication satellites, along with the associated ground support system, support a multi-billion-dollar business.

Estimates suggest, based on the number of launches made during the last half century, that the total number of satellites could be over 4 000, although it is likely that only a few hundred of them are still operational and active. The lifetime of satellites typically spans from just one year to ~ 10 years. Most of the satellites listed above are required to provide their respective services continuously, and therefore need to be replaced after their scheduled life. Predictions indicate that in order to improve global positioning systems and communication, more satellites will be expected to be launched. Space exploration is another area gaining prominence due to the associated technological, scientific and, often, political reasons. This will add to the demand for increased launch schedules and facilities. Therefore, the demand for launching is continuously growing and it also comprises a lucrative business proposition.

Satellites are placed in their specified orbits by specially designed launch vehicles. Usually, these launch vehicles are space rockets with suitable control and guidance systems. The only exception would be orbital spacecraft such as the NASA space shuttle [1]. Perhaps, until the full realization of space elevators, which are still at a very primitive stage, launch vehicles will remain the only mode of transporting men and equipment into space. Launch vehicles and space shuttles are susceptible to lightning threat not only during the launching stage but also during their transit in the lower atmosphere. The consequences of a lightning strike could be catastrophic. Extreme care must therefore be exercised to prevent such an exigency.

In order to have a clear picture, it is worthwhile to begin with a brief description of space rockets and their launch pads.

16.2 Structure of a rocket

The external shape of a rocket comprises a slender cylinder in order to minimize air drag at high velocities. To propel the given load to a specified velocity or height, it becomes necessary to construct rockets in multiple stages. Figure 16.1 gives the schematic of such a system and a brief description of the individual stages.

The stages are as follows.

1. *Payload fairing.* The conical structure at the nose of the launch vehicle shields the payload from aerodynamic loads, aerodynamic heating, high-speed air-stream and acoustic noises. It consists of two half-shells made typically of carbon–epoxy composites and aluminium honeycomb composite structures with external thermal insulation. The fairing jettisoning is carried out at altitudes of above 100 km by pyrotechnic devices.
2. *Payload.* This is the satellite or other scientific instrument that is to be put into space.
3. *Vehicle equipment bay.* This is a critical element of the launch vehicle housing onboard computers for flight guidance, the inertial measurement system feeding guidance and altitude data to the computers, the altitude control system and the telemetry transmitter.
4. *Upper stage (third stage).* This comprises a re-ignitable engine with stored propellants. It is responsible for placing the payload into its final orbit.
5. *Second stage.* This is the rocket ignited by the vehicle control during the flight after separation of the first stage. This stage may or may not be present.
6. *First or main stage.* This includes a main rocket along with boosters. In many designs, boosters contain solid propellants and the main rocket liquid or cryogenic propellants. These are ignited on the ground by the ground control.

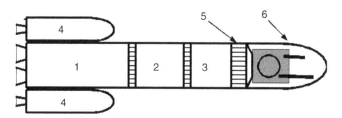

Figure 16.1 Schematic of the launch vehicle: 1, first/main stage; 2, second stage; 3, third stage; 4, boosters; 5, vehicle equipment bay; 6, payload/satellite

Typically launch vehicles for satellites weigh ∼300 to 1 000 tons, depending on the weight of the payloads. During takeoff, hot exhaust from the engine produce a thrust higher than the total weight. This exhaust will be at high temperatures in excess of 1 000 K. These extreme ambient conditions necessitate a highly robust structure to support the launch vehicle, both prior to and during the actual takeoff. This structure is called the launch pad.

The rocket and payload are transported to the launch complex and then assembled together to realize the launch vehicle. Of course, several mandatory health checks of various subsystems and systems are carried out prior to and after its assembly. There seem to be two distinct approaches adopted. In one, the vehicle is assembled on the launch pad with the help of a massive structure called the mobile service tower [2–4], which contains all the mechanical hoist arrangement, including the ultraclean room, for the payload assembly. After completing the task, this structure will retract from the launch pad. Alternatively, the assembly of the launch vehicle is carried out on stationary structures called the vehicle assembly building or the vehicle/horizontal integration building [5–8], which is typically 1–2.5 km away from the launch site. The latter design ensures safety to the assembly building in the event of an accidental vehicle explosion on the launch pad. The following two incidents provide some idea of the impact in such explosions. On 4 October 1990, at the Russian Baikonur launch site, the rocket Zenit's first stage failed five seconds after lift-off. The shockwave from the explosion lifted a 1 000-ton metal body of the launch structure as high as 20 m from the ground before crashing down. The debris from the rocket weighing 2–3 tons was scattered around the complex, 2–3 km from the epicentre [9]. Similarly, on 22 August 2003, at the Alcantara launch site, Brazil's launch vehicle exploded, causing the massive service tower to collapse [10]. It must be noted here that the rocket contains both the oxidants and the fuel. Also, for one particular reason (self-destruction on demand), electro-explosives are also present on the vehicle. Thus, any such accident at the launch site would be simply disastrous, including loss of life, time and money.

The assembly of the launch vehicle could be done vertically or horizontally and in the latter case the vehicle is lifted to the vertical position at the launch pad through hoists. When the vehicle is transported to the launch site on the launch pedestal or launch table, it will be exposed to the atmosphere for more than an hour. Generally, the vehicle is deemed to be safe from rain and lightning when it is within the mobile service tower or the vehicle assembly building. In one design [11], the service towers are integrated with the launch pad and these towers retract (on their bottom hinge) prior to the launch.

As far as the lightning threat to the launch pad and launch vehicle is considered, no distinction needs to be made with respect to space rockets and space shuttles.

16.3 Launch pad

Generally, this comprises an umbilical tower (a massive steel structure), a launch pedestal/table, and a concrete structure with flame duct/trenches supporting the launch pedestal.

The umbilical tower provides umbilicals to the launch vehicle for (i) the electrical connections necessary for providing ground command and power, (ii) air-conditioning

792 *Lightning Protection*

to the payload and (iii) depending on the scheme, the liquid propellant services for the launch vehicle and spacecraft. Owing to the limited literature available on this matter, only launch sites built on land will be considered in this chapter. It appears that there is also a launch complex built entirely on sea, owned by Sea Launch [12].

16.3.1 Launch campaign and duration of exposure

Typically, the launch vehicle will stay on the pad for a duration from a few hours (>7 h) to a few days. Functional checks of the onboard subsystem and systems will be performed to ensure a safe and successful operation. During the last 7–8 h, several important tasks are executed, including evacuation of personnel. Some of the typical tasks performed during the final countdown are as follows:

1. Filling of cryogenic fuel and topping until synchronized sequencing;
2. Launch vehicle Control System and Telemetry System power-on and functional checkout;
3. Cooling of the cryogenic engines;
4. Loading of the final flight program onto the onboard computer;
5. Disconnection of the fuel supply;
6. Switching the electrical supply from ground to the launch vehicle;
7. Disconnection of Control, Telemetry and Tracking system umbilicals.

   The following steps are irreversible:

8. Checking of onboard computer and activation of guidance, navigation and control functions (switching over to on board control);
9. Start of ignition sequence automatic system.

16.4 Lightning threat to launch vehicle

It is reported that lightning strikes to launch pads (that are not protected by any specific lightning protection scheme) have triggered sounding rockets and other similar rocket systems. The following two incidents are testimony to the possible damages to the launch vehicle caused by a lightning strike. The incidents were actually related to vehicle-triggered lightning, which occurred soon after successful lift-off and during the initial flight minutes [13,14].

Apollo 12 was launched on 14 November 1969 from Kennedy Space Center in Florida. A vehicle-triggered lightning was observed at 36.5 and 52 s into the mission. Nine non-essential sensors with solid-state circuits were permanently damaged. Temporary upsets included loss of communication, flashing and the sounding of various warning lights and alarms, disconnection of three fuel cells from the power bus, loss of altitude reference by the inertial platform, and disturbances to the timing system, clocks and other instruments. All critical system problems were subsequently corrected (by the crew) and the mission was completed successfully.

The Atlas-Centaur 67 rocket, carrying the FltSatCom (Fleet Satellite Communications) satellite, was launched on 26 March 1987 from Cape Canaveral.

Weather conditions were similar to those existing at the time of the Apollo 12 launch. No cloud-to-ground lightning had been observed within 9 km of the launch site in the 42 min prior to launch, and only one strike occurred within 18.5 km during this time. At launch, the intensity of the electric field at the field mill site closest to the launch site was -7.8 kV m^{-1}. After the launch, a lightning strike caused a memory upset in the vehicle guidance system. This upset then caused the vehicle to commence an unplanned yaw rotation. The stresses associated with this motion caused the vehicle to begin breaking apart. About 70 s after lift-off the range safety officer ordered the vehicle's self-destruction. Substantial portions of the fibreglass–honeycomb structure that covered the front 6–7 m of the vehicle were subsequently recovered from the Atlantic Ocean. These showed physical evidence of lightning attachment. These two incidents clearly showed that a lightning strike, whether natural or triggered, could be catastrophic to the launch vehicle. The increased use of digital systems in modern launch vehicles has further enhanced the vulnerability of the launch vehicle to lightning.

Launch pads are typically tall structures with heights in the range 60–100 m, built on a plain terrain. This makes them the most preferred targets for a cloud-to-ground flash in that region. Furthermore, most launch sites are situated near the seacoast (e.g. the John F. Kennedy Space Center, USA, the Ariane launch complex of the European Space Transportation System, French Guyana, Satish Dhawan Space Centre, Sriharikota, India), where lightning activity is very severe. Therefore, it can be concluded that launch pads are prone to lightning strikes. There have been several incidents during which lightning has struck the pad with the launch vehicle on the pad [13]. This has led to interruption of the testing, and postponement of the launch operation. There could have been similar events in other launch pads, which may not have been fully documented and analysed.

16.4.1 Limitations of present-day knowledge in quantifying the risk

In the event of a direct lightning hit to the launch pad, the resulting electromagnetic forces do not pose any serious threat. Similarly, the resulting thermal effects also cannot directly lead to any explosion. However, a direct hit to the launch vehicle can cause local damage to the thermal insulation on the spacecraft/fairing, and this certainly needs serious attention. If the stroke involves current of large magnitude and possible significant continuing currents, then structural damages to the launch vehicle cannot be ruled out and this would need to be evaluated.

With regard to the lightning threat arising from electromagnetic origins, it should be noted here that the resulting currents in the umbilical cables and in the vehicle equipment assembly bay can cause serious interference or damage, in spite of being four to five orders of magnitude lower than that of the incident lightning current. This is because they contain sensitive circuits that clearly have very low damage thresholds. Consequently, modelling of the system and the subsequent calculations (computed currents) must be of very high accuracy so as to precisely determine the current induction/distribution. This makes the problem extremely difficult to handle.

794 *Lightning Protection*

At present, possible threats due to a direct strike or a strike nearby are estimated to be (i) damage to the digitally controlled flight systems and other instrumentation, (ii) disruption of power and digital lines and (iii) possibly uncontrolled ignition of the fuel. Both natural and triggered lightning are safety threats.

Strictly speaking, even from the electromagnetic point of view, lightning is a non-linear electromagnetic initial-value problem with significant high-frequency components in the lightning current. The non-linearity arises due to the arc dynamics in the channel and soil ionization at the ground end. The true excitation mainly arises from the charge stored in the lightning channel and the grounded objects that are being struck.

Addressing this complex problem invariably necessitates making several simplifying assumptions. For example, in existing electromagnetic models for the channel, a lumped transient source is used to form a current source (or, a source similar to it) with reduced velocity of current propagation being artificially emulated. The actual complex dynamics of the channel are generally not amenable to any feasible solution, and in most cases, they do not need to be modelled. Such a modelling approach for the channel has provided satisfactory results (current and resulting field) for a strike to simple elevated slender objects [15,16], at least for the critical time regime. Even after such simplifications, extension of such models to the launch pad is rather difficult (due to its physical dimensions). Furthermore, the induced current to be computed is along paths of cross-sections that are orders of magnitude lower than the elements of the launch pad and they terminate inside the launch vehicle and the umbilical tower. So, with current modelling approaches, unless several simplifications are imposed, it appears quite difficult (if not impossible) to be able to quantify the induced currents circulating through the umbilicals. Whenever simplifications are adopted, it is also necessary to carefully review them for their ramifications regarding reliable simulation of the physical environment.

16.5 Lightning protection systems

16.5.1 External protection

For launch sites located at an altitude close to the mean sea level (which is the case for most of launch sites), significant upward lightning activities may not be expected for the prevailing range of launch pad heights. In view of this fact, design of the launch pad protection systems seems to be based only on downward flashes.

Similar to lightning protection schemes for other structures, the basic responsibility of the launch pad protection scheme can be categorized into the following stages:

1. Intercepting and diverting all dangerous flashes away from the launch pad;
2. Controlling the consequential so-called potential differences/rises to within safe limits;
3. Limiting the resulting electromagnetic field in the launch pad area to a minimum.

Obviously, the primary aim of the protection system would be to safely divert the prospective strokes away from the launch pad/complex. In order to build an economic protection system, and one that would permit smooth erection of the launch vehicle on the pad as well as the launch operation itself, it is mandatory to accept a small percentage of shielding failure flashes that may sneak through. In other words, the descending leader with prospective return stroke current lower than a specified value can sneak into the protected volume and could terminate on the launch pad. According to the present understanding of the attractive radii for strokes of different polarity, it can be said that the magnitude of the current in bypass strokes will be higher for positive flashes than that for negative flashes. Any quantification of the stroke parameters into what can be considered safe and what is unsafe for a given launch pad is a highly debatable issue. The reasons for this have already been mentioned. Fortunately, it can be stated with certainty that the maximum di/dt of the stroke current is the key factor, and governs the level of induction/interference with the system connected through the umbilicals. Therefore, one needs to concentrate on the front portion of the stroke current rather than the remaining portions.

Further, the interception efficacy of the lightning protection system (LPS) generally increases with the magnitude of the prospective return-stroke current for strokes of both the polarity. Therefore, the current in the inevitable bypass strokes would be towards the lower magnitude regime. As the probability of occurrence of strokes with lower magnitude (like that with higher magnitudes) is low, the situation naturally tends to be become more safe. At this point it would be worth recalling that the attractive radii of the ground-based structure increases with their height. Therefore, for any protective action, the air termination network of the protection system must stand taller than the launch pad. As might be expected, the air termination system of the launch pad LPS is realized by suitably placed set of masts and shield wires.

In principle, there are two distinct philosophies used in the design of LPSs for satellite launch pads. In many cases, the importance of a suitable LPS was realized only after the launch system was built and had been in service. As a consequence, different types of configurations have been adopted and schematics of the same are provided in Figure 16.2.

The lightning current through the down-conductor produces significant electromagnetic fields. Similarly, the current entering the soil at the earth termination will raise the ground potential typically to a few tens of kilovolts or sometimes even higher. Efforts to minimize such secondary effects lead to the philosophy of the first approach, which basically attempts to divert the current in the intercepted stroke far away from the protected volume. In order to realize this, the protection system is isolated from the supporting structures using insulating masts. The ground wires/catenoids that form the down-conductors run from the lightning rod to the remote ground. In the second approach, which is in line with the classical protection schemes, no such additional efforts are made. The supporting towers themselves form down-conductors.

In any LPS the ground termination will govern the soil potential rise and has significant influence on the so-called potential rise at other elements of the protection scheme. Driven rods along with radial counterpoises forming a mesh are generally used for the earth termination of the protection scheme.

In the following sections, the different protection systems that are currently being used will be briefly discussed.

16.5.1.1 Brief description of some of the present protection schemes

In Scheme 1, which employs the first approach mentioned above, a long insulating mast at the top of the umbilical/service tower supports the lightning rod (a metallic pipe/rod). The launch vehicle lies within the cone of protection (with an angle of 45°) provided by the lightning rod. This scheme is adopted in John F. Kennedy Space Center [13,17]. A 24.4-m (80-ft) fibre-glass mast sits at the top of the fixed service structure and carries a 1.22-m lightning rod at its top. Two 2.5-cm (1-in) diameter stainless steel cables from the lightning rod stretch ~304.8 m (1 000 ft) in either direction to the ground points.

In Scheme 2, which is inline with classical approaches, one or more tall towers without any aerial interconnection are used around the launch pad. The limited available data indicate that such an approach has been employed in the launch pads of Russia, Japan and China [18–21]. For example, in the Russian Energia launch pad complex, all three launch pads are surrounded by two 225-m lightning protection towers and two shorter lighting towers. All four towers are ~150 m away from the pad.

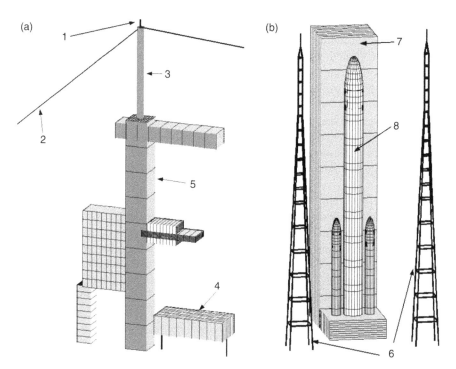

Figure 16.2   Schematic of the current protection schemes used at different places: (a) Scheme 1, (b) variant of Scheme 2, (c) Scheme 3 and (d) Scheme 4. (1, lightning rod; 2, catenary/ground wire; 3, insulating mast; 4, launch pedestal; 5, fixed service structure; 6, protection towers; 7, umbilical tower; 8, launch vehicle; 9, shield wires.)

Lightning protection of satellite launch pads 797

In the other schemes, much more effort is made to provide protection against a direct hit due to a descending leader approaching from any direction. This aspect becomes important for strokes with relatively lower current magnitudes. For this, the protection system is made to encircle or surround the launch pad complex.

Scheme 3 presents the first approach described in the previous section. The air termination network is generally formed by lightning rods interconnected by shield wires. The air termination network is insulated from the supporting tower by a long cylindrical insulating support. The catenoids or ground wires originate from the

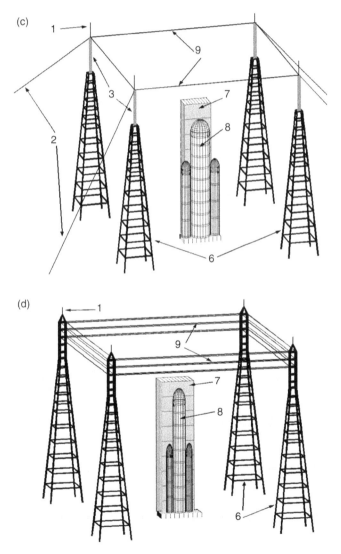

Figure 16.2   Continued

lightning rod and terminate to the ground end at remote earth points. Typically, four towers are used surrounding the launch pad. Further discussion on this will be made later. This scheme has been used in the Ariane 5 launch pad, where four towers surround the launch pad [22,23]. The net height of the lightning rods is ~90 m and the effective length of the insulating support is ~10 m. Similarly, the launch complexes SLC 40/41 of the Cape Canaveral Air Force Station in Florida also have four towers with an interconnected air termination system of ~106.7 m total height. The tower supports insulators of length close to 22.86 m [24]. The shield wires in this system form a double square over the launch pad, a structurally complicated arrangement to construct.

The protection system at Delta IV Launch Facilities, Cape Canaveral Air Force Station [25] comprises only two 116-m lightning protection towers standing on either side of the launch pad. The catenaries or ground wires run from the top of the insulating support, i.e. from the bottom of the lightning rods to remote ground.

Scheme 4, on the other hand, has lightning rods directly supported on the towers, which are suitably placed around the launch pad complex. Here, the towers play the role of down-conductors. The initial setup of the protection scheme for both of the Indian satellite launch pads used this scheme. The first launch pad involves three tower configurations [26] and the recent second launch pad utilizes four tower configurations [8,27]. In both cases, the towers are 120 m tall, supporting a 10-m lighting rod at the top. Six shield wires interconnect the towers at heights of 120, 115 and 110 m. Experience of more than five years with the protection system on the first pad has been quite satisfactory. Incidentally, some modifications of these systems are under consideration. The modified scheme would imbibe the combined design philosophies of both these approaches.

With the limited information available on a time-correlated record of events, it appears that all the schemes are working satisfactorily. This may be partly due to the launch commit criteria, which ensure that the operation is as safe as possible.

16.5.2 Principles used for the design of the external protection system

16.5.2.1 Air termination network

For the design of the abovementioned air termination schemes, the electrogeometric models invariably seem to have been used [22,28]. Using the same along with local lightning data, estimation of the shielding failure flashes has been made [29]. Some field data on power transmission lines, as well as indirect evaluation using the line trip-out rates, have provided strong support to the use of electrogeometric models and even the protection angle approach for the design of transmission lines. However, these deductions cannot be extended to objects of different geometries and specifically larger heights.

Attempts have also been made to verify the efficacy of the air termination scheme by using high-voltage impulses on a rod suspended above reduced-scale models [22,28]. The applicability of such tests has been questioned in several technical publications (e.g. see Reference 30). It must first be noted here that only the geometry of the protection system is scaled, while the pressure at the test area is not scaled.

Consequently, the physical phenomena leading to the final attachment of natural lightning to grounded objects are not fully and satisfactorily simulated in the laboratory tests.

16.5.2.2 Earth termination

Very little published literature provides details of the earth termination of launch pad protection systems. The accurate modelling of the earth termination would be of interest for deducing the soil step potential, potential rise at various nearby grounds (power and switching), touch potential and the temporal decay rate of induced currents in the protection system.

There are several issues in modelling the earth termination. The classical power system approach would be to employ a distributed ladder network comprising of series L and shunt R and C elements. This approach could consider the soil stratification [31] as well as, to a large extent, soil ionization [32]. However, the eddy current and wave propagation effects in the soil are not handled adequately. Owing to the symmetrical diffusion of the current into the soil, as well as high localization of the effective resistance of buried conductors, the eddy current effects in the soil may not assume any dominance.

16.5.2.3 Down-conductor system

The limited literature indicates that the lightning surge response of the system has been analysed using two different methods. The finite-difference time-domain (FDTD) method has been used [33] for analysing the response of Scheme 3, making use of simplified models for the channel, shield wires and catenoids. A distributed circuit model deduced from power transmission engineering has been attempted for the initial analysis of Scheme 4 [31], with multistorey tower models for the protection towers and transmission line models for shield wires. A detailed review of the above will be taken up later.

It must be noted here that the mechanical design of the protection system, apart from considering the wind and earthquake loads, must also take into account the pressure, thermal and acoustic load arising out of vehicle exhaust during the launch.

16.5.3 Internal protection

The basic philosophy of the protection is provided here. There could be certain exceptions and careful evaluation of the individual situation would then become necessary.

16.5.3.1 Launch vehicle

In the event of a bypass stroke terminating on the launch vehicle or the triggered lightning during the transit of the spacecraft, the nose or fairing of the vehicle forms the preferential attachment point for the stroke. This is basically due to the geometry of the spacecraft/rocket structure. To protect the vehicle and components, several lines of defence are used, starting with the vehicle's structure, bonding requirements and cable shielding [34–36]. The existing line of defence has not been verified to ensure a full level of safety against a natural or triggered lightning hit, and therefore they must be considered only as basic safety measures. At present, a launch exercise is not permitted if there is a possibility of a direct or a triggered lightning.

When the vehicle has all-metallic external surfaces, it will form an excellent conducting path, provided that adequate bonding is made between the various stages. If there are no external connections, this would assure good protection to the internal equipment. However, when the vehicle is on the launch pad, umbilicals form external conducting connections and therefore circulating/induced current can sneak in. With a composite skin for the external surface, a separate conducting path is needed and it should extend to the entire length of the vehicle. Also, to avoid tribo-electrifications, a conductive paint is necessary over the skin.

The bonding of all-electrical components (connections, metallic plumbing, and so on) is mandatory to achieve an equipotential environment [34]. Specifically, in the regions on the vehicle where critical electrical components are situated and in places where large conducting loops could be formed, the above aspect becomes very important. To ensure these, (i) all tank sections should be welded and bonded to achieve a low-impedance reference plane and (ii) all metallic parts of dimension larger than ~ 30 cm should have a discharge path to the structure.

In line with the general lightning protection principles, it is necessary to ensure the following: (i) the bonding region should have adequate cross-section and skin (surface) area; and (ii) the connection points should be free from any insulation material and foreign materials such as paint, oxide and corrosion. Further, as sharp bends cause large mechanical forces and contribute to high transient impedances, they should be minimized. They can also form potential arcing points.

Electrical wires and cables

For metallic conductive surfaces enclosing cables, direct coupling to internal cables would be negligible. However, when there are apertures, then corresponding coupling must be considered. For composite skin, such as in carbon-fibre structures, as the skin depth is greater than the structure thickness for higher frequencies present in lightning current spectrum [22], coupling to internal cables needs careful attention. The coupling mode on cables connected to different units inside the vehicle is mainly common mode in nature.

In order to reduce the magnitude of induction due to lightning, the internal electrical wiring/cabling should follow the general guidelines for electromagnetic shielding. This would limit the level of noise or interference throughout the system. The use of fibre-optic cables wherever possible would eliminate the susceptibility of the electrical connection to the electromagnetic effects of lightning. Such an approach could be possible for control and data transmission.

Broadly speaking the electrical wiring in the vehicle can be categorized into the following.

1. *Power and low frequency.* This is used for supplying electrical energy to various electrical devices. The basic d.c. supply is through onboard batteries.
2. *Radio-frequency circuits.* The launch vehicle invariably contains onboard radio equipment for the purpose of telemetry, tracking and tele-command. The

information sent through telemetry is very useful in monitoring and post-launch analysis of system performances and fault analysis.
3. *Digital data lines*. These serve for data acquisition system and control.
4. *Ordnance circuit*. Pyroelectric or electroexplosive techniques are employed for jettisoning the nose fairing and sometimes for separating used and/or burnt stages. They are also used for the destruction of the launch vehicle under violation of range safety or when certain anomalies are detected in its operation.

The general principles adopted for grounding and shielding of cables/wires inside the vehicle are as follows:

1. For power and low-frequency wiring, it is suggested that the wires must be twisted with its return and must be referenced to ground at both ends. When inductive loads such relay coils and motors are not provided with a free-wheeling diode or Transorb (diode suppression), they can be a significant source of electromagnetic noise during the period of current chopping. In such cases, shielded lines are suggested with the shield grounded at both ends.
2. The radio-frequency circuits or circuits susceptible to radio frequency should have the outer braid of the coaxial cable grounded at both ends [35] and at all points along the length of the shield as is practicable.
3. For digital data lines, the use of a shielded twisted pair with the shield being grounded at both ends is suggested.
4. For ordnance circuits, twisted separate shielded [37] wire should be used with the shields grounded at both ends.
5. In general, circuits that are sensitive to high frequencies should have shields grounded to the structure at both the ends. In addition to electromagnetic (EM) hardening of the electrical system, structural designs should avoid susceptibility to triboelectric/frictional charging [38].

16.5.3.2 Vehicle on launch pad

A vehicle is placed on the launch pad for carrying out the final ground operations, as have been described in the previous section. For the reasons already mentioned in Section 16.3.1, the tasks performed at this stage are very important and quite dangerous as well. Both the bypass (or shielding failure) stroke, which possesses smaller peak currents, and the nearby strokes can cause current flow in ground support equipment. With regard to the electrical systems, lightning (including the induced effects) and transient hardening are very essential. The umbilical connection should have a solid connection between the umbilical tower and the vehicle.

In order to limit the damage inflicted by lightning and high current transients, the following steps have been suggested [35].

All the ground support equipment (umbilical tower, service tower, etc.) must be sufficiently grounded and bonded. The individual equipment should be grounded to the facility structure when the ground support equipment is installed. In order to minimize the inductive impedance and hence the resulting potential, heavy-gauge grounding cables should be instituted, to ground external items to the major structural members.

All wires and components connecting ground support equipment with payloads should be appropriately grounded. Lightning surge suppression devices and appropriate measures should be employed at critical circuit interfaces and in current loop areas where potential differences can be substantial during direct and induced lightning strikes. When protecting the high-frequency circuits, the capacitance of the protective device needs to be considered while selecting the device. Use of twisted pairs with shields is recommended for umbilical cables. For cables connecting to critical systems, two insulated outer shields are suggested with the outer shield grounded at both ends. Owing to the length of connection between the ground support equipment and the umbilical tower, it would be necessary to ground the outer sheath/braid of the cable to the supporting the metallic structure at every 10–20 m interval. For cases, where many of the existing electrical connections do not have two shields, wires are without braids, and so on, separate grounding strips should be run parallel with strips grounded at regular intervals of, say, 10–20 m. These strips must be connected to the ground of the junction boxes at their ends. If the cable support structure does not possess sufficient ground connection, driven rods can be employed at 20 m intervals.

16.6 Weather launch commit criteria

The final countdown, commencing with the filling of cryogenic fuel to the rockets, is governed by the prevailing weather conditions. Suitable weather launch commit criteria have been developed and are used in practice. For example, temperature is specified by an upper and lower bound; similarly wind has an upper bound and no precipitation is acceptable in the zone. The lightning launch/flight commit criteria (LLCC) forms an important subset of the weather launch commit criteria [39,40].

The goal of the LLCC is to ensure that no lightning strike, either natural or triggered by the vehicle, should occur during the entire of phase of launching, including the transit phase. Tanking should not be started if the forecast predicts a greater than 20 per cent chance of lightning within 9.25 km (5 nautical miles) of the launch pad during the first hour of the tanking. Furthermore, the umbilical connections may even be taken out if the electric field in the area exceeds a certain level. During the process, if by chance the ground electric field in the pad area exceeds $5-10$ kV m^{-1}, the prevailing safety rules in certain cases demand that personnel should be evacuated from the launch site.

Only those rules directly pertaining to the launch rather than governing the flight path will be considered here [39,40]. These are as follows.

1. Do not launch (and fly) within 18.52 km (10 nautical miles) of any type of lightning or any cloud that has produced it within the past 30 min. An exception is allowed if the cloud has moved beyond 18.52 km and if the electric field within 9.25 km of the flight path is lower than 1 kV m^{-1} for the last 15 min.
2. Do not launch if the electric field within 9.25 km of the flight path has exceeded $1-1.5$ kV m^{-1} in the past 15 min. Exceptions are permitted for the rules under certain restricted conditions.

Apart from these, there are about six to ten rules related to the type and condition of the clouds around the flight path. Many of these rules appear to be rather conservative and at present they are deemed to be safe.

Therefore, the main line of defence against lightning is LLCC, which does not permit the launch operation to occur during any lightning activity in the area surrounding the launch pad or along the flight path. Because of these rules, the real efficacy of the protection system has not been subjected to a critical test. However, the prevailing internal records indicate that many of the systems have successfully intercepted a direct hit and have withstood the consequential fields. However, details of the associated current and the operation that was being performed on the pad during the event are not available.

16.7 Review of present status and suggested direction for further work

According to Reference 41, the space sector is subject to the strong influence of market forces. The basic criteria are services, their costs and quality. In this regard, the quality of the launch operation and the dependable launch on demand [42] would clearly play a vital role. Present estimates indicate that weather-related launch delays and scrubs, of which lightning is a major contributor, amount in many places to up to 30–40 per cent of the total delays. Furthermore, it is the one factor that cannot be controlled. However, in order to minimize its effects, it is necessary to improve our understanding and knowledge on the quantification of the interception efficacy, quantification of the consequences due to a shielding failure flash terminating on the pad, and ascertain the induction to sensitive systems during stroke interception and a nearby stroke.

In the following, the requirements for a reliable analysis of the efficacy of the protection system will be dealt with. This section will draw inferences from the available literature on lightning, some of which are not necessarily related to launch pad protection. Owing to the abundance of the literature, only the more recent references have been cited although an exhaustive list of the remaining literature is available in the other chapters in this book.

16.7.1 Attachment process

Lightning is recognized to be essentially a breakdown process in air, with an impulse-like input excitation. It is well known that the breakdown process, even under controlled conditions in a laboratory, is in itself so complicated that it is not fully amenable to theoretical evaluation. This being fact, one can easily visualize the degree of complexity that would be associated when natural lightning has to be modelled.

From the protection point of view, it is necessary to model the last stages of bridging during which the descending stepped leader produces a strong electric field near the ground. Because of their geometry, the net field at the tip of tall objects such as launch pads/launch vehicle is magnified. Consequently, significant discharge activities involving upward-moving streamers and leaders can be expected from them.

The process terminates with the bridging of the descending leader and the upward discharge (which is predominantly a leader discharge). A detailed description of the attachment process can be found in Chapter 4.

As pointed out in Chapter 4, electrogeometric models, which have been quite successfully employed for simple structures like transmission lines and ground-based structures with short height, do not differentiate between the streamer and leader mode of bridging. Also, different stages of bridging, *viz.* starting from inception, then subsequent propagation and final bridging of streamers ahead of the two leaders are not adequately dealt with. Therefore, their application to critical systems such as launch pads must be carefully examined.

For simulation involving final stages of bridging, it would be ideal to use physical models that deal with various gaseous processes such as ionization, attachment, velocity, energy of the molecules, their statistical distribution and so on. After suitable simplifications, a modified physical model was proposed in which equivalent avalanche and discharge channels were considered [43]. A further extension of this model for lightning upward leaders from slender grounded structures can be found in Reference 44. More recently, Becerra and Cooray [45] developed a model based on the physical principles developed in Reference 43 to study lightning attachment to structures and power lines. The predictions of this model and their comparison with available experimental data are presented in Chapter 4. The results of the simulations to be presented here are based on the model introduced by Rizk [46], which was the best tool available at the time this analysis was performed.

In parallel to the above, there are some relatively simple engineering models that attempt to simulate the macroscopic aspects of the discharge phenomena. The agreement between experimental results and those given by this approach has been very encouraging. Use of such models would definitely be far more accurate than the electrogeometric model currently employed for designs. Therefore, they can be considered as one big step towards the realistic modelling of the final attachment process. Development and application of such a model to the lightning attachment process to isolated tall towers and conductors can be found in Reference 46.

Based on Rizk's work, a modified inception criteria was developed for towers interconnected by shield wires, as well as for a set of parallel shield wires. It was successfully employed for arriving at a more reliable evaluation of the interception efficacy of LPS to Indian launch pads [26,47]. A detailed simulation of the final stages of bridging has been considered from inception of the upward connecting leaders to their subsequent propagation towards the descending leader. The charge simulation method was found to be best suited for field computation, and was therefore used. In this work, the ratio of velocities of descending and upward leaders has been taken to be unity. It is shown that the vertical descent of the main leader forms the worst possible scenario and, therefore, deductions are to be made with respect to it. A sample simulation result for the LPS to pad-I is provided in Figure 16.3a. The trajectory of upward connecting leaders and the final jump region can be clearly seen in the result. As only the deterministic part of the phenomena is modelled, all the three upward leaders are successful in bridging to the main descending leader. The interception efficacy and the maximum possible current in the shielding failure strokes have

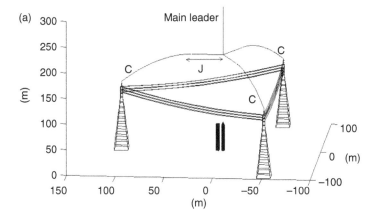

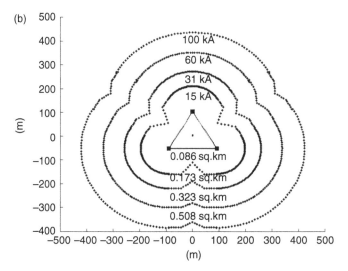

Figure 16.3 Simulation results for the protection scheme for Indian satellite launch pad-I [26]: (a) stroke interception (C → connecting leaders and J → final jump); (b) zone of protection for different stroke currents along with corresponding protected area

been deduced. For the negative strokes, the protective area covered by the protection systems is evaluated as a function of prospective return stroke current (see Figure 16.3b) and the probabilities of shielding failure and the number of strokes collected by the LPS are also evaluated. It is evident from Figure 16.3b that the collection area for strokes with higher return stroke is quite high. As compared to the towers, even in the interconnected system, shield wires possess relatively lower efficacy in launching upward leaders. This aspect has resulted in an inward kink in the locus of the protective area.

In summary, the important factors that need to be modelled for a reliable evaluation of the protective action are (i) the criterion for the determination of upward leader inception, (ii) tracing leader development for both descending and upward connecting leaders and (iii) determination of the final bridging. A macroscopic field-based model for the determination of upward leader inception from any arbitrary structure has been proposed [48]. This is an improvement over the inception criteria used in an earlier work [26]. It would be equally important to deduce the relative velocities of both leaders. Apart from the field data, knowledge on the dependency of the leader dynamics on the prevailing ambient field needs to be acquired. This velocity ratio will govern the attractive radius, the locus of the upward discharge and the attachment point. Also, when upward leaders develop from the protected system and the protection system, the velocity issue becomes very critical in deciding the interception. At present, due to the predominance of negative lightning (in most geographical locations), the engineering models concentrate mostly on negative lightning. This needs to be extended to positive lightning as well.

Both physical and engineering models attempt to dynamically trace the set of events. Therefore, a solution of the electrostatic field would be required for every step of the descending leader. This would be in addition to the simulation of the associated physical processes. Therefore, extensive computation would be necessary for the simulation of every single attachment event. The salient features of the problem suggest that the boundary-based electrostatic field computation methods are best suited to handling such a problem. If the objective is to also estimate the protective area as a function of current and the probabilities of shielding failure strokes per year, the number of strokes per year to protection system, and so on, then the associated computation at present would require much more effort and time.

The present modelling approaches are in effect aimed at tracing the definitive part of the actual attachment process. It is modulated by the randomness in the development of streamers ahead of the leaders, space charge pockets in air, humidity and velocity of air, the corona space charge around the air termination network and so on. The resulting stochastic nature of the problem is to be suitably considered and this would require significant progress in the understanding, subsequent modelling and collection of the field data on actual system. Now, even if the deterministic part of the phenomena were accurately simulated, it would be a great step ahead. The actual impact of the stochastic nature of the problem is not clearly known; however, the success of the classical methods indicates that it may not be very significant.

Recent studies have attempted to include most of the above features in self-consistent physically based models. Detailed descriptions of these models are given in Chapter 4.

16.7.2 Lightning surge response

16.7.2.1 Earth termination

When the earth termination is situated far away from the system to be protected, then the reflection produced at the junction between down-conductor and earth termination, which governs the voltage rise, is the only factor that is of concern. In a well-designed earth termination, the transient impedance would be lower by nearly an order of magnitude with respect to the 'surge impedance' of the down-conductors. In such circumstances, the actual surge response of the earth termination is of less importance. Perhaps, the decay rate of the successively reflected current waves, at later time periods, will be governed by the low-frequency impedance of the earth termination. In such remote earth terminations, a protective fencing would be necessary to avoid electrocution through dangerous step potentials prevailing during stroke interception. On the other hand, for earth terminations close to the launch pad (as in Schemes 2 and 4), the consequential soil potential rise becomes an important issue. In the following, the approaches that are currently employed for the analysis of the earth termination will be discussed.

The distributed circuit model is a detailed circuit-based model, and has been extensively used for the analysis of grounding systems [49]. Its validity in distributed earth termination would be limited by the retardation effects of the fields and higher modes of current propagation along the conductors. A general rule of thumb states that when the geometrical extension of the earth termination exceeds one-tenth of the wavelength along the conductor in the soil, quasi-static field approximation becomes invalid. Of course, the wavelength corresponding to the significant higher-frequency component of the current injected into the earth termination network must be considered. As an engineering approximation, transmission line-based modelling is suggested in Reference 50 in which the coupling across different elements is deduced from quasi-static fields. A further discussion on these factors will be made in conjunction with the down-conductor system.

16.7.2.2 Down-conductor system

The lightning surge response of down-conductors is important for the evaluation of the 'voltage rise' on the elements of the protection system, current waves that undergo successive reflections at various junctions, current flowing into the earth termination network and the resulting electromagnetic fields in the protected volume. Owing to the dominance of the transverse magnetic (TM) mode on the protection system during the initial portion of the stroke current, definition of the quasi-static voltage is not permissible. From the breakdown point of view, whatever experimental data are currently available and that which can be obtained in the high-voltage laboratory are only valid for the classical electrostatic regime. As there is no credible alternative, the line integral of the electric field (preferably along the shortest path between the protection system and the protected system) is to be employed as the definition of 'voltage rise' for estimating the possibility of flashover. Even for the quasi-static domain, the breakdown strength of the gaps required for the design of an LPS is currently not available.

For a rough estimation of the voltage rise, simplified modelling employing the techniques developed for the analysis of the lightning response of power transmission lines can be used. Accordingly, the protection tower, shield wires and catenary or ground wires are represented by suitable transmission line models. For example, in Reference 31, for the initial estimation, multistorey line models were used for towers and uniform line models for the interconnecting shield wires. The minimal efforts required for such an approach would be a major advantage, and they can be used for rough estimation.

Theoretical approach

In contrast, for a more reliable study, it will be more appropriate to use a field model where the governing electromagnetic fields are solved for the system. The FDTD method has been used for the analysis of the lightning surge response of the protection system for Ariane [??], but with some simplifications. In the following, a field-based approach for studying the lightning surge response of a protection system will be considered.

A full-wave solution of Maxwell's field equation over the whole protection system, including the lightning channel, would be ideal; however, it would be more complicated. Such an approach would correctly account for the coupling between the buried and aerial conductors. Apart from difficulties in making a pertinent representation of the lightning channel, there are several associated complexities that need to be first addressed. Only the more important ones will be discussed now.

The spatial extension of the protection system is quite large, spanning several wavelengths for significant higher-frequency components of the lightning current spectrum. The geometry of the protection system involves thin elements (shield and ground wires), as well as intricate interconnected structures. Junctions involving thin wires or thin wires with towers are to be accurately modelled. There will be considerable corona activity during lightning current propagation along the air termination network as well as the catenoid down-conductor systems.

The umbilical tower and the launch vehicle are of a considerable cross-section. The number of cables and pipes running within the umbilical tower as well as between the umbilical tower and the vehicle are extremely difficult to represent individually. More importantly, the magnitude of the resulting current in the umbilicals that could potentially damage many electronic systems is lower by almost four to five orders when compared to the magnitude of the lightning current.

The ground parameters vary with distance, depth and season. In fact, two- or multilayer soil stratification has been extensively used in power engineering. With a typical top dry sand layer, which is common in the coastal belts where most of the launch pads are situated, it would be necessary to consider the conducting bottom layer along with the top layer. Further, close to the pad, there are several deep foundations modulating the soil current flow pattern. This requires modelling of the concrete bed for the relevant frequency regime, as well as soil compaction. Furthermore, there are cable trenches and pipelines either buried in or at the surface of the soil.

Because of these complexities, the analysis is carried out with several simplifications of the original problem. The simplifications must be carried out with engineering judgement. In the following, salient points pertaining to different numerical approaches will be made.

Domain-based methods

Any domain-based method such as FDTD, finite element methods (FEM) and transmission line modelling (TLM) suffer from difficulties arising due to discretization, large matrix size (unless sparse matrix techniques are fully employed) and huge amounts of computational data arising due to accumulation of intermediate results while time stepping. They can also suffer from occasional numerical oscillations. However, they have distinct advantages in a time-domain approach, in that quantities are available directly in the time domain and there are, as such, no frequency-domain to time-domain inversion problems. Further, the non-linearity of the channel conductivity, and if required, the corona on shield and ground wires, can be represented. Irrespective of whether the frequency-domain or time-domain approach is used, modelling of ground with a good number of stratifications and even taking account of the cable trenches is in principle possible. These methods need artificial truncation of the problem geometry with appropriate boundary conditions, so as to have a finite spatial grid (or discretization). Therefore, the adequacy of channel representation for a feasible spatial discretization is to be carefully considered.

In the author's opinion, the following entities are sometimes overlooked (which is not acceptable). Shield wires, catenoids, umbilical cables, driven rods and counterpoises are of thin geometry and, as a result, the process of discretization considering them would be quite impractical. In view of this some efforts are made in which they are not explicitly represented, but their effects are considered to be of second order, which is strictly not permissible. Special approaches (in which their accurate representation is possible), have to be used for their modelling [51–55]. However, these methods are not intended for junctions of multiple slender conducting elements, such as the ones encountered in umbilical/service towers. With modelling of soil, the time step used for the simulation must be appropriate for the velocity of waves in soil, and not in air. The former will invariably be much lower than the latter. Following the trend in power engineering, the required simulations are most often carried out using a current source model for the lightning channel. With such practices in the time domain, in order to minimize the computational time, instead of an actual lightning current excitation, a Gaussian pulse is sometimes used. Then, by using Fourier techniques, the required transfer functions are evaluated. This would be acceptable provided that the significant frequency component of the pulse, a spatial resolution of 10 cells or element per wavelength, is employed and the time stepping is chosen corresponding to the velocity in soil.

Boundary-based methods
In principle, the boundary-based methods, like the method of moments (MoM) [56], have several advantages. As discretization is limited only to the channel and the structures, they need the least discretization. The resulting matrix will be small but fully dense. Apart from the numerical problems associated with inversion to the time domain and the tediousness in modelling the soil stratifications, at present, frequency-domain-based approaches also do not handle non-linearity in the channel and soil. Therefore, it will be difficult to emulate the physical phenomena associated with the transient change in channel conductivity. Any attempt to realize the slower propagation of the channel current will not model the change in the effective surge impedance of the channel with the current front. The actual excitation consists of charges present in the channel, system, and bridging streamers. At present, this has not been appropriately represented.

In the literature pertaining to the lightning surge response of tall towers and similar systems, only the frequency-domain-based MoM (considering perfectly conducting ground) has been used extensively. More specifically, it is a public domain software NEC (Numerical Electromagnetic Code) [57] that is mostly employed. This approach has provided acceptable results for the current in simple down-conductors and for the electromagnetic field produced by a direct strike [15,16]. The same code has also been used for the evaluation of induction to the launch complex by a remote stroke (which will be dealt with later). There are also limited efforts taking the ground modelled as a uniform semi-infinite media, and a direct time domain simulation is performed. In these works, a current source model is used for the channel.

There are publications dealing with the performance of power-frequency grounding systems for fast current pulses. A MoM approach that is currently used involves frequency-domain thin-wire approximation for the conductors [58,59] and a simplified image system for the air–earth interface [60]. Detailed modelling of the earth termination will have to be weighted against the accuracy of the soil parameters used in the simulation and their seasonal and spatial variation, the influence of other buried metallic objects in the vicinity, structural foundations and so on.

In summary, at present, studies on the lightning surge response of the protection system to launch pads seems to be possible with the following simplifications:

1. a current source model for excitation;
2. a linear model for the channel with loading for realizing the reduced velocity;
3. simplified geometry for the protection system, as well as the launch pad (where the simplification depends on the numerical approach employed);
4. a linear model for the ground and no corona or ionization in the system.

With the present state of the art in the field, any approach with the abovementioned simplifications would be a significant step ahead. Hence, it is evident that considerable progress is necessary for a reliable theoretical quantification of the surge response of the protection system to launch pads.

Experimental approach
As an alternative to the field theoretical approach, with the above set of simplifications, the volt–ampere based experimental approach on the scaled electromagnetic model of the system can be efficiently used.

Such an approach has been used in analysing lightning surge voltages at the top of power transmission line towers (e.g. in References 61 and 62). This can be seen to be analogous to the wind tunnel experiments for determining aerodynamic profiles. The principle of electromagnetic modelling [63] involves scaling of the geometry and time with the same scaling factor. In other words, the associated frequencies must be inversely scaled. The same principle has been used in Reference 64 for the study of the surge response of the following configurations: (i) isolated 120-m-tall protection tower, (ii) 120-m tower with neighbouring conducting objects, (iii) 120-m tower with ground wires, and (iv) insulated mast scheme involving a 120-m tower with many ground wires. This study used a frequency-domain characterization approach, and the required time-domain quantities were then deduced from discrete inverse Fourier transform. Direct time-domain experimentation, as well as some simulations using NEC have shown that the frequency-domain approach is quite adequate for simple down-conductors, as well as schemes with a single tower. Of course, the ground was modelled to be perfectly conducting and wideband current monitors were used for the required non-intrusive measurements [63].

For obtaining the lightning surge response characteristics of Indian satellite launch pad-I and pad-II protection schemes, experiments have been conducted in the frequency domain on reduced-scale models with a 40:1 scaling factor [65,73]. The actual cross-sections of the tower elements were rather difficult to reproduce in the model and hence a cylindrical approximation was made according to the geometric mean radii. All the intricacies, spanning much less than the wavelength corresponding to significant frequency components of the lightning current spectrum, have been neglected [63]. The model tower was made of copper. According to scale model theory [63], the conductivity of the model should be 40 times that of the actual tower. Although the magnetic permeability of the tower is not scaled exactly, the use of copper for the model attempts to very crudely respect the scale factor for the conductivity. In any case, these quantities have a very little influence on the response. Owing to the spatial extent of the experimental model, measuring the phase data was rather difficult. The phase data were artificially extracted using the principle of the Hilbert transform as described in Reference 64. In view of this, the parameters estimated from the frequency-domain approach can have errors in the range of 10–15 per cent. In this work, characteristics of three schemes have been investigated: (i) Scheme 4, (ii) Scheme 4 with ground wires for a given tower and (iii) Scheme 4 modified for an insulated mast scheme. The basic quantities that were addressed were the so-called 'voltage rise' at the struck point and the tower base currents. It is necessary to scale back the time axis to the original value.

Some sample results deduced for the protection system to launch pad-I are presented in Figure 16.4. It can be seen that the 'voltage rise' at the lightning rod is reduced when multiple ground wires are connected to the tower. Also, it results in a significant reduction of the tower base current. This aspect would be of special interest for protection towers situated close to ground support systems.

On the other hand, the insulated mast scheme for the same layout would develop significantly higher voltages. Further, the induced current in the supporting tower, which is insulated from the air termination network, is significant. The tower base current is shown to reach $\sim$30 per cent of the incident stroke current for a time to crest of 1 μs and $\sim$10 per cent for a time to crest of 5.5 μs. An important issue that is difficult to quantify with regard to the insulated mast scheme is briefly discussed in References 64 and 65. The supporting tower and the protection system have two critical clearances: one along the insulating support and other between the ground wire/catenary and the supporting tower. Experience in high-voltage engineering shows that the surface strength can drop below the air gap strength, and this is specifically true when the deposition of dirt, salinity and so on occurs on the surface. However, at present, surface-withstand strength for the oscillatory voltage, which develops across the surface, is not available. Similarly, withstand strength for the

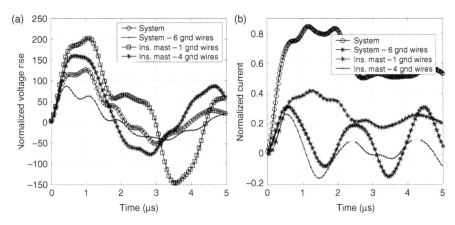

Figure 16.4 Simulation results for a stroke intercepted by the protection system to Indian satellite launch pad-I: (a) top voltage at the struck point; (b) tower base current at tower 1 (ins, insulated; gnd, ground; stroke parameters: time to crest = 1 μs; amplitude normalized to unity) (from Reference 65)

geometry and gap length prevailing between the ground wire and tower top is not available. This withstand strength will become an important issue for strokes with large currents and fast/moderate time to crest. In summary, for strokes with small and moderate amplitudes, the insulated mast scheme can provide smaller soil gradients and electromagnetic fields in the protected volume. However, depending on the length of the gaps (which in turn depends on the length of the insulator) and the surface condition of the insulating mast, there would be a certain current amplitude and time to crest for which a flashover along the surface cannot be ruled out. Although the probability of occurrence of strokes with higher currents is smaller, the collection area projected by the protection system (see Figure 16.3b) is larger and therefore needs considerable attention. Therefore further work is required in this direction.

In principle, neglecting the non-linearity in the channel and soil, the reduced geometric scale electromagnetic modelling of the protection system and launch pad can serve as a very useful technique for the evaluation of their lightning surge response and also the associated fields. This approach is not limited by the numerical computation problems (which are encountered many times with the high-frequency numerical field evaluation codes), has no ambiguity in the application and interpretation of results, and no restriction on the geometry of the system under study. However, this approach needs a spacious laboratory (of large height), which is preferably shielded, a suitably modelled ground plane, and importantly a pertinent source of good strength (especially for time-domain experimentation). Care would be necessary to minimize the disturbance introduced by the physical presence of a measuring system. If other factors permit, the required experimentation can also be conducted outdoors, provided that the source strength is sufficiently high and ambient electromagnetic noise arising due to telecommunication networks and broadcasting is within acceptable limits.

16.7.3 Weather launch commit criteria

It is reported that the threat due to natural or triggered lightning is a major source of launch delay. The launch delay results in considerable costs, overstressing of personnel due to an extended working period and the possible delay of subsequent launch operations [66]. As an increasing number of launches are scheduled, this could become a matter of serious concern.

It is opined that atmospheric electricity is one of the least matured meteorological sciences. It is recognized that the electric field in cloud is the primary threat, and needs to be measured and quantified [39]. A suitable aircraft that can measure three-dimensional electric fields inside the cloud would be useful for both quantification of the cloud electrification and to optimize LLCC.

16.8 Indirect effects

In the event of a nearby strike, the existing launch procedures in many cases require that the launch operation be suspended and all system-level tests be repeated to confirm that damage or upset to the payload or launch vehicle systems has not really occurred [67,68]. This is time-consuming, costly and in many instances might not really be required. In order to improve the situation, it is essential to deduce the quantitative relation between the stroke parameter, including its location with respect to the pad, and the corresponding induction level, at sensitive systems. Any meaningful study on the indirect effect should address the induction levels at the launch pad and supporting system. There are some efforts in the literature that deal with the electromagnetic impact of a nearby stroke on the system.

Attempts have been made in References 69 and 70 to evaluate the induced currents and resultant fields around the critical areas of the launch pad and launch/space vehicle at launch sites in Florida. A 30 kA stroke at 1 km distance and finite ground conductivities was considered. Using several simplifications, numerical simulations using NEC have been made. Comparisons with the actual measurements, which are made at selected locations on the pad, indicate that the numerical results are lower than those measured by a factor of 2 to 3.

In Reference 71, the resulting voltage on the shield wires of the protection system for the Indian launch pad due to a nearby stroke to ground is considered. As the system is designed for stroke interception, the voltage rise due to a strike nearby is not of any serious concern and hence it is practically irrelevant. Further, the electromagnetic response of the massive protection towers, as well as the launch pad (umbilical tower and launch vehicle) has been neglected. The number of shield wires and the earth termination system considered are different from the actual configuration. Hence, the results must be viewed cautiously.

Modelling for the indirect effect involves suitable modelling for the lightning channel and the system under study. In this case, modelling of the channel is basically for the field produced by the bridging and the return stroke phase. An appreciable induction can be expected only for the strokes within a kilometre, so finite conductivity of the soil may not assume any large significance. Modelling of the system would be as complex as modelling of the same for a direct lightning strike.

16.9 Protection of other supporting systems

There are several supporting systems around the launch pad. Depending on the type and nature of the structure, either an isolated protection system or a protection system that is supported on the structure to be protected is used for intercepting and safely diverting potentially dangerous strokes. Although, in principle, they may come within the protective zone of the launch pad protection system, local care is always preferred. This will ensure protection even for strokes with smaller currents. For buildings housing sensitive electronic systems, an internal metallic shield is preferred. All the cable trenches running between the electronics/control/electrical ground support equipment building to the umbilical tower are preferably shielded with interconnected metallic sheets or at least by interconnected metallic strips with appropriate grounding arrangements. Wherever possible, fibre-optic links are preferred. Appropriate surge protection devices such as gas discharge tubes and varistors (when the system permits) are used at the wiring/cable ends. From the lightning point of view, it is preferred that the various earths of the control building such as power, switching and signal are all brought together electrically. However, as this is not accepted in many cases, they need to be tied together through special semiconducting devices that act only for the transient duration for which the potential difference exceeds a certain value.

16.10 On-site measurements

A quantitative evaluation of the efficacy of protection for both direct strokes and nearby strokes would require a suitable measuring system. The basic quantities that are to be measured would be the induced currents and voltages in the elements of critical systems. However, accessibility and cost would limit such measurements only to umbilical cables and some cables running between the electronic/control/electrical ground support equipment building and the umbilical tower. As simultaneous measurement on a large number of connections would be quite tedious, it would be appropriate to measure the electric and magnetic fields at critical regions in the launch pad complex, which in turn would be a good indicator for the induction levels. Considering the rate of rise of lightning current and especially the associated fields, broadband sensors along with a high-speed data acquisition system would be essential for recording the data. Note that any system tuned for recording the quantities due to a stroke nearby would definitely go to saturation (or even destruction, unless suitably protected) during a direct hit to the system by a bypass stroke.

As mentioned earlier, an important application of such a measuring system would be to acquire quantitative data on the induced currents and voltages on a critical system during a nearby strike. The measurement system discussed above would serve the purpose, provided the processed data were available online. A description of such a system, which has been used at many launch sites in the United States, can be found in Reference 72. A complete knowledge of the stroke parameters and the corresponding induction levels, if made available, would save time, efforts and of course cost.

16.11 Summary

Launch pad complexes are massive, expensive establishments, and at present only a few countries possess them. They can be considered to be pinnacles of engineering excellence involving multidisciplinary efforts. Intensive care is necessary to provide safety and reliability. Launch exercise for satellites and exploratory space missions are expected to be a continuously growing activity and, therefore, launch pads will become more and more busy. Needless to say, lightning protection of the satellite launch pad as well as the launch vehicle is mandatory. Protection engineering should encompass threats posed due to a direct (natural or triggered) strike and a nearby stroke. At present, the protection philosophy involves a three-level defence:

1. adhere to lightning launch commit criteria;
2. use a suitable protection system to protect the launch pad from all dangerous strokes;
3. use EMC hardening techniques to protect the system inside the vehicle and that related to the launch system from the electromagnetic field due to lightning.

Present-day understanding and knowledge of the lightning attachment process as well as the response of the system during a lightning strike/interception is still incomplete. Considerable work is essential to make any further progress. The areas that need to be addressed span and include the breakdown of very long air gaps for both negative and positive polarity, lightning surge response of the launch pad, as well as the protection system under the transverse magnetic mode of current propagation. For this, both theoretical and suitable experimental approaches would be necessary. However, at present a pessimistic engineering evaluation seems to be possible with certain simplifications. For the evaluation of both direct and indirect effects, experiments on reduced-scale models of the system can serve as a very useful tool.

On-site measurements through a suitable measuring system would provide a strong support for further work. In particular, the acquisition of data on lightning-induced currents and voltages in critical systems during operating situations will be very beneficial.

In order to provide a more reliable and launch-on-demand capability, the lightning launch commit criteria (LLCC), which at present appears to be conservative, will necessarily have to be revised. At the same time, it is essential to make sure that small local discharges initiated during transit of the vehicle through clouds will not cause any significant disturbance/damage to the vehicle or any of its subsystems. All these call for a clear understanding of charge build-up mechanism in clouds and interaction of the lightning with the launch pad and launch vehicle. Further work is very essential in all these areas.

References

1. http://www.nasa.gov/columbia/home/ (January 2006).
2. http://www.boeing.com/defense-space/space/delta/flo_facilities_delta_IV.htm (January 2006).
3. http://www.isro.org/gslvd1/gslvd1.htm (January 2006).
4. http://www.sinodefence.com/space/facility/xichang.asp (January 2006).
5. http://www.arianespace.com/site/images/ARIANE5_tech_GB.pdf (January 2006).
6. http://www.nasa.gov/centers/kennedy/home/index.html (January 2006).
7. http://www.russianspaceweb.com/baikonur_energia_110.html (January 2006).
8. http://www.isro.org/Cartosat/INDEX.htm (January 2006).
9. http://www.russianspaceweb.com/baikonur_zenit.html (January 2006).
10. http://www.spacetoday.org/Rockets/Brazil/BrazilRockets.html (January 2006).
11. http://home.comcast.net/~rusaerog/boosters/R-7.html (January 2006).
12. http://www.sea-launch.com (January 2006).
13. NASA Facts AC 321/867-2468. *Lightning and Space Program*. August 1998.
14. Uman M.A., Rakov V.A. 'The interaction of lightning with airborne vehicles'. *Progr. Aerospace Sci.* 2003;**39**:61–81.
15. Baba Y., Ishi M. 'Numerical electromagnetic field analysis of lightning current in tall structures'. *IEEE Trans. Power Deliv.* 2001;**16**:324–28.
16. Kordi B., Moini R., Janischewskyj W., Hussein A.M., Shostak V.O., Rakov V.A. 'Application of the antenna theory model to a tall tower struck by lightning'. *J. Geophys. Res.* 2003;**108**:4542, ACL 7.
17. http://www-pao.ksc.nasa.gov/kscpao/nasafact/padsfss2.htm (January 2006).
18. http://www.geo.at/spacetravel/images/raketenstart.jpg (January 2006).
19. http://www.energia.ru/english/energia/iss/iss08/photo_archive-iss08-all.html (January 2006).
20. http://www.jaxa.jp/countdown/f4/rcc_e.html (January 2006).
21. http://www.sinodefence.com/space/facility/xichang.asp (January 2006).
22. Schaffar A., Lemeur P., Gobin V., Bertuol S. 'ARIANE 5 lightning verification plan'. *International Conference on Lightning and Static Electricity*; Toulouse, France, June 1999, paper no. 1999-01-2334.
23. http://www.esa.int/esaCP/SEMDAXXEM4E_index_0.html (January 2006).
24. http://www.globalsecurity.org/space/facility/ccas-lc-41-pics.htm (January 2006).
25. http://www.boeing.com/defensespace/space/delta/delta4/gallery/delta_iv_canaveral.html (January 2006).
26. Kumar U., Nelson T.J. 'Analysis of the air termination system of the lightning protection scheme for the indian satellite launch pad'. *Proc. IEE Sci. Meas. Technol.* 2003;**150**:3–10.
27. Kumar U., Nelson T.J. 'Evaluation of the protective action of LPS to Indian satellite launch pad-II'. *12th International Symposium on High Voltage Engineering*; August 2001, **1**, paper no. 2–2, pp. 91–94.

28. Kumar U., Nagabhushana G.R. 'Analysis of lightning protection system for India satellite launch pad'. *National Interagency Coordination Group Lightning Conference*; Orlando, Florida, 12 April 2000.
29. Johnson D.L., Vaughan W.W. *Analysis and Assessment of Peak Lightning Current Probabilities at the Kennedy Space Center*. NASA report, NASA/TM-2000-210131, May 1999.
30. Van Brunt R.J., Nelson T.L., Stricklett K.L. 'Early streamer emission lightning protection systems: an overview'. *IEEE DEIS Magazine* 2000;**16**:5–23.
31. Kumar U. *Final report Development of a Lightning Protection Scheme For Launch Pad – II at SHAR*. Sponsored Project by SDSC, ISRO, January 2001, pp. 116.
32. Cooray V., Zitnik M., Manyahi M., Montano R., Rahman M., Liu Y. 'Physical model of surge-current characteristics of buried vertical rods in the presence of soil ionisation'. *J. Electrostat.* 2004;**60**:193–202.
33. Schaffar A. 'ARIANE 5 lightning protection'. *International Conference on Lightning and Static Electricity*; 1992, Paper 2A-1.
34. Fisher F.A., Plumer J.A. *Lightning Protection of Aircraft, National Aeronautics and Space Administration*. NASA reference publication 1008, 1977.
35. NASA. *Design Considerations for Lightning Strike Survivability*. Preferred Reliability Practices, practice no. PD-ED-1231.
36. Yiming Li. 'Review of EMC practice for launch vehicle systems', *IEEE International Symposium on Electromagnetic Compatibility*; Seattle, WA, USA, 1988. pp. 459–64.
37. ECSS-E-30 Standard Part 6A. *Space Engineering, Mechanical Part – 6: Pyrotechnics*; 25 April 2000.
38. NASA. *Assessment and Control of Electrical Charges*, practice no. PD-ED-1210.
39. Krider E.P., Christian H.J., Dye J.E., Koons H.C., Madura J., Mercert F. *et al.* 'Natural and triggered lightning launch commit criteria'. *The 86th American Meteorological Society Annual Meeting, 12th Conference on Aviation Range and Aerospace Meteorology*; Atlanta, GA, USA, January 2006.
40. Roeder W.P., McNamara T.M. 'A survey of the lightning launch commit criteria'. *The 86th American Meteorological Society Annual Meeting, 2nd Conference on Meteorological Applications of Lightning Data*, Atlanta, GA, USA, January 2006.
41. Martínez A.A., Mura F., Dionisio C., Howes S., Slim R., Erickson P.D. *Future Satellite Services, Concepts and Technologies*. ESA Bulletin 95, August 1998.
42. Birkeland P., Johnson T. 'Kistler aerospace and the desirability of non-federal launch sites'. *AIAA Space 2001 Conference and Exposition*; Albuquerque, New Mexico, August 2001; paper no.AIAA-2001-4533.
43. Gallimberti I., Bacchiega G., Bondiou-Clergerie A., Lalande P. 'Fundamental processes in long air gap discharges'. *C.R. Physique* 2002;**3**:1335–59.
44. Lalande P., Bondiou-Clergerie A., Bacchiega G., Gallimberti I. 'Observations and modeling of lightning leaders'. *C.R. Physique* 2002;**3**:1375–92.

45. Becerra M., Cooray V. 'Time dependent evaluation of the lightning upward connecting leader inception'. *J. Phys. D: Appl. Phys.* 2006;**39**:4695–4702.
46. Rizk F. 'Modeling of lightning incidence to tall structures, part I & II'. *IEEE Trans. Power Deliv.* 1994;**9**:162–93.
47. Nelson T.J. *Analysis of the Protective Action of a Closely Distributed Lightning Protection System for a Satellite Launch Pad*. MSc (Eng.) thesis, Department of High Voltage Engineering, Indian Institute of Science, October 2000.
48. Kumar U., Prasanth Kumar B., Jagannath Padhi 'A macroscopic inception criterion for the upward leaders of natural lightning'. *IEEE Trans. Power Deliv.* 2005;**20**:904–11.
49. Mazzetti C. 'Principle of protection of structures against lightning' in V. Cooray (ed.). *The Lightning Flash*. London: IEE, 2003.
50. Liu Y., Zitnik M., Rajeev T. 'An improved transmission-line model of grounding system'. *IEEE Trans. Electromagn. Compat.* 2001;**43**:348–55.
51. Umashankar K.R., Teflove A., Berker B. 'Calculation and experimental validation of induced currents on coupled wires in an arbitrary shaped cavity'. *IEEE Trans. Antennas Propagation* 1987;**35**:1248–57.
52. Boonzaaier J.J., Pistonius C.W.I. 'Finite-difference time-domain field approximations for thin wires with a lossy coating'. *IEE Proc. Microw. Antennas Propag.* 1994;**141**:107–13.
53. Baba Y., Nagaoka N., Ametani A. 'Modeling of thin wires in a lossy medium for FDTD simulations'. *IEEE Trans. Electromagn. Compat.* 2005;**47**:54–60.
54. Edelvik F. 'A new technique for accurate and stable modeling of arbitrarily oriented thin wires in the FDTD method'. *IEEE Trans. Electromagn. Compat.* 2003;**45**:416–23.
55. Paul J., Christopoulos C., Thomas D.W.P., Liu X. 'Time-domain modeling of electromagnetic wave interaction with thin-wires using TLM'. *IEEE Trans. Electromagn. Compat.* 2005;**47**:447–55.
56. Harrington R.F. *Field Computation by Moment Method*. Malabar, FL: Krieger, 1983.
57. Bruke G.J., Poggio A.J. *Numerical Electromagnetic Code (NEC) Part I & III*. Technical document 116, Naval Ocean Systems Center, San Diego, 1980.
58. Xiong W., Dawalibi F.P 'Transient performance of substation grounding systems subjected to lightning and similar surge currents'. *IEEE Trans. Power Del.* 1994;**9**:1412–20.
59. Olsen R.G., Willis M.C. 'A comparison of exact and quasi-static methods for evaluating grounding system at higher frequencies'. *IEEE Trans. Power Deliv.* 1996;**11**:1071–81.
60. Vesna A., Grcev L. 'Electromagnetic analysis of horizontal wire in two-layered soil'. *J. Comput. Appl. Math.* 2004;**168**:21–29.
61. Kawai M. 'Studies of the surge response of a transmission line tower'. *IEEE Trans. Power Appar. Syst. J* 1964;**83**:30–34.
62. Baba Y., Ishii M. 'Numerical electromagnetic field analysis on measuring methods of tower surge response'. *IEEE Trans. Power Deliv.* 1999;**14**:630–35.

63. Anderson J.G., Hagenguth J.H. 'Magnetic fields around a transmission line tower'. *AIEE Trans. Power Appar. Syst.* 1959;**78**:1391–98.
64. Kumar U., Kumar P. 'Investigations on the voltages and currents in lightning protection schemes involving single tower'. *IEEE Trans. Electromagn. Compat.* 2005;**47**:543–51.
65. Kumar U. *Experimental Investigation with the Scaled-Down Models for the Post-stroke Potential Differences & Currents in UT/MST and LPS*. Final report. Sponsored Project by ISRO-IISc Space Technology Cell, 2002, pp. 104.
66. Roeder W.P., Sardonia J.E., Jacobs S.C., Hinson M.S., Guiffrida A.A., Madura J.T. 'Lightning launch commit criteria at the Eastern Range/Kennedy Space Center'. *37th AIAA Aerospace Sciences Meeting and Exhibit*; Teno, NV, USA, January 1999. Paper no. AIAA-99-0890.
67. Rivera A.F., Koons H.C., Walterschied R.L, Briet R. 'Protecting space systems from lightning'. *Crosslink, The Aerospace Corporation Magazine of Advances in Aerospace Industry*, Summer 2001.
68. Chai J.C., Monegutt J.J., de Russy S.D. 'Survey of CGLSS/SLC40 lightning data and retest criteria'. *IEEE International Symposium on Electromagnetic Compatibility*; Austin, Texas, USA, 18–22 August 1997, pp. 391–96.
69. Chai J.C., Monos S. 'Electromagnetic response of structures induced by nearby lightning events at launch sites'. *International Conference on Lightning Protection 2000*; Rhodes, Greece, 18–22 September 2000. Paper no. 3.9, pp. 278–83.
70. Chai J.C., Britting A.O., Jr. 'Does an overhead lightning protection system protect structures against nearby strikes?' *International Symposium on Electromagnetic Compatibility*; Beijing, China, 21–23 May 1997. pp. 480–83.
71. Kannu P.D., Thomas M.J. 'Lightning-induced voltages in a satellite launch-pad protection system'. *IEEE Trans. Electromagn. Compat.* November 2003;**45**: 644–51.
72. Sechi P.G., Adamo R.C. 'The developement, deployment and operation of an on-line lightning monitoring system (OLMS) for spacecraft launch support'. *International Conference on Lightning and Static Electricity*; Seattle, WA, USA, September 2001. Paper no. 2001-01-2901.
73. Kumar U., Hedge V., Darji P. 'Investigations on the voltages and currents in the lightning protection system of the Indian satellite launch pad-I during a stroke interception'. *IET Sci. Meas. Technol.* 2007;**1**:225–31.

Chapter 17
Lightning protection of structures with risk of fire and explosion
Arturo Galván Diego

17.1 Introduction

This chapter is devoted to the provision of guidance on the protection against lightning-related electric sparks to structures containing explosive or highly flammable materials that can generate an explosive atmosphere. These materials can take the form of solids, liquids, gases, vapours or dusts. In this chapter, 'structures' is the term used for vessels, tanks or other containers in which these materials are contained.

An ignition source is anything that can heat even a small portion of a fuel to its auto-ignition temperature. Owing to its high temperature, lightning and electric sparks (such as electrostatic discharges) can generate ignition in a volatile atmosphere, and thus cause fire and explosion. Ignitions with high pressure waves are termed 'explosions' and ignitions with minor pressure waves are known as 'flash-fires'.

Despite the great variety of types of structure involved, the main causes of flash-fire and explosion arising from lightning, which are common in many situations, are the following [1]:

- ignition of an explosive atmosphere;
- penetration of a metal enclosure;
- a temperature rise in a metal container;
- ignition of a gas/air mixture by a point-discharge current;
- mechanical impact caused by a direct strike; and
- earth currents.

It is worth mentioning that, even after all known precautions have been taken, the prevention of direct/indirect lightning-related spark effects cannot be absolutely assured. However, one can minimize the possibility of damaging accidents by implementing an adequate level and type of protection, and then regularly maintain and inspect the protection system.

Ignition is unlikely or impossible when the following conditions are fulfilled [2]:

- sparks may occur, but flammable vapours are always excluded by gas freeing the atmosphere from or inerting it in the area of discharge;
- product handling occurs in a closed system, and oxygen in that system is always below the minimum concentration required to support combustion, such as in the handling of liquefied petroleum gas (LPG);
- the flammable concentration is always maintained above the upper flammable limit (UFL)

17.2 Tanks and vessels containing flammable materials

17.2.1 General

There is no doubt that lightning strikes that hit storage or process vessels containing flammable materials can cause devastating accidents at refineries, bulk plants, processing sites and other facilities. Some of the registered accidents are listed in the references [3–7].

Chang and Lin [7] conducted a review of 242 accidents involving storage tanks that have occurred in industrial facilities since 1960. It was shown that 74 per cent of accidents took place in petroleum refineries, oil terminals or while in storage. Fire and explosions account for 85 per cent of these accidents and 80 of them (33 per cent) were lightning-related. Table 17.1 shows the percentage of lightning-related accidents in tanks.

It is well known that a small accident in installations with a flammable and hazardous chemicals content may lead to million-dollar property loss, some days of production interruptions and, in the worst case, loss of life.

There are two main mechanisms by which these lightning-related accidents may have occurred: (i) a direct strike and (ii) indirect or nearby strikes. Direct strikes account for the flow of a large amount of lightning current from the point of incidence through the tank body and towards the earth system. In contrast, the occurrence of lightning nearby, which is highly probable, may lead to dangerous potential differences between different parts of the tank.

Table 17.1 *Lightning-related accidents involving tanks over the last 40 years (adapted from Reference 7)*

Year	1960–1969	1970–1979	1980–1989	1990–1999	2000–2003	Total
LRA*	4	10	19	37	10	80
Total	17	36	53	85	51	242
% of total	**24**	**28**	**36**	**44**	**20**	**33**

*LRA, lightning-related accidents.

Table 17.2 Apparent susceptibility to lightning strikes and lightning damage to oil-field facilities (adapted from Reference 8)

Facilities most susceptible to a lightning hit* (from most to least susceptible)	Equipment sustaining the most lightning-related damage† (listed in decreasing order of damage [as determined from insurance claims])
Drilling rigs	Tanks
Pulling units	Electronics
Electric lines/transformers	Motors/controls
Pumping units	Transformers
Dehydration towers	Electric lines
Battery equipment	Pulling units
Treaters–tanks	Drilling rigs
Motors/controls	Pumping units

*Determined by considering the relative height and degree of isolation.
†From interviews with relevant personnel regarding the frequency of damage and the extent of the damage.

Table 17.2 shows that tanks are not the objects most susceptible to a lightning hit in a facility, but, unfortunately, they are prone to suffer the largest amount of lightning-related damage. One possible reason for this is the poor lightning protection measures adopted for tanks (in terms of grounding, bonding and air terminals). According to the information provided by Welker [9], documented tank explosions include accidents involving full tanks, empty tanks, partially full tanks, stock tanks, and metal and fibre-glass tanks. This means that tanks are prone to damage irrespective of the type and operation conditions once they are hit by lightning or subjected to the very strong electromagnetic influence of nearby lightning. Static electricity can also be created by the friction of dust or liquids against a non-conducting object, causing ignition of flammable vapours, which very often is blamed on lightning [8].

There is universal consensus in and technical support from lightning physics that lightning strikes cannot be eliminated, so sound engineering practices and accepted industry standards for the operation and maintenance of the equipment have to be used to greatly reduce the consequences of lightning [2]. However, as mentioned by Welker [8], 'There is some degree of randomness regarding lightning strikes and there are seldom first-hand witnesses to a hit or the resulting damage sequence. Consequently, the subject of lightning damage in the oilfield is surrounded by mystery and myth, opinions vary widely and the subject of what, if anything, can be done about it still leaves room for speculation.' Even worse, empirical evidence shows lightning protection to be the exception rather than the rule.

When considering the installation of a lightning protection system, one problem is the misconception about what comprises a technically sound lightning protection system for tanks and vessels containing flammable materials. In addition to this, knowing which concepts given in standards are relevant and the lack of knowledge about who should be responsible for implementing the lightning protection system ensure that making the appropriate decisions is not a foregone conclusion. Thus an absence of sound knowledge, guidelines and responsibility may generate confusion among those who seek reliable protection for their installation. The end result could be a solution that is universally questionable or, worse still, no protection at all on a sensitive structure.

Another classical situation is that lightning protection systems are installed only after catastrophic events, like that mentioned above [5], in which a Faraday cage was selected and designed for the protection of hydrocarbon storage tanks in the Luján de Cuyo refinery, Argentina, after the disaster had taken place.

There is a vast difference between the likelihood of a given tank actually being hit by lightning and the likelihood that the tank will sustain damage if it is hit [8]. Therefore, all those measures relating to the minimization of potential differences between different parts of a tank and between the parts of the tank and the surrounding elements should be taken as mandatory actions, irrespective of whether a decision is made to install air terminals or not for the interception of lightning currents to protect the tank from a strike.

Lightning-related damage can be caused by one or more of the following actions in the oil industry [8]:

- heat and possibly vaporization of materials from the extreme current flow;
- side-flashing or arcing to adjacent structures as a result of the high voltages;
- damage to equipment from induced current/voltage/frequency fluctuations;
- ignition of flammable/explosive vapours by the strike or side-flashing;
- ignition of flammable/explosive vapours by corona discharges;
- possible hydrolysis and subsequent ignition of water in storage vessels.

17.2.2 Risk assessment

Tables 17.1 and 17.2 show that lightning strikes are associated with 25–45 per cent of damage in tank accidents, and that tanks are more prone to lightning-related damage than any other piece of equipment in the oil industry. In general, the risk to life and property is so high when tanks and vessels containing flammable materials are concerned, that the provision of every means possible for protection from the consequences of a lightning discharge is essential, unless the tank or vessel has been specially designed and is situated in a place chosen specifically to limit the effects of a catastrophe. If an external lightning protection system incorporating air terminations is to be installed for tanks and vessels in the open field, in which the lightning current flow can cause a serious problem to the skin of the tank, the system should be isolated (because the thermal and explosive effects at the point of strike may cause damage to the tank, its content and the surroundings, as mentioned in Chapter 8); otherwise, a non-isolated system can be applied in which several air terminals are installed onto the structure, and bonded to the body. If an external lightning protection system with an air termination system is to be installed for installations containing tanks and vessels, the system might or might not have to be isolated according to the specific requirements.

17.2.3 Lightning protection measures

In general, tanks and vessels have numerous grounding and other protection systems. These include static discharge from fluid-movement bonding, lightning protection with external air terminals, equipment bonding, cathodic protection, stray-current control, power and external-line protection, instrumentation connection and instrumentation protection [9].

Although all these protection systems are mandatory, for obvious reasons, they are, unfortunately, only applied partially or not at all. In these cases there is a risk of ignition, fire and eventually explosion, with catastrophic effects. On the other hand, no single guideline covers all design and installation considerations.

A lightning protection system (LPS) for structures at risk of catching fire and exploding should be mandatory [10], and should consider both external lightning protection (ELPS) (comprised of air terminations and down-conductors) and internal lightning protection (ILPS) (in the form of bonding and grounding), combined with a clean environment surrounding the structure. This is because electric sparks are not only generated by direct lightning strikes, but also by nearby lightning flashes.

There is a common misconception among engineers and lightning protection designers, and even in some relevant standards, that the term lightning protection refers to an air terminals system alone, which is conceptually inadequate as a sole means of providing protection from lightning. Instead, as mentioned above, the lightning protection term refers to all measures and steps required to ensure adequate protection, including external and internal measures. It is important to mention that air terminals are not the panacea of lightning protection and, furthermore, when they are wrongly installed, they can even increase the vulnerability of a facility.

The consequences of a shielding failure of the external lightning protection system could have disastrous effects, ensuring that no cost is spared to prevent it happening [1]. However, almost all explosions and secondary effects attributable to lightning strikes come from an underestimation or a relaxation of the protective measures (against which there might be no protection at all), in the belief that nothing serious will happen. In this respect, it should be noted that, owing to the catastrophic effects of any incidence involving the kind of structures being discussed here, it is recommended that the criteria for the provision of integral protection, fully discussed in this chapter, be adhered to, irrespective of any devices or systems employed that are claimed to provide enhanced protection.

When facilities that handle flammable substances and have been damaged by lightning were investigated, it was found that little or no information relevant to lightning protection had been used. However, given what is currently known about lightning, many or all of these incidents could have been preventable. In the case of storage tanks containing flammable substances, one important thing to consider is that they may represent a special hazard in the event of a lightning strike because a spark, which might otherwise cause little or no damage, could ignite flammable vapours, resulting in a flash-fire or explosion. Releases of toxic substances can also occur [3–5].

The recommended lightning protection practices considered below can be divided into groups according to three main measures: (i) air terminations, (ii) equipotential bonding (including grounding systems) and (iii) maintaining a clean environment.

17.2.3.1 Air terminations

Air terminations are used to avoid direct lightning impact upon a tank's shell or protrusions from it. They could be part of an isolated or non-isolated ELPS. The air terminations take the form of vertical masts, overhead ground wires or a combination of the two. In general, two methods are applied in constructing lightning protection of installations at risk of fire and explosion: the rolling sphere method (RSM) and the protective angle method (PAM).

It is generally agreed that the locations of the air terminations should provide a theoretical shielding against lightning currents from currents as low as 5 kA (to those corresponding to a radius of 30 m for the RSM). It is worth mentioning that, in some lightning protection standards such as the Mexican [10] and Australian [11] standards, the application of a rolling sphere of radius 20 m, providing theoretical shielding against lightning currents of 3 kA and greater, is recommended in conjunction with the usage of the RSM. The PAM is the recommendation in other standards, such as the British standard [12], where the specification is an angle of 45° in the bounded space of two or more overhead grounding wires and 30° elsewhere.

Figure 17.1 shows the zone of protection provided by both the RSM (for a radius of 20 and 30 m) and the PAM for the protection of tanks and vessels using an isolated ELPS, as mentioned above. Note that the zone of protection given by the RSM is greater than that given by the PAM, the latter being more restrictive.

The vertical masts should provide multiple paths for the lightning current to reduce potential values [13] (Figure 17.2). On the other hand, it is recommended that the elements of the external lightning protection system should be located at a distance of at least 2 m from any element of the tank or vessel to be protected; however, the minimum distance might need to be increased if a safety distance analysis [10,16], considering $k_c = 1$ (partitioning coefficient), reveals 2 m to be inadequate.

Practical considerations impose some restrictions on the application of isolated ELPSs in tanks and vessels of huge dimensions. For example, observing a minimum separation of at least 2 m between the shell of the tank and the elements of the LPS, tanks of 30 m diameter would need to be protected by vertical masts of 25 m height if the tanks were 15 m high and the masts were erected in a four-vertical-mast arrangement, and the height of the vertical masts would have to be increased as soon as either the height or the diameter of the tank needed to be increased. Higher vertical masts are subject to mechanical and operative restrictions, which might limit their use in a protective scheme, for example, for tanks with huge diameters (from 40 to 85 m) commonly used to store less volatile liquids, like crude oil, for which a self-protecting scheme could be observed as the unique protective measure.

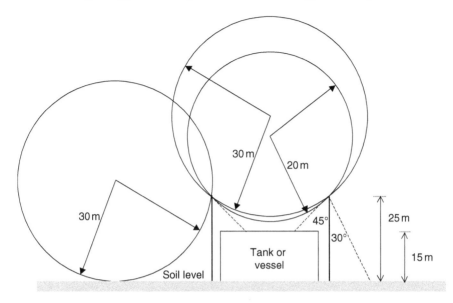

Figure 17.1 Rolling sphere method and protective angle method for the protection of tanks and vessels in an isolated external lightning protection system (ELPS). The protective angle is 45° in the bounded space of two overhead ground wires and 30° elsewhere (dashed lines). The diameter and height of the tank are 30 and 15 m, respectively. The height of the masts is 25 m.

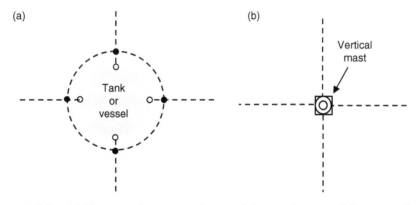

Figure 17.2 (a) The grounding system (top view) for a tank or vessel. Points at which the grounding system is bonded to the tank are marked by empty circles. The vertical ground rods are marked by full circles. The ground conductor buried at a depth of at least 0.6 m is marked by a dashed line. (b) The vertical mast (top view) should have multiple paths for lightning current near the connecting point in order to reduce potential rise [13].

17.2.3.2 Equipotential bonding

Equipotential bonding is relevant to two parts of the tank, divided as follows: (i) ground level and (ii) the other parts (namely, the walls, basement and roof top).

At ground level

At ground level the grounding system is essential. The general recommendation is that the grounding system should encircle the tank or vessel (at a distance of at least 1 m from the shell) with a single or double ring, and be interconnected with the grounding system of the other tanks or vessels and to the general grounding system of the facility (Figure 17.1). The resistance of the grounding system should not exceed 10 Ω. The number of points for connecting the tank or vessel to the grounding system will depend on the diameter of the tank, with at least four points being connected for diameters ≤ 30 m (Figure 17.1) and at least eight points for diameters above this. Some lightning protection standards recommend a smaller number of connections [10–12,14,17], but the benefit of adding extra connections with the grounding system is much greater than the cost of adding extra connections; it should be considered that the entire grounding system is essentially an insurance coverage, and, like insurance, it is much better to have it and not use it than need it and not have it. This is the case because the cost of lightning-related damage in tanks and vessels is always very high.

It is recommended that vertical ground rods be installed at each point of the grounding connections to the tank, with a length determined by the resistivity of the soil, obtained by field measurements (avoid using generic tables). In addition, it is strongly recommended that the joints be welded to the tank when making the connections to the grounding system to ensure that none of the connections will be loose connections. Pressure storage tanks, vessels and process equipment that are designed to contain flammable liquids or gas under pressure do not normally require lightning protection, because it is assumed that they are well grounded and thick enough not to be punctured by a direct lightning strike [14]. Nevertheless, the need for an ELPS should be considered, especially in geographical areas with high ground flash densities to ensure that, in those instances where a system is needed, the installation is not erroneously overlooked.

Different parts of the tank

All parts of the tank or vessel should be attached to the grounding system to ensure electrical continuity. The utmost care must be taken to bond any metal pipes or cables entering the structure to the tank, observing the cathodic protection system. One special case for bonding is the floating roof tank. As the floating roof is not in direct contact with the tank shell, high potential differences can arise under the influence of direct or nearby lightning strikes. One recommendation is that the roof should be electrically bonded to shoes of the seal through the most direct electrical path at intervals not greater than 3 m on the circumference of the tank. This bonding can be done by shunts made of stainless steel straps, but care must be taken to ensure that they have an adequate current-carrying capacity and a suitable corrosion resistance [14]. It is imperative that the metallic shoe stays in permanent contact with the tank shell through the shunts. Of course, the shunts are not required when tanks do not have a vapour space at the seal. However, it has been recognized that the shunts used in the oil industry can give rise to poor contact between the tank shell and the floating roof, so one supplementary bonding measure is required, especially for tanks of huge diameters where an isolated external lightning protection has mechanical restrictions – retractile grounding cable devices [20]. These are permanent, easy to handle and require hardly any maintenance. The main purpose of retractile grounding cables is to ensure a permanent equalization surface between the floating roof and the tank shell, thus avoiding the devastating effects of secondary arcs mainly generated by nearby lightning, which would otherwise be produced by the charge bridging the small air gap between the structures in a volatile atmosphere. It is important to emphasize that the only purpose of the retractile grounding cable device is to enhance the measures taken to ensure that the connection is sound, not to replace them, and the cable should be of low inductance. The number of devices will depend on special requirements, but it is good practice to install at least four devices, with one end in the highest part of the shell and the other end on the floating roof [20].

17.2.3.3 A clean environment

The main goal of this measure is to avoid the formation of a volatile atmosphere on and around the tank, under any circumstances. Such a volatile atmosphere could be formed from the gases or vapours within the tank, and therefore several measures are applied to ensure that this does not happen: (i) flammable air–vapour mixtures should be avoided and (ii) all appurtenances should be kept in good working order.

It is quite common to observe that, during venting processes in manufacturing, storage, compression, and chemical plants, explosive mixtures of gases or vapours and air are released into the atmosphere. Such a condition makes lightning protection almost impossible to achieve, because the distances over which such dangerous mixtures can persist after release from outlets can reach several tens of metres, which represents a serious condition as ignition (through flash-fires or explosions) can be generated when a lightning channel is located near to or passes through the outlet. For deliberate and unintentional venting processes, the following measures should be fulfilled to minimize the risk of flash-fire and explosion:

- As far as possible, making use of the venting process, particularly during active thunderstorms, should be avoided.
- Vents should be fitted with flame arresters to prevent eventual ignition in the air being transmitted through the vent pipe into the container or installation.
- Where possible, vents should be made from non-conducting material to reduce the probability of a lightning strike.

17.2.3.4 Self-protecting system

It is normally agreed that in self-protecting systems, metallic structures are electrically continuous, tightly sealed to prevent the escape of liquids, vapours or gases, of a thickness adequate to withstand direct strokes (that is, at least 5 mm thick), and encase highly dangerous material. When these conditions are met, other than ensuring adequate bonding and grounding, an ELPS based on an air termination system for lightning interception may not be required at all [12]. Tanks or vessels with non-metallic roofs cannot be considered to be self-protecting.

Irrespective of the above, it should be observed that the self-protecting criterion varies from one place to another as it depends on several parameters of a typical lightning flash in that region, such as the habitual variation in the charge of a discharge and the duration of an event. In this way, for a severe lightning discharge, 5 mm thickness might not be adequate to remain undamaged. Thus, the self-protecting criterion is always determined by local conditions and regulations.

17.2.3.5 Resumé for lightning protection

The general recommendations for the lightning protection of structures with a risk of fire and explosion are as follows.

1. In the case of an obvious risk to life and property, every means possible of providing protection should be installed.
2. A single or double ground ring should be used for the lightning protection system of each tank or vessel, interconnecting neighbouring structures.
3. It should be ensured that multiple paths exist for attaching the connecting points to the ground system of air terminations.
4. It should be ensured that bonding is done thoroughly and that the safety distance is observed to avoid arc flashing.
5. An isolated LPS should be installed when the lightning current flow poses a problem to the skin of the tank or vessel, having suspended air terminals at least 2 m apart from the structure to be protected, by using protection level I or II.
6. A non-isolated LPS should be installed when the lightning current flow does not pose a problem to the skin of the tank or vessel, by using protection level I or II.
7. It should be ensured that the environment in the immediate vicinity of the tank of vessel is free of gas or vapour and flametraps should be used for vents emitting flammable vapours or powders.
8. If dangerous materials are encased in metal of adequate thickness, other than ensuring that the bonding and grounding are sound, the additional measures provided by systems such as air terminations may not be required at all (i.e. the self-protecting condition is satisfied).
9. Tanks or vessels containing non-metallic materials cannot be considered to be self-protecting.

17.3 Offshore oil platforms

17.3.1 General

A lightning flash can subject offshore oil installations (OOI) to damage ranging from undesirable effects to emergencies, particularly when a direct strike occurs [19]. Lightning strikes that hit equipment and storage or process vessels containing flammable materials in OOIs can cause devastating accidents. On the other hand, lightning protection is becoming increasingly important, given the use of digital and low-voltage analogue systems in OOIs, in which lightning-related malfunction operations can cause loss of function and, therefore, loss of production. Given this, OOIs should have well designed, properly installed, and appropriately maintained lightning protection systems, to avoid the obvious catastrophic effects that could endanger personnel and the installation and its contents.

As discussed above for storage tanks and vessels onshore, an integral lighting protection system should be designed and installed, incorporating an ELPS (air terminals and down-conductors) and internal lightning protection (equipotential ring, grounding conductors, equipotential zones and surge protection devices, SPD). The approach adopted should be to consider each of the following points, ensuring that each is met if a complete and effective system is to be constructed [19]:

- external lightning protection
- general grounding system
- internal grounding system
- shielding effectiveness
- location of SPD

17.3.2 Relevant standards

The NORSOK standards have been developed by the Norwegian petroleum industry to ensure that adequate safety measures are adopted, and that value adding procedures are in place and cost-effectiveness ensured. Section 5.4.2 of Standard E-CR-001 [21] for OOIs states that, 'No additional installations will be required for the lightning protection, provided the unit consists of bolted and welded steelwork that will provide a continuous current path from the highest point of the unit to the main earth.'

Standard IEC 61892-6 [22], which deals with the electrical installation of mobile and fixed offshore units, establishes in Section 16 appropriate provisions for lightning protection against primary structural damage and secondary damage to the electrical system. Again, in Section 16.2.3, it is stressed that 'A protective system need not be fitted to a unit of metallic construction, where a low resistance path to earth will be inherently provided by bolted and welded steelwork from the highest point of the unit to earth.'

Det Norske Veritas (DNV), an independent foundation with the objective of safeguarding life, property and the environment, has also developed standards. In DNV-OS-D201 [23] for OOIs, Section I-600, it specifies that lightning protection with air terminals should be considered under certain conditions, for instance when masts over vessels are made of non-conductive material, and lays down the appropriate size of the air terminals (with a minimal height of 150 mm over the mast) and specifies that the down-conductors (terminated to the nearest point of the metal hull) intended to carry lightning current should be double the size of those specified in IEC 62305-3 [17] recommendations for land-based installations.

On the other hand, the IEC standards, produced by a worldwide organization producing universal standards comprising all national electrotechnical committees,

832 *Lightning Protection*

deal with this topic in several related standards. For instance, IEC 92 [24], recommends lightning conductors on ships only if all the masts are wooden. Like Reference 21, IEC 92 recommends double sized lightning conductors in comparison to those specified in IEC 62305-3 [17] recommendations for land-based installations and the same characteristics for their terminations or ultimate connection. Amendment 2 of IEC 60092-401 [24] establishes, in Section 51.2.3, that 'Vent outlets for flammable gases located at or near the say *top of masts* on tankships are to be protected by air terminals which extend at least 2 m above the vent outlet ...'; as can be observed, this protection element is compulsory irrespective of the thickness of the tankship. IEC 60092-502 [25], relating to electrical installations in ships, recommends in Section 5.6.1 that 'Account shall be taken of the risks due to lightning attachment', and in Section 5.6.2, 'Consideration should be given to the risk and effects of lightning attachment to high level gas or vapor vents, or adjacent structures'; however, it says nothing about 'how to provide the appropriate protection'.

17.3.3 Risk assessment

Owing to the nature of OOIs, the risk to life and property is so obvious that the provision of every means possible for protection from the consequences of a lightning discharge is essential. Unlike land-based installations, however, it is very difficult to accomplish an isolated ELPS with respect to the metallic structure of the OOI. Therefore, a non-isolated ELPS should be put in place.

17.3.4 Lightning protection measures

17.3.4.1 External lightning protection

When lightning strikes an OOI, the lightning currents and the spurious signal generated interact with the equipment by a variety of mechanisms: capacitive, magnetic and conductive coupling, direct arc-over and potential rise. The primary effects of direct lightning strikes on OOIs are shown in Figure 17.3. As can be observed, direct damage to equipment, containers and electrical systems occurs. In addition, there might be a loss of production, risk of fire and explosion, generation of spurious signals, and so on. Damage can be reduced or avoided through the installation of an ELPS.

The International Electrotechnical Commission (IEC) is the recognized entity in many countries around the world for lightning protection standards. Thus, IEC standards and others [10–12,15–17,22,24,25] give sound guidance on lightning protection and on the philosophies underlying the general principles. However, there is a lack of information concerning OOIs and other special installations.

In dealing with lightning protection of OOIs, two key assumptions are made: (i) it is a very risky installation and (ii) its structure is generally metallic. The first implies that a risk assessment is irrelevant as lightning protection should be provided to ensure that the maximum level of protection is given. The latter implies that much of the equipment and many devices could already be protected against direct lightning incidence by the metallic structure itself (which provides a 'natural' air terminal).

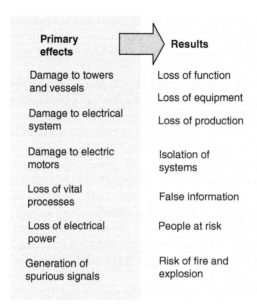

Figure 17.3 Primary effects of direct lightning incidence in offshore oil installations (OOIs)

However, apparatus or devices out of the protective zone of the 'natural' metallic air terminal formed by the structure itself should be considered.

Although it is done in many cases, lightning protection must not rely upon the metallic condition of the structure of the OOI. It is possible, by combining the information content of related standards with the experience gained from real situations, to establish the following protection scheme for the provision of an ELPS.

1. Verify all the equipment and devices are within the frame of the metallic structure of the OOI.
2. Verify that the metallic frame of the structure is electrically continuous. If not, install jumpers between major metallic parts or use another suitable measure.
3. If there are devices or some equipment is likely to be hit by lightning (due to their height and location), install external lightning protection based on air terminal(s) with a covering zone suitable for level I or II protection as recommended by IEC 62305-3 [17], by applying RSM over and around the volume of the OOI. Air terminations should be attached at the base of this to the frame of the metallic structure of the OOI, which will work as a 'natural' down-conductor.
4. Use air terminals and lightning conductors double the size of the IEC 62305-3 [17] recommendations for land-based installations.
5. When the electrical continuity of the path for the lightning current is not ensured, down-conductors intended to carry the lightning current should be installed. When such condutors are installed, they should be bonded to the main hull (or to main base pillars – not necessarily up to seawater level) of the OOI, taking the most direct route, and the cable should be bare.

6. Corrosion-resistant metals and materials resistant to seawater should be used and the possibility of galvanic action between dissimilar metallic elements should be avoided by careful choice of materials in an ELPS used at sea.

Furthermore, it should be borne in mind that, irrespective of the strike point of a direct lightning strike, the type of air terminal and the down-conductor, whether intentionally constructed or part of the structure itself, the entire metallic structure will supply a frame through which to conduct the lightning current. This condition makes it very hard (if not impossible) to apply an isolated lightning protection system. On the other hand, the thickness of the body of the processing towers and of the vessels located in the OOI that are exposed to lightning strikes generally satisfies the requirements for the flow of lightning current (5 mm). Such constructions could, however, have small pipes, valves, or vent outlets for flammable gases and signalling cables attached at the top, making it imperative to protect the constructions against direct hits. Many problems have been generated by overlooking or inadequately considering these topside elements, even though the body of the tower or vessels avoided primary damage.

Particular attention should be paid to the topsides of I&C, Electric Power and Control rooms. The walls and roof of these rooms are generally made of continuous steel sheets that satisfy the recommended thickness for direct lightning strikes, by acting as a Faraday cage. However, the situation changes when an air conditioning system is installed on the roof as it incorporates motors and ducts, thereby making the room vulnerable to lightning hazard. In this situation, then, air terminals should be used to avoid a direct impact to an air conditioning system. Finally, the metallic roofs of turbine sheds are not generally designed to withstand a lightning current, so they should be protected against a direct lightning strike by installing air terminals.

17.3.4.2 Grounding system and common bonding network

There are two main types of steel-framed installation for OOIs: floating (i.e. ships) and bottom-founded structures. So, strictly speaking, in this instance, earthing is the term used in OOI structures to refer to connection to the huge mass provided by the metallic structure. Protective earth conductors or safety earth conductors are connected to the structure in two ways: directly, by forming a connection onto the metallic structure, and by using a common bonding network (CBN) characterized by an equipotential ring conductor attached to the metallic structure (Figure 17.4).

For ships, IEC 92 [24] specifies, in Section 67, that one should '... verify that ... earthing leads are connected to the frames of apparatus and to the hull'. Section 52.2.1 of IEC 60092-401 [24] states that 'Metallic enclosures shall be earthed to the metal hull or to the protective system ...'. It is then clear that grounding leads should be connected to the hull, and it should be noted that the shortest route should be used for floating OOIs. What about bottom-founded structures? In this case, the grounding leads can be connected to the main pillars of the OOI by the CBN. In this scheme, the frame of all metallic enclosures should be radially connected to the CBN located beneath the main floor of the platform, as shown in Figure 17.4; this ring conductor,

Lightning protection of structures with risk of fire and explosion 835

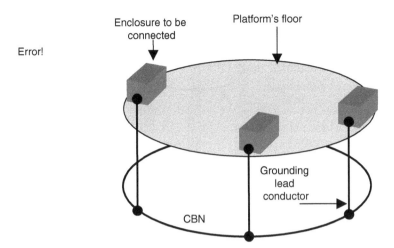

Figure 17.4   Radial connection from enclosures to the nearest CBN (equipotencial ring) of the platform's floor or level. The leads must follow the shortest distance, avoiding any abrupt turns.

in turn, should be joined to the main pillars, as shown in Figure 17.5. The key ingredient that makes these schemes work is that, for a land-based installation, it is more important that the grounding system extends under the entire structure than that the ground is low resistance. This is similar to the CBN used in several-storey structural steel buildings.

Section I-701 of the standard DNV-OS-D201 [23] specifies that 'Aluminium superstructures that are provided with insulating material between aluminium and steel in order to prevent galvanic action are to be earthed to the hull ... with wires or bands ... with a maximum connection distance of 10 m.' This is important because many problems encountered in grounding systems of OOIs are related to corrosion generated by galvanic action from the use of dissimilar metals. The distance should be reduced to below 10 m when radio interference is concerned. Standard E-CR-001 [21] refers to local recommendations for the grounding system.

Connections between CBN, grounding lead conductors and the main pillars of the OOI should be exothermic or of an irreversible compression type, able to meet the specific requirements of the task for which they are being installed, with corrosion-resistant layers or films at the point of connection.

17.3.4.3   Internal grounding system

The internal grounding system is related to the grounding of electrical distribution systems and devices related to the level of the power supply (motors, circuit breakers and transformers, and electronic I&C systems). The boundary, as far as lightning is concerned, depends upon the type of electrical supply. If the OOI generates its own

836 Lightning Protection

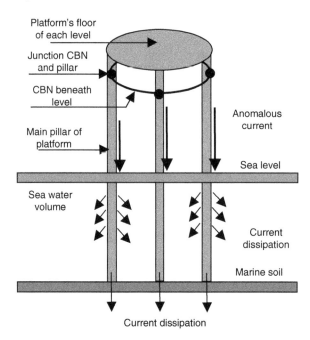

Figure 17.5   *Anomalous current path from CBN (equipotential rings) and main pillars, which are considered the lowest impedance elements in OOIs of the bottom founded structure type. The current is dissipated (to some extent) in the seawater volume and (predominantly) in marine soil.*

electrical supply, then generators will also be included in the scheme; if not, the boundary begins at the service entrance and ends at the digital and analogue system.

Equipment housed in the OOI structure is protected from direct lightning strikes. However, lightning-related transients can follow a variety of paths, reaching power-level devices and I&C sensitive devices. I&C devices are particularly vulnerable to transients, which can enter from power supply connections and sensor signal paths, or be induced directly by the lightning channel or metallic structures carrying a lightning current. This interaction arises because the way in which lightning will behave is not easy to predict.

Connections for the internal grounding scheme for low and medium power distribution systems, as well as electrical devices, must follow the safety rules laid out in IEC 61892-6 [22], in accordance with the type of earthing system. Further actions for lightning protection and for the TN–S system are as follows.

- Place a master bonding bar near the main distribution panel.
- Connect a grounded conductor to the master bonding bar. This must be applied to the motor control central.
- Secondary distribution panels must be provided with separate neutral and grounding bars.

- Install safety and reference grounding conductors.
- I&C cabinets must be provided with isolated bonding bars.
- A reference grounding grid should be supplied to telecommunication and control rooms (when necessary).
- Adequate grounding bars should be provided for surge suppression devices (SPDs).
- Care should be taken to monitor the effect of corrosion on metallic conductors and enclosures.

In the case of I&C circuits, devices and cabinets, Figure 17.6 shows the lightning protection scheme, including ELPSs (where necessary), the general grounding system and the internal grounding system. The protective zones, radial bonding arrangement and the shielding are as specified in IEC 62305-4 [18].

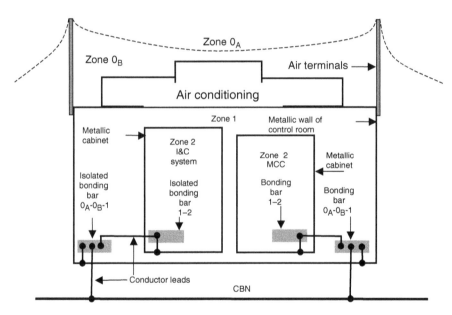

Figure 17.6 Typical lightning protection scheme for a control room containing motor control centre (MCC) and I&C devices and circuits in an OOI

Note that, in Figure 17.6, the following statements apply.

- Air terminals are installed only when air conditioning or other equipment is not shielded by any higher metallic element. In the case that there is no device or any system at risk on the roof of the control room, the ELPS can be omitted.
- A lightning current will flow through the air terminals and the walls of the control room, which will act as a Faraday cage.

- The bonding bars for I&C systems are isolated from the metallic walls of the cabinets.
- The bonding bars for the motor control centre (MCC) can be isolated or non-isolated from the metallic walls of cabinets and the room.
- All cabinets and the control room are made of continuous metallic sheets of adequate thickness to satisfy the requirements specified in standards for lightning current flow.
- Cabinets for I&C systems should be isolated from the metallic floor and walls of the control room

17.3.4.4 Shielding

Shielding of OOIs can be observed by considering the following parts separately: the air terminals for lightning attachment (in the event of direct impact), control of radiated lightning electromagnetic fields and their effects, control of conducted lightning-generated transients at the interface of protected zones, and bonding leads, bars and the CBN.

The shielding effectiveness of air terminals in a direct lightning strike will depend on the level of protection used, according to IEC 62305-3 [17].

The effectiveness of shielding against radiated lightning electromagnetic fields and the effect and the damage inflicted by such fields on sensitive electronic devices, sensors and cables inside enclosures will depend on the configuration of and the material used for the barrier. As far as lightning-induced induction is concerned, the typical frequency characterizing the magnetic field associated with the first stroke of a lightning strike is 25 kHz and that of subsequent strokes $\sim$1 MHz, as indicated in IEC 62305-4 [18].

Shields rely on two major electromagnetic phenomena: reflection from a conducting surface and absorption in a conductive volume. For plane waves (i.e. far fields), the combined effect of these losses, known as the attenuation (arising from reflection and absorption), determines the effectiveness of the shield. If the components in the enclosure are to be protected from external fields, then the material from which the enclosure is constructed should be selected to maximize the absorption and reflection losses. Reflection, however, is a surface-dependent effect and, is, therefore independent of the barrier thickness. It is a function of the material's conductivity and magnetic permeability and of the frequency of the field. Absorption is the transformation of wave energy to heat, in this case, in the shield, and is frequently defined by the term 'skin depth'; it is not directly related to near- or far-field conditions.

It is a fact that shielding effectiveness varies with frequency, shield geometry, the positioning of a victim, component or device within the shield, the type of field being attenuated, the direction of incidence and polarization. Moreover, the definition of shielding effectiveness for electric and magnetic fields will be identical if the fields concerned are uniform plane waves and the media on each side of the barrier are identical. For example, for 20 mil copper (1/2 mm), the combined attenuation is $\sim$140 dB (10^7:1) and for 20 mil steel, the combined attenuation is $\sim$180 dB (10^9:1) for a signal of 25 kHz (typical of a first stroke); these attenuation values are higher for 1 MHz signals (associated with subsequent strokes) [26]. Cases outlined in IEC 62305-4 [18] are only valid for gridlike spatial shields; however, they provide the basis for an analysis of the effectiveness of the shielding of metallic framed structures like the OOI itself, and others. For example, for a 1-mm width copper/aluminium gridlike spatial shield, the magnetic attenuation would be $\sim$80 dB for 25 kHz and for 1 MHz, but for 5-m width, the magnetic attenuation would only be $\sim$5 dB.

The above analysis indicates that perfectly welded sealed rooms are ideal for producing effective lightning shielding. However, all openings, such as doors, windows, side panels, I/O panels, and ventilation ports and cable exits that ingress/egress can endanger the integrity of the shielding, especially through the production of conducted signals. Thus, all openings should correctly be treated so that the effiency of shielding is not undermined.

17.3.4.5 Location of SPD

IEEE Std 1100 [27] addresses the filtering and grounding of service lines and other conductors that ingress/egress the LPS boundary zones, and IEC 62305-4 [18] specifies the appropriate arrangement of SPDs within lightning protection zones, particularly at transition points and with a special emphasis on the coordination of an SPD with the equipment to be protected. IEEE C62.41 [28] offers guidelines for the application of SPDs.

The following are some important factors to be considered when evaluating the need for and installing SPDs.

- SPDs appropriate for lightning protection of low-voltage installations are designed to handle only a portion of the total lightning current.
- The protection levels of the selected SPDs must meet the requirements for the insulation coordination and the immunity levels of the equiment to be protected (IEC 61000-4-5 [29]).
- SPDs should be located at well identified critical zone interfaces of the LPS, and in particular, at interfaces of voltage levels.
- SPDs should be located as close as possible to the equipment to be protected, with the ground conductor being as short and straight as possible.

References

1. Golde R.H. *Lightning Protection*. New York: Chemical Publishing Co.; 1973.
2. API 2003. *Protection Against Ignition[s] Arising out of Static, Lightning and Stray Currents*; 1998.
3. EPA. *Lightning Hazard to Facilities Handling Flammable Substances*. Report of Environmental Protection Agency, EPA 550/F/97/002c, May 1997.
4. Sirait K.T., Pakpahan P., Anggoro B., Soewono S., Iskanto E., Garniwa I., Rahardjo A. 'An analysis of the origin of internal sparks in kerosene tanks due to lightning strikes'. *Lightning and Mountains Conference* 1997; paper A.4.5.

5. Marianni E., Rodriguez M. 'Controlling the risk of fire caused by lightning hydrocarbon storage tanks'. *25th International Conference on Lightning Protection ICLP 2000*; Sep 2000, paper 8.9, pp. 786–91.
6. Lightfoot F.K. 'Lightning protection for metal production storage tanks'. *Professional Safety* March 1985, 21/23.
7. Chang J.I., Lin C.C. 'A study of storage tank accidents'. *Journal of the Loss Prevention Process Industries* 2006;**19**:51–59.
8. Welker A.J. 'Lightning – Its effects and some simple safeguards in regard to oilfield operations'. *Proceedings of Annual Southwestern Petroleum Short Course*; 1998, pp. 329–46.
9. Durham M.O., Durham R.A. 'Ground systems: design considerations for vessels'. *IEEE Industry Applications Magazine* Nov–Dec 2001, pp. 41–49.
10. NMX-J-549-ANCE-2005. *Protection Against Lightning – Specifications, Materials and Methods of Measure* (in Spanish). ANCE; 2006.
11. Australian Standard AS/NZS 1768. *Lightning Protection*; 2003.
12. British Standard BS-6651. *Code of Practice for Protection of Structures Against Lightning*; 1999.
13. Grcev L. 'Modelling of grounding systems for better protection of communication installations against effects from electric power systems and lightning'. *23rd Telecommunications Energy Conference, INTELEC 2001*; 14–18 October 2001, pp. 461–68.
14. NFPA 780. *Standard for the Installation of Lightning Protection Systems*; 2008.
15. IEC 62305-1. *Protection Against Lightning. Part 1: General Principles*; 2006.
16. IEC 62305-2. *Protection Against Lightning. Part 2: Risk Management*; 2006.
17. IEC 62305-3. *Protection Against Lightning. Part 3: Physical Damage to Structure and Life Hazard*; 2006.
18. IEC 62305-4. *Protection Against Lightning. Part 4: Electrical and Electronic Systems within Structures*; 2006.
19. Galván A. 'A technical basis for guidance of lightning protection for offshore oil installations'. *Journal of Lightning Research* 2007;**3**:1–9.
20. Galván A. 'Retractile grounding cable used for floating roof tanks: four-year experience'. Presented at *Proceedings of International Conference on Lightning Protection*; Kanazawa, Japan, 18–22 September 2006.
21. NORSOK Standard. *Electrical Installation of Offshore Installations*, E-CR-001, Rev. 3; 1997.
22. IEC 61892-6. Mobile and fixed offshore units – electrical installations – part 6: installation, 2nd edition, 2007.
23. DNV Offshore Code DNV-OS-D201. *Electrical Installations*; 1993.
24. IEC 92: 1980, *Electrical Installations in Ships. Part 401: Installation and Test of Completed Installation*; Amendment 2 – IEC 60092-401, 1997-04.
25. IEC 60092:1999-02. *Electrical Installations in Ships. Part 502: Tankers – Special Features.*

26. Clayton P.R. *Introduction to Electromagnetic Compatibility.* John Wiley & Sons, USA, 1992.
27. IEEE Std 1100. *Recommended Practice for Powering and Grounding Electronic Equipment*; 1999.
28. IEEE Std C62.41. *Recommended Practice on Surge Voltages in Low Voltage AC Power Circuits*; 1995.
29. IEC 61000-4-5. *Electromagnetic Compatibility (EMC): Part 4. Testing and Measurement Techniques, Section 5: Pulse Magnetic Field Immunity Test*; 1995.

Chapter 18

Lightning and trees

Mahendra Fernando, Jakke Mäkelä and Vernon Cooray

18.1 Introduction

Each year a significant number of lightning flashes should strike trees, as about a 30 per cent of land is covered by forest. According to DeCoursey and colleagues [1] (as cited by Price and Rind [2]), about half a million lightning strikes terminate on trees every day. This figure includes strikes in dense forests, which are hardly ever observed, except perhaps as forest fires, and also strikes in inhabited areas where trees are often the highest structures and hence the most likely objects to be struck. Remarkably, only a few studies have been conducted to understand the interaction of lightning with trees. As a result, the mechanisms by which lightning damages trees are understood only at a general level. It is also worth considering that trees might preserve a 'recording' of the lightning flash, and could therefore be used to study the very fine structure and physics of the lightning attachment processes [3]. On the other hand, there is a considerable body of literature on the long-term effects of lightning damage, especially related to forest fires [4,5].

Most of the books on lightning physics and lightning protection include a chapter on lightning and trees [6–9]. In general, the sources acknowledged in the literature are fragmented and, to a large extent, rely on individual case studies. There appear to have been only a few systematic large-scale attempts at data collection, such as those of Covert [10] (and references therein) and Taylor [11], as well as an ongoing study in Finland, whose first results have been published by Mäkelä *et al.* [3]. In this Chapter, an attempt is made to evaluate the present state of knowledge of the interaction between lightning and trees, and of lightning protection of trees.

Generally, damage caused by lightning to trees is limited to individual trees or to their immediate surroundings. At the level of practical impact, the largest effect of lightning on trees is the initiation of forest fires. In tropical forests, lightning hardly ever appears to initiate fires [12]; however, elsewhere, the estimated figures vary from 34 per cent in Central Siberia [13] and 35 per cent in Canada [14] to 90 per cent in Northern Siberia [13]. These percentages, however, are highly dependent on the presence of human habitation (which increases the incidence of forest fires from other causes such as campfires). In addition, the existence of a fire-control infrastructure decreases the effect of forest fires. Thus, in Finland it is estimated that lightning is responsible for the ignition of only 10 per cent of forest fires [15], despite it being the only natural means of ignition [16]. In addition to direct economic damage, forest fires can decrease the air quality over large regions [17]. However, studies have been conducted to determine whether forest fires can produce a positive feedback by increasing the number of lightning flashes as a result of increased pollutant levels; so far, the results have been inconclusive [18] and are assumed to be fairly short-term [19].

On the other hand, even though forest fires may be economically devastating, they may be ecologically useful, or even necessary. Lightning tends to strike the tallest object, which helps to remove the tallest and oldest trees in forests where they have been obstructing the growth of smaller younger ones. Furthermore, when lightning ignites forest fires, it can help to keep the ecosystem in balance by burning vegetation, avoiding the accumulation of insects and diseases, which is beneficial for the wellbeing of forests, keeping them healthy and, eventually, enabling them to spread. Ideally, rather than aiming for a total eradication of forest fires, fire protection authorities should try to balance this ecological role against the threat to security and property, as well as the cost of fire management [14].

The economic impact of lightning-induced damage to trees is even more difficult to quantify. For example, in August 1972, a group of citrus trees in Florida were struck by lightning and, after six months, 2 527 of the affected trees were found to have either died or were dying [20]. Very large and ancient trees may hold a cultural and spiritual value for communities; it is impossible to quantify such value economically other than by noting the amount of money communities are willing to pay to protect such trees. Aside from these special cases, it is noteworthy that there seems to be absolutely no literature on the cumulative economic effects of damage to individual trees. Holle and colleagues [21], analysing the data from small insurance claims, reported that the number of lightning-related insurance claims in the United States may exceed 300 000 annually, and lead to cumulative losses of more than 300 million USD. Most of the damage they reported was due to indirect effects by lightning overvoltages as a result of flashes on or near power lines, and as such is not related to trees. However, the presence of trees is known to affect the number of strikes to power lines, although it is not even clear whether the predominant effect is to shield the wires (as suggested by Mousa and Srivastava [22]) or to increase the number of strokes (as suggested by Sakae and colleagues [23]). Tree strikes may also result in flashover currents to nearby buildings.

18.2 Strike and damage probability of lightning to trees

The physical effects of lightning strikes on trees have been extensively documented and are described in Section 18.3. From the viewpoint of lightning protection, however, the probability of a tree being struck (known as the strike probability) would be the most interesting parameter. As a poor electrical conductor, wood does not provide a good path for a lighting channel. On the other hand, trees are often relatively tall in comparison to their surroundings and have many-pointed ends. Thus, they are able to provide the electric field intensification needed to launch an upward leader.

It is also necessary to distinguish between the strike probability and the probability of causing damage (known as the damage probability), because lightning flashes do not always have any visible impact on a tree [24]. When forests were searched by the volunteers working with Mäkelä and colleagues [3], guided by lightning location data, fewer than 10 per cent of the searches found a physically damaged tree. As pointed out by Mäkelä and colleagues [3], evaluation of the strike probability is seriously hampered by the difficulty of locating damaged trees. Similarly, attempts to identify specific tree types most likely to be hit [10] are hampered by lack of information on the distribution of trees in the given forests as well as information on the height of any trees struck [7]. Furthermore, because most of the results are based on observed damage, they may be skewed toward those flashes that produce more serious damage.

The importance of ground conductivity was already understood in the early pioneering study of Covert [10], which referred to even earlier studies. Although the analysis by Covert [10] does not make a clear distinction between the strike probability and the damage probability, the results still provide a good foundation on which to base further studies. The following claims are made by Covert [10].

1. Covert [10] claims that trees that stand well above the neighbouring trees are more vulnerable to lightning than their neighbours. This is in general a common-sense assumption, but one that does not take into consideration a possible interaction between conductivity and height. Although the height of a tree does play a role in the probability that it is struck, the preliminary results of Mäkelä and colleagues [3] suggest that height alone may not be the sole determining factor.
2. Lightning strikes are also being documented from trees growing in the open, either alone or in a small group. There is a high probability that these observations could bias the statistics somewhat because strikes and damage are more likely to be observed when they occur to such groups than when they occur in thick woods or deep forest. This observer bias has been discussed in detail by Mäkelä and colleagues [3].
3. Trees growing along avenues or in the border of woods are struck by lightning more often [10]. The observer effect is important in these cases too, because they are easily observed.
4. Trees growing in moist soil are suggested to be better conductors than other trees [10]. The resistance of the path taken by the current may impede the growth of connecting leaders and thereby lead to a change in the probability of lightning strikes, although it is not clear how this would affect the damage probability.
5. According to Covert [10], trees growing in loam or sandy soils are much more likely to be attacked by lightning than those in clay, marl and calcareous soils. This result has not been verified as far as we are aware by subsequent studies. Mäkelä and colleagues [3] suggest that ground moisture from rain could be the

dominant parameter that changes the ground conductivity, a result that is supported by the conductivity figures by Saraoja [25].

6. Covert [10] claims that starchy trees are better conductors than oily ones, with conifers lying in between. Although this may be the case, measurements by Simpson and TenWolde [26] show that the water content of the wood is the parameter that controls conductivity, with the resistance decreasing by four orders of magnitude as the moisture content changes from 10 to 20 per cent. The moisture content dominates over the differences in wood species where the conductivity is concerned. However, how much the moisture content varies in living trees remains unknown from scientific studies.

7. Healthy trees in general are less likely to be seriously damaged by lightning than rotten wood [10]. This is supported by case studies such as that of Heidler and colleagues [27] as well as the statistics of Mäkelä and colleagues [3].

18.3 Types of lightning damage

Although damage to trees brought about by lightning strikes has been studied extensively, this has been done mainly from a horticultural point of view. Lightning-induced damage to trees can mainly be divided into two categories according to whether the damage is evident on a macroscale or a microscale [20]. We examine damage more profoundly by considering these two categories in the following sections.

18.3.1 Microscale damage

Taylor's work [20] provides a useful framework in which to place microscale damage. Six categories were identified and a considerable number of case studies, both old and new, exist that support these observations.

1. When lightning strikes conifers, a shower of fine needle particles can be produced [28]. The diameter of these particles can vary from 0.05 to 3 mm.
2. Spherical and coloured beads of chemicals deposited in lightning-struck conifers are occasionally observed. The diameter of these chemical compositions were found to vary from $\sim$0.1 to 0.6 mm.
3. Strips of crushed inner bark are commonly observed in furrows on lightning-struck trees [29].
4. Lightning furrows have been observed in the inner bark at places where the visible furrow on the surface of the trunk is discontinuous.
5. Wood splinters with charred tips have been observed when lightning splits tree trunks. The diameter of these splinters can vary from 1 to 3 mm and their length can vary from 1 to 4 cm [30].
6. Grooves, cracks and hardened tissues appear in the inner bark [20].

18.3.2 Macroscale damage

Taylor [11] provides a useful framework to classify macroscale damage on trees. The observed damage was classified into three types: bark loss, wood loss and explosion. However, after Orville [24], a new category has to be added to this classification: trees with no visible damage. Thus the categories are as follows:

1. *No physical damage.* No reliable statistics exist to suggest how common this is. Mäkelä and colleagues [3] found damaged trees in fewer than 10 per cent of the cases where a lightning-location system placed a flash in a forest, but the result could be attributable to multiple factors, including inaccuracies in lightning-locating systems and dense forestation.
2. *Bark loss.* Only a strip of bark is removed from the tree. Such damage was observed in 38 of the 53 cases ($\sim$70 per cent) studied in detail by Taylor [11].
3. *Loss of wood.* This comprises a deeper scar that also removes wood from the tree. Taylor [11] observed this in 15 of 53 cases ($\sim$30 per cent).
4. *Explosive damage.* Such damage was observed in only 10 cases out of 1 000 by Taylor [11], implying an occurrence probability of $\sim$1 per cent. This statistic alone points to a very strong observer-related effect in most case studies, because the focus is on the dramatic cases, which are, in fact, highly exceptional.
5. *Fires.*

The possible mechanisms responsible for each type of damage are discussed in the subsections that follow. Unfortunately, the statement by Taylor [20] that 'the mechanism by which lightning disrupts trees is still largely an enigma' is still a valid statement today. Our understanding of the damage-inflicting mechanisms has not improved significantly over the last twenty years. Moreover, not many experiments have been carried out over this period that could confirm or negate any proposed mechanism.

18.3.2.1 No physical damage

As reported by Orville [24], most lightning discharges striking trees may not be reported or investigated owing to the fact that there is no any visible damage. Mäkelä and colleagues [3] describe a similar case in which a flash incident on a tree was photographed, without any ill effects being found, despite a thorough examination being conducted (Figure 18.1). In both studies, the flashes were preceded by a long period of rain, making both the ground and the tree surface wet. In this case, the simplest explanation is that the current was simply following the path of least resistance, which in this case would have been the wet surface of the tree (and perhaps even a film of water on the tree surface). This model does not, however, explain why there was no observable damage caused by the shock wave from the flash [31]. In both of the above cases, no long-term deleterious effects were observed, despite examinations being made over a significant period of time.

Figure 18.1   Lightning strike to a tree that did not produce any visible damage (photo from Niklas Montonen). From the dataset of Mäkelä and colleagues [3].

Figure 18.2   (a) A regular bark loss case. (b) An irregular bark loss case (photo from Eero Karvinen). From the dataset of Mäkelä and colleagues [3].

18.3.2.2 Bark-loss damage

Golde [6] developed a model that has not been seriously challenged by later researchers. The assumption made is that, when lighting strikes the top of a tree, the current initially travels along the cambium (the thin moist layer just inside the bark). When the potential gradient becomes large enough, a flashover occurs through the bark and into the ground. This would explain why the damage tends to follow the grain of the tree, as observed by Taylor [11], for example, rather than being completely vertical. On the other hand, this simple model does not adequately explain the difference between bark-loss and wood-loss damage [11] or the occasional presence of explosive damage [20]. Presumably, damage of this kind would be caused by explosive evaporation of the water in the cambium.

Bark damage differs from one type of tree to another (Figure 18.2). In coniferous trees with thick and irregular bark, the damage takes the form of a scar aligned with the grain of the tree [11]. In deciduous trees with smoother bark, the loss can be much more irregular [8]. In these cases, the bark may be removed in its entirety. The height of the scar may vary considerably, and no systematic pattern appears in the literature.

18.3.2.3 Wood-loss damage

Taylor [11] classifies wood-loss damage as damage in which part of the inner (bole) wood is blown away from the trunk in addition to the bark (Figures 18.3 and 18.4). This implies that the current has taken a deeper path within the wood rather than proceeding through the cambium, as in bark-loss cases. Wood-loss damage is mostly associated with old and large trees, which fits with the mechanism proposed by Taylor [20], which suggests there are voids and ruptures within the core wood that let the current penetrate deep into the core. The rapid instantaneous water vaporization within the wood would produce forces strong enough to make the tree explode. Mäkelä and colleagues [3] found that the two main parameters capable of predicting wood loss or explosive damage are dry ground (lack of rainfall preceding the strike) and large peak currents.

18.3.2.4 Explosive damage

A small number of trees are completely annihilated by lightning strikes. Taylor [11] estimates the number of such cases to be ~ 1 per cent of lightning damage. Mäkelä and colleagues [3] observed that 2 of the 27 trees exploded, implying a probability of less than 10 per cent. On the other hand, even a quick analysis of the case-study literature (both scientific and anecdotal) shows that such cases are reported disproportionately often. Heidler and colleagues [27] analysed two such cases for which the flash can be identified on a lightning identification system, and showed that they are associated with positive currents ($+35$ and $+112$ kA). The explosion was intense enough to throw a 10 kg piece of wood more than 20 m. However, Mäkelä and colleagues [3] did not find explosive cases to be systematically associated with anomalously large currents. The explosive damage shown in Figure 18.5 was due to a flash with negative peak current of only 12 kA. Although the dataset is small, the

850 Lightning Protection

Figure 18.3   Wood-loss damage [photos from Eero Karvinen (a, b) and Matti Mäkelä (c)]. From the dataset of Mäkelä and colleagues [3].

implication is that flash characteristics alone are not a key determinant for explosive damage.

18.3.2.5 Ignition

The effects of forest fires are well covered in the literature on ecology and forestry. Somewhat less well understood, however, is the physics that leads to some flashes

Figure 18.4 Unusual case of wood loss in one tree and bark loss on a neighbouring tree. The flash most likely propagated along the telephone wire located next to both trees (photo from Tero Pajala). From the dataset of Mäkelä and colleagues [3].

causing fires. Why do some flashes cause fires and others not? Granström [32] and Larjavaara [33] studied forest fires caused by lightning in Sweden and Finland, respectively, but unfortunately it is difficult to estimate from these datasets the percentage of lightning flashes that cause ignition. However, one can find values for the annual incidence of fires per unit area ranging from 0.01 to 1.5 per 100 km^2 per year in the literature, at least at high latitudes. Because lightning flash rates in the same areas are more than 1 km^{-2} yr^{-1}, less than one flash in a thousand appears to result in ignition. Various authors [19,20] suggest that, nearer to the equator, lightning is an insignificant cause of forest fires. The wet conditions prevailing in tropical

Figure 18.5 *(a) Explosive damage. (b) The surroundings have been partly cleared of tree fragments (photo from Mats Kommonen). From the dataset of Mäkelä and colleagues [3].*

rain forests could be the reason. The tropics may, however, become more vulnerable in the future if climate change continues [19]. Thus, regardless of what the situation may be today, a better understanding of the ignition mechanisms could help improve predictions.

Taylor [20] suggests that the presence of charred bark or wood splinters in almost all newly struck trees is an indication that most or all flashes are in fact igniting the material they strike. On the other hand, Heidler and colleagues [27] (and references therein) describe cases in which no scorch marks were observed, as do Mäkelä and colleagues [3]. This observation is therefore unverified at the moment. However, the spread of a fire initiated by lightning demands that there is a supply of suitably located fuel. According to Granström [32], the actual effect of an ignition depends on a number of factors, such as the presence of fuel, weather patterns and topography, so the spread of forest fires has more to do with meteorology and forestry than with the initial physical mechanism causing the ignition. This tends to be supported by existing laboratory and field experiments. It is generally thought that continuing currents are an important cause of ignition [34–36], but Larjavaara and colleagues [37] showed that, for forest fires in Finland, at least, continuing currents are not necessary. Furthermore, laboratory tests by Darveniza and Zhou [38] showed that a continuous current can significantly increase ignition, but even an impulsive current can cause ignition. It would therefore seem that the characteristics of the flash itself have only a marginal effect on ignition. In effect, the lightning flash simply produces an opportunistic spark, and the subsequent events are determined by biological characteristics and meteorological conditions.

18.3.3 Other damage scenarios

There are documented cases of damage that are not easy to explain by means of the simple physical models described earlier.

18.3.3.1 Long-term propagation of damage

The immediate macroscale damage from a lightning strike to a tree may or may not result in long-term damage to the tree. Fires aside, any individual long-term damage is mostly explained by biology, and falls outside the scope of a physical analysis. For example, although a weakened tree may be left standing, in some cases insects could more easily colonize it and therefore destroy it [39]. In an experiment to investigate long-term damage, Rykiel and colleagues [40] simulated a lightning strike by winding detonating cords around trees and setting off explosions. The damaged trees eventually became infested by bark beetles, with the infestation spreading to significantly larger areas. Therefore, the long-term damage cannot be predicted without conducting an in-depth analysis of the entire forest ecology.

18.3.3.2 Group damage to trees

A single lightning flash may cause damage to a large number trees, through mechanisms that are presently not well understood [41]. One possible mechanism is that of damage to the roots of nearby trees. Another mechanism is the spread of insects attacking the damaged tree then spreading across to nearby trees, creating insect epidemics [40].

18.3.3.3 Damage to ground

Lightning can cause a variety of furrows in the ground (Figure 18.6). It is not known whether the presence of furrows correlates with group deaths of trees. Furthermore, the

Figure 18.6 Ground damage to the roots of a tree. A hole was gouged, and the ground singed (photo from Ilkka Juga). From the dataset of Mäkelä and colleagues [3].

Figure 18.7 Ground damage caused by a lightning flash proceeding from a tree to a house almost 20 m away. The grass was thrown about, and the flash also shattered a concrete slab (photo from Kim Lund). From the dataset of Mäkelä and colleagues [3].

mechanism producing these furrows is not understood. Case studies include that of Payne [42]. The furrows can, under some conditions be tens of metres long and cause extensive ground damage, as seen in Figure 18.7.

18.3.3.4 Damage to vegetation

Lightning has been known to cause damage in low-lying vegetation as well as in trees, although the reports are highly anecdotal. Known reports include damage to potato fields [43], cotton plants [44] and setting fire to standing wheat [45]. No general studies of the phenomenon could be identified from a literature search.

18.4 Protection of trees

Although forest fires appear to be the most economically damaging aspect of lightning damage, there is no realistic way of protecting large areas of uninhabited forest from lightning. A more useful approach to lightning protection, however, is to improve the accuracy of lightning detection and radar measurements, and to use that information to forecast the probability of ignition [37,46]. Such forecasts can help guide observers and aid in prioritizing fire-prevention measures.

The protection of individual trees is difficult to justify at a macro-economic scale, but lightning protection measures have nevertheless been defined. Such measures are used to protect historically valuable trees and trees neighbouring structures. The National Arborists Association (NAA) of America has published its own lightning protection standards. These are based on the criteria issued by the National Fire

Protection Association and Lightning Protection Institute (NFPA 780:2004). In addition, the American National Systems Institute standard ANSI A300 provides guidelines to lightning protection for trees.

The protection of individual trees can also make economic sense if the trees are located less than a few metres from structures such as houses. In such cases, flashovers from a struck tree can cause damage even if the structure itself is protected in accordance with the standards. The standard components used for structural protection are not automatically usable for protecting trees, because wind-induced movement can alter or damage the lightning protection system. Instead, air terminals should be fixed to suitable branch tops, which are earthed with copper cables, the cables being interconnected at suitable points. Cables used in tree protection systems usually consist of a number of strands. The number of strands per cable and the gauge of a strand may be different for different standards. The ground cable is connected to the tree using special clamps and connectors. The grounding of the system should be done at a considerable distance from the tree to avoid possible damage to tree roots. It is advised that the depth of the grounding system be ~ 3 m. In addition, it is advised that all grounding systems, including any metallic pipes within 10 m, be connected with the tree grounding systems; however, in certain standards this is not a requirement.

18.5 Conclusions

The interactions between trees and lightning are understood sketchily, if at all. Most of the research conducted in this field focuses on the measurable after-effects of a lightning strike, in particular forest fires. The long-term effects of lightning strikes to trees are fundamentally biological and ecological, rather than physical, and they have little or no correlation with the characteristics of the flash that initiated the damage. However, given the fact that millions of lightning flashes terminate on trees each year, it could be important to understand the phenomenon better.

Acknowledgements

The authors would like to acknowledge the valuable additional comments provided by Niko Porjo. We thank the following people for the photographs: Ilkka Juga, Mats Kommonen, Kim Lund, Eero Karvinen, Matti Mäkelä and Tero Pajala.

References

1. DeCoursey D.G., Chameides W.L., McQuigg J., Frere M.H., Nicks A.D. 'Thunderstorms in agriculture and in forest management' in E. Kessler (ed.). *Thunderstorms in Human Affairs*. University of Oklahoma Press; 1983.
2. Price C., Rind D. 'The impact of a $2xCO_2$ climate on lightning-caused fires'. *J. Climate* 1994;7:1484–94.

3. Mäkelä J.S., Karvinen E., Porjo N., Mäkelä A., Tuomi T.J. 'Attachment of natural lightning flashes to trees: Preliminary statistical characteristics'. *J. Lightning Res.* 2009;**1**:9–21.
4. Latham D., Williams E. 'Lightning and forest fires' in E.D. Johnson, K. Miyanishi (eds.). *Forest Fires*. San Diego, CA: Academic Press; 2001. pp. 375–418.
5. Weber M.G., Stocks B.J. 'Forest fires and sustainability in the boreal forests of Canada'. *Ambio.* 1998;**27**:545–50.
6. Golde R.H. *Lightning Protection*. Edward Arnold; 1973.
7. Golde R.H, *Lightning, Volume 2: Lightning Protection*. Academic Press; 1977.
8. Rakov V.A., Uman M.A. *Lightning: Physics and Effects*. Cambridge University Press; 2003.
9. Uman M.A. *The Art and Science of Lightning Protection*. Cambridge University Press; 2008.
10. Covert R.A. 'Why an oak is often struck by lightning: A method of protection of trees against lightning'. *Monthly Weather Review.* October 1924, pp. 492–93.
11. Taylor A.R. 'Lightning damage to forest trees in Montana'. *Weatherwise.* April 1964, pp. 61–65.
12. Tutin C.E.G., White L.J.T., Mackanga-Missandzou A. 'Lightning strike burns large forest tree in the Lopé Reserve, Gabon'. *Global Ecology and Biogeography Lett.* 1996;**5**:36–41.
13. Ivanova G.A., Ivanov V.A. 'Fire regimes in Siberian forests'. *International Forest Fire News* 2005;**32**:67–69.
14. Podur J., Martell D.L., Csillag F. 'Spatial patterns of lightning-caused forest fires in Ontario, 1976–1998'. *Ecological Modelling* 2003;**164**:1–20.
15. Larjavaara M. *Climate and Forest Fires in Finland – Influence of Lightning-Caused Ignitions and Fuel Moisture*. Department of Forest Ecology, Faculty of Agriculture and Forestry; 2005.
16. Gromtsev A. 'Natural disturbance dynamics in the boreal forests of European Russia'. *Silva Fennica* 2002;**36**(1):41–55.
17. Wotawa G., Trainer M. 'The influence of Canadian forest fires on pollutant concentrations in the United States'. *Science* 2000;**288**(5464):324–28.
18. Smith J.A., Baker M.B., Weinman A. 'Do forest fires affect lightning?' *Q.J.R. Meteorol. Soc.* 2003;**129**:2651–70.
19. Price C., Rind D. 'The impact of a $2 \times CO_2$ climate on lightning-caused fires'. *J. Climate* 1994;**7**:1484–94.
20. Taylor A.R. 'Lightning and trees' in R.H. Golde (ed.). *Lightning, Volume 2: Lightning Protection*. Academic Press; 1977.
21. Holle R.L., López R.E., Arnold L.J., Endres J. 'Insured lightning-caused property damage in three Western states'. *J. Appl. Meteorology* 1996;**35**:1344–51.
22. Mousa A.M., Srivastava K.D. Effect of shielding by trees on the frequency of lightning strokes to power lines. *IEEE Trans. Power Deliv.* 1988;**3**(2):724–32.
23. Sakae M., Asakawa A., Ikesue K., Shindo T., Yokoyama S., Morooka Y. 'Investigation on lightning attachment manner by use of an experimental distribution line and a tree'. *Proceedings of IEEE/PES Transmission and Distribution Conference Exhibition 2002*; Asia Pacific, 2002. pp. 2217–22.

24. Orville R.E. 'Photograph of a close lightning flash'. *Science* 1968;**162**:666–67.
25. Saraoja E.K. 'Lightning earths' in R.H. Golde (ed.). *Lightning, Volume 2: Lightning Protection*. Academic Press; 1977.
26. Simpson W., TenWolde A. *Physical Properties and Moisture Relations of Wood, Wood Handbook – Wood as an Engineering Material*. US Department of Agriculture, Forest Service, Forest Products Laboratory; 1999.
27. Heidler F., Diendorfer G., Zischank W. 'Examples of trees severely destructed by lightning'. Presented at *International Conference on Lightning and Static Electricity (ICOLSE)*; Seattle, USA; 2005.
28. Taylor A.R. 'Ecological aspects of lightning in forests'. *Proceedings of the 13th Tall Timbers Fire Ecology Conference*; Tallahassee, Florida, 1974. pp. 455–82.
29. Taylor A.R. 'Diameter of lightning as indicated by tree scars'. *J. Geophys. Res.* 1965;**70**(22):5693–94.
30. Taylor A.R. 'Lightning effects on the forest complex'. *Proceedings of the Annual Tall Timbers Fire Ecology Conference*; Tall Timbers Research Station, Tallahassee, 1969. pp. 127–50.
31. Goyer G.G. On the mechanism of bark damage and forest fire inception by lightning discharge, 1966; quoted in Taylor [20].
32. Granström A. 'Spatial and temporal variation in lightning ignitions in Sweden'. *J. Vegetation Science* 1993;**4**:737–44.
33. Larjavaara M. 'Spatial distribution of lightning-ignited forest fires in Finland'. *Forest Ecology and Management* 2005;**208**:177–88.
34. Latham D.J., Schlieter J.A. *Ignition Probabilities of Wildland Fuels Based on Simulated Lightning Discharges*. United States Department of Agriculture, Forest Service. Research paper INT-411, 1989.
35. Fuquay D.M., Taylor A.R., Hawe R.G., Schmidt C.W., Jr. 'Lightning discharges that caused forest fires'. *J. Geophys. Res.* 1972;**77**:2156–58.
36. Fuquay D.M., Taylor A.R., Hawe R.G. 'Characteristics of seven lightning discharges that caused forest fires'. *J. Geophys. Res.* 1967;**72**:6371–73.
37. Larjavaara M., Pennanen J., Tuomi T.J. 'Lightning that ignites forest fires in Finland'. *Agricultural and Forest Meteorology* 2005;**132**:171–80.
38. Darveniza M., Zhou Y. 'Lightning-initiated fires: Energy absorbed by fibrous materials from impulse current arcs'. *J. Geophys. Res.* 1994;**99**(D5):10663–70.
39. Anderson N.H., Anderson D.B. 'Ips bark beetle attacks and brood development on a lightning-struck pine in relation to its physiological decline'. *The Florida Entomologist* 1968;**51**(1):23–30.
40. Rykiel J.R., Coulson R.N., Sharpe P.J.H., Allen T.F.H., Flamm R.O. 'Disturbance propagation by bark beetles as an episodic landscape phenomenon'. *Landscape Ecology* 1988;**1**(3):129–39.
41. Magnusson W.E., Lima A.P., Lima O.D. 'Group lightning mortality of trees in a Neotropical forest'. *J. Tropical Ecology* 1996;**12**(6):899–903.
42. Payne F.F. 'Ground markings by lightning'. *Monthly Weather Review* June 1928, p. 216.
43. Johnston E.S. 'Lightning injury in a potato field'. *Monthly Weather Review* August 1920, p. 452.

44. Jones L.R., Gilbert W.W. 'Lightning injury to cotton and potato plants'. *Monthly Weather Review* March 1915, p. 135.
45. Wells E.L. 'Standing wheat fired by lightning'. *Monthly Weather Review* August 1920, p. 452.
46. Rorig M.L., Ferguson S.A. 'Characteristics of lightning and wildland fire ignition in the Pacific Northwest'. *J. Appl. Meteorology* November 1999;**38**:1565–75.

Chapter 19
Lightning warning systems
Martin J. Murphy, Kenneth L. Cummins and Ronald L. Holle

19.1 Introduction

Previous chapters have dealt primarily with lightning protection of buildings, power systems and equipment, and principally from the point of view of minimizing or avoiding direct or induced damage from direct or nearby lightning strikes. In this chapter, we change the focus to discuss primarily lightning protection of people through providing advance notice of the threat of lightning. Although the primary focus of this chapter is on human safety in the presence of lightning, this material also has relevance in other areas of lightning protection and avoidance. For example, even though modern aircraft are capable of withstanding a triggered lightning discharge, other hazards posed by thunderstorms are a more serious threat. Examples are the severe thunderstorm downdrafts that may be encountered by aircraft upon take-off or landing and the icing and turbulence associated with the clouds aloft. In addition, the whole class of lightning protection methods known as 'active protection' [1] is based on shutting down sensitive systems or processes in advance of the presence of lightning.

A lightning warning system as defined in this chapter consists of two components: (i) a device or collection of devices for detecting lightning or a thunderstorm-related phenomenon and (ii) an application or algorithm that uses the data from the detection system to provide the lightning warning. It is critical to note that the detection technique can both set bounds on and be bounded by the application. That is, certain warning algorithms cannot be practised using some detection techniques, but at the same time, the desired application or warning requirement can and should guide the decision about the detection technique to be used. Relevant detection techniques are discussed briefly in Section 19.2. A more thorough discussion of lightning detection and location systems is given by Cummins and Murphy [63]. Sections 19.3 to 19.6 of this chapter deal primarily with the applications or algorithms component of the lightning warning system. Section 19.3 gives a discussion of the two general frames of reference in which lightning warning information can be derived. Section 19.4

discusses performance measures for lightning warning algorithms. Section 19.5 assesses the performance of a couple of different warning algorithms using the metrics described in Section 19.4. Finally, Section 19.6 deals with applying the concepts from earlier Sections in risk assessment and decision-making processes.

19.2 Thunderstorm lifecycle and associated detection methods

In this section, we briefly describe the lifecycle and electrical behaviour of a typical thunderstorm. We also provide some specific examples of the detection techniques used in lightning warning systems. The detection methods may be broken into two broad categories: stand-alone, or single-point, detection systems that typically have a fairly limited range (often by design), and systems with a much wider spatial coverage, which often involve networks of sensors at different locations. Wide-area, network-based lightning detection systems have been described in more detail in [63]. In this section, we describe single-point lightning detection techniques in more detail.

19.2.1 Thunderstorm life cycle

19.2.1.1 Convective development and electrification

The development of a thunderstorm cell typically begins with rising moisture forced by updrafts, the resulting separation of charge due to collisions between frozen precipitation (graupel and small hail) and ice crystals, and the organization of this separated charge due to the different sizes, and hence fall speeds, of the charged particles. Although the amount and polarity of charge transferred during these collisions depend on temperature and a number of other factors, in most ordinary thunderstorms, the charge separation process leads to a simple tripolar arrangement of charge with positive charge near the top of the cloud (typically 8–12 km in summer thunderstorms), a layer or region of negative charge in the middle at temperatures of -10 to -20 °C (typically 4–8 km in summer thunderstorms), and a smaller region of positive charge at or near freezing level [2].

Modern weather radar systems are capable of identifying the conditions necessary for charge separation, and therefore have the potential to provide early warning for thunderstorms. Dual-polarization radars are able to provide information about particle type, which relates more directly to the charging mechanism. A less direct radar-based correlate with electrification is the availability of sufficient precipitation at the appropriate temperature levels. This can be inferred from volumetric radar scans by observing high reflectivity levels (usually ≥ 30 dBZ) at or above the -10 °C altitude. However, all radar-based measurements are limited by the relative location of the radar with respect to the developing thunderstorm, and by the vertical resolution (number of 'tilts' in the radar volume scan [3]).

The most direct ground-based measure of electrification is obtained from devices that measure the near-d.c. electric field produced at the ground. The most common measuring device is the electric field mill (EFM). During charge separation, the electric field will begin to deviate from a low-level negative 'fair-weather' field of $100-200$ V m^{-1}, typically reversing polarity and increasing to levels greater than

1 000 V m^{-1} over several minutes before the first lightning flash occurs. The polarity of the field at the ground is defined by the superposition of fields produced by the distribution of charges as a function of height and the distances of those charges with respect to the EFM [4].

19.2.1.2 Early stages of lightning activity

The first lightning flash in a storm/cell follows the initial electrification, typically within ∼5 min. In most storms, the first flashes occur exclusively in the cloud, neutralizing charge between the dominant upper positive and negative charge regions. We refer to these as cloud discharges. These discharges are most readily detected by VHF 'total lightning' detection or mapping systems, which respond to VHF emissions produced by initial electrical breakdown or self-propagating electrical discharges in existing conductive channels [5]. Some pulse-like components of cloud flashes can be detected by lightning sensors operating in the VLF–LF frequency range, but very few of these pulses are large enough to be detected and located. (See [63] for more details about lightning detection systems.)

Cloud-to-ground (CG) flashes typically follow the first cloud flashes with a lag time ranging anywhere from a few minutes to over one hour (depending on storm type) and are associated with a more mature phase of the storm lifecycle. However, depending on geographic location, as many as one-quarter of all storms will produce a CG flash as the first lightning discharge [6]. CG flashes are easily detected using wide-area networks of VLF–LF sensors as well as by a single-point lightning sensor within a limited range.

19.2.1.3 Late stages of lightning activity

The late stages of a thunderstorm include the mature phase, where CG and cloud flash rates are usually at a maximum, and the dissipation phase, where lightning rates diminish rapidly and eventually cease. By the time a storm reaches the mature phase, the storm-onset element of lightning warning is over, and all protective activities should have been initiated. The remaining lightning warning task is to determine when the storm has dissipated and it is safe to return to normal activities. The rapidly diminishing lightning rate that occurs in the dissipation phase will frequently give people a false sense of safety. In fact, this period in the thunderstorm lifecycle has been shown to be just as deadly as all other phases [7], and determining the end of the storm requires the widest range of instrumentation and insight.

During the dissipation phase of an ordinary thunderstorm, charge is no longer being separated actively, but regions of electrified cloud can still exist. A certain class of storm called a mesoscale convective system (MCS) has a large area of stratiform cloud and precipitation, usually located behind a main line of thunderstorm cells (the 'convective line'). In the stratiform regions, active charge separation can take place for long periods of time. Large near-d.c. electric fields can sometimes exist under the charged stratiform regions, as long as the charge structure is not so complex that the fields due to multiple charge layers cancel themselves at the ground. These elevated fields can be detected using electric field mills, signalling the potential for additional lightning discharges. During the dissipating phase of ordinary storms and especially in the stratiform regions of mesoscale convective systems (MCSs), two unique types of lightning flashes co-exist. Long horizontal cloud discharges known as 'spider lightning' discharges can be seen travelling through and below the extensive stratiform clouds, sometimes having overall flash extents exceeding 100 km. These spider flashes are frequently associated with one or more isolated CG flashes with large peak currents in the range 50–200 kA.

19.2.2 Associated detection methods

19.2.2.1 Detection of initial electrification

Devices that detect cloud electrification by measuring the near-d.c. electric field produced at the ground, such as EFMs [4] and other field-change devices [8] are inherently omnidirectional. These devices respond to the vertical component of the field produced at the ground by the charge distribution aloft. The vertical component of the field, in turn, weights the charges inversely with the cube of distance because of the conducting lower boundary provided by the earth's surface (represented by image charges). For a distribution of charges in the clouds aloft, the field observed by an EFM is the superposition of the fields due to the individual charges and their corresponding image charges. Because of the strong distance dependence, the effective range of EFMs and other electrostatic field change devices is generally in the range 5–8 km, although these devices may respond to strong fields at distance of tens of kilometres. The effective range is also limited by instrument sensitivity.

Examples of commercially available EFMs are shown in Figure 19.1. Figure 19.1a has a classical upward-looking configuration, clearly showing the rotor and stator assemblies that are used to convert the d.c. field to a (measurable) a.c. current

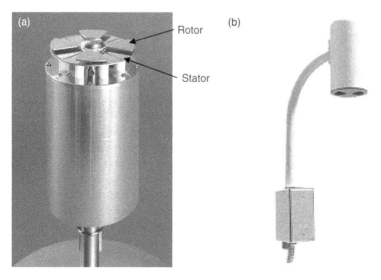

Figure 19.1 Electric field mills (EFMs). (a) Upward-looking model, showing locations of rotor and stators. (b) Downward-looking model.

[9, Appendix C]. Some modern EFMs have a downward-looking configuration, as shown in Figure 19.1b, which has the advantage of protecting the measured fields from direct contamination when charged raindrops strike the stators. EFMs are sometimes networked together over small regions such as the NASA Kennedy Space Center and can be used to identify regions of high charge density and to localize changes in the charge distribution produced by lightning flashes [10–12].

19.2.2.2 Single-point lightning detection sensors

There is a large class of single-point lightning sensors that detect fast transient field changes produced by CG strokes and cloud discharges. These devices employ various characteristics of these transient electric and/or magnetic fields to estimate the range from the sensor to the discharge. Most of the commercially available devices have an effective range of 50–100 km. The ranging accuracy varies with the specific device and method, but is generally between 10 and 40 per cent of the estimated range.

One example of a commercially available sensor is shown in Figure 19.2. This device operates in the VLF/LF frequency range and employs two orthogonally

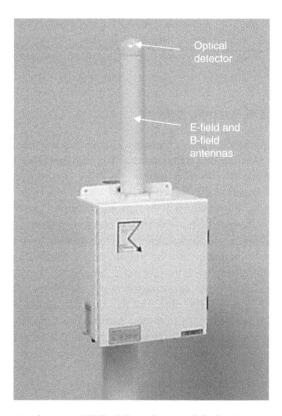

Figure 19.2 *A single-point VLF–LF and optical lightning sensor. The vertical antenna assembly contains both electric and magnetic field antennas. The optical sensor sits atop the antenna assembly.*

oriented magnetic loop antennas to determine the direction to the lightning discharge. It produces a range estimate using the vertical electric field. Non-lightning events are rejected by requiring time coincidence between the electric field signal and light pulses detected by an optical detector located at the top of the sensor.

19.2.2.3 Lightning detection networks

Network-based lightning detection systems have been discussed in detail in [63]. Briefly, networks operating in the LF and VLF bands typically have sensors separated by distances of 100–300 km and are designed to respond to the signals that propagate over the surface of the earth in the form of a ground wave. The largest signals in this band by far are produced by the return strokes in CG flashes, and therefore these are primarily what such networks are able to detect and locate. Other networks detect emissions in the VHF band. These propagate by line of sight, which restricts the maximum separation between sensors. The emissions in the VHF band, however, are numerous and are produced during all portions of a cloud discharge as well as the leaders and in-cloud components of CG flashes. These emissions allow for the detailed mapping of the spatial extent of lightning activity within the cloud. Because all flashes produce VHF emissions and because of the capability for spatial mapping, lightning detection networks that operate in the VHF band are often referred to as 'total lightning mapping systems'. As we will see later in this chapter, the capabilities of these systems have important applications in the lightning warning problem.

19.3 Examples of warning systems

Section 19.2 introduced the various measurement devices and methods in the context of describing the lifecycle of a thunderstorm. We now introduce the application, or algorithm, component of lightning warning systems. To be useful to consumers, warning information necessarily has to be point- or area-specific. However, broadly speaking, the warning information may be derived in either of two frames of reference, one that is fixed on the specific point or area of concern, and another that moves with cells or storms but produces, as its output, targeted warning information for specific locations or areas. Some techniques or algorithms are adaptable to either frame of reference. In the following sections, we refer broadly to the warning algorithms described as either fixed-point or storm-following algorithms, depending on their normal reference frame. We also point out the detection techniques best suited for operating in one or both of these reference frames. The limitations and challenges specific to the various warning algorithms described in this Section are discussed in Section 19.4.

19.3.1 Fixed-point warning applications

The simplest, and perhaps the most widely used, fixed-point lightning warning technique involves only human observation, with no automated meteorological observations of any kind. This method is known as the '30–30' rule and is highly recommended for personal or small-group safety applications [13]. The first 30 in the rule refers to lightning onset, and the second refers to lightning cessation. If an observer sees lightning, the method recommends that he or she monitor the time from the visible flash until thunder is heard. If that time is less than 30 s, then the lightning struck within 10 km (6 mi), and the observer is advised to seek a lightning-safe location. This first part of the 30–30 rule is based on the observation that successive flashes in storms are randomly distributed in space with a particular distribution of separation distances. A distance of 10 km bounds the 80–90th percentile value of the distribution of flash separation distances for isolated cells and about the 60th percentile value in one large MCS [14]. The second 30 in the 30–30 rule encourages observers to wait 30 min after hearing the last thunder before returning to outdoor activity. The motivation for this value is addressed in Section 19.4.2.2 of this chapter, where we talk about the lightning cessation problem.

An automated fixed-point warning algorithm that provides information for objectively assessing the threat of lightning can be implemented using a single-point lightning detection sensor. Figure 19.3 shows lightning flashes detected by a network-based lightning location system (the U.S. National Lightning Detection Network, NLDN) and a conventional spatial representation of the same activity by a VLF/LF single-point sensor that provides range and direction information. Figure 19.3a shows the lightning locations determined by the network in a 15-min interval in northern Texas, USA. Each 'dot' is the location of a CG flash determined by the network, and red lines represent the highway system in this area. Figure 19.3b shows the representation of the flashes produced by the single-point sensor. The star in the centre represents the location of the sensor. The effective detection range of the sensor is broken down into 17 sectors: eight azimuthal sectors in each of two different range rings with respect to the sensor (8–16 km, and 16–50 km) and one overhead sector (0–8 km). The lightning flash count in each sector is colour-coded, with green representing no lightning, yellow representing 'moderate' lightning, and red representing 'frequent' lightning. In this application, the count of flashes is also shown in each bin.

Automated fixed-point algorithms may also use network-based lightning detection information. In addition, they may involve more complex combinations of network-based lightning data with EFM and/or radar observations. When network-based lightning detection systems are used for lightning warning purposes, there is always some algorithm that determines when a warning begins based on when a threshold, or some combination of thresholds, is exceeded. The end of a warning is typically determined by waiting a certain time after the conditions for a warning are no longer satisfied. This end-of-warning time period is referred to here as the 'dwell time'. A common fixed-point algorithm using network-based lightning detection (sometimes with EFMs) involves establishing a central area of concern (AOC) around the fixed point of interest and at least one warning area (WA) surrounding the AOC. Figure 19.4 gives an illustration of the algorithm configuration with a hypothetical set of lightning events moving through the WA and approaching the AOC. In the context of automated fixed-point algorithms, we refer to a 'storm' not in terms of a physical thunderstorm cell or cluster of cells but rather in terms of a continuous period of lightning activity within

866   Lightning Protection

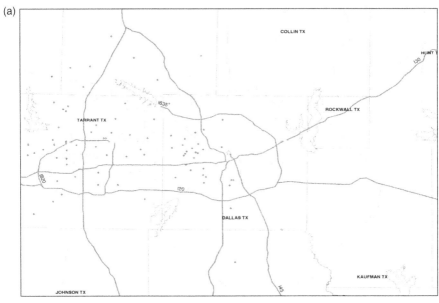

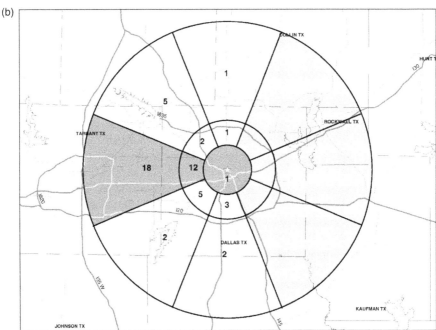

Figure 19.3 (a) Lightning locations determined by the US NLDN over a 15-min period. Map background is of northern Texas, USA (b) Representation of the same information by an application receiving information from a single-point LF/VLF lightning sensor.

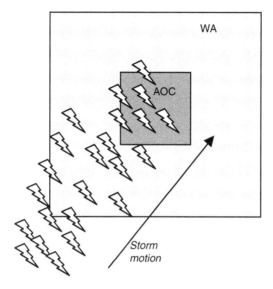

Figure 19.4 Illustration of algorithm configuration for a fixed-point warning method using network-based lightning detection data. AOC = area of concern. WA = warning area.

the AOC and WAs. The definition of 'continuous' is determined by the dwell time: if the interval between flashes is less than the dwell time, then the flashes are part of the same 'storm'. Because of the limitations inherent in using CG lightning data alone for warnings, the method is often augmented with the addition of EFMs, cloud lightning detection information, and/or radar data. Both EFMs and cloud lightning detection can be used to provide a warning in cases when a thunderstorm develops directly overhead. Further details are given in Sections 19.4 and 19.5 of this chapter.

19.3.2 Storm-following algorithms

Storm-following algorithms require the ability to track storms as they move through a region. These methods therefore require detection systems capable of observing thunderstorms over a large area, at least the size of a region that extends 100 km or more from any particular place where warning information is needed. For this reason, EFMs and other electrification-related detection devices, as well as short-range single-point lightning sensors, are not capable of operating in this regime. Network-based lightning detection systems and radars are the most commonly used tools for this type of warning algorithm.

There are many storm-following algorithms in use today that track radar-identified storms or cells, and in many cases, the ultimate objective is short-term thunderstorm forecasting. The simplest storm-following method involves extrapolating the positions of cells using a recent history of motion. Algorithms such as Storm Cell ID and Tracking (SCIT) used in the US National Weather Service's NEXRAD network

[15], Thunderstorm Information, Tracking, Analysis and Nowcasting (TITAN) [16], and the algorithm used as part of the ASPOC system in France [17] employ this methodology. SCIT, for example, shows the current and past positions of cells, and uses those to determine the recent cell motion vectors for a short-term extrapolation of cell positions. More sophisticated systems not only track cells, but they also monitor the cells for trends in the radar (and lightning, if available) characteristics to provide some short-term estimate of expected cell growth and decay. Such systems include the NCAR AutoNowcaster [18], a growth-decay tracking algorithm [19] that feeds the MIT-Lincoln Laboratory Tactical Convective Weather Forecast system [20], and the National Convective Weather Forecast [21].

Radar information may be used to estimate the onset of electrical activity within a cell. This is based on the fact that the dominant electrification mechanism requires precipitation-sized ice particles in regions of the cloud where the temperature is below 0 °C that also contain small ice crystals and supercooled liquid water droplets. Substantial reflectivity in this region of the cloud, known as the mixed-phase region, is usually an indicator of when the proper ingredients for electrification are present. A large number of rule-based algorithms have been developed in different parts of the world that involve looking for reflectivity to exceed some threshold (typically 20–40 dBZ) at or above some altitude corresponding to an environmental temperature level (typically between −10 and −20 °C) where a dominant negative charge layer is located in most ordinary thunderstorms [10,22]. Discussion of the specific performance of these methods is reserved for Section 19.5. Dual-polarization radar measurements and associated particle classification algorithms make it possible to determine specifically when the necessary precipitation-sized ice particles are present in the proper region of the cloud and in what quantity.

Lightning detection data can also be used together with, or even as a substitute for, radar data in a simple cell tracking algorithm, as has been demonstrated by Lojou and Cummins [23,24] and Kononov and Yusupov [25]. Even if radar data alone are used to do the cell tracking, lightning flashes may be assigned to cells, and information about which cells contain lightning and which do not can constitute a form of lightning warning information, although it is not specifically directed to the task. The use of storm-following methods specifically for the lightning warning problem has been reduced to practice by Conway *et al.* [26], Saxen and Mueller [27], and Brunza *et al.* [28]. The method described by Conway *et al.* [26] includes cell tracking with growth and decay [19] as well as model data to derive lightning threat potential forecasts at several fixed time intervals in the short term. The method of Saxen and Mueller [27] uses a fuzzy logic system to produce a new lightning potential forecast after each radar scan. No particular lead time is specified in the forecast, although a forecast is only considered successful in their analysis if it leads the first observed flash by at least 5 min. Brunza *et al.* [28] use a modified Monte Carlo method with lightning data alone to determine likely cell track positions for lightning warning.

Many lightning warning applications require only short lead times, perhaps up to 15 min or so. Extrapolations of storm positions based on cell-tracking using either radar or lightning detection data are typically valid over periods of 5–15 min for a localized thunderstorm cell, and therefore, simple storm-following algorithms are suitable for many lightning warning situations. Some applications, however, might require lead times of 1 h or more. Examples where longer lead times are useful or required include (i) large numbers of people outdoors, (ii) a large geographic area where people and/or equipment are distributed over the whole area, or (iii) a time-consuming protection procedure (e.g., shutting down a facility or complex operational process). For these situations, simple cell-tracking extrapolations are usually insufficient, and additional data are required. Surface weather observations of sufficient density (10–50 km between stations) can be used to determine when the conditions required for thunderstorm development are in progress. Watson et al. [29] and Forbes and Hoffert [30] have discussed such methods specifically in the context of lightning warning. Similar observations can also be obtained from radars, as long as there is sufficient density of radars, as has been proposed by McLaughlin et al. [31]. Radar techniques relevant to detecting pre-storm conditions include determining the low-altitude winds (Doppler techniques) and the distribution of available water vapour [32]. Mixed human–automated systems such as the NCAR AutoNowcaster [18] incorporate a variety of information to indicate areas where thunderstorms are expected to form. These methods for detecting the precursors of thunderstorm formation can extend the warning lead time to an hour or two. Beyond this time range, numerical weather prediction (NWP) is required for further extension of the lead time. The NCAR AutoNowcaster allows for combining short-term (3–6 h) NWP forecasts with the observational information discussed above. NWP model data may be used to generate forecasts of lightning out to a couple days, as demonstrated by Burrows et al. [33].

19.4 Warning system performance measures

In this section, we describe methods for measuring the performance of a lightning warning system. We begin by presenting the metrics and then describe how the metrics apply in both the fixed-point and storm-following frames of reference described in Section 19.3. Finally, this section describes some of the specific limitations of lightning warning algorithms at each stage of the warning problem.

19.4.1 Performance metrics

The essential element of verification for any warning system or method is a 2×2 contingency table [34] of the type shown in Table 19.1. Four main metrics are derived from the table:

1. The probability of detection (POD), computed as the ratio of correctly predicted cases of observed lightning to the total number of cases of observed lightning $[A/(A+B)]$;
2. The false alarm ratio (FAR), computed as the ratio of cases where lightning was predicted but did not occur to the total number of cases in which lightning was predicted $[C/(A+C)]$;

*Table 19.1 Sample of a 2 × 2 contingency table.
See text for description of statistics
derived from this table*

		Predicted	
		Yes	No
Observed	Yes	A	B
	No	C	–

3. The failure-to-warn rate (FTW), also known as 'false negative rate', which is the fraction of cases of observed lightning in which no lightning was predicted [B/(A + B)];
4. The critical success index (CSI), defined as the ratio of successful predictions to all cases of predicted or observed lightning [A/(A + B + C)].

In a standard contingency table, POD = 1 − FTW. The CSI is useful as an overall discriminator of best prediction accuracy, but because either a high FAR or a high FTW can lead to a poor CSI, we choose to focus on the FAR and FTW themselves in the following discussion.

Depending on the type of warning system or algorithm, there can be variations on how the contingency table entries are defined or slight modifications to the table itself. The single most important modification (for any of the algorithms discussed in Section 19.3) is usually to introduce the amount of lead time provided by a prediction of lightning occurrence. This modification makes the quantity A in Table 19.1 a function of lead time. The quantity B in its pure form is not a function of lead time but may become so if the pure failures to warn are augmented by adding cases in which the lead time was insufficient. Specific examples of how these metrics are applied and modified are given in the following paragraphs.

19.4.1.1 Performance metrics for fixed-point algorithms

Recall that Figure 19.4 illustrated a fixed-point lightning warning algorithm for which the corresponding detection method was a lightning detection network. In the following discussion, the lightning detection network is assumed to provide only CG lightning locations. In that case, the ideal situation is to have lightning begin within the WA prior to its onset within the AOC in order to have advance notice of the threat to the AOC. In practice, this does not always occur, and when it does occur, it occurs with varying amounts of lead time. Figure 19.5 shows a histogram and cumulative distribution of the time interval between the first flash in the WA and the first flash in the AOC for an analysis done with the US NLDN. These results are a composite over four different sites in the United States that were

Lightning warning systems 871

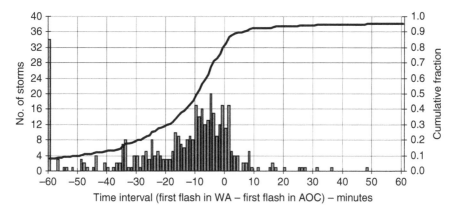

Figure 19.5 Distribution and cumulative distribution of the time interval between the first flash in the warning area (WA) and the first flash in the area of concern (AOC) for 417 storms in four locations in the United States as detected by the US NLDN

also analysed by Holle *et al.* [35]. For this particular set of calculations, the AOC and WA were concentric circles with the AOC having a radius of 3 km and the WA having an outer radius of 10 km. In Figure 19.5, negative values on the time axis indicate that the first flash in the WA occurred before the first flash in the AOC; that is, a negative value of time indicates that there was some advance notice of the onset of lightning in the AOC.

Tables 19.2 to 19.4 show three different contingency tables based on the data in Figure 19.5. In Table 19.2, we consider any lead time to be a successful prediction. In Table 19.3, successful predictions include only those where the lead time was at least 3 min, but the failures-to-warn (upper right-hand entry) are the same as in Table 19.2. That is, Table 19.3 applies a strict definition of a failure to warn (no lead time at all). Finally, Table 19.4 makes one further modification to Table 19.3, which is to consider all lead times less than 3 min to be failures to warn in addition to those storms with zero lead time. Table 19.5 summarizes the POD, FAR and

Table 19.2 Contingency table derived from Figure 19.5, considering all storms in which the WA provided some lead time to be successful predictions

		Predicted	
		Yes	No
Observed	Yes	345	72
	No	624	–

Table 19.3 Contingency table derived from Figure 19.5, considering all storms in which the WA provided at least 3 min of lead time to be successful predictions, but not modifying the unsuccessful prediction count

		Predicted	
		Yes	No
Observed	Yes	305	72
	No	624	–

Table 19.4 Contingency table derived from Figure 19.5, considering all storms in which the WA provided at least 3 min of lead time to be successful predictions and all storms where the WA provided less than 3 min of lead time to be unsuccessful predictions

		Predicted	
		Yes	No
Observed	Yes	305	112
	No	624	–

Table 19.5 Summary of POD, FAR and FTW derived from Tables 19.2 to 19.4

	Table 19.2	Table 19.3	Table 19.4
POD	$345/(345+72) = \mathbf{0.827}$	$305/(305+72) = \mathbf{0.809}$	$305/(305+112) = \mathbf{0.731}$
FAR	$624/(624+345) = \mathbf{0.644}$	$624/(624+305) = \mathbf{0.672}$	$624/(624+305) = \mathbf{0.672}$
FTW	$72/(345+72) = \mathbf{0.173}$	$72/(305+72) = \mathbf{0.191}$	$112/(305+112) = \mathbf{0.269}$

FTW values derived from Tables 19.2 to 19.4. Note that in Tables 19.2 and 19.4, the sum of the first row is 417 storms, which is the actual number of storms that produced lightning within the AOC. However, in Table 19.3, the sum of the first row is only 377 because storms with insufficient lead time (<3 min in this example) have been excluded. If we say that insufficient lead time is tantamount to a failure to warn, as in Table 19.4, then the higher FTW computed from Table 19.4 may be viewed as the appropriate value. As for FAR, the 40 storms that produced lead times less than 3 min were not false alarms. Lightning did occur within the AOC in those 40 cases, but they do not show up in the left-hand column of either Table 19.3 or 19.4. Therefore, the most appropriate determination of FAR comes from Table 19.2. Thus, we see that the most useful performance metrics may involve choosing from a combination of the possible modifications to the contingency table, depending on how performance is to be quantified.

19.4.1.2 Performance metrics for storm-following algorithms

Many storm-following algorithms depict a lightning threat area at some future time based on the cell or storm tracking algorithm. In this case, performance may be measured using the same basic contingency table method but with some modifications. One approach is to break the entire forecast area down into a series of small grid cells and create a contingency table based on the grid cells themselves. Figure 19.6 shows an example of this analysis. The grey, shaded area with downward cross-hatching represents the estimated threat area, and the red areas with upward cross-hatching represents those grid cells that had lightning at the valid time of the

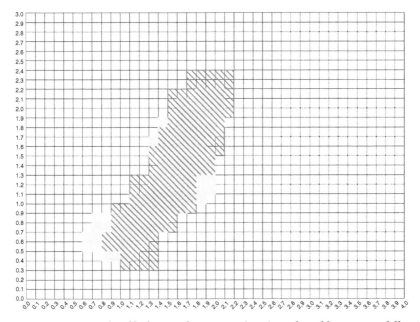

Figure 19.6 *Example of lightning threat area (grey) predicted by a storm-following algorithm for some future time and the actual lightning coverage area (cross-hatching) at that future time. Verification is done by setting up a grid and counting cells that fall into the threat area, verification area or both.*

Table 19.6 Contingency table for grid-based verification of the storm-following method in the example of Figure 19.6

		Predicted	
		Yes	No
Observed	Yes	93	17
	No	44	–

warning. Cells where the two overlap obviously represent successful predictions, while cells with cross-hatching only represent false alarms and cells with grey shading only represent failures to warn. The contingency table is thus constructed from the counts of these three categories of grid cells, and the results for the specific example shown in Figure 19.6 are given in Table 19.6. The POD in this example is 0.845, the FAR is 0.321 and the FTW is 0.155.

The foregoing method of verification based on a grid analysis has its strengths and limitations. The method can be quite good for indicating whether the predicted shape and size of the threat area are approximately correct as long as the placement of the forecast threat area coincides well with the observations. If shape, size and placement are all approximately correct, POD will be high and both FAR and FTW low. When the forecast area's placement is approximately correct but the size of the area is too large, then both POD and FAR are high and FTW is low. Conversely, if FTW is high but FAR is low or nearly zero, then the threat area is significantly smaller than the actual lightning occurrence area. If both FTW and FAR are high, then at a minimum, the placement of the threat area is poor, and its shape and size may be poor as well. Brown and colleagues [36] pointed out that this grid-based verification method falls apart completely when the forecast and observation areas do not overlap at all in their placement. In those cases, POD is exactly zero and FTW and FAR are exactly 1 regardless of how close in space and time the forecast came to reality, and the standard contingency table offers no way to measure the error in the forecast. Brown and colleagues introduced the idea of an object-based verification that gives metrics for a 'close' forecast. Given that the objective of lightning warning is usually to provide non-expert users with a binary decision, the concept of a 'close' forecast, although certainly applicable in many settings, may not be of much use.

Other possible modifications to the verification technique for storm-following algorithms involve applying either a temporal or spatial smoothing to the observed lightning data field. Temporal smoothing is designed mainly to eliminate an overly strict definition of a false alarm in the case of moving storms. Spatial smoothing primarily avoids an overly strict definition of a failure to warn, especially in situations where the rate-density of CG flashes is low relative to the grid size. These modifications, in turn, change the contingency table results. The 'closeness' issue discussed by Brown and colleagues [36] is addressed in their work by a fuzzy logic algorithm that attempts to determine the best match between forecasts and observations based on a variety of characteristics. The technique could also be implemented by computing spatial and/or temporal cross-correlations of forecast and observation grids valid at different times and then evaluating the contingency table at each spatial or temporal lag/offset value.

19.4.2 Specific challenges at different stages of the warning problem

Equipped with the basic performance metrics, we can now take a more detailed look at the warning problem. In this section, we consider two different aspects of the thunderstorm and lightning warning problem, lightning onset and lightning cessation, and their unique challenges. Each of these is described in general, and then we point out the specific issues in the context of the performance metrics discussed in the previous section.

19.4.2.1 Lightning onset

Storms that are already electrified and are moving towards an observer's location usually present little difficulty in terms of lightning onset, as long as they are within the valid extrapolation time period, meaning that rapid storm decay does not occur before the storm arrives at the AOC. The most difficult cases are the storms that develop overhead.

Lightning onset in a developing storm is necessarily preceded by the separation of charge in the cloud. Thus, the most reliable precursor to lightning onset is a measure of cloud electrification. Of the techniques available to do this, the best potential POD is provided by the measurement of static electric field because it is the direct result of this charge separation. Electrostatic field measurements (by EFMs typically) are appropriate for a fixed-point warning algorithm because, as noted above, the target issue is to detect storms that form directly overhead. Electrostatic field measurements can have many false alarms, caused in part by electrified clouds that produce no lightning, which can be a significant contributor under some conditions, and in part by other phenomena such as blowing dust, snow, surf spray and exhaust [8]. Normally, a warning based on electric field requires that the field cross a threshold that is indicative of significant electrification. The resulting POD can be quite high if the threshold is tuned well, but both POD and FAR are dependent on the threshold value. Pierce [8] advocated against the use of thresholds and for using the time trend of the field to identify an approaching storm. Nevertheless, thresholds are used operationally, and Hoeft and Wakefield [37] confirmed Pierce's inference of a high FAR, but they also observed a high POD for a commonly used threshold of 2 kV m^{-1}.

In storm-following algorithms, the initial electrification of the cloud is not measured directly but is rather inferred using radar. As discussed in Section 19.3, the radar has to sample the cold parts of the cloud aloft, where charge separation occurs. This indirect technique for inferring electrification presents a wide range of challenges. First, the sampling resolution of the radar in both space and time can limit its effectiveness. Clouds can become electrified and produce lightning within

5 min of the first appearance of precipitation-sized ice in the proper region of the cloud [38], so a radar volume scan time of 5 min or more can result in missing the onset of electrification and lightning. In terms of spatial sampling, operational radar scans typically only include low elevation angles, perhaps up to 20–25° or so. This leaves a 'cone of silence' over the radar where the development of precipitation aloft cannot be detected. At longer distances from the radar, the increased vertical separation between the beams can result in missing the first appearance of precipitation aloft. Additionally, radars operating in mountainous terrain face beam blockage by the terrain. Finally, as mentioned before, conventional reflectivity is not a direct measure of electrification, so it can produce lower POD and higher FAR than electric field measurements. Despite these limitations, several specific studies have explored using conventional reflectivity as a warning method for CG lightning, and their results are summarized in Section 19.5. Dual-polarization radar, with associated particle classification algorithms, can specifically identify when the proper ingredients for electrification are present in the cloud, thereby improving warning performance over conventional reflectivity.

In most storms, the first lightning flashes following the initial electrification process are cloud flashes. For these storms, any lightning detection system capable of detecting cloud flashes can provide advance warning of the threat of CG flashes. In some regions, however, there is evidence that as many as 20–25 per cent of cells produce a CG flash as the very first flash. In these cases, no lightning detection system can provide advance warning of the first flash if the storm develops directly overhead. However, for the majority of storms that develop elsewhere and move toward the AOC, a storm-following algorithm can still provide advance warning by tracking the storm as it moves toward the AOC.

19.4.2.2 Lightning cessation

Lightning cessation is a significant problem, both in simpler localized storm situations and in more organized large-scale MCSs that contain both a convective line, which is the predominant lightning producer, and an electrified region of stratiform precipitation. In the latter case, lightning frequency and density in the stratiform region are typically very low relative to the convective line, but the area of electrified cloud is usually significantly larger. An added complication is that some portions of the stratiform precipitation region may be producing lightning while others are not.

As in the case of lightning onset, the most direct information about whether the potential for lightning still exists is the observation of overhead electrification with EFMs. For small, localized storms, once the ingredients for electrification have fallen out of the mixed-phase part of the cloud, then the electric field and lightning activity fall off rather rapidly. As this occurs, however, there can be a period of high electric fields that swing gradually from one polarity to the other. This is referred to as the end-of-storm oscillation [39], and it can occur with little or no further lightning activity. As mentioned previously in Section 19.2, electric field measurements can be equally useful under the stratiform regions of MCSs, as long as the fields due to different charge layers do not cancel.

With regard to fixed-point algorithms, the time interval between successive lightning flashes within the AOC can be used as another means of assessing the end-of-storm lightning risk. Distributions of the time interval between successive lightning flashes have been computed for several locations in the United States using the NLDN. Figure 19.7 shows the distribution for Pittsburgh, in the northeastern United States. This graph shows the probability of having an interval between CG flashes that is longer than the value shown on the horizontal axis. These results are fairly representative of other areas in the NLDN. After a 10-min interval, there is typically a few percent probability of having another flash.

However, there is a long tail on the interval distribution, such that after 30 min, there is still typically a 0.5–1 per cent chance of having another flash. Statistics such as these were used to make the 30-min end-of-storm recommendation in the 30–30 rule for personal safety. In an automated warning algorithm, the relevant parameter is the dwell time. If the dwell time is too short, then warnings will often end too soon. If there is a subsequent flash after the dwell time has passed and the warning expires, it will essentially be a surprise, or failure to warn. Thus, the primary challenge for lightning cessation is to minimize the FTW, which can be quite high in places where MCSs are frequent because of the low density and rate of CG flashes in the stratiform region. At the same time, however, it is desirable to minimize the warning duration to allow resumption of activity. Techniques for handling lightning cessation problems can include (i) increasing the size of the AOC (which lowers the probability of successive flashes at all intervals), (ii) increasing the dwell time, and (iii) including additional datasets such as radar and total lightning mapping.

For storm-following purposes, total lightning mapping systems have been shown to provide a good indication of the electrified region because lightning discharges require the maintenance of an electric field at their tips in order to keep propagating. Where the charged area of the cell or storm system ends, so do the conditions for further propagation of discharges. Because of the low rate and density of CG flashes in an MCS

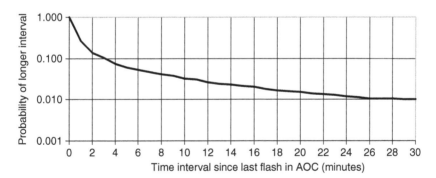

Figure 19.7 Cumulative distribution of the time interval between successive flashes in a $10 \times 10\ km^2$ AOC centred on the Pittsburgh International Airport in the United States. Eight years of NLDN data were used to produce the analysis.

stratiform region, if one observes only the return strokes from CG flashes using a VLF–LF lightning detection system, no clear spatial domain that contains these isolated CG flashes is obvious. However, simultaneous observations using total lightning mapping produce a clear picture of the threat domain. This effect is illustrated in Figure 19.8, which shows 5 min of lightning activity as detected by two different lightning detection networks in an MCS that moved from northwest to southeast across northern Texas, U.S. The grey dots in Figure 19.8 are from a VHF total lightning mapping system, and the black dots are CG flash locations from the United States NLDN. Most of the CG flashes are concentrated along the southeastern edge of the storm system. This corresponds to the convective line of the MCS. The remainder of the storm system is the stratiform region, and isolated CG flash locations are also seen in various places throughout this region. The VHF lightning mapping data show that these isolated CG flashes are clearly associated with extensive in-cloud lightning activity. The mapping of the in-cloud activity depicts a single coherent region over which isolated CG flashes can be expected. In Section 19.5, we will show how the coherent presentation of the lightning threat area by a VHF lightning mapping system improves lightning warning performance.

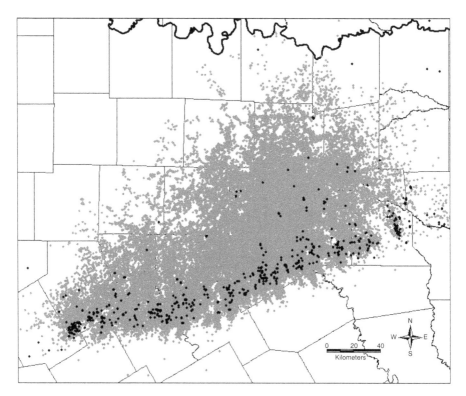

Figure 19.8 Five minutes of lightning activity in an MCS. Grey dots: VHF total lightning mapping data. Black dots: CG flash locations.

19.5 Application of performance measures to cloud-to-ground warning systems

Previously, we said that the choice of lightning detection technology or other observation method should be guided by the required level of warning performance, because it affects the values of the performance metrics (Section 19.4). The choice of detection method also determines whether a fixed-point or storm-following algorithm is the more appropriate choice for the warning algorithm, as well as the configuration of the algorithm. The objective of this Section is to show how the detection method and warning algorithm configuration interact as observed through the basic performance measures, POD, FAR and FTW, discussed in the previous section. For fixed-point algorithms, many examples of the interaction of detection method and warning algorithm exist, and in what follows, we show the influence of detection method on the algorithm and its performance. On the other hand, very few details about operational storm-following lightning warning algorithms exist in the literature, so we will concentrate primarily on what is known about the relevant observation techniques, particularly radar.

19.5.1 Assessment of a fixed-point warning algorithm

Throughout Section 19.5.1, we discuss a fixed-point warning algorithm that relies on information provided by lightning detection networks. The reader is referred back to Figure 19.4 and the associated discussion for a basic description of this type of warning algorithm. We will first discuss various lightning detection network technologies and how these impact the appropriate configuration of the algorithm, using POD, FAR and FTW to determine the best configuration in each case. Then, using a consistent detection technology, we will vary the configuration of the algorithm and show how that affects the resulting POD, FAR and FTW.

19.5.1.1 Effects of lightning detection technology

We employ two different lightning detection technologies in the examples discussed in this section. The first is the US NLDN, which consists of sensors operating in the LF/VLF frequency range. As described elsewhere in this book, the dominant signals in this frequency range are those due to the return strokes in CG flashes. The detection efficiency for CG return strokes in the interior of the NLDN is currently 65–70 per cent [40,41] and 90–95 per cent for CG flashes. Cloud discharges, on the other hand, produce much weaker signals in the LF band than CG strokes. For this reason, estimated cloud flash detection efficiency is considerably lower. However, for the examples discussed in this section, we take advantage of an enhanced test network of LF sensors embedded within the NLDN near Dallas–Fort Worth, Texas. This region of the network had an estimated detection efficiency of 20–30 per cent for cloud flashes during the case studies discussed in this section. The second major lightning detection network type is a VHF total lightning mapping system, also located in the Dallas–Fort Worth region. As discussed earlier, VHF total lightning mapping systems detect and map the spatial extent of all flashes, including cloud discharges and the in-cloud components associated with CG flashes. For all practical purposes, the system has a flash detection efficiency of 100 per cent.

Aside from the detection efficiency values stated above, there is a very significant difference between the representations of cloud discharges obtained with an LF detection network and a VHF total lightning mapping network. Figure 19.9 shows two images of a large MCS in a satellite-type projection over the south-central United States looking towards the northwest. The first image (Figure 19.9a) shows maximum column reflectivity from a single radar and points out the convective line and stratiform region of the MCS. In Figure 19.9b, two sets of lightning data are overlaid on the reflectivity data. The small grey dots are from a VHF total lightning mapping system, while the white squares are cloud discharge positions determined by an LF system. Note, how the vast majority of the LF cloud discharge positions are coincident with the convective line of the MCS. Very few of them appear outside this area. By contrast, the VHF lightning mapping system shows extensive lightning activity in the stratiform precipitation region in the northwestern portion of the MCS. This difference in the spatial representation of cloud discharges has important impacts on the warning application, as the following examples illustrate.

As mentioned above, we refer the reader to Figure 19.4 and the associated text for a discussion of the warning algorithm used in the following examples. We will use two different configurations of the algorithm, both coming from a study by Murphy and Holle [42]. In that study, the full width of the AOC was 20 km, and the full width of the WA was 40 km. The data in that study were from 20 storms in the Dallas–Fort Worth region in 2005, including a variety of different storm types. In the first example discussed here, we omit the WA and use only the AOC. If only CG flash data from the NLDN are available, this configuration provides no lead time at all, because the first CG flash within the AOC is assumed to represent a threat condition. As a result, the FTW by definition is 100 per cent, and the FAR and POD are both 0 per cent. However, it is instructive to compare this condition to the POD, FTW and FAR obtained under the same algorithm configuration using either the LF cloud discharge dataset or the VHF total lightning mapping dataset. The results are shown in Figure 19.10, which compares the POD (minimum lead time of 10 min), FTW and FAR, as well as the total duration of warnings during the 2005 storms. Figure 19.10 shows that, for CG data alone, the POD and FAR are 0 per cent and the FTW is 100 per cent.

When LF cloud discharge data are available, the POD is small but non-zero. Thus, by detecting cloud lightning even with low detection efficiency, an LF network can still provide advance warning of the onset of CG lightning in a small number of storms. The POD is significantly higher, and the FTW significantly lower, when the VHF lightning mapping system is used. The FAR is not significantly different between LF cloud discharge detection and VHF total lightning mapping (labelled 'TL' in Figure 19.10). The addition of total lightning mapping increases the total duration of warnings by ~50 per cent over LF cloud discharge detection, but that accompanies a significant reduction in FTW.

Obviously, on the basis of the results in Figure 19.10, one should conclude that the AOC-only configuration of the algorithm provides very poor results when the

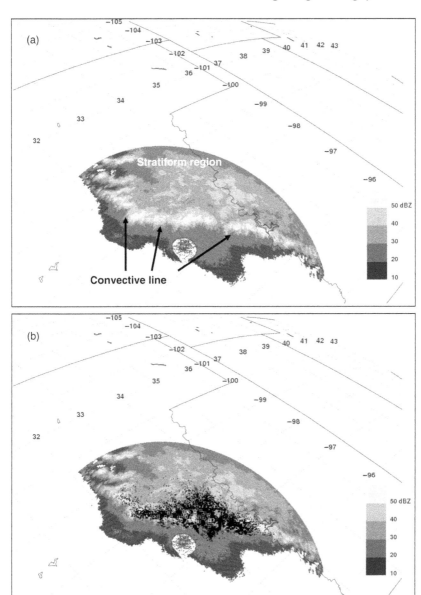

Figure 19.9 (a) Maximum column reflectivity from the National Weather Service Doppler radar at Fort Worth, Texas, USA, showing the convective line and stratiform region of a large MCS. The view is towards the northwest. The scale at the right is reflectivity in dBZ. The circular area cut out of the reflectivity near the bottom is close to the radar, and the beams do not intersect the cloud. (b) Same reflectivity image with LF cloud discharges (white squares) and VHF total lightning mapping information (grey dots) superimposed.

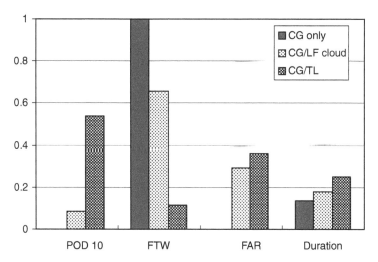

Figure 19.10 Graph of all warning performance statistics when the AOC only is used. CG flash data are always involved in the analysis as the predictand. The other datasets are (1) LF cloud discharge data, light grey, stippled bars, and (2) VHF total lightning mapping: checkerboard bars.

detection technique involves CG data alone or LF cloud discharge data. In these two cases, it is appropriate to use at least one WA in order to provide lead time before the first CG in the AOC. When we use the WA with CG data alone or with LF cloud discharge data, we obtain the results shown in Figure 19.11. When the WA is included, the CG flash data alone are sufficient to provide a POD of 50 per cent for a required lead time of at least 10 min. Under this configuration of the algorithm, the LF cloud discharge data provide only one noticeable benefit, a reduced FTW. Again, this is due to the fact that cloud discharge detection by an LF network provides advance notice of the onset of CG lightning in some storms, despite the low detection efficiency.

Finally, Figure 19.12 shows the comparison of POD, FAR, FTW and total warning duration corresponding to the algorithm configuration that is best suited to each type of detection technology. To reiterate, in the case of CG data alone and in the case of LF cloud discharge information, that means both the WA and AOC are used, whereas whenever VHF total lightning mapping data are available, only the AOC is used. Figure 19.12 shows a couple of distinct advantages to using the VHF total lightning mapping data. First, because we are able to eliminate the WA, the total duration of warnings is somewhat shorter overall. In addition, the FAR is significantly lower. When the WA is required, there is a significant probability that a storm will pass by the AOC, producing lightning in the WA but not in the AOC. These cases add to the FAR in the two cases where the WA is required. Finally, as mentioned before, the fact that the VHF mapping system captures in-cloud discharge activity that extends outside the area where CG flashes occur means that the FTW can be reduced significantly.

Lightning warning systems 883

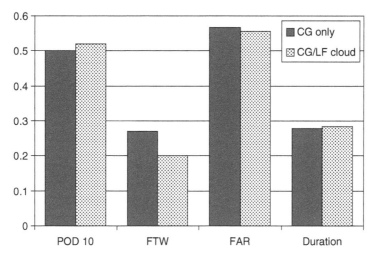

Figure 19.11   Warning performance metrics for the two cases where a WA is required in addition to the AOC. The bar styles are the same as in Figure 19.10.

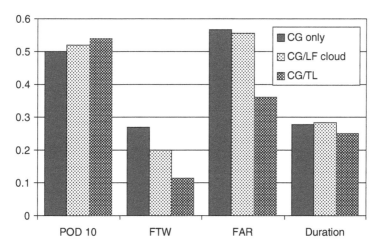

Figure 19.12   Warning performance metrics for best possible usage of the datasets. The bar styles are the same as in Figure 19.10. In this analysis, the CG only and CG plus LF cloud datasets require both the AOC and WA in the warning algorithm, while the example involving the VHF total lightning mapping system (with 'TL' in the legend) requires the AOC only.

19.5.1.2 Effects of algorithm configuration using a single detection technology

In this section, all examples use only CG lightning information from the US NLDN. The examples presented here are from a study by Holle and colleagues [35]. The warning algorithm is basically identical to Figure 19.14 except that the AOC and WA in the Holle and colleagues study were concentric circles. The AOC had a fixed radius of 3 km, and the outer radius of the WA varied between 4.5 and 16 km. The purpose of this analysis is to show how altering the configuration of the algorithm affects the performance metrics POD, FAR and FTW.

Figure 19.13 shows the results of the analysis in the form of a composite of the data presented by Holle and colleagues [35]. This figure shows how POD (for a required lead time of at least 3 min), FAR and FTW vary as a function of the outer radius of the WA. As the radius of the WA is increased, both POD and FAR increase. As POD increases, there is a corresponding decrease in FTW. These results are consistent with expectations. The larger the WA, the more like it is that there will be at least 3 min of lead time between the first flash in the WA and the first flash in the AOC (higher POD). At the same time, there is also a greater likelihood of finding storms that pass by the AOC, producing flashes in the WA only (higher FAR). The strong correlation between POD and FAR is a consistent feature of lightning warning systems.

19.5.2 Lightning cessation in MCS cases

The importance of VHF total lightning mapping for reducing FTW in the particularly difficult case of MCS stratiform regions was examined by Murphy and Holle [42]. The same analysis methods summarized above in Section 19.5.1 were used in this study, but no LF cloud discharge data were available. The results showed that FTW could be reduced by a factor of about four, from 23 per cent in the case of CG-based warnings

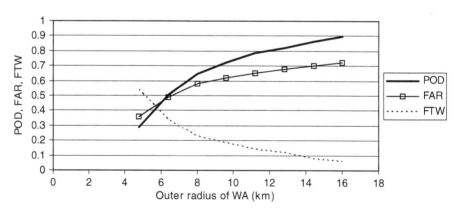

Figure 19.13   Variation of POD, FAR and FTW as a function of the outer radius of a circular WA centered on a circular AOC with a constant radius of 3 km. Analysis is a composite of several cases in the United States using data from the NLDN [35].

to 5 per cent in the combined radar–total lightning algorithm. This is a much greater reduction in FTW than that seen in the Murphy and Holle [43] study, which included many different types of storms, not MCSs alone. This result is to be expected because the total lightning mapping data maintains the continuity of warnings in the stratiform region where CG flashes are widely separated in space and time, as we showed previously (Section 19.4.2.2). Another major concern when MCS stratiform regions are involved is the duration of warnings. Murphy and Holle [42] found that the combined radar–total lightning method produced only a 19 per cent increase in total warning duration relative to CG-based warnings.

19.5.3 Radar applications for lightning onset in storm-following algorithms

As discussed earlier, in the storm-following reference frame, one common way of warning of the onset of lightning is to use radar reflectivity within the mixed-phase portion of the cloud where the dominant electrification mechanism operates. Among the many studies in which such a method has been used are Lhermitte and Krehbiel [44], Dye et al. [45], Buechler and Goodman [46], Michimoto [47], Hondl and Eilts [48], Zipser and Lutz [49], Petersen et al. [50], Rison et al. [51], MacGorman and Filiaggi [52], Hoffert [53], Forbes and Hoffert [30], Gremillion and Orville [54], Saxen and Mueller [27], and Vincent et al. [55]. Several of these studies include enough samples to permit an examination of the performance of the method. Table 19.7 summarizes these results, where possible, in the context of the contingency table statistics used throughout this Chapter. With the exception of Saxen and Mueller [27], none of these studies required a particular amount of lead time from the time the radar threshold was reached until the onset of lightning. With caution, we can compare these results with the fixed-point analysis discussed above, although we must note that a complete storm-following warning algorithm must translate the cell- or storm-level information into a lightning warning for a specific place, and that was not part of the studies cited above. Further, we must caution that the storms analysed in the above studies were chosen to avoid the radar 'cone of silence' and not to be at too great a distance from the radar, but otherwise the temporal and spatial sampling of the radar was that of normal operational conditions (except in Dye et al. [45]). With these caveats, we note that the PODs shown in Table 19.7 are comparable to/slightly greater than the PODs attained by the fixed-point warning algorithm employing total lightning mapping data at Dallas–Fort Worth when all lead times are taken into account. The FAR in Dallas–Fort Worth was 0.3, also consistent with the values shown in Table 19.7.

As mentioned in Section 19.2.1, dual-polarization radar provides information that can be used to discriminate between raindrops and various forms of precipitation-sized ice. Because of the requirement for frozen precipitation for electrification, these dual-polarization radar observations more directly address the eventual possibility of lightning. Goodman et al. [56], Jameson et al. [38], Carey and Rutledge [57,58] and others have examined dual-polarization radar data during storm electrification and lightning production and have found that the development of graupel (soft hail pellets) in the

Table 19.7 Summary of results using radar-based thresholds for the onset of lightning activity

Study	Location	Radar criterion	Lightning criterion	POD	FAR	Lead time
Forbes and Hoffert [30]	Kennedy Space Center, Florida, USA	30 dBZ at −10 °C	Onset of CG flashes	1.0	0.35	7.8 min (average)
Dye et al. [45]	Langmuir Lab, New Mexico, USA	40 dBZ at −10 °C	First lightning	1.0	0.22	17 min (average)
Saxen and Mueller [27]	WSMR, New Mexico, USA	Fuzzy logic including max dBZ above −20 °C	Onset of CG flashes	0.904	0.217	lead time >5 min required for success
Gremillion and Orville [54]	Kennedy Space Center, Florida, USA	40 dBZ at −10 °C	Onset of CG flashes	0.84	0.07	7.5 min (median)
Hondl and Eilts [48]	Kennedy Space Center, Florida, USA	10 dBZ at 0 °C	Onset of CG flashes	1.00	0.405	
Hoffert [53]	Kennedy Space Center, Florida, USA	30 dBZ at 7 km altitude	First lightning	1.00	0.35	7.8 min (average)
Vincent et al. [55]	North Carolina, USA	40 dBZ at −10 °C	Onset of CG flashes	1.00	0.37	14.7 min

CG, cloud-to-ground; POD, probability of detection; FAR, false alarm ratio.

−10 to −20 °C region of the cloud corresponds with the development of significant electric fields and lightning. Moreover, the mass of graupel in the mixed-phase part of the cloud has been shown in some cases to be correlated to either the rate of VHF emissions from lightning or the flash rate [57–60]. Much of this dual-polarization radar work is directly relevant to the lightning warning problem, but all of the studies mentioned above are case studies involving one or a few storms. A more systematic study specifically oriented to the lightning warning problem is warranted. We expect that dual-polarization data will likely reduce both FAR and FTW relative to conventional reflectivity observations because it provides information more directly related to the electrification process.

19.6 Assessing the risks

As discussed above, lightning warning systems are not perfect. There are failures-to-warn (false negative condition) and false alarms (false positive condition), and the trade-off between these two conditions is a function of the amount of warning lead-time required. In this section we provide a conceptual framework for evaluating and configuring lightning warning systems in the context of specific warning applications.

19.6.1 Decision making

The practical use of lightning warning systems requires either an implicit or explicit decision-making strategy – probabilities of occurrence are combined with the costs of being right or wrong, and decisions are made. In this section we provide a brief overview of objective decision making that will help in understanding the cases to be presented in Sections 19.6.2 and 19.6.3.

A fundamental overview of decision making can be provided in the context of Bayesian decision making, where one selects the decision (action) that minimizes a 'cost function' produced by the sum of the cost of each possible pair of actions and occurrences, weighted by the probability of that condition. Mathematically, this is expressed as

$$\min_{k \in K} \sum_{k=1}^{K} [c(a_k) + c(o_k)] p(o_k)$$

where a_k and o_k are the action and occurrence for condition k, $c(\cdot)$ is the cost associated with a specific action or occurrence, K is the number of possible conditions, and $p(o_k)$ is the known (or estimated) probability for occurrence k.

The cost function for a single-stage lightning warning system can be derived from the simple 2 × 2 contingency table discussed in Section 19.4. In this example, prediction of lightning in the AOC is equated with taking 'protective action'. The probabilities and costs for this example are shown in Table 19.8. The probabilities (P_{ij}) and costs (C_{ij}) are subscripted by <row,column> in the table. Note that the probability

Table 19.8 Probabilities and costs associated with a single-stage lightning warning method, with probabilities taken from the 2 × 2 contingency table (Section 19.4)

	No action	Protective action
No lightning in AOC	$C_{11} = 0$	$C_{12} = C_a$
	P_{11} = unknown	P_{12} = FAR
Lightning in AOC	C_{21}	$C_{22} = C_a$
	P_{21} = FTW	P_{22} = POD

AOC, area of concern.

of 'no action' coupled with 'no lightning in the AOC' (P_{11}) is unknown. This does not impact the estimate of cost because C_{11} is zero. Also note that $C_{12} = C_{22}$, which is the cost of taking action (C_a). This assumes that the protective action results in no damage or injury. The (average) cost in this example is therefore

$$\text{Cost} = C^*_{21}P_{21} + C^*_{12}P_{12} + C^*_{22}P_{22} = C^*_{21}\text{FTW} + C^*_a(\text{FAR} + \text{POD})$$

Some insight into the decision-making problem can be drawn from an analysis of the cost function shown above. First, note that the cost of action (C_a) scales linearly with the sum of POD and FAR. We also know from Section 19.5 that for realistic warning systems these probabilities increase or decrease together, and that they are inversely related to FTW. This insight, coupled with knowledge of C_{21} and C_a, leads to an objective way to select the right operating conditions. For example, if C_{21} (cost of failure-to-warn) is 10 times higher than the cost of taking the protective action (C_a), then the average cost can be re-written as

$$\text{Cost} = C^*_a[10^*\text{FTW} + (\text{FAR} + \text{POD})]$$

From this equation, it is clear that if an increase in (FAR + POD) can lead to more than a (FAR + POD)/10 decrease in FTW, the overall cost will decrease.

An alternative representation of conditions and their associated probabilities for the single-stage warning system is depicted by the Venn Diagram shown in Figure 19.14. The space defined by 'B' is the set of warning actions, and the space defined by 'A' is the set of lightning occurrences. The intersection of A and B (the set A ∩ B) is associated with the probability POD. The probability of A with no warning (the set of A − A ∩ B) is FTW. The probability of B with no lightning in the AOC (the set of B − A ∩ B) is FAR.

Although the Venn diagram depiction is not required to understand the simple single-stage warning system, it can be very helpful in the analysis of more complex situations. For example, a fairly general (two-stage) warning system has three actions: no action, preparative and protective (warning). Preparative can be viewed as a limited action that has low cost and will facilitate the protective action if it is required. To avoid complexity, we assume that the preparative action will NOT prevent damage or injury due to lightning in the AOC; it will only improve the

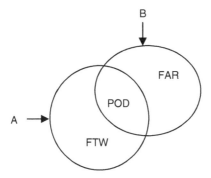

Figure 19.14 *Venn diagram depicting a single-stage lightning warning system. B is the set of warning actions and A is the set of lightning occurrences. The probabilities for each of the three interior regions (members of the two set) are FAR, POD and FTW, as discussed in Section 19.4.*

quality of the warning that results in the protective (warning) action, by providing more time to make the decision. That extra time is assumed to translate to a reduced FAR for the protective action. Now considering the possible occurrences, they can be broken down into no lightning (in AOC), lightning in AOC and damage (or injury). The damage occurrence reflects the poorly predictable relationship between a non-damaging strike in the AOC, which has no real cost, and the subset of these occurrences that have the potential to be costly. Figure 19.15 shows the Venn diagram for this general, two-stage warning system.

The space defined by set C is the set of preparative actions, and the space defined by set A is the set of lightning occurrences. The set of protective (warning) actions, B, is fully contained in C, and the set of lightning events capable of producing damage or injury, D, is fully contained in A. As in Figure 19.14, the probabilities are shown inside each set.

The set of flashes capable of damage or injury is broken down into three components (α, β, γ) corresponding to the fraction of those flashes where no action was taken (α), the fraction where only preparative action was taken (β), and the fraction where protective action was taken (γ). Note, however, that only portions α and β actually produce damage or injury, because we assume that once protective action is taken (the γ portion), no damage/injury results.

The fact that A ∩ C is larger than A ∩ B does not change the POD (defined by the set of A − A ∩ B), because we have assumed that preparative action does not result in protection. On the basis of this assumption, the failure-to-warn rates have been partitioned depending on whether no action was taken at all or whether preparative action was taken. These are both considered failures to warn because neither results in protection, in accordance with the assumption about preparative action stated previously. The area defined by A ∩ C − A ∩ B − β, which is also shown by the grey Xs in Figure 19.15, is defined as FTW_p, the failure-to-warn rate for protective action, while the area A ∩ not C is now defined as FTW_n, the failure-to-warn rate for no action.

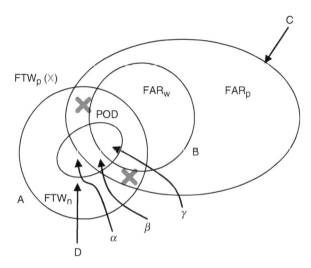

Figure 19.15 Venn diagram depicting two-stage lightning warning system. C is the set of preparative actions, and A is the set of lightning occurrences. B is the set of protective (warning) actions, which is fully contained in C. D is the set of flashes capable of damage or injury and is fully contained in A. The probabilities for each of the interior regions are shown. See text for details.

Table 19.9 shows this set of actions and occurrences, with their associated probabilities and costs. Note that the two conditions with no action and no damage/injury have no cost, irrespective of the probability of that condition. We make the assumption that protective action completely prevents damage or injury, so that $C_{13} = C_{23} = C_{33} = C_a$. This also leads us to set $P_{33} = 0$ because although there is a subset of flashes (γ) capable of producing damage/injury, no damage or injury is actually incurred under the protective action. Note that both the cost of preparative action (C_p) and the cost of damage/injury (C_d) are incurred for C_{32}, based on our definition of β.

The average cost in this example is therefore

$$\text{Cost} = C_d^*(\alpha + \beta) + C_p^*(\text{FAR}_p + \text{FTW}_p + \text{POD}) + C_a^*(\text{FAR}_w + \text{POD})$$

The term in this equation associated with C_d is equivalent to the C_{12} term in the cost function for single-stage decision making, and the other two terms are a re-partitioning of the cost of taking action.

By comparing the cost functions for the single-stage and two-stage warning systems, it is possible to determine when two-stage warning is financially beneficial. First, we note that according to our assumptions, POD is the same for both systems, and it is reasonable to assume that C_d and C_a are the same for both systems. Therefore, the average cost is reduced when the added (probability-weighted) cost of the preparative action is less than the savings from the reduced false-alarm rate that we attain by being

Table 19.9 Probabilities and costs associated with a two-stage lightning warning method. The probabilities are described by the Venn diagram in Figure 19.15.

	No action	Preparative action	Protective action
No lightning in AOC	$C_{11} = 0$ P_{11} = unknown	$C_{12} = C_p$ $P_{12} = FAR_p$	$C_{13} = C_a$ $P_{13} = FAR_w$
Lightning in AOC, no damage/injury	$C_{21} = 0$ $P_{21} = FTW_n - \alpha$	$C_{22} = C_p$ $P_{22} = FTW_p + POD - \beta$	$C_{23} = C_a$ $P_{23} = POD$
Lightning in AOC, with damage/injury	$C_{31} = C_d$ $P_{31} = \alpha$	$C_{32} = C_p + C_d$ $P_{32} = \beta$	$C_{33} = C_a$ $P_{33} = 0$
Column total probabilities	FTW_n	$FAR_p + FTW_p + POD$	$FAR_w + POD$

AOC, area of concern.

able to wait longer before taking protective action. Mathematically, the relationship between the total cost due to preparative action and the savings in protective action relative to the single-stage warning case is represented as

$$C_p^*(FAR_p + FTW_p + POD) < C_a^*(FAR - FAR_w)$$

which is also

$$C_p < C_a \frac{FAR - FAR_w}{FAR_p + FTW_p + POD}$$

where FAR is the probability of a false alarm for the single-stage case. This relation shows that the more the preparative action can reduce the FAR_w relative to the single-stage FAR, the greater the value of the preparative action. Conversely, the more the preparative action is taken unnecessarily (FAR_p) or fails to lead to protective action when necessary (FTW_p), the lower its value.

19.6.2 Equipment protection application

The following is a simplified example of Bayesian decision making applied to an equipment protection problem. Consider a manufacturing facility requiring 'dynamic' lightning protection due to power quality and reliability problems during nearby thunderstorms. The protective action is isolation of the facility from the power grid and running off of a local generation system. The preparative action is to start the large local generator motor, applying a very light load to the generator. We will assume that the cost of the protective action (C_a) is 1 per cent of the purchase

Table 19.10 A specific example of probabilities and costs associated with a two-stage lightning warning method

	No action	Start generator	Switch-over to generator
No power problem	0	$C_p = \$500$ $FAR_p = 0.6$	$C_a = \$5\,000$ $FAR_w = 0.3$
Power problem (action taken)		$C_p - \$500$ $POD + FTW_p = 0.92 + 0.03$	$C_a - \$5\,000$ $POD = 0.92$
Power problem (damage)	$C_d = \$100\,000$ $FTW_n = 0.05$	$C_d = \$100\,000$ $FTW_p = 0.03$	

price of the generator system and associated switches (\$500 000), and that the cost of the preparative action (C_p) is one-tenth of the protective action. The cost of failure to protect is the scrapping of all in-process material during a brown-out or power loss (C_d), taken to be \$100 000. In this example, we further simplify the problem by assuming that any time lightning is observed in the AOC it results in a power problem (damage). This is the equivalent of letting $\alpha = FTW_n$ and $\beta = FTW_p$. Table 19.10 shows the specific costs and probabilities associated with this example. For clarity, the second and third rows of Table 19.10 have been modified (relative to Table 19.9) to isolate the costs associated with action and damage, respectively.

For this example, the average cost is given by

$$\text{Cost} = C_d^*(FTW_n + FTW_p) + C_p^*(FAR_p + POD + FTW_p) + C_a^*(FAR_w + POD)$$
$$= \$8\,000 + \$775 + \$6\,100 = \$14\,875$$

Given the relative costs of damage, preparation and protection, it is clear that the most significant terms are those associated with damage and protective action. Therefore, this company would benefit most from even a slightly improved FTW_n and/or a protective action that leads to a low FAR_w (at the expense of FAR_p).

19.6.3 Trade-offs between performance and risks for cloud-to-ground warning in safety applications

The following subsections describe how warning decision making is impacted by performance of lightning warning systems in safety-related applications. The examples provided here are not intended to be specific recommendations, but rather a set of applications that were selected to represent the spectrum of 'costs' and performance issues.

19.6.3.1 Personal and small-group warning

Small groups can make quicker responses to the lightning threat than a large number of people. At a typical public venue, the evacuation time for individuals and small groups is considered to be less than 10 min [13]. In such safety applications, the cost of 'damage' is essentially infinite, and the cost of taking action (moving indoors) is usually very small. The one significant exception is when people are essentially trapped far from a safe location, for example when camping or hiking in remote areas. In situations where the cost of action is small and the cost of failure to warn is very large, a high FAR is acceptable and there is no need for a two-stage warning process.

Statistically speaking, very few of the flashes within the AOC have the potential to cause 'damage' (injury or death). In practice, this fraction of flashes cannot be determined *a priori*, but it can often be quantified using historical data in order to get a sense for the cost of providing warning. The following is a simplified analysis for a common small-group situation, an outdoor athletic activity with perhaps up to 20 participants. In this scenario, we assume the following.

- The site for the activities is a small stadium or field with sides of 0.2 km length.
- The annual CG flash density in the area is 5 flashes km^{-2} yr^{-1}.
- The field is occupied 20 per cent of the time, on average, throughout the year.
- There is no correlation between when lightning activity occurs and when the field is occupied.
- If a flash strikes the field, at least one person is injured or killed.

Under these assumptions, we may estimate the number of flashes that have the potential to cause injury or fatality. This corresponds to the set D shown in Figure 19.15. This set consists of (5 flashes km^{-2} yr^{-1}) * (0.2 yr) * (0.2 km)^2 = 0.04 flashes each year, or one flash every 25 years. Assuming further that the cost of a lightning warning data service or a small stand-alone warning device is ~$300 yr^{-1}, then the cost of warning per flash capable of causing injury or fatality is ($300 yr^{-1})/(0.04 flash yr^{-1}) = $7 500 per flash.

This cost should be compared with the potential cost incurred by using the 30–30 rule, which involves essentially no implementation cost but can be expected to have higher FTW than an automated method. The precise FTW of the 30–30 rule cannot be specified exactly because it is determined by the availability and reliability of a human observer and the noise environment that might prevent timely observation of thunder. In any case, it should be noted that the cost of providing some sort of warning (automated or 30–30 rule) is significantly less than the cost of a lightning injury. Lightning injuries are about ten times more common than lightning fatalities and have two sets of costs, those due to medical expenses and those due to loss of income as a result of the neurological disability that usually results from lightning injury [61]. These costs can easily exceed several hundred thousand dollars.

19.6.3.2 Large venue warning

Larger groups are less able to react quickly to lightning than a person or small group. The cost of taking action is significantly larger, involving the time required for evacuation of the site and travel to safe locations as well as any additional costs associated with the evacuation. Because of the higher cost of action, false alarms are much less tolerable than in the personal or small-group case, while failures to warn are equally intolerable because personal safety is at issue in both situations. Two examples of large-venue warning situations are large sporting events and airport ground operations. Spectators at a major sporting event may require up to an hour or more of advance warning for adequate evacuation [62], and the ideal lightning-safe location may be away from the site itself, for example, nearby buildings or buses that have to be brought to the site. In the latter case, there is an additional cost associated with providing the lightning-safe location. In the airport case, the necessary lead time is determined primarily by the time required for ground crews to stop work activities and move to the building. The additional costs associated with the evacuation are related to the loss of productivity and associated flight delays.

Because false alarms are very costly in large-venue situations, a two-stage warning approach is appropriate. Recall from Section 19.6.1 that one of the benefits of a two-stage warning is to reduce the false alarm ratio for the protective action step by being able to wait longer before taking protective action. In the large sporting event example, perhaps the preparative action would involve not allowing any more spectators to enter the stadium. This step reduces the number of people involved in an eventual evacuation and therefore reduces the cost of, and required lead time for, that evacuation. The preparative action would also include the preparations for a full evacuation of those who are already in the stadium. The protective action would then consist of a full evacuation of the stadium. In the airport case, the preparative action might be to stop fueling operations and disconnect communications cables between ground crew members and airplanes, while the protective action is to bring all ground crews indoors. In Section 19.6.1, we said that to be economically beneficial, the cost of the preparative action must be less than the savings realized by reducing the FAR for the protective action. Thus, the more the preparative action can reduce the FAR relative to a single-stage warning method, the higher the cost of the preparative action can be. For each situation, such as the examples above, these costs need to be evaluated in advance in order to determine the required performance of the lightning warning system.

Consistent with the low tolerance for false alarms, it is also highly undesirable to send people back outdoors at the end of a storm and then have to reverse the decision because of subsequent lightning activity in the area. At the end of the storm, it may be advisable to extend the duration beyond the 30 min recommended by the 30–30 rule in order to be sure that there will not be a subsequent re-evacuation. The cost of waiting somewhat longer once a large group is in a safe place is typically much less than the cost of another evacuation. Excessively long wait times may produce a negative reaction to lightning safety warnings among spectators at a sporting event, but false alarms and repeated evacuations are even more likely to have this effect.

References

1. Soulage A., Demetriades N., Murphy M., Hufnagel K., Dunn M., Cummins K. 'On the use of thunderstorm warning in active lightning protection'. *27th International Conference on Lightning Protection*; Avignon, France: SEE; 2004, pp. 202–07.
2. Williams E.R. 'The tripole structure of thunderstorms'. *J. Geophys. Res.* 1989;**94**:13151–67.

3. Maddox R.A., Zhang J., Gourley J.J., Howard K.W. 'Weather radar coverage over the contiguous United States'. *Weather and Forecasting* 2002;**17**:927–34.
4. Malan D.J. *Physics of Lightning*. London: English Universities Press, Ltd; 1963. p. 176.
5. Mazur V., Williams E., Boldi R., Maier L., Proctor D. 'Initial comparison of lightning mapping with operational time-of-arrival and interferometric systems'. *J. Geophys. Res.* 1997;**102**:11071–85.
6. MacGorman D., Apostopakopoulos I., Nierow A., Cramer J., Demetriades N., Krehbiel P. 'Improved timeliness of thunderstorm detection from mapping a larger fraction of lightning flashes'. Preprints. International Lightning Meteorology Conference; Tucson, AZ, USA: Vaisala, Inc.; (CD-ROM), 2006.
7. Holle R.L., López R.E., Ortiz R., Paxton C.H., Decker D.M., Smith D.L. 'The local meteorological environment of lightning casualties in central Florida'. Preprints. *17th Conference on Severe Local Storms and Conference on Atmospheric Electricity*; St. Louis, MO, USA: American Meteorological Society; 1993. pp. 779–84.
8. Pierce E.T. 'Lightning warning and avoidance' in R.H. Golde (ed.) *Lightning, Vol. 2 Lightning Protection*. London: Academic Press; 1977. pp. 876.
9. Uman M.A. *The Lightning Discharge*. Orlando: Academic Press; 1987.
10. Jacobson E.A., Krider E.P. 'Electrostatic field changes produced by Florida lightning'. *J. Atmos. Sci.* 1976;**33**:103–17.
11. Krider E.P. 'Electric field changes and cloud electrical structure'. *J. Geophys. Res.* 1989;**94**:13145–49.
12. Murphy M.J., Krider E.P., Maier M.W. 'Lightning charge analyses in small convection and precipitation/electrification (CaPE) experiment storms'. *J. Geophys. Res.* 1996;**101**:29615–26.
13. Holle R.L., Lopez R.E., Zimmerman C. 'Updated recommendations for lightning safety – 1998'. *Bull. Am. Meteorol. Soc.* 1999;**80**:2035–41.
14. Lopez R.E., Holle R.L. *The Distance between Successive Lightning Flashes*. NOAA Technical Memorandum NWS ERL NSSL-105, 1999.
15. Johnson J.T., MacKeen P.L., Witt A., Mitchell E.D., Stumpf G.J., Eilts M.D., Thomas K.W. 'The storm cell identification and tracking algorithm: an enhanced WSR-88D algorithm'. *Weather and Forecasting* 1998;**13**:263–76.
16. Dixon M., Wiener G. 'TITAN: thunderstorm identification, tracking, analysis and nowcasting – a radar-based methodology'. *J. Atmos. Oceanic Technol.* 1993;**10**:785–97.
17. Autones F., Carriere J-M., Girres S., Senesi S., Thomas P. 'ASPOC – a French project for a thunderstorm product designed for air traffic control'. *8th Conference on Aviation, Range and Aerospace Meteorology*; Dallas, Texas, USA: American Meteorological Society; 1999, pp. 235–39.
18. Mueller C.K., Saxen T., Roberts R., Wilson J., Betancourt T., Dettling S., Oien N., Yee J. 'NCAR Auto-Nowcast system'. *Weather and Forecasting* 2003;**18**:545–61.
19. Wolfson M.M., Forman B.E., Hallowell R.G., Moore M.P. 'The growth and decay storm tracker'. *8th Conference on Aviation, Range and Aerospace Meteorology*; Dallas, Texas, USA: American Meteorological Society; 1999, pp. 58–62.

20. Wolfson M.M., Forman B.E., Calden K.T., Boldi R.A., Dupree W.J., Johnson R.J., Wilson C., Bierenger P.E., Mann E.B., Morgan J.P. 'Tactical 0–2 hour convective weather forecasts for FAA'. *11th Conference on Aviation, Range and Aerospace Meteorology*. Hyannis, MA, USA: American Meteorological Society; paper 3.1 (CD-ROM), 2004.
21. Megenhardt D.L., Mueller C., Trier S., Ahijevych D., Rehak N. 'NCWF-2 probabilistic forecasts'. *11th Conference on Aviation, Range and Aerospace Meteorology*; Hyannis, MA, USA: American Meteorological Society; paper 5.2 (CD-ROM), 2004.
22. Koshak W.J., Krider E.P. 'Analysis of lightning field changes during active Florida thunderstorms'. *J. Geophys. Res.* 1989;**94**:1165–86.
23. Lojou J.-Y., Cummins K.L. 'On the representation of two- and three-dimensional total lightning information'. *Conference on Meteorological Applications of Lightning Data*; San Diego, CA, USA: American Meteorological Society; paper 2.4 (CD-ROM), 2005.
24. Lojou J.-Y., Cummins K.L. 'An assessment of total lightning mapping using both VHF interferometry and time-of-arrival techniques'. *19th International Lightning Detection Conference*; Tucson, AZ, USA: Vaisala, Inc.; (CD-ROM), 2006.
25. Kononov I.I., Yusupov I.E. 'Cluster analysis of thunderstorm development in relation to synoptic patterns'. *18th International Lightning Detection Conference*; Helsinki, Finland: Vaisala Inc.; paper 23 (CD-ROM), 2004.
26. Conway J.W., Thurston T.R., Mitchell E.D., Bassett G.M., Poon T.W., Eilts M.D. 'Automatically predicting lightning threat using integrated data sources'. *17th International Lightning Detection Conference*; Tucson, Arizona, USA: Vaisala, Inc.; paper 45, 2002.
27. Saxen T.R., Mueller C.K. 'A short-term lightning potential forecasting method'. *30th International Conference on Radar Meteorology*; Munich, Germany: American Meteorological Society; 2001, pp. 237–39.
28. Brunza S.J., Coleman C.R., Jakacky J.M. 'Using filtering theory to predict lightning strikes'. *16th International Conference on Interactive Information and Processing Systems (IIPS) for Meteorology, Oceanography and Hydrology*. Long Beach, CA, USA: American Meteorological Society; 2000, pp. 16–18.
29. Watson A.I., Holle R.L., Lopez R.E., Ortiz R., Nicholson J.R. 'Surface wind convergence as a short-term predictor of cloud-to-ground lightning at Kennedy Space Center'. *Weather and Forecasting* 1991;**6**:49–64.
30. Forbes G.S., Hoffert S.G. 'Studies of Florida thunderstorms using LDAR, LLP and single Doppler radar data'. *11th International Conference on Atmospheric Electricity*; Guntersville, AL, USA: NASA Conf. Pub. 1999-202261; 1999.
31. McLaughlin D.J., Chandrasekhar V., Droegemeier K., Frasier S., Kurose J., Junyent F., Philips B., Cruz-Pol S., Colom J. 'Distributed collaborative adaptive sensing (DCAS) for improved detection, understanding and prediction of atmospheric hazards'. *9th Symposium on Integrated Observing and Assimilation Systems for the Atmosphere, Oceans and Land Surface*; San Diego, CA, USA: Amer. Meteorol. Soc.; paper 11.3 (CD-ROM), 2005.

32. Weckwerth T.M., Parsons D.B., Koch S.E., Moore J.A., LeMone M.A., Demoz B.B., Flamant S., Geerts B., Wang J., Feltz W.F. 'An overview of the International H_2O Project (IHOP_2002) and some preliminary highlights'. *Bull. Am. Meteorol. Soc.* 2004;**85**:253–77.
33. Burrows W.R., Price C., Wilson L.J. 'Warm season lightning probability prediction for Canada and the northern United States'. *Weather and Forecasting.* 2005;**20**:971–88.
34. Wilks D.S. *Statistical Methods in the Atmospheric Sciences*. London: Academic Press; 1995. p. 467.
35. Holle R.L., Murphy M.J., Lopez R.E. 'Distances and times between cloud-to-ground flashes in a storm'. *International Conference on Lightning and Static Electricity*; Blackpool, England: BAE Systems and Royal Aeronautical Society; paper I03-79 KMI (CD-ROM), 2003.
36. Brown B.G., Bullock R.R., Davis C.A., Gotway J.H., Chapman M.B., Takacs A., Gilliland E., Manning K., Mahoney J.L. 'New verification approaches for convective weather forecasts'. *11th Conference on Aviation, Range and Aerospace Meteorology*; Hyannis, MA, USA: American Meteorological Society; paper 9.4 (CD-ROM), 2004.
37. Hoeft R., Wakefield C. 'Evaluation of the electric field mill as an effective and efficient means of lightning detection'. *1992 International Aerospace and Ground Conference on Lightning and Static Electricity*; Atlantic City, NJ, USA: National Interagency Coordinating Group; 1992. 4-1–4-13.
38. Jameson A.R., Murphy M.J., Krider E.P. 'Multiple-parameter radar observations of isolated Florida thunderstorms during the onset of electrification'. *J. Appl. Meteorol.* 1996;**35**:343–54.
39. Krehbiel P.R. 'The electrical structure of thunderstorms', in E.P. Krider *et al.* (eds) *The Earth's Electrical Environment*. Washington, DC: National Academy Press; 1986. p. 263.
40. Jerauld J., Rakov V.A., Uman M.A., Rambo K.J., Jordan D.M., Cummins K.L., Cramer J.A. 'An evaluation of the performance characteristics of the U.S. National Lightning Detection Network in Florida using rocket-triggered lightning'. *J. Geophys. Res.* 2005;**110**:doi:10.1029/2005JD005924.
41. Biagi C., Krider E.P., Cummins K.L. 'NLDN performance in southern Arizona, Texas and Oklahoma in 2003–2004'. *19th International Lightning Detection Conference*; Tucson, Arizona, USA: Vaisala, Inc., CD-ROM; 2006.
42. Murphy M.J., Holle R.L. 'A warning method for cloud-to-ground lightning based on total lightning and radar information'. *2005 International Conference on Lightning and Static Electricity*; Seattle, WA, USA: The Boeing Co.; paper LDM-36 (CD-ROM), 2005.
43. Murphy M.J., Holle R.L. 'Warnings of cloud-to-ground lightning hazard based on total lightning and radar information'. *2nd Conference on Meteorological Applications of Lightning Data*; Atlanta, GA, USA: American Meteorological Society; paper 2.2 (CD-ROM), 2006.
44. Lhermitte R., Krehbiel P. 'Doppler radar and radio observations of thunderstorms'. *IEEE Trans. Geosci. Electron.* 1979;**GE-17**:162–71.

45. Dye J.E., Winn W.P., Jones J.J., Breed D.W. 'The electrification of New Mexico thunderstorms. Part I: Relationship between precipitation development and the onset of electrification'. *J. Geophys. Res.* 1989;**94**:8643–56.
46. Buechler D.E., Goodman S.J. 'Echo size and asymmetry: impact on NEXRAD storm identification'. *J. Appl. Meteor.* 1990;**29**:962–69.
47. Michimoto K. 'A study of radar echoes and their relation to lightning discharge of thunderclouds in the Hokuriku District. Part I: Observation and analysis of 'single-flash' thunderclouds in midwinter'. *J. Meteorol. Soc. Jpn* 1991;**71**: 195–204.
48. Hondl K.D., Eilts M.D. 'Doppler radar signatures of developing thunderstorms and their potential to indicate the onset of cloud-to-ground lightning'. *Monthly Weather Rev.* 1994;**122**:1818–36.
49. Zipser E.J., Lutz K.R. 'The vertical profile of radar reflectivity of convective cells: A strong indicator of storm intensity and lightning probability?' *Monthly Weather Rev.* 1994;**122**:1751–59.
50. Petersen W.A., Rutledge S.A., Orville R.E. 'Cloud-to-ground lightning observations from TOGA COARE: Selected results and lightning location algorithms'. *Monthly Weather Rev.* 1996;**124**:602–20.
51. Rison W., Krehbiel P., Maier L., Lennon C. 'Comparison of lightning and radar observations in a small storm at Kennedy Space Center, Florida'. *10th International Conference on Atmospheric Electricity*. Osaka, Japan: Meteorological Soc. of Japan; 1996, pp. 196–99.
52. MacGorman D., Filiaggi T. 'Lightning ground flash rates relative to radar-inferred storm properties'. *28th International Conference on Radar Meteorology*; Austin, TX, USA: American Meteorological Society; 1997, pp. 143–44.
53. Hoffert S.G. 'Studies of lightning and non-lightning convective clouds over the John F. Kennedy Space Center'. *16th Conference on Weather Analysis and Forecasting*. Phoenix, AZ, USA: American Meteorological Society; 1998. pp. 129–31.
54. Gremillion M.S., Orville R.E. 'Thunderstorm characteristics of cloud-to-ground lightning at the Kennedy Space Center, Florida: a study of lightning initiation signatures as indicated by the WSR-88D'. *Weather and Forecasting* 1999; **14**:640–49.
55. Vincent B.R., Carey L.D., Schneider D., Keeter K., Gonski R. 'Using WSR-88D reflectivity for the prediction of cloud-to-ground lightning: a central North Carolina study'. *Conference on Meteorological Applications of Lightning Data*; San Diego, California: Amer. Meteorol. Soc.; paper 3.2 (CD-ROM), 2005.
56. Goodman S.J., Buechler D.E., Wright P.D., Rust W.D. 'Lightning and precipitation history of a microburst-producing storm'. *Geophys. Res. Lett.* 1988; **15**:1185–88.
57. Carey L.D., Rutledge S.A. 'The relationship between precipitation and lightning in tropical island convection: a C-band polarimetric radar study'. *Monthly Weather Review* 2000;**128**:2687–710.
58. Carey L.D. Rutledge S.A. 'The relationship between precipitation and lightning in tropical island convection: A C-band polarimetric radar study'. *Mon. Wea. Rev.* 2000;**128**:2687–710.

59. MacGorman D., Rust D., van der Velde O., Askelson M., Krehbiel P., Thomas R., Rison W., Hamlin T., Harlin J. 'Lightning relative to precipitation and tornadoes in a supercell storm'. *12th International Conference on Atmospheric Electricity*; Versailles, France: ONERA and others; paper TuA3-014 (CD-ROM), 2003.
60. Wiens K.C., Rutledge S.A., Tessendorf S.A. 'The 29 June 2000 supercell observed during STEPS. Part II: Lightning and charge structure'. *J. Atmos. Sci.* 2005;**62**:4151–77.
61. Cooper M.A. 'Disability, not death, is the main problem with lightning injury'. *National Weather Digest* 2001;**25**:43–47.
62. Gratz J., Noble E. 'Lightning safety and large stadiums'. *Bull. Am. Meteorol. Soc.* 2006;**87**:1187–94.
63. Cummins K.L., Murphy M.J. 'An overview of lightning locating systems: History, techniques, and data uses, with an in-depth look at the U.S. NLDN'. *IEEE Trans. Electromagn. Compat.* 2009 (in press).

Chapter 20
Lightning-caused injuries in humans
Vernon Cooray, Charith Cooray and Christopher Andrews

(Sections 20.1 to 20.3.9 and 20.4 were published previously by the same authors in *J. Electrostatics* **65**, 386–94, 2007. They are reproduced here by permission from *J. Electrostatics*.)

20.1 Introduction

Lightning is one of the most powerful and spectacular natural phenomena that mankind has ever encountered. It is both breathtakingly beautiful and treacherous and menacing at the same time. Every year around 3 billion lightning flashes occur around the world. In the tropical regions $\sim 10-20$ per cent of the lightning flashes strike the ground while the rest take place inside a cloud. In temperate regions the corresponding figure is ~ 50 per cent. From time to time lightning flashes striking the ground interact with humans, causing injuries and sometimes death. Statistics concerning the number of deaths caused by lightning are available only for a few countries. In the United Kingdom about three people are killed by lightning annually, whereas the number of people injured by lightning is ~ 50 [1]. With a 58.2 million population in the United Kingdom, the probability of being struck by lightning is one in 1.2 million and being struck and killed by lightning is 1 in 19 million. In the United States there were $\sim 2\,566$ deaths and 6 720 injuries due to lightning over the period 1959–1985 [2,3]. Thus ~ 100 people die of lightning-caused injuries annually in the United States. Indeed in the United States, lightning is the top storm killer from all natural phenomena after floods. In Switzerland 12 people died between 1988 and 1992 of lightning injuries (Swiss Federal Office of Statistics) and in Germany 19 people were killed between 1991 and 1993 (data from the Association of German Electrotechnicians). In South Africa ~ 1.5 per million deaths are caused by lightning in urban regions and the figure is ~ 9 per million in rural populations [4]. This also shows that the localized death rate caused by lightning is related not only to the number of thunderstorm days (variation between countries) but also on the population density and housing conditions (variation between regions). In Sweden the annual number of deaths due to lightning is about one [5]. Statistics from tropical regions are not available but the number of deaths caused by lightning is likely to be higher both due to the presence of lightning and the amount of time spent outside and in unprotected buildings. For example, in Zimbabwe 430 people died of lightning-caused injuries between 1965 and 1972 [6,7]. The available evidence suggests that the number of people killed annually by lightning in Sri Lanka is more than 50 (Fernando, personal communication, 2000). The global mortality rate could be $\sim$1 000 per year [8]. According to the data from the United Kingdom the risk of being killed by lightning is $\sim$1 in 15. However, if the accidents taking place outdoors alone are considered the ratio increases to 1 in 4 [1]. Indeed, the majority of people struck by lightning survive. However, many of those who survive have permanent injuries [9,10].

In this chapter, the different ways in which lightning can interact with humans and the medical implications of this are discussed.

20.2 The different ways in which lightning can interact with humans

There are seven different ways in which humans can be affected by a lightning strike: direct strike, side flash, touch voltage, step voltage, subsequent stroke, connecting leaders and shock waves. In the case of a direct strike the lightning channel terminates on the body, exposing it to the full lightning current. The channel will usually terminate on the head or the upper part of the body. It is thought that this accounts for the largest number of deaths [11].

When lightning strikes for example, a tree, the current injected by the lightning flash into the tree will flow along the trunk of the tree to ground. If a human stands close to this tree then, due to a potential gradient, a discharge path may be created between the tree and the human. A portion of the lightning current may flow along this discharge path and through the body to ground. Such an event is called a side flash. It is important to note here that more than 50 per cent of the lightning injuries that take place outdoors are caused by side flashes from trees while the tree is being used as a shelter from rain. This highlights the danger of this sort of practice. When lightning current flows along an object (a tree or a structure), a potential difference is created between the ground and any other point on the object. If a person happens to be holding an object that is struck by lightning then this potential causes a current to flow through his body from the contact point to the ground, causing injuries. This is called injury due to touch voltage.

During a lightning strike the current injected into the ground at the point of strike will flow radially outwards. This current flow will result in a potential difference between any two points located in the radial direction. If a person happens to be standing close to a lightning strike this potential difference, known as step voltage, appears between his two feet leading to a current surge through the lower body. The current will enter the body through one leg and leave from the other. In this case the current does not flow through the heart or the brain. The resulting injuries are usually not severe. However, if the person happens to be sitting or lying close to the point of strike the magnitude and the path of the current through the body may depend on the way in which the body contacts the ground. This is even more important for a four-footed animal where current may flow from front leg to back leg with the heart in the pathway.

In general a lightning flash consists of several strokes and the point of termination of different strokes may not be the same [12]. That is, the first stroke of the flash may strike the ground or any other object in the vicinity of a human and a subsequent flash may strike the person concerned directly. In this case the person will be exposed to the step voltage of the first stroke and the subsequent stroke will strike him directly.

Another way in which a person can receive injuries from a lightning flash, although only recently identified [13,14] in the literature, is through the connecting leader current. As a stepped leader reaches within about a few hundred metres of the ground several connecting leaders may rise from several grounded objects towards the down-coming stepped leader. Only one of these connecting leaders will make the connection between the stepped leader and the ground. The rest of the connecting leaders will be aborted the moment the charge on the stepped leader is neutralized by the return stroke. These are the aborted connecting leaders. For example, in the case of a lightning strike to a nearby object a connecting leader may arise from the head of a person who is located in the vicinity and cause injuries. Even though these leaders will be aborted almost simultaneously with the initiation of the return stroke, they may still support currents sufficient to injure a person. In the literature it is assumed that current flow through the body due to aborted upward leaders is in the order of 10 to 100 A and lasts a few tens to hundreds of microseconds [15,16]. However, these figures correspond to upward leaders from tall objects (higher than several tens of metres) and may not represent the values in aborted upward leaders originated from the human body. Becerra and Cooray [17] analysed this problem using a self-consistent lightning interception model. Their study shows that the currents in the aborted connecting leader may reach a peak value of $\sim$20 A for in the case of a 30 kA return stroke current. The duration of the current flow could be hundreds of microseconds. However, during the return stroke the electric field that drives the connecting leader disappears and the charges collected on the connecting leader flows back to ground within a time less than about a microsecond. During this phase the current flowing through the body may reach several kiloamperes even though the duration of the current is not more than a microsecond. Figure 20.1 depicts the attractive distance of a human body (1.75 m tall) for several postures as a function of the return stroke peak current. Note that one can reduce the probability of lightning strikes by lowering ones body, whereas one can increase it by extending it upwards, for example by raising a hand or holding an umbrella. Figure 20.2 shows the zones where the body is vulnerable to direct strikes and to aborted leaders. Note that in the case of severe lightning flashes a connecting leader may originate from a human body even when it is as far as 100 m from the path of the down-coming stepped leader. Note, that what is being calculated in the paper is the distance at which a thermalized leader is generated from the body. Several streamer bursts may be generated from the body before the inception of the leader. Because these streamer bursts may originate from the body in electric fields lower than the one required for the inception of a leader, one may experience an electric shock due to the streamer bursts at distances larger than the ones indicated by the dotted line in Figure 20.2.

904   *Lightning Protection*

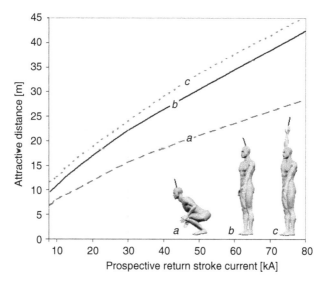

Figure 20.1   The lightning attractive distance of a human body in different postures as a function of prospective first return stroke current. (a) Head of a person in the squat position. (b) Head of a person standing. (c) Extended hand of a person (adapted from Reference 17).

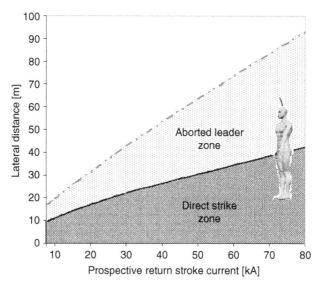

Figure 20.2   Predicted values of the maximum lateral distances to the stepped leader as a function of first return stroke peak current at which a 1.75 m tall human is vulnerable to direct strikes (solid line) and initiation of an aborted connecting leader (dot-dashed line) (adapted from Reference 17)

Finally, injuries can also be caused by shock waves created by the lightning channel. During a lightning strike, channel temperature will be raised to $\sim$25 000 K in a few microseconds, and as a result the pressure in the channel may increase to several atmospheres. The resulting rapid expansion of the air creates a shock wave. This shock wave can injure a human being located in the vicinity of the lightning flash. The pressure associated with the shock wave rapidly decreases with distance, so the shock wave can injure a human being located only in the very close vicinity of the lightning flash.

In the case of a lightning strike only a very small fraction of the current may generally flow through the body and the rest will flow over the body. As the current through the body increases a potential difference is created across the body due to its resistance and capacitance. This voltage increases as the lightning current increases. As this voltage builds up, a stage will be reached at which it increases beyond the voltage necessary to create an electric discharge in air along the skin of the body. When this happens a discharge channel is created along the outer surface of the body to ground. Because the resistance of this breakdown channel is much less than that of the body, most of the lightning current will follow this external path to ground, reducing the current flowing through the body to a small value [18]. For example, assume that the height of the victim is 1.8 m. In air the voltage necessary to create a discharge across a 1.8 m gap is $\sim$900 kV. The voltage needed to create a discharge across an insulating surface of similar length is less than the above value. In the case of human skin smeared with salt from sweat it would be even less. Assume therefore that the voltage needed to create surface breakdown along the human skin is $\sim$450 kV. Now, the resistance of the body is $\sim$1 000 Ω. Thus, when the current through the body reaches 450 A the voltage across the body reaches the surface flashover value, thus leading to a surface discharge. The surface discharge is created long before the lightning current reaches its peak value of $\sim$30 000 A. Now let us consider what happens after this event. The resistance of an arc channel in air is $\sim$1 Ω m^{-1}. Thus the resistance of the surface discharge across the body is $\sim$2 Ω. Thus, the lightning current will be divided between the body resistance of 1 000 Ω and the external resistance of 2 Ω. Therefore, at peak current, say 30 000 A, only 60 A will flow through the body and the rest will flow outside. If one assumes that the duration of the impulse current of a return stroke is $\sim$100 μs and the shape of the current is of triangular shape, the total electrical energy dissipated inside the body will be $\sim$120 J. For a 60 kg human the energy dissipation is $\sim$2 J kg^{-1}. The lethal electrical energy based on animal models is $\sim$62.6 J kg^{-1} [19]. Thus the effect of the surface discharge is to reduce drastically both the current flowing through the body and the energy dissipation inside the body. The surface discharge may cause burn injuries, however, and will be referred to later.

The lightning current flowing inside the body, although small, can cause various types of injuries by heating of tissue, electrolysis and by upsetting the electrical state of excitable tissue (i.e. depolarization). These effects are controlled by the way in which this current distributes itself inside the body. This in turn depends on the conductivity of body fluids and different types of tissues in the body. The current flowing outside can also cause injuries from heat and shock waves.

Table 20.1 *Lightning injuries to large groups. The number of deaths are given in the second column in parentheses (adapted partly from Reference 20)*

Number injured	Activity
10 soldiers [21]	On manoeuvers (0)
16 soldiers [22]	On manoeuvers (0)
38 children [23]	Playing soccer (1)
46 adults [24]	By concession stand (2)
28 children and adults [20]	Camping (4)
41 adults [25]	Mountain climbing (11)
11 teenagers [1]	Sheltering under a slide (0)
14 teenagers [1]	Camping (0)
17 children and adults [26]	Golf (0)
8 children and adults [27]	Shelter under a tree (0)

It is also of importance to note that a single lightning flash can injure several humans at the same time. Table 20.1 summarizes some of the available data on such incidences.

20.3 Different types of injuries

20.3.1 Injuries to the respiratory and cardiovascular system

Cardiopulmonary arrest is the major cause of death following a lightning strike. With appropriate first aid it is reversible in some cases. However, the mortality from lightning strike remains ~ 20 per cent. A small number of patients can be successfully resuscitated with external cardiac massage and expired air ventilation [28] after cardiac arrest due to lightning injury, demonstrating the importance of this as primary first aid. There is no support for the dogma [29] that individuals are capable of being resuscitated after a longer than normal period of cardiac arrest.

The function of the heart is controlled by the systematic and sequential electrical depolarization and subsequent contraction of different parts of the heart muscles (myocardium) (see Figure 20.3). The current flowing through the body during a lightning strike may depolarize the myocardium which may result in myocardial dysfunction including arrhythmias (conduction abnormalities that affect the electrical system of the heart muscle, producing abnormal heart rhythms which can cause the heart to pump less effectively), cardiac arrest either in complete standstill (asystole), or in an uncontrolled and unsynchronized contraction pattern of the myocardium known as ventricular fibrillation (VF). In both cases the forward pumping action of the heart is lost, and blood does not then perfuse vital organs. Probably asystole occurs more often than ventricular fibrillation [31].

The sequence of electrical activity within the heart occurs as follows. First, the electrical impulse leaves the sinus node (SA node) and travels to the right and left atria,

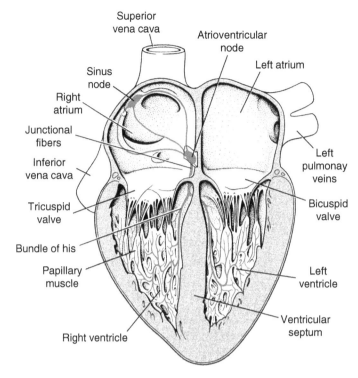

Figure 20.3   The anatomy of the heart (from Reference 30)

causing them to contract together. This takes 0.04 s. This electrical activity can be recorded from the surface of the body as a 'P' wave on the patient's EKG (or ECG) (electrocardiogram). The basics of an EKG is shown in Figure 20.4. The electrical impulse then moves to an area known as the atrioventricular node (AV) node. Here, the electrical impulse is held up for a brief period. This delay allows the right and left atrium to continue emptying their blood contents into the two ventricles. This delay is recorded as a 'PR interval'. The AV node thus acts as a 'relay station' delaying stimulation of the ventricles long enough to allow the two atria to finish emptying. Following the delay, the electrical impulse travels to the bundle of his, then divides into the right and left bundle branches where it rapidly spreads using Purkinje fibres to the muscles of the right and left ventricle, causing them to contract at the same time. The spread of electrical activity through the ventricular myocardium produces the QRS complex on the ECG. The T wave represents the repolarization of the ventricles. It is known that the heart is more sensitive to electrical shock during the early T wave [33–36]. This is the time, the 'vulnerable window', when the ventricles are repolarizing randomly after electrical depolarization, and are potentially at their most disorganized and in a vulnerable state. Any external electrical current that transgresses this portion of the cycle may produce the most deleterious effects. A lightning strike

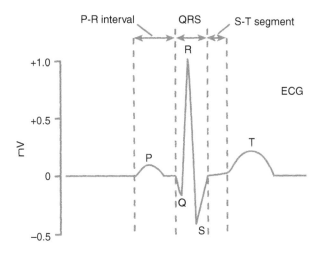

Figure 20.4 EKG (or ECG) waveform (from Reference 32)

during the vulnerable window may have more serious consequences on the function of the heart than at other times. Sometimes, although much more rarely than with industrial electrical shocks, the heart muscles can be permanently damaged due to a lightning strike and this may appear as a change in the EKG resembling a myocardial infarction, or 'heart attack'. EKG changes may also develop subsequently, although not apparent at the time of injury [37,38]. These changes, however, generally disappear over a long period of time.

It is possible to predict how much current is needed in an industrial electric shock to cause VF. This may be done by estimating the current flowing in a given path from the applied voltage and the resistance of the pathway. Our ability to quantify the injuring agent in a pulse as short as a lightning shock is markedly limited. Although for long-duration shocks, current seems to be the important parameter, for ultrashort-duration shocks it seem to be the charge transferred that is the important parameter for estimating injury thresholds.

The breathing action in a human is controlled by the respiratory centres in the brain stem, pons and medulla. They control respiration's rhythm, rate and depth. Current flow through this region may lead to a respiratory arrest (central apnoea). The blast associated with the lightning flash can also cause injuries to the respiratory system [39]. Usually, the cardiac arrest caused by depolarization of the myocardium may recover naturally after the cessation of the current flow through the body, because the heart has its own 'intrinsic' pacemaker. The respiratory apparatus does not, however, act similarly and remains at a standstill [40]. The persistence of the respiratory arrest may then deprive the myocardium of oxygen, leading to a second cardiac arrest. The lack of oxygen to the heart may lead to permanent damage of the myocardium, but more importantly the lack of oxygenated blood reaching the brain quickly leads to the death of brain tissue.

20.3.2 Injuries to the eye

In the case of a lightning strike both the current passing through the head and the strong radiation produced by the channel may cause a series of medical problems in the eye [41–43]. Figure 20.5 depicts the main parts of the human eye. Many eye problems develop over a long period, and so prolonged surveillance of a lightning strike survivor is necessary. The cataract is the most common long-term injury reported in lightning strikes. The first lightning-induced cataract was reported in 1722 [44]. A cataract is a clouding of the lens in the eye that affects vision. A cataract can occur in either or both eyes. The lens consists mostly of water and proteins. When the proteins clump up, it clouds the lens and reduces the light that reaches the retina. The cause of the cataract could be the heating of the lens fluids due to current flow or due to exposure of the eye to very strong optical radiation including ultraviolet light during a lightning strike. Indeed, the lightning channel is a very strong source of ultraviolet radiation and recently it has been shown that it gives rise to strong X-ray and gamma radiation [45]. In the case of lightning injuries the cataract may occur days or years after the injury [1]. Cataracts have been observed not only in the case of lightning strikes outdoors but also in cases of lightning accidents indoors associated with telephones [46].

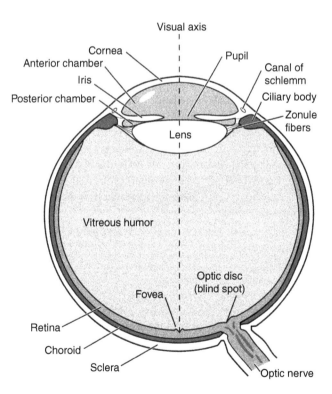

Figure 20.5   The anatomy of the eye (from Reference 32)

In addition to cataracts, the observed effects of lightning strikes on the ocular region of human beings are numerous. Indeed, lightning is known to have caused a multitude of ocular injuries [41].

The retina is the light-sensitive layer of tissue that lines the inside of the eye and sends visual messages through the optic nerve to the brain. The central region of the retina, which contains a high density of photoreceptors, is known as the macula. The macula provides the sharp, central vision we need for seeing fine detail. During lightning strikes a small break in the macula can occur acutely, causing blurred and distorted central vision. Such an injury is called a macular hole. The lightning injury may lead to a pulling or shifting of the retina from its normal position. Such damage is called retinal detachment. In addition to retinal detachment lightning can induce wrinkles in the retinal tissue in one or more areas. They cause small blind spots and are called retinal folds.

The vitreous humour is a clear jelly-like substance within the eye that takes up the space behind the lens and in front of the retina. The vitreous is attached to the retina, more strongly in some places than others. The lightning injury may cause the vitreous to come away from the retina, leading to vitreous detachment. Moreover, a lightning flash can also induce haemorrhages in the vitreous.

Lightning can also cause inflammation within the uveal tract (called uveitis) and in the iris (iritis). Uveitis may cause extreme sensitivity to light (photophobia) with changes of inflammation. During a lightning flash strong ultraviolet and high energetic radiation may enter the eye, causing eye injuries. The cornea is a layer of protective and light transparent tissue covering the iris on the front part of the eyeball. Indeed it is the cornea that takes the main part of the damage when eyes are exposed to energetic radiation. Some of these damages are corneal burns, swelling (oedema), corneal opacities, ulcers and punctuate keratitis. It may lead to changes in vision or complete loss of vision. Lightning can also lead to double vision (diplopia) and this is due to damage to the muscles controlling eye movement or their various nerve supplies. In this case the eyes do not track conjointly and this is a very troublesome visual disorder. The ability to read, walk and perform common activities is suddenly disrupted. In one reported case (Stig Lundquist, personal communication), after receiving a lightning strike a young girl experienced for some time inversion of the optical image, seeing the outside world upside down.

Lightning victims may exhibit fixed or dilated pupils but this does not suggest a bad prognosis [47].

20.3.3 Ear

The anatomy of the ear can be divided into three parts, the outer, inner and middle ear. Figure 20.6 shows the anatomy of the human ear. The outer ear includes the canal, which ends at the eardrum or tympanic membrane. The middle ear consists of a chamber in which there are three tiny bones (malleus, incus and stapes) called ossicles. The ossicles connect the tympanic membrane to the oval window on the opposite side of the middle ear. Their task is to transmit and amplify sound vibration from the external to inner ear. The inner ear contains the cochlea, housing thousands of hair

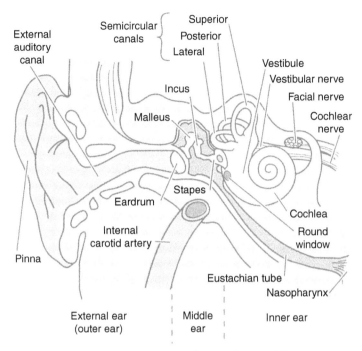

Figure 20.6   Anatomy of the human ear (from Reference 32)

cells and nerve endings. They mediate the conversion of vibration into nerve impulses, thus transmitting an image of sound to the brain. The inner ear also mediates the balance mechanism. About 20–50 per cent of lightning-injured victims suffer a ruptured tympanic membrane in the ear [48,49]. The cause for this could probably be the shock wave created by the lightning flash. During a direct lightning strike to the upper part of the body the ears can be located within a few centimetres of the lightning channel. Calculations by Hill [50] show that the over pressure within a few centimetres of the lightning channel can reach $\sim$10–20 atm. This over pressure is equivalent to a sound impulse of $\sim$200 dB (taking 20×10^{-6} Pa as the reference level). In the case of human hearing the pain threshold level is $\sim$120 dB. In some cases even if the tympanic membrane remains intact, the victims still may suffer from varying degrees of permanent hearing loss and 'ringing in the ear' (tinnitus). This is probably caused by the damage to the hair cells and nerves in the cochlea either from the shock wave or by the flow of current through it. The blast can also cause damage to ossicles that will result in conductive deafness, especially at high frequency. Lightning-induced skull fractures can also cause damage in the middle ear.

It is important to note that the special sense orifices in the cranium (eye sockets, ear canals, nasal and sinus passages) have been pointed out as entry points for electric current leading easily to body fluids such as cerebrospinal fluid (CSF) and blood [40].

20.3.4 Nervous system

The nervous system of a human can be divided into two parts: the central nervous system and the peripheral nervous system. The central nervous system consists of the brain and the spinal cord. The peripheral nervous system can be divided into two main parts: the somatic nervous system and the autonomic nervous system. The former sends sensory information to the central nervous system and receives instructional output to motor nerve fibres that project to skeletal muscles, inducing voluntary movement. The latter controls the unconscious activity of many internal organs, glands and other structures. The processing of pain input to the central nervous system is extremely complex and may mediate long-term pain syndromes, often seen after many physical injuries. During a lightning strike both the central and peripheral nervous systems are often affected. Indeed, the majority of sequelae following a lightning strike are neurological and they are found in 70 per cent of survivors [51].

In the nervous system the lightning-generated currents may cause acute traumatic injuries simply due to the trauma of the insult. These include various types of intra-cranial haemorrhages, swelling of the tissues (oedema) and neuronal injury [27,51,52]. These can cause prolonged or even permanent neurological symptoms. The nervous system can also be affected due to the lack of oxygen resulting from the cardio-respiratory arrest. Lightning can also cause intense vasospasm and constriction of blood vessels and restriction in blood flow (and thus oxygen) to a part of the body. The lack of oxygen to a tissue is termed tissue ischaemia and can cause further injuries to individual parts of the nervous system. A large current flowing through the brain can also lead to neuronal damage, which can lead to permanent brain damage.

In some cases one may also observe a delayed onset of neurological disturbances such as epileptic seizures, tremor, progressive hemiparesis (paralysis of half the body), malfunction of nerves and neurological defects in the central nervous system.

Of particular importance is the phenomenon of 'keraunoparalysis' [53]. It is a flaccid paralysis of an extremity in the path of the current. It is associated with the pulseless and ischaemic limbs. It is suggested that the latter may result from arterial wall constriction as the current flows along them. Facial nerve palsy [54] may also be an expression of this. Keraunoparalysis is thought to be caused by damage to the small blood vessels accompanying the nerves that control the muscles of the extremity involved, along with ischaemia of these muscles. It resolves spontaneously and requires no intervention.

The lightning victim may experience loss of consciousness for varying periods. If the spinal cord is damaged paraplegia may result. The lightning current can also affect the memory of the victim [27,31,36], producing 'amnesia'. Many do not have any recollection of the event, and in some cases the memory of the events a few days to a few weeks before and after the event could be affected. Lightning can cause other specific types of brain dysfunction, such as aphasia, an impairment of language expression. This may affect the production or comprehension of speech. The ability to read or write may also be affected.

In addition to keraunoparalysis, lightning victims may experience weakness, numbness and tingling feelings in muscles and tissues (paresthesias) that may last for several weeks to years.

One case report illustrates the case of growth arrest after a lightning strike [55]. The victim suffered a lightning strike and presented with asymmetric growth arrest two years after the accident. During the strike there was swelling and venous congestion below both knees, multiple blisters on all toes, third-degree burns over the right upper arm and first degree burns over the flank and abdomen. On arrival at the hospital the victim was conscious and oriented and examination showed no bony deformities.

20.3.5 Skin and burn injuries

It is a hallmark of lightning injury that burns are usually minor and require little treatment. This is in severe contrast with other electrical injuries. Lightning can cause burn injuries ranging from superficial burns to full-thickness burns [28,42,56]. The location of burns can be anywhere from head, neck, trunk, upper extremity, hands, lower extremity and legs. There are several ways in which lightning can cause burn injuries. When an electric discharge in air terminates on a solid body, a voltage difference of ~ 10 V is created across a thin layer of gas and vaporized solid matter. In the case of metal objects this is called a cathode fall and has a thickness of less than a millimetre. A similar 'electrode layer' may arise at the gas-to-solid interface of the entrance and exit points of the lightning current into and out of the body. The heat generated in this gas layer is proportional to the total charge passing through this layer. This heat can cause full-thickness burns in the body tissue in contact with it. In lightning-burn victims a characteristic burn pattern is often observed in the form of small, circular, full-thickness burns involving the sides of the soles of the feet and the tips of the toes [28,51,57]. These are probably caused as the lightning current exits from the body by creating an electric discharge between the feet and the ground.

As mentioned previously, as the lightning current passes through the body, it builds up a potential difference between the point of strike and the ground, leading to a surface discharge. This surface discharge may follow the surface of the skin. Any discharge in air may heat the discharge channel to several thousands of degrees, and this heat may cause burn injuries on the skin. Most probably these will be superficial due to the fact that this discharge channel may be isolated from full contact with the skin through a layer of vaporized moisture on the skin. On the other hand, if the victim is wearing any metal objects such as necklaces then the surface discharge may intercept the metal object and the full current may flow through it, causing it to melt. This molten metal can cause deep burns on the skin.

Many lightning strike victims also develop a skin discoloration that looks like redbrown feathery skin markings [20,27,31,58]. These marks, sometimes known as keraunographic marks or arborizations, are probably caused by the streamer-like electrical discharges connected to the main discharge channel propagating over the surface of the skin. This may be an inflammatory reaction that usually disappears within a day or two. Indeed, the pattern of discharge is very similar to the one that can be observed when electrical discharges are directed onto insulating photographic paper, i.e. Lichtenberg figures.

One has to keep in mind that the nature of lightning injuries depends not only on the parameters of the lightning flash but also on the physiology of the body and on the location of the victim during lightning strikes. For example, there is a case of a soldier who suffered full-thickness burns of the scalp and cranial bones extending down to the dura mater [59]. He, together with four other soldiers, took cover from rain using a thick nylon cover. The burn injuries were probably caused by the heating and vaporization of the water on the nylon cover, which was in contact with the head.

20.3.6 Psychological

It is usual that although physical injury can be marked, it is the psychological components of the injury that cause the most ongoing distress [60]. In addition to physical damage, lightning victims may experience a range of psychological problems. These include the fear of thunderstorms, anxiety, depression, disturbances in sleeping rhythm, panic attacks (a sudden rush of uncomfortable physical symptoms such as increased heart rate, dizziness or light-headedness, shortness of breath, inability to concentrate and confusion), disorders of memory, learning, concentration and higher mental facility. There has been at least one reported case in which the patient had to be transferred to a mental hospital [61]. Some lightning victims repeatedly re-experience the ordeal in the form of flashback episodes, memories, nightmares or frightening thoughts, especially when they are exposed to events or objects reminiscent of the trauma, such as thunderstorms or sudden bright lights. This may, in some, be part of a post-traumatic stress disorder. These problems may lead to altered bowel habits, constipation and gastric dilation, in which the stomach becomes excessively dilated with gas, causing it to expand.

20.3.7 Blunt injuries

During the lightning flash the channel temperature may increase to $\sim$30 000 K within a few microseconds. This rapid heating leads to the creation of a shock wave in the vicinity of the channel. As mentioned previously, the shock wave associated with the lightning flash may reach over-pressures of 10 to 20 atm in the vicinity of the channel. In addition to causing damage in the ears and eyes, this shock wave can also cause damage to other internal organs such as the spleen, liver, lungs and bowel tract [27]. Moreover, it may displace the victim suddenly from one place to another, causing head and other traumatic injuries. Indeed, as well as appraising a victim for specific lightning-caused injuries, one must always have in mind associated trauma. In one situation the victim received fractures of the facial bones during a lightning strike [62]. At the time of strike he was wearing a helmet and the damage may have been caused by the intense pressure created by a discharge that resulted during the passage of the lightning current from the helmet to the head across the layer of gas lying between the head and the helmet.

One can also receive blunt injuries from material ejected from the object that is being struck. For example, when lightning strikes trees, the trunk of the tree can explode and the splinters cause injuries in those standing in the vicinity. One can also receive blunt injuries from flying objects inside buildings. During a lightning strike to an unprotected building the central power distribution switches, television sets and antenna cables may explode, causing injuries. Trauma may also be associated with falls from a region (e.g. a cliff) in which a victim finds himself.

20.3.8 Disability caused by lightning

Even though the risk of being killed by lightning is very small and a major number of lightning victims survive a lightning strike, Cooper and Andrews [60] are of the view that the disability resulting from a lightning strike is a serious problem. The primary cause of these disabilities is injuries to the nervous system coupled with psychological disability. According to these authors, many survivors of lightning strikes will be unable to go back to their previous occupation, which may have devastating consequences both to the individual and the family. The primary areas of disability caused by lightning involve neurocognitive functions such as deficits in short-term memory, processing new information, personality changes, easy fatigability, decreased work capacity, chronic pain syndromes, sleep difficulties, dizziness and severe headache. Some patients, after a few months of injury may develop absence types of seizure activity where they stare into empty space or may perform some automatic activity without remembering. Because of these symptoms they may often find that they will not be able to carry out their duties and former functions. One serious problem is that survivors find themselves isolated because friends, family and physicians do not recognize their disability and may sometimes feel that they are faking. Inability to convince a medical practitioner of the underlying problem, a disabled person is likely to lose the possibility to get disability benefit insurance. This difficulty of convincing a medical practitioner of their disability can also stem from the fact that some of the lightning injuries are associated with remote effects (see the next section); these are increasingly being recognized in both electrical and lightning injuries.

20.3.9 Remote injuries

Several recent studies dealing mainly with electrical injuries have brought to focus a new aspect of these injuries [63,64]. In general it is assumed that the effects of injuries are connected to the pathway of current through the body. However, the above studies show that morbidity from electrical injuries is not confined to effects in the pathway of current flow. In other words, some of the injuries are associated with tissues not traversed by current. These are called remote injuries. They may include personality changes, emotional liability, poor concentration, forgetfulness, intolerance of bustle, intolerance of noise, irritability, increased sleep and general slowing and depression. A victim may find it easier to convince others of these symptoms if the lightning strike path involved the head. Unfortunately, according to the data, these symptoms may occur even if the current path is not through the head. According to Andrews [64], remote injuries are a common feature of electrical injury and may even be the norm.

Because most of these symptoms may not be associated with the current path, the exact cause for these symptoms is not known. However, Andrews [64,65] speculates that they could be chemical in nature. The above papers deal mainly with electrical injuries and more work is necessary to elucidate remote injuries associate with lightning encounters although they are quite similar in nature. Even though there is a great deal of difference between normal electrical injury and lightning injury, a comparison of the behavioural consequences of lightning and electrical injury as depicted in Table 20.2 shows many similarities [66]. In this table, Engelstatter [66] reviewed symptom checklists collected retrospectively from 100 lightning strike survivors and 65 electric shock survivors, all of whom presented with chronic sequelae two

Table 20.2 *Frequencies of common after effects reported by 65 electrical injury and 100 lightning injury survivors (adapted from Reference 66)*

Symptom	Electrical injury sample (%)	Lightning injury sample (%)
Neurobehavioural		
Sleep disturbance	74	44
Memory deficit	71	52
Attention deficit	68	41
Headaches	65	30
Irritability	60	34
Inability to cope	60	29
Reduced libido	55	26
Unable to work	54	29
Chronic fatigue	48	32
Dizziness	48	38
Easily fatigued	48	38
Communication problems	46	25
Incoordination	40	28
Confusion	38	25
Chronic pain	29	21
Weakness	25	29
Sensory		
Numbness	63	36
Paresthesias	60	40
Tinnitus	48	33
Photophobia	46	34
Hearing loss	31	25
Visual acuity reduced	25	20
Emotional		
Depression	63	32
Flashbacks	51	20

(*Continued*)

Table 20.2   Continued

Symptom	Electrical injury sample (%)	Lightning injury sample (%)
Agoraphobia	46	29
Emotional problems	38	24
Personality change	29	19
Storm phobia	7	29
Nightmares	26	12
Other		
Muscle spasms	63	34
External burns	54	32
Decreased grip strength	51	34
Stiff joints	48	35
Back problems	46	25
Inability to sit long	45	32
Arthritis	38	19
Hyperhidrosis	31	25
Internal burns	28	21
Bowel problems	25	23

or more years post-injury (mean interval 4.5 yr). The similarity between the two types of injuries suggests that several causes of disability associated with lightning injury could also be due to remote injuries. Understanding this nature of the injury is very important both for the families of the victims and the medical practitioner, because lack of knowledge concerning this mechanism of injury may convince a doctor that the symptoms of the patients do not have any underlying physical cause.

20.3.10   Lightning electromagnetic fields

One important question that has been raised in the lightning literature recently is the medical consequences of electromagnetic fields generated by close lightning flashes. Cherington and colleagues [67] suggested that the currents induced in the body through its interaction with the magnetic field generated by a nearby strong lightning flash could cause myocardial disturbances. This suggestion was based on an incident in which a man standing close to a tree suffered cardiac arrest when lightning struck the tree. This hypothesis has been reconsidered by Andrews and colleagues [68]. They showed that the magnetic fields of lightning flashes are not strong enough to cause myocardial disturbances in humans located in the vicinity of lightning strike points. In another study conducted recently, Cooray and Cooray [69] investigated whether the time derivative of the magnetic field of a close lightning flash could generate currents that can alter the electrical activity of the brain. The study was motivated by the strong similarity between the hallucinations associated with epileptic seizures of the occipital lobe and ball lightning observations. Because most ball lightning observations are associated with close lightning flashes, these authors investigated whether the magnetic field derivative of a nearby lightning flash is strong enough to stimulate nerves in the brain and hence to induce an epileptic seizure of the occipital lobe of a person with a lower seizure threshold located in the vicinity of a lightning flash. Because such a seizure can produce optical hallucinations, almost identical to the reported features of ball lightning, while the person is in a conscious state, these authors hypothesized that some of the ball lightning observations in reality are optical hallucinations.

In order to study this, Cooray and Cooray [69] calculated the time derivative of magnetic fields of lightning flashes with different currents located at different distances and compared the results with the threshold magnetic field time derivatives necessary to stimulate nerves in the brain. In the calculations the transmission line model is used together with a return stroke speed of 1.5×10^8 m s^{-1}. The current at the channel base of the return stroke was represented by the following waveform, which has been adopted in lightning protection standards:

$$i = \frac{I}{k} \cdot \frac{(t/\tau_1)^{10}}{1 + (t/\tau_1)^{10}} \cdot \exp(-t/\tau_2) \tag{20.1}$$

where I gives the peak current and $\tau_1 = 19.0$ μs, $\tau_2 = 485$ μs and $k = 0.93$. Calculations were conducted for peak currents of 50, 100 and 200 kA.

Time-varying magnetic fields are being used routinely in medical diagnosis such as magnetic resonance imaging (MRI) and transcranial magnetic stimulation (TMS). The safety precautions necessary in using these techniques have motivated scientists to study the strength of the magnetic field time derivatives necessary for the excitation of nerves and cardiac muscle [70,71]. From a combination of theory and experiment, they have come up with threshold levels for the magnetic time derivatives necessary for nerve and cardiac stimulation [71]. Figure 20.7 shows the rate of change of a magnetic field applied in the form of a ramp required to excite nerves and to induce cardiac stimulation. In this graph the horizontal axis gives the duration of the ramp and the vertical axis the rate of change of the magnetic field. The curve for the cardiac stimulation corresponds to the most sensitive population percentile, while the nerve stimulation corresponds to the mean population. Usually, the safe level for magnetic fields in MRI is set to be about three times less than the one corresponding to nerve stimulation [70]. Because the full width of the lightning-generated magnetic field derivative evaluated in the study is ~20 μs, Cooray and Cooray [69] superimposed the peak values of these magnetic field time derivatives in Figure 20.7 at 20-μs pulse width. Based on this figure they pointed out that the magnetic field derivative of a strong lightning flash striking close to a person could not induce cardiac stimulation. This is in agreement with the conclusions made by Andrews and colleagues [68]. They also pointed out that, depending on the distance to the current path, the peak values of magnetic time derivatives exceed the values required for nerve stimulation. Based on these results they conclude that a person located within a few metres of the path of a lightning current could be exposed to a magnetic field derivative that is large enough to stimulate neurons in the brain. Combining this information with the fact that

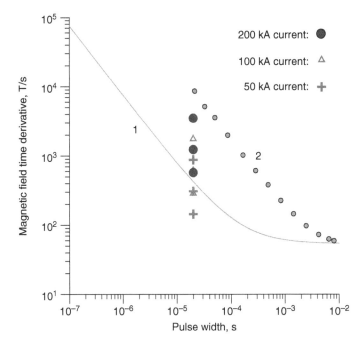

Figure 20.7 The rate of change of magnetic field applied in the form of a ramp required to excite the nerves (curve 1) and induce cardiac stimulation (curve 2) [70,71]. On the same figure, the peak magnetic field derivative for lightning flashes 10, 5 and 2 m from the channel is also depicted for currents of 50 kA, 100 kA and 200 kA.

intracranial magnetic stimulation, where the brain is exposed to strong magnetic field derivatives, can cause seizures in epileptic patients [72], Cooray and Cooray [69] inferred that the probability of a close lightning flash triggering an epileptic seizure of the occipital lobe of a person located close to lightning strikes is non-zero. Based on this study they suggested the possibility that some of the 'ball lightning observations' experienced by people located in the vicinity of lightning strikes could be hallucinations generated by seizures in the occipital lobe.

Cooray and Cooray [69] pointed out that the striking distance, i.e. the distance of attraction, of lightning flashes supporting large currents is larger than several tens of metres and the chances that lightning flashes of this magnitude will strike ground within 10 m of a human standing on open ground is rather small (see also Figures 20.1 and 20.2). In such cases the lightning flash would terminate on the human. However, they pointed out that there are several situations in which a person could be exposed to the magnetic fields generated by strong lightning flashes striking within 10 m. One such example is a person standing within 10 m of a tree or a high object struck by lightning. A similar scenario could also occur, for example, when lightning strikes a protected building. The current of the lightning flash flows along the down-conductors of the lightning protection system and the person could be standing within metres of such a conductor during a lightning strike. Thus, the probability that a human could be exposed to the magnetic field generated by a strong lightning flash within 10 m is not negligible, according to Cooray and Cooray [69].

20.4 Concluding remarks

The various types of injuries described in this paper are not limited to outdoor lightning victims. A person staying indoors can also receive injuries either through side flashes or by lightning surges travelling along telephone or electrical distribution lines. Indeed ~52 per cent of lightning accidents happen indoors [1]. Even though the magnitude of the current to which the body is exposed could be less than those of outdoor lightning injuries, almost all the injuries mentioned above can also happen indoors. Andrews [73] and Andrews and Darveniza [74] analysed over 300 cases of telephone-mediated lightning injuries and found that ~10 per cent of the victims were severely injured. This is less, however, than the 40–60 per cent for direct strike victims.

The information given here shows that an interaction with lightning strikes can have severe immediate as well as long-term consequences both for victims and their families. The best way to prevent being injured by lightning and suffering the resulting consequences is to take proper precautions during thunderstorms and to offer immediate medical assistance to those struck by lightning. Such precautions and advice are described in detail in References 75 and 76.

References

1. Elsom D.M. 'Deaths and injuries caused by lightning in the United Kingdom: analysis of two databases'. *Atmospheric Res.* 2000;**56**:325–34.
2. National Oceanic and Atmospheric Administration. Storm Data. 1985;**27**(12).
3. Bernstein T. *Lightning Death and Injury in the United States; Fourteen Years (1968–1981)*. Department of Electrical and Computer Engineering, University of Wisconsin, Madison, 1982.
4. Eriksson A.J., Smith M.A. 'A study of lightning fatalities and related incidents in Southern Africa'. *Trans. SA Inst. Elec. Eng.* 1986; 163–78.
5. Eriksson A., Örnehult L. 'Death by lightning'. *Am. J. Forens. Med. Pathol.* 1988;**9**(4):295–300.
6. Castle W.M., Kreft J.A. 'A survey of deaths in Rhodesia caused by lightning'. *Centr. Afr. J. Med.* 1974;**20**:93–95.
7. Virenque C., Laguerre J. 'Les accidents de la fulguration'. *Anesth. Analg. Réanim.* 1976;**33**:775–84.
8. Mackerras D. 'Occurrence of lightning death and injury'. *Lightning Injuries: Electrical, Medical and Legal Aspects*. Andrews *et al.* (eds.). CRC Press, London, 1992. pp. 47–70.

9. Sarma B.P., Sarma N. 'Epidemiology, morbidity, mortality and treatment of burn injuries: A study in a peripheral industrial hospital'. *Burns* 1994;**20**:253–55.
10. Duclos P.J., Sanderson L.M. 'An epidemiological description of lightning related deaths in the United States'. *Int. J. Epidemiol.* 1990;**19**:673–79.
11. Kitagawa N., Ohashi M., Ishikawa T. 'The lightning accidents that involve numerous people in the vicinities of struck points'. *Proceedings of the 26th International Conference on Lightning Protection*; Cracow, Poland, 2002.
12. Thottappillil R., Rakov V.A., Uman M.A., Beasley W.H., Master M.J., Shelukhin D.V. 'Lightning subsequent-stroke electric field peak greater than the first stroke peak and multiple ground terminations'. *J. Geophys. Res.* 1992;**97**:7503–509.
13. Mackerras D. 'Protection from lightning'. *Lightning Injuries: Electrical, Medical and Legal Aspects*. Andrews et al. (eds.). CRC Press, London, 1992.
14. Anderson R.B., Jandrell I.R., Nematswerani H.E. 'The upward streamer mechanism versus step potentials as a cause of injuries from close lightning discharges'. *Trans. S. Afr. Inst. Electr. Eng.* 2002; **93**(1):33–37.
15. Uman M.A. 'Physics of lightning phenomena'. *Lightning Injuries: Electrical, Medical and Legal Aspects.* C.J. Andres, M.A. Cooper, M. Darveniza, D. Mackerras (eds.). CRC Press, Boca Raton, FL, 1992. pp. 6–22.
16. Mackeras D. 'Protection from lightning'. *Lightning Injuries: Electrical, Medical and Legal Aspects.* C.J. Andres, M.A. Cooper, M. Darveniza, D. Mackerras (eds.). CRC Press, Boca Raton, FL, 1992. pp. 6–22.
17. Becerra M., Cooray V. 'On the physics of the interaction of aborted upward lightning connecting leaders with humans'. *Proceedings of the International Conference on Grounding and Earthing & 3rd International Conference on Lightning Physics and Effects*; Brazil, 2008.
18. Andrews C.J. 'Electrical aspects of lightning strikes to humans'. *The Lightning Flash*. V. Cooray (ed.). The Institute of Electrical Engineers, London, UK, 2003. pp. 549–64.
19. Ishikawa T., Ohashi M., Kitagawa N., Nagai Y., Miyazawa T. 'Experimental study on the lethal threshold value of multiple successive voltage impulses to rabbits simulating multi-strike lightning flash'. *Inst. J. Biometeorol.* 1985;**29**(2):157.
20. Carte A.E., Anderson R.B., Cooper M.A. 'A large group of children struck by lightning'. *Ann. Emerg. Med.* 2002;**39**(6):665–70.
21. Epperly T.D., Stewart J.R. 'The physical effects of lightning injury'. *J. Fam. Pract.* 1989;**29**:267–72.
22. Buechner H.A., Rothbaum J.C. 'Lightning stroke injury: A report of multiple casualties resulting from a single lightning bolt'. *Mil. Med* 1961;**153**: 755–62.
23. Dollinger S.J. 'Lightning strike disaster among children'. *Br. J. Med. Psychol.* 1985;**58**:375–83.
24. Anden G.P., Harrison S.H., Lister J. 'Lightning accident at Ascot'. *Br. Med. J.* 1956;**1**:1450–53.
25. Golde R.H. *Lightning Protection*. Edward Arnold Publishers, London, UK, 1973.

26. Webb J., Srinivasan J., Fahmy F., Frame J.D. 'Unusual skin injury from lightning'. *Lancet* 1996;**347**:321.
27. Graber J., Ummenhofer W., Herion H. 'Lightning accident with eight victims: Case report and brief review of the literature'. *J. Trauma* 1996;**40**:288–90.
28. Fahmy F.S., Brinsden M., Smith J., Frame J.D. 'Lightning: multisystem group injuries'. *J. Trauma* 1999;**46**(5):937–40.
29. Taussig H. '"Death" from lightning and the possibility of living again'. *Ann. Int. Med.* 1968;**68**:1345.
30. Eckert R., Randall D., Augustine G. *Animal Physiology*. Freeman and Company, New York, 1978.
31. Muehlberger P., Vogt M., Munster A.M. 'The long term consequences of lightning injuries'. *Burns* 2001;**27**:829–33.
32. Rhoades R.A., Tanner G.A. *Medical Physiology*. Lippincott Williams & Wilkins, New York, 2003.
33. Jackson S.H.D., Parry D.J. 'Lightning and heart'. *Br. Heart J*. 1980;**43**:454.
34. Eber B., Himmel G., Schubert B. 'Myokardiale schadinung nach blitzschlag'. *Z. Kardiol*. 1989;**78**:402.
35. Schwab M., Wiegand H., Bentsen P. 'Infrakt-EKG nach blitzunfall'. *Z. Kardiol* 1989;**78**:811.
36. Andrews C.J., Coorper M.A. 'Clinical presentation of the lightning victims'. *Lightning Injuries: Electrical, Medical and Legal Aspects*. Andrews *et al.* (eds.). CRC Press, London, 1992.
37. Read J.M. 'Man struck by lightning reveals marked ECG changes hours later'. *Med. Trib.* p3, March 28, 1966.
38. Zeana C.D. 'Acute transient myocardial ischaemia after lightning injury'. *Int. J. Cardiol*. 1984;**5**:207.
39. Moulson A.M. 'Blast injury of the lungs due to lightning'. *Br. Med. J.* 1984;**289**:1270.
40. Andrews C. 'Structural changes after lightning strike, with special emphasis on special sense orifices as portals of entry'. *Semin. Neurol.* 1995;**15**(3):296–303.
41. Norman M.E., Albertson D., Younge B.R. 'Ophthalmic manifestations of lightning stroke'. *Surv. Ophthalmol.* 2001;**46**(1):19–24.
42. Hunt L. 'Ocular injuries induced by lightning'. *Insight* **XXV**(2), 2000.
43. Espaillat A., Janigian R. K. 'To, cataracts, bilateral macular holes and rhegmatogenous retinal detachment induced by lightning'. *Am. J. Ophthalmol.* 1999;**127**(2):216–17.
44. Noel L.P., Clarke W.N., Addison D. 'Ocular complications of lightning'. *J. Pediatr. Ophthalmol. Strabismus* 1980;**17**:245–46.
45. Dwyer J.R. *et al*. 'Energetic radiation produced during rocket triggered lightning'. *Science* 2003;**299**:694–97.
46. Dinakaran S., Desai S.P., Elsom D.M. 'Telephone-mediated lightning injury causing cataract'. *Injury* 1998;**29**(8):645–46.
47. Abt J.L. 'The papillary response after being struck by lightning'. *JAMA* 1985;**254**:3312.

48. Coorper M.A. 'Lightning injuries: prognostic signs for death'. *Ann. Emerg. Med.* 1980;**9**:134–38.
49. Andrews C.J., Darvaniza M., Mackerras D. 'Lightning injury: A review of clinical aspects, pathophysiology and treatment'. *Adv. Trauma* 1989;**4**:241–87.
50. Hill R.D. 'Thunder'. *Lightning.* R.H. Golde (ed.). vol. 1, Academic Press, New York, 1977.
51. Lewis A.M. 'Understanding the principles of lightning injuries'. *J. Emerg. Nursing* 1997;**23**:535–41.
52. Mora-Magana I., Collado-Corona M.A., Toral-Martinon R., Cano A. 'Acoustic trauma caused by lightning'. *Int. J. Pediatr. Otorhinolaryngol.* 1996;**35**:59–68.
53. ten Duis H.J., Klasen H.J., Reenalda P.E. 'Keraunoparalysis, a 'specific' lightning injury'. *Burns Incl. Therm. Inj.* 1985;**12**(1):54–57.
54. Richards A. 'Traumatic facial palsy'. *Proc. R. Soc. Med.* 1973;**66**:28.
55. Lim J.-K., Lee E.-H., Chhem R.K. 'Physeal injury in a lightning strike survivor'. *J. Pediatric Orthopaedics* 2001;**21**(5):608–12.
56. Ohashi M., Kitagawa N., Ishikawa T. 'Lightning injury caused by discharges accompanying flashovers: a clinical and experimental study of death and survival'. *Burns* 1987;**13**:141–46.
57. Courtman S.P., Wilson P.M., Mok Q. 'Case report of a 13-year-old struck by lightning'. *Paediatric Anaesthesia* 2003;**13**:76–79.
58. ten Duis H.J. *et al.* 'Superficial lightning injuries – their 'fractal' shape and origin'. *Burns* 1987;**13**(2):141–46.
59. Celiköz B., Isik S., Turegun M., Selmanpakoglu N. 'An unusual case of lightning strike: full-thickness burns of the cranial bones'. *Burns* 1996;**22**(5):417–19.
60. Coorper M.-A., Andrews C.J. 'Disability, not death is the issue'. *International Conference on Lightning and Static Electricity*; Blackpool, United Kingdom, 2003.
61. Panse F. 'Electrical trauma'. *Handbook of Clinical Neurology*, Vinken *et al.* (eds.). vol. 23, ch. 34, Elsevier, Amsterdam, 1975.
62. Tibesar R.J., Saswata R., Hom D.B. 'Fracture from a lightning strike injury to the face'. *Otolaryngology – Head and Neck Surgery*, 647–49, November 2000.
63. Morse M.S., Berg J.S., TenWolde R.L. 'Diffuse electrical injury: A study of 89 subjects reporting long-term symptomatology that is remote to the electrical current pathway'. *IEEE Trans. Biomed. Eng.* 2004;**51**(8):1449.
64. Andrews C.J. 'Further documentation of remote effects of electrical injuries, with comments on the place of neuropsychological testing and functional scanning'. *IEEE Trans. Biomed. Eng.* 2006;**53**(10):2102–13.
65. Andrews C.J. 'Acceptance speech on the Award of the Nobu Kitagawa Medal'. *Proceedings of the International Conference on Lightning and Static Electricity (ICOLSE)*; Paris, 2006.
66. Engelstatter G.H. 'Neuropsychological and psychological sequelae of lightning and electric shock injuries'. Presented at *Fourth Annual Meeting of Lightning Strike and Electric Shock Victims International*; Maggie Valley, NC, May 1994.

67. Cherington M., Wachtel W., Yarnell P. 'Could lightning injury be magnetically induced?' *Lancet* 1998;**351**:1788.
68. Andrews C., Cooper M.A., Kotsos T., Kitagawa N., Mackerras D. 'Magnetic effect of lightning strokes on the human heart'. *J. Lightning Res.* 2007;**1**:158–65.
69. Cooray G., Cooray V. 'Could some ball lightning observations be hallucinations caused by epileptic seizures?' *Open Access Atmos. Sci. J.* 2008;**2**:101–105.
70. Reilly J.P. 'Peripheral nerve stimulation by induced electric currents: exposure to time-varying magnetic fields'. *Med. Biol. Eng. Comput.* 1989;**27**:101–10.
71. Schaefer D.J., Bourland J.D., Nyenhuis J.A. 'Review of patient safety in time varying gradient fields'. *J. Magn. Res. Imaging* 2000;**12**:20–29.
72. Tassinari C.A., Cincotta M., Zaccara G., Michelucci R. 'Transcranial magnetic stimulation and epilepsy'. *Clin. Neurophysiol.* 2003;**114**:777–98.
73. Andrews C.J. 'Telephone-related lightning injury'. *Med. J. Austr.* 1992;**157**:823–26.
74. Andrews C.J., Darvaniza M. 'Telephone-mediated lightning injury: An Australian survey'. *J. Trauma* 1989;**29**(5):665–71.
75. Zimmermann C., Coorper M.A., Holle R.L. 'Lightning safety guidelines'. *Ann. Emerg. Med.* 2002;**39**(6):660–65.
76. Lederer W., Wiedermann F.J., Cerchiari E., Baubin M.A. 'Electricity-associated injuries II: outdoor management of lightning induced casualties'. *Resuscitation* 2000;**43**:89–93.

Chapter 21
Lightning standards
Fridolin Heidler and E.U. Landers

21.1 Introduction

The international lightning standards are issued by the International Electrical Commission (IEC) and lain down in the IEC 62305 standard series developed by the Technical Committee TC 81 of IEC. The following parts of this series are in force.

- IEC 62305-1 [1]: Part 1 provides the principles of lightning protection.
- IEC 62305-2 [2]: Part 2 gives the procedures for the evaluation of risk due to lightning (see Chapter 8 of this book).
- IEC 62305-3 [3]: Part 3 deals with protection against physical damages and life hazard in a structure.
- IEC 62305-4 [4]: Part 4 provides the principles of lightning protection of electrical and electronic systems within structures.

The IEC 62305 standards are not mandatory, but all nations are encouraged to transfer them to national standards. In the European Union (EU) the standards are accepted and lain down in the EN 62305 standard series. These standards are mandatory for members of the EU.

The scope of the IEC 62305 series is restricted to immobile common structures located on earth, as buildings, towers and industrial facilities. These structures are typically provided by such services as water or gas pipes, electrical power lines, telecommunication lines, and so on.

Mobile systems such as vehicles, boats or aircrafts, and moveable systems such as tents or containers are not connected to any of such services or only to a minor number. These systems do not have earth termination systems comparable to immobile structures. Therefore, different or additional regulations are used for mobile and moveable systems. Different regulations are also necessary for special structures such as nuclear power plants, offshore installations or underground high-pressure pipelines. Moreover, the lightning protection of military equipment is fixed in military standards. Therefore, all these systems are outside the scope of the IEC 62305 standard series.

The IEC 62305 standard series provides the principles to be followed in the lightning protection of structures including their installations, services and contents as well as persons. This standard series does not cover protection against electromagnetic interference due to lightning, which may cause malfunctioning of electronic systems. Lightning protection against malfunctioning is included in the general rules of electromagnetic compatibility. These rules are given in the IEC 61000 standard series. In this series the immunity of susceptible devices against the electromagnetic environment is fixed, including the requirements of lightning protection.

The lightning protection concept is based on actual lightning data measured over years in observation stations. Such data give a continuously distributed probability function for each lightning current parameter in the shape of logarithmic normal distributions. It is obvious, for instance, that an industrial plant needs better lightning protection than a simple barrack. Therefore, in IEC 62305 four lightning protection levels (LPL) are defined, accepting that lightning currents may exceed the defined levels with a certain probability. The lightning protection of the structures has to be realized according to the chosen lightning protection level.

Lightning current is the primary source of damage. Depending on the point of strike the following four sources are considered in IEC 62305: flashes to the structure, near the structure, to the services and near the services connected to the structure. Whereas protection of electrical and electronic systems (IEC 62305-4) considers all these sources, the protection against physical damage and life hazard (IEC 62305-3) has to consider only flashes direct to the structure or direct to its services.

To protect against direct flashes IEC 62305-3 provides a lightning protection system (LPS) using the following measures (see Section 21.4): air termination system, down-conductor system, earth termination system, lightning equipotential bonding of all services at the entry point and a separation distance against dangerous sparking.

To protect even sensitive electrical and electronic systems IEC 62305-4 provides a LEMP protection measures system (LPMS) using, additionally (see Section 21.5), an equipotential bonding network, coordinated SPD protection, line routing and shielding, spatial shielding and a safety distance against too high magnetic fields.

Protection against the surges coming in from the services requires equipotential bonding at the entry point. Where galvanic bonding is not possible as for electrical live conductors, so-called surge protective devices (SPDs) are installed. The requirements, testing procedures and principles for their selection and application are given in the IEC 61643 standard series.

21.2 Standardized lightning currents

21.2.1 Threat parameters of the lightning current

Lightning current is the primary source for damages, disturbances and malfunctions. The lightning threat is associated with the following parameters defined in IEC 62305-1 [1]:

- peak current I
- maximum current steepness $\left(\dfrac{di}{dt}\right)_{max}$

- charge $Q = \int i \times dt$
- specific energy $W/R = \int i^2 \times dt$

Peak current is important for the design of the earth termination system. When the lightning current enters the earth, the current flowing through the earthing resistance produces a voltage drop. The peak current determines the maximum of this voltage.

Electronic devices are normally connected to different electrical services such as the mains supply and the data link. Depending on the line routing and grounding inside the structures, large open-loop networks are often built up. The maximum current steepness is responsible for the maximum of the magnetically induced voltages in such open loops.

The charge Q is responsible for melting effects at the attachment points of the lightning channel. The energy input at the arc root is given by the anode/cathode voltage drop $u_{a,c}$ multiplied by the charge Q. In IEC 62305-1 [1], a worst-case approach assumes that the total energy input is used only for melting. Following this assumption, the melted volume V is given by

$$V = \frac{u_{a,c} Q}{\gamma} \times \frac{1}{c_w (\vartheta_s - \vartheta_u) + c_s} \tag{21.1}$$

where γ denotes the material density, c_s the latent heat of melting and c_w the thermal capacity. The temperature rise is given by the difference between the melting temperature ϑ_s and the ambient temperature ϑ_u.

The specific energy W/R is responsible for the heating effects arising when the lightning current flows through a metallic conductor of cross-section q. The temperature rise can be evaluated as

$$\vartheta - \vartheta_u = \frac{1}{\alpha} \left[\exp \frac{W/R \times \alpha \times \rho_0}{q^2 \times \gamma \times c_w} - 1 \right] \tag{21.2}$$

where α denotes the temperature coefficient of the resistance and ρ_0 the specific ohmic resistance at ambient temperature. Typical materials for the down-conductors and air termination are copper, aluminium or steel. Table 21.1 presents the physical quantities of these materials according to IEC 62305-1 [1].

Table 21.1 Parameters of metallic materials used in lightning protection systems

Quantity	Unit	Material			
		Aluminium	Copper	Mild steel	Stainless steel
ρ_0	$\Omega\,m$	29×10^{-9}	17.8×10^{-9}	120×10^{-9}	0.7×10^{-6}
α	$1/K$	4.0×10^{-3}	3.92×10^{-3}	6.5×10^{-3}	0.8×10^{-3}
γ	$kg\,m^{-3}$	2 700	8 920	7 700	8×10^3
ϑ_s	°C	658	1 080	1 530	1 500
c_s	$J\,kg^{-1}$	397×10^3	209×10^3	272×10^3	–
c_w	$J\,kg^{-1}\,K^{-1}$	908	385	469	500

21.2.2 Current waveforms

For each of the four lightning protection levels (LPLs) a set of maximum lightning current parameters is fixed in IEC 62305-1 [1]. These maximum current values define the lightning threat for lightning protection components and for the equipment to be protected. Table 21.2 gives the probability that the lightning current exceeds the data fixed for the LPL. For instance, for LPL I it is accepted that 1 per cent of the anticipated currents exceeds one or more of the fixed current parameters.

Table 21.2 *Probability that the lightning current parameters exceed the data fixed for a particular lightning protection level (LPL)*

Lightning protection levels	I	II	III	IV
Probability (%)	1	2	3	3

The set of lightning currents comprises the first short stroke current, the subsequent short stroke current and the long stroke current. The first and subsequent short stroke currents are associated with the first and subsequent return strokes, while the long stroke current simulates the continuing current. For calculation purposes, the short stroke currents are defined by the formula

$$i = \frac{I}{k} \times \frac{(t/\tau_1)^{10}}{1 + (t/\tau_1)^{10}} \times \exp(-t/\tau_2) \quad (21.3)$$

where t denotes time and I the peak current. The correction factor k is necessary to achieve the correct peak current I. The waveform is fixed by the front time constant τ_1 and the tail time constant τ_2. The parameters are listed in Table 21.3 for the first and subsequent short stroke currents. Figures 21.1 and 21.2 show the current waveforms during the rise and the decay. The definitions of the rise time T_1 and the decay time T_2 are also given in these figures. With the fixed parameters, the current

Table 21.3 *Parameters of the short stroke current to be used for equation (21.3)*

Parameters	First short stroke current			Subsequent short stroke current		
	LPL			LPL		
	I	II	III–IV	I	II	III–IV
I (kA)	200	150	100	50	37.5	25
k	0.93	0.93	0.93	0.993	0.993	0.993
τ_1 (µs)	19.0	19.0	19.0	0.454	0.454	0.454
τ_2 (µs)	485	485	485	143	143	143

LPL, lightning protection level.

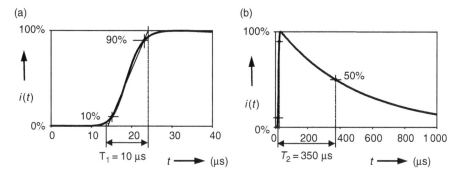

Figure 21.1 Waveform of the first short stroke current during the rise (a) and the decay (b)

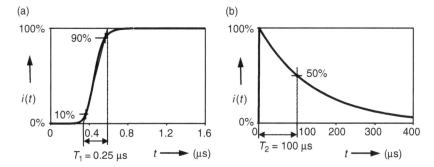

Figure 21.2 Waveform of the subsequent short stroke current during the rise (a) and the decay (b)

of the first short stroke results in the 10/350 μs waveform, and the subsequent short stroke is represented by the 0.25/100 μs waveform. Except for the peak current, these current waveforms are not altered for the different LPLs.

The current waveform of the long stroke current is not fixed by a mathematical formula. It is considered that the current can be represented by a rectangular waveform having an average current I and duration T_{long}. The long stroke charge Q_{long} results directly from the multiplication of I and T_{long}.

Table 21.4 gives an overview of the lightning parameters fixed for the four LPLs in IEC 62305-1 [1]. Compared to the subsequent short stroke current, the first short stroke current has much higher values of peak current, short stroke charge and the specific energy. Consequently, these quantities are taken into account only for the first short stroke current. Following this simplification, the flash charge is composed only of the charge of the first short stroke current and of the long stroke current. On the other hand, the current steepness of the subsequent short stroke is about one order of magnitude higher compared to the first short stroke. Therefore, the current steepness is only fixed for the subsequent short stroke current.

Table 21.4 Lightning parameters according to lightning protection level (LPL)

Current parameters	Symbol	Unit	LPL			
			I	II	III	IV
First short stroke						
Peak current	I	kA	200	150	100	
Short stroke charge	Q_{short}	C	100	75	50	
Specific energy	W/R	MJ Ω^{-1}	10	5.6	2.5	
Time parameters	T_1/T_2	µs/µs		10/350		
Subsequent short stroke						
Peak current	I	kA	50	37.5	25	
Average steepness	di/dt	kA µs^{-1}	200	150	100	
Time parameters	T_1/T_2	µs/µs		0.25/100		
Long stroke						
Long stroke charge	Q_{long}	C	200	150	100	
Time parameter	T_{long}	s		0.5		
Flash						
Flash charge	Q_{flash}	C	300	225	150	

21.2.3 Requirements for the current tests

Owing to the limitations of the test equipment, in laboratory tests the fixed stroke currents can be realized only within certain tolerances. The peak current I, the charge Q_{short}, and the specific energy W/R are tested only for the first short stroke current. The test parameters and their tolerances are listed in Table 21.5. The test parameters have to be obtained in the same test current.

Table 21.5 Test parameters of the first short stroke

Test parameter	Unit	LPL			Tolerance (%)
		I	II	III–IV	
Peak current, I	kA	200	150	100	±10
Charge, Q_{short}	C	100	75	50	±20
Specific energy, W/R	MJ Ω^{-1}	10	5.6	2.5	±35

LPL, lightning protection level.

For the long stroke current, the long stroke charge Q_{long} is fixed with a duration of 0.5 s. The test parameters and the tolerances are listed in Table 21.6.

The current steepness can be tested for the first short stroke and the subsequent short stroke. For these tests, there are no special requirements defined regarding the current decay. It is only necessary to meet the requirements of the current front

Table 21.6 Test parameters of the long stroke

Test parameter	Unit	LPL			Tolerance (%)
		I	II	III–IV	
Charge, Q_{long}	C	200	150	100	±20
Duration, T_{long}	s	0.5	0.5	0.5	±10

LPL, lightning protection level.

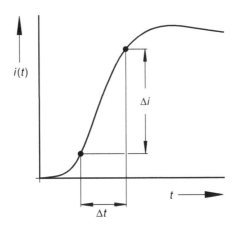

Figure 21.3 Definition of the current steepness for test purposes

Table 21.7 Test parameters to simulate the current steepness of the short strokes

Type of short stroke	Test Parameter	Unit	LPL			Tolerance (%)
			I	II	III–IV	
First	Δi	kA	200	150	100	±10
	Δt	µs	10	10	10	±20
Subsequent	Δi	kA	50	37.5	25	±10
	Δt	µs	0.25	0.25	0.25	±20

LPL, lightning protection level.

shown in Figure 21.3, where the current rise is fixed by the current value Δi and the time duration Δt. The associated test parameters are listed in Table 21.7.

21.3 Determination of possible striking points

21.3.1 Rolling sphere method

The rolling sphere method is suggested in IEC 62305-3 [3] for use in the detection of possible striking points. This is a universal method and there are no limitations

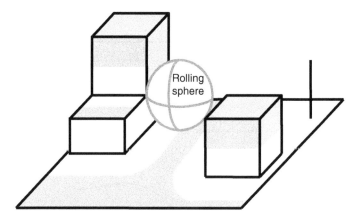

Figure 21.4 Application of the rolling sphere method to a structure to be protected

regarding the structure to be protected. Figure 21.4 shows an example of how to apply this method. A sphere with a certain radius r is rolled around and over the structure to be protected in all feasible directions. Lightning strikes are possible at any points touched by the sphere. The shaded areas are exposed to lightning interception and need lightning protection. Lightning strikes are excluded at all points not touched by the rolling sphere. In this way the protected volume is derived. The unshaded areas are within the protected volume.

Following the electro-geometric model the rolling sphere radius r is identical to the striking distance. The radius r (m) is correlated with the peak current I (kA) of the first short stroke and is given by the relation [1]

$$r = 10\, I^{0.65} \tag{21.4}$$

Table 21.8 contains the rolling sphere radii and the associated current peaks for the different LPL fixed in IEC 62305-1 [1]. The minimum values of the rolling sphere radius r define the interception efficiency of the LPS. Therefore, LPL I has the smallest rolling sphere radius, and LPL IV has the highest. The probability P denotes the percentage of lightning with a current peak lower than the current peak valid for the LPL. For these lightning, it cannot be excluded that they terminate inside the protected volume.

Table 21.8 Values of the rolling sphere radius, the associated current peak and the probability of possible lightning termination inside the protected volume

LPL	I	II	III	IV
Current peak, I (kA)	3	5	10	16
Rolling sphere radius, r (m)	20	30	45	60
Probability, P (%)	1	3	9	16

21.3.2 Mesh method

The mesh method can only be applied to air termination systems consisting of meshes with metal wires. The method is based on the rolling sphere method, which gives the sag between the meshed wires. The interspacing is reduced to such small distance that the sag is negligible. In IEC 62305-3 [3] it is considered that the structure covered by such a small-meshed air termination system is within the protected volume.

21.3.3 Protection angle method

Figure 21.5 shows the fundamental assumptions of the protection angle method considering a Franklin rod. According to the rolling sphere model the boundary of the volume protected is given by the segment of a circle. The protection angle α is chosen in such a way that area A_1 and area A_2 are equal. The resulting protected volume becomes the shape of a cone. The protection angle α is given by the formula

$$\alpha = \arctan\left[\frac{\sqrt{2rh - h^2}}{h} + \frac{r\sqrt{2rh - h^2}}{h^2} - \frac{r^2}{h^2}\arccos\left(1 - \frac{h}{r}\right)\right] \quad (21.5)$$

The protection angle α depends on the height h above a reference plane. Only for simple structures is the reference plane equal to the earth surface. For more

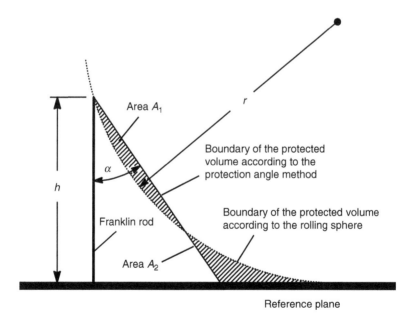

Figure 21.5 Application of the protection angle method for a Franklin rod of height h, where r is the rolling sphere radius and α the protection angle

complex structures, additional reference planes have to be defined. The protection angle method is therefore limited to simple structures. This method cannot be applied if the height h is greater than the rolling sphere radius r.

21.4 The lightning protection system (LPS)

According to IEC 62305-3 [3], the lightning protection system (LPS) consists of the external and internal lightning protection systems. The functions of the external lightning protection are to intercept a lightning flash, to conduct the lightning flash safely to earth and to disperse it into the earth. The function of the internal lightning protection is to prevent dangerous sparking within the structure. The LPS does not provide sufficient lightning protection for the electrical and electronic equipment installed inside the structure. It is allowed to include 'natural' components in the lightning protection concept. 'Natural' components are conductive parts not installed specially for the lightning protection, e.g. metal components of the roof construction. They can be used in addition to the LPS or in some cases can provide the function of one or more parts of the LPS.

In IEC 62305-3 [3] four classes of LPS (I, II, III, IV) are defined as a set of construction rules, based on the four corresponding LPLs. Each set includes level-dependent construction rules (e.g. mesh size of the air termination system) and construction rules specified independently from the protection level (e.g. cross-section of the down-conductors).

The LPS consists of the following five components [3]:

- air termination system (component of the external LPS)
- down conductor system (component of the external LPS)
- earth termination system (component of the external LPS)
- lightning equipotential bonding (component of the internal LPS)
- separation distance s (component of the internal LPS)

21.4.1 Air termination system

The air termination system is necessary to intercept the lightning flash. It can be composed of any combinations of wires, meshed conductors and rods, including 'natural' components. The air termination components have to withstand the minimum requirement of the specific energy. For a specific energy of 10 MJ Ω^{-1}, according to LPL I, the minimum cross-section to avoid melting is 16 mm^2 for copper, 25 mm^2 for aluminium and 50 mm^2 for steel and stainless steel. Table 21.9 gives an overview of materials and their configurations. For all materials the minimum cross-section is fixed at 50 mm^2, because some special configurations require an increase of the cross-section to a somewhat greater area.

The rolling sphere method is suggested as a general principle for the positioning of the air termination system. It is also allowable to use the protective angle method for simple shaped buildings up to heights equal to the radius of the rolling sphere. The mesh method is suitable only for plane surfaces to be protected by a mesh.

Table 21.9 Materials, configuration and cross-sections for air termination conductors and down-conductors

Material	Configuration	Minimum cross-section (mm²)
Copper, tin-plated copper	Solid tape, solid round, stranded	50
Aluminium	Solid tape	70
	Solid round, stranded	50
Aluminium alloy	Solid tape, solid round, stranded	50
Steel	Solid tape, solid round, stranded	50
Stainless steel	Solid tape, solid round	50
	Stranded	70

Table 21.10 Values of the rolling sphere radius, mesh size and protection angle corresponding to the class of lightning protection system (LPS)

Class of LPS	Rolling sphere method Radius r (m)	Mesh method Mesh size (m × m)	Protection angle method
I	20	5 × 5	Equation (21.5)
II	30	10 × 10	r, rolling sphere radius of the class of LPS
III	45	15 × 15	h, height above reference plane with $h \leq r$
IV	60	20 × 20	

Table 21.10 contains the values of the rolling sphere radius, mesh size and protection angle corresponding to the class of LPS.

21.4.2 Down-conductor system

The down-conductor system is necessary to conduct the lightning flash safely to earth. The downward conductors should be installed straight and oriented vertically such that they provide the shortest and most direct path to earth. The cross-sections are the same as for the air termination conductors listed in Table 21.9. Natural components can be used. In particular, the reinforcement of a building can be part of the down-conductor system.

21.4.3 Earth termination system

The earth termination system is necessary to disperse the current into the earth. There is no special value fixed in the standard, but in general a low earthing resistance of less than 10 Ω, measured at low frequency, is recommended. The dimension of each earth

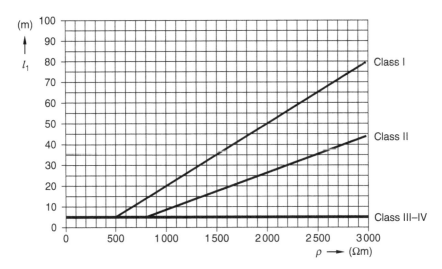

Figure 21.6 Reference length l_1 of an earth electrode as a function of soil resistivity ρ

electrode has to be chosen according to the reference length l_1 shown in Figure 21.6. For LPS I and LPS II the reference length varies depending on the soil resistivity ρ, whereas for the LPS III and LPS IV the reference length is fixed to a constant value of 5 m.

For the earth termination systems, two basic types of earth electrode arrangements apply, the type A arrangement and the type B arrangement. The type A arrangement comprises a set of horizontal or vertical earth electrodes connected to each down-conductor. The total number of earth electrodes should not be less than two. The minimum length is fixed to

$$l_{min} = l_1 \quad \text{for each horizontal electrode} \quad (21.6a)$$
$$l_{min} = 0.5 l_1 \quad \text{for each vertical electrode} \quad (21.6b)$$

The type B arrangement comprises either a ring conductor external to the structure or a foundation earth electrode. The mean radius r_e of the area enclosed by the ring conductor or the foundation earth electrode should not be less than the value of reference length l_1:

$$r_e \geq l_1 \quad (21.7)$$

If the radius r_e is smaller than the reference length l_1, additional horizontal or vertical electrodes should be added with individual length as follows:

$$l_r = l_1 - r_e \quad \text{for each horizontal electrode} \quad (21.8a)$$
$$l_v = 0.5 \, (l_1 - r_e) \quad \text{for each vertical electrode} \quad (21.8b)$$

The number of such additional electrodes should not be less than the number of down-conductors, with a minimum number of two.

The external ring electrode (type B arrangement) should be buried at a depth of at least 0.5 m and at a distance of ~ 1 m around the external walls. The earth electrodes of type A arrangement should be installed outside the structure to be protected at a depth of the upper end of at least 0.5 m.

21.4.4 Lightning equipotential bonding

The lightning equipotential bonding should avoid dangerous sparking between the external LPS and the other conductive parts. Equipotentialization is achieved by interconnecting the LPS using

- structural metal parts
- metal installations
- external conductive parts and lines connected to the structure
- electrical and electronic systems within the structure to be protected

Permanent interconnecting means are bonding conductors. These permanent connections cannot be applied for electrical lines under voltage in normal operation. In this case the conductors are bonded by surge protective devices (SPDs) making connections only during the duration of the lightning surge.

In the case of isolated external LPSs lightning equipotential bonding should be established at ground level only. For not-isolated external LPSs the lightning equipotential bonding should be installed at the basement or approximately at ground level. The bonding bar should be connected to the earth termination system. For large structures exceeding 20 m in length, more than one bonding bar can be installed.

Table 21.11 contains the minimum cross-sections required for the bonding conductors. The cross-sections depend on the material used and on the type of connection, but not on the class of LPS. The first type of connection is applied between different bonding bars or from bonding bars to the earth termination. These connections have relatively high cross-sections, because a remarkable fraction of the lightning current may flow through them. The second type of connection is applied between internal metal installations to the bonding bar. Their cross-sections are about a factor of three lower, because of the much lower currents expected.

Table 21.11 Minimum cross-sections (mm^2) of bonding conductors

Material	Connection between different bonding bars or from bonding bars to the earth termination	Connection between internal metal installations to the bonding bar
Copper	14	5
Aluminium	22	8
Steel	50	16

21.4.5 Separation distance

The electrical isolation between the air termination or the down-conductor system and the internal metallic installations can be achieved by a separation distance between the parts. Where the requirements of the isolation are not fulfilled, the lightning equipotential bonding is to apply to avoid dangerous sparking. The minimum separation distance is given by

$$s = k_i(k_c/k_m)l \tag{21.9}$$

where the coefficients k_i, k_c and k_m denote non-dimensional quantities. The length l has to be considered along the air termination or the down-conductor, from the point where the separation distance is to be considered to the nearest equipontential bonding point.

The values of the coefficients k_i, k_c and k_m are given in Tables 21.12 to 21.14. Because the coefficient k_i is based on the current steepness of the subsequent short stroke current, this coefficient varies depending on the LPL (Table 21.4). The coefficient k_c takes into account the current share through the downward conductors. This current share mainly depends on the number of down-conductors. Because the earth termination and the arrangement of the downward conductor and air termination system are also of influence, this value can be given only approximately. A somewhat

Table 21.12   Values of the coefficient k_i

Class of LPS	k_i
I	0.08
II	0.06
III and IV	0.04

Table 21.13   Values of the coefficient k_c

Number of down-conductors	k_c (approximate values)
1	1
2	0.66
4 and more	0.44

Table 21.14   Values of the coefficient k_m

Material	k_m
Air	1
Concrete, bricks	0.5

more precise method is given in Annex C of IEC 62305-3 [3]. The coefficient k_m considers the different spark-over processes in air and through concrete or bricks.

21.5 The LEMP protection measures system (LPMS)

The electrical and electronic devices inside buildings have to be protected against the lightning electromagnetic impulse (LEMP). In IEC 62305-4 [4], the item LEMP is not restricted to the electromagnetic field as considered usually. The LEMP involves all electromagnetic effects due to lightning current, including the conducted surges as well as the radiated impulse electromagnetic field effects.

The LPMS comprises the following protection measures [4]:

- the lightning protection zones concept
- earthing and bonding (including external LPS components and a bonding network)
- line routing and shielding (to minimize induction effects)
- coordinated SPD protection (for adequate limiting of surges)
- spatial magnetic shielding (to minimize the magnetic field)

21.5.1 The lightning protection zones (LPZ) concept

Protection against the LEMP is based on the lightning protection zones (LPZ) concept. The principle of LPZ requires forming nested zones of successively reduced values of the electromagnetic environment. This objective is achieved by (i) shielding to reduce the electromagnetic fields and (ii) equipotential bonding of all lines at the LPZ boundaries to limit the line-conducted surges (over-voltages and -currents). Figure 21.7 depicts the principles of the LPZ concept. The LPZ are assigned volumes of space where the LEMP severity is compatible with the immunity level of the internal systems enclosed. Successive zones are characterized by significant reduction of the LEMP severity.

The outer zone LPZ 0 is endangered by the unattenuated lightning field and by surges up to the full or partial level of the lightning current. LPZ 0 is subdivided into LPZ 0_A and LPZ 0_B. LPZ 0_A is endangered by direct lightning strikes, by the unattenuated lightning field and by the full lightning current surges. In contrast, direct lightning strikes are excluded from LPZ 0_B as well as from the inner zones of LPZ 1, 2 or higher. In LPZ 1 the surges are limited by current sharing and by SPDs at the boundary. The lightning fields can be attenuated by spatial shielding. In LPZ 2 the surges are further limited by current sharing and by SPDs at the boundary. The lightning fields are usually attenuated by spatial shielding. In special cases LPZ 3 or higher may be necessary for a further reduction of the surges and fields.

As shown in Figure 21.8 an LPS according to IEC 62305-3 (see Section 21.4) usually has only a single lightning protection zone LPZ 1. The primary electromagnetic source to harm the electrical and electronic systems is the lightning current I_0 and the associated magnetic field H_0. The LPS intercepts the lightning and protects against the penetration of the lightning current I_0. The SPD installed at the boundary

940 Lightning Protection

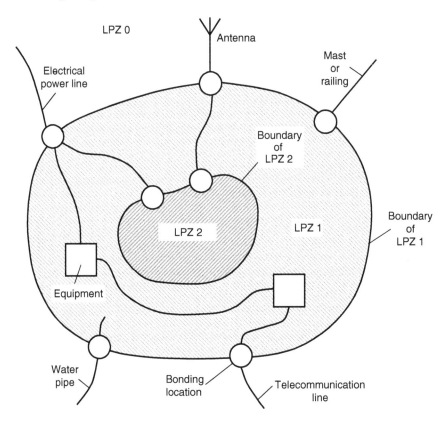

Figure 21.7 Principle of the LPZ concept

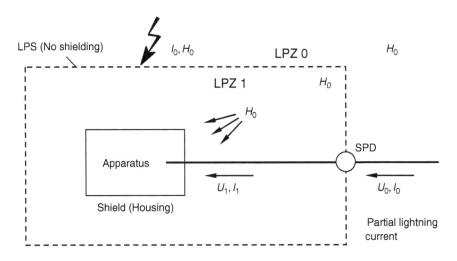

Figure 21.8 Example of a structure with LPS according to IEC 62305-3 (Section 21.4)

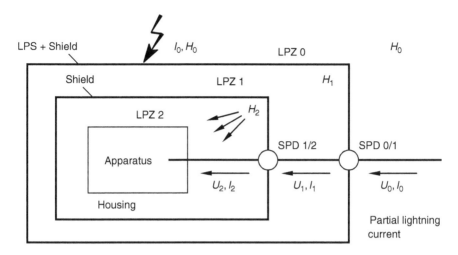

Figure 21.9 Example of a structure with a LEMP protection measures system (LPMS)

of the LPS limits the incoming conducted surges as the (partial) lightning current. These are the basic requirements to build up the LPZ 1. A higher-level LPZ cannot be achieved, because the magnetic field is not significantly attenuated by the LPS. Moreover, the incoming conducted surges (U_0, I_0) are limited only to a basic level (U_1, I_1), which commonly exceeds the immunity level of the electronic equipment. Therefore, the apparatus in Figure 21.8 has no effective protection either against conducted surges or against radiated magnetic fields.

The adequate protection of the apparatus therefore requires a LEMP protection measures system (LPMS) according to IEC 62305-4. The LPMS comprises the complete system of protection measures for internal systems against LEMP. Basic requirements of the LPMS are an adequate earth termination system and an appropriate bonding system. Figure 21.9 shows an example of an LPMS, where LPZ 1 is formed by the spatial shield of LPZ 1. This spatial shield reduces the magnetic field H_0 of LPZ 0 to the lower level magnetic field H_1 inside LPZ 1. The SPDs installed at the boundary of LPZ 1 reduce the incoming surges to (U_1, I_1). The successive LPZ 2 is formed by a second spatial shield, in which the magnetic field H_2 is significantly lower compared to LPZ 1. The LPZ 2 also requires that the conducted surges coming in from LPZ 1 are further reduced to (U_2, I_2) by the SPD at the boundary of LPZ 2.

21.5.2 Earthing system and bonding network

Of course, the LPMS also has to intercept, down-conduct and disperse direct lightning flashes to earth and requires therefore adequate external LPS components. However, to avoid dangerous potential differences between all the equipment inside the inner LPZ, a low-impedance bonding network is also needed. Moreover, such a bonding network also reduces the magnetic field. This can be realized by a meshed bonding network integrating any conductive parts of the structure or parts of the internal systems and by bonding all metal parts or conductive services at the boundary of each LPZ directly or by suitable SPD. It can be set up by a three-dimensional, meshed bonding network with a typical mesh width of 5 m. This requires multiple interconnections of all metal components in and on the structure. The bonding bars and magnetic shields of the LPZ should be similarly integrated.

21.5.3 Line routing and shielding

The electrical and electronic devices are interconnected by different networks for power and data. Induction effects in the loops of such networks can be reduced by minimizing loop areas by adequate line routing or the use of shielded cables.

21.5.4 Coordinated surge protection device application

Different types of SPDs are applied at the boundaries of the different LPZs. For unprotected lines coming in from LPZ 0_A, the SPD has to protect against the 10/350 μs current waveform according to LPL I–IV. For lines coming in from LPZs not exposed to direct strike, the SPD has to protect against surges with reduced levels. These surges are characterized by the 1.2/50 μs voltage waveform and/or the 8/20 μs current wave waveform. The maximum considered values depend on the requirements of the selected LPZ. Typically, the peak value of the 1.2/50 μs voltage is considered in the range of some kilovolts and the peak value of the 8/20 μs current waveform is considered in the range of some kiloamperes.

The principle of LPZs requires that the SPDs are installed subsequently in the same circuit. In this case the SPDs have to be energy coordinated to share the stress among them according to their energy-absorbing capability. For an effective coordination the characteristics of the individual SPD as published by the manufacturer, the threat at their installation point and the characteristics of the equipment to be protected must be considered. For example, as a general rule it has to be ensured that the SPD between LPZ 0 and LPZ 1 blocks the majority of the lightning current. Rules and examples for the coordination of SPDs are given in Annex D of the IEC 62305-4 standard [4]. The general requirements, testing procedures and principles for the selection and application of the SPDs are lain down in the IEC 61643 standard series.

21.5.5 Spatial magnetic shielding

A cost-effective method by which to form spatial electromagnetic shields is to use existing metallic structural components, such as the reinforcement elements of concrete. The shield of a building (shield around LPZ 1) can be part of an external LPS, and therefore the lightning currents will flow along it in the case of a direct lightning strike. This situation is shown in Figure 21.10 assuming that the lightning hits the structure at an arbitrary point of the roof. For such a grid-like shield, the magnetic field strength H_1 at an arbitrary point inside the LPZ is given by the formula

$$H_1 = 0.01 I_0 w / (d_w \times \sqrt{d_r}) \quad (\text{A m}^{-1}) \tag{21.10}$$

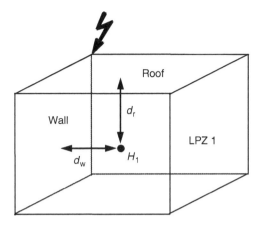

Figure 21.10 Magnetic field inside LPZ 1, when the grid-like shield is struck by direct lightning

where I_0 (A) denotes the lightning current in LPZ 0_A and w (m) the mesh width of the grid-like shield. The distance d_r (m) is the shortest distance between the point considered to the roof of the shielded LPZ 1 and the distance d_w (m) is the shortest distance between the point considered to the wall of the shielded LPZ 1. The magnetic fields are valid only inside the grid-like shield with a minimum safety distance $d_{s/1} = w$ to the shield.

In the case of a nearby stroke, the lightning channel is assumed to be perpendicular to the earth surface. For LPZ 0, the incident magnetic field is calculated with the formula

$$H_0 = I_0/(2\pi s_a) \quad (\text{A m}^{-1}) \tag{21.11}$$

where I_0 denotes the considered lightning current in LPZ 0_A and s_a is the horizontal distance between the striking point and the centre of the shielded volume. The magnetic field inside LPZ 1 can be derived from the relation

$$SF_1 = 20 \log (H_0/H_1) \tag{21.12}$$

where SF_1 denotes the shielding factor of the shield of LPZ 1.

If a LPZ 2 is established, it is assumed that no lightning current or only a negligible fraction of the lightning current flows through the screen of LPZ 2. For both a nearby strike and a direct strike, the magnetic field inside LPZ 2 is determined by

$$SF_2 = 20 \log (H_1/H_2) \tag{21.13}$$

Equation (21.13) has to be applied analogously if LPZ 3 or higher are also established. Inside the different LPZs the waveform of the magnetic field is assumed to be identical to the waveform of the lightning current. This means that inside the different

Table 21.15 Magnetic attenuation of grid-like spatial shields

Material	Shielding factor, SF (dB)	
	25 kHz (valid for the first stroke)	1 MHz (valid for the subsequent stroke)
Copper, aluminium	$20 \times \log \dfrac{8.5}{w}$	$20 \times \log \dfrac{8.5}{w}$
Steel	$20 \times \log \left[\dfrac{8.5}{w} \times \dfrac{1}{\sqrt{1 + \dfrac{18 \times 10^{-6}}{r^2}}} \right]$	$20 \times \log \dfrac{8.5}{w}$

w, mesh width of the grid-like shield (m); r, radius of a rod of the grid-like shield (m).

LPZs the magnetic field is given by the 10/350 μs waveform for the first short stroke and the 0.25/100 μs waveform for the subsequent short stroke. On the other hand, the threat of the magnetic field is more or less restricted to the fast rising front. Ignoring the field decay, this fast rising front is simulated by an equivalent sinusoidal magnetic field having the same peak value and the same rise time. The rise time from zero to peak is considered as 10 μs for the first short stroke and as 0.25 μs for the subsequent short stroke. Following this procedure, the threat of the magnetic field is expressed by a sinusoidal magnetic wave with a frequency of 25 kHz for the first short stroke and a second sinusoidal magnetic wave with a frequency of 1 MHz for the subsequent short stroke. For these frequencies the shielding factors of grid-like shields are given in Table 21.15. For copper and aluminium the shielding factors only depend on the mesh width, but for steel there is a minor influence of the frequency on the first stroke.

21.6 Conclusions

IEC 62305-1 [1] provides the scientific background, the basic data and the general rules for lightning protection. The standard postulates that the lightning current is the primary source of damage. Because this source is based on natural phenomena, the lightning currents vary very much in amplitude, in duration and in the number and kind of current components. From the variety of different current components three characterizing currents are identified: (i) first short stroke current, (ii) subsequent short stroke current and (iii) long stroke current. This set of current components represents the comprehensive lightning threat against which a structure is to be protected. In 62305-1, four sets of current components are fixed for four different lightning protection levels (LPLs). Each set is associated with a certain risk that the lightning current exceeds the fixed lightning current parameters (see Section 21.2).

IEC 62305-1 gives the general rules to avoid direct lightning strikes to the volume to be protected. For the design of air termination systems, the rolling sphere method is suggested as a universal method. For each LPL, the radius of the rolling sphere is fixed. The rolling sphere radius is associated with a minimum value of the peak current. Weak lightning with peak currents lower than the fixed value may penetrate the air termination system with a certain probability. Therefore, the sphere radius is successively reduced for higher lightning protection levels (see Section 21.3).

Risk management according to IEC 62305-2 [2] is given in Chapter 8 of this book. It is used first to determine the need for protection, then to select suitable protection measures, and finally to determine the residual risk for the protected structure.

Neither IEC 62305-1 nor IEC 62305-2 provide installation guides, but they do give the basic information for lightning protection and the related risk assessment. Installation guides are provided by IEC 62305-3 [3] and IEC 62305-4 [4].

To protect a structure against direct lightning strikes, IEC 62305-3 gives the rules for the installation of a lightning protection system (LPS) consisting of external and internal lightning protection systems (see Section 21.4). The installation of an LPS is not necessary in cases where direct lightning to the volume to be protected is excluded. The functions of the external lightning protection are to intercept a lightning flash, to conduct the lightning flash safely to earth and to disperse it into the earth. The function of the internal lightning protection is to prevent dangerous sparking within the structure. Four classes of LPSs are defined as a set of construction rules, based on the four corresponding LPLs. Each set includes level-dependent construction rules (e.g. mesh size of the air termination system) and construction rules specified independently from the protection level (e.g. cross-section of the down-conductors).

The LPS does not provide sufficient protection for the electrical and electronic equipment installed inside the structure. Adequate protection of the apparatus requires a LEMP protection measures system (LPMS) according to IEC 62305-4 (see Section 21.5). The protection against LEMP is based on the LPZ concept. The principle of LPZ requires the formation of nested zones of successively reduced values of the electromagnetic environment. This objective is achieved by shielding to reduce the electromagnetic fields and by equipotential bonding of all lines at the LPZ boundaries to limit the line-conducted surges (over-voltages and -currents). The LPZs are assigned volumes of space where the LEMP severity is compatible with the immunity level of the internal systems enclosed. Successive zones are characterized by significant reduction of LEMP severity.

The LPS uses four classes of predefined bundles for the protection system, and the LPMS offers a construction kit of optional protection measures, which can be freely chosen and then combined to a tailored protection system. The risk analysis helps to select the most cost-effective protection measures. For instance, to limit surges to the required level, an improved bonding network or a changed line routing or the installation of coordinated SPDs could be possible solutions. In any case, the final protection system should reduce the residual risk below the tolerable level.

References

1. IEC 62305-1:2006-01. *Protection Against Lightning – Part 1: General Principles.* January 2006.
2. IEC 62305-2:2006-01. *Protection Against Lightning – Part 2: Risk Management.* January 2006.
3. IEC 62305-3:2006-01. *Protection Against Lightning – Part 3: Physical Damage to Structures and Life Hazard.* January 2006.
4. IEC 62305-4:2006-01. *Protection Against Lightning – Part 4: Electrical and Electronic Systems Within Structures.* January 2006.

Chapter 22
High-voltage and high-current testing
Wolfgang Zischank

22.1 Introduction

Lightning involves both high potential differences between the cloud charge and ground and high currents. The potential difference between the lower boundary of the negative charge region and ground is in the range 50–100 MV [1]. Also, the potential difference between the downward-moving stepped-leader tip and ground is ~10 MV. The step length of an approaching stepped leader is in the order of tens of metres. Lightning currents flowing after the attachment process may reach peak values of ~100 kA, lowering charges of ~100 C to ground.

For several reasons it is not possible to duplicate lightning itself in testing: the highest voltages that can be generated in laboratories are ~6 MV. For outdoor equipment 10 MV is about the maximum that can be produced by air-insulated generators. Further limitations are imposed by laboratory size, preventing discharges much longer than 10 m or so.

High-current test generators, on the other hand, are well able to generate unidirectional currents similar to natural lightning, having peak values of ~100 kA and a charge transfer of ~100 C. However, the charging voltage of such high-current generators does not usually exceed ~100 kV. There is no equipment available that could produce the high voltages and high currents of natural lightning, simultaneously using the same machine.

Although it is not possible to duplicate lightning in its entirety, it is very possible to evaluate the effects of lightning by separately testing the effects relating to high voltages and high currents [2]. High-voltage tests help to identify possible lightning attachment points or to evaluate the efficiency of air termination systems. The injection of high currents into these previously determined attachment points allows the deleterious effects of lightning to be studied, including heating, burning, mechanical damage, and resistively or inductively coupled overvoltages.

The lightning effects to which structures, systems or equipments are exposed are often subdivided into direct and indirect effects [3]. Direct effects are related to the attachment of a lightning channel and/or the conduction of a lightning current. Indirect effects result from the interaction of the electromagnetic fields generated by lightning with electrical or electronic equipment. The coupling mechanism of indirect effects may be inductive, capacitive or resistive.

22.2 Lightning test equipment

The simulation of lightning effects related to high voltages is mainly performed with impulse voltage generators. Tests using a.c. or d.c. voltages are of minor importance and are not addressed here. Methods and equipment for the generation of high a.c. or d.c. voltages are described in almost any book on high-voltage techniques [4,5]. The generation of impulse voltages as well as the simulation of lightning direct and indirect current effects are described in the following sections. Methods for the measurement of high voltages and currents are discussed.

22.2.1 High-voltage impulse test generators

Impulse voltage tests are used to determine possible lightning attachment points and breakdown paths across or through non-conducting materials. Because there is a wide range of possible waveforms caused by natural lightning, three major waveforms have been established for simulation of the effects of slow, medium and fast rates of voltage rise.

Slow rates of rise, as may occur during the approach of a stepped leader, are simulated by the so-called 'switching impulse voltage'. This impulse voltage waveform originates from the switching phenomena in high-voltage transmission networks. International standards define such a waveform as having a time to crest $T_{cr} = 250 \pm 20\%$ µs and a decay time to half value of $T_2 = 2\,500$ µs $\pm 60\%$. The definitions of IEC 60060-1 [6] are reproduced in Figure 22.1.

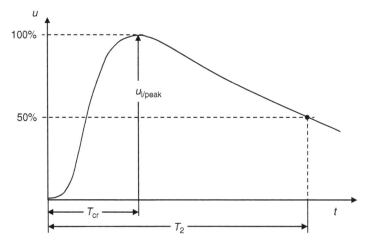

Figure 22.1 *General shape and definitions of the standard switching impulse voltage 250/2 500 µs*

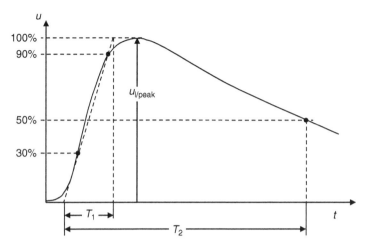

Figure 22.2 General shape and definitions of the standard lightning impulse voltage 1.2/50 μs

Impulse voltages having a medium rate of rise can be simulated by the so-called 'standard lightning impulse' waveform. Such waveforms were originally derived from measurements of lightning-induced overvoltages on power transmission lines. They are generally unidirectional, having front times of a few microseconds and decay times to half value of a few tens of microseconds. The rise time and the duration of lightning impulse voltages are thus significantly shorter compared to switching impulse voltages. International standard IEC 60060-1 [6] defines this as a waveform of 1.2/50 μs, characterized by a front time of $T_1 = 1.2$ μs ± 30% and a decay time to half value of $T_2 = 50$ μs ± 20%. The exact definitions can be seen in Figure 22.2.

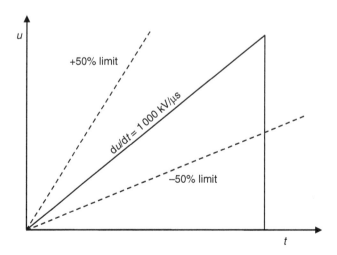

Figure 22.3 General shape and definitions of the steep front voltage waveform

This waveform may also be chopped by a disruptive discharge during the rise or decay portions, either intentionally by means of, for example, a parallel spark gap or by a break down at the object under test.

A fast rate of rise impulse voltage waveform used for the simulation of lightning effects is termed the 'steep front wave'. It rises in a mostly linear fashion at a rate of 1 000 kV µs^{-1} ± 50% until it is interrupted by puncture of, or flashover across, an object under test. This waveform is commonly used in aircraft lightning testing and is termed the 'Voltage Waveform A' in EUROCAE ED 84 [7]. A graph of the steep front wave is provided in Figure 22.3.

22.2.1.1 Single-stage impulse voltage circuits

Single-stage impulse voltage circuits are suitable for the generation of voltage peak values of up to ~200 kV. The two basic circuits for impulse generators are shown in Figure 22.4. The load capacitance C_ℓ is composed of a discrete capacitor and the capacitance of the measuring devices and equipment under test.

Initially, the surge capacitor C_s charges slowly from a d.c. voltage source up to the charging voltage U_{ch}. At time $t = 0$ the surge capacitor C_s discharges via a start switch

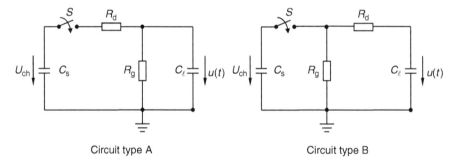

Figure 22.4 Basic circuit diagrams for impulse voltage generators

S (often a spark gap) and via a damping resistor R_d into a load capacitance C_ℓ. The load capacitance C_ℓ is usually much smaller than C_s. The damping resistor R_d and the load capacitance C_ℓ form a low-pass filter that controls the front time T_1

$$T_1 \propto R_d \cdot C_\ell$$

The decay of the voltage is determined by the time it takes to discharge the capacitors C_s and C_ℓ through the resistors R_g and R_d to ground. Because $C_s \gg C_\ell$ and $R_g \gg R_d$, the decay time to half value, T_2, is predominantly governed by

$$T_2 \propto R_g \cdot C_s$$

The main characteristic data for the impulse voltage generator are the charging voltage U_ch, the energy W stored in the surge capacitor C_s and the voltage efficiency η. The stored energy is given by

$$W = \frac{1}{2} \cdot C_\text{s} \cdot U_\text{ch}^2$$

The peak value of the output voltage u_peak appearing at the load capacitance C_ℓ is less than the charging voltage U_ch, because after closing of the start switch S the charge originally stored in C_s is distributed to both capacitor C_s and C_ℓ. The voltage efficiency η is defined as the ratio of the peak output voltage u_peak to the charging voltage U_ch

$$\eta = \frac{u_\text{peak}}{U_\text{ch}}$$

The two basic circuits of Figure 22.4, henceforth referred to as circuit 'type A' and circuit 'type B', differ only in the positioning of the discharge resistor R_g. In circuit type A the resistors R_d and R_g form a voltage-dividing system and, for a given ratio C_ℓ / C_s, the voltage efficiency of a type A circuit is somewhat lower compared to that of a type B circuit. Type B circuits are therefore more favourable for the generation of high voltages. An advantage of the type A circuit is that the grounding resistor R_g can be used as a voltage divider to measure the output voltage. However, this is applied only to generators of moderate voltages of ~ 10 kV.

The output voltage $u(t)$ for both types of circuits is given by the differential equation

$$\frac{d^2 u}{dt^2} + K_1 \cdot \frac{du}{dt} + K_0 \cdot u = 0$$

Evaluating this equation results in

$$u = \frac{U_\text{ch}}{K} \cdot \frac{\tau_1 \cdot \tau_2}{\tau_1 - \tau_2} \cdot \left(e^{-t/\tau_1} - e^{-t/\tau_2} \right)$$

The output voltage is therefore the superposition of two exponential functions containing the time constants τ_1 and τ_2. These time constants are given by

$$\tau_1 = \frac{2}{K_1 - \sqrt{K_1^2 - 4 \cdot K_0}} \quad \text{and} \quad \tau_2 = \frac{2}{K_1 + \sqrt{K_1^2 - 4 \cdot K_0}}.$$

The exact relationship of K, K_0 and K_1 to the circuit components R_d, R_g, C_s and C_ℓ are summarized in Table 22.1 for both circuit types, A and B. For most practical cases the decay time to half value T_2 is much greater than the front time T_1, and therefore $R_\text{g} C_\text{s} \gg R_\text{d} C_\ell$. The simplified relationships given in Table 22.2 can then be applied.

Table 22.1 Constants K, K_0 and K_1 as a function of circuit type and components

Circuit	Type A	Type B
K	$R_d \cdot C_\ell$	$R_d \cdot C_\ell$
K_0	$\dfrac{1}{R_d \cdot C_\ell \cdot R_g \cdot C_s}$	$\dfrac{1}{R_d \cdot C_\ell \cdot R_g \cdot C_s}$
K_1	$\dfrac{1}{R_d \cdot C_\ell} + \dfrac{1}{R_d \cdot C_s} + \dfrac{1}{R_g \cdot C_\ell}$	$\dfrac{1}{R_d \cdot C_\ell} + \dfrac{1}{R_d \cdot C_s} + \dfrac{1}{R_g \cdot C_s}$

Table 22.2 Time constants and voltage efficiency as a function of the circuit type and components for $R_g C_s \gg R_d \cdot C_\ell$

Circuit	Type A	Type B
τ_1	$(R_d + R_g) \cdot (C_s + C_\ell)$	$R_g \cdot (C_s + C_\ell)$
τ_2	$\dfrac{R_d \cdot R_g}{R_d + R_g} \cdot \dfrac{C_s \cdot C_\ell}{C_s + C_\ell}$	$R_d \cdot \dfrac{C_s \cdot C_\ell}{C_s + C_\ell}$
η	$\dfrac{R_g}{R_d + R_g} \cdot \dfrac{C_s}{C_s + C_\ell}$	$\dfrac{C_s}{C_s + C_\ell}$

The time constant τ_1 mainly determines the decay time T_2, whereas the front time T_1 is governed by τ_2. However, there is no simple relationship between these time constants and the times T_1 and T_2 as defined in the international standards. The relationships, therefore, must be computed numerically. Values for the standard waveforms 1.2/50 µs and 250/2 500 µs are given in Table 22.3. The time to crest value T_{cr} is given for both circuit types by

$$T_{cr} = \frac{\tau_1 \cdot \tau_2}{\tau_1 - \tau_2} \cdot \ln\left(\frac{\tau_1}{\tau_2}\right)$$

Table 22.3 Time constants τ_1 and τ_2 for the standard waveforms 1.2/50 µs and 250/2 500 µs

1.2/50 µs	250/2 500 µs
$\tau_1 = 68.5$ µs	$\tau_1 = 3\,200$ µs
$\tau_2 = 0.405$ µs	$\tau_2 = 62.3$ µs

22.2.1.2 Multistage impulse voltage circuits

Single-stage circuits are not suitable for the generation of impulse voltages greater than a few 100 kV, due to limitations in the physical size of the circuit components and the difficulties encountered with corona discharges at megavolt d.c. charging voltages. These difficulties can be overcome by the use of multistage impulse generators, as suggested by Marx in the early 1920s [8]. The functional principle is to slowly charge a group of capacitors in parallel through high-ohmic resistors and then to rapidly discharge them in series through spark gaps. Marx originally developed this generator principle for '... testing of insulators and other electrical apparatuses'.

Figure 22.5 shows the circuit design for a multistage impulse generator, often referred to as a Marx generator after its inventor. In essence, the circuit of each

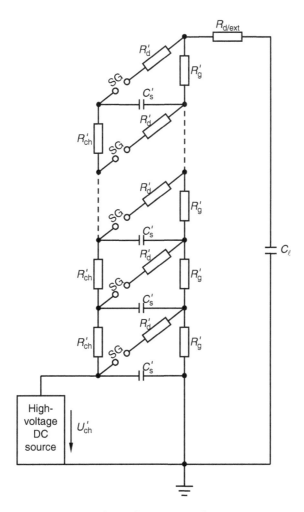

Figure 22.5   Circuit design of a multistage impulse generator

stage is similar to the single-stage generator of Figure 22.4, composed of a surge capacitor C'_s, a damping resistor R'_d and a grounding resistor R'_g. Usually, the load capacitance is not distributed to the single stages. Instead, an external load capacitor C_ℓ rated to the full output voltage of the generator is used. This external load capacitor is often constructed as a voltage divider for measurement of the impulse voltage. Also, a part of the damping resistance may be installed externally ($R_{d/ext}$) in front of the load capacitor in order to facilitate easy adjustment of the front time to various capacitances of the equipment under test.

The operation of a multistage impulse generator requires that all the spark gaps (SGs) operate almost simultaneously. The distances of the spark gaps are set to a value slightly above the natural breakdown at the charging voltage U'_{ch}. The spark gap of the lowest stage is then triggered either by moving the spheres together or by applying an impulse to a trigger electrode inserted into the first-stage sphere gap. A simple sparkplug may serve to induce breakdown of the first-stage gap.

Traditionally, it was thought that after triggering of the first stage a voltage of $2U'_{ch}$ would appear at the second stage, followed by $3U'_{ch}$ at the third stage, and so on. This, however, is not true. Assuming $R'_g \gg R'_d$ the voltage U'_{ch} appears at R'_g (labelled 'A' in Figure 22.6) after triggering of the first stage. As the load capacitance C_ℓ is still uncharged, the generator output terminal remains at ground potential. U'_{ch} therefore

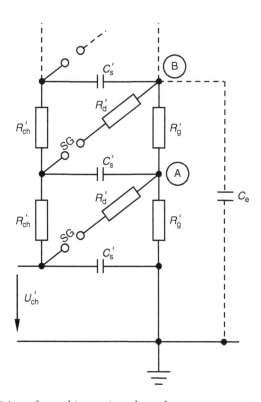

Figure 22.6   Firing of a multistage impulse voltage generator

appears at the serial connection of the remaining discharge resistors R'_g, and the overvoltage per stage is just $U'_{ch}(n-1)$, with n being the number of stages. For generators with a large number of stages ($n = 10-30$) this overvoltage will not be sufficient to reliably fire the rest of the spark gaps.

In the early days of Marx generators, laboratory engineers had already empirically found that moving the generator somewhat closer to the wall improved the firing quality. Later, when computers became available, detailed research work [9,10] proved the decisive role of stray capacitances for the firing performance of large multi-stage generators. Predominant among these is the stray capacitance C_e of the single stages to ground potential, that is, to the laboratory floor, walls and ceiling, as indicated in Figure 22.6 for the second stage. After triggering of the first stage, the voltage at point 'A' will quickly rise to U'_{ch}. The voltage at point 'B', however, will not follow immediately, but will be delayed due to the low-pass characteristic given by $R'_g C_e$. Therefore, for a short time of 10 ns or so, an overvoltage much higher than $U'_{ch}(n-1)$ will appear at the second-stage spark gap. This short-duration overvoltage is sufficient to reliably fire the spark gap. The process is then successively repeated at the other stages, so that all spark gaps will fire within a few 100 ns.

Figure 22.7 Example of a 12-stage impulse voltage generator (1.2 MV, 36 kJ)

For calculation purposes, the circuit of a multistage impulse voltage generator can be reduced to an equivalent single-stage circuit, as shown in Figure 22.4, using the following equations:

$$U_{ch} = nU'_{ch}$$
$$C_s = C'_s/n$$
$$R_g = nR'_g$$
$$R_d = nR'_d + R_{d/ext}$$

The impulse voltage waveform, then, can be calculated using the equations given in Section 22.2.1.1 for single-stage circuits. In general, multistage impulse voltage generators may be designed according to circuit type A or B. However, due to the higher voltage efficiency, type B circuits, as shown in Figure 22.5, are commonly preferred. As an example, the 12-stage, 1.2 MV generator at the high-voltage laboratory of the University of the Federal Armed Forces in Munich is shown in Figure 22.7.

22.2.2 High-current test generators

An entire lightning flash may last up to a second or so and may be composed of several different current components. The main current components found in natural lightning are

- the first return stroke
- subsequent return strokes
- long-duration continuing currents

Lightning currents are essentially unidirectional. Parameters for the simulation of the effects of these current components are internationally standardized by the IEC [11–14] for ground-based structures and in EUROCAE [7] for aircraft.

First return strokes typically have high current amplitudes of some 10 kA up to a few 100 kA, lasting up to several 100 μs. Subsequent return strokes are lower in amplitude and duration compared to first return strokes, but often have a significantly higher current steepness di/dt of up to a few 100 kA μs^{-1}. Long-duration continuing currents exhibit comparatively low current amplitudes of just a few 100 A, but last much longer (up to several 100 ms). As the waveforms of these three basic current components are quite different, testing calls for different generator designs.

Lightning currents in general are considered as impressed currents, which are not affected by the impedance of the struck object. High-current effects testing of equipment, systems or lightning protection components requires the injection of high currents into the object under test either through a small air gap at a spacing of a few centimetres or by direct connection.

The definition of current front time T_1 and decay time to half value T_2 for impulse currents according to IEC 60060-1 [6] is reproduced in Figure 22.8.

The following sections separately contain the description of test generators for the simulation of first return strokes, subsequent return strokes and long-duration currents. To approach natural lightning phenomena as closely as possible, it may be necessary for specific cases to combine different generator types in one test facility. Directions

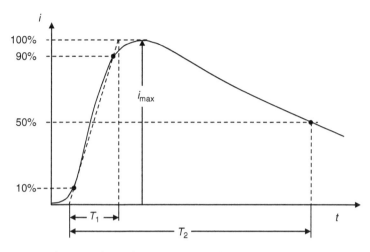

Figure 22.8 Definition of impulse current waveform parameters

are given for current injection into an object under test. Finally, testing of indirect effects is briefly addressed.

22.2.2.1 Simulation of first return stroke effects

Deleterious effects such as heating and mechanical damage are mainly related to first return strokes. The most important current parameters for testing are the current peak value i_{max}, the impulse charge Q_i and specific energy W/R. The front time T_1 is usually only of minor interest as it does not remarkably affect degradation or physical damage in most practical cases. Nevertheless, T_1 should have a typical value not exceeding a few tens of microseconds. The waveform should be unidirectional.

RLC circuits

High-current impulse generators usually consist of a set of large high-voltage capacitors $C_{s/1} \ldots C_{s/n}$ connected in parallel (Figure 22.9). As an example, Figure 22.10 shows

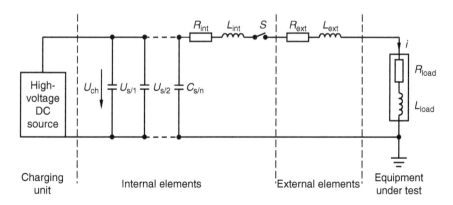

Figure 22.9 High-current impulse generator

Figure 22.10 Example of an impulse current generator

the impulse current generator at the high-voltage laboratory of the University of the Federal Armed Forces in Munich. The 24 individual surge capacitors $C_{s/v}$ are arranged in the form of a 'U' shape. The object under test can be placed in the centre of the 'U'.

The capacitor bank is slowly charged from a d.c. source to a high voltage U_{ch} (e.g. 100 kV) and then rapidly discharged via a starting switch S (usually a spark gap), external wave-forming elements R_{ext} and L_{ext} into the object under test with the load characteristics R_{load} and L_{load}. The connections inside the generator should be configured so as to minimize its internal resistance R_{int} and inductance L_{int}.

Such a generator is characterized by its maximum charging voltage U_{ch} and the energy W stored in the capacitor bank C_s:

$$W = \frac{1}{2} \cdot C_s \cdot U_{ch}^2$$

where C_s is the sum of all individual surge capacitors $C_s = \sum_1^n C_{s/v}$.

Essentially, the generator design shown in 9 forms an R–L–C circuit with the equivalent circuit components (Figure 22.11)

$$R = R_{int} + R_{ext} + R_{load}$$
$$L = L_{int} + L_{ext} + L_{load}$$
$$C_s = \sum_1^n C_{s/v}.$$

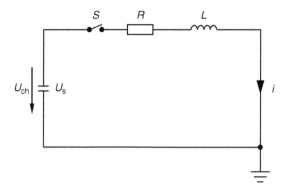

Figure 22.11 Equivalent R–L–C circuit of a high-current generator

The current i of such an R–L–C circuit is given by the differential equation [15]

$$\frac{d i^2}{dt^2} + \frac{R}{L}\cdot\frac{di}{dt} + \frac{1}{L\cdot C_s}\cdot i = 0$$

with the boundary conditions at closing of the start switch ($t = 0$) being

$$i = 0 \quad \text{and} \quad \frac{di}{dt} = \frac{U_{ch}}{L}.$$

Depending on the magnitude of the damping resistance R, three basic current waveforms may result from an R–L–C circuit (Figure 22.12):

- $0 < R < 2\sqrt{L/C_s}$: undercritically damped (damped oscillating) current
- $R = 2\sqrt{L/C_s}$: critically damped (unidirectional) current
- $R > 2\sqrt{L/C_s}$: overcritically damped (unidirectional) current

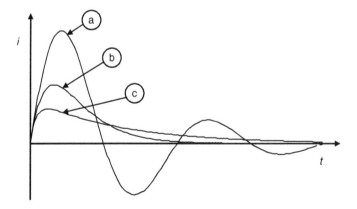

Figure 22.12 Basic current waveforms of an R–L–C circuit: (a) an undercritically damped circuit, (b) a critically damped circuit and (c) an overcritically damped circuit

Lightning Protection

Table 22.4 to 22.6 summarize the resulting waveforms and the parameters for these three basic types of R–L–C circuit currents. Note that in the case of the oscillating waveform the impulse charge Q_i is defined as the time integral of the absolute value of the current i, $Q_i = \int_0^\infty |i| \cdot dt$, because the positive as well as the negative half wave will cause damage at an object. The specific energy W/R is defined as the time integral of the square of the current i and is also referred to as the 'action integral':

$$W/R = \int_0^\infty i^2 \cdot dt.$$

To obtain maximum current output, impulse current generators have to be operated in an undercritically damped mode. Undercritically damping, however, means that the current waveform is oscillatory, contrary to the unidirectional currents associated with natural lightning strokes. It also means that the front time of the impulse current may become rather long.

To obtain a unidirectional waveform from an R–L–C circuit requires critical (or overcritical) damping. Critical damping is obtained by increasing the circuit

Table 22.4 Parameters of undercritically damped R–L–C circuits

Condition	$0 > R > 2\sqrt{L/C_s}$
Current waveform	$i = \dfrac{U_{ch}}{\omega L} \cdot \sin(\omega t) \cdot e^{-t/\tau}$
Angular frequency	$\omega = \sqrt{\dfrac{1}{LC_s} - \dfrac{1}{\tau^2}}$
Frequency of oscillation	$f = \dfrac{\omega}{2\pi} = \dfrac{1}{2\pi} \cdot \sqrt{\dfrac{1}{LC_s} - \dfrac{1}{\tau^2}}$
Time constant	$\tau = \dfrac{2L}{R}$
Time to crest value	$T_{cr} = \dfrac{\arctan(\omega\tau)}{\omega}$
Current peak value	$i_{max} = U_{ch} \cdot \sqrt{\dfrac{C_s}{L}} \cdot e^{-[\arctan(\omega\tau)/\omega\tau]}$
Maximum current steepness	$\left(\dfrac{di}{dt}\right)_{max} = \dfrac{U_{ch}}{L}$
Impulse charge	$Q_i = \dfrac{U_{ch}}{L \cdot (\omega^2 + 1/\tau^2)} \cdot \left(\dfrac{2}{1 - e^{-\pi/\omega\tau}} - 1\right)$
Specific energy	$W/R = \dfrac{U_{ch}^2}{4\omega^2 L^2} \cdot \dfrac{\tau}{1 + (1/\omega\tau)^2}$
Ratio of front/decay time	$0.263 > T_1/T_2 > 0.482$

Table 22.5 Parameters of critically damped R–L–C circuits

Condition	$R = 2\sqrt{L/C_s}$
Current waveform	$i = \dfrac{U_{ch}}{L} \cdot e^{-t/\tau} \cdot t$
Time constant	$\tau = 2L/R$
Time to crest value	$T_{cr} = \tau$
Current peak value	$i_{max} = \dfrac{2}{e^1} \cdot \dfrac{U_{ch}}{R} = 0.736 \cdot \dfrac{U_{ch}}{R}$
Maximum current steepness	$\left(\dfrac{di}{dt}\right)_{max} = \dfrac{U_{ch}}{L}$
Impulse charge	$Q_i = U_{ch} \cdot C_s$
Specific energy	$W/R = \dfrac{U_{ch}^2 \cdot C_s}{4} \cdot \sqrt{\dfrac{C_s}{L}}$
Ratio of front/decay time	$T_1/T_2 = 0.263$

Table 22.6 Parameters of overcritically damped R–L–C circuits

Condition	$R > 2\sqrt{L/C_s}$
Current waveform	$i = \dfrac{U_L}{\sqrt{R^2 - 4L/C_s}} \cdot \left(e^{-t/\tau_1} - e^{-t/\tau_2}\right)$
Decay time constant	$\tau_1 = \dfrac{1}{\dfrac{R}{2L} - \sqrt{\left(\dfrac{R}{2L}\right)^2 - \dfrac{1}{LC_s}}}$
Front time constant	$\tau_2 = \dfrac{1}{\dfrac{R}{2L} + \sqrt{\left(\dfrac{R}{2L}\right)^2 - \dfrac{1}{LC_s}}}$
Time to crest value	$T_{cr} = \dfrac{\tau_1 \cdot \tau_2}{\tau_1 - \tau_2} \cdot \ln\dfrac{\tau_1}{\tau_2}$
Current peak value	$i_{max} = \dfrac{U_L}{\sqrt{R^2 - 4L/C_s}} \cdot \left(e^{-T_{cr}/\tau_1} - e^{-T_{cr}/\tau_2}\right)$
Maximum current steepness	$\left(\dfrac{di}{dt}\right)_{max} = \dfrac{U_{ch}}{L}$
Impulse charge	$Q_i = \dfrac{U_L}{\sqrt{R^2 - 4L/C_s}} \cdot (\tau_1 - \tau_2) = U_L \cdot C_s$
Specific energy	$W/R = \dfrac{U_L^2/2}{R^2 - 4L/C_s} \cdot \dfrac{(\tau_1 - \tau_2)^2}{\tau_1 + \tau_2}$
Ratio of front/decay time	$0 < T_1/T_2 < 0.263$

resistance. This, however, means diminishing the current peak value and wasting most of the energy initially stored in the capacitor bank by heating the generator's damping resistors. Critically damped $R-L-C$ circuits, capable of generating peak currents of 100–200 kA with a charge transfer of 50–100 C would become rather large and expensive. Therefore, they are used in practice only for generating impulse currents with a charge transfer of up to 20 C or so.

Crowbar technologies
A very effective way to obtain a unidirectional current with a tolerable size of capacitor bank is the use of a crowbar device in an $R-L-C$ circuit. The basic circuit diagram of such a generator is shown in Figure 22.13.

The principle of operation of an impulse generator with crowbar device is illustrated in Figure 22.14 [16,17]. An external inductance L_{ext}, significantly higher than the internal inductance L_{int}, is inserted into the circuit. To obtain a high current peak value the generator is operated in a strong undercritically damped mode with low resistance. The discharge is initiated by a starting gap S at $t = 0$, while the crowbar device $S_{crowbar}$ remains open. At the instant of the crest value of the current ($t = T_{cr}$) the crowbar device $S_{crowbar}$ is closed. Most of the energy initially stored in the capacitor bank is at the instant T_{cr} transferred to the inductances L_{ext} and L_{load}. By shorting out the capacitor with the crowbar device the current is converted from an oscillatory to an exponentially decaying waveform having a decay time constant of

$$\tau = \frac{L_{ext} + L_{load} + L_{crowbar}}{R_{ext} + R_{load} + R_{crowbar}}$$

where $L_{crowbar}$ is the self-inductance and $R_{crowbar}$ the resistance of the crowbar branch.

For the calculation of an impulse generator with crowbar device the current has to be slip up into the section before and after T_{cr}. For $t \leq T_{cr}$ the equations for an

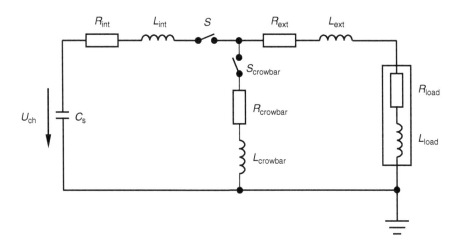

Figure 22.13   Circuit of an impulse current generator with crowbar device

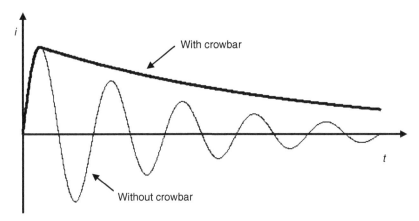

Figure 22.14 Principle of an impulse current generator

undercritically damped circuit (Table 22.4) apply, with $R = R_{int} + R_{ext} + R_{load}$ and $L = L_{int} + L_{ext} + L_{load}$. For $t \geq T_{cr}$ the current changes to an exponentially decaying waveform according to

$$i = i_{max} \cdot e^{-t/\tau}.$$

Assuming that the damping of the circuit is kept low, $R \ll 2\sqrt{L/C_s}$, the charge and the specific energy can be approximated by

$$Q_i \approx U_{ch} \cdot C_s + \frac{U_{ch}}{\omega \cdot L} \cdot \tau$$

and

$$W/R \approx \left(\frac{U_{ch}}{\omega \cdot L}\right)^2 \cdot \left(\frac{\pi}{4\omega} + \frac{\tau}{2}\right)$$

For more exact calculations, the non-linear behaviour of the crowbar device as a function of current has also to be taken into account.

The external resistance R_{ext} is usually composed of only the resistance of the copper bars used for connection and can therefore be quite low. The decay of the current is then proportional to the resistance of the object under test plus the resistance of the crowbar device. Thus, a major part (50 per cent and more) of the energy originally stored in the capacitor bank can be transferred into the object under test. Objects intended to handle high lightning currents inherently have to be of low resistance. Therefore, using the crowbar technique, high impulse currents with decay times to half value of several 100 μs can be obtained. An actual waveform is shown in Figure 22.15. The oscillations after the current peak arise from the internal inductance L_{int}: at the instant $t = T_{cr}$ some energy is also trapped in the internal inductance L_{int}, giving rise to an oscillatory current through the internal part (C_s, R_{int} and L_{int}) of

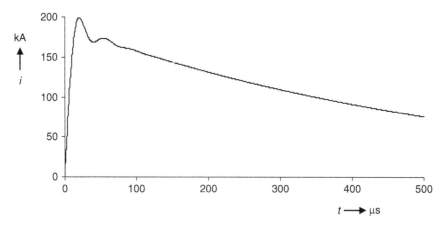

Figure 22.15 Actual waveform of an impulse current with crowbar device

the generator. This current then splits between the parallel connection of the crowbar branch and the branch containing the object under test. It is recommended to use some internal damping R_{int}, typically in the range of 100 mΩ. The oscillatory current through C_s then quickly dies out and the stress to the surge capacitors is reduced. In order to keep the internal inductance L_{int}, and with it the oscillations, low, the crowbar device should be placed as close as possible to the capacitor bank.

An advantage of the crowbar technique is that the inductance of the object under test is not a crucial factor. The external inductance L_{ext} can be partly or completely replaced by the self-inductance of the object under test.

The requirements for the crowbar device, however, are demanding. At $t = 0$, when it is still open, it has to withstand nearly the whole charging voltage U_{ch}. At the instant of the current peak, $t = T_{cr}$, most of the energy is transferred into the inductance L_{ext} and L_{load}. The voltage across the capacitor bank C_s, and thus across the crowbar device, is then near or at zero. Operation of the crowbar device near or at zero voltage calls for advanced technologies. Mechanical switches are usually too slow to operate at a microsecond timescale. More sophisticated crowbar techniques, however, have been developed and successfully applied since the 1980s.

Three-electrode spark gaps
The principle of three-electrode spark gaps in ambient air, under pressurized gases (e.g. SF_6) or vacuum is shown in Figure 22.16. An external high-voltage impulse is applied via a peaking circuit to the centre trigger electrode. The 'centre' electrode is located somewhat closer to the high-voltage main electrode of the crowbar gap, so that sparkover occurs here first. The circuit is then closed via the internal inductance L_{int} and the capacitor bank C_s. As the peaking circuit generates a high di/dt, the voltage drop across L_{int} can become high enough to initiate a second discharge in the crowbar gap between the centre electrode and the ground electrode. Thus, it is possible to achieve a complete ionized channel between the two main electrodes at zero voltage across the three-electrode crowbar gap [16].

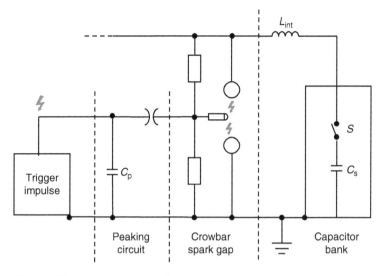

Figure 22.16   *Three-electrode crowbar spark gap*

High-power diodes

High-power diodes have been successfully applied as a crowbar device [18]. In order to handle the high charging voltage, many diodes need to be connected in series. Furthermore, several such columns of diodes in parallel are necessary to cope with the high current (Figure 22.17). The polarity of the diodes is chosen to not conduct during the current rise after triggering the start switch at $t = 0$. At the instant of the

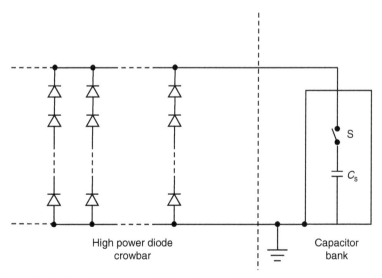

Figure 22.17   *High-power diode crowbar device*

current peak, $t = T_{cr}$, the voltage across C_s reverses polarity and the diodes become conductive. No external trigger impulse is necessary. However, the large number of diodes required makes such a crowbar device quite expensive. Care must be taken to protect the diodes from overvoltages.

Ignitrons

Ignitrons are a kind of mercury arc controlled rectifier dating from the 1930s. They usually consist of a container with a pool of mercury at the bottom, acting as a cathode (Figure 22.18). The anode is fixed above the pool by an insulating support. An igniting electrode is quickly pulsed to release electrons from the surface of the mercury, initiating a conducting arc through the tube between the cathode and

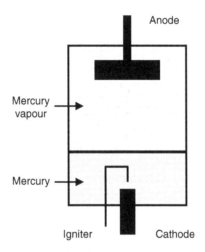

Figure 22.18   Functional principle of an ignitron

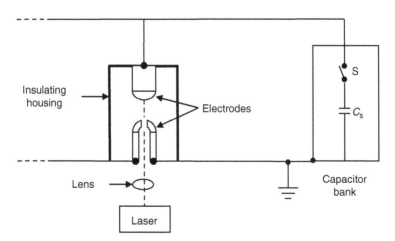

Figure 22.19   Laser-triggered crowbar spark gap

anode. The arc lasts until the voltage across the ignitron has been reduced to the point where the arc can no longer be sustained. Ignitrons can switch hundreds of kiloamperes and hold off as much as 50 kV. Auxiliary equipment is often needed for cooling.

Laser-triggered spark gaps
Two electrode spark gaps can be triggered by a laser beam [19]. The principle is shown in Figure 22.19. A laser beam (e.g. a UV laser) is fired axially through an opening in one of the electrodes into the interspace between the electrodes. Triggering at zero voltage is in principle not possible. However, depending on the laser power and characteristics, an arc can be established with just a few kilovolts across the gap.

22.2.2.2 Simulation of subsequent return stroke effects

Magnetically induced overvoltages are mainly related to the high current steepness di/dt during the rise time portion of a stroke, while the induction during the slower decay is less important. The highest current steepness is found in negative subsequent strokes with values up to 100 or 200 kA μs^{-1}. Peak current, charge and specific energy of subsequent return strokes are significantly lower than that of first impulse currents and are therefore considered here.

The maximum current steepness of an R–L–C circuit is equal to the ratio of the charging voltage U_{ch} to the total inductance L:

$$di/dt_{max} = \frac{U_{ch}}{L}$$

The generator design for simulating the high current steepness of negative subsequent strokes is greatly dependent on the inductance of the object to be tested. Given a certain load inductance, attempts to simply increase the charging voltage yield little benefit. Increasing the charging voltage requires more insulation spacing, which in turn increases the generator circuit internal inductance. Specific measures to keep the internal inductance low or to boost the current front therefore have to be applied.

Smaller objects such as surge protective devices having an inductance of just a few 100 nH can usually be tested with generators having charging voltages of a few hundred kilovolts. A quasi-coaxial return path arrangement may help to keep the inductance of the test circuit low. Typically, it consists of at least four return path conductors, symmetrically arranged around the object under test.

Larger test objects, like parts of an aircraft, as well as indirect effects testing where an object is exposed to the electromagnetic field radiated from a nearby down-conductor, usually involve load inductances of several μH. For such tests, sophisticated and expensive generator equipment is necessary. In the following, several techniques to reduce the front time of impulse current generators are presented.

Low-inductance megavolt generators
As stated above, attempts to simply increase the charging voltage yield little benefit, because the generator's internal inductance increases too. For air-insulated generators

a value of ~4 µH MV^{-1} can be assumed [20,21]. To significantly lower the internal inductance, the generators have to be embedded in a (metal) tank filled with a highly insulating medium such as oil or SF_6. Connections between the generator and the object under test preferably should be designed coaxially. An example of an oil-insulated 1 MV generator is given in Reference 22. Such a generator can produce a current of more than 200 kA with a steepness of 200 kA µs^{-1}.

Peaking circuits
Peaking circuits added to a conventional impulse current generator help to boost the current rise. The principle is shown in Figure 22.20. A low-inductance peaking capacitor C_p ($C_p \ll C_s$) and a peaking spark gap S_p are added to an $R-L-C$ circuit. After closing the start switch S at $t = 0$, the peaking capacitor is charged from the main capacitor bank C_s. The distance of the peaking gap is set to fire when the voltage across C_p reaches its maximum. As a result of oscillations after closing the start switch, C_p is charged to a voltage higher than U_{ch}. Because C_p is exposed only to a short-duration impulse voltage its insulation and the inductance L_p of the peaking circuit can be reduced compared to the main surge capacitor C_s, which has to be

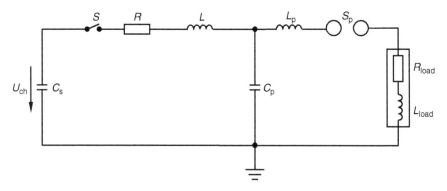

Figure 22.20 Impulse current generator with peaking circuit

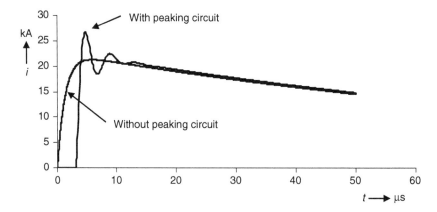

Figure 22.21 Example waveforms obtained from a peaking circuit

designed for d.c. voltage application. The dimensioning of the circuit elements is complicated and usually has to be optimized numerically or experimentally. The example of Figure 22.21 shows the current waveforms obtained from a generator with and without a peaking circuit.

Peaking circuits can be built by lumped, low-inductance capacitances [2,23]. For outdoor generators large parallel-plate capacitors have also been used [24]. Such a distributed capacitance can be formed from a large grid of meshed wires, for example with dimensions 50 m × 50 m, suspended on insulating poles located 10 m above ground. The object under test is place in the centre underneath the overhead grid. In order to minimize the peaking circuit inductance, a large-diameter conductor is used to connect the grid to the peaking gap.

Exploding wires
Another principle to increase the current steepness of an impulse current generator has been suggested in References 25 and 26 through the use of exploding wires [27]. A small-diameter wire and a peaking spark gap S_p are added to an $R-L-C$ circuit (Figure 22.22). After closing the start switch S at $t = 0$, the current through the wire rapidly raises its temperature until the superheated wire vaporizes with explosive violence. The abrupt explosion leads to a rapid increase of the wire resistance by several orders of magnitude. The circuit current then drops down to a very low rate. Because of the rapid change of current, a voltage remarkably higher than the charging voltage U_{ch} (e.g. ten times) appears across the inductance L. The distance of the peaking gap S_p is set to fire when the voltage reaches its maximum, driving a fast rising current into the object under test. The current through the load rises within about the same time interval necessary to explode the wire.

The wire length must be sufficient to withstand the voltage appearing across the wire path, a typical value being a few metres. The voltage withstand of the inductance L must also be designed accordingly. The front time of the current through the load is a function of the wire diameter. For a copper wire having a diameter of 0.35 mm, a front time of a few hundred nanoseconds has been measured [26]. Figure 22.23 shows an example of the current steepness di/dt and the resulting current i through the load for such a wire diameter.

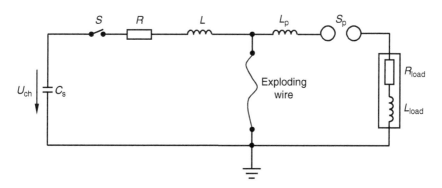

Figure 22.22   Impulse current generator with exploding wire

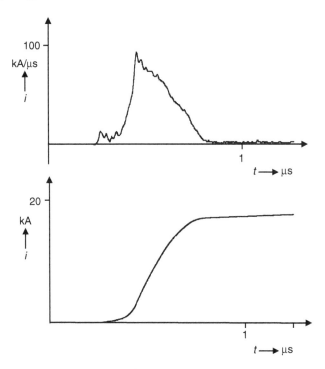

Figure 22.23 Current steepness di/dt and current i from an impulse generator using exploding wires

22.2.2.3 Generation of long-duration currents

Melting and burn-through at the lightning attachment point are predominantly related to continuing currents: their comparatively long duration of several hundreds of milliseconds enables a deep penetration of the melt front into the material, whereas for first return stroke currents, the finite thermal conductivity of the metals prevents the melt front from penetrating deep into the material during the shorter duration of a few hundreds of microseconds.

Long-duration currents are characterized by average currents of ~ 100 A lasting up to 500 ms and resulting in a charge transfer Q_l of a few hundreds of coulombs. International standards often define a rectangular waveform [11], whereas other standards [7] also allow unidirectional waveforms with exponential or linear decay.

If a test standard does not require a rectangular waveform, long-duration currents can be produced from a critically damped capacitor discharge. Rectangular waveforms are generated using a d.c. source (Figure 22.24), which is applied to the object under test via a resistor to adjust the required current amplitude. The d.c. source should have a minimum voltage of 500 V (better 1 kV) in order to maintain an arc of several centimetres at the injection point. Commonly used d.c. sources include

- storage batteries connected in series
- single-phase transformers with rectifiers and smoothing capacitor

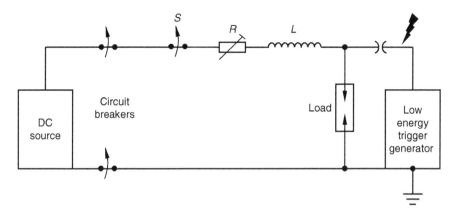

Figure 22.24 Basic circuit of a long-duration generator for a rectangular waveform

- three-phase transformers with rectifiers
- rotating machines, with or without rotating mass

The long-duration current flow can be initiated by either a low-energy trigger impulse voltage or by means of a fine metallic starter wire (diameter ≤ 0.1 mm), initially shunting the injection spark gap. Such a starter wire melts within a short period of time and usually does not affect the test results adversely. An inductance L of some mH in the circuit is intended to decouple the trigger generator from the d.c. source and helps to smooth the current. After the required duration of typically 500 ms, the current flow is interrupted.

In the case of storage batteries, interruption is performed by circuit breakers or by an additional switch S. Power semiconductors (e.g. insulated gate bipolar transistors, IGBT) are often used as switching device for normal operation, while the circuit breakers provides personnel safety. Transformer/rectifier units are best interrupted on the primary side of the transformer at zero crossing of the a.c. voltage and rotating machines by turning of the excitation.

22.2.2.4 Current injection for direct effects testing

Depending on the effects that are to be evaluated, lightning currents may either be coupled directly via a conductor into the object under test or injected via a spark gap.

Direct or conducted current injection is appropriate if only the damage along the current path through an object is to be evaluated. This includes physical damage, arcing and sparking at the connection point, magnetic force effects or heating of conductors.

If effects such as arc root damage, puncture of metal sheets or temperature rise in the vicinity of the attachment point are of importance, the current has to be injected via a spark gap simulating the lightning channel near and at the point of strike. Owing to the limited source voltages available in laboratory simulations of lightning currents, the injection electrode has to be placed relatively close to the object under test.

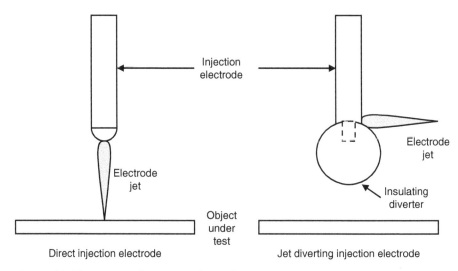

Figure 22.25 Types of injection electrodes

This represents a major departure from a natural lightning attachment, because strong jets of ions and vaporized material originate from the surface of the injection electrode (Figure 22.25). These electrode jets result from magnetic self-compression in the arc and the development of a longitudinal pressure gradient in the vicinity of the arc root on an electrode surface [28,29]. The electrode jets may cause overtesting when they blow out molten material from the attachment point. On the other hand, they may also give rise to undertesting when the severe turbulence of the jet causes the arc root to skip around erratically on the surface of the sample. The dispersion of the incident energy over multiple attachment spots then reduces the damage.

To avoid undesirable influences of the injection electrode on the object under test, a gap spacing of at least 50 mm should be maintained [3]. Further improvement can be achieved by using so-called jet-diverting injection electrodes (Figure 22.25). The tip of such an electrode is equipped with an insulating material, for example in the form of a sphere. Because electrode jets are primarily oriented perpendicular to the electrode surface, the jet is diverted away from the object under test.

If the charging voltage of a generator is not sufficient to reliably fire the current injection gap, a fine metallic or carbon initiating wire with a diameter not exceeding 0.1 mm [3] may be used to direct the arc from the injection electrode to the test object surface. Such a wire explodes within microseconds and does not usually adversely affect the test results.

22.2.3 Indirect effects testing

Indirect effects can be studied by injecting full threat level currents and/or current steepness into an object. Testing the indirect effects of lightning strikes, however, becomes quite demanding for objects or systems of considerable size [30], as discussed in Section 22.2.2.2. Tests are often performed at reduced current levels and the measured responses in the system are extrapolated to full threat.

The expected transients, then, are simulated

- by injecting conducted overvoltages and overcurrents directly to the equipment's interfaces (pin injection test) or
- by magnetically coupling to wires or wire bundles of a system or a major subsystem

The simulation at pin or wire/wire bundle level does not require the sophisticated generators that were necessary for full-level tests on whole systems. When testing complex equipment, often of unknown internal design, it is uncertain whether a voltage or a current wave will arise from the lightning-induced transient. In recognition of this problem, hybrid or combination wave testing has been proposed [31,32]. Combination wave generators are characterized by specifying both an open-circuit output voltage u_{oc} into a high-impedance load and a short-circuit output current i_{sc} into a low-impedance load. The ratio of peak open-circuit voltage $u_{oc/max}$ to peak short-circuit current $i_{sc/max}$ is defined as fictive impedance Z_f

$$Z_f = \frac{u_{oc/max}}{i_{sc/max}}$$

For any intermediate load impedance the equipment under test will form its specific voltage/current response in interaction with the combination wave generator, as it would in a real installation in interaction with an incoming surge.

A multitude of combination waves are in use for various applications. A widely used combination wave provides an open-circuit voltage waveform of 1.2/50 μs and a short-circuit current waveform of 8/20 μs, with a fictive impedance of $Z_f = 2\,\Omega$ [33]. The peak value of the open-circuit voltage does not usually exceed 10 kV. Although not representative for all the possible waveforms inside a real installation, for many years these waveforms have proven to be a reasonable stress, especially for equipment located in a lightning protection zone LPZ 1. An example circuit design for such a combination wave generator is given in Figure 22.26 [31].

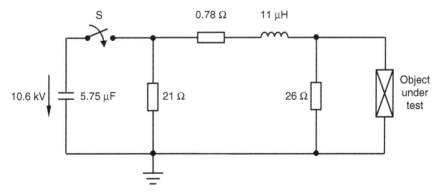

Figure 22.26 Example of a combination wave generator ($u_{oc/max} = 10\,kV$, $Z_f = 2\,\Omega$)

22.3 Measurement techniques

In this section, techniques to measure high impulse voltages and currents are briefly addressed. Furthers details and dimensioning criteria can be found in the landmark book by Schwab [34] and in many books on high-voltage techniques [4,5].

22.3.1 Measurement of impulse voltages

Originally, spark gaps were used for the measurement of high voltages. However, they can only determine the voltage peak value. To also record the waveform, voltage dividers are necessary. In general there are three types of voltage dividers:

- resistive dividers
- capacitive dividers
- damped capacitive dividers

Resistive dividers are suited only for d.c. measurements. Owing to their stray capacitance to ground they have a pronounced low-pass characteristic. Capacitive voltage dividers are appropriate mostly for a.c. measurements. At impulse voltages they tend to oscillate due to the inductance of the measuring circuit. Impulse applications are therefore limited to a few hundred kilovolts by the circuit inductance, which is related to physical size.

Damped capacitive dividers are especially suited for the measurement of high impulse voltages in the megavolt range, even for sub-microsecond front times. They consist of distributed capacitors and resistors connected in series (Figure 22.27). The oscillations are damped by the resistors. At high frequencies, when the impedance $1/(\omega C)$ is low, the divider ratio is given by the resistances and inversely at lower frequencies by the capacitances. A frequency-independent divider ratio is achieved when the time constants RC of the high- and low-voltage section of the divider are chosen to be equal:

$$R'_1 \cdot C'_1 = R_2 \cdot C_2.$$

For a divider with n distributed capacitors C'_1 and resistors R'_1, the equivalent capacitance C_1 and resistance R_1 of the high-voltage section are given by

$$C_1 = C'_1/n$$

and

$$R_1 = R'_1 \cdot n$$

The output voltage u_2 then becomes

$$u_2 = u_1 \cdot \frac{C_1}{C_1 + C_2}$$

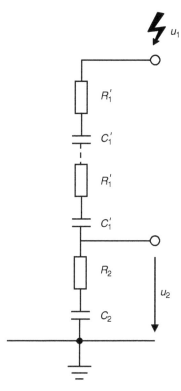

Figure 22.27   Damped capacitive impulse voltage divider

22.3.2   Measurement of impulse currents

For the measurement of impulse currents three basic methods are commonly used:

- resistive shunts
- Rogowski coils
- current monitors

Rogowski coils and current monitors are not suited for the measurement of d.c. or long-duration rectangular currents. The fact, however, that they make no ohmic contact with the circuit being measured is very useful with respect to problems arising from ground-loop currents or potential differences.

22.3.2.1   Resistive shunts

The voltage across a resistive shunt is basically composed of a resistive and an induction component:

$$u = R \cdot i + d\Phi/dt$$

976 Lightning Protection

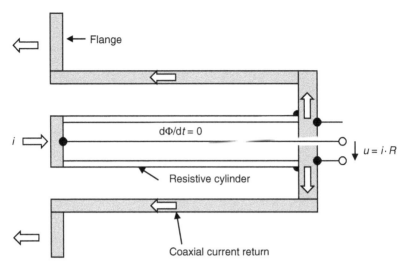

Figure 22.28 Principle of a tubular shunt

where $d\Phi/dt$ is the magnetic flux change in the loop formed by the resistor and the connecting leads. Simple shunts are not useful for the measurement of currents with microsecond or sub-microsecond front times, because the induction component becomes dominant. To overcome this drawback, tubular (or coaxial) shunts have been developed. A thin resistive metal sheet (e.g. a nickel chromium material) is formed into a cylindrical tube (Figure 22.28). The voltage drop along this tube is measured by a sensor wire running in the centre of the tube. The current return path to the flange, too, consists of a cylinder (e.g. steel). Because the magnetic field change $d\Phi/dt$ inside the resistive tube is zero the output voltage is proportional to the current. The upper bandwidth of a tubular shunt is limited by the current displacement (skin effect) in the resistive tube.

22.3.2.2 Rogowski coils

A Rogowski coil is basically an induction loop. It consists of a helical coil of wire. The whole assembly is then wrapped around the straight conductor for which the current is to be measured (Figure 22.29). The voltage induced in the coil is proportional to the derivative of current in the straight conductor:

$$u = M \, di/dt$$

where M is the mutual inductance between the coil and conductor. The output of a Rogowski coil is usually connected to an integrating network in order to provide an output signal proportional to the current.

One advantage of a Rogowski coil is that it can be made open-ended and flexible, allowing it to be wrapped around a conductor without disturbing it. Because a Rogowski coil has an air core, it has a low inductance and can respond to fast-changing

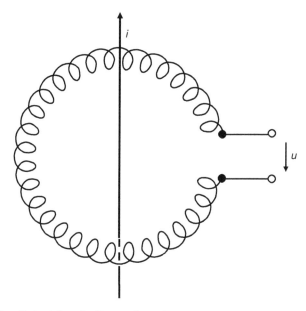

Figure 22.29 Principle of a Rogowski coil

currents. Also, because it has no iron core to saturate, it is highly linear even when subjected to large currents.

22.3.2.3 Current monitors

A current monitor is basically a current transformer where the primary winding is the conductor whose current is to be measured. Such a monitor consists of a

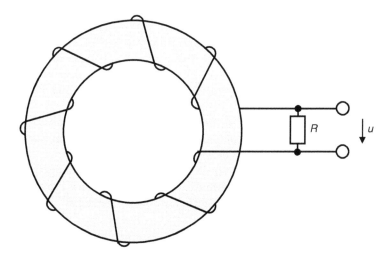

Figure 22.30 Principle of a current monitor

high-permeability magnetic core. The secondary winding is wrapped around the toroidal core and is terminated into a resistor R. The principle is shown in Figure 22.30. The magnetic field change produced by the current flowing in the primary conductor induces a voltage in the secondary winding proportional to the current change di/dt. The inductance L of the secondary winding together with the termination resistor R forms a low-pass filter with a cut-off angular frequency ω_c of

$$\omega_c = \frac{R}{L}$$

For angular frequencies $\omega \gg \omega_c$ (i.e. for impulse durations $T \ll 1/\omega_c$) the low-pass filter acts as an integrator and the output voltage becomes proportional to the current. The high-frequency response of a current transformer is determined by the inductance, resistance and stray capacitance of the winding. Internally distributed termination resistances extend the usable high-frequency limit.

Because the maximum magnetic flux is limited by core saturation, there is a corresponding limit on the charge transfer (or so-called $I \cdot t$ product) that a current monitor can handle. However, current monitors are commercially available rated for peak currents of several hundred of kiloamperes and a corresponding charge transfer in the range of 100 C.

References

1. Rakov V.A., Uman M.A. *Lightning: Physics and Effects*. Cambridge University Press; 2003.
2. Fisher F.A., Plumer J.A., Perala R.A. *Aircraft Lightning Protection Handbook*. Lightning Technologies, Inc.; 2005.
3. EUROCAE ED-105, April 2005: *Aircraft Lightning Test Method*.
4. Kuffel E., Zaengl W.S. *High Voltage Engineering: Fundamentals*. Butterworth-Heinemann Ltd.; 2000.
5. Beyer M., Boeck W., Möller K., Zaengl W. *Hochspannungstechnik: Theoretische und Praktische Grundlagen*. Berlin: Springer Verlag; 1998.
6. IEC 60060-1, Ed. 2.0, 1989-11: *High-Voltage Test Techniques. Part 1: General Definitions and Test Requirements*.
7. EUROCAE ED-84, 1997-8: *Aircraft Lightning Environment and Related Test Waveforms Standard*.
8. Marx E. *Verfahren zur Schlagpruefung von Isolatoren und anderen Elektrischen Vorrichtungen*. German patent no. DE000000455933A, application date 12.10.1923, publication date 13.02.1928.
9. Rodewald A. 'Ausgleichsvorgänge in Marxschen Vervielfachungsschaltungen nach der Zündung der ersten Schaltfunkenstrecke'. *Bulletin SEV* 1969;**60**:37–44.
10. Heilbronner F. 'Das Durchzünden mehrstufiger Stossgeneratoren'. *Elektrotechnische Zeitschrift-A* 1971;**92**:372–76.
11. IEC 62305-1 Ed. 1.0, 2006-01: *Protection Against Lightning – Part 1: General Principles*.

12. IEC 62305-2 Ed. 1.0, 2006-01: *Protection Against Lightning – Part 2: Risk Management*.
13. IEC 62305-3 Ed. 1.0, 2006-01: *Protection Against Lightning – Part 3: Physical Damage to Structures and Life Hazard*.
14. IEC 62305-4 Ed. 1.0, 2006-01: *Protection Against Lightning – Part 4: Electrical and Electronic Systems Within Structures*.
15. Hasse P., Wiesinger J., Zischank W. *Handbuch für Blitzschutz und Erdung*. 5th edn. München: Pflaum Verlag; 2006.
16. Zischank W. 'Simulation von Blitzströmen bei direkten Einschlägen'. *Elektrotechnische Zeitschrift* 1984;**105**:12–7.
17. Zischank W. 'A surge current generator with a double-crowbar sparkgap for the simulation of direct lightning stroke effects'. *5th International Symposium on High Voltage Engineering ISH*; Braunschweig, 1987. Paper 61.07.
18. Hourtane J.-L. 'DICOM: Current generator delivering the A, B, C, D waveform for direct lightning effects simulation on aircraft'. *International Aerospace and Ground Conference on Lightning and Static Electricity ICOLSE*; Bath, 1989. Paper 5A.5.
19. Landry M.J., Brigham W.P. 'UV laser triggering and crowbars used in the Sandia lightning simulator'. *International Aerospace and Ground Conference on Lightning and Static Electricity*; Orlando, 1984. pp. 46-1–46-13.
20. Modrusan M., Walther P. 'Aircraft testing with simulated lightning currents of high amplitude and high rate of rise time'. *International Aerospace and Ground Conference on Lightning and Static Electricity ICOLSE*; Paris, 1985. Paper 3A-5.
21. Modrusan M., Walther P. *Aircraft Testing with Simulated Lightning Currents of High Amplitude and High Rate of Rise Time*. Haefely Publication no. E5–23.
22. White R.A. 'Lightning simulator circuit parameters and performance for severe threat, high-action-integral testing'. *International Aerospace and Ground Conference on Lightning and Static Electricity*; Orlando, 1984. Paper 40-1.
23. Craven J.D., Knaur J.A., Moore T.W., Shumpert Th.H. 'A simulated lightning effects test facility for testing live and inert missiles and components'. *International Aerospace and Ground Conference on Lightning and Static Electricity*; Cocoa Beach, 1991. Paper 108-1.
24. Perala R.A., Rudolph T.H., McKenna P.M., Robb J.D. 'The use of a distributed peaking capacitor and a Marx generator for increasing current rise rates and the electric field for lightning simulation'. *International Aerospace and Ground Conference on Lightning and Static Electricity ICOLE*; Orlando, 1984. pp. 45-1–45-6.
25. Möller J. 'Kommutierung und aufsteilung von stoßströmen in funkenstrecken erhöhter schlagweite durch parallelen zünddraht'. *Elektrotechnische Zeitschrift-A* Bd.91(1970), H.6, pp.361–3.
26. Zischank W. 'Simulation of fast rate-of-rise lightning currents using exploding wires'. *21st International Conference on Lightning Protection, ICLP*; Berlin, 1992. Paper 5.01.

27. Salge J., Pauls N., Neumann K.-K. 'Drahtexplosionsexperimente in kondensatorentladekreisen mit großer induktivität'. *Zeitschrift für Angewandte Physik*. 1970;**29**: 339–43.
28. Zischank W.J., Drumm F., Fisher R.J., Schnetzer G.H., Morris M.E. 'Reliable simulation of metal surface penetration by lightning continuing currents'. *1995 International Aerospace and Ground Conference on Lightning and Static Electricity*; Williamsburg, Virginia, 26–28 September 1995.
29. Zischank W.J., Drumm F., Fisher R.J., Schnetzer G.H., Morris M.E. 'Simulation of lightning continuing current effects on metal surfaces'. *23rd International Conference on Lightning Protection, ICLP*; Firenze, Italy, 1996. 519–526.
30. Clifford D.W., Crouch K.E., Schulte E.H. 'Lightning simulation and testing'. *IEEE Trans. EMC* May 1982;**EMC-24(2)**:209–24.
31. Wiesinger J. 'Hybrid-Generator für die Isloaktionskoordination'. *Elektrotechnische Zeitschrift* 1983;**104(H.21)**:1102–05.
32. Richman P. 'Single output, voltage and current test waveform'. *IEEE Symposium on Electromagnetic Compatibility* 1983;**47**:413–18.
33. IEC 61000-4-5 Ed. 2.0, 2005-11: *Electromagnetic compatibility (EMC) – Part 4–5: Testing and Measurement Techniques – Surge Immunity Test*.
34. Schwab A.J. *High-Voltage Measurement Techniques*. Cambridge, MA: MIT Press; 1972.

Chapter 23
Return stroke models for engineering applications
Vernon Cooray

23.1 Introduction

From the point of view of an electrical engineer, the return stroke is the most important event in a lightning flash; it is the return stroke that causes most of the destruction and disturbance in electrical and telecommunication networks. In their attempts to provide protection, engineers seek the aid of return stroke models for three reasons. First, they would like to characterize and quantify the electromagnetic fields produced by return strokes at various distances to provide them with the input for mathematical routines that analyse the transient voltages and currents induced in electrical networks by these fields. This calls for return stroke models that are capable of generating electromagnetic fields similar to those created by natural return strokes. Second, their profession demands detailed knowledge of the effects of direct injection of lightning current. In a real situation this direct injection will be superimposed on currents and voltages induced by electromagnetic fields in the system under consideration. This necessitates the use of return stroke models that are capable of generating channel base currents similar to those in nature. Third, in order to evaluate the level of threat posed by lightning, engineers require statistical distributions of peak currents and peak current derivatives in lightning flashes. Even though the characteristics of return stroke currents can be obtained through measurements at towers equipped with current measuring devices or by utilizing rocket-triggering techniques, gathering statistically significant data samples in different regions and under different weather conditions is an exceptionally difficult enterprise. Accurate return stroke models can simplify this task to a large extent by providing the connection between the electromagnetic fields and the currents so that the latter can be extracted from the measured electromagnetic fields.

In the case of return strokes, a model is a mathematical formulate that is capable of predicting the temporal and spatial variation of the return stroke current, the variation of the return stroke speed, the temporal and spatial characteristics of optical radiation, the features of electromagnetic fields at different distances, and the signature of thunder. From the point of view of an engineer, the lightning parameters of particular

982 *Lightning Protection*

interest are the return stroke current and its electromagnetic fields, whence most of the return stroke models available today, especially the engineering models, are constructed to predict either one or both of these features. On the basis of the concepts and aims of return stroke models, they can be separated into four main groups: (i) the electro-thermodynamic models, (ii) the transmission line or *LCR*

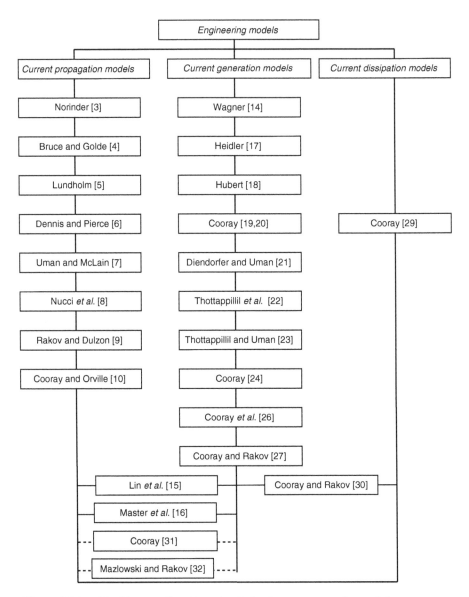

Figure 23.1   The history of engineering lightning return stroke models

models, (iii) antenna models, and (iv) engineering models [1,2]. Here, we concentrate on engineering models.

Even though the name 'engineering models' may give the impression that these models have down-played the physics completely, all these models are built using a valid physical concept as a foundation. However, the physics is somewhat neglected when selecting the input parameters necessary to build up the model. In many of these models input parameters are not evaluated using fundamental principles. Some of these parameters are obtained from available experimental data and the other parameters are selected, sometimes without any physical foundation, in such a way that the model predictions agree with experimental data. One problem with this procedure is the lack of uniqueness in the way in which the input parameters could be combined to generate the required result. The basic concept on which all the engineering models are built, whether it is stated as such or not, can be traced back to the physics of transmission lines. Using the basic physics of transmission line theory, three types of engineering models can be constructed. These are the current generation models, the current propagation models and the current dissipation models. Figure 23.1 shows a summary of the development of engineering return stroke models. At a later stage in this chapter we will show how these different types of return stroke models are related to each other.

In this chapter we will describe and discuss several engineering models that can be utilized either to evaluate electromagnetic fields from lightning flashes or to study the direct effects of lightning attachment to various structures including tall towers. We will start by describing the basic concepts of engineering return stroke models. This discussion will be followed by a description of various return stroke models and the equations necessary for the evaluation of electromagnetic fields using these return stroke models.

23.2 Current propagation models (CP models)

23.2.1 Basic concept

Consider a uniform and lossless transmission line. A current pulse injected into this line will propagate along the line with uniform speed without any change in the amplitude of the waveshape. The transmission line does not interfere with the current (of course this is not true in the case of a transmission line going into corona). It will only provide a path for the propagation of the current pulse from one location to another. This is the basis of the current propagation (CP) models. In these models it is assumed that the return stroke is a current pulse originating at ground level and propagating from ground to cloud along the transmission line created by the leader. The engineering models using this postulate as a base were constructed by Norinder [3], Bruce and Golde [4], Lundholm [5], Dennis and Pierce [6], Uman and McLain [7], Nucci and colleagues [8], Rakov and Dulzon [9] and Cooray and Orville [10]. The models differ from each other in the way they prescribe how the return stroke current varies as it propagates along the leader channel. For example, in the model introduced by Uman and McLain [7], popularly known as the transmission line model, the current is assumed to propagate along the channel without attenuation and with constant speed. In the model introduced by Nucci and colleagues [8] (MTLE – modified transmission line model with exponential current decay), the current amplitude decreases exponentially and in the one introduced by Rakov and Dulzon [9] (MTLL – modified transmission line model with linear current decay), the current amplitude decreases linearly. Cooray and Orville [10] introduced both current attenuation and dispersion while allowing the return stroke speed to vary along the channel.

23.2.2 Most general description

In order to generalize the description of the model let us assume that the return stroke process consists of two waves. The first one travels with speed u (which could very well be a function of height), preparing the channel for the transport of charge and current. The second one is the current wave and the associated charge travelling upwards with average speed v, depending again on the height. The current cannot be finite ahead of the plasma front that prepares the channel for conduction. Thus the current at level z is given by

$$I(z, t) = A(z) F(z, t - z/v) \qquad t > z/u \qquad (23.1)$$

In the above equation $A(z)$ is a function that represents the attenuation of the peak current and $F(z, t)$ describes the wave shape of the current at height z. Note also that $F(z, t) = 0$ for $t < 0$. One can define the function $F(z, t)$ as

$$F(z, t) = \int_0^t I_b(\tau) R(z, t - \tau) \, d\tau \qquad (23.2)$$

where $I_b(t)$ is the channel base current and $R(z, t)$ is a function that describes how the shape of the current waveform is being modified with height. However, this operation itself leads to the attenuation of the current and if we would like to represent the attenuation only by the factor $A(z)$, then we have to normalize this function to unity. Let t_p be the time at which the peak of the function defined in equation (23.2) is reached. Then the normalization can be carried out as follows:

$$F(z, t) = I_p \frac{\int_0^t I_b(\tau) R(z, t - \tau) \, d\tau}{\int_0^{t_p} I_b(\tau) R(z, t - \tau) \, d\tau} \qquad (23.3)$$

where I_p is the peak current at channel base. If the speeds of propagation of the pulses depend on z, then u and v in equation (23.1) have to be replaced by the average speeds. In Table 23.1 the expressions for the parameters of equations (23.1) to (23.3) pertinent

Table 23.1 Basic features of current propagation models

The model	v	u	$A(z)$	$R(z, t)$
Norinder [3]	∞		1.0	$\delta(t)$
Lundholm [5]	v varies as a function of peak current. The value of v is evaluated as a function of peak current using Wagner's equation [12]	$u = v$ or $u = \infty$	1.0	$\delta(t)$
Bruce and Golde [4]	$v = v_b e^{-\gamma_b t}$			
Dennis and Pierce [6]	$v = v_d e^{-\gamma_d t}$	Case 1: $u = v$ Case 2: $u = c$	1.0	$\delta(t)$
Uman and McLain [7]	For first: $v = v_d e^{-\gamma_d t}$ For subsequent: $v = v_u$	$u = v$	For first: $A(z) = I_p \dfrac{v}{v_d}$ For subsequent: $A(z) = 1.0$	$\delta(t)$ $\delta(t)$
Rakov and Dulzon [9]	$v = v_r$. Authors have also considered exponentially decaying speeds	$u = v$	$A(z) = \left(1 - \dfrac{z}{H}\right)$ $A(z) = \left(1 - \dfrac{z}{H}\right)^2$	$\delta(t)$
Nucci et al. [8]	$v = v_n$	$u = v$	$A(z) = e^{-z/\lambda}$	$\delta(t)$
Cooray and Orville [10]	$v = v_1 \{ a e^{-z/\lambda_1} + b e^{-z/\lambda_2} \}$	$u = v$	$A(z) = \{ g e^{-z/\lambda_3} + d e^{-z/\lambda_4} \}$	$R(z, t) = \dfrac{e^{-t/\tau(z)}}{\tau(z)}$ $\tau(z) = \eta z$ $\tau(z) = 1 - e^{-z/\lambda_5}$ $\tau(z) = 1 - e^{-(z/\lambda_6)^2}$

Parameters: $v_d = 8.0 \times 10^7$ m/s, $\gamma = 2 \times 10^4$ s^{-1}, $v_u = 8.0 \times 10^7$ m/s, $v_r = 1.5 \times 10^8$ m/s, I_p is the peak current selected in the model, $v_n = 1.1 \times 10^8$ m/s, $H = 7.5$ km, $v_1 = 2.2 \times 10^8$, $a = 0.5$, $b = 0.5$, $\lambda_1 = 100.0$ m, $g = 0.3$, $d = 0.7$, $\lambda_3 = 100$ m, $\lambda_4 = 3000$ m, $\lambda_5 = 500$ m, $\eta = 7 \times 10^{-10}$ s m^{-1}, $\lambda_6 = 500 - 250$ m, c is the speed of light.

to different return stroke models are presented. In analysing these models, the following points should be kept in mind.

1. The numerical values of the parameters given in the table are only of historical value. The important point is the way in which different scientists attempted to incorporate the observed facts into the models.
2. In all the models the channel base current is given as an input parameter. It varies from one return stroke model to another, but these variations are mainly due to the lack of information available in the literature concerning the return stroke current at the time of the creation of the model. Thus one can replace it with analytical forms created recently by scientists to represent the return stroke current. Several such examples are given in Chapter 2 and also in the later part of this chapter. One exception to this is the Lundholm model [5] in which, for reasons of mathematical simplicity, the channel base current is assumed to be a step.
3. The return stroke speed is another input parameter of these models. In some models the return stroke speed is assumed to be uniform, whereas in others it is assumed to decrease exponentially. Cooray and Orville [10] assumed a more complicated function containing two exponentials to represent the variation of return stroke speed with height. However, recent information shows that the speed can actually increase initially, reach a peak and then continue to decay [11].
4. Note that Cooray and Orville [10] changed the rise time of the current directly with height without first defining the function $R(z, t)$. The functions given in Table 23.1 are the ones that approximate this variation in the standard form written above. However, use of this function in the model will not only change the rise time but will also change the shape of the current waveform with height. On the other hand any physically reasonable dispersion process will affect not only the rise time but also the fast variations in any other part of the current.
5. In the case of the Lundholm model one can either use $u = c$ or $u = \infty$. This is the case because in this model the current waveform is replaced by a step function.

23.3 Current generation models (CG models)

23.3.1 Basic concept

If a transmission line goes into corona, the corona currents released at each line element will give rise to currents propagating along the line and an observer will be able to measure a current appearing at the base of the line. A similar scenario is used in the current generation (CG) models to describe the creation of the return stroke current. In these models the leader channel is treated as a charged transmission line and the return stroke current is generated by a wave of ground potential that travels along it from ground to cloud. The arrival of the wave front (i.e. return stroke front) at a given point on the leader channel changes its potential from cloud potential to ground potential, causing the release of bound charge on the central core and the corona sheath giving rise to the current in the channel (this is called the corona current in the literature). These models postulate that as the return stroke front propagates upwards, the charge stored on the leader channel collapses into the highly conducting core of the return stroke channel. Accordingly, each point on the leader channel can be treated as a current source that is turned on by the arrival of the return stroke front at that point. The corona current injected by these sources into the highly conducting return stroke channel core travels to ground with a speed denoted by v_c. As we will see later, in most of the return stroke models it is assumed that $v_c = c$, where c is the speed of light.

The basic concept of CG models was first introduced by Wagner [14]. He assumed that the neutralization of the corona sheath takes a finite time and therefore the corona current can be represented by a decaying exponential function. The decay time constant associated with this function is called the corona decay time constant. Wagner assumed, however, that the speed of propagation of the corona current down the return stroke channel is infinite. Lin and colleagues [15] introduced a model in which both CG and CP concepts are incorporated in the same model. In the portion of the current described by the CG concept, the corona current is represented by a double exponential function. The speed of propagation of the corona current down the channel is assumed to be the same as the speed of light. A modified form of this model is introduced by Master and colleagues [16], but in this modification the CG description remained intact. Heidler [17] constructed a model based on this principle in which the channel base current and the return stroke speed are assumed as input parameters. Furthermore, it was assumed that the neutralization of the corona sheath is instantaneous and hence the corona current generated by a given channel section can be represented by a Dirac delta function. The speed of propagation of the corona current down the return stroke channel is assumed to be equal to the speed of light. This model gives rise to a current discontinuity at the return stroke front, which, according to the author's understanding, is not physically reasonable. Hubert [18] constructed a current generation model rather similar to that of Wagner's model with the exception that the downward speed of propagation of the corona current is equal to the speed of light. He utilized this model to reproduce experimental data (both current and electromagnetic fields) obtained from triggered lightning. Cooray [19,20] introduced a model in which the distribution of the charge deposited by the return stroke (i.e. the sum of the positive charge necessary to neutralize the negative charge on the leader and the positive charge induced on the channel due to the action of the background electric field) and the decay time constant of the corona current are taken as input parameters, with the model predicting the channel base current and return stroke speed. Moreover, he took into consideration that the process of neutralization of the corona sheath takes a finite time in reality and, as a consequence, the corona current was represented by an exponential function with a finite duration. This is the first model in which the decay time constant of the corona current (and hence the duration of the corona current) is assumed to increase with height. Because the leader channel contains a hot core surrounded by a corona sheath, he also divided the corona current into two parts, one fast and the other slow. The fast one was associated with the neutralization of the core and the slow one with the neutralization of the corona sheath. Furthermore, by treating the dart leader as an arc and assuming that the electric field at the return stroke front is equal to the electric field that exists in this arc channel, he managed to derive the speed of the return stroke. Diendorfer and Uman [21] introduced a model in which the channel base current, return stroke speed and the corona decay time constant were assumed as input parameters. They also divided the corona current into two parts, one fast and the other slow. Thottappillil and colleagues [22] and Thottappillil and Uman [23] modified this model to include variable return stroke speed and a corona decay time constant that varies with height. Cooray [24] developed the ideas introduced in References 19 and 20 to create a CG model with channel base current as an input. Cooray [25] and Cooray and colleagues [26] extended the concept to include first return strokes with connecting leaders.

In CG models one has the choice of selecting the channel base current, $I_b(t)$, the distribution of the charge deposited by the return stroke along the channel, $\rho(z)$, the return stroke speed, $v(z)$, and the magnitude and variation of the corona discharge time constant with height, $\tau(z)$, as input parameters. Any set of three of these four input parameters will provide a complete description of the temporal and spatial variation of the return stroke current. Most of the CG models use $v(z)$ and either $\rho(z)$ or $\tau(z)$ in combination with $I_b(t)$ as input parameters. Recently, Cooray and Rakov [27] developed a model in which $\rho(z)$, $\tau(z)$ and $I_b(t)$ are selected as input parameters. The model could generate $v(z)$ as a model output.

23.3.2 Mathematical background

As mentioned above, a CG model needs three input parameters, which can be selected from a set of four parameters, that is, $\rho(z)$, $\tau(z)$, $I_b(t)$ and $v(z)$. Once three of these parameters are specified the fourth can be evaluated either analytically or numerically. Let us now consider the mathematics necessary to do this.

23.3.2.1 Evaluate $I_b(t)$ given $\rho(z)$, $\tau(z)$ and $v(z)$

Because the current at any given level on the channel is the cumulative effect of corona currents associated with channel elements located above that level, the return stroke current at any height in the return stroke channel $I(z, t)$ can be written as

$$I(z, t) = \int_z^{h_e} I_{cor}\{t - \xi/v_{av}(\xi) - (\xi - z)/v_c\} \, d\xi \qquad t > z/v_{av}(z) \qquad (23.4)$$

$$I_{cor}(z) = \frac{\rho(z)}{\tau(z)} \exp\{-(t - z/v_{av}(z))\} \qquad t > z/v_{av}(z) \qquad (23.5)$$

Note that $I_{cor}(z)$ is the corona current per unit length associated with a channel element at height z and $v_{av}(z)$ is the average return stroke speed over the channel section of length z with one end at ground level. The latter is given by

$$v_{av}(z) = \frac{z}{\int_0^z \frac{1}{v(z)} dz} \qquad (23.6)$$

The value of h_e can be obtained from the solution of

$$t = \frac{h_e}{v_{av}(h_e)} + \frac{h_e - z}{v_c} \tag{23.7}$$

The current at the channel base is given by

$$I_b(0, t) = \int_0^{h_0} I_{cor}(t - \xi/v_{av}(\xi) - \xi/v_c) \, d\xi \tag{23.8}$$

$$t = \frac{h_0}{v_{av}(h_0)} + \frac{h_0}{v_c} \tag{23.9}$$

23.3.2.2 Evaluate $\tau(z)$ given $I_b(t)$, $\rho(z)$ and $v(z)$

In most of the return stroke models corona current is represented by a single exponential function. An exponential function gives an instantaneous rise time to the corona current that is not physically reasonable. For this reason, in a few models it is represented by a double exponential function. In the analysis to be given below we therefore assume that the corona current is represented by a double exponential function. The corona current in this case is given by

$$I_{cor}(z) = \frac{\rho(z)}{\tau(z) - \tau_r}[\exp\{-(t - z/v_{av}(z))/\tau(z)\} - \exp\{-(t - z/v_{av}(z))/\tau_r\}] \, t > z/v_{av}(z) \tag{23.10}$$

where both $\rho(z)$ and τ_r are known, but $\tau(z)$ is unknown. Results pertinent to a corona current with a single exponential function can be obtained by letting $\tau_r = 0$. With this corona current the return stroke current at ground level is given by

$$I_b(0, t) = \int_0^{h_0} \frac{\rho(z) \, dz}{\tau(z) - \tau_r}[\exp\{-(t - z/v_{av}(z) - z/v_c)/\tau(z)\} - \exp\{-(t - z/v_{av}(z) - z/v_c)/\tau_r\}] \tag{23.11}$$

where h_0 can be extracted by the solution of equation (23.9). If we divide the channel into a large number of segments of equal length dz, the above integral can be written as a summation:

$$I_b(t_m) = \sum_{n=1}^{m} \frac{\rho_n}{\tau_n - \tau_r} \exp\left\{-\left(t_m - \frac{(n-1)\,dz}{v_{av,n}} - \frac{(n-1)\,dz}{v_c}\right)/\tau_n\right\}$$

$$- \sum_{n=1}^{m} \frac{\rho_n}{\tau_n - \tau_r} \exp\left\{-\left(t_m - \frac{(n-1)\,dz}{v_{av,n}} - \frac{(n-1)\,dz}{v_c}\right)/\tau_r\right\} \tag{23.12}$$

where ρ_n is the charge deposited per unit length on the nth section, τ_n is the decay time constant of the corona current of the nth section and $v_{\text{av},n}$ is the average return stroke speed over the channel section connecting the ground and the nth element. In this equation t_m is the time for the corona current released from the mth segment to reach the ground. This is given by the equation

$$t_m = \left[\left(m - \frac{1}{2}\right)\bigg/v_{\text{av},m} + \left(m - \frac{1}{2}\right)\bigg/v_c\right]dz \quad (23.13)$$

If the return stroke speed and the current at the channel base is known, then the value of the discharge time constant at different heights can be estimated progressively by moving from $m = 1$. For example when $m = 1$ the only unknown is τ_1. Once this is found one can consider the case $m = 2$. In the resulting equation the only unknown is the value of τ_2, and this can be obtained by solving that equation. In this way the values of discharge time constants up to the mth element can be obtained sequentially [22,24].

23.3.2.3 Evaluate $\rho(z)$ given $I_b(t)$, $\tau(z)$ and $v(z)$

Equations (23.12) and (23.13) can also be used to evaluate the discharge time constant when the other parameters are given as inputs. For example, in this case when $m = 1$ the only unknown is ρ_1. Once this is found one can consider the case $m = 2$. In the resulting equation the only unknown is the value of ρ_2 and this can be obtained by solving that equation. In this way the values of discharge time constants up to any mth element can be obtained sequentially.

23.3.2.4 Evaluate $v(z)$, given $I_b(t)$, $\rho(z)$ and $\tau(z)$

As before, we start with equation (23.12). Because $I_b(t)$, $\rho(z)$ and $\tau(z)$ are given, the only unknown parameter in these equations is $v_{\text{av},n}$, the average speed along the nth channel segment. Solving the equations as before one can observe that when $m = 1$ the only unknown is $v_{\text{av},1}$, the average speed over the first channel segment. Once this is found the value of $v_{\text{av},2}$ can be obtained by considering the situation of $m = 2$. In this way the average return stroke speed as a function of height can be obtained. It is important to point out that in this evaluation the value of dz in equation (23.12) should be selected in such a manner that it is reasonable to assume constant return stroke speed along the channel element. Once the average return stroke speed as a function of height is known, the return stroke speed as a function of height can be obtained directly from it.

23.3.3 CG models in practice

As mentioned in the introduction, several current generation models are available in the literature, and they differ from each other by the way in which input parameters are selected. In the sections to follow information is provided that is necessary to use CG models to calculate the spatial and temporal variation of the return stroke current.

23.3.3.1 Model of Wagner [14]

This is the first CG model to be introduced in the literature and therefore the credit for the creation of CG models goes to Wagner. He was the first scientist to come up with the concept of corona current and to treat the return stroke current as a sum of corona currents generated by channel elements located along the channel. The input parameters of Wagner's model are the distribution of the charge deposited by the return stroke, corona decay time constant and the return stroke speed. In the model Wagner assumed that the speed of propagation of the corona current down the return stroke channel is infinite. The parameters of Wagner's model are the following:

- *Channel base current*. The channel base current can be calculated using the parameters given in equation (23.4).
- *Corona current per unit length*:

$$I_c(t, z) = \frac{\rho(z)}{\tau} e^{-t/\tau}$$

  with $\tau = 6.66 \times 10^{-6}$ s
- *Speed of the corona current*:

$$v_c = \infty$$

- *Linear density of the charge deposited by the return stroke*:

$$\rho(z) = \rho_0 e^{-z\lambda}$$

  The peak current at ground level varies with ρ_0 and the latter can be selected to get the desired current peak at ground level, $\lambda = 10^3$ m.

- *Return stroke speed*:

$$v(z) = v_0 e^{-\gamma t}$$

  where $\gamma = 3 \times 10^4$ s^{-1} and v_0, which is a constant, is assumed to vary between $0.1c$ and $0.5c$, where c is the speed of light.

Note that the predicted current at ground level has a double exponential shape. The rise time of the current is determined by τ and the decay time by λ and γ.

23.3.3.2 Model of Heidler [17]

Heidler was the first scientist to introduce the channel base current as an input parameter in CG models and to connect the other parameters of the model to the channel base current. The input parameters of the model in addition to the channel base current are the return stroke speed and the corona discharge time constant. In the model Heidler assumed that the discharge time constant of the corona current is zero; that is, the discharge process takes place instantaneously. With these parameters the model can predict the distribution of the charge deposited by the return stroke along the channel. The parameters of the model are given below. Observe that the linear density of the charge deposited by the return stroke can be derived analytically.

- Channel base current:

$$I_b(0, t) = \frac{I_p}{\eta} \frac{k^n}{1+k^n} e^{-t/\tau_2}$$

with $k = t/\tau_1$, $n = 10$, $\tau_1 = 1.68$ μs, $\tau_2 = 20\text{--}150$ μs. η is the factor that has to be adjusted to obtain the exact current peak value.
- Corona current per unit length:

$$I_c(t, z) = \rho(z)\delta(t)$$

- Speed of the corona current:

$$v_c = c$$

- Linear density of the charge deposited by the return stroke:

$$\rho(z) = \frac{I_b(0, z/v + z/v_c)}{v^*}$$

with $(1/v^*) = (1/v) + (1/v_c)$.
- Return stroke speed. In the model v is assumed to be a constant.

23.3.3.3 Model of Hubert [18]

Hubert utilized a model based on the CG concept to generate a fit to the measured currents and electromagnetic fields of triggered lightning flashes. The values of the various model parameters were selected so that the predictions agree with experiment. The input parameters of the model are the distribution of the charge deposited by the return stroke, return stroke speed and the corona decay time constant. The parameters of the model are the following.

- Channel base current. The channel base current can be calculated using the parameters given in equation (23.4).
- Corona current per unit length:

$$I_c(t, z) = \frac{\rho(z)}{\tau} e^{-t/\tau}$$

with $\tau = 10^{-7}$ s.
- Speed of the corona current:

$$v_c = c$$

- Linear density of the charge deposited by the return stroke:

$$\rho(z) = \rho_0 e^{-z/\lambda}$$

where $\lambda = 15 \times 10^3$ m and the value of ρ_0 is selected to provide the required peak current at ground level.
- *Return stroke speed.* In the model v is assumed to be a constant equal to 1.0×10^8 m s^{-1}.

23.3.3.4 Model of Cooray [19,20]

Cooray introduced a CG model where for the first time the discharge time constant was assumed to increase with height. He also assumed that the neutralization process can be divided into two parts: one fast and the other slow. The fast one is assumed to be generated by the neutralization of the charge on the central core of the leader channel and the slow one by the neutralization of the cold corona sheath. The input parameters of the model are the corona discharge time constant and the distribution of the charge deposited by the return stroke on the channel. Cooray also attempted to evaluate the return stroke speed by connecting the electric field at the return stroke front to the potential gradient of the leader channel. The results showed that the return stroke speed increases initially, reaches a peak, and then decreases with increasing height. However, in using this model in engineering studies one can skip this iterative calculation and use it as a normal CG model by plugging in a speed profile similar to that predicted by the full model as an input parameter. An approximate for this speed profile is given by

$$v(z) = v_1 + (v_2/2)[1 - e^{-(z-1)/a} - e^{-(z-1)/b}] \quad 1.0 \leq z \leq 50\,\text{m} \quad (23.14)$$

$$v(z) = v_3 e^{-z/a_1} + v_4 e^{z/b_1} \quad z \geq 50\,\text{m} \quad (23.15)$$

with $v_1 = 1.02 \times 10^8$ m s^{-1}, $v_2 = 1.35 \times 10^8$ m s^{-1}, $v_3 = 7.11 \times 10^7$ m s^{-1}, $v_4 = 1.66 \times 10^8$ m s^{-1}, $a = 1.4$ m, $b = 7.4$ m, $a_1 = 400$ m, $b_1 = 2\,100$ m. The other model parameters are summarized below.

- *Channel base current.* Inserting the parameters given here in equation (23.4) one can calculate the return stroke current at any level along the channel.
- *Corona current per unit length of the hot corona sheath*:

$$I_{\text{hc}}(t, z) = \frac{\rho_\text{h}(z)}{\tau_\text{h} - \tau_\text{b}} [e^{-t/\tau_\text{h}} - e^{-t/\tau_\text{b}}]$$

with $\tau_\text{h} \approx 50\text{--}100$ ns and $\tau_\text{b} = 5$ ns.
- *Corona current per unit length of the cold corona sheath*:

$$I_{\text{cc}}(t, z) = \frac{\rho_\text{c}(z)}{\tau_\text{s} - \tau_\text{h}} [e^{-t/\tau_\text{s}} - e^{-t/\tau_\text{h}}]$$

where $\tau_\text{s} = \tau_\text{s0}[1 - e^{-z/\lambda_\text{s}}]$ and $\tau_\text{s0} = 1$ μs, $\lambda_\text{s} = 200$ m.

- Corona current per unit length:

$$I_c(t, z) = I_{hc}(t, z) + I_{cc}(t, z)$$

- Speed of corona current:

$$v_c = c$$

- Total linear density of the charge deposited by the return stroke on the leader channel:

$$\rho(z) = \rho_0 [0.3 \, e^{-z/\lambda_1} + 0.7 \, e^{-z/\lambda_2}]$$

with $\rho_0 = 0.0001$ C m^{-1} (for a typical subsequent stroke), $\lambda_1 = 600$ m and $\lambda_2 = 5\,000$ m.

- Linear density of charge deposited by the return stroke on the hot core:

$$\rho_h(z) = \rho_0 e^{-z/\lambda_c}$$

with $\lambda_c = 50$ m.

- Linear density of charge deposited by the return stroke on the corona sheath:

$$\rho_c(z) = \rho(z) - \rho_h(z)$$

- Return stroke speed. This is predicted by the model. The speed profile given by equations (23.14) and (23.15) can be used as an input parameter.

Comment: Observe that the peak value of the current at ground level varies linearly with ρ_0.

23.3.3.5 Model of Diendofer and Uman [21]

Diendorfer and Uman introduced a return stroke model where, similar to the model of Cooray [19,20] described above, the corona current is separated into two parts, one fast and the other slow. The fast corona current is assumed to be generated by the neutralization of the leader core and the slow one by the corona sheath. However, in contrast to the Cooray model, they utilized the channel base current as one of the input parameters. In the model, this current was separated into two parts. One part was assumed to be generated by the cumulative effects of the fast corona current and the other part by the cumulative effects of the slow corona currents. In addition to the channel base current the input parameters of the model are the return stroke speed and the discharge time constants. The input parameters of this model are given in the following.

- Channel base current:

$$I_b(0, t) = i_h(t) + i_c(t)$$

- Channel base current component associated with the leader core:

$$i_h(t) = \frac{I_{01}}{\eta_1} \frac{(t/\tau_{11})^2}{(t/\tau_{11})^2 + 1} e^{-t/\tau_{21}}$$

with $I_{01} = 13$ kA, $\eta_1 = 0.73$, $\tau_{11} = 0.15$ μs, $\tau_{21} = 3.0$ μs for typical subsequent strokes and $I_{01} = 28$ kA, $\eta_1 = 0.73$, $\tau_{11} = 0.3$ μs, $\tau_{21} = 6.0$ μs for typical first strokes.

- Channel base current component associated with the corona sheath:

$$i_c(t) = \frac{I_{02}}{\eta_2} \frac{(t/\tau_{12})^2}{(t/\tau_{12})^2 + 1} e^{-t/\tau_{22}}$$

with $I_{02} = 7$ kA, $\eta_2 = 0.64$, $\tau_{12} = 5$ μs, $\tau_{22} = 50$ μs for typical subsequent strokes and $I_{02} = 16$ kA, $\eta_2 = 0.53$, $\tau_{12} = 10$ μs, $\tau_{22} = 50$ μs for typical first strokes.

- Corona current per unit length from the leader core:

$$I_h(t, z) = \frac{\rho_h(z)}{\tau_h} e^{-t/\tau_h}$$

with $\tau_h = 0.6$ μs.

- Corona current per unit length from the corona sheath:

$$I_c(t, z) = \frac{\rho_c(z)}{\tau_c} e^{-t/\tau_c}$$

with $\tau_c = 5$ μs.

- Total corona current:

$$I_c(t, z) = I_h(t, z) + I_c(t, z)$$

- Linear density of charge deposited by the return stroke on the leader core:

$$\rho_h(z) = \frac{i_h(0, z/v^*) + \tau_h \frac{di_h(0, z/v^*)}{dt}}{v^*}$$

with $(1/v^*) = (1/v) + (1/v_c)$.

- Linear density of charge deposited by the return stroke on the corona sheath:

$$\rho_c(z) = \frac{i_c(0, z/v^*) + \tau_c \frac{di_c(0, z/v^*)}{dt}}{v^*}$$

with $(1/v^*) = (1/v) + (1/v_c)$.

- Total linear charge density deposited by the return stroke:

$$\rho(z) = \rho_h(z) + \rho_c(z)$$

- Speed of corona current:

$$v_c = c$$

- Return stroke speed. In the model v is assumed to be a constant equal to 1.3×10^8 m s^{-1}.

Comment: Note that the charge densities are not input parameters but could be derived once the channel base current, return stroke speed and the corona decay time constants are given.

23.3.3.6 First modification of the Diendofer and Uman model by Thottappillil et al. [22]

Thottappillil and colleagues modified the Diendorfer and Uman model to introduce a return stroke speed that varies with height. The parameters are given below.

- Channel base current:

$$I_b(0, t) = i_h(t) + i_c(t)$$

- Channel base current component associated with the leader core:

$$i_h(t) = \frac{I_{01}}{\eta_1} \frac{(t/\tau_{11})^2}{(t/\tau_{11})^2 + 1} e^{-t/\tau_{21}}$$

with $I_{01} = 13$ kA, $\eta_1 = 0.73$, $\tau_{11} = 0.15$ μs, $\tau_{21} = 3.0$ μs for typical subsequent strokes and $I_{01} = 28$ kA, $\eta_1 = 0.73$, $\tau_{11} = 0.3$ μs, $\tau_{21} = 6.0$ μs for typical first strokes.

- Channel base current component associated with the corona sheath:

$$i_c(t) = \frac{I_{02}}{\eta_2} \frac{(t/\tau_{12})^2}{(t/\tau_{12})^2 + 1} e^{-t/\tau_{22}}$$

with $I_{02} = 7$ kA, $\eta_2 = 0.64$, $\tau_{12} = 5$ μs, $\tau_{22} = 50$ μs for typical subsequent strokes and $I_{02} = 16$ kA, $\eta_2 = 0.53$, $\tau_{12} = 10$ μs, $\tau_{22} = 50$ μs for typical first strokes.

- Corona current per unit length from the leader core:

$$I_h(t, z) = \frac{\rho_h(z)}{\tau_h} e^{-t/\tau_h}$$

with $\tau_h = 0.6$ μs.

- Corona current per unit length from the corona sheath:

$$I_c(t, z) = \frac{\rho_c(z)}{\tau_c} e^{-t/\tau_c}$$

with $\tau_c = 5$ μs.
- Total corona current:

$$I_c(t, z) = I_h(t, z) + I_c(t, z)$$

- Linear density of charge deposited by the return stroke on the leader core:

$$\rho_h(z) = \frac{i_h[0, z/v_{av}(z) + z/v_c] + \tau_h \dfrac{di_h[0, z/v_{av}(z) + z/v_c]}{dt}}{\Gamma}$$

with

$$\frac{1}{\Gamma} = \frac{v_{av}(z) - z\dfrac{dv_{av}(z)}{dz}}{[v_{av}(z)]^2} + \frac{1}{v_c}$$

- Linear density of charge deposited by the return stroke on the corona sheath:

$$\rho_c(z) = \frac{i_c[0, z/v_{av}(z) + z/v_c] + \tau_c \dfrac{di_c[0, z/v_{av}(z) + z/v_c]}{dt}}{\Gamma}$$

- Total linear charge density deposited by the return stroke:

$$\rho(z) = \rho_h(z) + \rho_c(z)$$

- Speed of corona current:

$$v_c = c$$

- Return stroke speed:

$$v = v_0 e^{-z/\lambda}$$

with $v_0 = 1.3 \times 10^8$ m s^{-1} and λ varying between 1 000 and 3 000 m.

23.3.3.7 Second modification of the Diendofer and Uman model by Thottappillil and Uman [23]

In a subsequent publication Thottappillil and Uman modified the Diendorfer and Uman model to include a discharge time constant that increases with height. The input parameters of the model are the charge density, return stroke speed and the channel base current. These input parameters are shown in the following.

- *Channel base current*:

$$I_b(0, t) = i_h(t) + i_c(t)$$

$$i_h(t) = \frac{I_{01}}{\eta_1} \frac{(t/\tau_{11})^2}{(t/\tau_{11})^2 + 1} e^{-t/\tau_{21}}$$

with $I_{01} = 13$ kA, $\eta_1 = 0.73$, $\tau_{11} = 0.15$ μs, $\tau_{21} = 3.0$ μs for typical subsequent strokes and $I_{01} = 28$ kA, $\eta_1 = 0.73$, $\tau_{11} = 0.3$ μs, $\tau_{21} = 6.0$ μs for typical first strokes.

$$i_c(t) = \frac{I_{02}}{\eta_2} \frac{(t/\tau_{12})^2}{(t/\tau_{12})^2 + 1} e^{-t/\tau_{22}}$$

with $I_{02} = 7$ kA, $\eta_2 = 0.64$, $\tau_{12} = 5$ μs, $\tau_{22} = 50$ μs for typical subsequent strokes and $I_{02} = 16$ kA, $\eta_2 = 0.53$, $\tau_{12} = 10$ μs, $\tau_{22} = 50$ μs for typical first strokes.

- *Corona current per unit length*:

$$I_{cor}(t, z) = \frac{\rho(z)}{\tau(z)} e^{-t/\tau(z)}$$

$\tau(z)$ has to be evaluated from equation (23.12).

- *Linear density of charge deposited by the return stroke on the leader channel*:

$$\rho(z) = \frac{i_h(0, z/v^*) + \tau_h \frac{di_h(0, z/v^*)}{dt}}{v^*} + \frac{i_c(0, z/v^*) + \tau_c \frac{di_c(0, z/v^*)}{dt}}{v^*}$$

where $(1/v^*) = (1/v) + (1/v_c)$. The values of τ_h and τ_c are given in Section 23.3.3.5.

- *Speed of corona current*:

$$v_c = c$$

- *Return stroke speed*. In the model v is assumed to be a constant equal to 1.3×10^8 m s^{-1}.

Comment: Note that the charge distribution used as an input to the model is identical to that obtained in the original Diendorfer and Uman model.

23.3.3.8 Model of Cooray [24]

In constructing the model, Cooray used the same principles as the ones used in his original model, but utilized the channel base current as one of the input parameters. The other input parameter is the distribution of the charge deposited by the return stroke. Both the variation of the corona discharge time constant with height and the return stroke speed were extracted from the model. In order to obtain the return stroke speed he assumed that the electric field at the front of the return stroke is equal to the potential gradient of the leader channel. Again it is observed that the return stroke speed increases initially, reaches a peak and then continue to decrease. However, one can skip the additional numerical procedures by treating the return stroke speed as an input parameter. Then the model can be used as a normal CG model. The parameters of the model are given below.

- Channel base current:

$$I_b(0, t) = \frac{I_{01}}{\eta} \frac{(t/\tau_1)^2}{(t/\tau_1)^2 + 1} e^{-t/\tau_2} + I_{02}(e^{-t/\tau_3} + e^{-t/\tau_4})$$

For a typical subsequent return stroke: $I_{01} = 9.9$ kA, $\eta = 0.845$, $\tau_1 = 0.072$ μs, $\tau_2 = 5$ μs, $I_{02} = 7.5$ kA, $\tau_3 = 100$ μs and $\tau_4 = 6$ μs.

- Corona current per unit length:

$$I_{cor}(t, z) = \frac{\rho(z)}{\tau(z) - \tau_b} [e^{-t/\tau(z)} - e^{-t/\tau_b}]$$

$\tau_b = 5$ ns and $\tau(z)$ has to be evaluated from equation (23.12).
- Linear charge density deposited by the return stroke:

$$\rho(z) = \rho_0[1 - (z/H)]$$

The value of ρ_0 scales linearly with peak current with 100 μC m^{-1} for a 10 kA current. H is the height of the return stroke channel (assumed to be 9 km in the model).

- Speed of corona current:

$$v_c = c$$

- Return stroke speed. This is predicted by the model. However, one can use the model as a normal CG model by using the return stroke speed as an input parameter. Equations (23.14) and (23.15), which agree with the model prediction, provide a good approximation that can be used as an input.

23.3.3.9 Model of Cooray and Rakov [27]

Because the return stroke speed is one of the possible input parameters of CG models, Cooray and Rakov realized that if the charge deposited by the return stroke, corona decay time constant and channel base current are given as input parameters, one can utilize the return stroke model itself to predict the return stroke speed profile without any additional mathematics. A model capable of this was introduced by Cooray and Rakov. The model showed again that the return stroke speed increases initially, reaches a peak, and then decays. The input parameters of the model are given below.

- Channel base current:

$$I_b(0, t) = \frac{I_{01}}{\eta} \frac{(t/\tau_1)^2}{(t/\tau_1)^2 + 1} e^{-t/\tau_2} + I_{02}(e^{-t/\tau_3} + e^{-t/\tau_4})$$

For a typical subsequent return stroke, $I_{01} = 9.9$ kA, $\eta = 0.845$, $\tau_1 = 0.072$ μs, $\tau_2 = 5$ μs, $I_{02} = 7.5$ kA, $\tau_3 = 100$ μs and $\tau_4 = 6$ μs.

- Corona current per unit length:

$$I_{cor}(t, z) = \frac{\rho(z)}{\tau(z)} e^{-t/\tau(z)}$$

$\tau(z) = \tau_i + \mu z$ with $\tau_i = 10^{-8}$ s and $\mu = 10^{-9}$ s/m.
- Linear charge density along the channel:

$$\rho(z) = a_o I_p + I_p(a + bz)/(1 + cz + dz^2)$$

I_p is peak return stroke current, $a_o = 5.09 \times 10^{-6}$ (s/m), $a = 1.325 \times 10^{-5}$ (s/m), $b = 7.06 \times 10^{-6}$ (s/m²), $c = 2.089$ (m⁻¹), $d = 1.492 \times 10^{-2}$ (m⁻²). This is based on the results obtained in Reference 37.

- Speed of corona current:

$$v_c = c$$

- Return stroke speed. This is evaluated from the model using equation (23.12).

23.3.3.10 Model of Cooray, Rakov and Montano [26]

All the CG models described so far have been introduced to describe subsequent return strokes. Recently, Cooray, Rakov and Montano introduced a CG model to describe first return strokes with the channel base current as an input parameter. In the model they assumed that the return stroke is initiated when the connecting leader reaches the streamer region of the stepped leader, and the slow front, usually observed in the first return stroke current waveforms, is generated during the time of passage of the return stroke front through the streamer region of the stepped leader. The rapid rise in the current occurs when the upward moving return stroke front reaches the hot core of the leader. It is important to point out that the model assumes that the return stroke is initiated at the moment when the connecting leader enters into the streamer region of the stepped leader. In the case of tall towers the tower may initiate a connecting leader before the arrival of leader streamers at the top of the tower. In this case the point of initiation of the return stroke is located not at the tip of the tower but at some height above the tower, that is, at the extremity of the streamer region of the stepped leader. In the case of short structures the initiation of the connecting leader takes place when the streamers of the stepped leader reach the structure, and in such cases the point of initiation of the return stroke is located on the structure. The model parameters are given below.

- Channel base current:

$$I_b(0, t) = I_{01} \frac{(t/\tau_1)^n}{(t/\tau_1)^n + 1} + I_{02}[1 - e^{-(t/\tau_1)^3}](ae^{-t/\tau_2} + be^{-t/\tau_3})$$

- For a typical first return stroke current these authors suggested the following parameters: $I_{01} = 7.8$ kA, $\tau_1 = 5$ μs, $n = 100$, $I_{02} = 32.5$ kA, $\tau_2 = 4$ μs, $\tau_3 = 100$ μs, $a = 0.2$ and $b = 0.8$.
- *Corona charge per unit length*:

$$I_{\text{cor}}(t, z) = \frac{\rho(z)}{\tau(z)} e^{-t/\tau(z)}$$

$\tau(z)$ has to be evaluated from equation (23.12).
- *Speed of corona current*:

$$v_c = c$$

- *Speed of return stroke*:

$$\begin{aligned} v(z) &= v_0 e^{z/\lambda_c} & \text{for } z < l_c \\ v(z) &= v_0 e^{l_c/\lambda_c} e^{-(z-l_c)/\lambda_r} & \text{for } z > l_c \end{aligned}$$

where l_c is the length of the streamer region of the stepped leader. In the model it is assumed that $v_0 e^{l_c/\lambda_c} = 2.0 \times 10^8$ m s^{-1} and for a typical first stroke $l_c \approx 70$ m. Note that $v_0 e^{l_c/\lambda_c}$ is the speed of the return stroke front at the moment of its contact with the hot core of the leader channel. Using the concepts of CG one can show that the duration of the slow front time t_f is given by $t_f = [\lambda_c(1 - e^{-l_c/\lambda_c})/v_0] + l_c/c$ where t_f is the duration of the slow front in the current waveform. From this information l_c and v_0 are estimated.
- *Linear density of charge deposited by the return stroke*. The charge density is assumed to be uniform along the channel [i.e. $\rho(z) = \rho_0$] and for a typical first stroke $\rho_0 = 0.001$ C m^{-1}.

Comment: Observe that the model assumes that the return stroke speed increases exponentially as it moves along the streamer region of the leader channel, reaches a peak when it encounters the hot core and then decays as it proceeds further.

23.4 Current dissipation models (CD Models)

23.4.1 General description

As mentioned previously, if a current pulse is propagating without corona along a transmission line, it will travel along the line without any attenuation and modification of the current waveshape. This concept is used as a base in creating current propagation models. When the current amplitude is larger than the threshold current necessary for corona generation, each element of the transmission line acts as a corona current source. Half of the corona current generated by the sources travels downwards and the other half travels upwards. The upward moving corona currents interact with the front of the injected current pulse in such a way that the speed of the upward moving current pulse is reduced, and for a transmission line in air, this is reduced to a value less than the speed of light [28]. In a recent publication Cooray [29] showed that the upward moving corona current concept can also be used to create return stroke models. He coined the term 'current dissipation models' for the same. The basic features of the current dissipation (CD) models are depicted in Figure 23.2. The main assumptions of the current dissipation models are given in the following. The return stroke is initiated by a current pulse injected into the

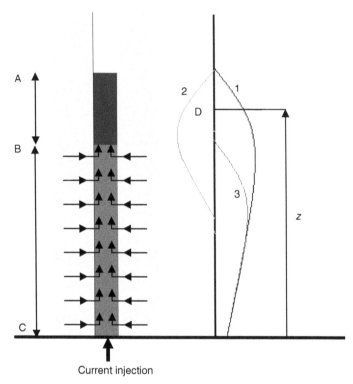

Figure 23.2 **Pictorial representation of the processes associated with a CD model at a given time t.** *The injected current (waveform 1 to the right) and the sum of corona currents (waveform 2 to the right) travel upwards with speed v_c. Point A is the front of these current waveforms. In the region A–B these two currents cancel each other, making the current above point B equal to zero. The cancellation is not complete below point B and therefore the net current below point B is finite (waveform 3 to the right). Thus point B is the front of the net current (i.e. return stroke front) moving upwards. Distance AC is equal to $v_c t$ and the distance BC is vt where v is the average speed of propagation of the net current front (i.e. return stroke front). Note that the current waveforms are not drawn to scale.*

leader channel from the grounded end. The arrival of the return stroke front at a given channel element will turn on a current source that will inject a corona current into the central core. It is important to stress here that the statement 'the arrival of the return stroke front at a given channel element' meant the onset of the return stroke current in that channel element (i.e. point B in Figure 23.2). Once in the core this corona current will travel upwards along the channel. In the case of negative return strokes the polarity of the corona current is such that it will deposit positive charge on the corona sheath and transport negative charge along the central core. Let us now incorporate mathematics into this physical scenario.

Assume that the return stroke is initiated by a current pulse injected into the leader channel at ground level. This current pulse propagates upward along the channel with speed v_c. When the return stroke front (i.e. the net current front) reaches a given channel element a corona source is turned on. This source will generate a corona current that will travel upwards along the central core with the same speed as the current pulse injected at the channel base (i.e. v_c). Note that the polarity of the upward moving corona current is opposite to that of the upward moving current injected at the channel base. For example, in the case of negative return stroke the current injected at the channel base carries positive charge upwards whereas the corona current transports negative charge upwards. According to this model the total current at a given point of the channel consists of two parts – an upward moving current pulse injected at the channel base and the total contribution of the upward moving corona currents. The upward moving corona current being of opposite polarity leads to the dissipation of the current pulse injected at the channel base.

23.4.2 Mathematical background

Consider the diagram to the right in Figure 23.2. This depicts a situation at any given time t. At this time the tip of the injected current is located at point A and the return stroke front is located at point B. The net current at any point above the return stroke front is zero. This is the case since the injected current and the cumulative effects of the corona current cancel each other above the return stroke front. Now, let us consider a point D located above the return stroke front. The height of this point from ground level is z. The net corona current at that point is given by

$$I_{\text{cor,total}}(z, t) = -\int_0^{h_d} I_c[\xi, t - \xi/v_{\text{av}}(z)(z - \xi)/v_c] \, d\xi \tag{23.16}$$

(Note that since the corona current is defined as positive here it has to be subtracted from the injected current). The value of h_d, the highest point on the channel whose corona current can reach point z at time t can be obtained by solving the equation

$$t - \frac{h_d}{v_{\text{av}}(h_d)} + \frac{z - h_d}{v_c} = 0 \tag{23.17}$$

The injected current at point z at time t is given by

$$I_{in}(z, t) = I_b(0, t - z/v_c) \tag{23.18}$$

Because the corona current annihilates the injected current at all points above the return stroke front we have

$$I_b(0, t - z/v_c) = \int_0^{h_d} I_c[\xi, t - \xi/v_{av}(\xi) - (z - \xi)/v_c] d\xi \tag{23.19}$$

Changing the variable we can write

$$I_b(0, t') = \int_0^{h_s} I_c(\xi, t' - \xi/v_{av}(\xi) + \xi/v_c) d\xi \tag{23.20}$$

with h_s given by

$$t' - \frac{h_s}{v_{av}(h_d)} + \frac{h_s}{v_c} = 0 \tag{23.21}$$

Now, a comparison of equations (23.20) and (23.21) with (23.8) and (23.9) shows that the only difference in the equations when moving from the CG concept to the CD concept is that v_c is replaced by $-v_c$. Moreover, as in the case of CG models, the input parameters of CD models are the charge deposited on the channel by the return stroke, corona decay time constant, return stroke speed and the channel base current. When three of these parameters are given, the fourth can be obtained in the same manner as was done in the CG models. However, in equations (23.4) to (23.12), v_c has to be replaced by $-v_c$ when using the equations in connection with current dissipation models (see [29]).

23.4.3 Cooray and Rakov model – a combination of current dissipation and current generation models [30]

In CG models the corona current generated by current sources located along the return stroke channel travels downwards with a speed equal to the speed of light (i.e. $v_c = c$). In general, it is assumed that this current waveform travelling down the channel will be completely absorbed by the ground. However, it is more correct to assume that the incident current would be reflected completely at ground level. If the current is reflected at ground level, it will propagate up with the speed of light, similarly to the injected current in the CD models. In the model, the incident current at ground level is represented by CG model principles and the propagation of the reflected current along the channel is represented by CD principles. Because the incident current is assumed to be completely reflected at ground level, the incident current component contributes to half of the channel base current, and the other half is caused by the reflected current component.

In treating the incident current at ground level, Cooray and Rakov [30] used the channel base current (i.e. half of the total channel base current), corona current and the return stroke speed as the input parameters. Because the return stroke speed and the discharge time constant are common for both current components (i.e. there is only one return stroke front), these, together with the channel base current (i.e. half of the total channel base current), are used as the input parameters of the CD model that simulated the reflected wave. The main features of the model are given below.

- Channel base current:

$$I_b(0, t) = i_i(t) + i_r(t)$$

- Incident component of the channel base current:

$$i_i(t) = \frac{1}{2}\left[\frac{I_{01}}{\eta}\frac{(t/\tau_1)^2}{(t/\tau_1)^2+1}e^{-t/\tau_2} + I_{02}(e^{-t/\tau_3} + e^{-t/\tau_4})\right]$$

For a typical subsequent return stroke, $I_{01} = 9.9$ kA, $\eta = 0.845$, $\tau_1 = 0.072$ µs, $\tau_2 = 5$ µs, $I_{02} = 7.5$ kA, $\tau_3 = 100$ µs and $\tau_4 = 6$ µs.
- Corona current per unit length associated with the incident current:

$$I_i(t, z) = \frac{\rho_i(z)}{\tau}e^{-t/\tau}$$

for $\tau = 0.1$ µs.
- Linear density of charge deposited by the incident current [29]:

$$\rho_i(z) = \frac{i_i(0, z/v^*) + \tau\dfrac{di_i(0, z/v^*)}{dt}}{v^*}$$

where $(1/v^*) = (1/v) + (1/v_c)$.
- Reflected component of the channel base current:

$$i_r(t) = \frac{1}{2}\left[\frac{I_{01}}{\eta}\frac{(t/\tau_1)^2}{(t/\tau_1)^2+1}e^{-t/\tau_2} + I_{02}(e^{-t/\tau_3} + e^{-t/\tau_4})\right]$$

For typical subsequent return stroke, $I_{01} = 9.9$ kA, $\eta = 0.845$, $\tau_1 = 0.072$ µs, $\tau_2 = 5$ µs, $I_{02} = 7.5$ kA, $\tau_3 = 100$ µs and $\tau_4 = 6$ µs.

- Corona current per unit length associated with the reflected current:

$$I_r(t, z) = \frac{\rho_r(z)}{\tau} e^{-t/\tau}$$

for $\tau = 0.1$ μs.
- Linear density of charge deposited by the reflected current:

$$\rho_r(z) = \frac{i_r(0, z/v^{**}) + \tau \frac{di_r(0, z/v^{**})}{dt}}{v^{**}}$$

where $(1/v^{**}) = (1/v) - (1/v_c)$.
- Speed of corona current:

$$v_c = c$$

- Speed of the return stroke. In the model v is assumed to be a constant equal to 1.3×10^8 m s^{-1}.

23.5 Generalization of any model to the current generation type

Cooray [31] showed that any return stroke model can be converted to a CG model by introducing an effective corona current. Here, we will illustrate the mathematical analysis that led to that conclusion.

Consider a channel element of length dz at height z and let $I(z, t)$ represent the temporal variation of the total return stroke current at that height. In the case of CG models, this current is generated by the action of corona current sources located above this height. Assume for the moment that the channel element does not generate any corona current. In this case the channel element will behave as a passive element that will just transport the current that is being fed from the top. In this case one can write

$$I(z + dz, t) = I(z, t + dz/v_c) \tag{23.22}$$

That is, the current injected at the top of the element will appear without any change at the bottom of the channel element after a time dz/v_c, which is the time taken by the current to travel from the top of the channel element to the bottom.

Now let us consider the real situation in which the channel element dz will also generate a corona current. As the current injected at the top passes through the channel element, the corona sources will add their contribution, resulting in a larger current appearing at the bottom than the amount injected at the top. The difference in these two quantities will give the corona current injected by the channel element.

Thus the average corona current generated by the element dz is given by

$$I_{cg}(z, t)dz = I(z, t + dz/v_c) - I(z + dz, t) \tag{23.23}$$

Using the Taylor expansion, the above equation can be rewritten as

$$I_{cg}(z, t)dz = I(z, t) - I(z + dz, t) + \frac{dz}{v_c}\frac{\partial I(z, t)}{\partial t} \tag{23.24}$$

Dividing both sides by dz and taking the limit d$z \to 0$, the corona current per unit length, $I_{cg}(z, t)$ injected into the return stroke channel at height z is given by

$$I_{cg}(z, t) = -\frac{\partial I(z, t)}{\partial z} + \frac{1}{v_c}\frac{\partial I(z, t)}{\partial t} \tag{23.25}$$

This equation can be used to transfer any return stroke model to a CG model with an equivalent corona current. It is important to stress here that even though the distribution of the return stroke current as a function of height remains the same during this conversion, there is a radical change in the corona current. If one attempts to extract the physics of the leader charge neutralization process using the temporal variation of the corona current as predicted by a return stroke model, the information one gathers will depend strongly on the way in which the return stroke model is formulated. This can easily be illustrated using the transmission line model (TL model) [7]. In the CP scenario of the TL model, the upward propagating current will not give rise to any corona and therefore the corona current is zero. On the other hand, if the same model is converted to a CG model then the equivalent corona current associated with the converted model [obtained from equation (23.25)] becomes bipolar [31]. The physics of the neutralization process pertinent to this equivalent corona current is the following. As the rising part of the upward moving current passes through a given channel element, the corona sheath located around that channel element will be neutralized by injection of positive charge into it. During the decaying part of the upward moving current all the deposited positive charge will be removed, bringing the corona sheath back to its original state. Thus the physics of corona dynamics in the two scenarios is completely different even though the longitudinal distribution of current and the charge along the channel at any given time is the same in the two formulations. This shows that conversion of a model from one type to another will change the underlying physics, even though both descriptions are identical from the point of view of the total current as a function of height. Thus one has to apply caution in deriving the physics of corona neutralization process using these models because the information extracted concerning it will be model dependent.

23.6 Generalization of any model to the current dissipation type

An analysis similar to the one presented in Section 23.5 was conducted by Cooray [29] for CD models. That analysis is presented below.

Consider a channel element of length dz at height z and let $I(z, t)$ represent the temporal variation of the total return stroke current at that height. Assume for the moment that the channel element does not generate any corona current. In this case the channel element will behave as a passive element that will just transport the current that is being fed from the top. In this case one can write

$$I(z + dz, t) = I(z, t - dz/v_c) \tag{23.26}$$

That is, the current injected at the bottom of the channel element will appear without any change at the top of the channel element after a time dz/v_c, which is the time taken by the current to travel from the bottom of the channel element to the top.

Now let us consider the real situation in which the channel element dz will also generate a corona current. As the current injected at the bottom passes through the channel element the corona sources will add their contribution, and because the polarity of the corona current is opposite to that of the injected current, this results in a smaller current appearing at the top than the amount of current injected at the bottom. The difference in these two quantities will give the corona current injected by the channel element. Thus the average corona current generated by the element dz is given by

$$I_{cd}(z, t)dz = I(z, t - dz/v_c) - I(z + dz, t) \tag{23.27}$$

Using the Taylor expansion, the above equation can be rewritten as

$$I_{cd}(z, t)dz = I(z, t) - I(z + dz, t) - \frac{dz}{v_c}\frac{\partial I(z, t)}{\partial t} \tag{23.28}$$

Dividing both sides by dz and taking the limit $dz \to 0$, the corona current per unit length, $I_{cd}(z, t)$, injected into the return stroke channel at height z is given by

$$I_{cd}(z, t) = -\frac{\partial I(z, t)}{\partial z} - \frac{1}{v_c}\frac{\partial I(z, t)}{\partial t} \tag{23.29}$$

Note that this equation is completely symmetrical to the one derived for the CG model [i.e. equation (23.25)], except that second term has a negative sign. This equation can be used to transfer any return stroke model to a current dissipation model with an equivalent corona current. The discussion given at the end of Section 23.5 is also applicable here.

23.7 Current dissipation models and the modified transmission line models

If the return stroke current associated with a current propagation model is assumed to decrease with height (as in the case of modified transmission line (MTL) models [8,9]), the conservation of charge requires deposition of charge along the channel as the return stroke front propagates upward. This leakage of charge from central core to the corona sheath can be represented by a radially flowing corona current. Recently, Maslowski and Rakov [32] showed that this corona current is given by

$$I_{cp}(z, t) = -\frac{\partial I(z, t)}{\partial t} - \frac{1}{v}\frac{\partial I(z, t)}{\partial t} \tag{23.30}$$

where $I_{cp}(z, t)$ is the corona current per unit length at height z, $I(z, t)$ is the longitudinal return stroke current at the same height as predicted by the return stroke model, and v is the speed of the return stroke front. Note that the direction of flow of the corona current is radial and, in contrast to the CG or CD models, it does not have a component flowing along the return stroke channel; that is, it is a stationary corona current. Maslowski and Rakov [32] showed that any return stroke model could be reformulated as a CP model with an equivalent stationary corona current given by equation (23.30).

Let us now go back to the CD models. Cooray [29] showed that in general the speed of propagation of the return stroke front in CD models is less than that of the injected current (i.e. v_c). However, he also showed that one can select the parameters of the corona current in such a way that the speed of the return stroke front remains the same as that of the injected current pulse and the corona current. When such a choice is made, CD models reduce to MTL models. This can be illustrated mathematically as follows. Let us represent the injected current at the channel base as $I_b(0, t)$. The injected current at height z is given by

$$I_b(z, t) = I_b(0, t - z/v_c) \tag{23.31}$$

Assume that the corona current per unit length at level z is given by

$$I_{cd}(z, t) = I_b(0, t - z/v_c)A(z) \tag{23.32}$$

where $A(z)$ is some function of z. According to this equation the corona current at a given height is proportional to the injected current at that height. Substituting this expression in equation (23.29), one finds that

$$I_b(0, t - z/v_c)A(z) = -\frac{\partial I(z, t)}{\partial z} - \frac{1}{v_c}\frac{\partial I(z, t)}{\partial t} \tag{23.33}$$

One can easily show by substitution that the solution of this equation is given by

$$I(z, t) = A'(z)I_b(0, t - z/v_c) \tag{23.34}$$

with

$$A'(z) = -\int A(z)\,dz \qquad (23.35)$$

Note that $I(z, t)$ in the above equation is the total current, that is the sum of the corona current and the injected current. According to equation (23.34), the total current propagates upwards with the same speed as that of the injected current and corona current. Moreover, it propagates upwards without any distortion while its amplitude varies with height according to the function $A'(z)$. Indeed, equation (23.34) describes an MTL model. In this special case equation (23.29) reduces to equation (23.30) as derived by Mazlowski and Rakov [32], because the return stroke speed v becomes equal to v_c. Thus, equation (23.30) is a special case of equation (23.29), and the latter reduces to the former in the case of MTL models. The above also demonstrates that all the CP models available in the literature are special cases of CD models.

23.8 Effect of ground conductivity

The way in which the ground conductivity can be incorporated into CG-type return stroke models and the effect of ground conductivity on the return stroke current has been described by Cooray and Rakov [30]. The procedure they have used to incorporate ground conductivity into return stroke models is described in the following.

Consider the physical process that leads to the formation of the corona current through the neutralization process. The neutralization process takes place when the channel core changes its potential from cloud to ground value. If this change is instantaneous, then in principle, the corona decay time constant could be very small and the shape of the corona current can be replaced by a Dirac delta function. However, when the ground is finitely conducting, the ground potential cannot be transferred to the channel faster than the relaxation time of the finitely conducting ground. Thus the relaxation time limits the rapidity with which the channel potential can be changed. In other words, the neutralization time and hence the corona decay time constant depend on the ground conductivity. Assume that the ground is perfectly conducting. The ground potential cannot then be transferred in a time less than ~ 10 ns, which is the time necessary for the heating and transfer of electron energy to the ions and neutrals in the central core. This sets a lower limit to the value of the corona decay time constant. When the ground is finitely conducting and the relaxation time is larger than the thermalization time, then the minimum value of the corona decay time constant, τ_0, is determined by the relaxation time. Thus one can write

$$\tau_0 = \tau_t \qquad \tau_r \leq \tau_t \qquad (23.36)$$
$$\tau_0 = \tau_r \qquad \tau_r \geq \tau_t \qquad (23.37)$$
$$\tau_r = \varepsilon\varepsilon_0/\sigma \qquad (23.38)$$

In the above equations ε is the relative permittivity of the ground, ε_0 the permittivity of air, σ the conductivity of soil, and τ_r is the relaxation time of soil. The value of τ_t (thermalization time) is 10 ns. The effects of ground conductivity on the return stroke current and return stroke current derivative as derived by Cooray and Rakov [30] are shown in Figure 23.3. Note that for typical ground conductivities (0.01–0.001 S m^{-1})

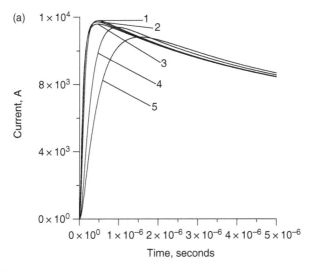

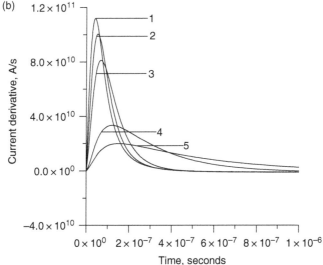

Figure 23.3 The effects of ground conductivity on the channel base current as predicted by the model of Cooray and Rakov [30] for (1) 0.01 S m^{-1}, (2) 0.002 S m^{-1}, (3) 0.001 S m^{-1}, (4) 0.0002 S m^{-1} and (5) 0.0001 S m^{-1}: (a) current; (b) current derivative.

the change in the peak current is insignificant, whereas it influences the return stroke peak current derivative significantly.

23.9 Equations necessary to calculate the electric and magnetic fields

Once the spatial and temporal distribution of the return stroke current is given, the electromagnetic fields can be obtained directly using Maxwell's equations. Without going into details of the derivations, we will give the equations that can be utilized for this purpose. First, we will give the electric and magnetic fields over finitely conducting ground based on Norton's analytical solution of Sommerfeld's equations. Cooray [33] showed that very close to the channel Norton's equations may produce significant errors if propagation effects are evaluated using them. However, in the case of close distances the overall propagation effects are not that significant and therefore these errors do not play a significant role in practice. We will then consider perfectly conducting ground and for this case the equations will be given in the time domain.

The geometry relevant for the calculation is given in Figure 23.4. Using the expressions for the electromagnetic fields of the vertical dipole as published by Norton [34] and Bannister [35] (see also [36]), the horizontal electric field, vertical

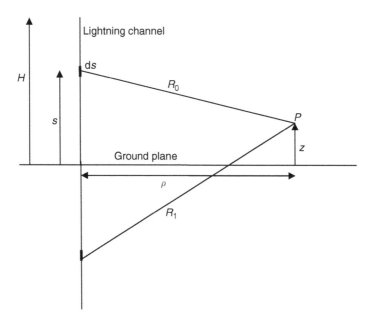

Figure 23.4   *Geometry relevant to the calculation of electromagnetic fields generated by the return stroke*

electric field and the azimuthal magnetic field at a height z from ground level and at a distance ρ from the return stroke can be expressed as

$$E_\rho(\rho, z, j\omega) = -\int_0^H \frac{jk_0 I(j\omega, s)\,ds\, Z_0}{4\pi}\left[\sin T_d \cos T_d \frac{\exp(-jk_0 R_0)}{R_0} + R_v \sin T_r \cos T_r\right.$$

$$\times \frac{\exp(-jk_0 R_1)}{R_1} - (1-R_v)\Delta_0 F(w) \sin T_r \frac{\exp(-jk_0 R_1)}{R_1}$$

$$-\sin T_r \cos T_r (1-R_v)\frac{\exp(-jk_0 R_1)}{jk_0 R_1^2} - \sin T_r \Delta_0 (1-R_v)\frac{\exp(-jk_0 R_1)}{2jk_0 R_1^2}$$

$$+ 3\sin T_d \cos T_d\left\{\frac{1}{jk_0 R_0} + \frac{1}{(jk_0 R_0)^2}\right\}\frac{\exp(-jk_0 R_0)}{R_0}$$

$$\left. + 3\sin T_r \cos T_r\left\{\frac{1}{jk_0 R_1} + \frac{1}{(jk_0 R_1)^2}\right\}\frac{\exp(-jk_0 R_1)}{R_1}\right] \quad (23.39)$$

$$E_z(\rho, z, j\omega) = -\int_0^H \frac{jk_0 I(j\omega, s)\,ds\, Z_0}{4\pi}\left[\sin^2 T_d \frac{\exp(-jk_0 R_0)}{R_0} + R_v \sin^2 T_r\right.$$

$$\times \frac{\exp(-jk_0 R_1)}{R_1} - (1-R_v)F(w)\sin^2 T_r \frac{\exp(-jk_0 R_1)}{R_1} + (1-3\cos^2 T_d)$$

$$\times \left\{\frac{1}{jk_0 R_0} + \frac{1}{(jk_0 R_0)^2}\right\}\frac{\exp(-jk_0 R_0)}{R_0} + (1-3\cos^2 T_r)$$

$$\left. \times \left\{\frac{1}{jk_0 R_1} + \frac{1}{(jk_0 R_1)^2}\right\}\frac{\exp(-jk_0 R_1)}{R_1}\right] \quad (23.40)$$

$$B_\phi(\rho, z, j\omega) = \int_0^H \frac{jZ_0 k_0 I(j\omega, s)\,ds}{4\pi c}\left[\sin T_d \frac{\exp(-jk_0 R_0)}{R_0} + R_v \sin T_r\right.$$

$$\times \frac{\exp(-jk_0 R_1)}{R_1} + (1-R_v)F(w)\sin T_r \frac{\exp(-jk_0 R_1)}{R_1} + \sin T_r(1-R_v)$$

$$\left. \times \frac{\exp(-jk_0 R_1)}{2jk_0 R_1^2} + \sin T_d \frac{\exp(-jk_0 R_0)}{jk_0 R_0^2} + R_v \sin T_r \frac{\exp(-jk_0 R_1)}{jk_0 R_1^2}\right]$$

$$\quad (23.41)$$

where $\sin T_d = \dfrac{\rho}{R_0}$, $\sin T_r = \dfrac{\rho}{R_1}$, $\cos T_d = \dfrac{z-s}{R_0}$, $\cos T_r = \dfrac{s+z}{R_1}$, $R_v = \dfrac{\cos T_r - \Delta_0}{\cos T_r + \Delta_0}$,

$Z_0 = \sqrt{\dfrac{\mu_0}{\varepsilon_0}}$, $k_0 = \omega\sqrt{\mu_0 \varepsilon_0}$, $\Delta_0 = \dfrac{\gamma_0}{\gamma}\sqrt{\left[1 - \dfrac{\gamma_0^2}{\gamma^2}\sin^2 T_r\right]}$, $\gamma_0 = \sqrt{\omega^2 \mu_0 \varepsilon_0}$,

$\gamma = \sqrt{-j\omega\mu_0(\sigma + j\omega\varepsilon_0 \varepsilon_r)}$, $w = -\dfrac{j\omega R_1}{2\sin^2 \beta}[\cos T_r + \Delta_0]^2$

and

$$F(w) = 1 - j(\pi w)^{1/2} e^{-w} \text{erfc}(jw^{1/2}).$$

In these equations erfc stands for the complementary error function, $I(j\omega, s)$ is the return stroke current at the element ds and $j = \sqrt{-1}$. Note that $I(j\omega, s)$, $E_\rho(\rho, z, j\omega)$ and $B_\phi(\rho, z, j\omega)$ are related to the time-domain quantities $i(t, s)$, $e_\rho(\rho, z, t)$ and $b_\phi(\rho, z, t)$ through the inverse Fourier transforms

$$I(j\omega, s) = \int_0^\infty i(t, s) e^{-j\omega t} \, dt \tag{23.42}$$

$$E_\rho(\rho, z, j\omega) = \int_0^\infty e_\rho(\rho, z, t) e^{-j\omega t} \, dt \tag{23.43}$$

$$B_\phi(\rho, z, j\omega) = \int_0^\infty b_\phi(\rho, z, t) e^{-j\omega t} \, dt \tag{23.44}$$

Now let us consider the perfectly conducting ground. In this case $\Delta_0 \to 0$ and $R_v \to 1$. The equations reduce to

$$E_\rho(\rho, z, j\omega) = -\int_0^H \frac{jk_0 I(j\omega, s) \, ds \, Z_0}{4\pi} \left[\sin T_d \cos T_d \frac{\exp(-jk_0 R_0)}{R_0} + \sin T_r \cos T_r \right.$$

$$\times \frac{\exp(-jk_0 R_1)}{R_1} + 3 \sin T_d \cos T_d \left\{ \frac{1}{jk_0 R_0} + \frac{1}{(jk_0 R_0)^2} \right\} \frac{\exp(-jk_0 R_0)}{R_0}$$

$$\left. + 3 \sin T_r \cos T_r \left\{ \frac{1}{jk_0 R_1} + \frac{1}{(jk_0 R_1)^2} \right\} \frac{\exp(-jk_0 R_1)}{R_1} \right] \tag{23.45}$$

$$E_z(\rho, z, j\omega) = -\int_0^H \frac{jk_0 I(j\omega, s) \, ds \, Z_0}{4\pi} \left[\sin^2 T_d \frac{\exp(-jk_0 R_0)}{R_0} + \sin^2 T_r \right.$$

$$\times \frac{\exp(-jk_0 R_1)}{R_1} + (1 - 3\cos^2 T_d) \left\{ \frac{1}{jk_0 R_0} + \frac{1}{(jk_0 R_0)^2} \right\} \frac{\exp(-jk_0 R_0)}{R_0}$$

$$\left. + (1 - 3\cos^2 T_r) \left\{ \frac{1}{jk_0 R_1} + \frac{1}{(jk_0 R_1)^2} \right\} \frac{\exp(-jk_0 R_1)}{R_1} \right] \tag{23.46}$$

$$B_\phi(\rho, z, j\omega) = \int_0^H \frac{jZ_0 k_0 I(j\omega, s) \, ds}{4\pi c} \left[\sin T_d \frac{\exp(-jk_0 R_0)}{R_0} + \sin T_r \frac{\exp(-jk_0 R_1)}{R_1} \right.$$

$$\left. + \sin T_d \frac{\exp(-jk_0 R_0)}{jk_0 R_0^2} + \sin T_r \frac{\exp(-jk_0 R_1)}{jk_0 R_1^2} \right] \tag{23.47}$$

These equations can be directly transformed into the time domain, resulting in

$$E_\rho(\rho, z) = -\frac{1}{4\pi\varepsilon_0} \int_0^H \left\{ \sin T_\text{d} \cos T_\text{d} \frac{1}{R_0 c^2} \frac{\partial i(s, t - R_0/c)}{\partial t} + \sin T_\text{r} \cos T_\text{r} \right.$$

$$\times \left. \frac{1}{R_1 c^2} \frac{\partial i(s, t - R_1/c)}{\partial t} \right\} + \left\{ 3 \sin T_\text{d} \cos T_\text{d} \frac{1}{R_0^2 c} i(s, t - R_0/c) \right.$$

$$+ 3 \sin T_\text{r} \cos T_\text{r} \frac{1}{R_1^2 c} i(s, t - R_1/c) \right\} + \left\{ 3 \sin T_\text{d} \cos T_\text{d} \frac{1}{R_0^3} \int_0^t i(s, \tau - R_0/c)\, d\tau \right.$$

$$+ 3 \sin T_\text{r} \cos T_\text{r} \frac{1}{R_1^3} \int_0^t i(s, \tau - R_1/c)\, d\tau \bigg\} ds \qquad (23.48)$$

$$E_z(\rho, z) = -\frac{1}{4\pi\varepsilon_0} \int_0^H \left\{ \sin^2 T_\text{d} \frac{1}{R_0 c^2} \frac{\partial i(s, t - R_0/c)}{\partial t} + \sin^2 T_\text{r} \frac{1}{R_1 c^2} \frac{\partial i(s, t - R_1/c)}{\partial t} \right\}$$

$$+ \left\{ (1 - 3\cos^2 T_\text{d}) \frac{1}{R_0^2 c} i(s, t - R_0/c) + (1 - 3\cos^2 T_\text{r}) \frac{1}{R_1^2 c} i(s, t - R_1/c) \right\}$$

$$+ \left\{ (1 - 3\cos^2 T_\text{d}) \frac{1}{R_0^3} \int_0^t i(s, \tau - R_0/c)\, d\tau \right.$$

$$+ (1 - 3\cos^2 T_\text{d}) \frac{1}{R_1^3} \int_0^t i(s, \tau - R_1/c)\, d\tau \bigg\} ds \qquad (23.49)$$

$$B_\phi(\rho, z) = \frac{\mu_0}{4\pi} \int_0^H \left\{ \sin T_\text{d} \frac{1}{cR_0} \frac{\partial i(s, t - R_0/c)}{\partial t} + \sin T_\text{r} \frac{1}{cR_1} \frac{\partial i(s, t - R_1/c)}{\partial t} \right\}$$

$$+ \left\{ \sin T_\text{d} \frac{1}{R_0^2} \frac{i(s, t - R_0/c)}{\partial t} + \sin T_\text{r} \frac{1}{R_1^2} \frac{i(s, t - R_1/c)}{\partial t} \right\} ds \qquad (23.50)$$

These equations define the electric field at any point in space over perfectly conducting ground.

23.10 Concluding remarks

In this chapter we have presented the basic principles underlying engineering return stroke models together with the information necessary to use available return stroke models to evaluate the spatial and temporal variation of the return stroke current and to use that information to calculate the electromagnetic fields generated by return strokes.

It is important to note here that any new return stroke model that is introduced into the scientific literature should be able to present a new way of studying the return stroke process. On the other hand, the model parameters should be considered as information that should or could be changed when more experimental data become available concerning the return stroke process. Unfortunately, some scientists give more emphasis to the model parameters and by doing so lose the important message that a model builder is trying to convey to the scientific establishment. This incorrect way of looking at models also leads to the creation of 'new models' by changing one or two parameters of an existing model.

References

1. Gomes C., Cooray V. 'Concepts of lightning return stroke models'. *IEEE Trans. Electromagn. Compat.* February 2000;**42**(1):18617–30.
2. Rakov V., Uman M.A. 'Review and evaluation of lightning return stroke models including some aspects of their application'. *IEEE Trans. Electromagn. Compat.* 1998;**40**:403–26.
3. Norinder H. *Quelques essais recents relatifs à la determination des surten sions indirectes*. CIGRE Session 1939, 29 June–8 July, Paris, p. 303.
4. Bruce C.E.R., Golde R.H. 'The lightning discharge'. *J. Inst. Elec. Eng.* 1941;**88**:487–520.
5. Lundholm PhD dissertation. KTH, Stockholm, Sweden, 1957.
6. Dennis A.S., Pierce E.T. 'The return stroke of lightning flash to earth as a source of atmospherics'. *Radio Sci.* 1964;777–94.
7. Uman M.A., McLain D.K. 'Magnetic field of lightning return stroke'. *J. Geophys. Res.* 1969;**74**:6899–910.
8. Nucci C.A., Mazzetti C., Rachidi F., Ianoz M. 'On lightning return stroke models for LEMP calculations'. Presented at the *19th International Conference on Lightning Protection*; Graz, Austria, 1988.
9. Rakov V.A., Dulzon A.A. 'A modified transmission line model for lightning return stroke field calculation'. *Proceedings of the 9th International Symposium on EMC*; Zurich, Switzerland, 1991. 44H1, pp. 229–35.
10. Cooray V., Orville R.E. 'The effects of variation of current amplitude, current risetime and return stroke velocity along the return stroke channel on the electromagnetic fields generated by return strokes'. *J. Geophys. Res.* October 1990; **95**(D11).
11. Olsen R.C., Jordan D.M., Rakov V.A., Uman M.A., Grimes N. 'Observed one-dimensional return stroke propagation speeds in the bottom 170 m of a rocket-triggered lightning channel'. *Geophys. Res. Lett.* 2004;**31**(L16107):4pp, doi: 10.1029/2004GL020187.

12. Wagner C.F. 'Relation between stroke current and velocity of return stroke'. *Trans. Inst. Elec. Electron. Eng. Power App. Syst.* 1963;609–17.
13. Cooray V., Theethayi N. 'Effects of corona on pulse propagation along transmission lines with special attention to lightning return stroke models and return stroke velocity'. *Proceedings of the 8th International Symposium on Lightning Protection*; Brazil, 2005. Invited paper.
14. Wagner C.F. 'A new approach to the calculation of the lightning performance of transmission lines'. *AIEE Trans.* 1956;**75**:1233–56.
15. Lin Y.T., Uman M.A., Standler R.B. 'Lightning return stroke models'. *J. Geophys. Res.* 1980;**85**:1571–83.
16. Master M, Lin Y.T., Uman M.A., Standler R.B. 'Calculations of lightning return stroke electric and magnetic fields above ground'. *J. Geophys. Res.* 1981;**86**: 12127–32.
17. Heidler F. 'Traveling current source model for LEMP calculation'. *Proceedings of the 6th International Symposium on EMC*; Zurich, Switzerland, 1985. 29F2, pp. 157–62.
18. Hubert P. 'New model of lightning return stroke – confrontation with triggered lightning observations'. *Proceedings of the 10th International Aerospace and Ground Conference on Lightning and Static Electricity*; Paris, 1985. pp. 211–15.
19. Cooray V. 'A return stroke model'. *Proceedings of the International Conference on Lightning and Static Electricity*; University of Bath, September 1989.
20. Cooray V. 'A model for the subsequent return strokes'. *J. Electrostat.* 1993;**30**:343–54.
21. Diendorfer G., Uman M.A. 'An improved return stroke model with specified channel base current'. *J. Geophys. Res.* 1990;**95**:13621–44.
22. Thottappillil R., McLain D.K., Uman M.A., Diendorfer G. 'Extension of Diendorfer–Uman lightning return stroke model to the case of a variable upward return stroke speed and a variable downward discharge current speed'. *J. Geophys. Res.* 1991;**96**:17143–50.
23. Thottappillil R., Uman M.A. 'Lightning return stroke model with height-variable discharge time constant'. *J. Geophys. Res.* 1994;**99**:22773–80.
24. Cooray V. 'Predicting the spatial and temporal variation of the current, the speed and electromagnetic fields of subsequent return strokes'. *IEEE Trans. Electromagn. Compat.* 1998;**40**:427–35.
25. Cooray V. 'A model for negative first return strokes in negative lightning flashes'. *Phys. Scripta* 1997;**55**:119–28.
26. Cooray V., Montano R., Rakov V. 'A model to represent first return strokes with connecting leaders'. *J. Electrostat.* 2004;**40**:97–109.
27. Cooray V., Rakov V. 'A current generation type return stroke model that predicts the return stroke velocity'. *J. Lightning Res.* 2007;**1**:32–39.
28. Cooray V., Theethayi N. 'Pulse propagation along transmission lines in the presence of corona and their implication to lightning return strokes'. *IEEE Trans. Antennas Propagat.* 2008;**56**(7):1948–59, doi: 10.1109/TAP.2008.924678, 2008.

29. Cooray V. 'A novel procedure to represent lightning strokes – current dissipation return stroke models'. *IEEE Trans. on Electromagnetic Compatibility,* in press, 2009. (Also presented at the *29th International Conference on Lightning Protection*; Uppsala, Sweden, 2008.)
30. Cooray V., Rakov V.A. '"Hybrid current source" return stroke model', *Proceedings of the 9th International Symposium on Lightning Protection*; Brazil, 2007.
31. Cooray V. 'On the concepts used in return stroke models applied in engineering practice'. *Trans. IEEE (EMC)* 2003;**45**:101–108.
32. Maslowski G., Rakov V.A. 'Equivalency of lightning return stroke models employing lumped and distributed current sources'. *Trans. IEEE (EMC)* 2007;**49**:123–32.
33. Cooray V. 'On the accuracy of several approximate theories used in quantifying the propagation effects on lightning generated electromagnetic fields'. *IEEE Trans. Antennas Propagat.* 2008;**56**(7):1960–67, doi: 10.1109/TAP.2008.924680, 2008.
34. Norton K.A. 'Propagation of radio waves over the surface of Earth and in the upper atmosphere. II'. *Proc. IEEE* 1937;**25**:1203–36.
35. Bannister P.R. *Extension of Finitely Conducting Earth-Image-Theory Results to Any Range.* Technical report, Naval Underwater Systems Center, January 1984.
36. Maclean T.S.M, Wu Z. *Radio Wave Propagation over Ground.* London: Chapman & Hall; 1993.
37. Cooray V., Rakov V., Theethayi N. 2004. 'Lightning striking distance – revisited'. *J. Electrostatics*, 2007;**65**:296–306.

Index

Action integral 48
Air discharges 15
Air voids 532, 545
Air-termination system 308–320, 434–435, 798–799, 934–935
  construction of 316–319
    metal sheet thicknesses 319
  location on structure 308–316
    mesh method 311–313
    method comparison 316
    protection angle method 313–315
    rolling sphere method 310–311
  non-conventional 320
Air-terminations 826–827
  protective angle method 826
  rolling sphere method 826
Airport runway lighting system 144–149
  cable currents 149
  counterpoise 147–148
  damage 149
  vertical ground rods 148–149
Airports, lightning damage to 360–361
Altitude triggering 103–105, 109–110
  positive lightning 105
  processes sequence 103–104
Andersson's measurements 53
Antenna support structure 726–729
  potential equalization 761
Armstrong and Whitehead 180–182
Arresters 417–421
  BLITZDUCTOR CT 417–421
Attempted leaders 17
Attractive radius expression 66
Autonomous power supply 747–748
Auxiliary power circuits 709–710

Backup surge protection 410–417
  installation protection 416–417

Bark-loss damage 849
Basic impulse insulation level (BIL) 646–648
Bazelyan
  empirical leader model 195–196
  leader inception model 209
Bearings 704–705
Becerra 197
  leader progression model 227–239
Berger's measurements 50, 53, 58
BIL: see basic impulse insulation level
Blades 695–702
BLITZDUCTOR CT installation 417–421
  cut-off frequency 421
  electrical isolation 422–423
  gas discharge tubes 418–420
  limiting voltage 420
  nominal current 420–421
  nominal discharge current 420
  voltage protection levels 418
Blunt injuries 914–915
Blunt tip lightning conductor 210
Bonding network 941–942
Boundaries, lightning protection zones and 373–380
  coordination of 385–388
Boundary-based methods 809–810
  method of moments 809–810
Branch components 16
Building equipment, potential equalization 761–763
Bunched cables vs. open wire line 597–598
Burn injuries 913–914

Cable connection to transformer, lightning damage to 734
Cable currents 149

1020 Index

Cable shielding 369–370
  double-ended 369
  earthing 369
  low-impedance 370
  single-ended 369–370
Camp Blanding, triggering 105, 106
Cantilevered construction 325–326
Capacitor 3, 295–296, 653
Carbon fibre reinforced plastic (CFRP) 696
Carbon fibre turbine blades 699–700
Cardiovascular system injuries 906–908
CFO: see critical impulse flashover
CFRP: see carbon fibre reinforced plastic
CG: see cloud-to-ground lightning
Channel terminations, ground flashes
     and 31–34
Charge transferred 274–275
Chokes 89
Circuit breakers 288–289
Circuit screening, external 763–765
Circuits, high-frequency groundings
     and 503–506
City areas, lightning damage to 357–360
Classical triggering 98–103, 107–109
  initial continuous current 100
  negative lightning 102
  positive lightning 102
  upward positive leader 100
Close lightning electromagnetic
     environment 128–131
  dart-leader electric field change 130
Cloud discharges 570–571
  cloud to air flashes 570
  cloud to cloud 570
  intracloud 570
Cloud flashes 15
  air discharges 15
  intercloud discharges 15
  intracloud 15
Cloud to air flashes 570
Cloud to cloud discharges 570
Cloud-to-ground (CG) lightning 274
Cluster of needles 254–256
CM: see common mode
Collection volume/field intensification
     method (CVM/FIM) 214–215
Collinson 5
Column of charge 15
  stepped leader 15

Common bonding network 834–835
Common mode (CM) voltages 270–273
Common-mode transfer 786
Communication systems, protection
     of 711–712
Conductivity, ionized region 536–537
Conductor penetration 279–280
  shield 279–280
Connecting leaders 16
  currents in 45–47
  lengths of 403–405
  speed of 44–45
Connections, down-conductors and 326–328
Continuing current pulses (CCP) 72
Continuing currents 16, 84–88
  long continuing 85
  protection of 710–712
  short continuing 85
  surge protectiona and 421
Cooray 179–180, 197, 545–549, 993–994,
     998–999
  air voids 545
  analytical expression 72–74
  leader progression model 227–239
  mathematical theories 546–548
  modelling parameters 548–549
Coordinated SPD protection system 766–768
Corona 19, 21, 23
Corona current, cluster of needles 254–256
Corona screening, electric field rate of
     change 253–254
Corrosion 753
Cost-benefit analysis, wind turbines
     and 688–691
Counterpoise, current delay 147–148
Critical electric field 534–536
Critical impulse flashover (CFO)
     voltage 646–648
Critical radius concept 186–190
Critical streamer length concept 194–195
Critical success index (CSI) 870
Cross-sectional areas 410–417
Crowbar technologies 962–964
CSI: see critical success index
CT: see current transformers
Current delay 147–18
Current dissipation model 1001–1006, 1008,
     1009–1010
  modified transmission line 1009–1010

Index   1021

Current distribution in down-conductors 330
Current flow damage 729–730
Current generation 970–971
Current generation model 986–1001
  mathematical concepts 988–990
  use of 990–1011
    Cooray 993–994, 998–999
    Diendofer and Uman 994–998
    Heidler 991–992
    Hubert 992–903
    Montano 1000
    Rakov 999–1000
    Thottappillil 996–997
    Wagner 991
Current generation type 1006–1007
Current injection 540
  deionization 540
  direct effects testing and 971–972
  ionization 540
  negative ionization 540
  sparking 540
Current limiters 288–290
  circuit breakers 288–289
  fuses 288–289
  inductors 289–290
  positive temperature coefficient
    devices 290
Current monitors 977–978
Current propagation model 983–986
  features 985
Current transformers (CT) 136
Current waveforms 67–70, 107–110
  altitude triggering 109–110
  classical triggering 107–109
  initial current variation 107
  mathematical representation 67
  negative rocket-triggered
    lightning 113–115
  return-stroke 67–70
Currents in
  connecting leaders 45–47
  vertical ground rods 148–149
Cut-off frequency 421
CVM/FIM: *see* collection volume/field
  intensification method

D-STATCOM: *see* distribution static
  compensator
d.c. line protection 436–437

Dalibard, Thomas-François 1, 5
Damage from lightning 355–361
Damage probabilities 452–457
  approach to 457–458
  evaluation 456
Dart leaders 17
  electric field
    change 130
    generation 40–44
  speed of 37–39
Dart-stepped leader 17
Darveniza 537–540
DAS: *see* dissipation array systems
Data processing systems 437–438
De Buffon, Comte 5
Deionization 540
Delay switches 775
Delayed fuses 774–776
Delfino, analytical expression 72
Dellera 179, 215–217
Delta configurations 558–559
Diendorfer 994–998
  analytical expression 71
Differential mode (DM) currents 289
Digital data lines 801
Diodes 286–288
Direct effects testing, current
  injection 971–972
Direct lightning strikes 568–570,
  603–604
  expected number 638–639
  overvoltages 636, 640–642
Disability, human injuries and 915
Discharge current 766–767
Dissipation array systems (DAS) 251–259
  experimental contradictions to 253–259
    corona current 254–256
    corona screening 253–254
    space charge 257–259
Distribution lines
  lightning performance 666–667
  lightning strikes to 636–640
Distribution static compensator
  (D-STATCOM) 294
Distribution systems, TN 394
Distribution transformer 558–564, 591–593,
  612–614
  delta configurations 558–559
DM: *see* differential mode

1022 *Index*

Domain-based methods 808–809
  FDTD 808–809
  finite element methods 808
  transmission line modelling 808–809
Double-ended cable earthen shielding 369
Down-conductor connections 326–328
Down-conductor current distribution 330
Down-conductor joints 326–328
Down-conductor system 320–330, 434–435, 743–745, 799, 807–812, 935
  boundary-based methods 809–810
  construction of 322–325
  domain-based 808–809
  location of 320–322
  positioning of 320–322
  ring conductors 321
Downward negative ground flash 15–17
  attempted leaders 17
  branch components 16
  column of charge 16
  connecting leaders 16
  continuing currents 16
  dart leader 17
  dart-stepped leader 17
  junction processes 17
  K-changes 17
  M-components 17
  partially conducting stage 17
  preliminary breakdown 15
  recoil streamer 17
  return stroke 16
  striking distance 16
  subsequent return stroke 17
Downward positive ground flash 15, 17
DVR: *see* dynamic voltage restorer
Dynamic leader inception evaluation 202–204
Dynamic voltage restorer (DVR) 293–294

Ear injuries 910–911
Early streamer emission (ESE)
  experimental contradictions to 241–243
  rods 243–244
  theoretical contradictions to 243–251
    amplitude of voltage pulses 249–250
    claimed time advantage 250–251
    early streamer emission rods 243–244
    lightning-like electric fields 246–249

    switching impulse voltages 244–246
Earth electrodes 332–335
Earth resistance 594–595
  artificial decrease of 342–343
  measurement of 348
Earth-termination system 330–348, 730–731, 745–752, 799, 806–807, 935–937
  autonomous power supply 747–748
  earth resistance 342, 348
  earthing arrangements 332–335
  external power supply 749–752
  large area coverage 341–342
  materials 348–352
  potential equalization 757–761
  principles 330–331
  rocky soils 341
  sandy soils 341
  soil ionization 343
  soil resistivity 345–347
  step voltages 344–345
  touch voltages 343
Earthing 693–695
Earthing arrangements 332–335
  small structures 335–341
  software for 338
Earthing cable shielding 369
Earthing improvement 753–754
Earthing reference point (ERP) 372
Earthing system 435–436, 594–595, 941
  IT 556
  TN 556
  TT 556
EFMs: *see* electric field mills
Electric field equations 1012–1015
Electric field generation 19–23
Electric field generation
  dart leaders 40–44
  stepped leaders 39–40
Electric field mills (EFMs) 862
Electric field peak value, observations on 77
Electric field rate of change, corona screening and 253–254
Electric fields
  lightning-like 246–249
  strokes from 74–81
    sea strikes 78
Electric fields generation, space charge layer 21

Electrical systems 706–710, 726
  auxiliary power circuits 709–710
  generator circuit 707–709
  medium-voltage 709
Electrical wires
  grounding and shielding of 801
  internal protection of 800–801
  low frequency 800
  ordnance circuit 801
  power 800
  radio-frequency 800–801
Electrification 860–861
  detection of 862–864
    electric field mills 862
Electro-geometrical method 168–170
Electrodes
  grounding arrangement and 479–481
  lightning currents to, resistance and 482–487
  location of 533
Electrodynamic effects 726
Electromagnetic compatibility 278–280
Electromagnetic field
  amplitude 79–80
  generation of 19
  remote sensing 48
Electromagnetic Transient Program (EMTP) 675
Electronic propagation 506–508
Electronic systems 726
Elevated structures, striking distance to 182–212
EMTP: *see* Electromagnetic Transient Program
Energy balance equation 543
Equipotential bonding 328–329, 363–365, 693–695, 828
  information technology systems 364–365
  low-voltage system 364
Equipotential bonding network 371–373
  boundaries 373–380
  earthing reference point 372
  information technology installation 378–380, 382–383, 384–385
  metal installations 373–375, 383
  power supply installations 375–378, 380–382, 383–384
Eriksson 177–179
Eriksson's measurements 51–52, 53

ERM: *see* Extended Rusck Model
ERP: *see* earthing reference point
ESE: *see* early streamer emission
Exploding wires 969
Explosion 415
Explosive tree damage 849
Extended Rusck Model (ERM) 573
External lightning protection system (LPS) 307–352
  air-termination system 308
  cantilevered construction 325–326
  down-conductor system 320–330
  earth-termination system 330–348
  lightning equipotential bonding 328–329
  offshore oil platforms and 832–834
External power supply 749–752
External zone lightning protection 366
Eye injuries 909–910

Failure-to-warn (FTW) 870
False alarm ratio (FAR) 869
FAR: *see* false alarm ratio
Fault current limiters 294
  high-temperature superconductors 294
FDTD: *see* finite-difference time-domain method
FEM: *see* finite element methods
Ferrites 289
Filter protection, parasitic elements of components 294–302
Filters 291–293, 294–302
  capacitors 295–296
  inductors 297–298
  resistors 298–302
Finite difference time-domain method (FDTD) 575, 808–809
Finite element methods (FEM) 808
First strokes
  correlation coefficients 57–60
  electric fields from 74–81
  peak electric radiation 81–83
Fixed-point
  algorithms 870–873
  warning algorithm 879–884
  warning system 864–867
Flammable materials
  air terminations 826–827
  equipotential bonding 828
  lightning protection 825–830

1024 *Index*

Flammable materials (*Continued*)
  risk assessment 824–825
  self-protecting system 829–830
  storage of 830
  tanks containing 822–830
  venting 829
  vessels containing 822–830
Flash
  polarity 15
  subsequent strokes 470–471
Flat ground, striking distance to 176–182
Follow current
  current limitation 393
  extinguishing capability 392–393
Fork lightning 31
Foundation earth electrode 754–757
Franklin, Benjamin 1–13
  capacitor 3
  legacy of 12–13
  Leyden jar capacitor 3
  lightning 3
  lightning rod
    early designs 7–9
    improved versions 9–12
    origins of 4
  thunderstorms, study of 6–7
Frequency dependent
    behaviour 514–520
  non-linear grounding vs. 526–527
FTW: *see* failure-to-warn
Fuses 288–289

Gaisberg tower 73
Garbagnati 179, 215–217
Gas discharge tubes (GDT) 282–285,
    418–420
GDT: *see* gas discharge tubes
Gears 704–705
Generator circuit 707–709
GFRP: *see* glass fibre reinforced plastics
Glass fibre reinforced plastics (GFRP) 695
Golde 177
GPR: *see* grounding potential rise
Grabagnati and Piparo's measurements 51
Grid size 518–520
Ground conditions, return-stroke current peak
    and 122–128
Ground conductivity 1010–1012
Ground flashes 15

channel terminations 31–34
connecting leaders 44, 45–47
density 23–25
  thunderstorm days 23–25
downward negative 15
downward positive 15
fork lightning 31
multiple terminations 33
structure interaction 18–19
  electromagnetic field generation 19
  injected current 18–19
two branches 31
upward negative 15
Grounding components 475–476
Grounding high-frequency 503–527
Grounding lightning protection
    systems 494–496
Grounding overhead distribution
    lines 496–497
Grounding potential rise (GPR) 476
Grounding resistance 476–478
  electrode arrangement 479–481
  electrodes to lightning currents 482–487
  low value 497
Grounding substations 490–494
Grounding systems 834–835
Grounding transmission lines 487–490

Hazardous areas, lightning damage
    to 357–360
Heidler 992
High-current test generators 956–972
  direct effects testing 971–972
  long-duration current generation 970–971
  return stroke effect 957–967
  subsequent return stroke effects 967–969
High-frequency groundings 503–527
  behaviour of 521–525
  circuit concepts 503–506
  electronic propagation 506–508
  frequency dependence 514–520
  frequency-dependent behaviour 526–527
  grid size 518–520
  modelling of 511–514
  soil characteristics 509–510
High-power diodes 965–966
High-temperature superconductors
    (HTS) 294
High-voltage impulse test generators

Index   1025

multistage impulse voltage
      circuits 953–956
   single-stage impulse voltage
      circuits 950–952
High-voltage stations 771–772
High-voltage testing 947–978
Horizontal configuration distribution
      line 136–142
   current transformers 136
HTS: see high-temperature superconductors
Hub 702–703
Hubert 992–993
Human injuries 901–920
   blunt 914–915
   burn 913–914
   cardiovascular system 906–908
   disability 915
   ear 910–911
   eye 909–910
   lightning electromagnetic
      fields 917–920
   lightning strike type 902–906
   nervous system 912–913
   psychological 914
   remote 915–917
   respiratory system 906–908
   skin 913–914
Hydraulic systems 705–706

ICC: see initial continuous current
ICV: see initial current variation
IEC: see International Electrotechnical
      Commission
Ignition, tree damage and 850–852
Ignitrons 966–967
Impulse current 392
   explosion 415
   measurement 975–978
   melting 412–414
   monitors 977–978
   resistive shunts 975–976
   Rogowski coils 976–977
Impulse voltages, measurement 974–978
Impulse withstand voltage (IEC) 646
Indirect effects
   modelling of 813–814
   testing 972–973
Indirect lightning strikes, overvoltages 636
Indirect strikes 571–603, 605–611

lightning-induced voltages 573–578
   sensitivity analysis 578–603
      bunched cables vs. open wire
         lines 597–598
      distribution transformer 591–593
      earth resistance 594–595
      earthing systems 594–595
      lightning channel 583
      lightning stroke point 587–588
      line height 591
      line length 588–590
      low-voltage power installations 594
      number of services 596
      rural lines 599–601
      soil electrical parameters 601–603
      stroke current
         magnitude 583–585
         propagation velocity 585–586
      stroke location 587
      topology 588–590
      urban lines 599–601
      waveform 583–585
Induced overvoltages 642–645
Induction protective distance 768
Inductors 289–290, 297–298
   chokes 289
   differential mode currents 289
   ferrites 289
Information technology
   installations 378–380, 382–383,
      384–385, 387–388
Information technology systems, equipotential
      bonding and 364–365
Initial continuous current (ICC) 100
Initial current variation (ICV) 107
Injected current 18–19
Installation protection 416–417
Instrument towers, measuring,
      Viscaro 54, 55
Instrumented towers 48
   measuring 50–55
      Andersson 53
      Eriksson 51–52
      Grabagnati 51
      Takami and Okabe 53, 54
Intercloud discharges 15
Internal grounding system 835–838
Internal lightning protection system 355–472
   concept of 365–366

Internal lightning protection system
(*Continued*)
  equipotential bonding 363–365
  surge protection 389–417
Internal protection, launch vehicle 799–801
  electrical wires 800–801
Internal zone lightning protection
  cable shielding 369–370
  equipotential bonding network 371
  magnetic shielding 366–369
International Electrotechnical Commission
  (IEC) analytical form 70
Interstroke interval 30–31
  worldwide statistics 31
Intracloud discharges 570
Intracloud flashes 15
Ionization 540
Ionized region, conductivity of 536–537
Isolation 422–423
  devices 290–291
  optocouplers 422–423
IT earthing systems 556
IT system 394, 401

Joints 326–328
Junction processes 17

K-changes 17

Lalande's stabilization field
  equation 196–197
Laser-triggered spark gaps 967
Launch vehicle
  grounding 801–802
  internal protection of 799–801
  lightning threat 792–794
Leader branches, effects of 226
Leader inception criterion 224–225
Leader inception models 186–212
  Bazelyan and Raizer's empirical leader
    model 195
  critical radius concept 186–190
  critical streamer length concept 194–195
  Lalande's stabilization field
    equation 196–197
  Rizk's generalized leader inception
    equation 190–194
  self-consistent leader inception
    model 197–212

Leader progression model 212–239
  assumptions of 223
  inception criterion 224–225
  leader branches 226
  orientation of the stepped
    leader 223–224
  thundercloud electric field 226–227
  tortuosity 226
  upward connecting leader 225–226
  Becerra and Cooray 227
    self-consistent lightning interception
      model 231–239
    theory 227–231
    validity 231
  collection volume/field intensification
    method 214–215
  concept of 212
  Dellera and Garbagnati 215–217
  Eriksson 212–215
  Rizk 217–220
  validation of 220–223
LEMP protection measures system
  (LPMS) 939–944
  bonding network 941–942
  earthing system 941–942
  lightning protection zone 939–941
  line routing 942
  shielding 942
  spatial magnetic shielding 942–944
  surge protection device 942
LEMP (lightning electromagnetic pulse)
Lengths of connecting leads 401–410
  parallel connections 403–405
  phase-side connecting cables 406–410
  protective earth neutral 405–406
  series connections 402
Leyden jar capacitor 3
Liew and Darveniza 537–540
  single driven rod 538–540
Lightning 3
  early stages of 861
  frequency 424–425
  interaction with objects 131–150
    airport runway lighting
      system 144–149
    miscellaneous tests 149–150
    overhead power distribution
      lines 131–142
    power transmission lines 143–144

residential buildings 144
    underground cables 143
  late stages 861–862
Lightning cessation 876–878, 884–885
Lightning channel 583
Lightning conductor
  radii of 211–212
  shape of 210–212
  shape of, blunt tip 210
Lightning current parameters, negative first return-stroke 54
Lightning currents
  standardized 926–931
  striking points determination 931–934
  threat parameters 926–929
Lightning-current parameters
  current measurements 54–55
  definition 49
  measuring of 48
    electromagnetic field remote sensing 48
    instrumented towers 48
    triggered lightning 48
  statistical representation 55–67
    Berger 58
    correlation issues 57–58
    current waveforms 67
    first strokes 59
    peak current derivatives 56
    positive strokes 58
    Takami 61
    tower height 60–67
    Viscaro 59
Lightning damage 355–361
  airports 360–361
  city areas 357–360
  consequences of 361
  hazardous areas 357
Lightning detection networks 864
Lightning discharges
  cloud flashes 15
  ground flashes 15
  upward 15
Lightning electric fields 276
Lightning electromagnetic fields 917–920
Lightning electromagnetic pulse (LEMP) 365
Lightning equipotential bonding 436, 937

Lightning flashes
  effects of 726–731
    antenna support structure 726–729
    current flow 729–730
    earth-termination segments 730–731
    electromagnetic field amplitude 79–80
    interstroke interval 30–31
    number of strokes 25, 28–30
    time interval between 25
Lightning-induced voltages
  calculations of 573–578
    finite-difference time-domain method 575
  Extended Rusck Model 573
Lightning interception
  dissipation array systems 251–259
  early streamer emission 240–251
  electro-geometrical method 168–170
  leader progression model 212–239
  mesh method 175
  non-conventional systems 239–259
  protection angle method 166–168
  rolling sphere method 170–175
  striking distance to
    elevated structures 182–212
    flat ground 176–182
Lightning launch/flight commit criteria (LLCC) 802–803
Lightning leaders 36–47
  dart leaders 37–39
  stepped leaders 36–37
Lightning-like electric fields 246–249
Lightning magnetic pulse protection
  inspection 388–389
  maintenance of 388–389
Lightning onset 875–876
Lightning protection 611–622
  distribution transformers 612–614
  external 307–352
  flammable materials and 825–830
  power installations 614–622
Lightning protection system (LPS) 307–352, 494–496, 934–939
  air termination 743, 934–935
  applications of 423–441
    photovoltaic systems 428–434
    wind turbines 423–428
  corrosion 753
  down-conductor 743–745, 935
  earth termination 745–752, 935–937

Lightning protection system (LPS) (*Continued*)
  earthing improvement 753–754
  failure reduction 757–763
    antenna support structure 761
    earth-termination potential equalization 757–761
    external building equipment 763
    internal building equipment 761–763
    foundation earth electrode 754–757
  internal 355–72
  launch pads 794–802
  lightning equipotential bonding 937
  separation distance 938
  video surveillance systems 438–441
Lightning protection zone (LPZ) 354–389, 939–941
  boundaries 373–380
    coordination of 385–388
  concepts 365–366, 691–692
    external zones 366
    internal 366
    lightning electromagnetic pulse 365
Lightning rod
  early designs 7–9
  experiments in France 5
  improved versions 9–12
  origins of 4
Lightning sensors 716
Lightning standards 926–945
  LEMP protection measures system 939–944
  protection system 934–939
Lightning strikes
  direct 636
    expected number 638–639
  distribution lines 636–640
    shielding 639–640
  human injuries and 902–906
  indirect 636
  location 447–452
Lightning stroke point 587–588
Lightning surge protection 269–303
Lightning surge response, earth termination 806–807
Lightning surges
  direct strikes 568–570
  low-voltage networks and 568–611
  medium-voltage line transfer 603–611
Lightning test equipment 948–973

high-current 956–972
high-voltage impulse test generators 948–956
indirect effects 972–973
Lightning threat, launch vehicles and 792–794
Lightning transients 273–278
  charge transferred 274–275
  negative first return stroke 277
  peak current 274
  prospective energy 275
  waveshape 276–278
Lightning warning systems 859–894
  performance metrics 869–875
    critical success index 870
    failure-to-warn 870
    false alarm ratio 869
    fixed-point algorithms 870–873
    lightning cessation 876–878
    lightning onset 875–876
    probability of detection 869
    storm-following algorithms 873–875
  performance of 869–878, 879–887
    fixed-point warning algorithm 879–884
    lightning cessation 884–885
    single detection technology 884–885
    storm-following algorithms 885–887
  risk assessment 887–894
    decision making 887–891
    equipment protection 891–892
    group warnings 892–893
  types 864–869
    fixed-point 864–867
    storm-following algorithms 867–869
Limiting voltage 420
Line height 591
Line routing 942
Lining, Dr. John 9
LIOV code 670–675
  Electromagnetic Transient Program (EMTP) 675
LIOV-Monte Carlo procedure 668–670
LLCC: *see* lightning launch/flight commit criteria
Long continuing currents 85
Long-duration current generation 970–971
Low-frequency grounding 476–499
Low-impedance cable shielding 370

Index 1029

Low-inductance megavolt
  generators 967–968
Low-voltage networks 554–568
  cloud discharges 570
  configurations 555–557
  distribution transformers 558–564
  earthing practices 554
  IEC nomenclature 556
  indirect strikes 571–603
  lightning protection 611–622
    distribution transformers 612–614
    power installations 614–622
  lightning surges 568–611
    direct strikes 568–570
    medium-voltage line transfer 603–611
  power installations 564–568
Low-voltage power installation (LVPI)
  network 270–273
  common mode voltages 270–273
Low-voltage power installations 594
Low-voltage power supply 434
Low-voltage stations 773–776
  delay switches 775
  delayed fuses 774–776
  reclosing switches 775–776
Low-voltage system, equipotential bonding
  and 364
LPMS: *see* LEMP protection measures system
LPS: *see* lightning protection system
LPZ: *see* lightning protection zone
LVPI: *see* low-voltage power installation

M-components 17, 87, 88
Macroscale tree damage 847
  bark-loss 849
  explosive 849
  ignition 850–852
  physical damage 847
  wood-loss 849
Magnetic field 276
  equations 1012–1015
Magnetic shielding 366–369
Maximum continuous voltage 391
MCS: *see* meoscale convective system
Measuring system 421–423
  satellite launch pads and 814–815
Medium voltage (MV) lines
  distribution lines lightning
    performance 666–667

lightning protection 635–675
LIOV code 670–675
LIOV-Monte Carlo procedure 668–670
protection of 645–653
  basic impulse insulation level 646–648
  critical impulse flashover
    voltage 646–648
  protective devices 650–653
  shield wire 648–649, 653–655,
    662–665
  soil resistivity 662
  surge arresters 655–660, 666
Medium-voltage electrical systems 709
Medium-voltage line transfer 603–611
  direct strikes 603–604
  indirect strikes 605–611
Melting, impulse currents and 412–414
Meoscale convective system (MCS) 861
Mesh method 175, 311–313, 933
Metal installations 383
  equipotential bonding and 373–375
Metal oxide varistor (MOV) 270
Metal sheet thickness 319
Method of moments (MoM) 809–810
Microscale tree damage 846
Modified transmission line 1009–1010
MoM: *see* method of moments
Montano 1000
MOV: *see* metal oxide varistor
Multiple terminations 33
Multistage impulse voltage
  circuits 953–956
MV: *see* medium voltage

Nacelle 703
Negative first return stroke 54, 277
Negative lighting 102
Negative rocket-triggered lightning, current
  waveforms 113–115
Nervous system injuries 912–913
Nollet, Abbe Jean-Antoine 5
Nominal current 420–421
Nominal discharge current 392, 420
Non-conventional air-termination system 320
Non-linear grounding behaviour, frequency-
  dependent vs. 526–527
Nucci, analytical expression 71
Number of strokes per flash 25, 28–30
  worldwide statistics 29

Offshore oil installation (OOI) 830
Offshore oil platforms 830–839
  common bonding network 834–835
  design standards 831–832
  external lightning protection 832–834
  grounding system 834–835
  internal grounding system 835–838
  risk assessment 832
  shielding 838–839
  SPD location 839
Off-shore wind turbines 714–716
Okabe measurements 53, 54
OOI: *see* offshore oil installation
Open wire lines vs. bunched
    cables 597–598
Optocouplers 422–423
Ordnance circuit 801
Overhead distribution lines 496–497
Overhead lines 769–770
Overhead power distribution lines 131–142
  direct strikes 135–136
  horizontal configuration 136–142
  vertical configuration 142
Overhead power lines, lightning 731–733
  effect 731
  cable damage 734
  near station 733–734
  transformer 734
  flashes to ground 733
Overvoltages 777–778, 786–787
  common-mode transfer 786
  differential-mode transfer 786
  direct lightning 640
    strike and 636
  indirect lightning and 636
  induced 642–645
  transfer across transformer 735–736

PAM: *see* protective angle method
Parallel connections 403–405
  two-conductor terminals 403
Partially conducting stage 17
Peak current 274
  cloud-to-ground lightning 274
  correlation of 61
  derivatives 56
Peak electric radiation 81–83
Peaking circuits 968–969
Peissenberg tower 72

PEN: *see* protection earth neutral
Phase-side connecting cables 406–410
Photovoltaic (PV) systems 428–434
  d.c. line protection 436–437
  low-voltage power supply 434
  protection necessity of 429
  with LPS 431–434
  without LPS 429–431
Piparo's measurements 51
POD: *see* probability of detection
Polarity, flash 15
Positive leader discharges 183–186
Positive lightning 102
Positive strokes 58
Positive temperature coefficient (PTC)
    devices 290
Power circuits 709–710
Power distribution network surge
    protection 293–294
  distribution static compensator 294
  dynamic voltage restorer 293–294
  fault current limiters 294
  solid-state breaker 293
  solid-state transfer switch 293
Power installations, protection from
    lightning 614–622
Power lines, direct lightning strike 269–270
Power supply circuits 771–776
  high voltage stations 771–772
  low voltage 773–776
  self-contained power supplies 776
Power supply installations 375–378, 380–382,
    383–384, 385–386
Power supply 747–748, 749–752
Power supply system
  lightning effect on 731
    overhead lines 731
    overvoltage transfer across
      transformer 735–736
    underground cable 735
  surge protection for 389–417
    backup 410–417
    coordination 393
    cross-sectional areas 410–417
    distribution systems 393–396
    follow current extinguishing
      capability 392–393
    follow current limitation 393
    impulse current 392

IT 394
lengths of connecting leads 401–410
maximum continuous voltage 391
nominal discharge current 392
short-circuit withstand capability 392
technical characteristics 391–393
temporary overvoltage 393
TT 394
voltage protection level 392
Power transmission lines 143–144
Power wires 800
Preliminary breakdown 15
Probability of detection (POD) 869
Prospective energy 275
Protection angle method 166–168, 313–315, 933–934
Protection earth neutral (PEN) 405–406
Protective angle method (PAM) 826
Protective devices 650–653
  capacitors 653
  spark gaps 651
  surge arresters 652–653
Protective distance calculation 767–768
Psychological injuries 914
PTC: see positive temperature coefficient
PV: see photovoltaic systems

Radii, lightning conductor and 211–212
Radio frequency wiring 800–801
Raizer's empirical leader model 195–196
Rakov 999–1000
RCD: see residual current device
Reclosing switches 775–776
Recoil streamer 17
Relative loss assessment 458–459
Remote injuries 915–917
Residential buildings 144
Residual current device (RCD) 397
Resistance, grounding and 476–478
Resistive shunts 975–976
Resistors 298–302
  behaviour of 299–302
Respiratory system injuries 906–908
Return-stroke 16, 47–55
  action integral 48
  Berger's measurements 50
  parameters of 47–55
Return-stroke current peak

ground conditions vs. 122–128
surface arcing 127
Return-stroke current waveform 67–70, 110–121
  Cooray 72–74
  Delfino 72
  Diendorfer 71
  Gaisberg tower 73
  International Electrotechnical Commission 70
  Nucci 71
  Peissenberg tower 73
  Uman 71
  upward-initiated flashes 72
Return-stroke effect simulation 957–967
  crowbar technologies 962–964
  high-power diodes 965–966
  ignitrons 966–967
  laser triggered spark gaps 967
  three-electrode spark gap 964
Return-stroke modelling 981–1016
  current dissipation 1001–1006, 1008, 1009–1010
  current generation 986–1001
  current propagation 983–986
  electric field equations 1012–1015
  ground conductivity 1010–1012
  magnetic field equations 1012–1015
Richmann, Georg Wilhelm, death of 5
Ring conductors, down-conductor system and 321
Risk analysis 443–499
  concept of 445–447
  damage probabilities 452–457
  flash, subsequent strokes 470–471
  lightning strikes 447–452
  relative loss assessment 458–459
Risk assessment 688–691, 887–883
  decision making 887–891
  flammable materials and 824–825
  offshore oil platform and 832
Risk components 459–460
  evaluation of 464–468
  reduction in 468–470
Risk reduction 468–470
Rizk 217–220
  generalized leader inception equation 190–194
Rocket lightning, objects interaction 131–150
Rocket structure 790–791

Rocket structure (*Continued*)
  electrical wires 800–801
    digital data lines 801
    low frequency 800
    ordnance circuit 801
    power 800
    radio-frequency 800–801
Rocket-triggered lightning 97–150
  close lightning electromagnetic
      environment 128–131
  current waveforms 107–110
  ground conditions 122–128
  history of 98
  return-stroke
    current peak 122–128
    current waveforms 110
  techniques 105–106
  triggering techniques 98–106
Rocky soil, earth-termination system and 341
Rogowski coils 976–977
Rolling sphere method (RSM) 170–175,
    310–311, 826, 931–932
  air terminal positioning 173
RSM: *see* rolling sphere method
Rural lines 599–601

Sand soil, earth-termination system 341
Satellite launch pads
  design of 791–792
  final countdown tasks 792
  launch vehicle 792–794
  lightning protection systems 794–802
    air termination network 798–799
    down-conductor system 799
    down-conductor systems 807–812
    earth termination 799
    external protection 794–795
    internal protection 799–803
    lightning surge response 806–813
    modelling 803–806
    on-site measurements 814–815
    presently used 790–798
    supporting systems 814
    weather launch commit
        criteria 802–803
  lightning protection 789–815
  rocket structure 790–791
  weather criteria 813
Screened cables 770–771

Sea strikes 78–79
Sekioka 543–545
  energy balance equation 543
Self-consistent leader inception
    model 197–212
  Bazelyan leader inception model 209
  Becerra 197
  comparisons 205–212
  Cooray 197
  dynamic leader inception
      evaluation 202–204
  laboratory based 205
  lightning conductor
    radii of 211–212
    shape of 210–212
  rocket-triggered lightning
      techniques 105–106
  static leader evaluation 198–202
  time-dependent leader inception 207
  value parameters 200
Self-consistent lightning interception model
    (SLIM) 231–239
Self-protecting system 829–830
Sensitivity analysis
  bunched cables vs. open wire
      lines 597–598
  distribution transformer 591–593
  earth resistance 594–595
  earthing systems 594–595
  lightning channel 583
  lightning stroke point 587–588
  line height 591
  length 588–590
  low-voltage power installations 594
  number of services 596–597
  rural lines 599–601
  soil electrical parameters 601–603
  stroke current magnitude 583–585
  stroke location 587
  topology 588–590
  urban lines 599–601
  waveform 583–585
Separation distance 938
Series connections 402
Services, number of 596–597
Shield wire 648–649, 653–655, 662–665
Shielding 426, 639–640, 838–839, 942
  conductor penetration 279–280
Short-circuit withstand capability 392

Short continuing currents 85
Single detection technology 884–885
Single driven rod 538–540
Single-ended shielding 369–370
Single-point lightning detections
    sensors 863–864
Single-stage impulse voltage
    circuits 950–952
Skin injuries 913–914
SLIM: *see* self-consistent lightning
    interception model
Soil electrical parameters 601–603
Soil ionization 343, 531–550
    air voids 532
    critical electric field 534–536
    electrode location 533
    modelling of 532–533, 536–548
        Cooray 545–549
        ionized region 536–537
        Liew and Darveniza 537–540
        Sekioka 543–545
        Wang 540–543
Soil resistivity 345–347, 662
Soil, frequency dependent characteristic
    of 509–510
Solar power plants 434–438
    air-termination system 434–435
    data processing systems 437–438
    down-conductor system 434–435
    earthing system 435–436
    lightning equipotential bonding 436
Solid-state breaker (SSB) 293
Solid-state transfer switch (SSTS) 293
Space charge layer 21
Space charge, connecting leaders 257–259
Spark gaps 282–285, 651
Spatial magnetic shielding 942–944
SPD: *see* surge protection device
Surge protection device 766–768
    discharge current 766–767
    installation 767–768
        induction protective distance 768
        protective distance 767–768
    location of 766–767
Speed of connecting leaders 44–45
Speed of dart leaders 37–39
Speed of stepped leaders 36–37
SSB: *see* solid-state breaker
SSTS: *see* solid-state transfer switch

Standardized lighting currents 926–931
    test requirements 930–931
Static leader inception evaluation 198–202
Step voltages 344–345, 742
Stepped leader 15
    electric field generation 39–40
    orientation of 223–224
    speed of 36–37
Storm following algorithms 867–869,
    873–875, 885–887
Striking distance 16
    modelling of 66
Striking distance to elevated
    structures 182–212
    leader inception models 186–212
    positive leader discharges 183–186
Striking distance to flat ground 176–182
    coefficients 181
    researchers
        Armstrong and Whitehead 180–182
        Dellera and Garbagnati 179
        Cooray 179–180
        Eriksson 177–179
        Golde 177
Striking points determination 931–934
    mesh method 933
    protection angle method 933–934
    rolling sphere method 931–932
Stroke current magnitude 383–585
Stroke current propagation
    velocity 585–586
Stroke location 587
Strokes, subsequent 470–471
Structures, upward ground flash and 17–19
Subsequent return stroke effects 967–969
    exploding wires 969
    low-inductance megavolt
        generators 967–968
    peaking circuits 968–969
Subsequent return stroke 17
Subsequent stroke, peak electric
    radiation 81–83
Substations 490–494
Surface arcing 35, 127
Surface discharge 34–36
Surface flash over
    surface arcing 35
    surface discharge 34–36
Surge arresters 652–653, 655–660, 666

Surge protection 269–303
  component properties 283
  coordination of 302–303
  current limiters 288–290
  diodes 286–288
  direct strike to power
      lines 269–270
    metal oxide varistor 270
  filters 291–293, 294–302
  gas discharge tubes 282–285
  isolation devices 290–291
  lightning transients 273–278
  low-voltage power installation
      network 270–273
  parasitic elements of
      components 294–302
  philosophy of 278–294
    conductor penetration 279–280
    electromagnetic compatibility 278–280
  power distribution networks 293–294
  power supply systems 389–417
  principles of 280–294
  spark gaps 282–285
  telecommunication systems 417–423
  thyristors 286–288
  varistors 285–286
Surge protection device (SPD) 942
Surge testing 776–787
  case study 779–784
    overvoltages 786–787
Surge, simulation of 776–779
  overvoltages 777–778
Switches
  delay 775
  reclosing 775
Switching impulse voltages 244–246

Takami measurements 53, 54
  peak current correlation 61
Tanks, flammable materials and 822–830
Telecommunication systems, surge
    protection 417–423
Telecommunication towers
  coordinated SPD protection
      system 766–768
  failure reduction 763–765
    circuit screening 763–765
    electrical systems 726
    external circuits 764–765

lightning damage
  electrodynamic effects 726
  electronic systems 726
  injury to people 725
  thermal effects 725–726
  touch-voltage 725
lightning flashes effect 726–731
  power supply system 731
lightning protection 723, 787
lightning protection system 743–757
  air-termination 743
  corrosion protection 753
  down-conductor 743–745
  earth-termination system 745–752
  earthing improvement 753–754
  failure reduction 757–763
    antenna support structure 761
    earth-termination 757–761
    external building equipment 763
    internal building equipment 761–763
  foundation earth electrode 754–757
  overhead lines 769–770
  screened cables 770–771
  power supply circuits 771–776
  protection level 736–737
    importance 737
    repair cost 737
  protection measures 736–741
    implementation 739–742
    selection of 739
    step voltages 742
    touch-voltages 742
  surge protection 417–423
    arresters 417–421
    control systems 421
    measuring systems 421–423
  surge testing 776–787
    case study 779–784
Temporary overvoltage 393
Thermal effects 725–726
Thottappillil 996–997
Threat parameters 926–929
Three-electrode spark gap 964
Thundercloud electric field 226–227
  generation 19–23
    corona 19, 21, 23
    space charge layer 21
Thunderstorm days 23–25
  worldwide 25

Index 1035

Thunderstorm life cycle 860–864
  convective development 860
  detection methods 862–864
    networks 864
    single-point 863–864
  electrification 860–861
  lightning
    early stages of 861
    late stages of 861–862
  meoscale convective system 861
Thyristors 286–288
Time advantage, early streamer emission and 250–251
Time intervals, lightning flashes and 25, 28–30
Time-dependent leader inception 207
TLM: see transmission line modelling
TN earthing systems 556
TN system 394, 396–397
  residual current device 397
Topology 588–590
Tortuosity, effects of 226
Touch voltages 343–344, 725, 742
Tower height
  effect of 60–67
    attractive radius expression 66
    attractive radius of the tower 64
    striking distance model 66
Tower measurements, Berger's measurements 50
Tower, wind turbine 703–704
Transformer, overvoltage transfer over 735–736
Transmission line modelling (TLM) 808–809
Transmissions lines 487–490
Trees 843–855
  lightning damage probability 844–846
  lightning damage to 843–844
  lightning damage types 846–854
    long-term 853
    macroscale 847
    microscale damage 846
  protection of 854–855
Triggered lightning 48
  summary of 99
Triggering techniques 98–106
  altitude 103–105
  Camp Blanding 105–196
  classical 98–103

TT
  earthing systems 556
  system 394, 398–401
Turbine blades 695–702
  carbon fibre 699–700
    reinforced plastic 696
  documentation 700–702
  glass fibre reinforced plastics 695
Two branches 31
Two-conductor terminals 403

Uman 994–998
  analytical expression 71
Underground cable 143, 735
UPL: see upward positive leader
Upward connecting leader 225–226
Upward ground flash 17–19
  structure interaction 17–19
Upward lightning flashes 15
Upward negative ground flashes 15
Upward positive leader (UPL) 100
Upward-initiated flashes, continuing current pulses 72
Upward-initiated flashes, Cooray's analysis of 72–74
Upward-initiated lightning flashes, pulses from 76
Urban lines 599–601

Varistors 285–286
Vascaro's measurements 54, 55
Venting 829
Vertical configuration distribution line 142
Vertical ground rods 148–149
  currents 148–149
Vessels, flammable materials and 822–830
Video surveillance systems 438–441
  installed external LPS 439–441
  no external LPS 441
Viscaro's measurements 59
Voltage protection level 392, 418
Voltage pulses, amplitude of 249–250

Wagner 991
Wang 540–543
  current injection 540
  deionization 540
  ionization 540

Wang (*Continued*)
  negative ionization 540
  sparking 540
Waveform 583–585
Waveshape 276–278
  lightning electric fields 276
  magnetic fields 276
Weather criteria, satellite launches and 813
Weather launch commit criteria, lightning
    launch/flight commit criteria 802–803
Whitehead 180–182
Wind farms, location of 713–714
Wind turbines 423–428, 681–718
  actual lightning damage to 687–688
  construction safeguards 717–718
  cost-benefit evaluation 688–691
  design 684–685
  earth-termination system 426–428
  earthing 693–695
  equipotential bonding 693–695
  lightning
    frequency 424–425
    protection zoning concept 691–692
  sensors 716
  threat to 686–687
  location of 713–714
  off-shore 714–716
  potential damage to 424
  protection measures 425–426
  protection of 695–712
    bearings and gears 704–705
    blades 695–702
    communication systems 711–712
    control systems 711–712
    electrical systems 706–710
    hub 702–703
    hydraulic systems 705–706
    nacelle 703
    tower 703–704
  protection standardization of 425
  risk assessment 688–691
  shielding 426
  usage of 424
Windmills 681–682
Wire, exploding 969
Wood-loss damage 849

Lightning Source UK Ltd.
Milton Keynes UK
UKHW01n0210280918
329634UK00004B/217/P